T0212484

Lecture Notes in Computer Science 9890

Commenced Publication in 1973
Founding and Former Series Editors:
Gerhard Goos, Juris Hartmanis, and Jan van Leeuwen

More information about this series at http://www.springer.com/series/7407

Vladimir P. Gerdt · Wolfram Koepf
Werner M. Seiler · Evgenii V. Vorozhtsov (Eds.)

Computer Algebra in Scientific Computing

18th International Workshop, CASC 2016
Bucharest, Romania, September 19–23, 2016
Proceedings

Springer

Editors

Vladimir P. Gerdt
Laboratory of Information Technologies
Joint Institute of Nuclear Research
Dubna
Russia

Werner M. Seiler
Institut für Mathematik
Universität Kassel
Kassel
Germany

Wolfram Koepf
Institut für Mathematik
Universität Kassel
Kassel
Germany

Evgenii V. Vorozhtsov
Institute of Theoretical and Applied
    Mechanics
Russian Academy of Sciences
Novosibirsk
Russia

ISSN 0302-9743   ISSN 1611-3349 (electronic)
Lecture Notes in Computer Science
ISBN 978-3-319-45640-9   ISBN 978-3-319-45641-6 (eBook)
DOI 10.1007/978-3-319-45641-6

Library of Congress Control Number: 2016949104

LNCS Sublibrary: SL1 – Theoretical Computer Science and General Issues

Printed on acid-free paper

This Springer imprint is published by Springer Nature
The registered company is Springer International Publishing AG
The registered company address is: Gewerbestrasse 11, 6330 Cham, Switzerland

Preface

One of the main goals of the International Workshops on Computer Algebra in Scientific Computing, which started in 1998 and since then have been held annually, is to highlight cutting-edge advances in all major disciplines of Computer Algebra (CA). And the second goal of the CASC workshops is to bring together both the researchers in theoretical computer algebra and the engineers as well as other allied professionals applying CA tools for solving problems in industry and in various branches of scientific computing.

This year the 18th CASC conference was held in Bucharest (Romania). Computer Algebra is popular among scientists in Romania. Researchers from many institutions, such as the University of Bucharest, the Institute of Mathematics "Simion Stoilow" of the Romanian Academy, the West University of Timişoara, the University "Al. I. Cuza" of Iaşi, the Institute of Computing "Tiberiu Popoviciu" from Cluj-Napoca, the "Horia Hulubei" National Institute for Research and Development in Physics and Nuclear Engineering (Bucharest–Măgurele), and "Ovidius" University in Constanţa, are working on subjects such as numerical simulation using computer algebra systems, symbolic–numeric methods for polynomial equations and inequalities, algorithms and complexity in computer algebra, application of computer algebra to natural sciences and engineering, polynomial algebra, and real quantifier elimination. In Romania there are several international conferences on Computer Algebra and related topics such as the International Symposium on Symbolic and Numeric Algorithms for Scientific Computing, SYNASC, in Timişoara, or the series of conferences on commutative algebra and computer algebra held in Constanţa and Bucharest.

The above has affected the choice of Bucharest as a venue for the CASC 2016 workshop.

This volume contains 30 full papers submitted to the workshop by the participants and accepted by the Program Committee after a thorough reviewing process with usually three independent referee reports. Additionally, the volume includes two invited talks.

Polynomial algebra, which is at the core of computer algebra, is represented by contributions devoted to improved algorithms for computing the Janet and Pommaret bases, the dynamic Gröbner bases computation, the algorithmic computation of polynomial amoebas, refinement of the bound of Lagrange for the positive roots of univariate polynomials, computation of characteristic polynomials of matrices whose entries are integer coefficient bivariate polynomials, finding the multiple eigenvalues of a matrix dependent on a parameter, the application of a novel concept of a resolving decomposition for the effective construction of free resolutions, enhancing the extended Hensel construction with the aid of a Gröbner basis, a hybrid symbolic-numeric method for computing a Puiseux series expansion for every space curve that is a solution of a polynomial system, numerical computation of border curves of bi-parametric real polynomial systems, and the application of sparse interpolation in Hensel lifting and pruning algorithms for pretropisms of Newton polytopes. Polynomial algebra also plays a central role in contributions concerned with elimination algorithms for sparse matrices over finite fields, new algorithms for computing sparse representations of systems of parametric rational fractions, and quadric arrangement in classifying rigid motions of a 3D digital image.

Several papers are devoted to using computer algebra for the investigation of various mathematical and applied topics related to ordinary differential equations (ODEs): application of the Julia package `Flows.jl` for the analysis of split-step time integrators of nonlinear evolution equations, the use of the CAS Maple 18 for the derivation of operator splitting methods for the numerical solution of evolution equations, and the complexity analysis of operator matrices transformations as applied to systems of linear ODEs.

Three papers deal with applications of symbolic and symbolic-numeric computations for investigating and solving partial differential equations (PDEs) and ODEs in mathematical physics and fluid mechanics: the construction of a closed form solution to the kinematic part of the Cosserat partial differential equations describing the mechanical behavior of elastic rods, symbolic-numeric solution with Maple of a second-order system of ODEs arising in the problem of multichannel scattering, and symbolic-numeric optimization of the preconditioners in a numerical solver for incompressible Navier–Stokes equations.

Applications of CASs in mechanics and physics are represented by the following themes: qualitative analysis of the general integrable case of the problem of motion of a rigid body in a double force field, investigation of the influence of aerodynamic forces on satellite equilibria with the aid of the Gröbner basis method, and generation of irreducible representations of the point symmetry groups in the rotor + shape vibrational space of a nuclear collective model in the intrinsic frame.

The first invited talk by Th. Hahn focuses on the application of computer algebra in high-energy physics, in particular, the *Mathematica* packages FeynArts and FormCalc. The second invited talk by C.S. Calude and D. Thompson deals with the problems of incompleteness and undecidability. These are important problems related to the foundations of mathematics. The authors discuss some challenges proof-assistants face in handling undecidable problems. Several example problems, including the automated proofs, are presented. The authors briefly describe the computer program Isabelle, which they use as the proof assistant.

CASC 2016 features for the first time a full blown Topical Session. In this fairly new feature of the CASC series, up to six talks around a common theme are invited. The authors have an extended page limit, but their submissions are refereed according to the same principles and with the same rigour as normal submissions. This time the topic was *Satisfiability Checking and Symbolic Computation (SC^2)* and the session also marks the beginning of a European FET-CSA project with the same title (see http://www.sc-square.org for more information about this project and its objectives). There is a large thematic overlap between the fields of satisfiability checking (traditionally more a subject in computer science) and of symbolic computation (nowadays mainly studied by mathematicians). However, the corresponding communities are fairly disjoint and each has its own conference series. The central goal of the SC^2 project consists of bridging this gap and of bringing together people from both sides.

Thus, the 2016 Topical Session intends to familiarize the CASC participants with the many interesting problems in this domain. It was well organised by E. Abraham, J. Davenport, P. Fontaine, and Th. Sturm and comprises five talks. One is a one-hour survey talk by D. Monniaux on satisfiability modulo theory. The other four talks concern the investigation of the complexity of cylindrical algebraic decomposition with respect to polynomial degree, efficient simplification techniques for special real quantifier elimination, the description of a new SAT + CAS verifier for combinatorial conjectures, and a generalized branch-and-bound approach in SAT modulo nonlinear integer arithmetic.

The CASC 2016 workshop was hosted and supported by the University of Bucharest and the Romanian Mathematical Society. We appreciate that they provided free accommodation for a number of participants. The five speakers in the Topical Session received financial support from funds of the FET-CSA project *Satisfiability Checking and Symbolic Computation*.

Our particular thanks are due to the members of the CASC 2016 Local Organizing Committee at the University of Bucharest, Doru Ştefănescu, Luminiţa Dumitrică, Mihaela, Miruleţ, and Silviu Vasile, who ably handled all the local arrangements in Bucharest. Furthermore, we would like to thank all the members of the Program Committee for their thorough work. We are grateful to Matthias Seiß (Kassel University) for his technical help in the preparation of the camera-ready manuscript for this volume. Finally, we are grateful to the CASC publicity chair Andreas Weber (Rheinische Friedrich-Wilhelms-Universität Bonn) and his assistant Hassan Errami for the design of the conference poster and the management of the conference web page http://www.casc.cs.uni-bonn.de.

July 2016

Vladimir P. Gerdt
Wolfram Koepf
Werner M. Seiler
Evgenij V. Vorozhtsov

Organization

CASC 2016 was organized jointly by the Institute of Mathematics at Kassel University, the University of Bucharest, and the Romanian Mathematical Society.

Workshop General Chairs

Vladimir P. Gerdt	Dubna
Werner M. Seiler	Kassel

Program Committee Chairs

Wolfram Koepf	Kassel
Evgenii V. Vorozhtsov	Novosibirsk

Program Committee

Moulay Barkatou	Limoges
François Boulier	Lille
Jin-San Cheng	Beijing
Victor F. Edneral	Moscow
Matthew England	Coventry
Jaime Gutierrez	Santander
Sergey A. Gutnik	Moscow
Jeremy Johnson	Philadelphia
Victor Levandovskyy	Aachen
Marc Moreno Maza	London, CAN
Veronika Pillwein	Linz
Alexander Prokopenya	Warsaw
Georg Regensburger	Linz
Eugenio Roanes-Lozano	Madrid
Valery Romanovski	Maribor
Doru Ştefănescu	Bucharest
Thomas Sturm	Nancy
Elias Tsigaridas	Paris
Jan Verschelde	Chicago
Stephen M. Watt	W. Ontario
Kazuhiro Yokoyama	Tokyo

Additional Reviewers

Alkis Akritas
Carlos Beltrán
Nikolaj Bjorner
Paola Boito
Charles Bouillaguet
Martin Bromberger
Changbo Chen
Pascal Fontaine
Vijay Ganesh
Andrzej Góźdź
Gavin Harrison
Vadim Isaev
Hidenao Iwane
Manuel Kauers
Kinji Kimura
François Lemaire
Scott Mccallum

Marc Mezzarobba
Bernard Mourrain
Hiroshi Murakami
Clément Pernet
Marko Petkovšek
Gerhard Pfister
Clemens G. Raab
Anca Rădulescu
Marc Rybowicz
Yosuke Sato
Arne Storjohann
Yao Sun
Stefan Takacs
Thorsten Theobald
Tristan Vaccon
Bican Xia

Local Organization

Doru Ştefănescu
Luminiţa Dumitrică

Mihaela Mireţ
Silviu Vasile

Publicity Chair

Andreas Weber   Bonn

Website

http://www.casc.cs.uni-bonn.de/2016
(Webmaster: Hassan Errami)

Contents

On the Differential and Full Algebraic Complexities of Operator Matrices Transformations

Sergei A. Abramov$^{(\boxtimes)}$

Dorodnitsyn Computing Centre,
Federal Research Center Computer Science
and Control of Russian Academy of Sciences,
Vavilova, 40, Moscow 119333, Russia
sergeyabramov@mail.ru

Abstract. We consider $n \times n$-matrices whose entries are scalar ordinary differential operators of order $\leqslant d$ over a constructive differential field K. We show that to choose an algorithm to solve a problem related to such matrices it is reasonable to take into account the complexity measured as the number not only of arithmetic operations in K in the worst case but of all operations including differentiation. The algorithms that have the same complexity in terms of the number of arithmetic operations can though differ in the context of the full algebraic complexity that includes the necessary differentiations. Following this, we give a complexity analysis, first, of finding a superset of the set of singular points for solutions of a system of linear ordinary differential equations, and, second, of the unimodularity testing for an operator matrix and of constructing the inverse matrix if it exists.

1 Introduction

In this paper, we discuss some algorithms which use operations on elements of a differential field. A complexity analysis of such algorithms is based sometimes on considering the complexity as the number of arithmetic operations in K in the worst case (the *arithmetic complexity*). This approach is not always productive. First, the differentiations are not for free, and there is no reason to believe that, e.g., a differentiation is much cheaper than an addition or a multiplication when the differential field is $\mathbb{Q}(x)$ with the standard differentiation by x. Second, we may face a situation where two algorithms have the same arithmetic complexity. However, it may be that the complexities of those two algorithms are different, if we compute the number of differentiations in the worst case (the *differential complexity*). Therefore, consideration of not only arithmetic but also differential complexity seems reasonable. This is similar to the situation with sorting

S.A. Abramov—Supported in part by the Russian Foundation for Basic Research, project no. 16-01-00174.

V.P. Gerdt et al. (Eds.): CASC 2016, LNCS 9890, pp. 1–14, 2016.
DOI: 10.1007/978-3-319-45641-6_1

algorithms, when we consider separately the complexity as the number of comparisons and, resp., the number of swaps. (For example, in [19], an upper bound on the number of differentiations of equations in a differential system sufficient for testing its compatibility is established. Actually, such a bound is an estimate for the complexity of algorithms for the compatibility testing.)

We will also consider the *full* complexity as the total number of all operations in the basic differential field in the worst case, when differentiations are included. This complexity will be considered in the context of *algebraic complexity theory*: the complexity is measured as the number of operation in K in the worst case without taking into account the possible growth of "sizes" of the elements computed by algorithms (similarly, say, to the complexity $\Theta(n^{\log_2 7})$ of Strassen's algorithm [27] for multiplying square $n \times n$-matrices).

Below, we discuss two well known algorithms for transforming operator matrices, i.e., square matrices whose entries are scalar differential operators with coefficients in the basic differential field. Earlier, only arithmetic complexity of those algorithms was investigated, and it was established that asymptotically, their arithmetic complexities agree. However, we show that their differential complexities are different. Note that the functionalities of the algorithms are also slightly different.

In Sect. 4, we discuss two computational problems, for which the above-mentioned algorithms for transforming operator matrices are useful. The first problem is related to finding singular points of solutions of the corresponding system of linear ordinary differential equations. The second one is the problem of testing unimodularity (i.e., invertibility) of an operator matrix and of constructing the inverse matrix if it exists. Invertibility testing is a classical mathematical problem, whose specifics depend on a field or a ring containing the matrix entries. The question of unimodularity of such operator matrices arises, in particular, in connection with the existence problems for solutions of differential systems [23]. The "binary" testing (with the output *yes* or *no*, without constructing the inverse when it exists) can also be considered as a problem which is of independent interest. A careful complexity analysis allows one to make an informed choice of a transformation algorithm as an adequate auxiliary tool for solving each of these problems.

2   Preliminaries

2.1   Operator Matrices

Let K be a differential field of characteristic 0 with a derivation $\partial =\,'$. For a non-negative integer n, the ring of $n \times n$-matrices with entries belonging to a ring R is denoted by $\mathrm{Mat}_n(R)$. The ring of scalar differential operators with coefficients in K is denoted by $K[\partial]$; the order of an operator $l \in K[\partial]$ which is denoted by $\mathrm{ord}\, l$ is equal to the degree of the corresponding non-commutative polynomial from $K[\partial]$. Any non-zero operator matrix $L \in \mathrm{Mat}_n(K[\partial])$ can be represented as a differential operator with matrix coefficients in $\mathrm{Mat}_n(K)$:

$$L = A_d\partial^d + A_{d-1}\partial^{d-1} + \cdots + A_0, \tag{1}$$

where $A_0, A_1, \ldots, A_d \in \mathrm{Mat}_n(K)$, and the matrix A_d (the *leading matrix* of L) is non-zero. The number d is the *order* of L; we write $d = \mathrm{ord}\, L$. The order of a row of L is the biggest order of operators from $K[\partial]$ belonging to the row. Thus, the order of an operator matrix coincides with the biggest order of all rows of the operator matrix. A matrix $L \in \mathrm{Mat}_n(K[\partial])$ is of *full rank* if its rows are linearly independent over $K[\partial]$.

An operator matrix L is invertible in $\mathrm{Mat}_n(K[\partial])$ and M is its inverse, if $LM = ML = I_n$ where I_n is the identity $n \times n$-matrix. We write L^{-1} for M. Invertible operator matrices are also called *unimodular* matrices.

In [23], the following example of a unimodular matrix and the inverse is given $(K = \mathbb{Q}(x),\ \partial = \frac{d}{dx})$:

$$\begin{pmatrix} x^2/2 & -(x/2)\partial + 1 \\ -x\partial - 3 & \partial^2 \end{pmatrix}^{-1} = \begin{pmatrix} \partial^2 & (x/2)\partial \\ x\partial + 1 & x^2/2 \end{pmatrix}. \tag{2}$$

2.2 The Dimension of the Solutions Space

Let the constant field $Const(K) = \{c \in K \mid \partial c = 0\}$ of K be algebraically closed. We denote by Λ a fixed *universal Picard–Vessiot differential extension field* of K (see [25, Sect. 3.2]). This is a differential extension Λ of K with $Const(\Lambda) = Const(K)$ such that any differential system $\partial y = Ay$ with $A \in \mathrm{Mat}_n(K[\partial])$ has a solution space of dimension n over the constants. For arbitrary operator matrix L of the form (1), we denote by V_L the linear space over $Const(\Lambda)$ of solutions of L (i.e., solutions of the equation $L(y) = 0$) belonging to Λ^n. Its dimension will be denoted by $\dim V_L$.

Suppose that $\mathrm{Const}(K)$ is not algebraically closed. It is not difficult to see that for any differential field K of characteristic 0 there exists a differential extension whose constant field is algebraically closed. Indeed, this is the algebraic closure $\bar{K}$ with the derivation obtained by extending the derivation of K in the natural way. In this case, $\mathrm{Const}(\bar{K}) = \overline{\mathrm{Const}(K)}$ (see [25, Exercises 1.5, 2:(c),(d)], [26, Sect. 3]). In this case, V_L is the linear space over $Const(\bar{K})$ of solutions of L whose components belong to the the universal differential extension of $\bar{K}$.

We use the notation $M_{i,*}$, $1 \leqslant i \leqslant n$, for the $1 \times n$-matrix which is the ith row of an $n \times n$-matrix M. Let a full rank operator matrix L be of the form (1). If $1 \leqslant i \leqslant n$ then define $\alpha_i(L)$ as the maximal integer k, $1 \leqslant k \leqslant d$, such that $(A_k)_{i,*}$ is a nonzero row. So, $\alpha_i(L) = \mathrm{ord}\, L_{i,*}$.

The matrix $F \in \mathrm{Mat}_n(K)$ such that $F_{i,*} = (A_{\alpha_i(L)})_{i,*}$, $i = 1,\ldots,n$, is the *frontal matrix* of L.

The group of unimodular matrices from $\mathrm{Mat}_n(K[\partial])$ will be denoted by Υ_n.

We formulate a theorem which is a consequence of statements proven in [2,3] (the equivalence (iii) can also be proven using [23, Theorem III]).

Theorem 1. *Let $L \in Mat_n(K[\partial])$ be of full rank. In this case*

(i) If L' is the result of differentiating of a row of L then $\dim V_{L'} = \dim V_L + 1$.

(ii) If the frontal matrix of $L \in Mat_n(K[\partial])$ is invertible then

$$\dim V_L = \sum_{i=1}^{n} \alpha_i(L).$$

(iii) $L \in \Upsilon_n \Longleftrightarrow V_L = 0$.

We suppose in the sequel that the field K is constructive, in particular that there exists a procedure for recognizing whether a given element of K is equal to 0.

2.3 Algorithm EG (EG-Eliminations)

Given $L \in Mat_n(K[\partial])$ of full rank, algorithm EG ([2,4–6]) constructs an *embracing* operator $\widehat{L} \in Mat_n(K[\partial])$ such that

- ord $\widehat{L}$ = ord L,
- the leading matrix of $\widehat{L}$ is invertible,
- $\widehat{L} = QL$ for some $Q \in Mat_n(K[\partial])$, thus $V_L \subseteq V_{\widehat{L}}$.

If L is not of full rank then this algorithm reports this.

This algorithm is based on alternation of reductions and differentiations. First, explain how the reduction works. It is checked whether the rows of the leading matrix are linearly dependent over K. If they are, coefficients of the dependence $p_1, \ldots, p_n \in K$ are found. From the rows of L corresponding to nonzero coefficients, we select one. Let it be the ith row. This row is replaced by the linear combination of the rows of L with the coefficients $p_1, \ldots, p_n$. As a result, the ith row of the leading matrix vanishes. This operation is called *reduction*.

Let the ith row of the leading matrix be zero. Then we differentiate the ith row of the operator matrix, i.e., replace each entry $L_{ij} \in K[\partial]$ by the composition of ∂ and L_{ij}, $j = 1, \ldots, n$ (this operation is called *row differentiation*).

A version of algorithm EG is as follows.

If the rows of the leading matrix of L are linearly dependent over K then the reduction is performed. Suppose that this makes the ith row of the leading matrix zero. Then, we perform the differentiation of the ith row of the operator matrix and continue the process of alternated reductions and differentiations until the leading matrix becomes nonsingular, and, therefore, we get the output matrix $\widehat{L}$.

If at some moment a zero row appears in the operator matrix or the number of row differentiations becomes bigger than nd then L is not of full rank.

2.4 Algorithm RR

This algorithm is based originally on the algorithm FF [11]. A simplified version of FF is RowReduction [9]; for short, we will use the abbreviation RR in the sequel.

For a given $L \in Mat_n(K[\partial])$ of full rank, algorithm RR constructs $\check{L} \in Mat_n(K[\partial])$ such that

– ord $\check{L} \leqslant$ ord L,
– the frontal matrix of $\check{L}$ is invertible,
– $\check{L} = UL$ for some $U \in \Upsilon_n$, thus $V_L = V_{\check{L}}$.

A comment on the matrix U will be given later in this section. A version of algorithm RR is as follows.

Let the rows of the frontal matrix of L be linearly dependent over K and coefficients of the dependence be $p_1, \ldots, p_n \in K$. From the rows of L corresponding to nonzero coefficients, we select one having the highest order (if there are few such rows then take any of them). Let it be the ith row. Then we replace $L_{i,*}$ of the operator matrix by

$$\sum_{j=1}^{n} p_j \partial^{\alpha_i(L) - \alpha_j(L)} L_{j,*} \tag{3}$$

and continue this process until the frontal matrix becomes nonsingular, and, therefore, we get the output matrix $\check{L}$.

If at some moment, a zero row appears in the operator matrix then L is not of full rank.

Remark 1. *In addition, algorithm RR allows (if it is required) to construct such operator matrices $U_1, \ldots, U_l \in Mat_m(K[\partial])$, that $U_1, \ldots, U_l$ are unimodular and*

$$\check{L} = U_l \ldots U_1 L. \tag{4}$$

Each U_j is of the form

$$i : \begin{pmatrix} 1 & & & & & & & & \\ & \ddots & & & & & & & \\ & & 1 & & & & & & \\ p_1 \partial^{\alpha_i - \alpha_1} & \cdots & p_{i-1} \partial^{\alpha_i - \alpha_{i-1}} & p_i & p_{i+1} \partial^{\alpha_i - \alpha_{i+1}} & \cdots & p_n \partial^{\alpha_i - \alpha_n} \\ & & & & 1 & & & & \\ & & & & & \ddots & & \\ & & & & & & 1 \end{pmatrix} \tag{5}$$

with $1 \leqslant i \leqslant n$, $p_1, \ldots, p_n \in K$, $p_i \neq 0$ (this matrix corresponds to the replacement $L_{i,*}$ by (3)).

A matrix of the form (5) will be called *elementary*. Each an elementary matrix is unimodular: to obtain the inverse, one can replace in (5) its ith row by

$$\left(-\frac{p_1}{p_i} \partial^{\alpha_i - \alpha_1} \quad \cdots \quad -\frac{p_{i-1}}{p_i} \partial^{\alpha_i - \alpha_{i-1}} \quad \frac{1}{p_i} \quad -\frac{p_{i+1}}{p_i} \partial^{\alpha_i - \alpha_{i+1}} \quad \cdots \quad -\frac{p_n}{p_i} \partial^{\alpha_i - \alpha_n} \right).$$

The list

$$U_1, \ldots, U_l \tag{6}$$

of the elementary matrices involved into (4) can be constructed in the course of executing RR with no extra cost.

3 Differential and Full Complexities of Algorithms for Operator Matrices Transformation

3.1 Diversity of Algebraic Complexities

Besides the complexity as the number of arithmetic operations (the *arithmetic complexity*) one can consider the number of differentiations in the worst case (the *differential complexity*).

We will also discuss the *full complexity* as the total number of all operations (the differentiations are included) in the field K in the worst case. In [3,9], when the complexity of algorithms EG and RR was considered, the differentiations were ignored (the same concerning algorithm which we will denote by ExtRR and will discuss in Sect. 3.3). We will denote such a kind of complexity by $F_{xx}(n, d)$, where xx is the name of an algorithm under consideration, for example, xx $\in$ {RR, EG}. It is worthy to note that if we ignore the differentiations then the arithmetic complexity can be in wrong values, since differentiating a scalar differential operator requires to execute also arithmetic operations (additions). Therefore, $F_{xx}(n, d)$ is only a visible (apparent) complexity (F = "at First sight"). We will consider also the following functions of n, d:

- $B_{xx}(n, d)$ — the number of row differentiations in the worst case,
- $\tilde{T}_{xx}(n, d)$ — the number of differentiations of elements of K in the worst case (the differential complexity),
- $T_{xx}(n, d)$ — the number of all operations in K in the worst case (the full complexity).

Along with O-notation we use the Θ-notation which is very common in complexity theory ([21]). Recall that $f(n, d) = \Theta(g(n, d))$ is equivalent to

$$f(n, d) = O(g(n, d)) \ \& \ g(n, d) = O(f(n, d)).$$

If $f(n, d) = \Theta(g(n, d))$ then we call $\Theta(g(n, d))$ a *sharp* bound for $f(n, d)$.
If xx $\in$ {RR, EG} then

$$\tilde{T}_{xx}(n, d) = \Theta(B_{xx}(n, d)nd) \tag{7}$$

for the the differential complexity $\tilde{T}$. We have also

$$T_{xx}(n, d) = \Theta(F_{xx}(n, d) + B_{xx}(n, d)nd) \tag{8}$$

for the full complexity T, since the case of a big number of differentiations is concurrently the case when the number of arithmetic operations is big and vice versa.

Asymptotic relation similar to (8) holds for the arithmetic complexity since each differentiation of a row of L uses in the worst case besides nd of differentiations of elements of K also the same number of arithmetic operations (additions) in K.

Searching for coefficients $p_1, \ldots, p_n$ of a linear dependence for rows of a matrix from $\mathrm{Mat}_n(K)$ is equivalent to solving a homogeneous system of linear algebraic equations with coefficients in K. The complexity of solving such a system is $\Theta(n^\omega)$, where ω is the matrix multiplication exponent, $2 < \omega \leqslant 3$. We have $F_{\mathrm{EG}}(n, d) = \Theta(n^{\omega+1}d + n^3 d^2)$, $B_{\mathrm{EG}}(n, d) = \Theta(nd)$.

Proposition 1. *The differential and the full complexity of EG allow the asymptotic estimates*

$$\tilde{T}_{EG}(n, d) = \Theta(n^2 d^2), \quad T_{EG}(n, d) = \Theta(n^{\omega+1}d + n^3 d^2). \tag{9}$$

Proof. We have mentioned that by Theorem 1(i, ii) each differentiation of a row increases the dimension of the solutions space of an operator matrix by 1, and in our case that dimension does not exceed nd. This implies the first estimate from (9).

Each reduction step uses in the worst case $\Theta(n^\omega + n^2 d)$ arithmetic operations. The number of such steps is nd in the worst case. Together with the first estimate from (9) this gives the second estimate. (The differentiation operations do not affect significantly the full complexity of EG.)

Proposition 2. *The differential and the full complexity of RR admit the asymptotic estimates*

$$\tilde{T}_{RR}(n, d) = \Theta(n^3 d^3), \quad T_{RR}(n, d) = \Theta(n^{\omega+1}d + n^3 d^3). \tag{10}$$

Proof. We have $F_{\mathrm{RR}}(n, d) = \Theta(n^{\omega+1}d + n^3 d^2)$ and $B_{\mathrm{RR}}(n, d) = \Theta(n^2 d^2)$. This implies the claim.

We see that $B_{\mathrm{RR}}(n, d)$ grows faster than $B_{\mathrm{EG}}(n, d)$. In the case of RR, the differentiating operations increase the full complexity.

3.2 Algorithms $\triangle$EG and $\triangle$RR

Let the ith row r of the frontal matrix of $L \in \mathrm{Mat}_n(K)$ have the form

$$\underbrace{(0, \ldots, 0}_{k-1}, a, \ldots, b),$$

$1 \leqslant k \leqslant n$, $a \neq 0$. Then k is the *pin index* of the ith row of L.

If all rows of L have distinct pin indices then the frontal matrix of L is nonsingular. Suppose that two rows r_1, r_2 of L have the same pin index k. Set $d_1 = \mathrm{ord}\, r_1$, $d_2 = \mathrm{ord}\, r_2$. Let $d_1 \leqslant d_2$. There exists a v in K such that the difference

$$r_2 - v \partial^{d_2 - d_1} r_1 \tag{11}$$

either has the pin index which is bigger than k or has the order which is less than d_2. This can be used[1] instead of a search for a linear dependency of the

---

[1] For the difference case, this was used in first versions [1] of algorithm EG. In a discussion related to the differential case, A. Storjohann drew the author's attention to the fact that the complexity of this approach is less than of one which uses solving of linear algebraic systems (see also [24]).

rows of the frontal matrix of L (in the case of EG, on key moments, the leading and the frontal matrices coincide, and $d_2 - d_1 = 0$ in (11)). If L is of full rank then the frontal matrix after the transformation is in triangular form.

This leads to modified versions of EG and RR. We will denote them as $\triangle$EG and, resp., $\triangle$RR.

Proposition 3. *The differential and the full complexities of $\triangle EG$ and $\triangle RR$ admit the asymptotic estimates*

$$\tilde{T}_{\triangle EG}(n, d) = \Theta(n^2 d^2), \quad T_{\triangle EG}(n, d) = \Theta(n^3 d^2) \tag{12}$$

and

$$\tilde{T}_{\triangle RR}(n, d) = \Theta(n^3 d^3), \quad T_{\triangle RR}(n, d) = \Theta(n^3 d^3). \tag{13}$$

Proof. The replacement of r_2 by (11) is a unimodular operation on L. This operation has the complexity $\Theta(nd)$. A row can have its pin index increased at most n times before the order of the row is decreased. Thus, $F_{xx}(n, d) = \Theta(nd \cdot n \cdot nd) = \Theta(n^3 d^2)$ for $xx \in \{\triangle EG, \triangle RR\}$. Concerning the differential complexity, for $\triangle$EG and, resp. $\triangle$RR it is the same as for EG and RR.

We emphasize that the difference between two estimates $T_{\triangle RR} = \Theta(n^3 d^3)$ and $T_{\triangle EG}(n, d) = \Theta(n^3 d^2)$ is due to the differential component: if we ignored the operation of differentiation, then we would have the estimate $\Theta(n^3 d^2)$ for both complexities.

3.3   Extended RR: Computing U Along with $\check{L}$

To compute along with $\check{L}$ the multiplier U such that $\check{L} = UL$, one can apply the following algorithm (we call it ExtRR) to L:

Apply RR to L (this gives $\check{L}$), and repeat in parallel all the operations for the matrix which is originally equal to the identity matrix I_n (this gives U).

The algorithm was presented in [9, Sect. 4]. Evidently, $B_{\text{ExtRR}}(n, d) = 2B_{\text{RR}}(n, d)$. The following proposition is useful for estimating $F_{\text{ExtRR}}(n, d)$, $\tilde{T}_{\text{ExtRR}}(n, d)$ and $T_{\text{ExtRR}}(n, d)$:

Proposition 4. *Let algorithm RR compute step-by-step the unimodular matrices of the form (5) for $L \in Mat_n(K[\partial])$, $\text{ord } L = d$ (see Remark 1). Then $\text{ord}(U_k \dots U_1) = O(nd)$ for all $k = 1, \dots, l$.*

Proof. It follows from [9, Proposition 1].

Proposition 4 and estimates (10) imply the estimates $F_{\text{ExtRR}}(n, d) = O(n^{\omega+1} d + n^4 d^2) = O(n^4 d^2)$, and

$$\tilde{T}_{\text{ExtRR}}(n, d) = O(n^4 d^3), \quad T_{\text{ExtRR}}(n, d) = O(n^4 d^2 + n^4 d^3) = O(n^4 d^3). \tag{14}$$

These estimates can be sharpened.

Proposition 5. *Let L be unimodular. Then*

$$\operatorname{ord} L^{-1} \leqslant (n-1)d \tag{15}$$

and (15) is the tight bound: for all integer n, d such that $d \geqslant 0$, $n \geqslant 2$ there exists $L \in Mat_n(K[\partial])$ such that $\operatorname{ord} L^{-1} = (n-1)d$.

Proof. The bound (15) follows from some estimates related to computing the Hermite form given in [18, Theorem 5.5]. (The Hermite form of a unimodular matrix is the identity, the transformation matrix U is the inverse.) Let us prove that this bound is tight. Indeed, the $n \times n$ operator matrix of order d

$$\begin{pmatrix}
1 & \partial^d & 0 & 0 & \ldots & \ldots & 0 & 0 \\
0 & 1 & \partial^d & 0 & \ldots & \ldots & 0 & 0 \\
 & & \ddots & & & & & \\
 & & & \ddots & & & & \\
 & & & & \ddots & & & \\
0 & 0 & 0 & 0 & \ldots & \ldots & 1 & \partial^d \\
0 & 0 & 0 & 0 & \ldots & \ldots & 0 & 1
\end{pmatrix} \tag{16}$$

has the inverse of order $(n-1)d$:

$$\begin{pmatrix}
1 & -\partial^d & \partial^{2d} & -\partial^{3d} & \ldots & (-1)^{n-2}\partial^{(n-2)d} & (-1)^{n-1}\partial^{(n-1)d} \\
0 & 1 & -\partial^d & \partial^{2d} & \ldots & (-1)^{n-3}\partial^{(n-3)d} & (-1)^{n-2}\partial^{(n-2)d} \\
 & & \ddots & & & & \\
 & & & \ddots & & & \\
 & & & & \ddots & & \\
0 & 0 & 0 & 0 & \ldots & 1 & -\partial^d \\
0 & 0 & 0 & 0 & \ldots & 0 & 1
\end{pmatrix}. \tag{17}$$

If algorithm ExtRR is applied to matrix (16) then this yields two matrices U and I_n where U is equal to (17). Note in addition to (14) that applying RR to matrix (16) yields $U_1, \ldots, U_{n-1}$ (see (4), where $l = n-1$ in this case) such that $\operatorname{ord}(U_k \ldots U_1) = kd$, $k = 1, \ldots, n-1$. Since $\sum_{k=1}^{n-1} k^3 d^3 = \Theta(n^4 d^3)$, we obtain

$$\tilde{T}_{\text{ExtRR}}(n,d) = \Theta(n^4 d^3), \quad T_{\text{ExtRR}}(n,d) = \Theta(n^4 d^2 + n^3 d^3). \tag{18}$$

Remark 2. *As a consequence of (15) we obtain that if $L \in Mat_2(K[\partial])$ is unimodular then $\operatorname{ord} L^{-1} = \operatorname{ord} L$. (Miyake's example (2) illustrates this.)*

Note finally that to show the correctness of using $\triangle$RR instead of RR for ExtRR we have to prove an analog of the statement of Proposition 4. It is not clear whether such a statement holds. In addition, the replacement of RR by $\triangle$RR does not improve estimate (18) due to the term $n^4 d^3$, which replaces $n^3 d^3$ in $T_{\triangle \text{RR}}(n,d)$ since the number of elements in a row is now $(n-1)nd$.

3.4   When Differentiated Rows are Stored

One can store all the results of row differentiations. In this case, some upper estimates for the total number of differentiations can easily be obtained.

Proposition 6. *The number of row differentiations without repetitions (when the result of each differentiation is stored, i.e., when we collect all such results) executed by algorithms RR and $\triangle$RR is $O(nd^2)$ and, resp., $O(n^2d^2)$ in the worst case; as a consequence, the number of differentiations of elements of K is $O(n^2d^3)$ and, resp., $O(n^3d^3)$.*

Proof. Let a row r be changed in the course of RR or $\triangle$RR performance. Let the order of r after the changing be $d_0 < d$. In this case, one can compute and store $d - d_0$ rows

$$\partial r, \quad \partial(\partial r), \quad \ldots, \quad \underbrace{\partial(\partial \ldots (\partial r) \ldots)}_{d - d_0 \text{ differentiations}} .$$

After this, when a row of the form $\partial^m r$, $m \leqslant d - d_0$ is needed for following eliminations, pick the needed row from the collection of the stored rows. Thus, we get modified versions of RR and $\triangle$RR whose numbers of row differentiations are not less than the analogous numbers for the original versions. It is easy to see that for the modified version of RR, this number is not bigger than nd^2, and not bigger than n^2d^2 for $\triangle$RR. The claim follows.

However, the estimates $O(n^2d^3)$ and $O(n^3d^3)$ for the number of differentiations do not allow to decrease the exponent of d in (10), (13). Based on Proposition 6, we cannot draw the conclusion that the storage of the results of all the differentiations decreases significantly the full complexity of RR and $\triangle$RR. Similarly, using the upper bound $O(n^3d^3)$ we cannot decrease the exponent of d in (18). But the space complexity will go up when we store all the results of differentiating.

Remark 3. *It is not clear whether, say, the upper bound $O(nd^2)$ for the number of row differentiations by RR is sharp. If for this number, the estimate $O(nd)$ holds then we would have the estimate $O(n^4d^2)$ for the full complexity of the version of ExtRR such that if a row r is differentiated m times then the rows $\partial r, \ldots, \partial^m r$ are stored for potential later uses.*

4   Two Computational Problems

4.1   Singularities of Systems

If, for example, $K = \mathbb{Q}(x)$, $\partial = \frac{d}{dx}$ and we are interested in singular points of solutions of a system $L(y) = 0$, $L \in \mathrm{Mat}_n(K)$, $\mathrm{ord}\, L = d$, then by each of algorithms EG, RR, $\triangle$EG, DRR we can find a polynomial whose roots form a finite superset of the set of such points [6–8]. The basic idea is that if the leading matrix of L is invertible in $\mathrm{Mat}_n(K)$ then we can take the (square-free factorized) determinant of the leading matrix.

Similarly, we can use the frontal matrix, if it is invertible. The fact is that if $\alpha_1, \ldots, \alpha_n$ are the row orders of $\check{L}$, $d = \operatorname{ord} \check{L} = \max\{\alpha_1, \ldots, \alpha_n\}$ and $D = \operatorname{diag}(\partial^{d-\alpha_1}, \ldots, \partial^{d-\alpha_n})$ then the leading matrix of $D\check{L}$ coincides with the frontal matrix of $\check{L}$ (we do not need to compute $D\check{L}$).

Therefore, for example, algorithms $\triangle$EG, $\triangle$RR can be used to compute the desirable polynomial. The distinction between complexities (12) and (13) shows that at least when n and d are large enough algorithm $\triangle$EG is probably better. The full complexity (in the meaning of this paper) is $\Theta(n^3 d^2)$.

4.2 The Unimodularity Testing

Applying algorithm ExtRR to L we transform L into $\check{L}$. Theorem 1(ii, iii) implies that L is unimodular if and only if $\check{L}$ is invertible in $\operatorname{Mat}_m(K)$. In this case, $(\check{L})^{-1}UL = I_n$, where U is unimodular. Therefore, $(\check{L})^{-1}U$ is the inverse for L.

The matrices U and $\check{L}$ can be constructed by ExtRR. By (14) the full complexity of the computation of the inverse is

$$\Theta(n^4 d^3). \tag{19}$$

The multiplication of $\check{L}^{-1}$ and U does not change this estimate (recall that $\operatorname{ord} \check{L}^{-1} \leqslant (n-1)d$).

In the case when we want only to test whether L is unimodular without constructing L^{-1}, then we can use $\triangle$EG; as we have mentioned, L is unimodular if and only if the number of differentiations has to be exactly equal to nd, and the leading matrix is invertible in K after those differentiations. Such testing has the complexity

$$\Theta(n^3 d^2). \tag{20}$$

The formulated computational problems related to unimodularity can be solved by different algorithms. For example, algorithms to construct the Jacobson and Hermite forms of a given operator matrix can be used. A polynomial-time deterministic algorithm for constructing the Jacobson form of L was proposed in [22]. Its complexity is considered in [22] as a function of three variables, and two of them are our n, d (in [22], another notation is used). The value of the third variable is in the worst case nd, and for the complexity as a function of the variables n, d one can derive the estimate $\Theta(n^9 d^9)$. As we have mentioned in the proof of Proposition 5, the Hermite form of a unimodular matrix is the identity, the transformation matrix U is the inverse. The complexity estimate for the algorithm given in [18, Theorem 5.5] is $O(n^7 d^3 \log(nd))$ (in our notation). It looks like this estimate is tight. (Of course, the algorithms from [18,22] solve more general problems, and the algorithm given above in this section has some advantages only for recognizing invertibility of an operator matrix and computing the inverse matrix.)

The author is unaware of the algorithms which solve the testing unimodularity problem with a complexity which is less than (20). Search in the literature gave no positive result, but of course it is possible that such algorithms exist. The author makes no attempt to offer a champion algorithm for solving this problem. Perhaps, for example, using the ideas of the fast matrix multiplication over a field [16, 27, 28], as well as the fast multiplication algorithm for scalar differential operators [12, 13, 20], one can propose an algorithm for fast multiplication operator matrices and then get out of it the appropriate algorithm for solving the unimodularity testing problem.

5   Conclusions

The author's goal is to show that to choose an algorithm to solve a problem over a differential field K it is reasonable to take into account the complexity measured as not only the number of arithmetic operation in K but all operations including the operation of differentiation. The algorithms that have the equivalent complexity as the number of arithmetic operations in the worst case can differ in the context of the full algebraic complexity that includes needed differentiations.

It is worthy to note that some versions of algorithms EG and RR are used quite successfully, for example, for finding singular points of differential systems, and we mentioned this in Sect. 4.1. We can expect that by these algorithms, the unimodularity testing will be performed in practice in a reasonable time.

From the current work, new questions arise.

First, the question formulated in Sect. 3.4: suppose that we store all the results of differentiations; does it allow to decrease the complexities (13), (14), (19) (it would be desirable to get d^2 instead of d^3)?

Second. It is unclear whether there exists an algorithm for unimodularity testing whose complexity is $O(n^\alpha d^\beta)$, where α, β are real numbers and $\alpha < 3$. For matrices whose entries are commutative polynomials from $K[x]$, there is an algorithm [17] for constructing the inverse matrix with complexity $O(n^3 \rho)$, where ρ is the maximal degree of entries of given matrices (strictly speaking, algorithm from [17] is for the case of "generic matrix inversion" only). It is unclear whether the problem of constructing the inverse operator matrix is reducible to the problem of the operator matrix multiplication, similarly to the case when entries of matrices belong to a field [15, Sect. 16.4]. Going back to matrix with polynomial entries, note that there exists a matrix multiplication algorithm [14] with complexity $O(n^\omega \rho f(\log n \log \rho))$, where f is a polynomial. However, an algorithm with a similar complexity for the matrix inversion does not probably exists. It looks like that the problem of constructing the inverse matrix is not reducible to the problem of the operator matrix multiplication neither for polynomial matrices nor for operator matrices.

Third, much recent work has focused on properly dealing with the growth in the size of coefficients from K, for example, when $K = \mathbb{Q}(x)$ ([11, 18] etc.). It would be valuable to investigate the bit complexity of the unimodularity testing algorithms. Another way is to consider the complexity as a function of three variables: n, d and ρ, where ρ is such that all the polynomials involved into L as numerators and denominators of coefficients of entries of L are of degree which is ρ at most. The complexity itself is then the number of operations in $\mathbb{Q}$ in the worst case. The algorithms should allow control over coefficients growth. It is reasonable to continue to investigate this line of enquiry.

Acknowledgments. The author is thankful to M. Barkatou and A. Storjohann for interesting discussions, and to anonymous referees for useful comments.

References

1. Abramov, S.: EG-eliminations. J. Differ. Equ. Appl. **5**, 393–433 (1999)
2. Abramov, S., Barkatou, M.: On the dimension of solution spaces of full rank linear differential systems. In: Gerdt, V.P., Koepf, W., Mayr, E.W., Vorozhtsov, E.V. (eds.) CASC 2013. LNCS, vol. 8136, pp. 1–9. Springer, Heidelberg (2013)
3. Abramov, S., Barkatou, M.: On solution spaces of products of linear differential or difference operators. ACM Comm. Comput. Algebra **4**, 155–165 (2014)
4. Abramov, S., Bronstein, M.: On solutions of linear functional systems. In: ISSAC 2001 Proceedings, pp. 1–6 (2001)
5. Abramov, S., Bronstein, M.: Linear algebra for skew-polynomial matrices. Rapport de Recherche INRIA RR-4420, March 2002. http://www.inria.fr/RRRT/RR-4420.html
6. Abramov, S., Khmelnov, D.: On singular points of solutions of linear differential systems with polynomial coefficients. J. Math. Sci. **185**(3), 347–359 (2012). (Translated from: Fundamentalnaya i prikladnaya matematika **17**(1), 3–21 (2011/2012))
7. Abramov, S., Khmelnov, D., Ryabenko, A.: Procedures for searching local solutions of linear differential systems with infinite power series in the role of coefficients. Program. Comput. Softw. **42**(2), 55–64 (2016)
8. Barkatou, M., Cluzeau, T., El Bacha, C.: Simple form of higher-order linear differential systems and their application in computing regular solutions. J. Symb. Comput. **46**(6), 633–658 (2011)
9. Barkatou, M., El Bacha, C., Labahn, G., Pflügel, E.: On simultaneous row and column reduction of higher-order linear differential systems. J. Symb. Comput. **49**(1), 45–64 (2013)
10. Beckermann, B., Labahn, G.: Fraction-free computation of matrix rational interpolants and matrix GCDs. SIAM J. Matrix Anal. Appl. **77**(1), 114–144 (2000)
11. Beckermann, B., Cheng, H., Labahn, G.: Fraction-free row reduction of matrices of Ore polynomials. J. Symb. Comput. **41**, 513–543 (2006)
12. Benoit, A., Bostan, A., van der Hoeven, J.: Quasi-optimal multiplication of linear differential operators. In: FOCS 2012 Proceedings, pp. 524–530 (2012)
13. Bostan, A., Chyzak, F., Le Roix, N.: Products of ordinary differential operators by evaluating and interpolation. In: ISSAC 2008 Proceedings, pp. 23–30 (2008)
14. Bostan, A., Schost, E.: Polynomial evaluation and interpolation on special sets of points. J. Complex. **21**(4), 420–446 (2005)
15. Bürgisser, P., Clausen, M., Shokrollahi, M.A.: Algebraic Complexity Theory. Grundlehren der mathematischen Wissenschaften, vol. 315. Springer, Heidelberg (1997)
16. Coppersmith, D., Winograd, S.: Matrix multiplication via arithmetic progressions. J. Symb. Comput. **9**(3), 251–280 (1990)
17. Jeannerod, C.-P., Villard, G.: Essentially optimal computation of the inverse of generic polynomial matrices. J. Complex. **21**(1), 72–86 (2005)

18. Giesbrecht, M., Sub Kim, M.: Computation of the Hermite form of a matrix of Ore polynomials. J. Algebra **376**, 341–362 (2013)
19. Gustavson, R., Kondratieva, M., Ovchinnikov, A.: New effective differential Nullstellensatz. Adv. Math. **290**, 1138–1158 (2016)
20. van der Hoeven, J.: FFT-like multiplication of linear differential operators. J. Symb. Comput. **33**, 123–127 (2002)
21. Knuth, D.E.: Big omicron and big omega and big theta. ACM SIGACT News **8**(2), 18–23 (1976)
22. Middeke, J.: A polynomial-time algorithm for the Jacobson form for matrices of differential operators. Technical report No. 08–13 in RISC Report Series (2008)
23. Miyake, M.: Remarks on the formulation of the Cauchy problem for general system of ordinary differential equations. Tôhoku Math. J. **32**(2), 79–89 (1980)
24. Mulders, T., Storjohann, A.: On lattice reduction for polynomial matrices. J. Symb. Comput. **37**(4), 485–510 (2004)
25. van der Put, M., Singer, M.F.: Galois Theory of Linear Differential Equations. Grundlehren der mathematischen Wissenschaften, vol. 328. Springer, Heidelberg (2003)
26. Rosenlicht, M.: Integration in finite terms. Amer. Math. Mon. **79**(9), 963–972 (1972)
27. Strassen, V.: Gaussian elimination is not optimal. Numer. Math. **13**, 354–356 (1969)
28. Vassilevska Williams, V.: Multiplying matrices faster than Coppersmith-Winograd. In: STOC 2012 Proceedings, pp. 887–898 (2012)

Resolving Decompositions for Polynomial Modules

Mario Albert[(✉)] and Werner M. Seiler

Institut für Mathematik, Universität Kassel, 34132 Kassel, Germany
{albert,seiler}@mathematik.uni-kassel.de

Abstract. We introduce the novel concept of a resolving decomposition of a polynomial module as a combinatorial structure that allows for the effective construction of free resolutions. It provides a unifying framework for recent results of the authors for different types of bases.

1 Introduction

The determination of free resolutions for polynomial modules is a fundamental task in computational commutative algebra and algebraic geometry. Free resolutions are needed for derived functors like Ext and Tor and many important homological invariants like the projective dimension or the Castelnuovo-Mumford regularity are defined via the minimal resolution. Furthermore, the Betti numbers contain much geometric and topological information.

Unfortunately, it is rather expensive to compute a resolution. As a rough rule of thumb one may say that computing a resolution of length ℓ corresponds to computing ℓ Gröbner bases. In many cases one needs only partial information about the resolution like the Betti numbers simply measuring its size. However, all classical algorithms require to determine always a full resolution.

In [1] we provided a novel approach by combining the theory of Pommaret bases and algebraic discrete Morse theory (together with an implementation in the CoCoALib). It allows for the first time to determine Betti numbers—even individual ones—without computing a full resolution and thus is for most problems much faster than classical approaches (see [1,3] for detailed benchmarks). Furthermore, it scales much better and can be easily parallelised.

Because of these good properties it is of great interest to generalise this approach to other situations. In [3], we extended it to Janet bases. While the proofs remained essentially the same, the use of another involutive division required the adaption of a number of technical points. Currently we are working on extensions to some alternative bases which do not necessarily come from an involutive division but provide similar combinatorial decompositions. Again this would require a number of minor modifications of the same proofs. In a different line of work [2], we introduced recently modules marked on quasi-stable submodules. Again one obtains combinatorial decompositions based on multiplicative variables defined by the Pommaret division, but this time no term order is used. Nevertheless, we could show that many results of [1] remain true.

© Springer International Publishing AG 2016
V.P. Gerdt et al. (Eds.): CASC 2016, LNCS 9890, pp. 15–29, 2016.
DOI: 10.1007/978-3-319-45641-6_2

The main objective of the current paper is the development of an axiomatic framework that unifies all the above works. We introduce the novel concept of a *resolving decomposition* which is defined via several direct sum decompositions. It implies in particular the existence of standard representations and normal forms. We then show that such a decomposition allows for the explicit determination of a free resolution and of Betti numbers.

The point of such a unification is *not* that it leads to any new algorithms. Indeed, we will not present a general algorithm for the construction of resolving decomposition. Instead one should see our results as a "meta-machinery" which given any concept of a basis that induces a resolving decomposition delivers automatically an effective syzygy theory for this kind of basis. For the concrete case of the resolving decompositions induced by Janet or Pommaret bases, an implementation of this effective theory is described (together with benchmarks) in [1,3]. For other types of underlying bases only fairly trivial modifications of this implementation would be required.

2   Resolving Decompositions

Let $\Bbbk$ be an algebraic closed field and $\mathcal{P} = \Bbbk[\mathbf{x}]$ with $\mathbf{x} = (x_0, \ldots, x_n)$. Let $\mathcal{P}_\mathbf{d}^m = \bigoplus_{i=1}^m \mathcal{P}(-d_i)\mathbf{e}_i$ be a finitely generated free $\mathcal{P}$-module with grading $\mathbf{d} = (d_1, \ldots, d_m)$ and free generators $\mathbf{e}_1, \ldots, \mathbf{e}_m$. A module $U \subseteq \mathcal{P}_\mathbf{d}^m$ is called *monomial module* if it is of the form $\oplus_{k=1}^m J^{(k)}\mathbf{e}_k$ with $J^{(k)}$ is a monomial ideal in $\mathcal{P}$. A *module term (with index i)* is a term of the form $x^\mu \mathbf{e}_i$. For a monomial ideal $J \subseteq \mathcal{P}$ we define $\mathcal{N}(J) \subseteq \mathbb{T}$ as the set of terms in $\mathbb{T}$ not belonging to J. For a monomial module U we define $\mathcal{N}(U) = \bigcup_{k=1}^m \mathcal{N}(J^{(k)})\mathbf{e}_k$. For an element $\mathbf{f} \in \mathcal{P}_\mathbf{d}^m$ we define $\operatorname{supp}(\mathbf{f})$ as the set of module terms appearing in $\mathbf{f}$ with a non-zero coefficient: $\mathbf{f} = \sum_{x^\alpha \mathbf{e}_{i_\alpha} \in \operatorname{supp}(\mathbf{f})} c_\alpha x^\alpha \mathbf{e}_{i_\alpha}$. If $\mathcal{B}$ is a set of homogeneous elements of degree s in $\mathcal{P}_\mathbf{d}^m$, we write $\langle \mathcal{B} \rangle$ for the $\Bbbk$-vector space generated by $\mathcal{B}$ in $(\mathcal{P}_\mathbf{d}^m)_s$. For a module $U \subseteq \mathcal{P}_\mathbf{d}^m$ we denote by $\operatorname{pd}(U)$ the *projective dimension* and by $\operatorname{reg}(U)$ the *(Castelnuovo-Mumford) regularity* of U.

Let $U \subseteq \mathcal{P}_\mathbf{d}^m$ be a finitely generated graded submodule and $\mathcal{B} = \{\mathbf{h}_1, \ldots, \mathbf{h}_s\}$ a finite homogeneous generating set of U. For every $\mathbf{h}_i \in \mathcal{B}$ we choose a term $x^{\mu_i}\mathbf{e}_{k_i} \in \operatorname{supp} \mathbf{h}_i$ denoted by $\operatorname{hm} \mathbf{h}_i$ and call it *head module term*. In addition to that we define the *head module terms of* $\mathcal{B}$, $\operatorname{hm}(\mathcal{B}) := \{\operatorname{hm}(\mathbf{h}) \mid \mathbf{h} \in \mathcal{B}\}$ and the *head module of* U, $\operatorname{hm}(U) = \langle \operatorname{hm} \mathcal{B} \rangle$. Note that $\operatorname{hm}(U)$ depends on the choice of $\mathcal{B}$ and on the choice of the head module terms in $\mathcal{B}$.

Definition 1. *We define a* resolving decomposition *of the submodule* U *as a quadruple* $(\mathcal{B}, \operatorname{hm}(\mathcal{B}), X_\mathcal{B}, \prec_\mathcal{B})$ *with the following five properties:*

(i) $U = \langle \mathcal{B} \rangle$.

(ii) *Let* $\mathbf{h} \in \mathcal{B}$ *be an arbitrary generator. Then, for every module term* $x^\mu \mathbf{e}_k \in \operatorname{supp}(\mathbf{h}) \setminus \{\operatorname{hm}(\mathbf{h})\}$, *we have* $x^\mu \mathbf{e}_k \notin \operatorname{hm}(U)$.

(iii) *We assign a set of* multiplicative variables $X_\mathcal{B}(\mathbf{h}) \subseteq \mathbf{x}$ *to every head module term* $\operatorname{hm}(\mathbf{h})$ *with* $\mathbf{h} \in \mathcal{B}$ *such that we have direct sum decompositions of both the head module*

$$\mathrm{hm}(U) = \bigoplus_{\mathbf{h} \in \mathcal{B}} \Bbbk[X_{\mathcal{B}}(\mathbf{h})] \cdot \mathrm{hm}(\mathbf{h}) \tag{1}$$

and of the module itself

$$U \stackrel{\prime}{=} \bigoplus_{\mathbf{h} \in \mathcal{B}} \Bbbk[X_{\mathcal{B}}(\mathbf{h})] \cdot \mathbf{h} . \tag{2}$$

(iv) $(\mathcal{P}_{\mathbf{d}}^m)_r = U_r \oplus \langle \mathcal{N}(\mathrm{hm}(U))_r \rangle$ for all $r \geq 0$.

(v) Let $\{\mathbf{f}_1, \dots, \mathbf{f}_s\}$ denote the standard basis of the free module $\mathcal{P}^s$. Given an arbitrary term $x^\delta \in \mathbb{T}$ and an arbitrary generator $\mathbf{h}_\alpha \in \mathcal{B}$, we find for every term $x^\epsilon \mathbf{e}_i \in \mathrm{supp}(x^\delta \mathbf{h}_\alpha) \cap \mathrm{hm}(U)$ a unique $\mathbf{h}_\beta \in \mathrm{hm}(\mathcal{B})$ such that $x^\epsilon \mathbf{e}_i = x^{\delta'} \mathrm{hm}(\mathbf{h}_\beta)$ with $x^{\delta'} \in \Bbbk[X_{\mathcal{B}}(\mathbf{h}_\beta)]$ by (iii). Then the term order $\prec_{\mathcal{B}}$ on $\mathcal{P}^s$ must satisfy $x^\delta \mathbf{f}_\alpha \succeq_{\mathcal{B}} x^{\delta'} \mathbf{f}_\beta$.

In the sequel, we will always assume that $(\mathcal{B}, \mathrm{hm}(\mathcal{B}), X_{\mathcal{B}}, \prec_{\mathcal{B}})$ is a resolving decomposition of the finitely generated module $U = \langle \mathcal{B} \rangle \subseteq \mathcal{P}_{\mathbf{d}}^m$. In addition to the multiplicative variables, we define for $\mathbf{h} \in \mathcal{B}$ the *non-multiplicative variables* as $\overline{X}_{\mathcal{B}}(\mathbf{h}) = \{x_0, \dots, x_n\} \setminus X_{\mathcal{B}}(\mathbf{h})$.

Remark 1. Resolving decompositions may be considered as a refinement of Stanley decompositions. Indeed, (1) gives us a Stanley decomposition of the head module of U and (2) of U itself. Consequently, it is easy to compute the Hilbert functions of $\mathrm{hm}(U)$ and of U, respectively. Because of the identical structure of the two decompositions, these two Hilbert functions are trivially the same (which may be considered as a term order free version of Macaulay's theorem in the theory of Gröbner bases). In addition, (iii) gives us for every $\mathbf{f} \in U$ a unique *standard representation*

$$\mathbf{f} = \sum_{\alpha=1}^{s} P_\alpha \mathbf{h}_\alpha$$

with $P_\alpha \in \Bbbk[X_{\mathcal{B}}(\mathbf{h}_\alpha)]$. Condition (iv) implies the existence of unique *normal forms* for all homogeneous elements $\mathbf{f} \in \mathcal{P}^m$. Due to this condition, we find unique $P_\alpha \in \Bbbk[X_{\mathcal{B}}(\mathbf{h}_\alpha)]$ for every $\mathbf{h}_\alpha \in \mathcal{B}$ such that $\mathbf{f}' = \mathbf{f} - \sum_{\alpha=1}^{s} P_\alpha \mathbf{h}_\alpha$ and $\mathbf{f}' \in \langle \mathcal{N}(\mathrm{hm}(U)) \rangle$. Another important consequence of the definition of a resolving decomposition is that (1) implies that every generators in $\mathcal{B}$ has a different head module term.

While for the purposes of this work the mere existence of normal forms is sufficient, we note that (v) implies that they can be effectively computed. The choice of head terms and multiplicative variables in a resolving decomposition induces a natural reduction relation. If $\mathbf{f} \in \mathcal{P}_{\mathbf{d}}^m$ contains a term $x^\epsilon \mathbf{e}_i \in \mathrm{hm}(U)$, then there exists a unique generator $\mathbf{h} \in \mathrm{hm}(\mathcal{B})$ such that $x^\epsilon \mathbf{e}_i = x^\delta \mathrm{hm}(\mathbf{h})$ with $x^\delta \in \Bbbk[X_{\mathcal{B}}(\mathbf{h})]$ and we have a possible reduction $\mathbf{f} \xrightarrow{\mathcal{B}} \mathbf{f} - c x^\delta \mathbf{h}$ for a suitably chosen coefficient $c \in \Bbbk$.

Lemma 1. *For any resolving decomposition* $(\mathcal{B}, \mathrm{hm}(\mathcal{B}), X_{\mathcal{B}}, \prec_{\mathcal{B}})$ *the transitive closure* $\xrightarrow{\;\mathcal{B}\;}{}^*$ *of* $\xrightarrow{\;\mathcal{B}\;}$ *is Noetherian and confluent.*

Proof. It is sufficient to prove that for every term $x^\gamma \mathbf{e}_k$ in $\mathrm{hm}(U)$, there is a unique $g \in \mathcal{P}_d^m$ such that $x^\gamma \mathbf{e}_k \xrightarrow{\;\mathcal{B}\;}{}^* g$ and $g \in \langle \mathcal{N}(\mathrm{hm}(U)) \rangle$.

Since $x^\gamma \mathbf{e}_k \in \mathrm{hm}(U)$, there exists a unique $x^\delta \mathbf{h}_\alpha \in U$ such that $x^\delta \mathrm{hm}(\mathbf{h}_\alpha) = x^\gamma \mathbf{e}_k$ and $x^\delta \in X_{\mathcal{B}}(\mathbf{h}_\alpha)$. Hence, $x^\gamma \mathbf{e}_k \xrightarrow{\;\mathcal{B}\;} x^\gamma \mathbf{e}_k - c x^\delta \mathbf{h}_\alpha$ for a suitably chosen coefficient $c \in \Bbbk$. Denoting again the standard basis of $\mathcal{P}^s$ by $\{\mathbf{f}_1, \ldots, \mathbf{f}_s\}$, we associated the term $x^\delta \mathbf{f}_\alpha$ with this reduction step. If we could proceed infinitely with further reduction steps, then the reduction process would induce a sequence of terms in $\mathcal{P}^s$ containing an infinite chain which, by condition (v) of Definition 1, is strictly descending for $\prec_{\mathcal{B}}$. But this is impossible, since $\prec_{\mathcal{B}}$ is a well-ordering. Hence $\xrightarrow{\;\mathcal{B}\;}{}^*$ is Noetherian. Confluence is immediate by the uniqueness of the element that is used at each reduction step. □

Furthermore, every resolving decomposition $(\mathcal{B}, \mathrm{hm}(\mathcal{B}), X_{\mathcal{B}}, \prec_{\mathcal{B}})$ induces naturally a directed graph. Its vertices are given by the elements in $\mathcal{B}$. If $x_j \in \overline{X}_{\mathcal{B}}(\mathbf{h})$ for some $\mathbf{h} \in \mathcal{B}$, then, by definition, $\mathcal{B}$ contains a unique generator $\mathbf{h}'$ such that $x_j \, \mathrm{hm} \, \mathbf{h} = x^\mu \, \mathrm{hm} \, \mathbf{h}'$ with $x^\mu \in \Bbbk[X_{\mathcal{B}}(\mathbf{h}')]$. In this case we include a directed edge from $\mathbf{h}$ to $\mathbf{h}'$. We call the thus defined graph the $\mathcal{B}$-graph.

Lemma 2. *The $\mathcal{B}$-graph of a resolving decomposition* $(\mathcal{B}, \mathrm{hm}(\mathcal{B}), X_{\mathcal{B}}, \prec_{\mathcal{B}})$ *is always acyclic.*

Proof. Assume the $\mathcal{B}$-graph was cyclic. Then we find generators $\mathbf{h}_{k_1}, \ldots, \mathbf{h}_{k_t} \in \mathcal{B}$, which are pairwise distinct, variables $x_{i_1}, \ldots, x_{i_t}$ such that $x_{i_j} \in \overline{X}_{\mathcal{B}}(\mathrm{hm}(\mathbf{h}_{k_j}))$ for all $j \in \{1, \ldots, t\}$ and terms $x^{\mu_1}, \ldots, x^{\mu_t}$ such that $x^{\mu_j} \in \Bbbk[X_{\mathcal{B}}(\mathrm{hm}(\mathbf{h}_{k_j}))]$ for all $j \in \{1, \ldots, t\}$ satisfying:

$$x_{i_1} \mathrm{hm}(\mathbf{h}_{k_1}) = x^{\mu_2} \mathrm{hm}(\mathbf{h}_{k_2}),$$
$$x_{i_2} \mathrm{hm}(\mathbf{h}_{k_2}) = x^{\mu_3} \mathrm{hm}(\mathbf{h}_{k_3}),$$
$$\vdots$$
$$x_{i_t} \mathrm{hm}(\mathbf{h}_{k_t}) = x^{\mu_1} \mathrm{hm}(\mathbf{h}_{k_1}).$$

Multiplying with some variables, we obtain the following chain of equations:

$$x_{i_1} \cdots x_{i_t} \mathrm{hm}(\mathbf{h}_{k_1}) = x_{i_2} \cdots x_{i_t} x^{\mu_2} \mathrm{hm}(\mathbf{h}_{k_2})$$
$$= x_{i_3} \cdots x_{i_t} x^{\mu_2} x^{\mu_3} \mathrm{hm}(\mathbf{h}_{k_3})$$
$$\vdots$$
$$= x_{i_t} x^{\mu_2} \cdots x^{\mu_t} \mathrm{hm}(\mathbf{h}_{k_t})$$
$$= x^{\mu_1} \cdots x^{\mu_t} \mathrm{hm}(\mathbf{h}_{k_1})$$

which implies that $x_{i_1} \cdots x_{i_t} = x^{\mu_1} \cdots x^{\mu_t}$. Furthermore, condition (v) of Definition 1 implies in $\mathcal{P}^s$ the following chain:

$$x_{i_1} \cdots x_{i_t} \mathbf{f}_{k_1} \succeq_{\mathcal{B}} x_{i_2} \cdots x_{i_t} x^{\mu_2} \mathbf{f}_{k_2} \succeq_{\mathcal{B}} \cdots \succeq_{\mathcal{B}} x^{\mu_1} \cdots x^{\mu_t} \mathbf{f}_{k_1} \,.$$

Because of $x_{i_1} \cdots x_{i_t} = x^{\mu_1} \cdots x^{\mu_t}$, we must have throughout equality entailing that $k_1 = \cdots = k_t$ which contradicts our assumptions. □

Example 1. Let $\prec$ be a term order on the free module $\mathcal{P}_{\mathbf{d}}^m$, L a continuous involutive division ([7, Definition 2.1]) and $\mathcal{B}$ a finite, L-involutively autoreduced set ([7, Definition 5.8]) which is a strong L-involutive basis ([7, Definition 5.1]) of the submodule $U \subseteq \mathcal{P}_{\mathbf{d}}^m$ it generates. Then $\mathcal{B}$ induces a resolving decomposition of U with $\mathrm{hm}(\mathcal{B}) = \{\mathrm{lt}(\mathbf{h}_1), \ldots, \mathrm{lt}(\mathbf{h}_s)\}$. The multiplicative variables $X_{\mathcal{B}}$ are assigned according to the involutive division L and we take as $\prec_{\mathcal{B}}$ the Schreyer order induced by $\mathcal{B}$ and $\prec$. Condition (i) of Definition 1 follows from [7, Corollary 5.5], condition (ii) is a consequence of the fact that $\mathcal{B}$ is involutively autoreduced and condition (iii) follows from [7, Lemma 5.12]. According to [7, Proposition 5.13] every $\mathbf{f} \in \mathcal{P}_{\mathbf{d}}^m$ possesses a unique normal form. In Remark 1 we have seen that this is equivalent to the fourth condition in our definition. Finally, (v) is satisfied because of [8, Lemma 5.5] and the existence of an L-ordering. Hence an autoreduced involutive basis always induces a resolving decomposition.

Example 2. Another example for resolving decompositions are the marked modules introduced in [2]. Marked modules are only defined for quasi-stable modules. The construction of a marked basis is a bit different from the usual construction of Gröbner bases. We start with a quasi-stable monomial module $V \subseteq \mathcal{P}_{\mathbf{d}}^m$ which is generated by a monomial Pommaret basis $\mathcal{H} = \{x^{\mu_1} \mathbf{e}_{k_1}, \ldots x^{\mu_s} \mathbf{e}_{k_s}\}$. Then we define a marked basis $\mathcal{B} = \{\mathbf{h}_1, \ldots, \mathbf{h}_s\}$ such that $\mathrm{hm}(\mathbf{h}_i) = x^{\mu_i} \mathbf{e}_{k_i}$ and $\mathrm{supp}(\mathbf{h}_i - x^{\mu_i} \mathbf{e}_{k_i}) \subseteq \langle \mathcal{N}(V)_{\deg(x^{\mu_i} \mathbf{e}_{k_i})} \rangle$. Furthermore it is required that $\mathcal{N}(V)_r$ induces a $\Bbbk$-basis of $(\mathcal{P}_{\mathbf{d}}^m)_r / \langle \mathcal{B} \rangle_r$ for all degrees r, which implies that $(\mathcal{P}_{\mathbf{d}}^m)_r = \langle \mathcal{B} \rangle_r \oplus \langle \mathcal{N}(V)_r \rangle$ for all r. The multiplicative variables $X_{\mathcal{B}}$ are assigned according to the multiplicative variables of the Pommaret basis $\mathcal{H}$ (for a detailed treatment see section two in [2]). We see immediately that the conditions (i), (ii) and (iv) are satisfied. The first part of condition (iii) follows from the fact that $\mathcal{H} = \mathrm{hm}(\mathcal{B})$ is a Pommaret basis and the second part follows from the uniqueness of the reduction process [2, Lemma 5.1]. Finally, we take for $\prec_{\mathcal{B}}$ the TOP lift of the lexicographic order; condition (v) then follows from [2, Lemma 3.6].

Example 3. Even in the case of a monomial module, not every Stanley decomposition can be extended to a resolving decomposition. For $m = 1$, $n = 4$ and the standard grading, we take as U the homogeneous maximal ideal in $\mathcal{P}$. A Stanley decomposition of U is then given by the set

$$\mathcal{B} = \{\mathbf{h}_1 = x_0, \mathbf{h}_2 = x_1, \mathbf{h}_3 = x_2, \mathbf{h}_4 = x_3, \mathbf{h}_5 = x_4, \mathbf{h}_6 = x_0 x_1 x_3, \mathbf{h}_7 = x_0 x_2 x_3,$$
$$\mathbf{h}_8 = x_0 x_2 x_4, \mathbf{h}_9 = x_1 x_2 x_4, \mathbf{h}_{10} = x_1 x_3 x_4, \mathbf{h}_{11} = x_0 x_1 x_2 x_3 x_4\}$$

with multiplicative variables

$$X_{\mathcal{B}}(\mathbf{h}_1) = \{x_0, x_1, x_2\}, \qquad\qquad X_{\mathcal{B}}(\mathbf{h}_2) = \{x_1, x_2, x_3\}$$
$$X_{\mathcal{B}}(\mathbf{h}_3) = \{x_2, x_3, x_4\}, \qquad\qquad X_{\mathcal{B}}(\mathbf{h}_4) = \{x_0, x_3, x_4\}$$
$$X_{\mathcal{B}}(\mathbf{h}_5) = \{x_0, x_1, x_4\}, \qquad\qquad X_{\mathcal{B}}(\mathbf{h}_6) = \{x_0, x_1, x_2, x_3\}$$
$$X_{\mathcal{B}}(\mathbf{h}_7) = \{x_0, x_2, x_3, x_4\}, \qquad\qquad X_{\mathcal{B}}(\mathbf{h}_8) = \{x_0, x_1, x_2, x_4\}$$
$$X_{\mathcal{B}}(\mathbf{h}_9) = \{x_1, x_2, x_3, x_4\}, \qquad\qquad X_{\mathcal{B}}(\mathbf{h}_{10}) = \{x_0, x_1, x_3, x_4\}$$
$$X_{\mathcal{B}}(\mathbf{h}_{11}) = \{x_0, x_1, x_2, x_3, x_4\}\,.$$

It is not possible to find a term order $\prec_{\mathcal{B}}$ which makes this Stanley decomposition to a resolving one, as the corresponding $\mathcal{B}$-graph contains a cycle (note that here obviously $\mathrm{hm}(\mathbf{h}_i) = \mathbf{h}_i$):

$$x_3\mathbf{h}_1 = x_0\mathbf{h}_4\,, \quad x_1\mathbf{h}_4 = x_3\mathbf{h}_2\,, \quad x_0\mathbf{h}_2 = x_1\mathbf{h}_1\,.$$

3 Syzygy Resolutions via Resolving Decompositions

Let $\mathcal{P}^m_{\mathbf{d}_0}$ be a graded free polynomial module with standard basis $\{\mathbf{e}^{(0)}_1, \ldots, \mathbf{e}^{(0)}_m\}$ and grading $\mathbf{d}_0 = (d^{(0)}_1, \ldots d^{(0)}_m)$. Furthermore, let $(\mathcal{B}^{(0)}, \mathrm{hm}(\mathcal{B}^{(0)}), X_{\mathcal{B}^{(0)}}, \prec_{\mathcal{B}^{(0)}})$ be a resolving decomposition of a finitely generated graded module $U \subseteq \mathcal{P}^m_{\mathbf{d}_0}$ with $\mathcal{B}^{(0)} = \{\mathbf{h}_1, \ldots, \mathbf{h}_{s_1}\}$. Our first goal is now to construct a resolving decomposition of the syzygy module $\mathrm{Syz}(\mathcal{B}^{(0)}) \subseteq \mathcal{P}^{s_1}$ which may be considered as a refined version of the well-known Schreyer theorem for Gröbner bases.

For every non-multiplicative variable x_k of a generator $\mathbf{h}_\alpha$, we have a standard representation $x_k\mathbf{h}_\alpha = \sum_{\beta=1}^{s_1} P^{(\alpha;k)}_\beta \mathbf{h}_\beta$ and thus a syzygy

$$\mathbf{S}_{\alpha;k} = x_k\mathbf{e}^{(1)}_\alpha - \sum_{\beta=1}^{s_1} P^{(\alpha;k)}_\beta \mathbf{e}^{(1)}_\beta \tag{3}$$

where $\{\mathbf{e}^{(1)}_1, \ldots, \mathbf{e}^{(1)}_{s_1}\}$ denotes the standard basis of the free module $\mathcal{P}^{s_1}_{\mathbf{d}_1}$ with grading $\mathbf{d}_1 = (\deg(\mathbf{h}_1), \ldots, \deg(\mathbf{h}_s))$. Let $\mathcal{B}^{(1)}$ be the set of all these syzygies.

Lemma 3. Let $S = \sum_{l=1}^{s_1} S_l\mathbf{e}^{(1)}_l$ be an arbitrary syzygy of $\mathcal{B}^{(0)}$ with coefficients $S_l \in \mathcal{P}$. Then $S_l \in \Bbbk[X_{\mathcal{B}^{(0)}}(\mathbf{h}_l)]$ for all $1 \leq l \leq s_1$ if and only if $S = 0$.

Proof. If $S \in \mathrm{Syz}(\mathcal{B}^{(0)})$, then $\sum_{l=1}^{s_1} S_l\mathbf{h}_l = 0$. Each $\mathbf{f} \in U$ can be uniquely written in the form $\mathbf{f} = \sum_{l=1}^{s_1} P_l\mathbf{h}_l$ with $\mathbf{h}_l \in \mathcal{B}^{(0)}$ and $P_l \in \Bbbk[X_{\mathcal{B}^{(0)}}(\mathbf{h}_l)]$. In particular, this holds for $0 \in U$. Thus $0 = S_l \in \Bbbk[X_{\mathcal{B}^{(0)}}(\mathbf{h}_l)]$ for all l and hence $S = 0$. $\qquad\square$

For $\mathbf{h}_\alpha \in \mathcal{B}^{(0)}$ we denote the non-multiplicative variables by $\{x_{i^\alpha_1}, \ldots, x_{i^\alpha_{r_\alpha}}\}$ with $i^\alpha_1 < \cdots < i^\alpha_{r_\alpha}$. Thus $\mathcal{B}^{(1)} = \cup_{j=1}^{s_1}\{\mathbf{S}_{j;i^j_k} \mid 1 \leq k \leq i^j_{r_j}\}$.

Theorem 1. *For every syzygy $S_{\alpha;i_k^\alpha} \in \mathcal{B}^{(1)}$ we set*

$$\mathrm{hm}(S_{\alpha;i_k^\alpha}) = x_{i_k^\alpha} \mathbf{e}_\alpha^{(1)}$$

and

$$X_{\mathcal{B}^{(1)}}(\mathbf{S}_{\alpha;i_k^\alpha}) = \{x_0, \ldots x_n\} \setminus \{x_{i_1^\alpha}, \ldots, x_{i_{k-1}^\alpha}\}.$$

Furthermore, we define $\prec_{\mathcal{B}^{(1)}}$ as the Schreyer order associated to $\mathcal{B}^{(0)}$ and $\prec_{\mathcal{B}^{(0)}}$. Then the quadruple $(\mathcal{B}^{(1)}, \mathrm{hm}(\mathcal{B}^{(1)}), X_{\mathcal{B}^{(1)}}, \prec_{\mathcal{B}^{(1)}})$ is a resolving decomposition of the syzygy module $\mathrm{Syz}(\mathcal{B}^{(0)})$.

Proof. We first show that $(\mathcal{B}^{(1)}, \mathrm{hm}(\mathcal{B}^{(1)}), X_{\mathcal{B}^{(1)}}, \prec_{\mathcal{B}^{(1)}})$ is a resolving decomposition of $\langle \mathcal{B}^{(1)} \rangle$. In a second step, we show that $\langle \mathcal{B}^{(1)} \rangle = \mathrm{Syz}(\mathcal{B}^{(0)})$.

The first condition of Definition 1 is trivially satisfied. By construction it is obvious to see that

$$\mathrm{hm}(\langle \mathcal{B}^{(1)} \rangle) = \bigoplus_{i=1}^{s_1} \langle \overline{X}_{\mathcal{B}^{(0)}}(\mathbf{h}_i) \rangle \mathbf{e}_i^{(1)}. \tag{4}$$

A term $x^\mu \mathbf{e}_l^{(1)} \in \mathrm{supp}(\mathbf{S}_{\alpha;k} - x_k \mathbf{e}_\alpha^{(1)})$ must satisfy by (3) that $x^\mu \in \Bbbk[X_{\mathcal{B}^{(0)}}(\mathbf{h}_l)]$ and hence $x^\mu \mathbf{e}_l^{(1)} \notin \mathrm{hm}(\langle \mathcal{B}^{(1)} \rangle)$ which implies condition (ii). The first part of condition (iii) is again easy to see. It is obvious that

$$\langle \overline{X}_{\mathcal{B}^{(0)}}(\mathbf{h}_\alpha) \rangle \mathbf{e}_\alpha^{(1)} = \bigoplus_{k=1}^{r_\alpha} \Bbbk[X_{\mathcal{B}^{(1)}}(\mathbf{S}_{\alpha;i_k^\alpha})] x_{i_k^\alpha} \mathbf{e}_\alpha^{(1)}.$$

If we combine this equation with (4) the first part of the third condition follows.

The second part of this condition is a bit harder to prove. We take an arbitrary $\mathbf{f} \in \langle \mathcal{B}^{(1)} \rangle$ and construct a standard representation for this module element. We construct this representation according to $\mathrm{hm}(\langle \mathcal{B}^{(1)} \rangle)$. We take the biggest term $x^\mu \mathbf{e}_\alpha^{(1)} \in \mathrm{supp}(\mathbf{f}) \cap \mathrm{hm}(\mathcal{B}^{(1)})$ with respect to the order $\prec_{\mathcal{B}^{(0)}}$. There must be a syzygy $\mathbf{S}_{\alpha;i}$, such that $x_i \mid x^\mu$ and $x^\mu/x_i \in \Bbbk[X_{\mathcal{B}^{(1)}}(\mathbf{S}_{\alpha;i})]$. We reduce $\mathbf{f}$ by this element and get

$$\mathbf{f}' = \mathbf{f} - c\frac{x^\mu}{x_i}\mathbf{S}_{\alpha;i}$$

for a suitable constant $c \in \Bbbk$ such that the term $x^\mu \mathbf{e}_\alpha^{(1)}$ is no longer in the support of $\mathbf{f}'$. Every term $x^\lambda \mathbf{e}_\beta^{(1)}$ newly introduced by $\frac{x^\mu}{x_i}\mathbf{S}_{\alpha;i}$ which also lies in $\mathrm{hm}(\mathcal{B}^{(1)})$ is strictly less than $x^\mu \mathbf{e}_\alpha^{(1)}$ according to condition (v) of Definition 1 and Eq. (3) defining the syzygies $\mathbf{S}_{\alpha;i}$. Now we repeat this procedure until we arrive at an $\mathbf{f}''$ such that $\mathrm{supp}(\mathbf{f}'') \cap \mathrm{hm}(\langle \mathcal{B}^{(1)} \rangle) = \emptyset$. It is clear that we reach such an $\mathbf{f}''$ in a finite number of steps, since the terms during the reduction decrease with respect to $\prec_{\mathcal{B}^{(0)}}$ which is a well-order. We know that now all $x^\epsilon \mathbf{e}_\alpha^{(1)} \in \mathrm{supp}(\mathbf{f}'')$ have the property that $x^\epsilon \in X_{\mathcal{B}^{(0)}}(\mathbf{h}_\alpha)$. Therefore we get that $\mathbf{f}'' = 0$ due to Lemma 3, which finishes the proof of this condition.

The above procedure provides us with an algorithm to compute arbitrary normal forms and hence condition (iv) of Definition 1 follows immediately.

For the last condition we note that now each head term $x_i e_\alpha^{(1)}$ is actually the leading term of $\mathbf{S}_{\alpha;i}$ with respect to the order $\prec_{\mathcal{B}^{(0)}}$. Hence the corresponding Schreyer order satisfies the last condition of Definition 1. $\square$

As with the usual Schreyer theorem, we can iterate this construction and derive this way a free resolution of U. By contrast to the classical situation, it is however now possible to make precise statements about the *shape* of the resolution (even if we do not obtain explicit formulae for the differentials).

Theorem 2. *Let $\beta_{0,j}^{(k)}$ be the number of generators $\mathbf{h} \in \mathcal{B}^{(0)}$ of degree j having k multiplicative variables and set $d = \min\{k \mid \exists j : \beta_{0,j}^{(k)} > 0\}$. Then U possesses a finite free resolution*

$$0 \to \bigoplus \mathcal{P}(-j)^{r_{n+1-d,j}} \to \cdots \to \bigoplus \mathcal{P}(-j)^{r_{1,j}} \to \bigoplus \mathcal{P}(-j)^{r_{0,j}} \to U \to 0 \quad (5)$$

of length $n + 1 - d$ where the ranks of the free modules are given by

$$r_{i,j} = \sum_{k=1}^{n+1-i} \binom{n+1-k}{i} \beta_{0,j-i}^{(k)}.$$

Proof. According to Theorem 1, $(\mathcal{B}^{(1)}, \mathrm{hm}(\mathcal{B}^{(1)}), X_{\mathcal{B}^{(1)}}, \prec_{\mathcal{B}^{(1)}})$ is a resolving decomposition for the module $\mathrm{Syz}_1(U)$. Applying the theorem again, we can construct a resolving decomposition of the second syzygy module $\mathrm{Syz}_2(U)$ and so on. Recall that for every index $1 \leq l \leq m$ and for every non-multiplicative variable $x_k \in \overline{X}_{\mathcal{B}^{(0)}}(\mathbf{h}_{\alpha(l)})$ we have $|\overline{X}_{\mathcal{B}^{(1)}}(S_{l;k})| < |\overline{X}_{\mathcal{B}^{(0)}}(\mathbf{h}_{\alpha(l)})|$.

If D is the minimal number of multiplicative variables for a head module term in $\mathcal{B}^{(0)}$, then the minimal number of multiplicative variables for a head term in $\mathcal{B}^{(1)}$ is $D + 1$. This observation yields the length of the resolution (5). Furthermore $\deg(\mathbf{S}_{k;i}) = \deg(\mathbf{h}_k) + 1$, e. g. from the jth to the $(j+1)$th module the degree from the basis element to the corresponding syzygies grows by one.

The ranks of the modules follow from a rather straightforward combinatorial calculation. Let $\beta_{i,j}^{(k)}$ denote the number of generators of degree j of the i-th syzygy module $\mathrm{Syz}_i(U)$ with k multiplicative variables according to the head module terms. By definition of the generators, we find

$$\beta_{i,j}^{(k)} = \sum_{t=1}^{k-1} \beta_{i-1,j-1}^{(n+1-t)}$$

as each generator with less multiplicative variables and degree $j-1$ in the resolving decomposition of $\mathrm{Syz}_i(\mathcal{B}^{(0)})$ contributes one generator with k multiplicative variables. A lengthy induction allows us to express $\beta_{i,j}^{(k)}$ in terms of $\beta_{0,j}^{(k)}$:

$$\beta_{i,j}^{(k)} = \sum_{t=1}^{k-i} \binom{k-l-1}{i-1} \beta_{0,j-i}^{(t)}.$$

Now we are able to compute the ranks of the free modules via

$$r_{i,j} = \sum_{k=1}^{n+1} \beta_{i,j}^{(k)} = \sum_{k=1}^{n+1} \sum_{t=1}^{k-i} \binom{k-t-1}{i-1} \beta_{0,j-i}^{(t)} = \sum_{k=1}^{n+1-i} \binom{n+1-k}{i} \beta_{0,j-i}^{(k)}.$$

The last equality follows from a classical identity for binomial coefficients.   □

Theorem 2 allows us to construct recursively resolving decompositions for the higher syzygy modules. In the sequel, we denote the corresponding resolving decomposition of the syzygy module $\mathrm{Syz}_j(U)$ by $(\mathcal{B}^{(j)}, \mathrm{hm}(\mathcal{B}^{(j)}), X_{\mathcal{B}^{(j)}}, \prec_{\mathcal{B}^{(j)}})$. To define an element of $\mathcal{B}^{(j)}$, we consider for each generator $\mathbf{h}_\alpha \in \mathcal{B}^{(0)}$ all ordered integer sequences $\mathbf{k} = (k_1, \ldots, k_j)$ with $0 \le k_1 < \cdots < k_j \le n$ of length $|\mathbf{k}| = j$ such that $x_{k_i} \in \overline{X}_{\mathcal{B}^{(0)}}(\mathbf{h}_\alpha)$ for all $1 \le i \le j$. We denote for any $1 \le i \le j$ by $\mathbf{k}_i$ the sequence obtained by eliminating k_i from $\mathbf{k}$. Then the generator $\mathbf{S}_{\alpha;\mathbf{k}}$ arises recursively from the standard representation of $x_{k_j}\mathbf{S}_{\alpha;\mathbf{k}_j}$ according to the resolving decomposition $(\mathcal{B}^{(j-1)}, \mathrm{hm}(\mathcal{B}^{(j-1)}), X_{\mathcal{B}^{(j-1)}}, \prec_{\mathcal{B}^{(j-1)}})$:

$$x_{k_j}\mathbf{S}_{\alpha;\mathbf{k}_j} = \sum_{\beta=1}^{s_1} \sum_{\mathbf{l}} P_{\beta;\mathbf{l}}^{(\alpha;\mathbf{k})} \mathbf{S}_{\beta;\mathbf{l}}. \tag{6}$$

The second sum is over all ordered integer sequences $\mathbf{l}$ of length $j-1$ such that for all entries ℓ_i the variables x_{ℓ_i} is non-multiplicative for the generator $\mathbf{h}_\beta \in \mathcal{B}^{(0)}$. Denoting the free generators of the free module which contains the jth syzygy module by $\mathbf{e}_{\alpha,\mathbf{l}}^{(j)}$, such that $\alpha \in \{1, \ldots, s_1\}$ and $\mathbf{l}$ is an ordered subset of $\overline{X}_{\mathcal{B}^{(0)}}(\mathbf{h}_\alpha)$ of length $j-1$ we get the following representation for $\mathbf{S}_{\alpha,\mathbf{k}}$:

$$\mathbf{S}_{\alpha;\mathbf{k}} = x_{k_j}\mathbf{e}_{\alpha;\mathbf{k}_j}^{(j)} - \sum_{\beta=1}^{s_1} \sum_{\mathbf{l}} P_{\beta;\mathbf{l}}^{(\alpha;\mathbf{k})} \mathbf{e}_{\beta;\mathbf{l}}^{(j)}.$$

Corollary 1. *In the situation of Theorem 2, set $d = \min\{k \mid \exists j : \beta_{0,j}^{(k)} > 0\}$ and $q = \deg(\mathcal{B}^{(0)}) = \max\{\deg(\mathbf{h}) \mid \mathbf{h} \in \mathcal{B}^{(0)}\}$. Then we obtain the following bounds for the projective dimension, the Castelnuovo-Mumford regularity and the depth, respectively, of the submodule U:*

$$\mathrm{pd}(U) \le n+1-d, \qquad \mathrm{reg}(U) \le q, \qquad \mathrm{depth}(U) \ge d.$$

Proof. The first estimate follows immediately from the resolution (5) induced by the resolving decomposition $(\mathcal{B}^{(0)}, \mathrm{hm}(\mathcal{B}^{(0)}), X_{\mathcal{B}^{(0)}}, \prec_{\mathcal{B}^{(0)}})$ of U. The last estimate is a simple consequence of the first one and the graded form of the Auslander-Buchsbaum formula. Finally, the ith module of this resolution is obviously generated by elements of degree less than or equal to $q+i$. This observation implies that U is q-regular and thus the second estimate.   □

Remark 2. The resolving decomposition $(\mathcal{B}^{(1)}, \mathrm{hm}(\mathcal{B}^{(1)}), X_{\mathcal{B}^{(1)}}, \prec_{\mathcal{B}^{(1)}})$ constructed in Theorem 1 is always a Janet basis of the first syzygy module with respect to the term order $\prec_{\mathcal{B}^{(0)}}$. This is simply due to the fact that the choice of the multiplicative variables in the resolving decomposition of the syzygy module made in Theorem 1 is actually inspired by what happens for the Janet division. Hence in the special case that the resolving decomposition is induced by a Pommaret or a Janet basis, it is easy to see that also the resolving decompositions of the higher syzygy modules are actually induced by Pommaret or Janet bases for a Schreyer order constructed as in Theorem 1. Since a Janet basis which only consists of variables is simultaneously an involutive basis for the alex division (see [5] for the definition), the same is true for resolving decompositions induced by alex bases.

At this point, one can also see some advantages of our general framework. Our previous results require that the used involutive division is of Schreyer type. This assumption ensures that we obtain at each step again an L-involutive basis for the syzygy module with respect to a Schreyer order. In our new approach, we automatically obtain Janet basis, as we can choose the head terms and the multiplicative variables as we like. Consequently, we can now use an involutive basis $\mathcal{B}$ for an arbitrary involutive division L as starting point for the construction of a resolution, provided its L-graph is acyclic (which is always the case if L is continuous). The construction will not necessarily lead to L-involutive bases of the syzygy modules, but for most applications this fact is irrelevant.

4   An Explicit Formula for the Differential

As in Sect. 3 let $\mathcal{P}_{\mathbf{d}_0}^m$ be a graded free module with free generators $\mathbf{e}_1^{(0)}, \dots \mathbf{e}_m^{(0)}$ and grading $\mathbf{d}_0 = (d_1^{(0)}, \dots d_m^{(0)})$. We always work with a finitely generated graded module $U \in \mathcal{P}_{\mathbf{d}_0}^m$ with a resolving decomposition $(\mathcal{B}^{(0)}, \mathrm{hm}(\mathcal{B}^{(0)}), X_{\mathcal{B}^{(0)}}, \prec_{\mathcal{B}^{(0)}})$ where $\mathcal{B}^{(0)} = \{\mathbf{h}_1, \dots, \mathbf{h}_{s_1}\}$.

First we give an alternative description of the complex underlying the resolution (5). Let $\mathcal{W} = \bigoplus_{\alpha=1}^{s_1} \mathcal{P}\mathbf{w}_\alpha$ and $\mathcal{V} = \bigoplus_{i=0}^{n} \mathcal{P}\mathbf{v}_i$ be two free $\mathcal{P}$-modules whose ranks are given by the size of the resolving decomposition $(\mathcal{B}^{(0)}, \mathrm{hm}(\mathcal{B}^{(0)}), X_{\mathcal{B}^{(0)}}, \prec_{\mathcal{B}^{(0)}})$ and by the number of variables in $\mathcal{P}$, respectively. Then we set $\mathcal{C}_i = \mathcal{W} \otimes_{\mathcal{P}} \Lambda_i \mathcal{V}$ where $\Lambda_\bullet$ denotes the exterior product. A $\mathcal{P}$-linear basis of $\mathcal{C}_i$ is provided by the elements $\mathbf{w}_\alpha \otimes \mathbf{v}_{\mathbf{k}}$ where $\mathbf{v}_{\mathbf{k}} = \mathbf{v}_{k_1} \wedge \cdots \wedge \mathbf{v}_{k_i}$ for an ordered sequence $\mathbf{k} = (k_1, \dots, k_i)$ with $0 \leq k_1 < \cdots < k_i \leq n$. Then the free subcomplex $\mathcal{S}_\bullet \subset \mathcal{C}_\bullet$ generated by all elements $\mathbf{w}_\alpha \otimes \mathbf{v}_{\mathbf{k}}$ with $\mathbf{k} \subseteq \overline{X}_{\mathcal{B}^{(0)}}(\mathbf{h}_\alpha)$ corresponds to (5) upon the identification $\mathbf{e}_{\alpha;\mathbf{k}}^{(i+1)} \leftrightarrow \mathbf{w}_\alpha \otimes \mathbf{v}_{\mathbf{k}}$. Let $k_{i+1} \in \overline{X}_{\mathcal{B}^0}(\mathbf{h}_\alpha) \setminus \mathbf{k}$, then the differential comes from (6),

$$d_{\mathcal{S}}(\mathbf{w}_\alpha \otimes \mathbf{v}_{\mathbf{k}, k_{i+1}}) = x_{k_{i+1}} \mathbf{w}_\alpha \otimes \mathbf{v}_{\mathbf{k}} - \sum_{\beta, \mathbf{l}} P_{\beta; \mathbf{l}}^{(\alpha; \mathbf{k}, k_{i+1})} \mathbf{w}_\beta \otimes \mathbf{v}_{\mathbf{l}} \,,$$

and thus requires the explicit determination of all the higher syzygies (6).

In this section we present a method to directly compute the differential without computing higher syzygies. It is based on ideas of Sköldberg [9,10] and generalises the theory which we developed in [1,3] for the special case of a resolution induced by a Pommaret or a Janet basis for a given term order.

Definition 2. *A graded polynomial module U has* head linear syzygies, *if it possesses a finite presentation*

$$0 \longrightarrow \ker \eta \longrightarrow \mathcal{W} = \bigoplus_{\alpha=1}^{s} \mathcal{P}\mathbf{w}_\alpha \xrightarrow{\eta} U \longrightarrow 0 \qquad (7)$$

with a finite generating set $\mathcal{H} = \{\mathbf{h}_1, \ldots, \mathbf{h}_t\}$ of $\ker \eta$ where one can choose for each generator $\mathbf{h}_\alpha \in \mathcal{H}$ a head module term $\mathrm{hm}(\mathbf{h}_\alpha)$ *of the form $x_i \mathbf{w}_\alpha$.*

Sköldberg's construction begins with the following two-sided Koszul complex $(\mathcal{F}, d_\mathcal{F})$ defining a free resolution of U. Let $\mathcal{V}$ be a $\Bbbk$-linear space with basis $\{\mathbf{v}_0, \ldots, \mathbf{v}_n\}$ (with $n + 1$ still the number of variables in $\mathcal{P}$) and set $\mathcal{F}_j = \mathcal{P} \otimes \Lambda_j \mathcal{V} \otimes U$ which obviously yields a free $\mathcal{P}$-module. Choosing a $\Bbbk$-linear basis $\{m_a \mid a \in A\}$ of U, a $\mathcal{P}$-linear basis of $\mathcal{F}_j$ is given by the elements $1 \otimes v_\mathbf{k} \otimes m_a$ with ordered sequences $\mathbf{k}$ of length j. The differential is now defined by

$$d_\mathcal{F}(1 \otimes \mathbf{v}_\mathbf{k} \otimes m_a) = \sum_{i=1}^{j}(-1)^{i+1}\big(x_{k_i} \otimes \mathbf{v}_{\mathbf{k}_i} \otimes m_a - 1 \otimes \mathbf{v}_{\mathbf{k}_i} \otimes x_{k_i}m_a\big). \qquad (8)$$

Here it should be noted that the second term on the right hand side is not yet expressed in the chosen $\Bbbk$-linear basis of U. For notational simplicity, we will drop in the sequel the tensor sign $\otimes$ and leading factors 1 when writing elements of $\mathcal{F}_\bullet$.

Sköldberg uses a specialisation of head linear terms. He requires that for a given term order $\prec$ the leading module of $\ker \eta$ in the presentation (7) must be generated by terms of the form $x_i \mathbf{w}_\alpha$. In this case he says that U has *initially linear syzygies*. Our definition is term order free.

Under the assumption that the module U has initially linear syzygies via a presentation (7), Sköldberg [10] constructs a Morse matching leading to a smaller resolution $(\mathcal{G}, d_\mathcal{G})$. He calls the variables

$$\mathrm{crit}\,(\mathbf{w}_\alpha) = \{x_j \mid x_j\mathbf{w}_\alpha \in \mathrm{lt}\ \ker \eta\}\,;$$

critical for the generator $\mathbf{w}_\alpha$; the remaining *non-critical* ones are contained in the set $\mathrm{ncrit}\,(\mathbf{w}_\alpha)$. Then a $\Bbbk$-linear basis of U is given by all elements $x^\mu \mathbf{h}_\alpha$ with $\mathbf{h}_\alpha = \eta(\mathbf{w}_\alpha)$ and $x^\mu \in \Bbbk[\mathrm{ncrit}\,(\mathbf{w}_\alpha)]$.

According to [9] we define $\mathcal{G}_j \subseteq \mathcal{F}_j$ as the free submodule generated by those vertices $\mathbf{v}_\mathbf{k}\mathbf{h}_\alpha$ where the ordered sequences $\mathbf{k}$ are of length j and such that every entry k_i is critical for $\mathbf{w}_\alpha$. In particular $\mathcal{W} \cong \mathcal{G}_0$ with an isomorphism induced by $\mathbf{w}_\alpha \mapsto \mathbf{v}_\emptyset \mathbf{h}_\alpha$.

The description of the differential $d_\mathcal{G}$ is based on reduction paths in the associated Morse graph (for a detailed treatment of these notions, see [1,9] or [6]) and expresses the differential as a triple sum. If we assume that, after expanding the right hand side of (8) in the chosen $\Bbbk$-linear basis of U, the differential of the complex $\mathcal{F}_\bullet$ can be expressed as

$$d_\mathcal{F}(\mathbf{v}_\mathbf{k}\mathbf{h}_\alpha) = \sum_{\mathbf{m},\mu,\gamma} Q^{\mathbf{k},\alpha}_{\mathbf{m},\mu,\gamma}\mathbf{v}_\mathbf{m}(x^\mu\mathbf{h}_\gamma)\,,$$

then $d_{\mathcal{G}}$ is defined by

$$d_{\mathcal{G}}(\mathbf{v_k h_\alpha}) = \sum_{\mathbf{l},\beta} \sum_{\mathbf{m},\mu,\gamma} \sum_p \rho_p\big(Q^{\mathbf{k},\alpha}_{\mathbf{m},\mu,\gamma}\mathbf{v_m}(x^\mu \mathbf{h_\gamma})\big) \tag{9}$$

where the first sum ranges over all ordered sequences $\mathbf{l}$ which consists entirely of critical indices for $\mathbf{w}_\beta$. Moreover the second sum may be restricted to all values such that a polynomial multiple of $\mathbf{v_m}(x^\mu \mathbf{h_\gamma})$ effectively appears in $d_{\mathcal{F}}(\mathbf{v_k h_\alpha})$ and the third sum ranges over all reduction paths p going from $\mathbf{v_m}(x^\mu \mathbf{h_\gamma})$ to $\mathbf{v_l h_\beta}$. Finally ρ_p is the reduction associated with the reduction path p satisfying

$$\rho_p\big(\mathbf{v_m}(x^\mu \mathbf{h_\gamma})\big) = q_p \mathbf{v_l h_\beta}$$

for some polynomial $q_p \in \mathcal{P}$.

It turns out that Sköldberg uses the term order $\prec$ only for distinguishing the critical and non-critical variables. Therefore it is straightforward to see that his construction also works for modules which have head linear syzygies. We simply replace the definition of critical and non-critical variables. We define

$$\mathrm{crit}\,(\mathbf{w}_\alpha) = \{x_j \mid x_j \mathbf{w}_\alpha \in \mathrm{hm}(\mathcal{H})\}\,,$$

where $\mathcal{H}$ is chosen as in Definition 2. Again the remaining variables are contained in the set $\mathrm{ncrit}(\mathbf{w}_\alpha)$.

In the sequel we will show that for a finitely generated graded module U with resolving decomposition $(\mathcal{B}^{(0)}, \mathrm{hm}(\mathcal{B}^{(0)}), X_{\mathcal{B}^{(0)}}, \prec_{\mathcal{B}^{(0)}})$ the resolution constructed by Sköldberg's method is isomorphic to the resolution which is induced by the resolving decomposition if we choose the head linear syzygies properly. Firstly we obtain the following trivial assertion.

Lemma 4. *If the graded submodule $U \subseteq \mathcal{P}^{s_1}_{d_0}$ possesses a resolving decomposition $(\mathcal{B}^{(0)}, \mathrm{hm}(\mathcal{B}^{(0)}), X_{\mathcal{B}^{(0)}}, \prec_{\mathcal{B}^{(0)}})$, then it has head linear syzygies. More precisely, we can set $\mathrm{crit}(\mathbf{w}_\alpha) = \overline{X}_{\mathcal{B}^{(0)}}(\mathbf{h}_\alpha)$, i.e. the critical variables of the generator $\mathbf{w}_\alpha$ are the non-multiplicative variables of $\mathbf{h}_\alpha = \eta(\mathbf{w}_\alpha)$.*

The lemmata which we subsequently cite from [1] are formulated for a Pommaret basis, which is an involutive basis. Nevertheless we can apply them directly in our setting, if not stated otherwise, because their proofs remain applicable for resolving decompositions. The reason for this is that they only need the existence of unique standard representations and the division of variables into multiplicative and non-multiplicative ones. Some of the proofs in [1] explicitly use the class of a generator in $\mathcal{B}^{(0)}$, a notion arising in the context of Pommaret bases. When working with resolving decompositions, one has to replace it by the maximal index of a multiplicative variable.

The reduction paths can be divided into elementary ones of length two. There are essentially three types of reductions paths [1, Sect. 4]. The elementary reductions of *type 0* are not of interest [1, Lemma 4.5]. All other elementary reductions paths are of the form

$$\mathbf{v_k}(x^\mu \mathbf{h_\alpha}) \longrightarrow \mathbf{v_{k\cup i}}(\frac{x^\mu}{x_i}\mathbf{h_\alpha}) \longrightarrow \mathbf{v_l}(x^\nu \mathbf{h_\beta})\,.$$

Here $\mathbf{k} \cup i$ is the ordered sequence which arises when i is inserted into $\mathbf{k}$; likewise $\mathbf{k} \setminus i$ stands for the removal of an index $i \in \mathbf{k}$.

Type 1: Here $\mathbf{l} = (\mathbf{k} \cup i) \setminus j$, $x^\nu = \frac{x^\mu}{x_i}$ and $\beta = \alpha$. Note that $i = j$ is allowed. We define $\epsilon(i; \mathbf{k}) = (-1)^{|\{j \in \mathbf{k}|j > i\}|}$. Then the corresponding reduction is

$$\rho(\mathbf{v_k} x^\mu \mathbf{h}_\alpha) = \epsilon(i; \mathbf{k} \cup i)\epsilon(j; \mathbf{k} \cup i) x_j \mathbf{v}_{(\mathbf{k} \cup i)\setminus j}\left(\frac{x^\mu}{x_i}\mathbf{h}_\alpha\right).$$

Type 2: Now $\mathbf{l} = (\mathbf{k} \cup i) \setminus j$ and $x^\nu \mathbf{h}_\beta$ appears in the involutive standard representation of $\frac{x^\mu x_j}{x_i}\mathbf{h}_\alpha$ with the coefficient $\lambda_{j,i,\alpha,\mu,\nu,\beta} \in \Bbbk$. In this case, by construction of the Morse matching, we have $i \neq j$. The reduction is

$$\rho(\mathbf{v_k} x^\mu \mathbf{h}_\alpha) = -\epsilon(i; \mathbf{k} \cup i)\epsilon(j; \mathbf{k} \cup i)\lambda_{j,i,\alpha,\mu,\nu,\beta}\mathbf{v}_{(\mathbf{k} \cup i)\setminus j}(x^\nu \mathbf{h}_\beta).$$

These reductions follow from the differential (8): The summands appearing there are either of the form $x_{k_i}\mathbf{v}_{k_i}m_a$ or of the form $\mathbf{v}_{k_i}(x_{k_i}m_a)$. For each of these summands, we have a directed edge in the Morse graph $\Gamma^A_{\mathcal{F}_\bullet}$. Thus for an elementary reduction path

$$\mathbf{v_k}(x^\mu \mathbf{h}_\alpha) \longrightarrow \mathbf{v}_{\mathbf{k} \cup i}\left(\frac{x^\mu}{x_i}\mathbf{h}_\alpha\right) \longrightarrow \mathbf{v_l}(x^\nu \mathbf{h}_\beta),$$

the second edge can originate from summands of either form. For the first form we then have an elementary reduction path of type 1 and for the second form we have type 2.

To show that the resolution induced by a resolving decomposition is isomorphic to the resolution constructed via Sköldberg's method we need a classical theorem concerning the uniqueness of free resolutions.

Theorem 3. [4, Theorem 1.6] *Let U be a finitely generated graded $\mathcal{P}^m_{\mathrm{d}}$-module. If $\mathcal{F}$ is the graded minimal free resolution of U and $\mathcal{G}$ an arbitrary graded free resolution of U, then $\mathcal{G}$ is isomorphic to the direct sum of $\mathcal{F}$ and a trivial complex.*

Assume that we have two graded free resolutions $\mathcal{F}$, $\mathcal{G}$ of the same module U with the same shape (which means that the homogeneous components of the free modules in the two resolutions have always the same dimensions: dim $(\mathcal{F}_i)_j =$ dim $(\mathcal{G}_i)_j$). Then Theorem 3 implies that the two resolutions are isomorphic. For the next theorem, we note the following important observation. The bases of the free modules in the resolution $\mathcal{G}$ of Sköldberg are given by the generators $\mathbf{v_k}\mathbf{h}_\alpha$ with $\mathbf{k} \subseteq \overline{X}_{\mathcal{B}^{(0)}}(\mathbf{h}_\alpha)$.

Theorem 4. *Let $\mathcal{F}$ be the graded free resolution which is induced by the resolving decomposition $(\mathcal{B}^{(0)}, \mathrm{hm}(\mathcal{B}^{(0)}), X_{\mathcal{B}^{(0)}}, \prec_{\mathcal{B}^{(0)}})$ and $\mathcal{G}$ the graded free resolution which is constructed by the method of Sköldberg when the head linear syzygies are chosen such that $\mathrm{crit}(\mathbf{h}_\alpha) = \overline{X}_{\mathcal{B}^{(0)}}(\mathbf{h}_\alpha)$ for every $\mathbf{h}_\alpha \in \mathcal{B}^{(0)}$. Then the resolutions $\mathcal{F}$ and $\mathcal{G}$ are isomorphic.*

Proof. According to the observation made above, it is obvious that the two resolutions $\mathcal{F}$ and $\mathcal{G}$ have the same shape. Together with Theorem 3, the claim follows then immediately. □

For completeness, we repeat some simple results from [1]. They will show us, that the differentials of both resolutions are very similar. In fact we show for the resolution constructed via Sköldberg's method, that we can find head module terms in the higher syzygies which are equal to the head module terms of the resolving decompositions of the higher syzygies of the induced free resolution.

Lemma 5. [1, Lemma 4.3] *For a non-multiplicative index[1] $i \in \mathrm{crit}(\mathbf{h}_\alpha)$ let $x_i \mathbf{h}_\alpha = \sum_{\beta=1}^{s_1} P_\beta^{(\alpha;i)} \mathbf{h}_\beta$ be the standard representation. Then we have $d_\mathcal{G}(\mathbf{v}_i \mathbf{h}_\alpha) = x_i \mathbf{v}_\emptyset \mathbf{h}_\alpha - \sum_{\beta=1}^{s_1} P_\beta^{(\alpha;i)} \mathbf{v}_\emptyset \mathbf{h}_\beta$.*

The next result states that if one starts at a vertex $\mathbf{v}_i(x^\mu \mathbf{h}_\alpha)$ with certain properties and follows through all possible reduction paths in the graph, one will never get to a point where one must calculate an involutive standard representation. If there are no critical (i. e. non-multiplicative) variables present at the starting point, then this will not change throughout any reduction path. In order to generalise this lemma to higher homological degrees, one must simply replace the conditions $i \in \mathrm{ncrit}(\mathbf{h}_\alpha)$ and $j \in \mathrm{ncrit}(\mathbf{h}_\beta)$ by ordered sequences $\mathbf{k}, \mathbf{l}$ with $\mathbf{k} \subseteq \mathrm{ncrit}(\mathbf{h}_\alpha)$ and $\mathbf{l} \subseteq \mathrm{ncrit}(\mathbf{h}_\beta)$.

Lemma 6. [1, Lemma 4.4] *Assume that $i \cup \mathrm{supp}(\mu) \subseteq \mathrm{ncrit}(\mathbf{h}_\alpha)$. Then for any reduction path $p = \mathbf{v}_i(x^\mu \mathbf{h}_\alpha) \rightarrow \cdots \rightarrow \mathbf{v}_j(x^\nu \mathbf{h}_\beta)$ we have $j \in \mathrm{ncrit}(\mathbf{h}_\beta)$. In particular, in this situation there is no reduction path $p = \mathbf{v}_i(x^\mu \mathbf{h}_\alpha) \rightarrow \cdots \rightarrow \mathbf{v}_k \mathbf{h}_\beta$ with $k \in \mathrm{crit}(\mathbf{h}_\beta)$.*

The next corollary asserts that we can choose in Sköldberg's resolution head module terms in such a way that there is a one-to-one correspondence to the head terms of the syzygies contained in the free resolution induced by the resolving decomposition. This corollary is a direct consequence of Lemma 6.

Corollary 2. *Let $(k_1, \ldots, k_j) = \mathbf{k} \subseteq \mathrm{crit}\,\mathbf{h}_\alpha$, then*

$$x_{k_l} \mathbf{v}_{\mathbf{k} \setminus k_l} \mathbf{h}_\alpha \in \mathrm{supp}(d_\mathcal{G}(\mathbf{v}_\mathbf{k} \mathbf{h}_\alpha)).$$

In [1,3] we show a method to effectively compute graded Betti numbers via the induced free resolution of Janet and Pommaret bases and the method of Sköldberg. We show that we can compute the graded Betti numbers with computing only the constant part of the resolution. With this method it is also possible to compute only a single Betti number without compute the complete constant part of the free resolution. The reason for that is that Sköldberg's formula allows to compute a differential in the free resolution independently of the rest of the free resolution. Furthermore the theorem about the induced free resolution gives us a formula to compute the ranks of these resolution. These methods are also applicable for an arbitrary resolving decomposition due to the fact that we proved Theorem 2 and the form of the differential (9).

---

[1] For notational simplicity, we will identify sets X of variables with sets of the corresponding indices and thus simply write $i \in X$ instead of $x_i \in X$.

References

1. Albert, M., Fetzer, M., Sáenz-de Cabezón, E., Seiler, W.: On the free resolution induced by a Pommaret basis. J. Symb. Comp. **68**, 4–26 (2015)
2. Albert, M., Bertone, C., Roggero, M., Seiler, W.M.: Marked bases over quasi-stable modules. Preprint arXiv:1511.03547 (2015)
3. Albert, M., Fetzer, M., Seiler, W.M.: Janet bases and resolutions in CoCoALib. In: Gerdt, V.P., Koepf, W., Seiler, W.M., Vorozhtsov, E.V. (eds.) CASC 2015. LNCS, pp. 15–29. Springer, Switzerland (2015)
4. Eisenbud, D.: The Geometry of Syzygies: A Second Course in Algebraic Geometry and Commutative Algebra. Graduate Texts in Mathematics. Springer, New York (2005)
5. Gerdt, V.P., Blinkov, Y.A.: Involutive division generated by an antigraded monomial ordering. In: Gerdt, V.P., Koepf, W., Mayr, E.W., Vorozhtsov, E.V. (eds.) CASC 2011. LNCS, vol. 6885, pp. 158–174. Springer, Heidelberg (2011)
6. Jöllenbeck, M., Welker, V.: Minimal resolutions via algebraic discrete Morse theory. Mem. Amer. Math. Soc. **197** (2009). AMS
7. Seiler, W.: A combinatorial approach to involution and δ-regularity I: involutive bases in polynomial algebras of solvable type. Appl. Alg. Eng. Comm. Comp. **20**, 207–259 (2009)
8. Seiler, W.: A combinatorial approach to involution and δ-regularity II: structure analysis of polynomial modules with Pommaret bases. Appl. Alg. Eng. Comm. Comp. **20**, 261–338 (2009)
9. Sköldberg, E.: Morse theory from an algebraic viewpoint. Trans. Amer. Math. Soc. **358**, 115–129 (2006)
10. Sköldberg, E.: Resolutions of modules with initially linear syzygies. Preprint arXiv:1106.1913 (2011)

Setup of Order Conditions for Splitting Methods

Winfried Auzinger[1], Wolfgang Herfort[1], Harald Hofstätter[1]([✉]),
and Othmar Koch[2]

[1] Technische Universität Wien, Vienna, Austria
{w.auzinger,w.herfort}@tuwien.ac.at, hofi@harald-hofstaetter.at
[2] Universität Wien, Vienna, Austria
othmar@othmar-koch.org,
www.asc.tuwien.ac.at/∼winfried
www.asc.tuwien.ac.at/∼herfort
www.harald-hofstaetter.at
www.othmar-koch.org

Abstract. For operator splitting methods, an approach based on Taylor expansion and the particular structure of its leading term as an element of a free Lie algebra is used for the setup of a system of order conditions. Along with a brief review of the underlying theoretical background, we discuss the implementation of the resulting algorithm in computer algebra, in particular using Maple 18 (Maple is a trademark of MapleSoft[TM].). A parallel version of such a code is described, and its performance on a computational node with 16 threads is documented.

Keywords: Evolution equations · Splitting methods · Order conditions · Local error · Taylor expansion · Parallel processing

1 Introduction

The construction of higher order discretization schemes of one-step type for the numerical solution of evolution equations is typically based on the setup and solution of a system of polynomial equations for a number of unknown coefficients. Classical examples are Runge-Kutta methods, and their various modifications, see e.g. [9].

To design particular schemes, we need to understand:

(i) how to generate a system of algebraic equations for the coefficients of the higher-order method sought,
(ii) how to solve the resulting system of polynomial equations.

Here we focus on (i) which depends on the particular class of methods one is interested in.[1] We consider *operator splitting methods*, which are based on the idea of approximating the exact flow of an evolution equation by compositions of (usually two) separate subflows which are easier to evaluate. Splitting methods

---

[1] The aspect (ii) enters the discussion in [2].

© Springer International Publishing AG 2016
V.P. Gerdt et al. (Eds.): CASC 2016, LNCS 9890, pp. 30–42, 2016.
DOI: 10.1007/978-3-319-45641-6_3

represent a very useful class of one-step methods for certain types of evolution equations, as for instance Schrödinger type equations, and if the operator splitting is done in an appropriate way, they have very good stability properties (possibly with complex instead of real coefficients for the case of parabolic equations). The more difficult problem is to find coefficients such that a higher order of accuracy is obtained, i.e., coping with (i) and (ii),

Computer algebra is an indispensable tool for dealing with (i). Typically there is a tradeoff between 'manual' a priori analysis and machine driven automatization. For operator-splitting methods, a well-known approach is based on a cumbersome recursive application of the Baker-Campbell-Hausdorff (BCH) formula, see [8]. Here we are advocating another approach, namely implementation of an algorithm for (i) which runs in a fully automatic mode. This approach is described in [1,2]; it is based on Taylor expansion and a theoretical result concerning the structure of the leading term in this expansion. This has the advantage that explicit knowledge of the BCH coefficients is not required. It may be called a generic, 'brute-force' approach. The efficiency of such a general algorithm cannot be optimal in an overall sense; on the other hand, it is easy to implement with optimal speedup on a parallel computer. Moreover, it can be easily adapted to special cases like coefficient symmetries, to operator splitting into more than two parts, and to pairs of splitting schemes akin to embedded Runge-Kutta methods.

In the present paper we focus on the implementation aspect, in particular in a parallel environment. The preparation of a parallel version was motivated by the computational complexity which strongly grows when the desired order is increased. Our parallel code scales in computation time at an (almost) optimal rate, and this speedup is of great practical advantage when trying out different variants, especially for more complex higher-order cases. This may also be viewed as a hardness test for parallelization in Maple.

Topic (ii) is not discussed in this paper. Details concerning the theoretical background and a discussion concerning concrete results and optimized schemes obtained are given in [2], and a collection of optimized schemes can be found at [3]. We note that a related approach has recently also been considered in [5].

In the rest of this introductory section we describe the mathematical background. In Sect. 2 we review our algorithm introduced in [1,2] based on Taylor expansion of the local error. A parallel implementation is described in Sect. 3. Modifications and extensions are indicated in Sect. 4, and performance measures are documented in Sect. 5 by means of some examples.

1.1   Splitting Methods for the Integration of Evolution Equations

In many applications, the right hand side $F(u)$ of an evolution equation

$$\partial_t u(t) = F(u(t)) = A(u(t)) + B(u(t)), \quad t \geq 0, \quad u(0) \text{ given,} \qquad (1)$$

splits up in a natural way into two terms $A(u)$ and $B(u)$, where the separate integration of the subproblems

$$\partial_t u(t) = A(u(t)), \qquad \partial_t u(t) = B(u(t))$$

is much easier to accomplish than for the original problem.

Example 1. The solution of a linear ODE system with constant coefficients,

$$\partial_t u(t) = (A + B)\, u(t),$$

is given by

$$u(t) = e^{t(A+B)}\, u(0).$$

The simplest splitting approximation ('Lie-Trotter'), starting at some initial value u and applied with a time step of length $t = h$, is given by

$$S(h, u) = e^{hB}\, e^{hA}\, u \approx e^{h(A+B)} u.$$

This is not exact (unless $AB = BA$), but it satisfies

$$\|(e^{hB}\, e^{hA} - e^{h(A+B)})u\| = \mathcal{O}(h^2) \quad \text{for} \quad h \to 0,$$

and the error of this approximation depends on behavior of the commutator $[A, B] = AB - BA$.   □

A general splitting method takes steps of the form[2]

$$S(h, u) = S_s(h, S_{s-1}(h, \ldots, S_1(h, u))) \approx \Phi_F(h, u), \tag{2a}$$

with

$$S_j(h, v) = \Phi_B(b_j\, h, \Phi_A(a_j\, h, v)), \tag{2b}$$

where the (real or complex) coefficients a_j, b_j have to be found such that a certain desired order of approximation for $h \to 0$ is obtained.

The local error of a splitting step is denoted by

$$S(h, u) - \Phi_F(h, u) =: \mathcal{L}(h, u). \tag{3}$$

For our present purpose of finding asymptotic order conditions it is sufficient to consider the case of a linear system, $F(u) = F\, u = A\, u + B\, u$ with linear operators A and B. We denote

$$A_j = a_j\, A, \; B_j = b_j\, B, \quad j = 1 \ldots s.$$

Then,

$$S(h, u) = S(h)u, \quad S(h) = S_s(h)\, S_{s-1}(h) \cdots S_1(h) \approx e^{hF}, \tag{4a}$$

with

$$S_j(h) = e^{hB_j}\, e^{hA_j}, \quad j = 1 \ldots s. \tag{4b}$$

For the linear case the local error (3) is of the form $\mathcal{L}(h)u$ with the linear operator $\mathcal{L}(h) = S(h) - e^{hF}$.

---

[2] By Φ_F we denote the flow associated with the equation $\partial_t u = F(u)$, and Φ_A, Φ_B are defined analogously.

1.2   Commutators

Commutators of the involved operators play a central role. For formal consistency, we call A and B the 'commutators of degree 1'. There is (up to sign) one non-vanishing[3] commutator of degree 2,

$$[A, B] := A B - B A,$$

and there are two non-vanishing commutators of degree 3,

$$[A, [A, B]] = A [A, B] - [A, B] A, \quad [[A, B], B] = [A, B] B - B [A, B],$$

and so on; see Sect. 2.2 for commutators of higher degrees.

2   Taylor Expansion of the Local Error

2.1   Representation of Taylor Coefficients

Consider the Taylor expansion, about $h = 0$, of the local error operator $\mathcal{L}(h)$ of a consistent one-step method (satisfying the basic consistency condition $\mathcal{L}(0) = 0$),

$$\mathcal{L}(h) = \sum_{q=1}^{p} \frac{h^q}{q!} \frac{d^q}{dh^q} \mathcal{L}(h) \bigg|_{h=0} + \mathcal{O}(h^{p+1}). \tag{5}$$

The method is of asymptotic order p iff $\mathcal{L}(h) = \mathcal{O}(h^{p+1})$ for $h \to 0$; thus the conditions for order $\geq p$ are given by

$$\frac{d}{dh} \mathcal{L}(h) \bigg|_{h=0} = \ldots = \frac{d^p}{dh^p} \mathcal{L}(h) \bigg|_{h=0} = 0. \tag{6}$$

The formulas in (6) need to be presented in a more explicit form, involving the operators A and B. For a splitting method (4), a calculation based on the Leibniz formula for higher derivatives shows[4] (see [2])

$$\frac{d^q}{dh^q} \mathcal{L}(h) \bigg|_{h=0} = \sum_{|\boldsymbol{k}|=q} \binom{q}{\boldsymbol{k}} \prod_{j=s\ldots 1} \sum_{\ell=0}^{k_j} \binom{k_j}{\ell} B_j^\ell A_j^{k_j - \ell} - (A + B)^q, \tag{7}$$

with $\boldsymbol{k} = (k_1, \ldots, k_s) \in \mathbb{N}_0^s$.

Representation of (7) in Maple. The non-commuting operators A and B are represented by symbolic variables A and B, which can be declared to be non-commutative making use of the corresponding feature implemented in the package **Physics**. Now it is straightforward to generate the sum (7), with unspecified coefficients a_j, b_j, using standard combinatorial tools; for details see Sect. 3.

---

[3] 'Non-vanishing' means non-vanishing in general (generic case, with no special assumptions on A and B).

[4] If A and B commute, i.e., $AB = BA$, then all these expressions vanish.

2.2   The Leading Term of the Local Error Expansion

Formally, the multinomial sums in the expressions (7) are multivariate homogeneous polynomials of total degree q in the variables $a_j, b_j, j = 1 \ldots s$, and the coefficients of these polynomials are power products of total degree q composed of powers of the non-commutative symbols A and B.

Example 2 ([2]). For $s = 2$ we obtain

$$\frac{\mathrm{d}}{\mathrm{d}h} \mathcal{L}(h) \Big|_{h=0} = (a_1 + a_2) A + (b_1 + b_2) B - (A + B),$$

$$\frac{\mathrm{d}^2}{\mathrm{d}h^2} \mathcal{L}(h) \Big|_{h=0} = ((a_1 + a_2)^2) A^2 + (2 a_2 b_1) A B$$

$$+ (2 a_1 b_1 + 2 a_1 b_2 + 2 a_2 b_2) B A + ((b_1 + b_2)^2) B^2$$

$$- (A^2 + A B + B A + B^2).$$

The consistency condition for order $p \geq 1$ reads $\frac{\mathrm{d}}{\mathrm{d}h} \mathcal{L}(h) \big|_{h=0} = 0$, which is equivalent to $a_1 + a_2 = 1$ and $b_1 + b_2 = 1$.

At first sight, for order $p \geq 2$ we need 4, or (at second sight) 2 additional equations to be satisfied, such that $\frac{\mathrm{d}^2}{\mathrm{d}h^2} \mathcal{L}(h) \big|_{h=0} = 0$. However, assuming that the conditions for order $p \geq 1$ are satisfied, the second derivative $\frac{\mathrm{d}^2}{\mathrm{d}h^2} \mathcal{L}(h) \big|_{h=0}$ simplifies to the commutator expression

$$\frac{\mathrm{d}^2}{\mathrm{d}h^2} \mathcal{L}(h) \Big|_{h=0} = (2 a_2 b_1 - 1) [A, B],$$

giving the single additional condition $2 a_2 b_1 = 1$ for order $p \geq 2$. Assuming now that a_1, a_2 and b_1, b_2 are chosen such that all conditions for $p \geq 2$ are satisfied, the third derivative $\frac{\mathrm{d}^3}{\mathrm{d}h^3} \mathcal{L}(h) \big|_{h=0}$, which now represents the leading term of the local error, simplifies to a linear combination of the commutators $[A, [A, B]]$ and $[[A, B], B]$, of degree 3, namely

$$\frac{\mathrm{d}^3}{\mathrm{d}h^3} \mathcal{L}(h) \Big|_{h=0} = (3 a_2^2 b_1 - 1) [A, [A, B]] + (3 a_2 b_1^2 - 1) [[A, B], B]. \qquad \square$$

Remark 1. The classical second-order Strang splitting method corresponds to the choice $a_1 = \frac{1}{2}, b_1 = 1, a_2 = \frac{1}{2}, b_2 = 0$, or $a_1 = 0, b_1 = \frac{1}{2}, a_2 = 1, b_2 = \frac{1}{2}$.

The observation from this simple example generalizes as follows:

Proposition 1. *The leading term $\frac{\mathrm{d}^{p+1}}{\mathrm{d}h^{p+1}} \mathcal{L}(h) \big|_{h=0}$ of the Taylor expansion of the local error $\mathcal{L}(h)$ of a splitting method of order p is a Lie element, i.e., it is a linear combination of commutators of degree $p + 1$.*

Proof. See [1,8]. $\qquad \square$

Example 3. Assume that the coefficients $a_j, b_j, \; j = 1 \ldots s$ have been found such that the associated splitting scheme is of order $p \geq 3$ (this necessitates $s \geq 3$). This means that

$$\frac{\mathrm{d}}{\mathrm{d}h} \mathcal{L}(h) \Big|_{h=0} = \frac{\mathrm{d}^2}{\mathrm{d}h^2} \mathcal{L}(h) \Big|_{h=0} = \frac{\mathrm{d}^3}{\mathrm{d}h^3} \mathcal{L}(h) \Big|_{h=0} = 0,$$

and from Proposition 1 we know that

$$\frac{\mathrm{d}^4}{\mathrm{d}h^4} \mathcal{L}(h) \Big|_{h=0} = \gamma_1 \, [A, [A, [A, B]]] + \gamma_2 \, [A, [[A, B], B]] + \gamma_3 \, [[[A, B], B], B]$$

holds, with certain coefficients γ_k depending on the a_j and b_j. Here we have made use of the fact that there are three independent commutators of degree 4 in A and B. $\square$

Targeting for higher-order methods one needs to know a *basis of commutators* up to a certain degree. The answer to this question is known, and a full set of independent commutators of degree q can be represented by a set of words of length q over the alphabet $\{A, B\}$. A prominent example is the set of *Lyndon-Shirshov words* (see [6]) displayed in Table 1. A combinatorial algorithm due to Duval [7] can be used to generate this table.

Here, for instance, the word AABBB represents the commutator

$$[A, [[[A, B], B], B]] =$$
$$A^2 B^3 - 3ABAB^2 + 3AB^2AB - 2AB^3A + 3BAB^2A - 3B^2ABA + B^3A^2,$$

with leading power product $A^2 B^3 = AABBB$ (w.r.t. lexicographical order).

Table 1. L_q is the number of words of length q.

q	L_q	Lyndon-Shirshov words over the alphabet $\{A, B\}$
1	2	A, B
2	1	AB
3	2	AAB, ABB
4	3	AAAB, AABB, ABBB
5	6	AAAAB, AAABB, AABAB, AABBB, ABABB, ABBBB
6	9	AAAAAB, AAAABB, AAABAB, AAABBB, AABABB, AABBAB, AABBBB, ABABBB, ABBBBB
7	18	...
8	30	...
9	56	...
10	99	...
⋮	⋮	⋱

2.3   The Algorithm: Implicit Recursive Elimination

On the basis of Proposition 1, and with a table of Lyndon-Shirshov words available, we can build up a set of conditions for order $\geq p$ for a splitting method with s stages in the following way (recall the notation $A_j := a_j A$, $B_j = b_j B$). This procedure corresponds to [1, Algorithm 2]:

   For $q = 1 \ldots p$:

- *Generate the symbolic expressions (7) in the indeterminate coefficients a_j, b_j and the non-commutative variables A and B.*
- *Extract the coefficients of the power products (of degree q) represented by all Lyndon-Shirshov words of length q, resulting in a set of L_q polynomials $P_{q,k}(a_j, b_j)$ of degree q in the coefficients a_j and b_j.*

The resulting set of $\sum_{q=1}^{p} L_q$ multivariate polynomial equations

$$P_{q,k}(a_j, b_j) = 0, \quad k = 1 \ldots L_q, \quad q = 1 \ldots p \tag{8}$$

represents the desired conditions for order p.

   We call this procedure *implicit recursive elimination,* because the equations generated in this way are correct in an 'a posteriori' sense (cf. Example 2):

- For $q = 1$, the basic consistency equations

$$
\begin{aligned}
P_{1,1}(a_j, b_j) &= a_1 + \ldots + a_s - 1 = 0, \\
P_{1,2}(a_j, b_j) &= b_1 + \ldots + b_s - 1 = 0,
\end{aligned} \tag{9a}
$$

   are obtained.
- *Assume* that (9a) is satisfied. Then, due to Proposition 1, the additional (quadratic) equation (note that $L_2 = 1$)

$$P_{2,1}(a_j, b_j) = 0, \tag{9b}$$

   represents one additional condition for a scheme of order $p = 2$.
- *Assume* that (9a) and (9b) are satisfied. Then, due to Proposition 1, the additional (cubic) equations (note that $L_3 = 2$)

$$P_{3,1}(a_j, b_j) = P_{3,2}(a_j, b_j) = 0, \tag{9c}$$

   represent two additional conditions for a scheme of order $p = 3$.
- The process is continued in the same manner.

If we (later) *have found* a solution $S = \{a_j, b_j, \ j = 1 \ldots s\}$ of the resulting system

$$(8) = \{(9a), (9b), (9c), \ldots\}$$

of multivariate polynomial equations, this means that

- S satisfies (9a) $\Rightarrow$
   S represents a solution of order $q = 1$ at least;

- S satisfies (9a) and (9b) $\Rightarrow$
  S represents a solution of order $q = 2$ at least;
- S satisfies (9a), (9b), and (9c) $\Rightarrow$
  S represents a solution of order $q = 3$ at least;

and so on. By induction we conclude that the whole procedure indeed results in a solution S representing a method of the desired order p. See [2] for a more detailed exposition of this argument.

Remark 2. In addition, it makes sense to generate the additional conditions for order $p + 1$. Even if we do not solve for these conditions, they represent the leading term of the local error, and this can be used to search for optimized solutions for order p, where the coefficients in $\frac{d^{p+1}}{dh^{p+1}} \mathcal{L}(h)\big|_{h=0}$ become minimal in size.

3   A Parallel Implementation

In our Maple code, a table of Lyndon-Shirshov words up to a fixed length (corresponding to the maximal order aimed for; see Table 1) is included as static data. The procedure `Order_conditions` displayed below generates a set of order conditions using the algorithm described in Sect. 2.3.

- First of all, we activate the package `Physics` and declare the symbols `A` and `B` as non-commutative.
- For organizing the multinomial expansion according to (7) we use standard functions from the packages `combinat` and `combstruct`.
- The number of terms during each stage rapidly increases as more stages are to be computed. Therefore we have implemented a parallel version based on the package `Grid`. Parallelization is taken into account as follows:
  - On a multi-core processor, all threads execute the same code. Each thread identifies itself via a call to `MyNode()`, and this is used to control execution. Communication between the threads is realized via message passing.
  - Thread 0 is the master thread controlling the overall execution.
  - For $q = 1 \ldots p$:
    * Each of the working threads generates symbolic expressions of the form (recall $A_j = a_j\, A$, $B_j = b_j\, B$)

$$\Pi_{\boldsymbol{k}} := \binom{q}{\boldsymbol{k}} \prod_{j=s\ldots 1} \sum_{\ell=0}^{k_j} \binom{k_j}{\ell} B_j^{\ell}\, A_j^{k_j - \ell}, \quad \boldsymbol{k} \in \mathbb{N}_0^s,$$

      appearing in the sum (7). Here the work is equidistributed over the threads, i.e., each of them generates a subset of $\{\Pi_{\boldsymbol{k}},\ \boldsymbol{k} \in \mathbb{N}_0^s\}$ in parallel.
    * For each of these expressions $\Pi_{\boldsymbol{k}}$, the coefficients of all Lyndon-Shirshov monomials of degree q are computed, and the according subsets of coefficients are summed up in parallel.

* Finally, the master thread 0 sums up the results received from all the working threads. This results in the set of multivariate polynomials representing the order conditions at level q.
– The Maple code displayed below is, to some extent, to be read as pseudo-code. For simplicity of presentation we have ignored some technicalities, e.g., concerning the proper indexing of combinatorial tupels, etc. The original, working code is available from the authors.

```
> with(combinat)
> with(combstruct)
> with(Grid)
> with(Physics)
> Setup(noncommutativeprefix={A,B})

> Order_conditions := proc()
  global p,s,OC,    # I/O parameters via global variables
        Lyndon      # assume that table of Lyndon monomials is available
  this_thread := MyNode()    # each thread identifies itself
  max_threads := NumNodes() # number of available threads
  for j from 1 to s do
      A_j[j] := a[j]*A
      B_j[j] := b[j]*B
      term[-1][j] := 1
  end do
  OC=[0$p]
  for q from 1 to p do
    if this_thread>0 then  # working threads start computing
                           # master thread 0 is waiting
      Mn := allstructs(Composition(q+2),size=2)
      for j from 1 to s do
          term[q-1][j] := 0
          for mn from 1 to nops(Mn) do
              term[q-1][j] :=
                term[q-1][j] +
                  multinomial(q,Mn[mn])*B_j[j]^Mn[mn][2]*A_j[j]^Mn[mn][1]
          end do
      end do
      k := iterstructs(Composition(q+s),size=s)
      OC_q_this_thread := [0$nops(Lyndon[q])]
      while not finished(k) do # generate expansion (7) term by term
        Ms := nextstruct(k)
        if get_active(this_thread) then # get_active:
                                        #   auxiliary Boolean function
                                        #   for equidistributing workload
            Pi_k := 1
            for j from s to 1 by -1 do
                Pi_k := Pi_k*term[Ms[j]-1][j]
            end do
            Pi_k := multinomial(q,Ms)*expand(Pi_k)
            OC_q_this_thread := # compare coefficients of Lyndon monomials
              OC_q_this_thread +
                [seq(coeff(Pi_k,Lyndon[q][l]),l=1..nops(Lyndon[q]))]
        end if
```

```
        end do
        Send(0,OC_q_this_thread) # send partial sum to master thread
    else # master thread 0 receives and sums up
        #    partial results from working threads
        OC[q] := [(-1)$nops(Lyndon[q])] # initialize sum
        for i_thread from 1 to max_threads-1 do
            OC[q] := OC[q] + Receive(i_thread)
        end do
    end if
end do
end proc
```

```
> # Example:
> p := 4
> s := 4
> Launch(Order_conditions,imports=["p","s"],exports=["OC"])  # run
```

```
> OC[1]
  [a[1]+a[2]+a[3]+a[4]-1,
   b[1]+b[2]+b[3]+b[4]-1]
```

```
> OC[2]
  [2*a[2]*b[1]+2*a[3]*b[1]+2*a[3]*b[2]
   +2*a[4]*b[1]+2*a[4]*b[2]+2*a[4]*b[3]-1]
```

```
> OC[3]
  [3*a[2]^2*b[1]+6*a[2]*a[3]*b[1]+6*a[2]*a[4]*b[1]
   +3*a[3]^2*b[1]+3*a[3]^2*b[2]+6*a[3]*a[4]*b[1]+6*a[3]*a[4]*b[2]
   +3*a[4]^2*b[1]+3*a[4]^2*b[2]+3*a[4]^2*b[3]-1,
   3*a[2]*b[1]^2+3*a[3]*b[1]^2+6*a[3]*b[1]*b[2]
   +3*a[3]*b[2]^2+3*a[4]*b[1]^2+6*a[4]*b[1]*b[2]+6*a[4]*b[1]*b[3]
   +3*a[4]*b[2]^2+6*a[4]*b[2]*b[3]+3*a[4]*b[3]^2-1]
```

```
> OC[4]
  [4*a[2]^3*b[1]+12*a[2]^2*a[3]*b[1]+12*a[2]^2*a[4]*b[1]
   +12*a[2]*a[3]^2*b[1]+24*a[2]*a[3]*a[4]*b[1]+12*a[2]*a[4]^2*b[1]
   +4*a[3]^3*b[1]+4*a[3]^3*b[2]+12*a[3]^2*a[4]*b[1]
   +12*a[3]^2*a[4]*b[2]+12*a[3]*a[4]^2*b[1]+12*a[3]*a[4]^2*b[2]
   +4*a[4]^3*b[1]+4*a[4]^3*b[2]+4*a[4]^3*b[3]-1,
   6*a[2]^2*b[1]^2+12*a[2]*a[3]*b[1]^2+12*a[2]*a[4]*b[1]^2
   +6*a[3]^2*b[1]^2+12*a[3]^2*b[1]*b[2]+6*a[3]^2*b[2]^2
   +12*a[3]*a[4]*b[1]^2+24*a[3]*a[4]*b[1]*b[2]+12*a[3]*a[4]*b[2]^2
   +6*a[4]^2*b[1]^2+12*a[4]^2*b[1]*b[2]+12*a[4]^2*b[1]*b[3]
   +6*a[4]^2*b[2]^2+12*a[4]^2*b[2]*b[3]+6*a[4]^2*b[3]^2-1,
   4*a[2]*b[1]^3+4*a[3]*b[1]^3+12*a[3]*b[1]^2*b[2]
   +12*a[3]*b[1]*b[2]^2+4*a[3]*b[2]^3+4*a[4]*b[1]^3
   +12*a[4]*b[1]^2*b[2]+12*a[4]*b[1]^2*b[3]+12*a[4]*b[1]*b[2]^2
   +24*a[4]*b[1]*b[2]*b[3]+12*a[4]*b[1]*b[3]^2+4*a[4]*b[2]^3
   +12*a[4]*b[2]^2*b[3]+12*a[4]*b[2]*b[3]^2+4*a[4]*b[3]^3-1]
```

For practical use some further tools have been developed, e.g. for generating tables of polynomial coefficients for further use, e.g., by numerical software other than Maple. This latter job can also be parallelized.

3.1 Special Cases

Some special cases are of interest:

– *Symmetric schemes* are characterized by the property $\mathcal{S}(-h, \mathcal{S}(h, u)) = u$. Here, either $a_1 = 0$ or $b_s = 0$, and the remaining coefficient sets (a_j) and (b_j) are palindromic. Symmetric schemes have an even order p, and the order conditions for even orders need not be included; see [8]. Thus, we use a special ansatz and generate a reduced set of equations.
– *Palindromic schemes* were introduced in [2] and characterized by the property $\mathcal{S}(-h, \check{\mathcal{S}}(h, u)) = u$, where, $\check{\mathcal{S}}$ denotes the scheme $\mathcal{S}$ with the role of A and B interchanged. In this case, the full coefficient set

$$(a_1, b_1, \ldots, a_s, b_s)$$

is palindromic. As for symmetric schemes, this means that a special ansatz is used, and again it is sufficient to generate a reduced set of equations, see [2].

Apart from these modifications, the basic algorithm remains unchanged.

4 Modifications and Extensions

4.1 Splitting into More Than Two Operators

Our algorithm directly generalizes to the case of splitting into more than two operators. Consider evolution equations where the right-hand side splits into three parts,

$$\partial_t u(t) = F(u(t)) = A(u(t)) + B(u(t)) + C(u(t)), \tag{10}$$

and associated splitting schemes,

$$\mathcal{S}(h, u) = \mathcal{S}_s(h, \mathcal{S}_{s-1}(h, \ldots, \mathcal{S}_1(h, u))) \approx \Phi_F(h, u), \tag{11a}$$

with

$$\mathcal{S}_j(h, v) = \Phi_C(c_j\, h, \Phi_B(b_j\, h, \Phi_A(a_j\, h, v))), \tag{11b}$$

see [4]. Here the linear representation (7) generalizes as follows, with $A_j = a_j\, A$, $B_j = b_j\, B$, $C_j = c_j\, C$, and $\boldsymbol{k} = (k_1, \ldots, k_s) \in \mathbb{N}_0^s$, $\boldsymbol{\ell} = (\ell_A, \ell_B, \ell_C) \in \mathbb{N}_0^3$:

$$\frac{\mathrm{d}^q}{\mathrm{d}h^q} \mathcal{L}(h)\Big|_{h=0} = \sum_{|\boldsymbol{k}|=q} \binom{q}{\boldsymbol{k}} \prod_{j=s\ldots 1} \sum_{|\boldsymbol{\ell}|=k_j} \binom{k_j}{\boldsymbol{\ell}} C_j^{\ell_C} B_j^{\ell_B} A_j^{\ell_A} - (A + B + C)^q.$$

$$\tag{12}$$

On the basis of these identities, the algorithm from Sect. 2.3 generalizes in a straightforward way. The Lyndon basis representing independent commutators now corresponds to Lyndon words over the alphabet $\{\mathtt{A}, \mathtt{B}, \mathtt{C}\}$, see [2]. Concerning special cases (symmetries etc.) and parallelization, similar considerations as before apply.

4.2   Pairs of Splitting Schemes

For the purpose of adaptive time-splitting algorithms, the construction of (optimized) pairs of schemes of orders $(p, p+1)$ is favorable. Generating a respective set of order conditions can also be accomplished by a modification of our code; the difference lies in the fact that some coefficients are chosen a priori (corresponding to a given method of order $p+1$), but apart from this the generation of order conditions for an associated scheme of order p works analogously as before. Finding optimal schemes is then accomplished by tracing a large set of possible solutions; see [2].

5   Computational Performance; Conclusions

The following computations were performed on a node consisting of two processors of type AMD Opteron 6132 HR (2.2 GHz) with 8 cores. This means that together with a master thread up to 15 working threads can be used. An ample memory of 32 GB is available.

Beginning with order $p = 6$, the computational effort becomes significant (and strongly increases with higher orders). We consider two different parameter settings (without assuming any symmetries for the setup of order conditions):

(i) AB – splitting, 10-stage scheme ($s = 10$), desired order $p = 6$,
(ii) ABC – splitting, 15-stage scheme ($s = 15$), desired order $p = 6$.

Timing data are specified in the format [d]:[hh]:mm:ss.

For case (i) we compare the performance for the fully parallelized version including 15 working threads with a restricted, essentially sequential version where only 1 working thread is used.

(i) • 15 active working threads:
     wall clock time = 00:45,
     total CPU time = 09:48.
     This amounts to an overall processor utilization of about 85 %.

   • 1 active working thread:
     wall clock time = 07:46,
     total CPU time = 07:58.
   We approximately observe the expected linear speedup of running (wall clock) time with the number of threads used. The slightly increased cost (in terms of total CPU time) of the fully parallel version is to be attributed to communication between the threads.

(ii) • 15 active working threads:
     wall clock time = 01:46:03,
     total CPU time = 1:01:29:47.
     This amounts to an efficient overall processor utilization of about 90 %.
     For this case we have not performed a run with a single working thread.

For the ABC case the absolute timing data are significantly larger due to the fact that the number of terms in the Taylor expansion of the local error grows much faster.[5]

Especially in this latter case, the poor computational performance of a general-purpose symbolic system like Maple (including the `Physics` package) becomes evident. Here, parallelization is essential to reduce wall-clock times as much as possible. The algorithm presented here could also be implemented in a 'slimmer' language as for instance C or Julia, but of course at the expense of implementing many auxiliary components like various combinatorial functions and, in particular, handling of expressions involving non-commutative variables. In this sense, our implementation is a pragmatic one: Make use of a readily available software package and gain performance via parallelization, a strategy which may be relevant also for other kinds of symbolic codes.

Acknowledgements. This work was supported by the Austrian Science Fund (FWF) under grant P24157-N13, and by the Vienna Science and Technology Fund (WWTF) under grant MA-14-002. Computational results based on the ideas in this work have been achieved in part using the Vienna Scientific Cluster (VSC).

References

1. Auzinger, W., Herfort, W.: Local error structures and order conditions in terms of Lie elements for exponential splitting schemes. Opuscula Math. **34**, 243–255 (2014)
2. Auzinger, W., Hofstätter, H., Ketcheson, D., Koch, O.: Practical splitting methods for the adaptive integration of nonlinear evolution equations. Part I: Construction of optimized schemes and pairs of schemes. To appear in BIT Numer. Math
3. Auzinger, W., Koch, O.: Coefficients of various splitting methods. www.asc.tuwien.ac.at/ winfried/splitting/
4. Auzinger, W., Koch, O., Thalhammer, M.: Defect-based local error estimators for high-order splitting methods involving three linear operators. Numer. Algorithms **70**, 61–91 (2015)
5. Blanes, S., Casas, F., Farrés, A., Laskar, J., Makazaga, J., Murua, A.: New families of symplectic splitting methods for numerical integration in dynamical astronomy. Appl. Numer. Math. **68**, 58–72 (2013)
6. Bokut, L., Sbitneva, L., Shestakov, I.: Lyndon-Shirshov words, Gröbner-Shirshov bases, and free Lie algebras. In: Non-Associative Algebra and Its Applications, chap 3. Chapman & Hall / CRC, Boca Raton (2006)
7. Duval, J.P.: Géneration d'une section des classes de conjugaison et arbre des mots de Lyndon de longueur bornée. Theoret. Comput. Sci. **60**, 255–283 (1988)
8. Hairer, E., Lubich, C., Wanner, G.: Geometric Numerical Integration, 2nd edn. Springer, Heidelberg (2006)
9. Ketcheson, D., MacDonald, C., Ruuth, S.: Spatially partitioned embedded Runge-Kutta methods. SIAM J. Numer. Anal. **51**, 2887–2910 (2013)

---

[5] There are special cases of practical interest where this growth is much more moderate; we do not discuss such details here.

Symbolic Manipulation of Flows of Nonlinear Evolution Equations, with Application in the Analysis of Split-Step Time Integrators

Winfried Auzinger[1], Harald Hofstätter[1(✉)], and Othmar Koch[2]

[1] Technische Universität Wien, Vienna, Austria
w.auzinger@tuwien.ac.at, hofi@harald-hofstaetter.at
[2] Universität Wien, Vienna, Austria
othmar@othmar-koch.org
www.asc.tuwien.ac.at/~winfried
www.harald-hofstaetter.at
www.othmar-koch.org

Abstract. We describe a package realized in the Julia programming language which performs symbolic manipulations applied to nonlinear evolution equations, their flows, and commutators of such objects. This tool was employed to perform contrived computations arising in the analysis of the local error of operator splitting methods. It enabled the proof of the convergence of the basic method and of the asymptotical correctness of a defect-based error estimator. The performance of our package is illustrated on several examples.

Keywords: Nonlinear evolution equations · Time integration · Splitting methods · Symbolic computation · Julia language

1 Problem Setting

We are interested in the solution to nonlinear evolution equations

$$\partial_t u(t) = A(u(t)) + B(u(t)) = H(u(t)), \quad u(0) = u_0, \tag{1}$$

on a Banach space X, where A and B are general nonlinear, unbounded operators defined on a subset $D \subset X$, the solution is denoted by $\mathcal{E}_H(t, u_0)$, and analogously for the two sub-flows associated with A and B. The structure of the vector fields often suggests to employ additive splitting methods to separately propagate the two subproblems defined by A and B,

$$u(t_1) \approx u_1 := \mathcal{S}(h, u_0) = \mathcal{E}_B(b_s h, \cdot) \circ \mathcal{E}_A(a_s h, \cdot) \circ \ldots \circ \mathcal{E}_B(b_1 h, \cdot) \circ \mathcal{E}_A(a_1 h, u_0), \tag{2}$$

where the coefficients $a_j, b_j, j = 1 \ldots s$ are determined according to the requirement that a prescribed order of consistency is obtained [5].

Both in the a priori error analysis and for a posteriori error estimation, a defect-based approach has been introduced in [4], which serves both to derive theoretical error bounds and as a basis for adaptive step-size selection.

© Springer International Publishing AG 2016
V.P. Gerdt et al. (Eds.): CASC 2016, LNCS 9890, pp. 43–57, 2016.
DOI: 10.1007/978-3-319-45641-6_4

In the analysis of this error estimate, extensive symbolic manipulation of the flows defined by the operators in (1), their Fréchet derivatives and arising commutators is indispensable. As such calculations imply a high effort of formula manipulation and are highly error prone, we have implemented an automatic tool in the Julia programming language to verify the calculation. The present paper focusses on this tool, since in [4] no details on the implementation are given. We are not aware of available tools providing the functionality required for our task in established computer algebra systems, whence we decided on an implementation from scratch. Julia is a high-level, high-performance dynamic programming language for technical computing, see [1], which appeared convenient for our purpose.

By defining a suitable set of substitution rules, the algorithm can check the equivalence of expressions built from the objects mentioned. On this basis, it was possible to ascertain the correctness of all steps in the proofs given in [4].

2   Local Error, Defect, and Error Estimator

We describe the background arising from the application of splitting methods (2) to the solution of evolution equations (1), see [4]. The defect of the splitting approximation is defined as

$$\mathcal{D}(t, u) = \mathcal{S}^{(1)}(t, u) = \partial_t \mathcal{S}(t, u) - H(\mathcal{S}(t, u)),$$

while the local error is given by

$$\mathcal{L}(t, u) = \mathcal{S}(t, u) - \mathcal{E}_H(t, u) = \int_0^t \mathcal{F}(\tau, t, u) \, d\tau, \tag{3}$$

with

$$\mathcal{F}(\tau, t, u) = \partial_2 \mathcal{E}_H(t - \tau, \mathcal{S}(\tau, u)) \cdot \mathcal{S}^{(1)}(\tau, u).$$

The Lie-Trotter Method

We illustrate our analysis of the local error for the simplest Lie-Trotter splitting method,

$$\mathcal{S}(t, u_0) = \mathcal{E}_B(t, \mathcal{E}_A(t, u_0)),$$

see [4]. This involves some nontrivial crucial identities, namely (4)–(9) below, which will be verified using our Julia package in Sect. 4.

Our aim is to show

$$\mathcal{S}^{(1)}(t, u) = \mathcal{O}(t),$$

and thus

$$\mathcal{L}(t, u) = \mathcal{O}(t^2).$$

$\mathcal{S}^{(1)}(t, u)$ can be represented in the form

$$\mathcal{S}^{(1)}(t, u) = \partial_t \mathcal{S}(t, u) - H(\mathcal{S}(t, u)) = \tilde{\mathcal{S}}^{(1)}(t, \mathcal{E}_A(t, u)), \tag{4a}$$

with

$$\tilde{S}^{(1)}(t,v) = \partial_2 \mathcal{E}_B(t,v) \cdot A(v) - A(\mathcal{E}_B(t,v)). \tag{4b}$$

$\tilde{S}^{(1)}(t,v)$ satisfies

$$\partial_t \tilde{S}^{(1)}(t,v) = B'(\mathcal{E}_B(t,v)) \cdot \tilde{S}^{(1)}(t,v) + [B,A](\mathcal{E}_B(t,v)),$$
$$\tilde{S}^{(1)}(0,v) = 0, \tag{5}$$

where

$$[B,A](u) = B'(u)A(u) - A'(u)B(u)$$

denotes the commutator of the two vector fields.

From

$$\partial_t \mathcal{E}_F(t,u) = F(\mathcal{E}_F(t,u)) \quad \Rightarrow \quad \partial_t \partial_2 \mathcal{E}_F(t,u) \cdot v = F'(\mathcal{E}_F(t,u)) \cdot \partial_2 \mathcal{E}_F(t,u) \cdot v$$

it follows that $\partial_2 \mathcal{E}_F(t,u)$ is a fundamental system of the linear differential equation

$$\partial_t X(t,u) = F'(\mathcal{E}_F(t,u)) \cdot X(t,u)$$

which satisfies

$$\partial_2 \mathcal{E}_F(t,u)^{-1} = \partial_2 \mathcal{E}_F(-t, \mathcal{E}_F(t,u)).$$

The solution of an inhomogenous system like (5) of the form

$$\partial_t X(t,u) = F'(\mathcal{E}_F(t,u)) \cdot X(t,u) + R(t,u),$$
$$X(0,u) = X_0(u)$$

can be represented by the variation of constant formula,

$$X(t,u) = \partial_2 \mathcal{E}_F(t,u) \cdot \left(X_0(u) + \int_0^t \partial_2 \mathcal{E}_F(-\tau, \mathcal{E}_F(\tau,u)) \cdot R(\tau,u)\, d\tau \right).$$

Hence, the term $\tilde{S}^{(1)}(t,v)$ defined in (4b) satisfies

$$\tilde{S}^{(1)}(t,v) = \partial_2 \mathcal{E}_B(t,v) \cdot \int_0^t \partial_2 \mathcal{E}_B(-\tau, \mathcal{E}_B(\tau,v)) \cdot [B,A](\mathcal{E}_B(\tau,v))\, d\tau.$$

From this integral representation it follows

$$\mathcal{D}(t,u) = \tilde{S}^{(1)}(t, \mathcal{E}_A(t,u)) = \mathcal{O}(t),$$

and

$$\mathcal{L}(t,u) = \int_0^t \partial_2 \mathcal{E}_H(t - \tau, \mathcal{S}(\tau,u)) \cdot \mathcal{D}(\tau,u)\, d\tau = \mathcal{O}(t^2).$$

As a basis for adaptive time-stepping, we define an a posteriori local error estimator by numerical evaluation of the integral representation (3) of the local error by the trapezoidal rule, yielding

$$\widehat{\mathcal{L}}(t,u) = \tfrac{1}{2} t\, \mathcal{F}(t,t,u) = \tfrac{1}{2} t\, \mathcal{D}(t,u) = \tfrac{1}{2} t\, \mathcal{S}^{(1)}(t,u).$$

To analyze the deviation of this error estimator from the exact error, we use the Peano representation

$$\widehat{\mathcal{L}}(t, u) - \mathcal{L}(t, u) = \int_0^t K_1(\tau, t) \, \partial_\tau \mathcal{F}(\tau, t, u) \, d\tau,$$

with the kernel

$$K_1(\tau, t) = \tau - \tfrac{1}{2} t = \mathcal{O}(t).$$

To infer asymptotical correctness of the error estimator, we wish to show that

$$\widehat{\mathcal{L}}(t, u) - \mathcal{L}(t, u) = \mathcal{O}(t^3).$$

To this end, we compute

$$\begin{aligned}
\partial_\tau \mathcal{F}(\tau, t, u) &= \partial_2 \mathcal{E}_H(t - \tau, \mathcal{S}(\tau, u)) \cdot \mathcal{S}^{(2)}(\tau, u) \\
&\quad + \partial_2^2 \mathcal{E}_H(t - \tau, \mathcal{S}(\tau, u))(\mathcal{S}^{(1)}(\tau, u), \mathcal{S}^{(1)}(\tau, u)) \qquad (6) \\
&= \partial_2 \mathcal{E}_H(t - \tau, \mathcal{S}(\tau, u)) \cdot \mathcal{S}^{(2)}(\tau, u) + \mathcal{O}(t),
\end{aligned}$$

where

$$\begin{aligned}
\mathcal{S}^{(2)}(t, u) &= \partial_t \mathcal{S}^{(1)}(t, u) - H'(\mathcal{S}(t, u)) \cdot \mathcal{S}^{(1)}(t, u) \\
&= \tilde{\mathcal{S}}^{(2)}(t, \mathcal{E}_A(t, u)), \qquad\qquad\qquad\qquad\qquad (7a)
\end{aligned}$$

with

$$\tilde{\mathcal{S}}^{(2)}(t, v) = \partial_2 \tilde{\mathcal{S}}^{(1)}(t, v) \cdot A(v) - A'(\mathcal{E}_B(t, v)) \cdot \tilde{\mathcal{S}}^{(1)}(t, v) + [B, A](\mathcal{E}_B(t, v)). \ (7b)$$

$\tilde{\mathcal{S}}^{(2)}(t, v)$ satisfies

$$\begin{aligned}
\partial_t \tilde{\mathcal{S}}^{(2)}(t, v) &= B'(\mathcal{E}_B(t, v)) \cdot \tilde{\mathcal{S}}^{(2)}(t, v) \\
&\quad + B''(\mathcal{E}_B(t, v))(\tilde{\mathcal{S}}^{(1)}(t, v), \tilde{\mathcal{S}}^{(1)}(t, v)) \\
&\quad - [B, [B, A]](\mathcal{E}_B(t, v)) - [A, [B, A]](\mathcal{E}_B(t, v)) \qquad (8) \\
&\quad + 2[B, A]'(\mathcal{E}_B(t, v)) \cdot \tilde{\mathcal{S}}^{(1)}(t, v), \\
\tilde{\mathcal{S}}^{(2)}(0, v) &= [B, A](v).
\end{aligned}$$

This implies the integral representation

$$\begin{aligned}
\tilde{\mathcal{S}}^{(2)}(t, v) &= \partial_2 \mathcal{E}_B(t, v) \cdot [B, A](v) + \partial_2 \mathcal{E}_B(t, v) \cdot \int_0^t \partial_2 \mathcal{E}_B(-\tau, \mathcal{E}_B(\tau, v)) \cdot \\
&\quad \Big(B''(\mathcal{E}_B(\tau, v))(\tilde{\mathcal{S}}^{(1)}(\tau, v), \tilde{\mathcal{S}}^{(1)}(\tau, v)) \\
&\quad - [B, [B, A]](\mathcal{E}_B(\tau, v)) - [A, [B, A]](\mathcal{E}_B(\tau, v)) \\
&\quad + 2[B, A]'(\mathcal{E}_B(\tau, v)) \cdot \tilde{\mathcal{S}}^{(1)}(\tau, v) \Big) \, d\tau.
\end{aligned}$$

Thus,
$$\mathcal{S}^{(2)}(\tau, u) = \partial_2 \mathcal{E}_B(\tau, \mathcal{E}_A(\tau, u)) \cdot [B, A](\mathcal{E}_A(\tau, u)) + \mathcal{O}(t),$$

and altogether

$$\widehat{\mathcal{L}}(t, u) - \mathcal{L}(t, u) = \int_0^t K_1(\tau, t)\, \partial_\tau \mathcal{F}(\tau, t, u)\, \mathrm{d}\tau$$

$$= \int_0^t K_1(\tau, t)$$
$$\cdot\, \partial_2 \mathcal{E}_H(t - \tau, \mathcal{S}(\tau, u)) \cdot \partial_2 \mathcal{E}_B(\tau, \mathcal{E}_A(\tau, u)) \cdot [B, A](\mathcal{E}_A(\tau, u))\, \mathrm{d}\tau + \mathcal{O}(t^3)$$

$$= \int_0^t K_2(\tau, t)$$
$$\cdot\, \partial_\tau \Big(\partial_2\, \mathcal{E}_H(t - \tau, \mathcal{S}(\tau, u)) \cdot \partial_2\, \mathcal{E}_B(\tau, \mathcal{E}_A(\tau, u)) \cdot [B, A](\mathcal{E}_A(\tau, u)) \Big)\, \mathrm{d}\tau + \mathcal{O}(t^3),$$

where $K_2(\tau, t) = \frac{1}{2}\tau(t - \tau) = \mathcal{O}(t^2)$ by partial integration. Here,

$$\partial_\tau \Big(\partial_2 \mathcal{E}_H(t - \tau, \mathcal{S}(\tau, u)) \cdot \partial_2 \mathcal{E}_B(\tau, \mathcal{E}_A(\tau, u)) \cdot [B, A](\mathcal{E}_A(\tau, u)) \Big) = \mathcal{O}(1), \quad (9\mathrm{a})$$

because this derivative can be expressed as

$$\Big[\partial_2 \mathcal{E}_H(t - \tau, \mathcal{E}_B(\tau, v)) \cdot \partial_2 \mathcal{E}_B(\tau, v) \cdot [[B, A], A](v)$$
$$+ \partial_2 \mathcal{E}_H(t - \tau, \mathcal{E}_B(\tau, v)) \cdot \partial_2 \tilde{\mathcal{S}}^{(1)}(\tau, v) \cdot [B, A](v)$$
$$+ \partial_2^2 \mathcal{E}_H(t - \tau, \mathcal{E}_B(\tau, v))\big(\tilde{\mathcal{S}}^{(1)}(\tau, v), \partial_2 \mathcal{E}_B(\tau, v) \cdot [B, A](v) \big) \Big]_{v = \mathcal{E}_A(t, u)}. \quad (9\mathrm{b})$$

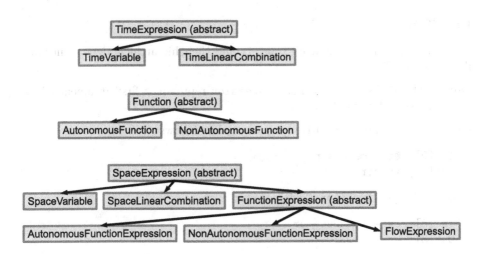

Fig. 1. Data type hierarchy of `Flows.jl`.

The Strang Splitting Method

For the Strang splitting method

$$\mathcal{S}(t, u_0) = \mathcal{E}_B(\tfrac{t}{2}, \mathcal{E}_A(t, \mathcal{E}_B(\tfrac{t}{2}, u_0))),$$

our aim is to show

$$\mathcal{S}^{(1)}(t, u) = \mathcal{O}(t^2),$$

and thus

$$\mathcal{L}(t, u) = \mathcal{O}(t^3).$$

In this case the necessary manipulations become significantly more complex than for the Lie-Trotter case, in particular concerning the analysis of a defect-based a posteriori error estimator. The analysis is based on multiple application of variation of constant formulas in a general nonlinear setting. This was the main motivation for the development of our computational tool; in particular, the theoretical results from [4], which are not repeated here, have been verified using this tool.

3   The Julia Package Flows.jl

The package described in the following is available from [2]. It consists of approximately 1000 lines of Julia code and is essentially self-contained, relying only on the Julia standard library but not on additional packages. A predecessor written in Perl has been used for the preparation of [4]. In a Julia notebook, the package is initialized as follows:

```
In [1]: using Flows
```

Data Types

The data types in the package Flows.jl and their hierarchical dependence are illustrated in Fig. 1.

Objects of the abstract type TimeExpression represent a first argument t in a flow expression like $\mathcal{E}_H(t, u)$.

Objects of type TimeVariable are generated as follows:

```
In [2]: @t_vars t s r
Out[2]: (t,s,r)
```

Objects of type TimeLinearCombination:

```
In [3]: ex = t - 2s + 3r
Out[3]: 3r + t - 2s
```

Similarly, objects of the abstract type SpaceExpression represent a second argument u in a flow expression like $\mathcal{E}_H(t, u)$.

Objects of type `SpaceVariables`:

```
In [4]: @x_vars u v w
Out[4]: (u,v,w)
```

In order to construct objects of type `AutonomousFunctionExpression` or `FlowExpression` like $A(u)$ or $\mathcal{E}_A(t, u)$ we first need to declare a symbol for A of type `AutonomousFunction`.[1]

```
In [5]: @funs A
Out[5]: (A,)
```

Now we can generate objects of types `AutonomousFunctionExpression` and `FlowExpression`:

```
In [6]: ex1 = A(u)
Out[6]: A(u)
In [7]: ex2 = E(A,t,u)
Out[7]: EA(t,u)
```
$$\text{Out}[6]: A(u)$$
$$\text{Out}[7]: \mathcal{E}_A(t, u)$$

Additional arguments in such expressions represent arguments for Fréchet derivatives with respect to a `SpaceVariable` u. Note that the order of the derivative is implicitly determined from the number of arguments.

```
In [8]: ex1 = A(u,v,w)
Out[8]: A''(u)(v,w)
In [9]: ex2 = E(A,t,u,v)
Out[9]: ∂2EA(t,u) · v
```
$$\text{Out}[8]: A''(u)(v, w)$$
$$\text{Out}[9]: \partial_2 \mathcal{E}_A(t, u) \cdot v$$

Objects of type `SpaceLinearCombination` can be built from such expressions:

```
In [10]: ex = -2E(A,t,u,v,w)+2u+A(v,w)
Out[10]: -2∂2²EA(t,u)(v,w) + 2u + A'(v) · w
```
$$\text{Out}[10]: -2\partial_2^2 \mathcal{E}_A(t, u)(v, w) + 2u + A'(v) \cdot w$$

Methods

differential

This method generates the Fréchet derivative of an expression with respect to a `SpaceVariable` applied to an expression.

```
In [11]: ex = A(B(u)) + E(A,t,u,v)
Out[11]: A(B(u)) + ∂2EA(t,u) · v
In [12]: differential(ex,u,B(w))
Out[12]: A'(B(u)) · B'(u) · B(w) + ∂2²EA(t,u)(v,B(w))
```
$$\text{Out}[11]: A(B(u)) + \partial_2 \mathcal{E}_A(t, u) \cdot v$$
$$\text{Out}[12]: A'(B(u)) \cdot B'(u) \cdot B(w) + \partial_2^2 \mathcal{E}_A(t, u)(v, B(w))$$

t_derivative

This method generates the derivative of an expression with respect to a `TimeVariable`.

---

[1] Likewise for objects of type `NonAutonomousFunction`.

```
In [13]:  ex = E(A,t-2s,u+E(B,s,v))
```
Out[13]: $\mathcal{E}_A(t - 2s, u + \mathcal{E}_B(s, v))$

```
In [14]:  t_derivative(ex,s)
```
Out[14]: $\partial_2 \mathcal{E}_A(t - 2s, u + \mathcal{E}_B(s, v)) \cdot B(\mathcal{E}_B(s, v)) - 2A(\mathcal{E}_A(t - 2s, u + \mathcal{E}_B(s, v)))$

expand

A (higher-order) Fréchet derivative is a (multi-)linear map. The method `expand` transforms the application of such a (multi-)linear map to a linear combination of expressions into the corresponding linear combination of (multi-)linear maps.

```
In [15]:  ex1 = E(A,t,u,2v+3w)
```
Out[15]: $\partial_2 \mathcal{E}_A(t, u) \cdot (2v + 3w)$

```
In [16]:  expand(ex1)
```
Out[16]: $2\partial_2 \mathcal{E}_A(t, u) \cdot v + 3\partial_2 \mathcal{E}_A(t, u) \cdot w$

```
In [17]:  ex2 = A(u,v+w,v+w)
```
Out[17]: $2A''(u)(v + w, v + w)$

```
In [18]:  expand(ex2)
```
Out[18]: $2A''(u)(v, w) + A''(u)(v, v) + A''(u)(w, w)$

substitute

Different variants of substitutions are implemented. The most sophisticated one is substitution of an object of type `Function` by an expression. For example, this allows to define the double commutator $[A, [A, B]](u)$ by substituting B in $[A, B](u)$ by $[A, B](u)$:

```
In [19]:  C_AB = A(u,B(u))-B(u,A(u))
```
Out[19]: $A'(u) \cdot B(u) - B'(u) \cdot A(u)$

```
In [20]:  substitute(C_AB,B,C_AB,u)
```
Out[20]: $-A''(u)(B(u), A(u)) + B'(u) \cdot A'(u) \cdot A(u) + B''(u)(A(u), A(u))$
$\qquad -A'(u) \cdot B'(u) \cdot A(u) + A'(u) \cdot (-B'(u) \cdot A(u) + A'(u) \cdot B(u))$

commutator

This method generates expressions involving commutators $[A, B]$ and double commutators $[A, [B, C]]$.

```
In [21]:  ex1 = commutator(A,B,u)
```
Out[21]: $A'(u) \cdot B(u) - B'(u) \cdot A(u)$

```
In [22]:  ex2 = commutator(A,B,C,u)
```
Out[22]: $C'(u) \cdot B'(u) \cdot A(u) + C''(u)(B(u), A(u)) - B'(u) \cdot C'(u) \cdot A(u)$
$\qquad - B''(u)(C(u), A(u)) + A'(u) \cdot (B'(u) \cdot C(u) - C'(u) \cdot B(u))$

Example: We verify the Jacobi identity $[A, [B, C]] + [B, [C, A]] + [C, [A, B]] = 0$:

```
In [23]:  ex3 = expand(commutator(A,B,C,u)+commutator(B,C,A,u)
                  +commutator(C,A,B,u))
```
Out[23]: 0

FE2DEF, DEF2FE

These methods substitute expressions according to the fundamental identity

$$A(\mathcal{E}_A(t,u)) = \partial_2 \mathcal{E}_A(t,u) \cdot A(u),$$

see [4].

In [24]: `ex1 = A(E(A,t,u))`
Out[24]: $A(\mathcal{E}_A(t,u))$

In [25]: `ex2 = FE2DEF(ex1)`
Out[25]: $\partial_2 \mathcal{E}_A(t,u) \cdot A(u)$

In [26]: `DEF2FE(ex2)`
Out[26]: $A(\mathcal{E}_A(t,u))$

reduce_order

This method constitutes the essential manipulation needed for the verification of the identities (4)–(9) in Sect. 2. By repeated differentiation of the fundamental identity

$$A(\mathcal{E}_A(t,u)) - \partial_2 \mathcal{E}_A(t,u) \cdot A(u) = 0$$

we obtain

$$A'(\mathcal{E}_A(t,u)) \cdot \partial_2 \mathcal{E}_A(t,u) \cdot v - \partial_2^2 \mathcal{E}_A(t,u)(A(u),v) - \partial_2 \mathcal{E}_A(t,u) \cdot A'(u) \cdot v = 0,$$

$$A''(\mathcal{E}_A(t,u))(\partial_2 \mathcal{E}_A(t,u) \cdot v, \partial_2 \mathcal{E}_A(t,u) \cdot w) + A'(\mathcal{E}_A(t,u)) \cdot \partial_2^2 \mathcal{E}_A(t,u)(v,w)$$
$$-\partial_2^3 \mathcal{E}_A(t,u)(A(u),v,w) - \partial_2^2 \mathcal{E}_A(t,u)(A'(u) \cdot v, w)$$
$$-\partial_2^2 \mathcal{E}_A(t,u)(A'(u) \cdot w, v) - \partial_2 \mathcal{E}_A(t,u) \cdot A''(u)(v,w) = 0,$$

and so on. The method **reduce_order** transforms expressions of the form of the highest order derivative in these identities by means of these very identities:

In [27]: `ex1 = E(A,t,u,A(u))`
Out[27]: $\partial_2 \mathcal{E}_A(t,u) \cdot A(u)$

In [28]: `reduce_order(ex1)`
Out[28]: $A(\mathcal{E}_A(t,u))$

In [29]: `ex2 = E(A,t,u,A(u),v)`
Out[29]: $\partial_2^2 \mathcal{E}_A(t,u)(A(u),v)$

In [30]: `reduce_order(ex2)`
Out[30]: $A'(\mathcal{E}_A(t,u)) \cdot \partial_2 \mathcal{E}_A(t,u) \cdot v - \partial_2 \mathcal{E}_A(t,u) \cdot A'(u) \cdot v$

In [31]: `ex3 = E(A,t,u,A(u),v,w)`
Out[31]: $\partial_2^3 \mathcal{E}_A(t,u)(A(u),v,w)$

In [32]: `reduce_order(ex3)`
Out[32]: $-\partial_2^2 \mathcal{E}_A(t,u)(A'(u) \cdot w, v) - \partial_2^2 \mathcal{E}_A(t,u)(A'(u) \cdot v, w)$
$$-\partial_2 \mathcal{E}_A(t,u) \cdot A''(u)(v,w) + A'(\mathcal{E}_A(t,u)) \cdot \partial_2^2 \mathcal{E}_A(t,u)(v,w)$$
$$+ A''(\mathcal{E}_A(t,u))(\partial_2 \mathcal{E}_A(t,u) \cdot v, \partial_2 \mathcal{E}_A(t,u) \cdot w)$$

Similarly for higher derivatives of analogous form.

4   Verification of Crucial Identities

We describe a Julia notebook which implements the verification of the identities (4)–(9) of Sect. 2.

Check (4)

We verify identity (4),

$$\partial_t \mathcal{S}(t, u) - H(\mathcal{S}(t, u)) = \Big[\partial_2 \mathcal{E}_B(t, v) \cdot A(v) - A(\mathcal{E}_B(t, v)) \Big]_{v = \mathcal{E}_A(t, u)}.$$

```
In [1]: using Flows
In [2]: @t_vars t
Out[2]: (t,)
In [3]: @x_vars u v
Out[3]: (u,v)
In [4]: @funs A B
Out[4]: (A,B)
In [5]: E_Atu = E(A,t,u)
Out[5]: ℰ_A(t, u)
In [6]: E_Btv = E(B,t,v)
Out[6]: ℰ_B(t, v)
In [7]: Stu = E(B,t,E(A,t,u))
Out[7]: ℰ_B(t, ℰ_A(t, u))
In [8]: S1tv = differential(E(B,t,v),v,A(v))-A(E(B,t,v))
Out[8]: ∂₂ℰ_B(t, v) · A(v) − A(ℰ_B(t, v))
In [9]: S1tu = substitute(S1tv,v,E(A,t,u))
        ex1 = S1tu
Out[9]: −A(ℰ_B(t, ℰ_A(t, u))) + ∂₂ℰ_B(t, ℰ_A(t, u)) · A(ℰ_A(t, u))
In [10]: ex2 = t_derivative(Stu,t)-(A(Stu)+B(Stu))
Out[10]: −A(ℰ_B(t, ℰ_A(t, u))) + ∂₂ℰ_B(t, ℰ_A(t, u)) · A(ℰ_A(t, u))
In [11]: ex1-ex2
Out[11]: 0
```

Check (5)

We verify identity (5),

$$\partial_t \tilde{S}^{(1)}(t, v) = B'(\mathcal{E}_B(t, v)) \cdot \tilde{S}^{(1)}(t, v) + [B, A](\mathcal{E}_B(t, v)).$$

```
In [12]: ex1 = B(E(B,t,v),S1tv)+commutator(B,A,E(B,t,v))
Out[12]: −A'(ℰ_B(t, v)) · B(ℰ_B(t, v)) + B'(ℰ_B(t, v)) · (∂₂ℰ_B(t, v) · A(v)
         −A(ℰ_B(t, v))) + B'(ℰ_B(t, v)) · A(ℰ_B(t, v))
In [13]: ex2 = t_derivative(S1tv,t)
Out[13]: −A'(ℰ_B(t, v)) · B(ℰ_B(t, v)) + B'(ℰ_B(t, v)) · ∂₂ℰ_B(t, v) · A(v)
In [14]: reduce_order(expand(ex1-ex2))
Out[14]: 0
```

Check (6)

We verify identity (6),

$$\partial_\tau \big(\partial_2 \mathcal{E}_H(t - \tau, \mathcal{S}(\tau, u)) \cdot \mathcal{S}^{(1)}(\tau, u) \big)$$
$$= \partial_2 \mathcal{E}_H(t - \tau, \mathcal{S}(\tau, u)) \cdot \mathcal{S}^{(2)}(\tau, u) + \partial_2^2 \mathcal{E}_H(t - \tau, \mathcal{S}(\tau, u))(\mathcal{S}^{(1)}(\tau, u), \mathcal{S}^{(1)}(\tau, u)).$$

```
In [15]: @nonautonomous_funs S
Out[15]: (S,)

In [16]: @funs H
Out[16]: (H,)

In [17]: S1tu = t_derivative(S(t,u),t)-H(S(t,u))
```
Out[17]: $-H(S(t, u)) + \partial_1 S(t, u)$

```
In [18]: S2tu = t_derivative(S1tu,t)-H(S(t,u),S1tu)
```
Out[18]: $-H'(S(t, u)) \cdot \partial_1 S(t, u) + \partial_1^2 S(t, u) - H'(S(t, u)) \cdot (-H(S(t, u))$
$\quad\quad + \partial_1 S(t, u))$

```
In [19]: @t_vars T
Out[19]: (T,)
```

In the following, a trailing semicolon in the input suppresses the display of the corresponding output.

```
In [20]: ex1 = E(H,T-t,S(t,u),S2tu)+E(H,T-t,S(t,u),S1tu,S1tu);

In [21]: ex2 = t_derivative(E(H,T-t,S(t,u),S1tu),t);

In [22]: reduce_order(expand(ex1-ex2))
Out[22]: 0
```

Check (7)

We verify identity (7),

$$\partial_t \mathcal{S}^{(1)}(t, u) - H'(\mathcal{S}(t, u)) \cdot \mathcal{S}^{(1)}(t, u)$$
$$= \Big[\partial_2 \tilde{\mathcal{S}}^{(1)}(t, v) \cdot A(v) - A'(\mathcal{E}_B(t, v)) \cdot \tilde{\mathcal{S}}^{(1)}(t, v) + [B, A](\mathcal{E}_B(t, v)) \Big]_{v = \mathcal{E}_A(t, u)}.$$

```
In [23]: S2tu = t_derivative(S1tu,t)-A(Stu,S1tu)-B(Stu,S1tu)
         ex1 = S2tu;

In [24]: S2tv = (differential(S1tv,v,A(v))-A(E(B,t,v),S1tv)
                  +commutator(B,A,E(B,t,v)))
         ex2 = substitute(S2tv,v,E(A,t,u));

In [25]: expand(ex1-ex2)
Out[25]: 0
```

Check (8)

We verify identity (8),

$$\partial_t \tilde{S}^{(2)}(t,v) = B'(\mathcal{E}_B(t,v)) \cdot \tilde{S}^{(2)}(t,v)$$
$$+ B''(\mathcal{E}_B(t,v))(\tilde{S}^{(1)}(t,v), \tilde{S}^{(1)}(t,v))$$
$$- [B,[B,A]](\mathcal{E}_B(t,v)) - [A,[B,A]](\mathcal{E}_B(t,v))$$
$$+ 2[B,A]'(\mathcal{E}_B(t,v)) \cdot \tilde{S}^{(1)}(t,v).$$

```
In [26]: ex1 = (B(E(B,t,v),S2tv) + B(E(B,t,v),S1tv,S1tv)
                -commutator(B,B,A,E(B,t,v))
                -commutator(A,B,A,E(B,t,v))
                +2*substitute(differential
                (commutator(B,A,w),w,S1tv),w,E(B,t,v)));

In [27]: ex2 = t_derivative(S2tv,t);

In [28]: expand(ex1-ex2)
Out[28]: 0
```

Check (9)

We verify identity (9),

$$\partial_\tau \Big(\partial_2 \mathcal{E}_H(t-\tau, \mathcal{S}(\tau,u)) \cdot \partial_2 \mathcal{E}_B(\tau, \mathcal{E}_A(\tau,u)) \cdot [B,A](\mathcal{E}_A(\tau,u)) \Big) =$$
$$\Big[\partial_2 \mathcal{E}_H(t-\tau, \mathcal{E}_B(\tau,v)) \cdot \partial_2 \mathcal{E}_B(\tau,v) \cdot [[B,A],A](v)$$
$$+ \partial_2 \mathcal{E}_H(t-\tau, \mathcal{E}_B(\tau,v)) \cdot \partial_2 \tilde{S}^{(1)}(\tau,v) \cdot [B,A](v)$$
$$+ \partial_2^2 \mathcal{E}_H(t-\tau, \mathcal{E}_B(\tau,v))(\tilde{S}^{(1)}(\tau,v), \partial_2 \mathcal{E}_B(\tau,v) \cdot [B,A](v)) \Big]_{v=\mathcal{E}_A(t,u)}.$$

```
In [29]: ex1 = t_derivative(E(H,T-t,Stu,E(B,t,E(A,t,u),
                commutator(B,A,E(A,t,u)))),t);

In [30]: ex2 = (substitute(-E(H,T-t,E(B,t,v),E(B,t,v,
                commutator(A,B,A,v)))+E(H,T-t,E(B,t,v),
                differential(S1tv,v,commutator(B,A,v)))
                +E(H,T-t,E(B,t,v),S1tv,E(B,t,v,
                commutator(B,A,v))),v,E(A,t,u)));

In [31]: diff = ex1-ex2
         diff = FE2DEF(diff)
         diff = substitute(diff,H,A(v)+B(v),v)
         diff = expand(reduce_order(diff))
Out[31]: 0
```

5   Elementary Differentials

To further demonstrate the functionality and correctness of our package, we consider the elementary differentials obtained by repeated differentiation of a differential equation,

$$y'(t) = F(y(t)),$$
$$y''(t) = F'(y(t)) \cdot F(y(t)),$$
$$y'''(t) = F''(y(t))(F(y(t), F(y(t)) + F'(y(t)) \cdot F'(y(t)) \cdot F(y(t)),$$

$$\vdots$$

The number of terms in these expressions are available from the literature, see [6, Table 2.1] or [3]. The following notebook generates the elementary differentials and counts their number:

```
In [1]: using Flows
```

```
In [2]: @t_vars t;
        @x_vars u;
        @funs F;
```

```
In [3]: ex = E(F,t,u)
Out[3]: $\mathcal{E}_F(t, u)$
```

```
In [4]: ex = t_derivative(ex,t)
Out[4]: $F(\mathcal{E}_F(t, u))$
```

```
In [5]: ex = t_derivative(ex,t)
Out[5]: $F'(\mathcal{E}_F(t, u)) \cdot F(\mathcal{E}_F(t, u))$
```

```
In [6]: ex = t_derivative(ex,t)
```
Out[6]: $F'(\mathcal{E}_F(t, u)) \cdot F'(\mathcal{E}_F(t, u)) \cdot F(\mathcal{E}_F(t, u))$
$\quad + F''(\mathcal{E}_F(t, u))(F(\mathcal{E}_F(t, u)), F(\mathcal{E}_F(t, u)))$

```
In [7]: ex = t_derivative(ex,t)
```
Out[7]: $3F''(\mathcal{E}_F(t, u))(F'(\mathcal{E}_F(t, u)) \cdot F(\mathcal{E}_F(t, u)), F(\mathcal{E}_F(t, u)))$
$\quad + F'''(\mathcal{E}_F(t, u))(F(\mathcal{E}_F(t, u)), F(\mathcal{E}_F(t, u)), F(\mathcal{E}_F(t, u)))$
$\quad + F'(\mathcal{E}_F(t, u)) \cdot (F'(\mathcal{E}_F(t, u)) \cdot F'(\mathcal{E}_F(t, u)) \cdot F(\mathcal{E}_F(t, u))$
$\quad + F''(\mathcal{E}_F(t, u))(F(\mathcal{E}_F(t, u)), F(\mathcal{E}_F(t, u))))$

This is not yet fully expanded. It is a linear combination consisting of 3 terms:

```
In [8]: length(ex.terms)
Out[8]: 3
```

If we expand it, we obtain a linear combination of 4 terms, corresponding to the 4 elementary differentials (Butcher trees) of order 4:

```
In [9]:   ex = expand(ex)
```
Out[9]: $3F''(\mathcal{E}_F(t,u))(F'(\mathcal{E}_F(t,u)) \cdot F(\mathcal{E}_F(t,u)), F(\mathcal{E}_F(t,u)))$
$\quad + F'(\mathcal{E}_F(t,u)) \cdot F'(\mathcal{E}_F(t,u)) \cdot F'(\mathcal{E}_F(t,u)) \cdot F(\mathcal{E}_F(t,u))$
$\quad + F'''(\mathcal{E}_F(t,u))(F(\mathcal{E}_F(t,u)), F(\mathcal{E}_F(t,u)), F(\mathcal{E}_F(t,u)))$
$\quad + F'(\mathcal{E}_F(t,u)) \cdot F''(\mathcal{E}_F(t,u))(F(\mathcal{E}_F(t,u)), F(\mathcal{E}_F(t,u)))$

```
In [10]:  length(ex.terms)
```
Out[10]: 4

```
In [11]:  ex = expand(t_derivative(ex,t));
```

```
In [12]:  length(ex.terms)
```
Out[12]: 9

```
In [13]:  ex = expand(t_derivative(ex,t));
```

```
In [14]:  length(ex.terms)
```
Out[14]: 20

```
In [15]:  ex = expand(t_derivative(ex,t));
```

```
In [16]:  length(ex.terms)
```
Out[16]: 48

```
In [17]:  ex = E(F,t,u)
          ex = expand(t_derivative(ex,t))
          println("order\t#terms")
          println("---------------")
          println(1,"\t",1)
          ex = expand(t_derivative(ex,t))
          println(2,"\t",1)
          for k=3:16
              ex = expand(t_derivative(ex,t))
              println(k,"\t",length(ex.terms))
          end
```

```
order    #terms
---------------
1        1
2        1
3        2
4        4
5        9
6        20
7        48
8        115
9        286
10       719
11       1842
12       4766
13       12486
```

14	32973
15	87811
16	235381

Acknowledgments. This work was supported in part by the projects P24157-N13 of the Austrian Science Fund (FWF) and MA14-002 of the Vienna Science and Technology Fund (WWTF).

References

1. http://julialang.org
2. https://github.com/HaraldHofstaetter/Flows.jl
3. The On-line Encyclopedia of Integer Sequences. https://oeis.org/A000081
4. Auzinger, W., Hofstätter, H., Koch, O., Thalhammer, M.: Defect-based local error estimators for splitting methods, with application to Schrödinger equations, part III: the nonlinear case. J. Comput. Appl. Math. **273**, 182–204 (2015)
5. Hairer, E., Lubich, C., Wanner, G.: Geometric Numerical Integration, 2nd edn. Springer, Heidelberg (2006)
6. Hairer, E., Nørsett, S., Wanner, G.: Solving Ordinary Differential Equations I, 2nd edn. Springer, Heidelberg (1993)

Improved Computation of Involutive Bases

Bentolhoda Binaei[1], Amir Hashemi[1,2($\boxtimes$)], and Werner M. Seiler[3]

[1] Department of Mathematical Sciences, Isfahan University of Technology,
84156-83111 Isfahan, Iran
h.binaei@math.iut.ac.ir, Amir.Hashemi@cc.iut.ac.ir
[2] School of Mathematics, Institute for Research in Fundamental Sciences (IPM),
19395-5746 Tehran, Iran
[3] Institut für Mathematik, Universität Kassel, Heinrich-Plett-Straße 40,
34132 Kassel, Germany
seiler@mathematik.uni-kassel.de

Abstract. In this paper, we describe improved algorithms to compute Janet and Pommaret bases. To this end, based on the method proposed by Möller et al. [20], we present a more efficient variant of Gerdt's algorithm (than the algorithm presented in [16]) to compute *minimal* involutive bases. Furthermore, by using an *involutive version* of the Hilbert driven technique along with the new variant of Gerdt's algorithm, we modify the algorithm given in [23] to compute a linear change of coordinates for a given homogeneous ideal so that the new ideal (after performing this change) possesses a finite Pommaret basis. All the proposed algorithms have been implemented in MAPLE and their efficiency is discussed via a set of benchmark polynomials.

1 Introduction

Gröbner bases are one of the most important concepts in computer algebra for dealing with multivariate polynomials. A Gröbner basis is a special kind of generating set for an ideal which provides a computational framework to determine many properties of the ideal. The notion of Gröbner bases was originally introduced in 1965 by Buchberger in his Ph.D. thesis and he also gave the basic algorithm to compute it [2,3]. Later on, he proposed two criteria for detecting superfluous reductions to improve his algorithm [1]. In 1983, Lazard [19] developed a new approach by relating Gröbner bases computations and linear algebra. In 1988, Gebauer and Möller [10] reformulated Buchberger's criteria in an efficient way to improve Buchberger's algorithm. Furthermore, Möller et al. [20] proposed an improved version of Buchberger's algorithm by using the syzygies of the constructed polynomials to detect useless reductions (this algorithm may be considered as the first *signature-based* algorithm to compute Gröbner bases). Relying on the properties of the Hilbert series of an ideal, Traverso [26] described the so-called *Hilbert-driven Gröbner basis algorithm* to improve Buchberger's

A. Hashemi—The research of the second author was in part supported by a grant from IPM (No. 94550420).

V.P. Gerdt et al. (Eds.): CASC 2016, LNCS 9890, pp. 58–72, 2016.
DOI: 10.1007/978-3-319-45641-6_5

algorithm by discarding useless critical pairs. In 1999, Faugère [6] presented his F_4 algorithm to compute Gröbner bases which stems from Lazard's approach [19] and uses fast linear algebra techniques on sparse matrices (this algorithm has been efficiently implemented in MAPLE and MAGMA). In 2002, Faugère presented the famous F_5 algorithm for computing Gröbner bases [7]. The efficiency of this signature-based algorithm benefits from an incremental structure and two new criteria, namely F_5 *and IsRewritten criteria* (nowadays known respectively as signature and syzygy criteria). We remark that several authors have studied signature-based algorithms to compute Gröbner bases and for novel approaches in this directions we refer to e.g. [8,9].

Involutive bases may be considered as an extension of Gröbner bases (w.r.t. a restricted monomial division) which include additional combinatorial properties. The origin of involutive bases theory must be traced back to the work of Janet [18] on a constructive approach to the analysis of linear and certain quasi-linear systems of partial differential equations. Then Janet's approach was generalized to arbitrary (polynomial) differential systems by Thomas [25]. Based on the related methods developed by Pommaret in his book [21], the notion of *involutive polynomial bases* was introduced by Zharkov and Blinkov in [27]. Gerdt and Blinkov [13] introduced a more general concept of *involutive division* and *involutive bases* for polynomial ideals, along with algorithmic methods for their construction. An efficient algorithm was devised by Gerdt [12] (see also [15]) for computing involutive and Gröbner bases using the involutive form of Buchberger's criteria (see http://invo.jinr.ru for the efficiency analysis of the implementation of this algorithm). In this paper, we refer to this algorithm as *Gerdt's algorithm*. Finally, Gerdt et al. [16] described a signature-based algorithm (with an incremental structure) to apply the F_5 criterion for deletion of unnecessary reductions. Some drawbacks of this algorithm are as follows: Due to its incremental structure (in order to apply the F_5 criterion), the selection strategy should be the POT module monomial ordering (which may not be efficient in general). Furthermore, to respect the signature of computed polynomials, the reduction process may be not accomplished and (since this may increase the number of intermediate polynomials) this may significantly affect the efficiency of the computation. Finally, the involutive basis that this algorithm returns may be not minimal.

The aim of this paper is to provide an effective method to calculate *Pommaret bases*. These bases introduced by Zharkov and Blinkov in [27] are a particular form of involutive bases respecting many combinatorial properties of the ideals they generate, see e.g. [22–24] for a comprehensive study of Pommaret bases. They are not only of interest concerning computational aspects in algebraic geometry (e.g. by providing deterministic approaches to transform a given ideal into some classes of generic positions [23]), but they also serve for theoretical aspects of algebraic geometry (e.g. by providing simple and explicit formulas to read off many invariants of an ideal like dimension, depth and Castelnuovo-Mumford regularity [23]).

Relying on the method developed by Möller et al. [20], we give a new signature-based variant of Gerdt's algorithm to compute *minimal* involutive bases. In particular, the experiments show that the new algorithm is more efficient than algorithm of Gerdt et al. [16]. On the other hand, [23] proposes an algorithm to compute *deterministically* a linear change of coordinates for a given homogeneous ideal so that the transformed ideal possesses a finite Pommaret basis (note that in general a given ideal does not have a finite Pommaret basis). In doing so, one computes iteratively the Janet bases of certain polynomial ideals. By applying the involutive version of Hilbert driven technique to the new variant of Gerdt's algorithm, we modify this algorithm to compute Pommaret bases. We have implemented all the algorithms described in this article and we assess their performance on a number of test examples.

The rest of the paper is organized as follows. In the next section, we will review the basic definitions and notations which will be used throughout this paper. Section 3 is devoted to the description of the new variant of Gerdt's algorithm. In Sect. 4, we present the improved algorithm to compute a linear change of coordinates for a given homogeneous ideal so that the new ideal has a finite Pommaret basis. We analyze the performance of the proposed algorithms in Sect. 5. Finally, in Sect. 6 we conclude the paper by highlighting the advantages of this work and discussing future research directions.

2   Preliminaries

In this section, we review the basic definitions and notations from the theory of Gröbner bases and involutive bases that will be used in the rest of the paper. Throughout this paper we assume that $\mathcal{P} = \Bbbk[x_1, \ldots, x_n]$ is the polynomial ring (where $\Bbbk$ is an infinite field). We consider also *homogeneous* polynomials $f_1, \ldots, f_k \in \mathcal{P}$ and the ideal $\mathcal{I} = \langle f_1, \ldots, f_k \rangle$ generated by them. We denote the total degree of and the degree w.r.t. a variable x_i of a polynomial $f \in \mathcal{P}$ by $\deg(f)$ and $\deg_i(f)$ respectively. Let $\mathcal{M} = \{x_1^{\alpha_1} \cdots x_n^{\alpha_n} \mid \alpha_i \geq 0, 1 \leq i \leq n\}$ be the monoid of all monomials in $\mathcal{P}$. A monomial ordering on $\mathcal{M}$ is denoted by $\prec$ and throughout this paper we shall assume that $x_n \prec \cdots \prec x_1$. The leading monomial of a given polynomial $f \in \mathcal{P}$ w.r.t. $\prec$ will be denoted by $\mathrm{LM}(f)$. If $F \subset \mathcal{P}$ is a finite set of polynomials, we denote by $\mathrm{LM}(F)$ the set $\{\mathrm{LM}(f) \mid f \in F\}$. The leading coefficient of f, denoted by $\mathrm{LC}(f)$, is the coefficient of $\mathrm{LM}(f)$. The leading term of f is defined to be $\mathrm{LT}(f) = \mathrm{LM}(f)\,\mathrm{LC}(f)$. A finite set $G = \{g_1, \ldots, g_k\} \subset \mathcal{P}$ is called a *Gröbner basis* of $\mathcal{I}$ w.r.t $\prec$ if $\mathrm{LM}(\mathcal{I}) = \langle \mathrm{LM}(g_1), \ldots, \mathrm{LM}(g_k) \rangle$ where $\mathrm{LM}(\mathcal{I}) = \langle \mathrm{LM}(f) \mid f \in \mathcal{I} \rangle$. We refer e.g. to [4] for more details on Gröbner bases.

Let us recall the definition of Hilbert function and Hilbert series of a homogeneous ideal. Let $X \subset \mathcal{P}$ and s a positive integer. We define the degree s component X_s of X to be the set of all homogeneous elements of X of degree s.

Definition 1. *The* Hilbert function *of $\mathcal{I}$ is defined by* $\mathrm{HF}_{\mathcal{I}}(s) = \dim\ (\mathcal{P}_s/\mathcal{I}_s)$ *where the right hand side denotes the dimension of $\mathcal{P}_s/\mathcal{I}_s$ as a $\Bbbk$-linear space.*

It is well-known that the Hilbert function of $\mathcal{I}$ is the same as that of $\mathrm{LT}(\mathcal{I})$ (see e.g. [4, Proposition 4, page 458]) and therefore the set of monomials not contained in LT($\mathcal{I}$) forms a basis for $\mathcal{P}_s/\mathcal{I}_s$ as a k-linear space (Macaulay's theorem). This observation is the key idea behind the Hilbert-driven Gröbner basis algorithm. Roughly speaking, suppose that $\mathcal{I}$ is a homogeneous ideal and we want to compute a Gröbner basis of $\mathcal{I}$ by Buchberger's algorithm in increasing order w.r.t. the total degree of the S-polynomials. Assume that we know beforehand $\mathrm{HF}_{\mathcal{I}}(s)$ for a positive integer s. Suppose that we are at the stage where we are looking at the critical pairs of degree s. Consider the set P of all critical pairs of degree s. Then, we compare $\mathrm{HF}_{\mathcal{I}}(s)$ with the Hilbert function at s of the ideal generated by the leading terms of all already computed polynomials. If they are equal, we can remove P.

Below, we review some definitions and relevant results on involutive bases theory (see [12] for more details). We recall first involutive divisions based on partitioning the variables into two subsets of the variables, the so-called *multiplicative* and *non-multiplicative variables*.

Definition 2. *An* involutive division $\mathcal{L}$ *is given on* $\mathcal{M}$ *if for any finite set* $U \subset \mathcal{M}$ *and any* $u \in U$, *the set of variables is partitioned into the subset of multiplicative* $M_{\mathcal{L}}(u, U)$ *and non-multiplicative variables* $NM_{\mathcal{L}}(u, U)$ *such that the following three conditions hold where* $\mathcal{L}(u, U)$ *denotes the monoid generated by* $M_{\mathcal{L}}(u, U)$:

1. $v, u \in U$, $u\mathcal{L}(u, U) \cap v\mathcal{L}(v, U) \neq \emptyset \Rightarrow u \in v\mathcal{L}(v, U)$ or $v \in u\mathcal{L}(u, U)$,
2. $v \in U$, $v \in u\mathcal{L}(u, U) \Rightarrow \mathcal{L}(v, U) \subset \mathcal{L}(u, U)$,
3. $V \subset U$ and $u \in V \Rightarrow \mathcal{L}(u, U) \subset \mathcal{L}(u, V)$.

We shall write $u \mid_{\mathcal{L}} w$ *if* $w \in u\mathcal{L}(u, U)$. *In this case,* u *is called an* $\mathcal{L}$-involutive *divisor of* w *and* w *an* $\mathcal{L}$-involutive *multiple of* u.

We recall the definitions of the Janet and the Pommaret division, respectively.

Example 1. Let $U \subset \mathcal{P}$ be a finite set of monomials. For each sequence $d_1, \ldots, d_n$ of non-negative integers and for each $1 \leq i \leq n$ we define the subsets

$$[d_1, \ldots, d_i] = \{u \in U \mid d_j = \deg_j(u), \ 1 \leq j \leq i\}.$$

The variable x_1 is Janet multiplicative (denoted by $\mathcal{J}$-multiplicative) for $u \in U$ if $\deg_1(u) = \max\{\deg_1(v) \mid v \in U\}$. For $i > 1$ the variable x_i is Janet multiplicative for $u \in [d_1, \ldots, d_{i-1}]$ if $\deg_i(u) = \max\{\deg_i(v) \mid v \in [d_1, \ldots, d_{i-1}]\}$.

Example 2. For $u = x_1^{d_1} \cdots x_k^{d_k}$ with $d_k > 0$ the variables $\{x_k, \ldots, x_n\}$ are considered as Pommaret multiplicative (denoted by $\mathcal{P}$-multiplicative) and the other variables as Pommaret non-multiplicative. For $u = 1$ all the variables are multiplicative. The integer k is called the *class* of u and is denoted by $\mathrm{cls}(u)$.

The Pommaret division is called a *global division*, because the assignment of the multiplicative variables is independent of the set U. In order to avoid repeating notations let $\mathcal{L}$ always denote an involutive division.

Definition 3. *The set* $F \subset \mathcal{P}$ *is called* involutively head autoreduced *if for each* $f \in F$ *there is no* $h \in F \setminus \{f\}$ *with* $\mathrm{LM}(h) \mid_{\mathcal{L}} \mathrm{LM}(f)$.

Definition 4. *Let $I \subset \mathcal{P}$ be an ideal. An $\mathcal{L}$-involutively head autoreduced subset $G \subset \mathcal{I}$ is an $\mathcal{L}$-involutive basis for $\mathcal{I}$ (or simply either an involutive basis or $\mathcal{L}$-basis) if for every $0 \neq f \in \mathcal{I}$ there exists $g \in G$ so that $\mathrm{LM}(g) \mid_{\mathcal{L}} \mathrm{LM}(f)$.*

Example 3. Let $\mathcal{I} = \{x_1^2 x_3, x_1 x_2, x_1 x_3^2\} \subset \Bbbk[x_1, x_2, x_3]$. Then, $\{x_1^2 x_3, x_1 x_2, x_1 x_3^2, x_1^2 x_2\}$ is a Janet basis for $\mathcal{I}$ and $\{x_1^2 x_3, x_1 x_2, x_1 x_3^2, x_1^2 x_2, x_1^{i+3} x_2, x_1^{i+3} x_3 \mid i \geq 0\}$ is a (infinite) Pommaret basis for $\mathcal{I}$. Indeed, Janet division is Noetherian, however Pommaret division is non-Noetherian (see [13] for more details).

Gerdt [12] proposed an efficient algorithm to construct involutive bases based on a completion process where prolongations of generators by non-multiplicative variables are reduced. This process terminates in finitely many steps for any Noetherian division.

Definition 5. *Let $F \subset \mathcal{P}$ be a finite. Following the notations in [23], the involutive span generated by F is denoted by $\langle F \rangle_{\mathcal{L}, \prec}$.*

Thus, if a set $F \subset \mathcal{I}$ is an involutive basis for $\mathcal{I}$ then we have $\mathcal{I} = \langle F \rangle_{\mathcal{L}, \prec}$.

Definition 6. *Let $F \subset \mathcal{I}$ be an involutively head autoreduced set of homogeneous polynomials. The involutive Hilbert function of F is defined by $\mathrm{IHF}_F(s) = \dim (\mathcal{P}_s / (\langle F \rangle_{\mathcal{L}, \prec})_s)$.*

Since F is involutively head autoreduced, one easily recognizes that $\langle F \rangle_{\mathcal{L}, \prec} = \bigoplus_{f \in F} \Bbbk[M_{\mathcal{L}}(\mathrm{LM}(f), \mathrm{LM}(F))] \cdot f$. Thus using the well-known combinatorial formulas to count the number of monomials in certain variables, we get

$$\mathrm{IHF}_I(s) = \binom{n + s - 1}{s} - \sum_{f \in F} \binom{s - \deg(f) + k_f - 1}{s - \deg(f)}$$

where k_f is the number of multiplicative variables of f (see e.g. [12]). We remark that an involutively head autoreduced subset $F \subset \mathcal{I}$ is an involutive basis for $\mathcal{I}$ if and only if $\mathrm{HF}_{\mathcal{I}}(s) = \mathrm{IHF}_F(s)$ for each s.

3   Using Syzygies to Compute Involutive Bases

We now propose a variant of Gerdt's algorithm [12] by using the intermediate computed syzygies to compute involutive bases and especially Janet bases. For this, we recall briefly the signature-based variant of the algorithm of Möller et al. [20] to compute Gröbner bases (the practical results are given in Sect. 5).

Definition 7. *Let us consider $F = (f_1, \ldots, f_k) \in \mathcal{P}^k$. The (first) syzygy module of F is defined to be $\mathrm{Syz}(F) = \{(h_1, \ldots, h_k) \mid h_i \in \mathcal{P}, \sum_{i=1}^{k} h_i f_i = 0\}$.*

Schreyer in his master thesis proposed a slight modification of Buchberger's algorithm to compute a Gröbner basis for the module of syzygies of a Gröbner basis. The construction of this basis relies on the following key observation (see [5]): Let $G = \{g_1, \ldots, g_s\}$ be a Gröbner basis. By tracing the dependency of each

$\text{SPoly}(g_i, g_i)$ on G we can write $\text{SPoly}(g_i, g_j) = \sum_{k=1}^{s} a_{ijk} g_k$ with $a_{ijk} \in \mathcal{P}$. Let $\mathbf{e}_1, \ldots, \mathbf{e}_s$ be the standard basis for $\mathcal{P}^s$ and $m_{ij} = lcm(\text{LT}(g_i), \text{LT}(g_j))$. Set

$$\mathbf{s}_{ij} = m_{i,j}/\text{LT}(g_i).\mathbf{e}_i - m_{i,j}/\text{LT}(g_j).\mathbf{e}_j - (a_{ij1}\mathbf{e}_1 + a_{ij2}\mathbf{e}_2 + \cdots + a_{ijs}\mathbf{e}_s).$$

Definition 8. *Let $G = \{g_1, \ldots, g_s\} \subset \mathcal{P}$. Schreyer's module ordering is defined by $x^\beta \mathbf{e}_j \prec_s x^\alpha \mathbf{e}_i$, if $\text{LT}(x^\beta g_j) \prec \text{LT}(x^\alpha g_i)$, and breaks ties by $i < j$.*

Theorem 1 (Schreyer's Theorem). *For a Gröbner basis $G = \{g_1, \ldots, g_s\}$ the set $\{\mathbf{s}_{ij} \mid 1 \le i < j \le s\}$ forms a Gröbner basis for $\text{Syz}(g_1, \ldots, g_s)$ w.r.t. $\prec_s$.*

Example 4. Let $F = \{xy - x, x^2 - y\} \subset \Bbbk[x, y]$. The Gröbner basis of F w.r.t. $x \prec_{dlex} y$ is $G = \{g_1 = xy - x, g_2 = x^2 - y, g_3 = y^2 - y\}$ and the Gröbner basis of $\text{Syz}(g_1, g_2, g_3)$ is $\{(x, -y + 1, -1), (-x, y^2 - 1, -x^2 + y + 1), (y, 0, -x)\}$.

Algorithm 1. GRÖBNERBASIS

Input: A set of polynomials $F \subset \mathcal{P}$; a monomial ordering $\prec$
Output: A Gröbner basis G for $\langle F \rangle$
$G := \{\}$ and $syz := \{\}$
$P := \{(F[i], \mathbf{e}_i) \mid i = 1, \ldots, |F|\}$
while $P \neq \emptyset$ **do**
  select (using normal strategy) and remove $p \in P$
  if $\nexists\, s \in syz$ s.t. $s \mid \text{Sig}(p)$ **then**
    $f := \text{Poly}(p)$
    $h := \text{NORMALFORM}(f, G)$
    $syz := syz \cup \{\text{Sig}(p)\}$
    if $h \neq 0$ **then**
      $j := |G| + 1$
      for $g \in G$ **do**
        $P := P \cup \{(r.h, r.\mathbf{e}_j)\}$ s.t. $r.\text{LM}(h) = \text{LCM}(\text{LM}(g), \text{LM}(h))$
        $G := G \cup \{h\}$
        $syz := syz \cup \{\text{LM}(g).\mathbf{e}_j \mid \text{LM}(h) \text{ and } \text{LM}(g) \text{ are coprime}\}$
      end for
    end if
  end if
end while
return (G)

According to this observation, Möller et al. [20] proposed a variant of Buchberger's algorithm by using the syzygies of the constructed polynomials to remove superfluous reductions. Algorithm 1 corresponds to it with a slight modification to derive a signature-based algorithm to compute Gröbner bases. We associate to each polynomial f the pair $p = (f, m\mathbf{e}_i)$ where $\text{Poly}(p) = f$ is the polynomial part of f and $\text{Sig}(p) = m\mathbf{e}_i$ is its signature. Furthermore, the function $\text{NORMALFORM}(f, G)$ returns a remainder of the division of f by G. If $\text{Sig}(p) = m\mathbf{e}_i$ in the first step of the reduction process, we must not use $f_i \in G$.

Algorithm 2. INVOLUTIVEBASIS

Input: A finite set $F \subset \mathcal{P}$; an involutive division $\mathcal{L}$; a monomial ordering $\prec$
Output: A minimal $\mathcal{L}$-basis for $\langle F \rangle$
$F := \text{sort}(F, \prec)$
$T := \{(F[1], F[1], \emptyset, \mathbf{e}_1)\}$
$Q := \{(F[i], F[i], \emptyset, \mathbf{e}_i) \mid i = 2, \ldots, |F|\}$
$syz := \{\}$
while $Q \neq \emptyset$ **do**
  $Q := \text{sort}(Q, \prec_s)$
  $p := Q[1]$
  if $\nexists s \in syz$ s.t $s \mid \text{Sig}(p)$ with non-constant quotient **then**
    $h := \text{INVOLUTIVENORMALFORM}(p, T, \mathcal{L}, \prec)$
    $syz := syz \cup \{h[2]\}$
    if $h = 0$ and $\text{LM}(\text{Poly}(p)) = \text{LM}(\text{Anc}(p))$ **then**
      $Q := \{q \in Q \mid \text{Anc}(q) \neq \text{Poly}(p)\}$
    end if
    if $h \neq 0$ and $\text{LM}(\text{Poly}(p)) \neq \text{LM}(h)$ **then**
      for $q \in T$ with proper conventional divisibility $\text{LM}(\text{Poly}(h)) \mid \text{LM}(\text{Poly}(q))$
      do
        $Q := Q \cup \{q\}$
        $T := T \setminus \{q\}$
      end for
      $j := |T| + 1$
      $T := T \cup \{(h, h, \emptyset, \mathbf{e}_j)\}$
    else
      $T := T \cup \{(h, \text{Anc}(p), \text{NM}(p), \text{Sig}(p))\}$
    end if
    for $q \in T$ and $x \in NM_{\mathcal{L}}(\text{LM}(\text{Poly}(q)), \text{LM}(\text{Poly}(T)) \setminus \text{NM}(q))$ **do**
      $Q := Q \cup \{(x.\text{Poly}(q), \text{Anc}(q), \emptyset, x.\text{Sig}(q))\}$
      $\text{NM}(q) := \text{NM}(q) \cup NM_{\mathcal{L}}(\text{LM}(\text{Poly}(q)), \text{LM}(\text{Poly}(T))) \cup \{x\}$
    end for
  end if
end while
return $(\text{Poly}(T))$

We show now how to apply this structure to improve Gerdt's algorithm [15]. [23, Theorem 5.10] contains an involutive version of Schreyer's theorem replacing S-polynomials by non-multiplicative prolongations and using involutive division. Algorithm 2 represents the new variant of Gerdt's algorithm for computing involutive bases using involutive syzygies. For this purpose, we associate to each polynomial f the quadruple $p = (f, g, V, m.\mathbf{e}_i)$ where $f = \text{Poly}(p)$ is the polynomial itself, $g = \text{Anc}(p)$ is its ancestor, $V = \text{NM}(p)$ is the list of non-multiplicative variables of f which have already been processed in the algorithm and $m.\mathbf{e}_i = \text{Sig}(p)$ is the signature of f. If P is a set of quadruples, we denote by $\text{Poly}(P)$ the set $\{\text{Poly}(p) \mid p \in P\}$. The functions $\text{sort}(X, \prec)$ and $\text{sort}(X, \prec_s)$ sort X by increasing $\text{LM}(X)$ w.r.t. $\prec$ and $\{\text{Sig}(p) \mid p \in X\}$ w.r.t. $\prec_s$, respectively. The involutive normal form algorithm is given in Algorithm 3.

Algorithm 3. INVOLUTIVENORMALFORM

Input: A quadruple p; a set of quadruples T; an involutive division $\mathcal{L}$; a monomial ordering $\prec$
Output: An $\mathcal{L}$-normal form of p modulo T, and the corresponding signature, if any
$S := \{\}$ and $h := \text{Poly}(p)$ and $G := \text{Poly}(T)$
while h has a monomial m which is $\mathcal{L}$-divisible by G **do**
    select $g \in G$ with $\text{LM}(g) \mid_{\mathcal{L}} m$
    if $m = \text{LM}(\text{Poly}(p))$ and $(m/\text{LM}(g).\text{Sig}(g) = \text{Sig}(p)$ or CRITERIA$(h,g))$ **then**
      return $(0, S)$
    end if
    if $m = \text{LM}(\text{Poly}(p))$ and $m/\text{LM}(g).\text{Sig}(g) \prec_s \text{Sig}(p)$ **then**
      $S := S \cup \{\text{Sig}(p)\}$
    end if
    $h := h - cm/\text{LT}(g).g$ where c is the coefficient of m in h
end while
return (h, S)

Furthermore, we apply the involutive form of Buchberger's criteria from [12]. We say that CRITERIA(p, g) holds if either $C_1(p, g)$ or $C_2(p, g)$ holds where $C_1(p, g)$ is true if $\text{LM}(\text{Anc}(p)).\text{LM}(\text{Anc}(g)) = \text{LM}(\text{Poly}(p))$ and $C_2(p, g)$ is true if $\text{LCM}(\text{LM}(\text{Anc}(p)), \text{LM}(\text{Anc}(g)))$ properly divides $\text{LM}(\text{Poly}(p))$.

Remark 1. We note that, due to the second **if**-loop in Algorithm 3, if $m_i \mathbf{e}_i$ is added to syz, then there exists an involutive representation of the form $m_i g_i = \sum_{j=1}^{\ell} h_j g_j + h$ where $T = \{g_1, \ldots, g_\ell\} \subset \mathcal{P}$ is the output of the algorithm, h the $\mathcal{L}$-normal form of p modulo T and $\text{LM}(h_j)\mathbf{e}_j \prec_s m_i \mathbf{e}_i$ for each j.

In the next proof, by an abuse of notation, we refer to the signature of a quadruple as the signature of its polynomial part.

Theorem 2. INVOLUTIVEBASIS *terminates in finitely many steps (if $\mathcal{L}$ is a Noetherian division) and returns a minimal involutive basis for its input ideal.*

Proof. The termination and correctness of the algorithm are inherited from those of Gerdt's algorithm [12] provided that we show that any polynomial removed using syzygies is superfluous. This happens in both algorithms. Let us deal first with Algorithm 2. Now, suppose that for $p \in Q$ there exists $s \in syz$ so that $s \mid \text{Sig}(p)$ with non-constant quotient. Suppose that $\text{Sig}(p) = m_i \mathbf{e}_i$ and $s = m_i' \mathbf{e}_i$ where $m_i = u m_i'$ with $u \neq 1$. Let $T = \{g_1, \ldots, g_\ell\} \subset \mathcal{P}$ be the output of the algorithm and $m_i' g_i = \sum_{j=1}^{\ell} h_j g_j + h$ be the representation of $m_i' g_i$ with $g_j \in T, h, h_j \in P$ and h the involutive remainder of the division of $m_i' g_i$ by T. Then, from the structure of both algorithms, we find that $\text{LM}(h_j g_j) \prec \text{LM}(m_i' g_i)$. In particular, we have $\text{LM}(h_j)\mathbf{e}_j \prec_s m_i' \mathbf{e}_i$ for each j. It follows that $\text{LM}(uh_j)\mathbf{e}_j \prec_s u m_i' \mathbf{e}_i = m_i \mathbf{e}_i$ for each j. On the other hand, if $h \neq 0$ then, again by the structure of the algorithm, uh has a signature less than $m_i \mathbf{e}_i$. For each j and for each term t in h_j we know that the signature of utg_j is less than $m_i \mathbf{e}_i$ and by the selection strategy used in the algorithm which is based on Schreyer's ordering, utg_j should be studied before $m_i'g_i$ and therefore it has an involutive representation in terms of T. Furthermore, the same holds also for uh provided that $h \neq 0$. These arguments show that $m_i'g_i$ is unnecessary and it can be omitted. Now we turn to Algorithm 3. Let $p \in Q$ and $g \in T$ so that $\mathrm{LM}(h) = u\,\mathrm{LT}(g)$ and $\mathrm{Sig}(p) = u\,\mathrm{Sig}(g)$ where $h = \mathrm{Poly}(p)$ and u is a monomial. Using the above notations, let $\mathrm{Sig}(p) = m_i\mathbf{e}_i$ and $\mathrm{Sig}(g) = m_i'\mathbf{e}_i$ where $m_i = um_i'$. Furthermore, let $m_i'g_i = \sum_{j=1}^{\ell} h_j g_j + g$ be the representation of $m_i'g_i$ with $\mathrm{LM}(h_j)\mathbf{e}_j \prec_s m_i'\mathbf{e}_i$ for each j. It follows from the assumption that $\mathrm{LM}(h_j g_j) \prec \mathrm{LM}(m_i'g_i) = \mathrm{LM}(g)$ for each j. We can write $um_i'g_i = \sum_{j=1}^{\ell} uh_j g_j + ug$. Since $\mathrm{LM}(uh_j)\mathbf{e}_j \prec_s um_i'\mathbf{e}_i = m_i\mathbf{e}_i$ for each j then, by repeating the above argument, we deduce that $uh_j g_j$ for each j has an involutive representation. Therefore, $um_i'g_i$ has a representation since u is multiplicative for g. Thus h has a representation and it can be removed. □

4   Hilbert Driven Pommaret Bases Computations

The Pommaret division is not Noetherian and therefore, a given ideal may not have a finite Pommaret basis. However, if the ideal is in *quasi-stable position* (see Definition 9) it has a finite Pommaret basis and a generic linear change of variables transforms any ideal into such a position. [23] proposes a deterministic algorithm to compute such a linear change by computing repeatedly the Janet basis of the last transformed ideal. We now improve it by incorporating an involutive version of the Hilbert driven strategy.

Algorithm 4. HDQUASISTABLE

Input: A finite set $F \subset \mathcal{P}$ and a monomial ordering $\prec$
Output: A linear change Φ so that $\langle \Phi(F) \rangle$ has a finite Pommaret basis
$\Phi := \emptyset$ and $J :=$ INVOLUTIVEBASIS$(F, \mathcal{J}, \prec)$ and $A :=$ TEST$(\mathrm{LM}(J), \prec)$
while $A \neq true$ **do**
   $G :=$ substitution of $\phi := A[3] \mapsto A[3] + cA[2]$ in J for a random choice of $c \in K$
   $Temp :=$ HDINVOLUTIVEBASIS$(G, \mathcal{J}, \prec)$
   $B :=$ TEST$(\mathrm{LM}(Temp))$
   if $B \neq A$ **then**
      $\Phi := \Phi, \phi$ and $J := Temp$ and $A := B$
   end if
end while
return (Φ)

It is worth noting that in [23] it is proposed to perform a Pommaret head autoreduction on the calculated Janet basis at each iteration. However, we do not need to perform this operation because each computed Janet basis is *minimal* and by [11, Corollary 15] each minimal Janet basis is Pommaret head autoreduced. All the used functions are described below. By the structure of the algorithm, we first compute a Janet basis for the input ideal using INVOLUTIVE-BASIS algorithm. From this basis, one can read off easily the Hilbert function of

---

Algorithm 5. TEST

Input: A finite set U of monomials
Output: True if any element of U has the same number of Pommaret and Janet
multiplicative variables, and false otherwise
if $\exists u \in U$ s.t. $M_{\mathcal{P},\prec}(u,U) \neq M_{\mathcal{J},\prec}(u,U)$ **then**
$\quad V := M_{\mathcal{J},\prec}(u,F) \setminus M_{\mathcal{P},\prec}(u,F)$
$\quad$**return**$(false, V[1], x_{\text{cls}(u)})$
end if
return $(true)$

---

the input ideal. Furthermore, the Hilbert function of an ideal is invariant under
a linear change of variables. Thus we can exploit this Hilbert function in the
next Janet bases computations as follows. The algorithm has the same structure
as the INVOLUTIVEBASIS algorithm and so we omit the similar lines. We add
the written lines in the HDINVOLUTIVEBASIS algorithm between $p := Q[1]$ and
the first **if**-loop in the INVOLUTIVEBASIS algorithm.

Algorithm 6. HDINVOLUTIVEBASIS

Input: A set of monomials F; an involutive division $\mathcal{L}$; a monomial ordering $\prec$
Output: A minimal $\mathcal{L}$-involutive basis for $\langle F \rangle$

$\vdots$

$d := \deg(p)$
while $\text{HF}_{\langle F \rangle}(d) = \text{IHF}_T(d)$ **do**
$\quad$ remove from Q all $q \in Q$ s.t. $\deg(\text{Poly}(q)) = d$
$\quad$**if** $Q = \emptyset$ **then**
$\quad\quad$**return** $(\text{Poly}(T))$
$\quad$**else**
$\quad\quad p := Q[1]$
$\quad\quad d := \deg(p)$
$\quad$**end if**
end while

$\vdots$

Theorem 3. HDQUASISTABLE *algorithm terminates in finitely many steps and
it returns a linear change of variables for a given homogeneous ideal so that the
changed ideal (after performing the change on the input ideal) possesses a finite
Pommaret basis.*

Proof. Let $\mathcal{I}$ be the ideal generated by F, the input of the HDQUASISTABLE
algorithm. The termination of this algorithm follows, from one side, from the
termination of the algorithms to compute Janet bases. From the other side,
[23, Proposition 2.9] shows that there exists a Zariski open set U of $\Bbbk^{n \times n}$ so
that for each linear change of variables, say Φ, corresponding to an element of

U the transformed ideal $\Phi(\mathcal{I})$ has a finite Pommaret basis. Moreover, he proved that the process of finding such a linear change terminates in finitely many steps (see [23, Remark 9.11]). Taken together, these arguments show that the HDQUASISTABLE algorithm terminates.

For the correctness, using the notations of the HDINVOLUTIVEBASIS algorithm, we shall prove that any $p \in Q$ removed by the Hilbert driven strategy reduces to zero. In this direction, we recall that any change of variables is a linear automorphism of $\mathcal{P}$, [17, page 52]. Thus, for each i, the dimension over $\Bbbk$ of components of degree i of $\mathcal{I}$ and that of $\mathcal{I}$ after the change remains stable. This yields that the Hilbert function of $\mathcal{I}$ does not change after a linear change of variables. Let J be the Janet basis computed by INVOLUTIVEBASIS. One can readily observe that $\mathrm{HF}_{\mathcal{I}}(d) = \mathrm{IHF}_J(d)$ for each d, and therefore from the first Janet basis one can derive the Hilbert function of $\mathcal{I}$ and use it to improve the next Janet bases computations. Now, suppose that F is the input of the HDINVOLUTIVEBASIS algorithm, $p \in Q$ and $\mathrm{HF}_{\mathcal{I}}(d) = \mathrm{IHF}_T(d)$ for $d = \deg(\mathrm{Poly}(p))$. It follows that $\dim\ (\langle F \rangle_d) = \dim\ (\langle \mathrm{Poly}(T) \rangle_d)$ and therefore the polynomials of $\mathrm{Poly}(T)$ generate involutively the whole set $\langle F \rangle_d$ and this shows that p is superfluous which ends the proof.   □

Remark 2. We assumed that the input of the algorithms INVOLUTIVEBASIS and HDQUASISTABLE was homogeneous. However, the former algorithm works also for non-homogeneous ideals. Furthermore, the latter algorithm also may be applied to non-homogeneous ideals provided that we consider the affine Hilbert function for such ideals; i.e. $\mathrm{HF}_{\mathcal{I}}(s) = \dim\ (\mathcal{P}_{\leq s}/\mathcal{I}_{\leq s})$.

[23] provides a number of equivalent characterizations of the ideals which have finite Pommaret bases. Indeed, a given ideal has a finite Pommaret basis if only if the ideal is in *quasi stable position* (or equivalently if the coordinates are δ-regular) see [23, Proposition 4.4].

Definition 9. *A monomial ideal $\mathcal{I}$ is called* quasi stable *if for any monomial $m \in \mathcal{I}$ and all integers i, j, s with $1 \leq j < i \leq n$, $s > 0$ and $x_i^s \mid m$ there exists an integer $t \geq 0$ such that $x_j^t m/x_i^s \in \mathcal{I}$. A homogeneous ideal $\mathcal{I}$ is in* quasi stable position *if $\mathrm{LT}(\mathcal{I})$ is quasi stable.*

Example 5. The ideal $\mathcal{I} = \langle x_2^2 x_3, x_2^3, x_1^3 \rangle \subset \Bbbk[x, y, z]$ is a quasi stable monomial ideal and its Pommaret basis is $\{x_2^2 x_3, x_2^3, x_1^3, x_1 x_2^2 x_3, x_1 x_2^3, x_1^2 x_2^2 x_3, x_1^2 x_2^3\}$.

5   Experiments and Comparison

We have implemented both algorithms INVOLUTIVEBASIS and HDQUASISTABLE in MAPLE 17[1]. We now compare the behavior of the algorithms INVOLUTIVEBASIS and HDQUASISTABLE, respectively, with the algorithms by Gerdt et al. [16]

---

[1] The MAPLE code of the implementations of our algorithms and examples are available at http://amirhashemi.iut.ac.ir/softwares.

and QUASISTABLE [23], respectively (we remark that QUASISTABLE has the same structure as the HDQUASISTABLE, however, to compute Janet bases we use Gerdt's algorithm). For this purpose, we used some well-known examples from the computer algebra literature. All computations were done over $\mathbb{Q}$ and for the degree reverse lexicographic monomial ordering. The results are shown in the following tables where the time and memory columns indicate, respectively, the consumed CPU time in seconds and the amount of used memory in megabytes. The C_1 and C_2 columns show, respectively, the number of polynomials removed by C_1 and C_2 criteria in the corresponding algorithm. The sixth column shows the number of polynomials eliminated by the new criterion related to syzygies applied in the INVOLUTIVEBASIS and INVOLUTIVENORMALFORM algorithms. The F_5 and S columns show the number of polynomials removed by the F_5 and the super-top-reduction criterion, respectively,. The last three columns represent the number of reductions to zero, the number and the maximum degree of polynomials in the final involutive basis, respectively (we note that for the algorithm by Gerdt et al. the number of polynomials is the size of the basis after the minimization process). The computations in this paper are performed on a personal computer with 2.70 GHz Intel(R) Core(TM) i7-2620M CPU, 8 GB of RAM, 64 bits under the Windows 7 operating system.

Liu	time	memory	C_1	C_2	Syz	F_5	S	redz	poly	deg
INVOLUTIVEBASIS	1.09	37.214	4	3	2	-	-	25	19	6
Gerdt et al.	2.901	41.189	7	39	-	25	0	1	19	7

Noon	time	memory	C_1	C_2	Syz	F_5	S	redz	poly	deg
INVOLUTIVEBASIS	3.822	43.790	4	15	6	-	-	69	51	10
Gerdt et al.	45.271	670.939	8	107	-	49	3	17	51	10

Haas3	time	memory	C_1	C_2	Syz	F_5	S	redz	poly	deg
INVOLUTIVEBASIS	8.424	95.172	0	20	24	-	-	203	73	13
Gerdt et al.	41.948	630.709	1	88	-	88	16	68	73	13

Sturmfels-Eisenbud	time	memory	C_1	C_2	Syz	F_5	S	redz	poly	deg
INVOLUTIVEBASIS	22.932	255.041	28	103	95	-	-	245	100	6
Gerdt et al.	2486.687	30194.406	29	1379	-	84	11	40	100	9

Lichtblau	time	memory	C_1	C_2	Syz	F_5	S	redz	poly	deg
INVOLUTIVEBASIS	24.804	391.3	0	5	6	-	-	19	35	11
Gerdt et al.	205.578	3647.537	0	351	-	18	0	31	35	19

Eco7	time	memory	C_1	C_2	Syz	F_5	S	redz	poly	deg
INVOLUTIVEBASIS	40.497	473.137	51	21	30	-	-	201	45	6
Gerdt et al.	1543.068	25971.714	63	1717	-	175	8	18	45	11

Katsura5	time	memory	C_1	C_2	Syz	F_5	S	redz	poly	deg
INVOLUTIVEBASIS	46.956	630.635	21	0	2	-	-	68	23	12
Gerdt et al.	42.416	621.551	62	73	-	114	1	21	23	8

Katsura6	time	memory	C_1	C_2	Syz	F_5	S	redz	poly	deg
INVOLUTIVEBASIS	81.526	992.071	43	0	4	-	-	171	43	8
Gerdt et al.	608.325	795.196	77	392	-	209	1	41	43	11

As one can observe, INVOLUTIVEBASIS is a signature-based variant of Gerdt's algorithm which has a structure closer to Gerdt's algorithm and it is more efficient than the algorithm by Gerdt et al. Moreover, we can see the detection of criteria and the number of reductions to zero by the algorithms are different.

Indeed, this difference is due to the selection strategy used in each algorithm. More precisely, in the algorithm by Gerdt et al. the set of non-multiplicative prolongations is sorted by POT ordering whereas in INVOLUTIVEBASIS it is sorted using Schreyer ordering. However, one needs to implement it efficiently in C/C++ to be able to compare it with GINV software[2].

The next tables illustrate an experimental comparison of HDQUASISTABLE and QUASISTABLE algorithms. In these tables, the HD column shows the number of polynomials removed by the Hilbert driven strategy in the corresponding algorithm. Further, the chen column shows the number of linear changes that one needs to transform the corresponding ideal into quasi stable position. The deg column represents the maximum degree of the output Pommaret basis (which is the Castelnuovo-Mumford regularity of the ideal, see [23]). The remaining columns show the number of detections of the corresponding criteria for all computed Janet bases. Finally, we shall remark that in the next tables we use the homogenization of the generating set of the test examples used in the previous tables. In addition, the computation of Janet basis of an ideal generated by a set F and the one of the ideal generated by the homogenization of F are not the same. For example, the CPU time to compute the Janet basis of the homogenization of the Lichtblau example is 270.24 s.

Liu	time	memory	C_1	C_2	Syz	HD	redz	chen	deg
HDQUASISTABLE	4.125	409.370	4	3	2	93	56	4	6
QUASISTABLE	9.56	1067.725	14	3	-	-	151	4	6

Katsura5	time	memory	C_1	C_2	Syz	HD	redz	chen	deg
HDQUASISTABLE	67.234	9191.288	44	3	6	185	168	2	8
QUASISTABLE	145.187	26154.263	86	29	-	-	359	2	8

Weispfenning94	time	memo	C_1	C_2	Syz	HD	reds	chen	deg
HDQUASISTABLE	110.339	6268.532	0	1	9	45	170	1	15
QUASISTABLE	243.798	16939.468	0	2	-	-	85	1	15

Noon	time	memory	C_1	C_2	Syz	HD	redz	chen	deg
HDQUASISTABLE	667.343	66697.995	4	25	6	325	119	4	11
QUASISTABLE	1210.921	205149.994	16	35	-	-	450	4	11

Sturmfels-Eisenbud	time	memory	C_1	C_2	Syz	HD	redz	chen	deg
HDQUASISTABLE	1507.640	125904.515	86	308	440	1370	1804	12	8
QUASISTABLE	843.171	96410.344	218	1051	-	-	3614	12	8

Eco7	time	memory	C_1	C_2	Syz	HD	redz	chen	deg
QUASISTABLE	2182.296	241501.340	298	98	373	1523	1993	8	11
QUASISTABLE	2740.734	500857.600	547	725	-	-	3889	8	11

Haas3	time	memory	C_1	C_2	Syz	HD	redz	chen	deg
HDQUASISTABLE	5505.375	906723.699	0	0	91	84	255	1	33
QUASISTABLE	10136.718	1610753.428	1	120	-	-	430	1	33

Lichtblau	time	memory	C_1	C_2	Syz	HD	redz	chen	deg
HDQUASISTABLE	16535.593	2051064.666	0	44	266	217	265	2	30
QUASISTABLE	18535.625	2522847.256	0	493	-	-	751	2	30

---

[2] See http://invo.jinr.ru.

6   Conclusion and Perspective

In this paper, a modification of Gerdt's algorithm [15] which is a signature-based version of the involutive algorithm [12,15] to compute minimal involutive bases is suggested. Additionally, we present a Hilbert driven optimization of the proposed algorithm, to compute (finite) Pommaret bases. In doing so, the proposed algorithm computes iteratively Janet bases by using the modified form of Gerdt's algorithm and use them, in accordance to the ideas of [23], to perform the variable transformations. The new algorithms have been implemented in Maple and they are compared with Gerdt's algorithm and with the algorithm presented in [23] in terms of the CPU time and used memory, and several other criteria. For all considered examples, the Maple implementations of the new algorithms are shown to be superior over the existing ones. One interesting research direction might be to develop a new variant of the proposed signature-based version of the involutive algorithm by incorporating the advantages of the algorithm in [15], in particular of the Janet trees [14]. Furthermore, it would be of interest to study the behavior of different possible techniques to improve the computation of Pommaret bases.

References

1. Buchberger, B.: A criterion for detecting unnecessary reductions in the construction of Gröbner-bases. In: Ng, K.W. (ed.) EUROSAM 1979 and ISSAC 1979. LNCS, vol. 72. Springer, Heidelberg (1979)
2. Buchberger, B.: Ein Algorithmus zum Auffinden der Basiselemente des Restklassenringes nach einem nulldimensionalen Polynomideal. Univ. Innsbruck, Mathematisches Institut (Diss.), Innsbruck (1965)
3. Buchberger, B.: Bruno Buchberger's PhD thesis 1965: an algorithm for finding the basis elements of the residue class ring of a zero dimensional polynomial ideal. J. Symb. Comput. $41(3–4)$, 475–511 (2006). Translation from the German
4. Cox, D., Little, J., O'Shea, D.: Ideals, Varieties, and Algorithms. An Introduction to Computational Algebraic Geometry and Commutative Algebra. Undergraduate Texts in Mathematics, 3rd edn. Springer, New York (2007)
5. Cox, D.A., Little, J., O'Shea, D.: Using Algebraic Geometry. Graduate Texts in Mathematics, vol. 185, 2nd edn. Springer, New York (2005)
6. Faugère, J.C.: A new efficient algorithm for computing Gröbner bases (F_4). J. Pure Appl. Algebra $139(1–3)$, 61–88 (1999)
7. Faugère, J.C.: A new efficient algorithm for computing Gröbner bases without reduction to zero (F_5). In: Proceedings of the International Symposium on Symbolic and Algebraic Computation, ISSAC 2002, Lille, France, 07–10 July, pp. 75–83 (2002)
8. Gao, S., Guan, Y., Volny, F.: A new incremental algorithm for computing Groebner bases. In: Proceedings of the International Symposium on Symbolic and Algebraic Computation, ISSAC 2010, Munich, Germany, 25–28 July, pp. 13–19 (2010)
9. Gao, S., Volny, F.I., Wang, M.: A new framework for computing Gröbner bases. Math. Comput. $85(297)$, 449–465 (2016)
10. Gebauer, R., Möller, H.: On an installation of Buchberger's algorithm. J. Symb. Comput. $6(2–3)$, 275–286 (1988)

11. Gerdt, V.P.: On the relation between Pommaret and Janet bases. In: Ganzha, V.G., Mayr, E.W., Vorozhtsov, E.V. (eds.) Computer Algebra in Scientific Computing, pp. 167–181. Springer, Heidelberg (2000)
12. Gerdt, V.P.: Involutive algorithms for computing Gröbner bases. In: Computational Commutative and Non-commutative Algebraic Geometry, Proceedings of the NATO Advanced Research Workshop, Chisinau, Republic of Moldova, 6–11 June 2004, pp. 199–225. IOS Press, Amsterdam (2005)
13. Gerdt, V.P., Blinkov, Y.A.: Involutive bases of polynomial ideals. Math. Comput. Simul. **45**(5–6), 519–541 (1998)
14. Gerdt, V.P., Blinkov, Y.A., Yanovich, D.: Construction of Janet bases I. monomial bases. In: Ganzha, V.G., Mayr, E.M., Vorozhtsov, E.V. (eds.) Computer Algebra in Scientific Computing, CASC 2001, pp. 233–247. Springer, Berlin (2001)
15. Gerdt, V.P., Hashemi, A., M.-Alizadeh, B.: A variant of Gerdt's algorithm for computing involutive bases. Bull. PFUR Ser. Math. Inf. Sci. Phys. **2**, 65–76 (2012)
16. Gerdt, V.P., Hashemi, A., M.-Alizadeh, B.: Involutive bases algorithm incorporating F_5 criterion. J. Symb. Comput. **59**, 1–20 (2013)
17. Herzog, J., Hibi, T.: Monomial Ideals. Springer, London (2011)
18. Janet, M.: Sur les systèmes d'équations aux dérivées partielles. C. R. Acad. Sci. Paris **170**, 1101–1103 (1920)
19. Lazard, D.: Gröbner bases, Gaussian elimination and resolution of systems of algebraic equations. In: van Hulzen, J.A. (ed.) Computer Algebra, EUROCAL 1983. LNCS, vol. 162, pp. 146–156. Springer, Heidelberg (1983)
20. Möller, H., Mora, T., Traverso, C.: Gröbner bases computation using syzygies. In: Proceedings of the International Symposium on Symbolic and Algebraic Computation, ISSAC 1992, Berkeley, CA, USA, 27–29 July, pp. 320–328 (1992)
21. Pommaret, J.: Systems of Partial Differential Equations and Lie Pseudogroups, vol. 14. Gordon and Breach Science Publishers, New York (1978). With a preface by Andre Lichnerowicz
22. Seiler, W.M.: A combinatorial approach to involution and δ-regularity. I: involutive bases in polynomial algebras of solvable type. Appl. Algebra Eng. Commun. Comput. **20**(3–4), 207–259 (2009)
23. Seiler, W.M.: A combinatorial approach to involution and δ-regularity. II: structure analysis of polynomial modules with Pommaret bases. Appl. Algebra Eng. Commun. Comput. **20**(3–4), 261–338 (2009)
24. Seiler, W.M.: Involution. The Formal Theory of Differential Equations and Its Applications in Computer Algebra. Algorithms and Computation in Mathematics, vol. 24. Springer, Berlin (2010)
25. Thomas, J.M.: Differential Systems, IX. 118 p. American Mathematical Society (AMS), New York (1937)
26. Traverso, C.: Hilbert functions and the Buchberger algorithm. J. Symb. Comput. **22**(4), 355–376 (1996)
27. Zharkov, A., Blinkov, Y.: Involution approach to investigating polynomial systems. Math. Comput. Simul. **42**(4), 323–332 (1996)

Computing All Space Curve Solutions
of Polynomial Systems by Polyhedral Methods

Nathan Bliss[✉] and Jan Verschelde

Department of Mathematics, Statistics, and Computer Science,
University of Illinois at Chicago, 851 S. Morgan Street (m/c 249),
Chicago, IL 60607-7045, USA
{nbliss2,janv}@uic.edu

Abstract. A polyhedral method to solve a system of polynomial equations exploits its sparse structure via the Newton polytopes of the polynomials. We propose a hybrid symbolic-numeric method to compute a Puiseux series expansion for every space curve that is a solution of a polynomial system. The focus of this paper concerns the difficult case when the leading powers of the Puiseux series of the space curve are contained in the relative interior of a higher dimensional cone of the tropical prevariety. We show that this difficult case does not occur for polynomials with generic coefficients. To resolve this case, we propose to apply polyhedral end games to recover tropisms hidden in the tropical prevariety.

Keywords: Newton polytope · Polyhedral end game · Polyhedral method · Polynomial system · Puiseux series · Space curve · Tropical basis · Tropical prevariety · Tropism

1 Introduction

In this paper we consider the application of polyhedral methods to compute series for all space curves defined by a polynomial system. Polyhedral methods compute with the Newton polytopes of the system. The *Newton polytope* of a polynomial is defined as the convex hull of the exponents of the monomials that appear with a nonzero coefficient.

If we start the development of the series where the space curve meets the first coordinate plane, then we compute Puiseux series. Collecting for each coordinate the leading exponents of a Puiseux series gives what is called a *tropism*. If we view a tropism as a normal vector to a hyperplane, then we see that there are hyperplanes with this normal vector that touch every Newton polytope of the system at an edge or at a higher dimensional face. A vector normal to such a hyperplane is called a *pretropism*. While every tropism is a pretropism, not every pretropism is a tropism.

This material is based upon work supported by the National Science Foundation under Grant No. 1440534.

V.P. Gerdt et al. (Eds.): CASC 2016, LNCS 9890, pp. 73–86, 2016.
DOI: 10.1007/978-3-319-45641-6_6

In this paper we investigate the application of a polyhedral method to compute all space curve solutions of a polynomial system. The method starts from the collection of all pretropisms, which are regarded as candidate tropisms. For the method to work, we focus on the following questions.

Problem Statement. Given that only the space curves are of interest, can we ignore the higher dimensional cones of pretropisms? In particular, if some tropisms lie in the interior of higher dimensional cones of pretropisms, is it then still possible to compute Puiseux series solutions for all space curves?

Related Work. In symbolic computation, new elimination algorithms for sparse systems with positive dimensional solution sets are described in [7]. Tropical resultants are computed in [13]. Related polyhedral methods for sparse systems can be found in [10,15]. Conditions on how far a Puiseux series should be expanded to decide whether a point is isolated are given in [8]. The authors of [12] propose numerical methods for tropical curves. Polyhedral methods to compute tropical varieties are outlined in [4] and implemented in Gfan [14]. The background on tropical algebraic geometry is in [16].

Algorithms to compute the tropical *prevariety* are presented in [21]. For *pre*processing purposes, the software of [21] is useful. However, the focus on this paper concerns the tropical variety for which Gfan [14] provides a tropical basis. Therefore, our computational experiments with computer algebra methods are performed with Gfan and not with the software of [21].

Organization and Contributions. In the next section we illustrate the advantages of looking for Puiseux series as solutions of polynomial systems. Then we motivate our problem with some illustrative examples. Relating the tropical prevariety to a recursive formula to compute the mixed volume characterizes the generic case, in which the tropical prevariety suffices to compute all space curve solutions. With polyhedral end games we can recover the tropisms contained in higher dimensional cones of the tropical prevariety. Finally we give some experimental results and timings.

2   Puiseux Series

When we work with Puiseux series we apply a hybrid method, combining exact and approximate calculations. Figure 1 shows the plot, in black, of Viviani's curve, defined as the intersection of the sphere $f = x_1^2 + x_2^2 + x_3^2 - 4 = 0$ and the cylinder $g = (x_1 - 1)^2 + x_2^2 - 1 = 0$.

There is one pretropism $\mathbf{v} = (2, 1, 0)$, which defines the initial forms of f and g respectively as $x_3^2 - 4$ and $x_2^2 - 2x_1$. For traditional Puiseux series, one would choose to set $x_1 = 1$, obtaining the four solutions $(1, \pm\sqrt{2}, \pm 2)$ and leading terms $(t^2, \pm\sqrt{2}t, \pm 2)$. If we instead use $x_1 = 2$, we obtain rational coefficients and the following partial expansion:

$$\begin{bmatrix} x_1 \\ x_2 \\ x_3 \end{bmatrix} = \begin{bmatrix} 2t^2 \\ 2t - t^3 - \frac{1}{4}t^5 - \frac{1}{8}t^7 - \frac{5}{64}t^9 \\ 2 - t^2 - \frac{1}{4}t^4 - \frac{1}{8}t^6 - \frac{5}{64}t^9 \end{bmatrix}. \tag{1}$$

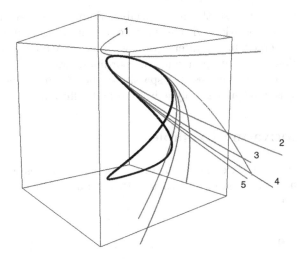

Fig. 1. Viviani's curve with improving Puiseux series approximations, labelled with the number of terms used to plot each one.

The plot of several Puiseux approximations to Viviani's curve is shown in gray in Fig. 1.

If we shift the Viviani example so that its self-intersection is at the origin, we obtain the following:

$$\mathbf{f}(\mathbf{x}) = \begin{cases} x_1^2 + x_2^2 + x_3^2 + 4x_1 = 0 \\ x_1^2 + x_2^2 + 2x_1 = 0 \end{cases} \tag{2}$$

An examination of the first few terms of the Puiseux series expansion for this system, combined with the On-Line Encyclopedia of Integer Sequences [17] and some straightforward algebraic manipulation, allows us to hypothesize the following exact parameterization of the variety:

$$\begin{bmatrix} x_1 \\ x_2 \\ x_3 \end{bmatrix} = \begin{bmatrix} -2t^2 \\ 2\dfrac{t^3}{1+\sqrt{1-t^2}} - 2t \\ -2t \end{bmatrix}. \tag{3}$$

We can confirm that this is indeed right via substitution. While this method is of course not possible in general, it does provide an example of the potential usefulness of Puiseux series computations for some examples.

3   Assumptions and Setup

Our object of study is space curves, by which we mean 1-dimensional varieties in $\mathbb{C}^n$. Because Puiseux series computations take one variable to be a free variable, we require that the curves not lie inside $V(\langle x_i \rangle)$ for some i; without loss of generality we choose to use the first variable. Some results require that the curve be in Noether position with respect to x_1, meaning that the degree of the variety is preserved under intersection with $x_1 = \lambda$ for a generic $\lambda \in \mathbb{C}$. It is of course possible to apply a random coordinate transformation to obtain Noether position, but we then lose the sparsity of the system's exponent support structure, which is what makes polyhedral methods effective.

4   Some Motivating Examples

In this section we illustrate the problem our paper addresses with some simple examples, first in 3-space, and then with a family of space curves in any dimensional space.

4.1   In 3-Space

Our first running example is the system

$$\mathbf{f}(\mathbf{x}) = \begin{cases} x_1x_3 - x_2x_3 - x_3^2 + x_1 = 0 \\ x_3^3 - x_1x_2 - x_2x_3 - x_3^2 - x_1 = 0 \end{cases} \tag{4}$$

which has an irreducible quartic and the second coordinate axis $(0, x_2, 0)$ as its solutions. Because the line lies in the first coordinate plane $x_1 = 0$, the system is not in Noether position with respect to the first variable. Therefore, our methods will ignore this part of the solution set. The algorithms of [6] can be applied to compute components inside coordinate planes. Computing a primary decomposition yields the following alternative, which lacks the portion in the first coordinate plane:

$$\tilde{\mathbf{f}}(\mathbf{x}) = \begin{cases} x_1x_3 - x_2x_3 - x_3^2 + x_1 \\ x_1x_2 - x_2^2 - x_2x_3 + x_3^2 + x_1 - 2x_2 - 2x_3 \\ x_3^3 - x_2^2 - 2x_2x_3 - 2x_2 - 2x_3 \end{cases} \tag{5}$$

The tropical prevariety contains the rays $(2, 1, 1)$, $(1, 0, 0)$, and $(1, 0, 1)$; because our Puiseux series start their development at $x_1 = 0$, rays that have a zero or negative value for their first coordinate have been discarded. The tropical variety however contains the ray $(3, 1, 1)$ instead of $(2, 1, 1)$, leading to the Puiseux expansion

$$\begin{bmatrix} x_1 \\ x_2 \\ x_3 \end{bmatrix} = \begin{bmatrix} 108t^3 \\ t - 3t^2 - 15t^3 + 27t^4 + 36t^5 \\ -t - 3t^3 - 18t^4 + 18t^5 + 162t^6 \end{bmatrix}. \tag{6}$$

This ray is a positive combination of $(2, 1, 1)$ and $(1, 0, 0)$. In other words, it is possible for the 1-dimensional cones of the tropical prevariety to fail to be in the tropical variety, and for rays in the tropical variety to "hide" in the higher-dimensional cones of the prevariety.

4.2   In Any Dimensional Space

This problem can also occur in arbitrary dimensions, as seen in the class of examples

$$
\mathbf{f}(\mathbf{x}) = \begin{cases}
x_1^2 - x_1 + x_2 + x_3 + \cdots + x_n = 0 \\
x_2^2 + x_1 + x_2 + x_3 + \cdots + x_n = 0 \\
x_3^2 + x_1 + x_2 + x_3 + \cdots + x_n = 0 \\
\quad\quad\quad\quad\quad\quad\vdots \\
x_{n-1}^2 + x_1 + x_2 + x_3 + \cdots + x_n = 0.
\end{cases}
\tag{7}
$$

The ray $(1, 1, 1, \ldots, 1)$ is a 1-dimensional cone of its prevariety since it is normal to a facet of each polytope, namely the linear portion of each polynomial. It is not, however, in the tropical variety, since the initial form system (as it will be defined in Sect. 5) contains the monomial x_1.

5   The Generic Case

This hiding of tropisms in the higher dimensional cones of the prevariety is problematic, as finding the tropical variety may require more expensive symbolic computations. For a comparison between various approaches see Sect. 9. Fortunately, this problem does not occur in general, as the next result will show. But first, a few definitions.

Definition 1. We write a polynomial f with support set A as

$$
f(\mathbf{x}) = \sum_{\mathbf{a} \in A} c_{\mathbf{a}} \mathbf{x}^{\mathbf{a}}, \quad c_{\mathbf{a}} \in \mathbb{C}^*, \mathbf{x}^{\mathbf{a}} = x_1^{a_1} x_2^{a_2} \cdots x_n^{a_n}.
\tag{8}
$$

The *initial form of f with respect to* $\mathbf{v}$ is then

$$
\mathrm{in}_{\mathbf{v}} f(\mathbf{x}) = \sum_{\mathbf{a} \in \mathrm{in}_{\mathbf{v}} A} c_{\mathbf{a}} \mathbf{x}^{\mathbf{a}},
\tag{9}
$$

where $\mathrm{in}_{\mathbf{v}} A = \{\, \mathbf{a} \in A \mid \langle \mathbf{a}, \mathbf{v} \rangle = \min_{\mathbf{b} \in A} \langle \mathbf{b}, \mathbf{v} \rangle \,\}$.

The initial form of a tuple of polynomials is the tuple of the initial forms of the polynomials in the tuple.

Definition 2. For $f \in \mathbb{C}[\mathbf{x}]$, I an ideal in $\mathbb{C}[\mathbf{x}]$ and $\mathbf{v} \in \mathbb{R}^n$, we define the *initial ideal* $\mathrm{in}_{\mathbf{v}}(I)$ as the ideal generated by $\{\mathrm{in}_{\mathbf{v}}(f) : f \in I\}$.

Definition 3. For $I = \langle f_1, \ldots, f_m \rangle \subset \mathbb{C}[\mathbf{x}]$ an ideal, *the tropical prevariety* is the set of $\mathbf{v} \in \mathbb{R}^n$ for which $\mathrm{in}_{\mathbf{v}}(f_i)$ is not a monomial for any i. The *tropical variety* is the set of $\mathbf{v} \in \mathbb{R}^n$ for which $\mathrm{in}_{\mathbf{v}}(f)$ is not a monomial for any $f \in I$.

Proposition 1. *For n equations in $n + 1$ unknowns with generic coefficients, the set of ray generators of the tropical prevariety contains the tropical variety.*

It is important to note that our notion of generic here refers to the coefficients, and not to generic tropical varieties as seen in [18] which are tropical varieties of ideals under a generic linear transformation of coordinates.

Proof. The tropical prevariety always contains the tropical variety. We simply want to show that all of the rays of the tropical variety show up in the prevariety as ray generators, and not as members of the higher-dimensional cones. Let $I = \langle p_1, \ldots, p_n \rangle \subseteq \mathbb{C}[x_0, \ldots, x_n]$, and let $\mathbf{w}$ be a ray in the tropical prevariety but not one of its ray generators. We want to show that $\mathbf{w}$ is not in the tropical variety, or equivalently that $\mathrm{in}_{\mathbf{w}}(I)$ contains a monomial. We will do so by showing that $I_{\mathbf{w}} := \langle \mathrm{in}_{\mathbf{w}}(p_1), \ldots, \mathrm{in}_{\mathbf{w}}(p_n) \rangle$ contains a monomial, which suffices since this ideal is contained in $\mathrm{in}_{\mathbf{w}}(I)$.

Suppose $I_{\mathbf{w}}$ contains no monomial. Then $(x_0 x_1 \cdots x_n)^k \notin I_{\mathbf{w}}$ for any k. By Hilbert's Nullstellensatz $V := \mathbb{V}(I_{\mathbf{w}}) \not\subseteq \mathbb{V}(x_0 x_1 \cdots x_n)$, i.e. V is not contained in the union of the coordinate hyperplanes. Then there exists $a = (a_0, \ldots, a_n) \in V$ such that all coordinates of a are all nonzero. Since $\mathbf{w}$ lies in the interior a cone of dimension at least 2, the generators of $I_{\mathbf{w}}$ are homogeneous with respect to at least two linearly independent rays $\mathbf{u}$ and $\mathbf{v}$. Thus $(\lambda^{u_0} \mu^{v_0} a_0, \ldots, \lambda^{u_n} \mu^{v_n} a_n) \in V$ for all $\lambda, \mu \in \mathbb{C} \setminus \{0\}$ where the $\mathbf{u}_i, \mathbf{v}_i$ are the components of $\mathbf{u}$ and $\mathbf{v}$, and V contains a toric surface. If we intersect with a random hyperplane, by Bernstein's theorem B [3] the result is a finite set of points, with the possibility of additional components that must be contained in the coordinate planes. Hence V can contain no surface outside of the coordinate planes, and we have a contradiction.

6   Polyhedral Methods

We will show that the tropical prevariety provides an upper bound for the degree of the solution curve. The inner product of a point $\mathbf{a}$ with a vector $\mathbf{v}$ is denoted as $\langle \mathbf{a}, \mathbf{v} \rangle = a_1 v_1 + a_2 v_2 + \cdots + a_n v_n$.

Lemma 1. *Consider an $(n-1)$-tuple of Newton polytopes $\mathcal{P} = (P_1, P_2, \ldots, P_{n-1})$ in n-space. Let E be the edge spanned by $(1, 0, \ldots, 0)$ and $(0, 0, \ldots, 0)$. The mixed volume of $(\mathcal{P}, E)$ equals*

$$V_n(\mathcal{P}, E) = \sum_{\mathbf{v}} v_1 V_{n-1}(\mathrm{in}_{\mathbf{v}} \mathcal{P}), \tag{10}$$

where $\mathbf{v}$ ranges over all rays in the tropical prevariety of $\mathcal{P}$ with $v_1 > 0$, normalized so that $\gcd(\mathbf{v}) = \gcd(v_1, v_2, \ldots, v_n) = 1$, and $\mathrm{in}_{\mathbf{v}} \mathcal{P} = (\mathrm{in}_{\mathbf{v}} P_1, \mathrm{in}_{\mathbf{v}} P_2, \ldots, \mathrm{in}_{\mathbf{v}} P_{n-1})$, where $\mathrm{in}_{\mathbf{v}} P_k$ is the face with support vector $\mathbf{v}$, formally expressed as

$$\mathrm{in}_{\mathbf{v}} P_k = \{ \, \mathbf{a} \in P_k \mid \langle \mathbf{a}, \mathbf{v} \rangle = \max_{\mathbf{a} \in P_k} \langle \mathbf{a}, \mathbf{v} \rangle \, \}. \tag{11}$$

Proof. We apply the following recursive formula [20] for the mixed volume

$$V_n(\mathcal{P}, E) = \sum_{\substack{\mathbf{v} \in \mathbb{Z}^n \\ \gcd(\mathbf{v}) = 1}} p_E(\mathbf{v}) \, V_{n-1}(\mathrm{in}_{\mathbf{v}} \mathcal{P}), \tag{12}$$

where p_E is the support function of the edge E:

$$p_E(\mathbf{v}) = \max_{\mathbf{e} \in E} \langle \mathbf{e}, \mathbf{v} \rangle \qquad (13)$$

and $\mathrm{in}_{\mathbf{v}} \mathcal{P} = (\mathrm{in}_{\mathbf{v}} P_1, \mathrm{in}_{\mathbf{v}} P_2, \ldots, \mathrm{in}_{\mathbf{v}} P_{n-1})$, where

$$\mathrm{in}_{\mathbf{v}} P_k = \{\ \mathbf{a} \in P_k \mid \langle \mathbf{a}, \mathbf{v} \rangle = p_k(\mathbf{v})\ \}, \qquad (14)$$

with p_k the support function of the polytope P_k.

Because the edge E contains $(0, 0, \ldots, 0)$: $p_E(\mathbf{v}) \geq 0$ and $p_E(\mathbf{v}) = 0$ when $v_1 \leq 0$. Only those rays for which $v_1 > 0$ contribute to $V_n(\mathcal{P}, E)$. We have then $p_E(\mathbf{v}) = v_1$.

The mixed volume of a tuple of polytopes equals zero if one of the polytopes consists of only one vertex. The rays in the tropical prevariety contain all vectors for which $\mathrm{in}_{\mathbf{v}}(P_k)$ is an edge or a higher dimensional face. These are the rays $\mathbf{v}$ for which $V_{n-1}(\mathrm{in}_{\mathbf{v}} \mathcal{P}) > 0$.

The application of Lemma 1 leads to a bound on the number of generic points on the space curve. Denote $\mathbb{C}^* = \mathbb{C} \setminus \{0\}$.

Lemma 2. *Consider the system $\mathbf{f}(\mathbf{x}) = \mathbf{0}$, $\mathbf{f} = (f_1,\ f_2,\ \ldots,\ f_{n-1})$ with $\mathcal{P} = (P_1, P_2, \ldots, P_{n-1})$ where P_k is the Newton polytopes of f_k. If the system is in Noether position with respect to x_1, then the degree of the space curve defined by $\mathbf{f}(\mathbf{x}) = \mathbf{0}$ is bounded by $V_n(\mathcal{P}, E)$.*

This result is a version of Lemma 2.3 from [15].

Proof. The proof of the lemma follows from the application of Bernshtein's theorem [3] to the system

$$\begin{cases} \mathbf{f}(\mathbf{x}) = \mathbf{0} \\ x_1 = \gamma, \quad \gamma \in \mathbb{C}^*. \end{cases} \qquad (15)$$

By the assumption of Noether position, there will be as many solutions to this system as the degree of the space curve defined by $\mathbf{f}(\mathbf{x}) = \mathbf{0}$. The theorem of Bernshtein states that the mixed volume bounds the number of solutions in $(\mathbb{C}^*)^n$.

Formula (12) appears in the constructive proof of Bernshtein's theorem [3] and was implemented in the polyhedral homotopies of [25]. For systems with coefficients that are sufficiently generic, the mixed volumes provide an exact root count.

Theorem 1. *Let $\mathbf{f}(\mathbf{x}) = \mathbf{0}$ be a polynomial system of $n - 1$ equations in n unknowns, with sufficiently generic coefficients. Assume the space curve defined by $f(\mathbf{x}) = \mathbf{0}$ is in Noether position with respect to the first variable. Then all rays $\mathbf{v}$ with $v_1 > 0$ in the tropical prevariety of $\mathbf{f}$ lead to Puiseux series expansions for the space curve defined by $\mathbf{f}(\mathbf{x}) = \mathbf{0}$. Moreover, the degree of the space curve is the sum of the degrees of the Puiseux series.*

We illustrate the application of polyhedral methods to the motivating examples.

Example 1. As a verification on the first motivating example (4), we consider the rays $(2, 1, 1)$, $(1, 0, 0)$, and $(1, 0, 1)$ of its tropical prevariety. The initial form of $\mathbf{f}$ in (4) w.r.t. to the ray $(2, 1, 1)$ is

$$\text{in}_{(2,1,1)}\mathbf{f}(\mathbf{x}) = \begin{cases} -x_2 x_3 - x_3^2 + x_1 = 0 \\ -x_2 x_3 - x_3^2 - x_1 = 0. \end{cases} \tag{16}$$

To count the number of solutions of $\text{in}_{(2,1,1)}\mathbf{f}(\mathbf{x}) = \mathbf{0}$ we apply a unimodular coordinate transformation, $\mathbf{x} = \mathbf{y}^U$:

$$U = \begin{bmatrix} 2 & 1 & 1 \\ 1 & 0 & 0 \\ 0 & 1 & 0 \end{bmatrix} \quad \begin{cases} x_1 = y_1^2\, y_2 \\ x_2 = y_1 \quad y_3 \\ x_3 = y_1 \end{cases} \tag{17}$$

which leads to the system

$$\text{in}_{(2,1,1)}\mathbf{f}(\mathbf{y}) = \begin{cases} -y_1^2 y_3 - y_1^2 + y_1^2 y_2 = 0 \\ -y_1^2 y_3 - y_1^2 - y_1^2 y_2 = 0. \end{cases} \tag{18}$$

After removing the common factor y_1^2, we see that this system has one solution for generic choices of the coefficients. As 2 is the first coordinate of $(2, 1, 1)$, this ray contributes two branches and adds two to the degree of the solution curve. The other rays $(1, 0, 0)$ and $(1, 0, 1)$ each contribute one to the degree, and so we recover the degree four of the solution curve.

Example 2. For the family of systems in (7), consider the curve in 4-space:

$$\mathbf{f}(\mathbf{x}) = \begin{cases} x_1^2 - x_1 + x_2 + x_3 + x_4 = 0 \\ x_2^2 + x_1 + x_2 + x_3 + x_4 = 0 \\ x_3^2 + x_1 + x_2 + x_3 + x_4 = 0. \end{cases} \tag{19}$$

For the tropism $\mathbf{v} = (2, 1, 1, 1)$, the initial form is

$$\text{in}_\mathbf{v}\mathbf{f}(\mathbf{x}) = \begin{cases} x_2 + x_3 + x_4 = 0 \\ x_2 + x_3 + x_4 = 0 \\ x_2 + x_3 + x_4 = 0. \end{cases} \tag{20}$$

This tropism is in the interior of the cone in the tropical prevariety spanned by $v_1 = (1, 1, 1, 1)$ and $v_2 = (1, 0, 0, 0)$. Using the same techniques as in the previous example, we find $\text{in}_{v_1}(I)$ has a mixed volume of one and $\text{in}_{v_2}(I)$ has a mixed volume of three, so for generic coefficients we again recover the degree of the solution curve.

7   Current Approaches

In [4] a method is given for computing the tropical variety of an ideal I defining a curve. It involves appending witness polynomials from I to a list of its generators such that for this new set, the tropical prevariety equals the tropical variety. Such a set is called a *tropical basis*. Each additional polynomial rules out one of the cones in the original prevariety that does not belong in the tropical variety. As stated in [4] only finitely many additional polynomials are necessary, since the prevariety has only finitely many cones.

The algorithm runs as follows. For each cone C in the tropical prevariety, we choose a generic element $\mathbf{w} \in C$. We check whether $\text{in}_\mathbf{w}(I)$ contains a monomial by saturating with respect to m, the product of ring variables; the initial ideal contains a monomial if and only if this saturation ideal is equal to (1). If $\text{in}_\mathbf{w}(I)$ does not contain a monomial, the cone C belongs in our tropical variety. If it does, we check whether $m^i \in I$ for increasing values of i until we find a monomial $m' \in \text{in}_\mathbf{w}(I)$. Finally, we append $m' - h$ to our list of basis elements, where h is the reduction of m with respect to a Gröbner basis of I under any monomial order that refines $\mathbf{w}$. For $\mathbf{w}$ to define a global monomial order, and thus allow a Gröbner basis, it may be necessary to homogenize the ideal first.

Bounding the complexity of this algorithm is beyond the scope of this paper, but for each cone it requires computing a Gröbner basis of I as well as another (possibly faster) basis when calculating the saturation to check if the initial ideal contains a monomial. In some cases we may only be concerned about tropisms hiding in a particular higher-dimensional cone of the prevariety, such as with our running example (7). Here it is reasonable to perform only one step of this algorithm, namely looking for a witness for a single cone, which could be significantly faster. However, this has the disadvantage of introducing more 1-dimensional cones into the prevariety. More details, including some timing comparisons, will be given in Sect. 9.

8   Polyhedral End Games

A polyhedral end game [11] applies extrapolation methods to numerically estimate the winding number of solution paths defined by a homotopy. The leading exponents of the Puiseux series are recovered via differences of the logarithms of the magnitudes of the coordinates of the solution paths. Even in the case – as in our illustrative example – where the given polynomials contain insufficient information to compute all tropisms only from the prevariety, a polyhedral end game manages to compute all tropisms. The setup is similar to that of [23], arising in a numerical study of the asymptotics of a space curve, defined by the system $\mathbf{f}(\mathbf{x}) = \mathbf{0}$:

$$\begin{cases} \mathbf{f}(\mathbf{x}) = \mathbf{0} \\ tx_1 + (1-t)(x_1 - \gamma) = 0, & \gamma \in \mathbb{C} \setminus \{0\}, \end{cases} \tag{21}$$

as t moves from 0 to 1, the hyperplane $x_1 = \gamma$ moves the coordinate plane perpendicular to the first coordinate axis.

As t moves from 0 to 1, it is important to note that t will actually never be equal to one. In the polyhedral end games of [11], to estimate the winding number via extrapolation methods, the step size decreases in a geometric ratio. In particular, denoting the winding number by ω, for $t = 1 - s^\omega$, and $0 < r < 1$, we consider the solutions for $s_k = s_0 r^k$, $k = 0, 1, \ldots$, starting at some $s_0 \approx 0$.

The constant γ in (21) is a randomly generated complex number. This implies that for $x_1 = \gamma$, the polynomial system in (21) for $t = 0$ has as many isolated solutions (generic points on the space, eventually counted with multiplicities) as the degree of the projection of the space curve onto the first coordinate plane. As long as $t < 1$, the points remain generic, although the numerical condition numbers are expected to blow up as t approaches one.

The deteriorating numerical ill conditioning can be mitigated by the use of multiprecision arithmetic. For example, condition numbers larger than 10^8 make results unreliable in double precision. In double double precision, much higher condition numbers can be tolerated, typically up to 10^{16}, and this goes up to 10^{32} for quad double precision. As we interpret the inverse of the condition number as the distance to a singular solution, with multiprecision arithmetic we can compute more points more accurately as needed in the extrapolation to estimate winding numbers.

An additional difficulty arises when a path diverges to infinity, which manifests itself by a tropism with negative coordinates. A reformation of the problem in a weighted projective space corresponds to a unimodular coordinate transformation which uses the computed direction of the solution path. Towards the end of the path, this direction coincides with the tropism.

The a posteriori verification of a polyhedral end game is similar to computing a Puiseux expansion starting at a pretropism.

9   Computational Experiments

In this section we focus on the family of systems (7) with a tropism hidden in a higher dimensional cone of pretropisms. Classical families such as the cyclic n-roots problems appear not to have such hidden pretropisms, at least not for the cases computed in [1,2] and [19].

9.1   Symbolic Methods

To substantiate the claim that finding the tropical variety is computationally expensive, we calculated tropical bases of the system (7) for various values of n. The symbolic computations of tropical bases was done with Gfan [14]. Times are displayed in Table 1. The computations were executed on an Intel Xeon E5-2670 processor running RedHat Linux. As is clear from the table, as the dimension grows for this relatively simple system, computation time becomes prohibitively large.

As mentioned in Sect. 7, an alternative to computing the tropical basis is to only calculate the witness polynomial for a particular cone of the tropical

Table 1. Execution times, in seconds, of the computation of a tropical basis for the system (7); averages of 3 trials.

n	3	4	5	6	7
time	0.052	0.306	2.320	33.918	970.331

prevariety. We implemented this algorithm in Macaulay2 [5] and applied it to (7) to cut down the cone generated by the rays $(1, 1, \ldots, 1)$ and $(1, 0, 0, \ldots, 0)$. In all the cases we tried, the new prevariety contained the ray $(2, 1, \ldots, 1)$, as we expected.

From Table 2 it is clear that this has a significant speed advantage over computing a full tropical basis. However, it has the disadvantage of introducing many more rays into the prevariety. The number can vary depending on the random ray chosen in the cone, so the third column lists some of the values we obtained over several trials. We only computed up through dimension 10 because the prevariety computations were excessive for higher dimensions.

Table 2. Execution times in seconds of the computation of a witness polynomial for the cone generated by $(1, 1, \ldots, 1), (1, 0, \ldots, 0)$ of the system (7); averages of 3 trials. The third column lists the number of rays in the fan obtained by intersecting the original prevariety with the normal fan of the witness polynomial; since this can vary with the choice of random ray, we list values from several tries.

dim	time	#rays in the new fan
3	0.004	4, 5
4	0.011	10, 11
5	0.004	13, 14
6	0.009	27, 49
7	0.033	13, 25, 102
8	0.170	124, 401, 504
9	0.963	758, 1076
10	10.749	514, 760, 1183, 2501
11	131.771	
12	1131.089	

9.2 Our Approach

The polyhedral end game was done with version 2.4.10 of PHCpack [22], upgraded with double double and quad double arithmetic, using QDlib [9]. Polyhedral end games are also available via the Python interface of PHCpack, since version 0.4.0 of phcpy [24].

Table 3. Execution times on tracking d paths in n-space with a polyhedral end game. The reported time is the elapsed CPU user time, in seconds. The last column represents the average time spent on one path.

n	d	time	time/d
4	4	0.012	0.003
5	8	0.035	0.006
6	16	0.090	0.007
7	32	0.243	0.010
8	64	0.647	0.013
9	128	1.683	0.016
10	256	4.301	0.017
11	512	7.507	0.015
12	1024	27.413	0.027

For the first motivating example (4) in 3-space, there are four solutions when $x_1 = \gamma$. The tropism $(3, 1, 1)$, with winding number 3, is recovered when running a polyhedral end game, tracking four solution paths. Even in quad double precision (double precision already suffices), the running time is a couple of hundred milliseconds.

Table 3 shows execution times for the family of polynomial systems in (7). The computations were executed on one core of an Intel Xeon E5-2670 processor, running RedHat Linux.

All directions computed with double precision at an accuracy of 10^{-8}. For this family of systems, double precision sufficed to accurately compute the tropism $(2, 1, \ldots, 1)$. Although the total number of paths grows exponentially, every path has the same direction, so tracking only one path suffices. Clearly, these times are significantly smaller than the time required to compute a tropical basis.

10   Conclusions

The tropical prevariety provides candidate tropisms for Puiseux series expansions of space curves. As shown in [1,2] on the cyclic n-root problems, the pretropisms may directly lead to series developments for the positive dimensional solution sets. In this paper we studied cases where tropisms are in the relative interior of higher-dimensional cones of the tropical prevariety. If the tropical prevariety contains a higher dimensional cone and Puiseux series expansion fails at one of the cone's generating rays, then a polyhedral end game can recover the tropisms in the interior of that higher dimensional cone of pretropisms. As our example shows, this takes drastically less time than computing the tropical variety via a tropical basis, especially as dimension grows. It is also faster than finding a witness polynomial for just that particular cone, and avoids the issue of adding rays to the tropical prevariety.

References

1. Adrovic, D., Verschelde, J.: Computing Puiseux series for algebraic surfaces. In: van der Hoeven, J., van Hoeij, M. (eds.) Proceedings of the 37th International Symposium on Symbolic and Algebraic Computation (ISSAC 2012), pp. 20–27. ACM (2012)
2. Adrovic, D., Verschelde, J.: Polyhedral methods for space curves exploiting symmetry applied to the cyclic n-roots problem. In: Gerdt, V.P., Koepf, W., Mayr, E.W., Vorozhtsov, E.V. (eds.) CASC 2013. LNCS, vol. 8136, pp. 10–29. Springer, Heidelberg (2013)
3. Bernshteĭn, D.: The number of roots of a system of equations. Funct. Anal. Appl. **9**(3), 183–185 (1975)
4. Bogart, T., Jensen, A., Speyer, D., Sturmfels, B., Thomas, R.: Computing tropical varieties. J. Symbolic Comput. **42**(1), 54–73 (2007)
5. Grayson, D., Stillman, M.: Macaulay2, a software system for research in algebraic geometry. http://www.math.uiuc.edu/Macaulay2/
6. Herrero, M., Jeronimo, G., Sabia, J.: Affine solution sets of sparse polynomial systems. J. Symbolic Comput. **51**(1), 34–54 (2012)
7. Herrero, M., Jeronimo, G., Sabia, J.: Elimination for generic sparse polynomial systems. Discrete Comput. Geom. **51**(3), 578–599 (2014)
8. Herrero, M., Jeronimo, G., Sabia, J.: Puiseux expansions and non-isolated points in algebraic varieties. Commun. Algebra **44**(5), 2100–2109 (2016)
9. Hida, Y., Li, X., Bailey, D.: Algorithms for quad-double precision floating point arithmetic. In: 15th IEEE Symposium on Computer Arithmetic (Arith-15 2001), 11–17, Vail, CO, USA, pp. 155–162. IEEE Computer Society (2001). Shortened version of Technical Report LBNL-46996, software at http://crd.lbl.gov/~dhbailey/mpdist/qd-2.3.9.tar.gz
10. Huber, B., Sturmfels, B.: A polyhedral method for solving sparse polynomial systems. Math. Comput. **64**(212), 1541–1555 (1995). http://www.jstor.org/stable/2153370
11. Huber, B., Verschelde, J.: Polyhedral end games for polynomial continuation. Numer. Algorithms **18**(1), 91–108 (1998)
12. Jensen, A., Leykin, A., Yu, J.: Computing tropical curves via homotopy continuation. Exp. Math. **25**(1), 83–93 (2016)
13. Jensen, A., Yu, J.: Computing tropical resultants. J. Algebra **387**, 287–319 (2013)
14. Jensen, A.: Computing Gröbner fans and tropical varieties in Gfan. In: Stillman, M., Takayama, N., Verschelde, J. (eds.) Software for Algebraic Geometry. The IMA Volumes in Mathematics and its Applications, vol. 148, pp. 33–46. Springer, Heidelberg (2008)
15. Jeronimo, G., Matera, G., Solernó, P., Waissbein, A.: Deformation techniques for sparse systems. Found. Comput. Math. **9**(1), 1–50 (2008). http://dx.doi.org/10.1007/s10208-008-9024-2
16. Maclagan, D., Sturmfels, B.: Introduction to Tropical Geometry, Graduate Studies in Mathematics, vol. 161. American Mathematical Society, Providence (2015)
17. OEIS Foundation Inc.: The on-line encyclopedia of integer sequences (2016). http://oeis.org. Accessed 03 Nov 2015
18. Römer, T., Schmitz, K.: Generic tropical varieties. J. Pure Appl. Algebra **216**(1), 140–148 (2012). http://www.sciencedirect.com/science/article/pii/S0022404911001290

19. Sabeti, R.: Numerical-symbolic exact irreducible decomposition of cyclic-12. LMS J. Comput. Math. **14**, 155–172 (2011)
20. Schneider, R.: Convex Bodies: The Brunn-Minkowski Theory, Encyclopedia of Mathematics and its Applications, vol. 44. Cambridge University Press, Cambridge (1993)
21. Sommars, J., Verschelde, J.: Pruning algorithms for pretropisms of Newton polytopes. In: Gerdt, V.P., Koepf, W., Seiler, W.M., Vorozhtsov, E.V. (eds.) CASC 2016. LNCS, vol. 9890, pp. 489–503 (2016)
22. Verschelde, J.: Algorithm 795: PHCpack: a general-purpose solver for polynomial systems by homotopy continuation. ACM Trans. Math. Softw. **25**(2), 251–276 (1999)
23. Verschelde, J.: Polyhedral methods in numerical algebraic geometry. In: Bates, D., Besana, G., Di Rocco, S., Wampler, C. (eds.) Interactions of Classical and Numerical Algebraic Geometry, Contemporary Mathematics, vol. 496, pp. 243–263. AMS (2009)
24. Verschelde, J.: Modernizing PHCpack through phcpy. In: de Buyl, P., Varoquaux, N. (eds.) Proceedings of the 6th European Conference on Python in Science (EuroSciPy 2013), pp. 71–76 (2014)
25. Verschelde, J., Verlinden, P., Cools, R.: Homotopies exploiting Newton polytopes for solving sparse polynomial systems. SIAM J. Numer. Anal. **31**(3), 915–930 (1994)

Algorithmic Computation of Polynomial Amoebas

D.V. Bogdanov[1], A.A. Kytmanov[2], and T.M. Sadykov[1](✉)

[1] Plekhanov Russian University, Stremyanny 36, Moscow 125993, Russia
Sadykov.TM@rea.ru
[2] Siberian Federal University, Svobodny 79, Krasnoyarsk 660041, Russia
aakytm@gmail.com

Abstract. We present algorithms for computation and visualization of polynomial amoebas, their contours, compactified amoebas and sections of three-dimensional amoebas by two-dimensional planes. We also provide a method and an algorithm for the computation of polynomials whose amoebas exhibit the most complicated topology among all polynomials with a fixed Newton polytope. The presented algorithms are implemented in computer algebra systems Matlab 8 and Mathematica 9.

Keywords: Amoebas · Newton polytope · Optimal algebraic hypersurface · The contour of an amoeba · Hypergeometric functions

1 Introduction

The amoeba of a multivariate polynomial in several complex variables is the Reinhardt diagram of its zero locus in the logarithmic scale with respect to each of the variables [17]. The term "amoeba" has been coined in [5, Chap. 6] where two competing definitions of the amoeba of a polynomial have been given: the affine and the compactified versions. Both definitions are only interesting in dimension two and higher since the amoeba of a univariate polynomial is a finite set that can be explored by a variety of classical methods of localization of polynomial roots.

The geometry of the amoeba of a polynomial carries much information on the zeros of this polynomial and is closely related to the combinatorial structure of its Newton polytope (see Theorem 1). Despite loosing half of the real dimensions, the image of the zero locus of a polynomial in the amoeba space reflects the relative size of some of its coefficients.

From the computational point of view, the problem of giving a complete geometric or combinatorial description of the amoeba of a polynomial is a task of formidable complexity [8,11], despite the substantial recent progress in this direction [12–14]. The number of connected components of an amoeba complement as a function of the coefficients of the polynomial under study is still to be explored by means of the modern methods of computer algebra. In particular,

© Springer International Publishing AG 2016
V.P. Gerdt et al. (Eds.): CASC 2016, LNCS 9890, pp. 87–100, 2016.
DOI: 10.1007/978-3-319-45641-6_7

the conjecture by M. Passare on the solidness of maximally sparse polynomials (see Definition 5) remains open for a long time.

Amoebas can be computed and depicted by means of a variety of approaches and methods (see [3,6,9,11,13,14] and the references therein). The present paper is meant to move forward the art of computing and drawing complex amoebas of algebraic hypersurfaces. A special attention paid to the geometrically most interesting and computationally most challenging case of optimal hypersurfaces (see Definition 4). We expose methods and algorithms for computation and visualization of amoebas of bivariate polynomials, their contours and compactified versions. The developed algorithms are used in higher dimensions for depicting sections of amoebas of polynomials in three variables. The main focus of the paper is on polynomials whose amoebas have the most complicated topological structure among all polynomials with a given Newton polytope (see Definition 4). We provide an algorithm for explicit construction of such polynomials.

The presented algorithms are implemented in the computer algebra systems Matlab 8 (64-bit) and Wolfram Mathematica 9 (64-bit). All examples in the paper have been computed on Intel Core i5-4440 CPU clocked at 3.10 GHz with 16 Gb RAM under MS Windows 7 Ultimate SP1.

2 Convex Polytopes, Cones and Amoebas: Definitions and Preliminaries

Let $p(x)$ be a polynomial in n complex variables:

$$p(x_1,\ldots,x_n) = \sum_{\alpha \in A} c_\alpha x^\alpha = \sum_{\alpha \in A} c_{\alpha_1 \ldots \alpha_n} x_1^{\alpha_1} \cdot \ldots \cdot x_n^{\alpha_n},$$

where $A \subset \mathbb{Z}^n$ is a finite set.

Definition 1. *The Newton polytope $\mathcal{N}_{p(x)}$ of a polynomial $p(x)$ is the convex hull of the set A of its exponent vectors.*

Definition 2. *The recession cone of a convex set M is the set-theoretical maximal element in the family of convex cones whose shifts are contained in M.*

Definition 3. *The amoeba $\mathcal{A}_{p(x)}$ of a polynomial $p(x)$ is the image of its zero locus under the map* $\mathrm{Log} : (x_1,\ldots,x_n) \longmapsto (\ln|x_1|,\ldots,\ln|x_n|)$.

The connected components of the complement to the amoeba $\mathcal{A}_{p(x)}$ are convex and in bijective correspondence with the expansions of the rational function $1/p(x)$ into Laurent series centered at the origin [4]. The next statement shows how the Newton polytope $\mathcal{N}_{p(x)}$ is reflected in the geometry of the amoeba $\mathcal{A}_{p(x)}$ [4, Theorem 2.8 and Proposition 2.6].

Theorem 1 (See [4]). *Let $p(x)$ be a Laurent polynomial and let $\{M\}$ denote the family of connected components of the amoeba complement $^c\mathcal{A}_{p(x)}$. There exists an injective function $\nu : \{M\} \to \mathbb{Z}^n \cap \mathcal{N}_{p(x)}$ such that the cone that is* dual to the polytope $\mathcal{N}_{p(x)}$ at the point $\nu(M)$ coincides with the recession cone of M. In particular, the number of connected components of $^c\mathcal{A}_{p(x)}$ cannot be smaller than the number of vertices of the polytope $\mathcal{N}_{p(x)}$ and cannot exceed the number of integer points in $\mathcal{N}_{p(x)}$.

Thus, the amoeba of polynomial $p(x)$ in $n \geq 2$ variables is a closed connected unbounded subset of $\mathbb{R}^n$ whose complement consists of a finite number of convex connected components. Besides, a two-dimensional amoeba has "tentacles" that go off to infinity in the directions that are orthogonal to the sides of the polygon $\mathcal{N}_{p(x)}$ (see Figs. 2 and 3).

The two extreme values for the number of connected components of an amoeba complement are of particular interest.

Definition 4 (See [4, Definition 2.9]). An algebraic hypersurface $\mathcal{H} \subset (\mathbb{C}^*)^n$, $n \geq 2$, is called *optimal* if the number of connected components of the complement of its amoeba $^c\mathcal{A}_\mathcal{H}$ equals the number of integer points in the Newton polytope of the defining polynomial for $\mathcal{H}$. We will say that a polynomial (as well as its amoeba) is optimal if its zero locus is an optimal algebraic hypersurface.

In other words, an algebraic hypersurface is optimal if the topology of its amoeba is as complicated as it could possibly be under the condition that the Newton polytope of the defining polynomial is fixed. The other extreme case of the topologically simplest possible amoeba is defined as follows.

Definition 5 (See [10]). An algebraic hypersurface $\mathcal{H} \subset (\mathbb{C}^*)^n$, $n \geq 2$, is called *solid*, if the number of connected components of its amoeba complement $^c\mathcal{A}_\mathcal{H}$ equals the number of vertices of the Newton polytope of its defining polynomial $\mathcal{H}$.

Thus, the solid and the optimal amoebas are the endpoints of the spectrum of amoebas of polynomials with a given Newton polytope. Of course, there exist plenty of optimal (as well as solid) amoebas defined by polynomials with a given Newton polytope. In fact, both sets of amoebas regarded as subsets in the complex space of coefficients of defining polynomials, have nonempty interior.

In the bivariate case, an amoeba is solid if and only if all of the connected components of its complement are unbounded and no its tentacles are parallel. A two-dimensional optimal amoeba has, on the contrary, the maximal possible number of bounded connected components in its complement and the maximal possible number of parallel tentacles.

The functional dependency of the topological type of the amoeba $\mathcal{A}_{p(x)}$ on the coefficients of its defining polynomial $p(x)$ is complex and little understood at the present moment. A sufficient condition for the amoeba of a polynomial to be optimal is that it satisfies a "natural" system of partial differential equations of hypergeometric type [1,2] while the support of the polynomial in question is complex enough [15]. In Sect. 4, we expose an algorithm for the computation of the hypergeometric polynomial with the prescribed Newton polytope.

Example 1. Let $\mathcal{N}$ denote the convex hull of the set of lattice points $\{(0,0),(1,0),(0,2),(2,1)\}$, see Fig. 1. This polygon will appear in several examples that follow and has been chosen as one of the simplest polygons that contain an inner integer point as well as an integer point in the relative interior of its edge.

Fig. 1. The support of the polynomials in Example 1

Figure 2 shows the amoebas of the four bivariate polynomials $p_1(x,y) = 1 + x + y + xy + y^2 + x^2y$, $p_2(x,y) = 1+x+3y+xy+y^2+x^2y$, $p_3(x,y) = 1+x+y+4xy+y^2+x^2y$, $p_4(x,y) = 1+x+3y+4xy+y^2+x^2y$ whose Newton polygons coincide with $\mathcal{N}$. The complement of the solid amoeba in Fig. 2(a) consists of the four unbounded connected components with two-dimensional recession cones. The complement to the optimal amoeba in Fig. 2(d) comprises six connected components: the four unbounded components with the two-dimensional recession cones, one unbounded component between the tentacles with the one-dimensional recession cone, and the bounded component. The other two amoebas depicted in Figs. 2(b) and (c) exhibit five connected components in their respective complements and topologically assume an intermediate position between the solid and the optimal amoebas defined by polynomials with the Newton polygon $\mathcal{N}$.

The existing analytic methods [4,12] do not in general allow one to predict the topological type of a polynomial with generic coefficients. From the computational point of view, the tasks of depicting the amoebas in Figs. 2(a) and (d) are rather similar. Yet, to predict the existence of a bounded connected component of a given order [4] in an amoeba complement by means of analytic methods is in general a task of formidable complexity [11].

3   Computing Two-Dimensional Amoebas

Definition 6. We will call the *"carcase"* of an amoeba $\mathcal{A}$ any subset of $\mathcal{A}$, such that the number of connected components of the complement to the intersection $\mathcal{A} \cap B$ for a sufficiently large ball B is as big as it could possibly be (that is, equal to the number of connected components in the complement of $\mathcal{A}$ in $\mathbb{R}^n$).

We remark that the carcase of an amoeba is not uniquely defined. However, the topology of its complement in a sufficiently big ball is well-defined and is as complex as possible. When speaking of depicting an amoeba we will mean depicting its suitable carcase.

In this section, we consider bivariate polynomials of the form $p(x,y) = \sum c_{ij}x^iy^j$ and provide an algorithm for computing their amoebas.

Theorem 1 yields that the geometry of the amoeba $\mathcal{A}_p$ is closely related to the properties of the Newton polytope $\mathcal{N}_p$ of the polynomial p. Yet, the coefficients of p also play a role and determine the size of the carcase of the amoeba in question. In what follows, the boundary of the domain where the carcase of an amoeba is depicted has been determined experimentally.

Algorithm 1. Algorithm for computing the amoeba $\mathcal{A}_p$ of a bivariate polynomial

---

Require: List of the polynomial coefficients cx_list, in x by all monomials y^k, $k = 0, \ldots, \deg_y p$; the boundaries of the rectangular domain in the logarithmic amoeba space a, b; the number of values of the absolute value n_r and the angle n_φ of the complex variable.

Ensure: List of coordinates of points that belong to the amoeba carcase z_list, w_list.

 1: **procedure** AMOEBA2D(cx_list, a, b, n_r, n_φ)
 2:     $z\_list :=$ empty list
 3:     $w\_list :=$ empty list
 4:     $d :=$ the number of elements in $cx\_list - 1$
 5:     $1d := (1, \ldots, 1) \in \mathbb{R}^d$ $\triangleright$ the vector with d units
 6:     $h_r := (\exp(b) - \exp(a)) / (n_r - 1)$
 7:     $h_\varphi := 2\pi / (n_\varphi - 1)$
 8:     **for** $r = \exp(a) : \exp(b) : h_r$ **do**
 9:         **for** $\varphi = 0 : 2\pi : h_\varphi$ **do**
10:             $x := r * \exp\left(\sqrt{-1} * \varphi\right)$
11:             $y := \text{roots}(cx\_list)$ $\triangleright$ the vector of zeros of the polynomial with the coefficients cx_list
12:             Add $\text{Log}(|x| * 1d)$ to z_list
13:             Add $\text{Log}(|y|)$ to w_list
14:         **end for**
15:     **end for**
16:     **return** $\{z\_list, w\_list\}$
17: **end procedure**

The next straightforward algorithm which computes the amoeba of a bivariate polynomial has been used by numerous authors in various forms [3,6,9]. We include it for the sake of completeness and future reference. We refer to [8] for a detailed discussion of computational complexity of polynomial root finding algorithms.

To obtain a picture of good quality the steps of the Algorithm 1 are repeated with the variables x and y interchanged. The points with the computed coordinates are depicted in the same figure.

Example 2. For the polynomial $p_4(x, y)$ in Example 1, the lists of polynomial coefficients in the variables x and y are as follows: $cx\_list=\{1, 3 + 4x + x^2, 1 + x\}$ and $cy\_list=\{y, 1 + 4y, 1 + 3y + y^2\}$. The (carcase of the) amoeba is depicted in the rectangle $[-5, 5] \times [-5, 5]$. The number of values of the absolute value of a variable is $n_r = 2000$ while the number of values of its argument is $n_\varphi = 180$. Figure 2 features the amoebas of the polynomials in Example 1 computed by means of Algorithm 1.

The topologically more involved amoeba of the polynomial in Example 4 has been computed in a similar way.

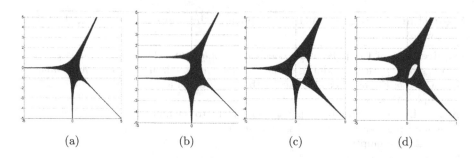

(a)   (b)   (c)   (d)

Fig. 2. The amoebas of the polynomials $p_1\,(x,y)$, $p_2\,(x,y)$, $p_3\,(x,y)$ and $p_4\,(x,y)$

4   Generating Optimal Polynomials

In this section, we employ the notion of a hypergeometric polynomial for the purpose of constructive generation of optimal amoebas. We will need the following auxiliary definition.

Definition 7. A formal Laurent series

$$\sum_{s\in\mathbb{Z}^n}\varphi\,(s)\,x^s \tag{1}$$

is called *hypergeometric* if for any $j = 1,\ldots,n$ the quotient $\varphi(s + e_j)/\varphi(s)$ is a rational function in $s = (s_1,\ldots,s_n)$. Throughout the paper, we denote this rational function by $P_j(s)/Q_j(s+e_j)$. Here $\{e_j\}_{j=1}^n$ is the standard basis of the lattice $\mathbb{Z}^n$. By the *support* of this series we mean the subset of $\mathbb{Z}^n$ on which $\varphi(s) \neq 0$.

A *hypergeometric function* is a (multi-valued) analytic function obtained by means of analytic continuation of a hypergeometric series with a nonempty domain of convergence along all possible paths [5,10].

Theorem 2 (Ore, Sato, cf [16]). The coefficients of a hypergeometric series are given by the formula

$$\varphi\,(s) = t^s\,U\,(s)\prod_{i=1}^m \Gamma\left(\langle \mathbf{A}_i, s\rangle + c_i\right), \tag{2}$$

where $t^s = t_1^{s_1}\ldots t_n^{s_n}$, $t_i, c_i \in \mathbb{C}$, $\mathbf{A}_i = (A_{i,1},\ldots,A_{i,n}) \in \mathbb{Z}^n$, $i = 1,\ldots,m$ and $U\,(s)$ is the product of a certain rational function and a periodic function $\phi\,(s)$ such that $\phi\,(s + e_j) = \phi\,(s)$ for every $j = 1,\ldots,n$.

Given the above data $(t_i, c_i, \mathbf{A}_i, U\,(s))$ that determines the coefficient of a multivariate hypergeometric Laurent series, it is straightforward to compute the rational functions $P_i\,(s)/Q_i\,(s+e_i)$ using the Γ-function identity $\Gamma(z + 1) = z\Gamma(z)$ (see e.g. [16]).

Algorithm 2. Generation of a bivariate hypergeometric polynomial and its defining system of equations

---

Require: List of vertices N_list of a convex integer polygon P.
Ensure: List of coefficients c_list of the hypergeometric polynomial $p(x, y)$ whose Newton polygon is P.

1: **procedure** HYPERPOLY2D(N_list)
2:     $B = $ list of outer normals to the sides of P
3:     $\varphi(s, t) = 1/\prod_{j=1}^{q} \Gamma(1 - \langle B_j, (s,t) \rangle) - c_j)$
4:     $c_list = $ list of the coefficients of the hypergeometric polynomial
5:     $R_1 = $ FunctionExpand$[\varphi(s+1, t)/\varphi(s, t)]$
6:     $R_2 = $ FunctionExpand$[\varphi(s, t+1)/\varphi(s, t)]$
7:     $P_1 = $ Numerator$[R_1]$
8:     $P_2 = $ Numerator$[R_2]$
9:     $Q_1 = $ Denominator$[R_1]$
10:     $Q_2 = $ Denominator$[R_2]$
11:     $\theta_x = xp'_x$
12:     $\theta_y = yp'_y$
13:     $p = $ the polynomial defined by c_list
14:     **if** $xP_1(\theta)p = Q_1(\theta)p$ **and** $yP_2(\theta)p = Q_2(\theta)p$ **then**
15:         **return** $\{c\_list\}$
16:     **end if**
17: **end procedure**

Definition 8. A (formal) Laurent series $\sum_{s \in \mathbb{Z}^n} \varphi(s) x^s$ whose coefficient satisfies the relations $\varphi(s + e_j)/\varphi(s) = P_j(s)/Q_j(s + e_j)$ is a (formal) solution to the following system of partial differential equations of hypergeometric type:

$$x_j P_j(\theta) f(x) = Q_j(\theta) f(x), \quad j = 1, \ldots, n. \tag{3}$$

Here $\theta = (\theta_1, \ldots, \theta_n)$, $\theta_j = x_j \frac{\partial}{\partial x_j}$.

The system (3) will be referred to as *the Horn hypergeometric system defined by the Ore–Sato coefficient* $\varphi(s)$ (see [2]) and denoted by Horn (φ).

For a convex integer polytope $\mathcal{N}$, we construct the list B of the outer normals to the faces of $\mathcal{N}$. We denote the length of this list by q. We assume the elements of B to be normalized so that the coordinates of each outer normal are integer and relatively prime. Define the Ore–Sato coefficient

$$\varphi(s) = \frac{1}{\prod_{j=1}^{q} \Gamma(1 - \langle B_j, s \rangle - c_j)}. \tag{4}$$

The hypergeometric system (3) defined by the Ore–Sato coefficient $\varphi(s)$ admits a polynomial solution with several interesting properties. Under the additional assumption that it cannot be factored in the ring of Puiseux polynomials $\mathbb{C}[x_1^{1/d}, \ldots, x_n^{1/d}]$ for any $d \in \mathbb{N}$, this polynomial turns out to have an optimal amoeba. We will employ Algorithm 2 to generate optimal polynomials of hyper-geometric type.

Example 3. The outer normals (normalized as explained above) to the sides of the polygon shown in Fig. 1 are as follows: $(0, -1)$, $(-1, 0)$, $(1, 2)$, $(1, -1)$. Using (4) we obtain

$$\varphi(s, t) = (\Gamma(s+1)\,\Gamma(t+1)\,\Gamma(-s-2t+5)\,\Gamma(-s+t+2))^{-1}.$$

By Definition 8 the above Ore–Sato coefficient $\varphi(s, t)$ gives rise to the polynomials

$$P_1(s, t) = (-s - 2t + 4)(-s + t + 1), \quad Q_1(s, t) = s + 2,$$
$$P_2(s, t) = (s + 2t - 4)(s + 2t - 3), \quad Q_2(s, t) = (t + 2)(-s + t + 3).$$

The corresponding hypergeometric system is defined by the linear partial differential operators

$$\begin{cases} x\,(-\theta_x - 2\theta_y + 4)\,(-\theta_x + \theta_y + 1) - (\theta_x + 2), \\ y\,(\theta_x + 2\theta_y - 4)\,(\theta_x + 2\theta_y - 3) - (\theta_y + 2)\,(-\theta_x + \theta_y + 3). \end{cases} \tag{5}$$

It is straightforward to check that the hypergeometric polynomial $p(x, y) = 1 + 4x + 6y + 24xy + 12x^2y + 2y^2$ (whose coefficients are found in accordance with Step 3 of Algorithm 2 and normalized to be relatively prime integers) belongs to the kernels of operators (5).

Example 4. Using Algorithms 1 and 2 we compute the coefficients of the hypergeometric polynomial supported in the polygon $\mathcal{N}_{p_h(x,y)}$ shown in Fig. 3(a):
$p_h(x, y) = 2x^2 + 20xy + 72x^2y + 20x^3y + 5y^2 + 160xy^2 + 450x^2y^2 + 160x^3y^2 + 5x^4y^2 + 12y^3 + 300xy^3 + 800x^2y^3 + 300x^3y^3 + 12x^4y^3 + 5y^4 + 160xy^4 + 450x^2y^4 + 160x^3y^4 + 5x^4y^4 + 20xy^5 + 72x^2y^5 + 20x^3y^5 + 2x^2y^6$. The amoeba $\mathcal{A}_{p_h(x,y)}$ of this polynomial is depicted in Fig. 3(a)(right).

   A more involved hypergeometric polynomial is given by $p_{h2}(x, y) = -456456x^3 + 488864376x^2y - 28756728x^3y + 25420947552x^2y^2 - 244432188x^3y^2 + 3003x^4y^2 - 119841609888xy^3 + 127104737760x^2y^3 - 465585120x^3y^3 + 6006x^4y^3 + 1396755360y^4 - 508418951040xy^4 + 139815211536x^2y^4 - 232792560x^3y^4 + 1729x^4y^4 + 4190266080y^5 - 355893265728xy^5 + 41611670100x^2y^5 - 29628144x^3y^5 + 57x^4y^5 + 698377680y^6 - 58663725120xy^6 + 3328933608x^2y^6 - 705432x^3y^6 - 2327925600xy^7 + 55023696x^2y^7 - 16930368xy^8$. Its defining Ore–Sato coefficient equals $\varphi(s, t) = \Gamma(s + t - 4)\Gamma(-4s + t - 16)\Gamma(-3s - 2t - 5)\Gamma(3s - t - 3)\Gamma(2s + t - 5)$. The amoeba of this polynomial is shown in Fig. 3(b).

5   Computing Contours of Amoebas of Bivariate Polynomials

The boundary of an amoeba admits a rich analytic structure that is revealed in the following definition.

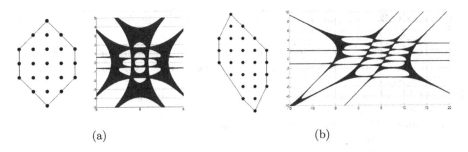

(a)   (b)

Fig. 3. The Newton polytope and the amoeba of the hypergeometric polynomial $p_h(x, y)$ and polynomial $p_{h2}(x, y)$

Definition 9. The *contour* of the amoeba $\mathcal{A}_{p(x)}$ is the set $\mathcal{C}_{p(x)}$ of critical points of the logarithmic map Log restricted to the zero locus of the polynomial $p(x)$, that is, the map Log : $\{p(x) = 0\} \rightarrow \mathbb{R}^n$.

The structure of the contour of an amoeba can be described in terms of the logarithmic Gauss map which is defined as follows.

Definition 10. Denote by $\mathrm{Gr}(n, k)$ the Grassmannian of k-dimensional subspaces in $\mathbb{C}^n$. The *logarithmic Gauss map* [7] is defined to be the map $\gamma : \mathcal{H} \longmapsto \mathrm{Gr}(n, k)$ that maps a regular point $x \in \mathrm{reg}\,\mathcal{H}$ into the normal subspace $\gamma(x)$ to the image $\log \mathcal{H}$.

The boundary of an amoeba $\partial \mathcal{A}_{p(x)}$ is necessarily a subset of the contour $\mathcal{C}_{p(x)}$ but is in general different from it.

Knowing the structure of the contour of an amoeba is important for describing the topological structure of the amoeba complement. Experiments show that a cusp of the contour inside the "body" of the corresponding amoeba is a counterpart of a missing connected component in its complement.

One of the ways to draw the contour of an amoeba is to depict the solutions to the system of algebraic equations

$$\begin{cases} p(x, y) = 0, \\ x\frac{\partial p(x,y)}{\partial x} - uy\frac{\partial p(x,y)}{\partial y} = 0. \end{cases} \tag{6}$$

Here $u \in \mathbb{R} \cup \{\infty\}$ is a real parameter that encodes the slope of the normal line to the contour of the amoeba.

Eliminating the variables x and y out of system (6) we obtain polynomials $s(y, u)$ and $t(x, u)$ that can be used to depict the contour of the amoeba by means of the following algorithm.

Remark 1. The lists of coordinates that are obtained at each iteration of the cycle in Algorithm 3 might contain additional elements that do not correspond to the amoeba contour. They are sorted out by checking against the system of equations (6). All visualization algorithms are linear in the lattice parameters h_r, h_φ, h_u.

Algorithm 3. Computing the contour C_p of the amoeba of a bivariate polynomial

---

Require: List of polynomial coefficients cx_list, cy_list depending on u, by all monomials x^k, $k = 0, \ldots, \deg_x t$, y^m, $m = 0, \ldots, \deg_y s$; initial value u_1, final value u_n and the step h_u.

Ensure: List of coordinates of points z_list, w_list that belong to the contour of the amoeba.

1: **procedure** CONTOUR2D(cx_list, cy_list, u_1, u_n, h_u)
2:     $z\_list :=$ the empty list
3:     $w\_list :=$ the empty list
4:     $dx :=$ the number of elements in $cx\_list - 1$
5:     $dy :=$ the number of elements in $cy\_list - 1$
6:     **for** $u = u_1 : u_n : h_u$ **do**
7:         $x :=$ roots(cx_list)   ▷ the vector of zeros of the polynomial with the coefficients cx_list
8:         $y :=$ roots(cy_list)   ▷ the vector of zeros of the polynomial with the coefficients cy_list
9:         Add Log($|x|$) to z_list
10:        Add Log($|y|$) to w_list
11:    **end for**
12:    **return** $\{z\_list, w\_list\}$
13: **end procedure**

Example 5. The system of equations (6) associated with the first polynomial in Example 1 has the form

$$\begin{cases} 1 + x + y + xy + y^2 + x^2 y = 0, \\ x - uy + xy - uxy + 2x^2 y - ux^2 y - 2uy^2 = 0. \end{cases}$$

Eliminating variables x and y yields the equations $u - 2uy + u^2 y - 4uy^2 + 2u^2 y^2 - 2uy^3 - u^2 y^3 + 3uy^4 + 4uy^5 + u^2 y^5 = 1 - y - 4y^2 - 9y^3 - 7y^4 - 4y^5$ and $-3u^2 - 3ux - 5u^2 x - 2ux^2 + u^2 x^2 + 3ux^3 + 5u^2 x^3 + 2ux^4 + 3u^2 x^4 + ux^5 + u^2 x^5 = x^2 + 2x^3 + 5x^4 + 2x^5$, respectively. The lists of the coefficients that depend on u are $cx\_list = \{-1+u, 1-2u+u^2, 4-4u+2u^2, 9-2u-u^2, 7+3u, 4+4u+u^2\}$ and $cy\_list = \{-3u^2, -3u-5u^2, -1-2u+u^2, -2+3u+5u^2, -5+2u+3u^2, -2+u+u^2\}$. The parameter u assumes values in the interval $[-120, 120]$ with the step $h_u = 0.001$. The obtained contours of the amoebas of the polynomials in Example 1 have nontrivial structure and are shown in Fig. 4.

6 Computing Two-Dimensional Compactified Amoebas

Definition 11. *The compactified amoeba $\overline{A}_{p(x)}$ of a polynomial $p(x)$ is defined to be the image of its zero locus under the mapping*

$$\mu(x) = \frac{\sum\limits_{\alpha \in A} \alpha \cdot |x^\alpha|}{\sum\limits_{\alpha \in A} |x^\alpha|} = \frac{\sum\limits_{(\alpha_1, \ldots, \alpha_n) \in A} (\alpha_1, \ldots, \alpha_n) \cdot |x^{\alpha_1} \cdot \ldots \cdot x^{\alpha_n}|}{\sum\limits_{(\alpha_1, \ldots, \alpha_n) \in A} |x^{\alpha_1} \cdot \ldots \cdot x^{\alpha_n}|}. \tag{7}$$

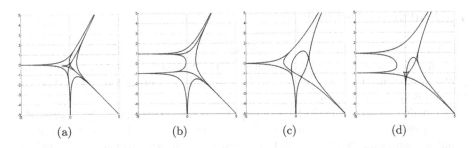

(a)   (b)   (c)   (d)

Fig. 4. Contours of the amoebas of the polynomials $p_1(x,y)$, $p_2(x,y)$, $p_3(x,y)$ and $p_4(x,y)$

Numeric computation of two-dimensional compactified amoebas is similar to that of their affine counterparts (see Algorithm 1). The main computational issue is dealing with the moment map (7) instead of the logarithmic map into the affine space.

Example 6. Applying (7) to the polynomials in Example 1 yields the mapping

$$(x, y) \longmapsto \mu(x, y) =$$

$$\frac{(1 \cdot (0,0) + |x| \cdot (1,0) + |y| \cdot (0,1) + |xy| \cdot (1,1) + |y^2| \cdot (0,2) + |x^2y| \cdot (2,1))}{(|1| + |x| + |y| + |xy| + |y^2| + |x^2y|)} =$$

$$\frac{(|x| + |xy| + 2|x^2y|, |y| + |xy| + 2|y^2| + |x^2y|)}{(|1| + |x| + |y| + |xy| + |y^2| + |x^2y|)}.$$

The corresponding compactified amoebas inside the Newton polygon of their defining polynomials are shown in Fig. 5.

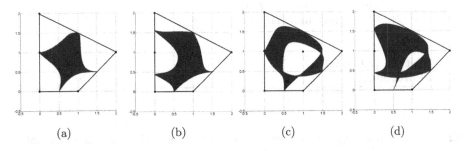

(a)   (b)   (c)   (d)

Fig. 5. Compactified amoebas of the polynomials $p_1(x,y)$, $p_2(x,y)$, $p_3(x,y)$ and $p_4(x,y)$ inside their Newton polygon

7 Multivariate Outlook

Depicting three-dimensional amoebas represents a substantial computational challenge due to the complex geometry of their shape. We will now treat the problem of computing sections of amoebas of polynomials in three variables by two-dimensional hyperplanes.

Consider a polynomial in three complex variables $p(x, y, z) = \sum c_{ijk} x^i y^j z^k$. To compute the section of its amoeba by the plane $|z| = \text{const}$ we fix the absolute value of z and modify Algorithm 1 by adding a cycle with respect to the values of the argument $\varphi_z \in [0; 2\pi]$.

Algorithm 4. Computing the section of the amoeba $\mathcal{A}_p$ of a polynomial in three variables by a two-dimensional plane

---

Require: List of polynomial coefficients cxz_list, depending on x and z by all monomials y^k, $k = 0, \ldots, \deg_y p$; bounds a, b for the rectangular domain in the amoeba space where the amoeba section is depicted; the numbers n_r and n_φ of values of the absolute value and the argument of the variable $z \in \mathbb{C}$, respectively.

Ensure: List of points that belong to the amoeba section z_list, w_list.

  1: **procedure** AMOEBA3D(cxz_list, a, b, n_r, n_φ, z)
  2:     $z\_list :=$ empty list
  3:     $w\_list :=$ empty list
  4:     $d :=$ the number of elements in $cxz\_list - 1$
  5:     $1d := (1, \ldots, 1) \in \mathbb{R}^d$ ▷ the vector of d units
  6:     $h_r := (\exp(b) - \exp(a))/(n_r - 1)$
  7:     $h_\varphi := 2\pi/(n_\varphi - 1)$
  8:     **for** $r = \exp(a) : \exp(b) : h_r$ **do**
  9:         **for** $\varphi = 0 : 2\pi : h_\varphi$ **do**
10:             $x := r * \exp\left(\sqrt{-1} * \varphi\right)$
11:             $y := \text{roots}(cxz\_list)$ ▷ the vector of zeros of the polynomial with the coefficients cxz_list
12:             Add $\text{Log}(|x| * 1d)$ to z_list
13:             Add $\text{Log}(|y|)$ to w_list
14:         **end for**
15:     **end for**
16:     **return** $\{z\_list, w\_list\}$
17: **end procedure**

Example 7. Using Algorithm 4 we compute a section of the amoeba of the polynomial $p(x, y, z) = 1 + 3y + y^2 + 6xy + x^2y + xyz + xyz^2$. This polynomial is one of the simplest polynomials with tetrahedral Newton polytopes that contain integer points in the interior as well as in the relative interior of faces of all positive dimensions (see Fig. 7). The section of the amoeba $\mathcal{A}_{p(x,y,z)}$ by the plane $\log |z| = 5$ is shown in Fig. 7(right).

Example 8. We now consider a computationally more challenging optimal hypergeometric polynomial in three variables supported in a regular integer octahedron. Due to symmetry it suffices to consider its part that belongs to the positive orthant and is shown in Fig. 6(left). We use a three-dimensional version of Algorithm 2 to compute the corresponding (uniquely determined up to scaling) hypergeometric polynomial: $p_h(x, y, z) = x^2 y^2 + 36 x^2 yz + 36 xy^2 z + 256 x^2 y^2 z + 36 x^3 y^2 z + 36 x^2 y^3 z + x^2 z^2 + 36 xyz^2 + 256 x^2 yz^2 + 36 x^3 yz^2 + y^2 z^2 + 256 xy^2 z^2 + 1296 x^2 y^2 z^2 + 256 x^3 y^2 z^2 + x^4 y^2 z^2 + 36 xy^3 z^2 + 256 x^2 y^3 z^2 + 36 x^3 y^3 z^2 + x^2 y^4 z^2 + 36 x^2 yz^3 + 36 xy^2 z^3 + 256 x^2 y^2 z^3 + 36 x^3 y^2 z^3 + 36 x^2 y^3 z^3 + x^2 y^2 z^4$. Figure 6(right) shows the intersection of the amoeba $\mathcal{A}_{p_h(x,y,z)}$ with the three coordinate hyperplanes as well as with the hyperplanes $|x| = 3$, $|x| = 8$, $|y| = 6$, $|z| = 4$ in the logarithmic amoeba space. The figure space is orthogonal to the vector $(1, 1, 1)$.

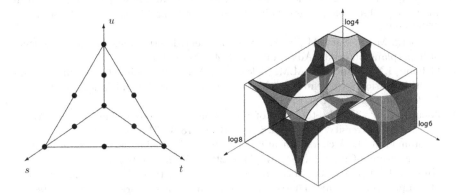

Fig. 6. The part of the support of $p_h(x, y, z)$ that belongs to the positive orthant and the carcase of the corresponding part of the amoeba

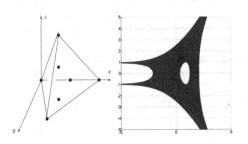

Fig. 7. The Newton polytope $\mathcal{N}_{p(x,y,z)}$ and the section of the amoeba $\mathcal{A}_{p(x,y,z)}$ by the plane $\log |z| = 5$

We denote by (s, t, u) the coordinates in the amoeba space of Fig. 6(right). The bold segments join the sections of the amoeba by the planes $t = 0$, $t = \log 6$, $u = 0$, $u = \log 4$, and the section of the amoeba by the plane $s = \log 3$. The bold curve is the "top of the box," i.e., the intersection of the amoeba with the plane $u = \log 4$.

Acknowledgements. This research was supported by the state order of the Ministry of Education and Science of the Russian Federation for Siberian Federal University (task 1.1462.2014/K), by grant of the Government of the Russian Federation for investigations under the guidance of the leading scientists of the Siberian Federal University (contract No. 14.Y26.31.0006) and by the Russian Foundation for Basic Research, projects 15-31-20008-mol_a_ved and 16-41-240764-r_a.

References

1. Dickenstein, A., Sadykov, T.M.: Algebraicity of solutions to the Mellin system and its monodromy. Dokl. Math. **75**(1), 80–82 (2007)
2. Dickenstein, A., Sadykov, T.M.: Bases in the solution space of the Mellin system. Sbornik Math. **198**(9–10), 1277–1298 (2007)
3. Forsberg, M.: Amoebas and Laurent Series. Doctoral Thesis presented at Royal Institute of Technology, Stockholm, Sweden. Bromma Tryck AB, ISBN 91-7170-259-8 (1998)
4. Forsberg, M., Passare, M., Tsikh, A.K.: Laurent determinants and arrangements of hyperplane amoebas. Adv. Math. **151**, 45–70 (2000)
5. Gelfand, I.M., Kapranov, M.M., Zelevinsky, A.V.: Discriminants, Resultants, and Multidimensional Determinants. Mathematics, Theory & Applications. Birkhäuser Boston Inc., Boston (1994)
6. Johansson, P.: On the topology of the coamoeba. Doctoral Thesis presented at Stockholm University, Sweden. US AB, ISBN 978-91-7447-933-1 (2014)
7. Kapranov, M.M.: A characterization of A-discriminantal hypersurfaces in terms of the logarithmic Gauss map. Math. Ann. **290**, 277–285 (1991)
8. Kim, M.-H.: Computational complexity of the Euler type algorithms for the roots of complex polynomials. Thesis, City University of New York, New York (1985)
9. Nilsson, L.: Amoebas, discriminants and hypergeometric functions. Doctoral Thesis presented at Stockholm University, Sweden. US AB, ISBN 978-91-7155-889-3 (2009)
10. Passare, M., Sadykov, T.M., Tsikh, A.K.: Nonconfluent hypergeometric functions in several variables and their singularities. Compos. Math. **141**(3), 787–810 (2005)
11. Purbhoo, K.: A Nullstellensatz for amoebas. Duke Math. J. **141**(3), 407–445 (2008)
12. Rullgård, H.: Stratification des espaces de polynômes de Laurent et la structure de leurs amibes (French). Comptes Rendus de l'Academie des Sciences - Series I: Mathematics **331**(5), 355–358 (2000)
13. Theobald, T.: Computing amoebas. Experiment. Math. **11**(4), 513–526 (2002)
14. Theobald, T., de Wolff, T.: Amoebas of genus at most one. Adv. Math. **239**, 190–213 (2013)
15. Sadykov, T.M., Tsikh, A.K.: Hypergeometric and Algebraic Functions in Several Variables (Russian). Nauka (2014)
16. Sadykov, T.M.: On a multidimensional system of hypergeometric differential equations. Siberian Math. J. **39**(5), 986–997 (1998)
17. Viro, T.: What is an amoeba? Not. AMS **49**(8), 916–917 (2002)

Sparse Gaussian Elimination Modulo p: An Update

Charles Bouillaguet[1]([envelope]) and Claire Delaplace[1,2]

[1] Univ. Lille, CNRS, Centrale Lille, UMR 9189 - CRIStAL -
Centre de Recherche en Informatique Signal et Automatique de Lille,
59000 Lille, France
charles.bouillaguet@univ-lille1.fr
[2] Université de Rennes-1/IRISA, Rennes, France
claire.delaplace@irisa.fr

Abstract. This paper considers elimination algorithms for sparse matrices over finite fields. We mostly focus on computing the rank, because it raises the same challenges as solving linear systems, while being slightly simpler.

We developed a new sparse elimination algorithm inspired by the Gilbert-Peierls sparse LU factorization, which is well-known in the numerical computation community. We benchmarked it against the usual right-looking sparse gaussian elimination and the Wiedemann algorithm using the Sparse Integer Matrix Collection of Jean-Guillaume Dumas.

We obtain large speedups (1000× and more) on many cases. In particular, we are able to compute the rank of several large sparse matrices in seconds or minutes, compared to days with previous methods.

1 Introduction

There are essentially two families of algorithms to perform the usual operations on sparse matrices (rank, determinant, solution of linear equations, etc.): *direct* and *iterative* methods.

Direct methods (such as gaussian elimination, LU factorization, etc.) generally produce an echelonized version of the original matrix. This process often incurs *fill-in*: the echelonized version has more non-zero entries than the original. Fill-in increases the time and space needed to complete the echelonization. As such, the time and space requirements of direct methods are usually unpredictable; they may fail if not enough storage is available, or become excruciatingly slow.

Iterative methods such as the Wiedemann algorithm [29] only perform matrix-vector products and only need to store one or two vectors in addition to the matrix. They do not incur any fill-in. On rank-r matrices, the number of matrix-vector products that must be performed is $2r$. Also, when a matrix M has $|M|$ non-zero entries, computing the matrix-vector product $\mathbf{x} \cdot M$ requires $\mathcal{O}(|M|)$ operations. The time complexity of iterative methods is thus $\mathcal{O}(r|M|)$, which is fairly easy to predict, and the space complexity is essentially that of

V.P. Gerdt et al. (Eds.): CASC 2016, LNCS 9890, pp. 101–116, 2016.
DOI: 10.1007/978-3-319-45641-6_8

keeping the matrix in memory. These methods are often the only option for very large matrices, or for matrices where fill-in makes direct methods impractical.

In a paper from 2002, Dumas and Villard [15] surveyed and benchmarked algorithms dedicated to rank computations for sparse matrices modulo a small prime number p. In particular, they compared the efficiency of sparse gaussian elimination and the Wiedemann algorithm on a collection of benchmark matrices that they collected from other researchers and made available on the web [16]. They observed that while iterative methods are fail-safe and can be practical, direct methods can sometimes be much faster. This is in particular the case when matrices are almost triangular, so that gaussian elimination barely has anything to do.

It follows that both methods are worth trying. In practical situations, a possible workflow could be: *"try a direct method; if there is too much fill-in, abort and restart with an iterative method"*.

We concur with the authors of [15], and strengthen their conclusion by developing a new sparse elimination algorithm which outperforms all other techniques in some cases, including several large matrices which could only be processed by an iterative algorithm.

Lastly, it is well-known that both methods can be combined: performing one step of elimination reduces the size of the "remaining" matrix (the Schur complement) by one, while increasing the number of non-zeros. This may decrease the time complexity of running an iterative method on the Schur complement. This strategy can be implemented as follows: *"While the product of the number of remaining rows and remaining non-zeros decreases, perform an elimination step; then switch to an iterative method"*. For instance, one phase of the record-setting factorization of a 768-bit number [20] was to find a few vectors on the kernel of a 2 458 248 361 × 1 697 618 199 very sparse matrix over $\mathbb{F}_2$ (with about 30 non-zero per row). A first pass of elimination steps (and discarding "bad" rows) reduced this to a roughly square matrix of size 192 796 550 with about 144 non-zero per row. A parallel implementation of the block-Wiedemann [7] algorithm then finished the job. The algorithm presented in this paper lends itself well to this hybridization.

1.1   Our Contribution

Our original intention was to check whether the conclusions of [15] could be refined by using more sophisticated sparse elimination techniques used in the numerical world. To do so, we developed the **SpaSM** software library (SPArse Solver Modulo p). Its code is publicly available in a repository hosted at: https://github.com/cbouilla/spasm

Our code is heavily inspired by **CSPARSE** (*"A Concise Sparse Matrix Package in C"*), written by Davis and abundantly described in his book [9]. We modified it to work row-wise (as opposed to column-wise), and more importantly to deal with non-square or singular matrices. At its heart lies a sparse LU factorization algorithm. It is capable of computing the rank of a matrix, but also of solving linear systems (and, with minor adaptations, of computing the determinant, finding a basis of the kernel, etc.).

We used this as a playground to implement the algorithms described in this paper —as well as some other, less successful ones. This was necessary to test their efficiency in practice. We benchmarked them using matrices from Dumas's collection [16], and compared them with the algorithms used in [15], which are publicly available inside the LinBox library. This includes a right-looking sparse gaussian elimination, and the Wiedemann algorithm.

There are several cases where the algorithm described in Sect. 3 achieve a $1000\times$ speedup compared to previous algorithms. In particular, it systematically outperforms the right-looking sparse gaussian elimination implemented in LinBox.

It is capable of computing the rank of several of the largest matrices from [16], where previous elimination algorithms failed. In these cases, it vastly outperforms the Wiedmann algorithm. In a striking example, two computations that required two days with the Wiedemann algorithm could be performed in 30 min and 30 s respectively using the new algorithm. More complete benchmark results are given in Sect. 4.

We relied on three main ideas to obtain these results. First, we built upon GPLU [19], a left-looking sparse LU factorization algorithm. We then used a simple pivot-selection heuristic designed in the context of Gröbner basis computation [18], which works well with left-looking algorithms. On top of these two ideas, we designed a new, hybrid, left-and-right looking algorithm.

Our intention is not to develop a competitor sparse linear algebra library; we plan to contribute to LinBox, but we wanted to check the viability of our ideas first.

1.2   Related Work

Sparse Rank Computation mod p. Our starting point was [15], where a right-looking sparse gaussian elimination and the Wiedemann algorithm are compared on many benchmark matrices. [23,24] consider the problem of large dense matrices of small rank modulo a very small prime, while [25] shows that most operations on extremely sparse matrices with only two non-zero entries per row can be performed in time $\mathcal{O}(n)$. [17] discusses sparse rank computation of the largest matrices of our benchmark collection by various methods. The largest computation were performed with a parallel block-variant of the Wiedemann algorithm.

Direct Methods in the Numerical World. A large body of work has been dedicated to sparse direct methods by the numerical computation community. Direct sparse numerical solvers have been developed during the 1970's, so that several software packages were ready-to-use in the early 1980's (MA28, SPARSPAK, YSMP, ...). For instance, MA28 is an early "right-looking" (cf. Sect. 2) sparse LU factorization code described in [12,13]. It uses Markowitz pivoting [22] to choose pivots in a way that maintains sparsity.

Most of these direct solvers start with a symbolic analysis phase that ignores the numerical values and just examines the pattern of the matrix. Its purpose is to predict the amount of fill-in that is likely to occur, and to pre-allocate data structures to hold the result of the numerical computation. The complexity of this step often dominated the running time of early sparse codes. In addition, an important step was to choose *a priori* an order in which the rows (or columns) were to be processed in order to maintain sparsity.

Sparse linear algebra often suffers from poor computational efficiency, because of irregular memory accesses and cache misses. More sophisticated direct solvers try to counter this by using *dense* linear algebra kernels (the BLAS and LAPACK) which have a much higher FLOP per second rate.

The supernodal method [6] does this by clustering together rows (or columns) with similar sparsity pattern, yielding the so-called *supernodes*, and processing them all at once using dense techniques. Modern supernodal codes include CHOLDMOD [5] (for Cholesky factorization) and SuperLU [11]. The former is used in Matlab on symmetric positive definite matrices.

In the same vein, the multifrontal method [14] turns the computation of a sparse LU factorization into several, hopefully small, *dense* LU factorizations. The starting point of this method is the observation that the elimination of a pivot creates a completely dense submatrix in the Schur complement. Contemporary implementations of this method are UMFPACK [8] and MUMPS [1]. The former is also used in Matlab in non-symmetric matrices.

Finally, "left-looking" algorithms are those that do not explicitly compute the Schur complement during the sparse factorization. This is for instance the case of SPARSPAK cited earlier. A very interesting algorithm, referred to as GPLU [19], computes an LU factorization in time proportionnal to the number of arithmetic operations needed to compute the product $L \times U$ (assuming the zeros are not stored). In particular, the symbolic part of the factorization does not dominate the numerical part asymptotically. This algorithm is implemented in Matlab, and is used for very sparse unsymmetric matrices. It is also the heart of the specialized library called KLU [10], dedicated to circuit simulation.

Direct Methods Modulo p. The world of exact sparse direct methods is much less populated. Besides LinBox, we are not aware of many implementations of sparse gaussian elimination capable of computing the rank of a matrix modulo p. According to its handbook, the MAGMA [2] computer algebra system computes the rank of a sparse matrix by first performing sparse gaussian elimination with Markowitz pivoting, then switching to a dense factorization when the matrix becomes dense enough. The Sage [28] system uses LinBox.

Some specific applications rely on exact sparse linear algebra. All competitive factoring and discrete logarithms algorithms work by finding a few vectors in the kernel of a large sparse matrix. Some controlled elimination steps are usually performed, which makes the matrix smaller and denser. This has been called "structured gaussian elimination" [21]. The process can be continued until the matrix is fully dense (after which it is handled by a dense solver), or stopped earlier, when the resulting matrix is handled by an iterative algorithm.

The current state-of-the-art factoring codes, such as CADO-NFS [26], seem to use the sparse elimination techniques described in [4].

Modern Gröbner basis algorithms work by computing the reduced row-echelon form of particular sparse matrices. An ad hoc algorithm has been designed, exploiting the mostly triangular structure of these matrices to find pivots without performing any computation [18]. An open-source implementation, GBLA [3], is available.

2   Sparse LU Factorization

In this section we discuss algorithms to compute a sparse LU factorization of a matrix A of size $n \times m$ over a finite field. The techniques described in this paper could in principle work over any field; for the sake of simplicity we focus on the case of integers modulo a prime p that fits into a machine integer.

2.1   Definitions and Notations

We recall here some usefull definitions. A is an n-by-m matrix and its (unknown) rank is denoted by r.

Because A is rectangular, we consider the $PLUQ$ factorization of A where L is n-by-r and lower-trapezoidal with a unit diagonal, U is r-by-m and upper-trapezoidal with a non-zero diagonal and P (resp. Q) is a permutation over the rows (resp. columns) of A. Computing a PLUQ factorization essentially amounts to performing gaussian elimination. We recall the usual (dense) algorithm: at each step i, choose a coefficient $a_{ik} \neq 0$ called the *pivot* and "eliminate" every coefficients under it by adding suitable multiples of row i to the rows below. This method is said to be *right-looking*, because at each step it accesses the data stored in the bottom-right of A (shaded aera in Fig. 1a). This contrasts with *left-looking* algorithms (or the *up-looking* row-by-row equivalent described in Sect. 2.3), that accesses the data stored in the left of L (respectively in the top of U) represented by the shaded aera in Fig. 1b (c).

If we assume that we have performed some steps of the PLUQ decomposition, then we have

$$PAQ = \begin{pmatrix} A_{00} & A_{01} \\ A_{10} & A_{11} \end{pmatrix},$$

such that A_{00} is square, nonsingular and can be factored as $A_{00} = L_{00} \cdot U_{00}$. It leads to the following factorization of A:

$$PAQ = \begin{pmatrix} L_{00} & \\ L_{10} & I \end{pmatrix} \cdot \begin{pmatrix} U_{00} & U_{10} \\ & S_{11} \end{pmatrix},$$

where $L_{10} = A_{10}U_{00}^{-1}$, $U_{01} = L_{00}^{-1}A_{01}$ and $S_{11} = A_{11} - A_{10}A_{00}^{-1}A_{01}$ is the *Schur Complement* of A_{11} in A. It is represented by S in Fig. 1a.

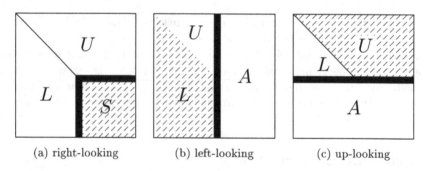

(a) right-looking   (b) left-looking   (c) up-looking

Fig. 1. Data access pattern of several PLUQ factorization algorithms. The dark area is the echelonization front. While the algorithm processes, they only access data in the shaded area.

2.2   The Classical Right-Looking Algorithm

LinBox implements a sparse gaussian elimination that only computes U (and not L) which is a straightforward adaptation of the classical (dense) algorithm to sparse data structures: find a pivot, permute the rows and columns to move the pivot to the diagonal, eliminate all the entries below the pivot, repeat. In LinBox's gaussian elimination code, a sparse matrix is an array of sparse vectors; a sparse vector is a dynamic array of (column index, coefficient) pairs, sorted by column indices. This dynamic array is essentially a `std::vector` object from the C++ STL.

This is very similar to the early MA28, except that in MA28, sparse vectors are unordered. Also, MA28 implements full Markowitz pivoting (greedily choosing the pivot that minimises fill-in in the Schur complement), while LinBox implements a faster relaxation thereof. Note that choosing the next pivot takes time $\Omega(n)$, so that the complexity of the whole process is lower-bounded by $\Omega(n^2)$. This is suboptimal in some cases (e.g. tridiagonal matrices).

2.3   The Left-Looking GPLU Algorithm

The GPLU Algorithm was introduced by Gilbert and Peierls in 1988 [19]. It is also abundantly described in [9], up to implementation details. During the k-th step, the k-th column of L and U are computed from the k-th column of A and the previous columns of L. The algorithm thus only accesses data on the left of the echelonization front, and the algorithm is said to be "left-looking". It does not compute the Schur complement at each stage. We implemented a row-by-row (as opposed to column-by-column) variant of this method which is then "up-looking".

The main idea behind the algorithm is that the next row of L and U can both be computed by solving a triangular system. If we ignore row and column permutations, this can be derived from the following 3-by-3 block matrix expression:

$$\begin{pmatrix} L_{00} & & \\ l_{10} & 1 & \\ L_{20} & l_{21} & L_{22} \end{pmatrix} \begin{pmatrix} U_{00} & \mathbf{u}_{01} & U_{02} \\ & u_{11} & \mathbf{u}_{12} \\ & & U_{22} \end{pmatrix} = \begin{pmatrix} A_{00} & \mathbf{a}_{01} & A_{02} \\ a_{10} & a_{11} & a_{12} \\ A_{20} & a_{21} & A_{22} \end{pmatrix}.$$

Here $(a_{10}\ a_{11}\ a_{12})$ is the k-th row of A. L is assumed to have unit diagonal. Assume we have already computed $(U_{00}\ \mathbf{u}_{01}\ U_{02})$, the first k rows of U. Then we have:

$$\begin{pmatrix} l_{10} & u_{11} & \mathbf{u}_{12} \end{pmatrix} \cdot \begin{pmatrix} U_{00} & \mathbf{u}_{01} & U_{02} \\ & 1 & \\ & & Id \end{pmatrix} = \begin{pmatrix} \mathbf{a}_{01} & a_{11} & \mathbf{a}_{21} \end{pmatrix}. \tag{1}$$

Thus, the whole PLUQ factorization can be performed by solving a sequence of n sparse triangular systems.

Sparse Triangular Solving. To make the above idea work, we need to solve efficiently $\mathbf{x} \cdot U = \mathbf{b}$, where U is a sparse matrix stored by rows, and $\mathbf{b}$ is a sparse vector (a row of A). The main trick of the GPLU algorithm is that it is possible to determine the sparsity pattern of $\mathbf{x}$ without performing any kind of numerical computation.

Theorem 1 (Gilbert and Peierls [19]). *Define the directed graph $G_U = (V, E)$ with nodes $V = \{1 \dots m\}$ and edges $E = \{(i,j)|u_{ij} \neq 0\}$. Let $\mathrm{Reach}_U(i)$ denote the set of nodes reachable from i via paths in G_U and for a set $\mathcal{B}$, let $\mathrm{Reach}_U(\mathcal{B})$ be the set of all nodes reachable from any nodes in $\mathcal{B}$. The non-zero pattern $\mathcal{X} = \{j|x_j \neq 0\}$ of the solution $\mathbf{x}$ to the sparse linear system $\mathbf{x} \cdot U = \mathbf{b}$ is given by $\mathcal{X} = \mathrm{Reach}_U(\mathcal{B})$, where $\mathcal{B} = \{i|b_i \neq 0\}$, assuming there is no numerical cancellation.*

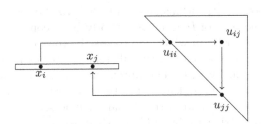

Fig. 2. The presence of x_i and of u_{ij} entails that of x_j in $\mathbf{x}$.

This is due to the fact that x_j can only be non-zero if either (1) b_j is non-zero, or (2) there is a node i such that $x_i \neq 0$ and $u_{ij} \neq 0$ as shown in Fig. 2. Thus, $\mathcal{X}$ can be determined by performing a graph traversal in G_U starting from each node of $\mathcal{B}$. If we perform a *depth-first search*, then we will naturally obtain $\mathcal{X}$ sorted in *topological order*: if $u_{ij} \neq 0$, then i comes up before j in $\mathcal{X}$. This enables the following efficient procedure to compute $\mathbf{x}$:

1: Scatter **b** into a (dense) working array **w**, initially filled with 0.
2: **for all** $i \in \mathcal{X}$ in topological order **do**
3:     $w_i \leftarrow w_i / u_{ii}$
4:     **for all** $j > i$ such that $u_{ij} \neq 0$ **do**
5:         $w_j \leftarrow w_j - w_i u_{ij}$
6: Gather **x** (get w_i for $i \in \mathcal{X}$), and reset **w**.

A striking feature of this triangular solver is that it may terminate in *constant time* in some cases (for instance if U is bidiagonal, and **b** has only a constant number of entries). In addition, the complexity of finding $\mathcal{X}$ is of the same order as that of the numerical computation of **x** once $\mathcal{X}$ is known: each iteration of line 5 in the above pseudo-code corresponds to the crossing of an edge in G_U during the construction of $\mathcal{X}$. This holds well in our implementation: computing $\mathcal{X}$ consistenty requires 25 % of the time needed to compute **x**.

Selecting the Pivots. For the triangular systems to have a solution, we need to make sure that the diagonal coefficient of U (the pivots) are non-zero. However, while solving the system (1), we may find $u_{11} = 0$. There are two possible cases: (1) If $u_{11} = 0$ and $\mathbf{u}_{12} \neq \mathbf{0}$, then we can permute the column j_0 where u_{11} is with another one j_1, such that $j_0 < j_1$. (2) If $u_{11} = 0$ and $\mathbf{u}_{12} = \mathbf{0}$, then there is no way we can permute the columns of U to bring a non-zero coefficent on the diagonal. This means that the current row of A is a linear combinations of the previous rows. When the factorization is done, the number of non-empty rows of U is the rank r of the matrix A.

Useful Heuristics. Right-looking algorithms have the possibility to choose the next pivot by exploiting knowledge of the Schur complement (this is typically what Markowitz pivoting is about). On the other hand, the only information available to left-looking algorithms such as GPLU to choose pivots is the original matrix, and the part of U that has already been computed. It is possible to choose an *a priori* order in which to process the rows of the matrix in order to keep U as sparse as possible. Many different strategies have been designed to do so in the numerical world: the (approximate) minimum degree algorithm, nested dissection, etc. The problem is that these algorithms are mostly adapted to symmetric matrices, and a fortiori to square matrices. They are usually retrofitted to the unsymmetric case by applying them to $A^T A$, which is square and symmetric. However, when A is very rectangular, this becomes completely dense and provide little useful information.

We instead implemented a much simpler heuristic inspired from an echelonization procedure dedicated to Gröbner basis computation by Faugére and Lachartre [18]. We map each row to the column of its leftmost coefficient. When several rows have the same leftmost coefficient, we select the sparsest row. We then move the selected rows before the others and sort them by increasing position of the left-most coefficient. As a result, the leftmost coefficient of each row cannot occur in any selected row below it. It follows that the selected rows are copied as-is into U, with zero fill-in. This is particularly well-suited to the

GPLU algorithm, because this keeps U as sparse as possible; this in turn makes the triangular solver fast.

Also, we observed that it can be beneficial to compute the rank of the *transpose* of the matrix. Indeed, U has size $r \times m$, so when m is much larger than n, transposing the matrix before starting the computation ensures than U is smaller. More often than not this decreases the running time of the factorization (on the matrices from [16]). Lastly, we note that the running time of the right-looking algorithm implemented in LinBox is usually not the same on A and A^T.

We also implemented a simple early abort test which is performed when sufficiently many rows have been processed without finding any new pivot. A random linear combination of the remaining rows is computed; if it belongs to the row-space of U, then no new pivot will be found in the remaining rows with probability greater than $1 - 1/p$. This test can be repeated to increase its success probability exponentially.

2.4 Left or Right?

The two algorithms presented in this section are incomparable. We illustrate this by exhibiting two situations where each one consistently outperforms the other. We generated two random matrices with different properties (situations A and B in Table 1). The entries of matrix A are identically and independently distributed. There are zero with probability 99 %, and chosen uniformly at random modulo p otherwise. GPLU terminates very quickly in this case because it only process the first ≈ 1000 rows before stopping, having detected that the matrix has full column-rank.

In matrix B, one row over 1000 is random (as in matrix A), and the 999 remaining ones are sparse linear combination of a fixed set of 100 other random sparse rows (as in matrix A). The right-looking algorithm discovers these linear combinations early and quickly skip over empty rows in the Schur complement. On the other hand, GPLU has to work its way thoughout the whole matrix. Section 4 contains further "real" examples where an algorithm clearly outperforms the other.

Table 1. Running time (in seconds) of the two algorithms on extreme cases.

Matrix	Rows	Columns	Non-zero	Rank	Right-Looking (LinBox)	GPLU (SpaSM)
A	100 000	1 000	995 076	1000	3.0	0.02
B	100 000	1 000	4 776 477	200	1.7	9.5

3 A New Hybrid Algorithm

The right-looking algorithm enjoys a clear advantage in terms of pivot selection, because it has the possibility to explore the Schur complement. The short-term objective of pivot selection is to keep the Schur complement sparse.

In GPLU pivot selection has to be done "in the dark", with the short-term objective to keep U sparse (since this keeps the triangular solver fast). However, GPLU performs the elimination steps very efficiently.

In this section, we present a new algorithm that combines these two strong sides. It usually outperforms both the classical right-looking elimination and GPLU. The new algorithm works as follows:

1. Use the Faugére-Lachartre heuristic to find as many pivots as possible in A.
2. Compute the Schur complement S with respect to these pivots, using GPLU.
3. Compute the rank of S by any mean (including recursively).

Pivot Selection. As in the right-looking algorithm, we begin by exploring the matrix looking for pivots. However, instead of choosing only one, we try to pick as many as possible. For this, we use the Faugére-Lachartre heuristic described in Sect. 2.3: it finds a sequence of k rows such that the leftmost coefficient of each rows does not appear in any subsequent row. It can thus be chosen as a pivot. Those specific rows are copied as-is into U without any arithmetic operation. The cost of this step is that of iterating over the entries of A.

Schur Complement. We then compute the Schur complement S with respect to the chosen pivots. This amounts to eliminating every entries below the chosen pivots. The left-looking side of the combination is that we use the GPLU algorithm to compute the Schur complement.

If we denote by P the permutation of the rows of A that "pushes" this well-chosen set of linearly independant rows at the top of A and if we ignore the permutation over the columns of A, the PLU factorization of A can be represented by the following 2-by-2 block matrix expression:

$$PA = \begin{pmatrix} U_{00} & U_{01} \\ A_{10} & A_{11} \end{pmatrix} = \begin{pmatrix} Id & \\ L_{10} & L_{11} \end{pmatrix} \cdot \begin{pmatrix} U_{00} & U_{01} \\ & U_{11} \end{pmatrix},$$

where $(U_{00}\ U_{01})$ is the part of U provided by the Faugére-Lachartre heuristic. What we need is $S = A_{11} - A_{10}U_{00}^{-1}U_{01}$, the Schur Complement of A with respect to U_{00}. We compute S row-by-row. To obtain the ith row $\mathbf{s}_i$ of S, denote by $(\mathbf{a}_{i0}\ \mathbf{a}_{i1})$ the ith row of $(A_{10}\ A_{11})$ and consider the following system:

$$(\mathbf{x}_0\ \mathbf{x}_1) \cdot \begin{pmatrix} U_{00} & U_{01} \\ & Id \end{pmatrix} = (\mathbf{a}_{i0}\ \mathbf{a}_{i1})$$

We get $\mathbf{x}_1 = \mathbf{a}_{i1} - \mathbf{x}_0 U_{01} = \mathbf{a}_{i1} - \mathbf{a}_{i0}U_{00}^{-1}U_{01}$. From the definition of S, it follows that $\mathbf{x}_1 = \mathbf{s}_i$. Thus S can be computed by a sequence of sparse triangular solve. Because we chose U to be as sparse as possible, this process is fast. It can also be parallelized efficiently: all the rows of S can be computed independently.

Computing the Rank of S. Once the Schur complement S has been computed, it remains to find its rank. Several options are possible. If S is sparse, we can either use the same technique recursively, or switch to GPLU, or switch to the Wiedemann algorithm. If S is small and dense, we switch to dense gaussian elimination. If S is big and dense, we abort and report failure.

In the first case, where several options are possible, some guesswork is required to find the best course of action. By default, we found that allowing only a limited number of recursive calls (usually less than 10, often 3) and then switching to GPLU yields good results.

Tall and Narrow Schur Complement. It is beneficial to consider a special case in the above procedure, when S has much more rows than columns (we transpose S if it has much more columns than rows). This happens in particular in the favorable situation where the Faugére-Lachartre heuristic finds almost all possible pivots, and only very few remain to be found.

This situation can be detected as soon as the pivots have been selected, because we known that S has size $(n - k) \times (m - k)$. In this case where S is very tall and very narrow, it is wasteful to compute S entirely. Many well-known techniques can be applied to obtain its rank by looking only at a fraction of its entries (see [24]). For instance, a naïve solution consists in choosing a small constant ϵ, building a dense matrix of size $(m - k + \epsilon) \times (m - k)$ with random (dense) linear combinations of the rows of S, and computing its rank using dense linear algebra. A linear combination of the rows of S can be formed by taking a random linear combinations of the rows of A, and then solving a triangular system, just like before.

4   Implementation and Results

All the experiments we carried on an Intel core i7-3770 with 8 GB of RAM. Only one core was ever used. We used J.-G. Dumas's Sparse Integer Matrix Collection [16] as benchmark matrices. We restricted our attention to the 660 matrices with integer coefficients. Most of these matrices are small and their rank is easy to compute. Some others are pretty large. In all cases, their rank is known, as it could always be computed using the Wiedemann algorithm. We fixed $p = 42013$ in all tests.

Our implementation is quite straightforward. Matrices are stored in Compressed Sparse Row format. Coefficients are stored in `int` variables, and are reduced modulo p after each multiplication.

Sparse Elimination: Right-Looking vs GPLU. We first compare the efficiencies of the right-looking gaussian elimination algorithm implemented in Lin-Box and our implementation of the GPLU algorithm. LinBox uses its Markowitz-like pivot selection, while SpaSM uses its default setting: transposing the matrix if it has more columns than rows, and using the Faugére-Lachartre heuristic to select pivots before actually starting the factorization.

We quickly observed that no algorithm is always consistently faster than the other; one may terminate instantly while the other may run for a long time and vice-versa. In order to perform a systematic comparison, we decided to set an arbitrary threshold of 60 s, and count the number of matrices that could be processed in this much time. LinBox could dispatch 579 matrices, while SpaSM processed 606. Amongst these, 568 matrices could be dealt with by both algorithms in less than 60s. This took 1100 s to LinBox and 463 s to SpaSM. LinBox was faster 112 times, and SpaSM 456 times. These matrices are "easy" for both algorithms, and thus we will not consider them anymore.

Table 2 shows the cases where one algorithm took less than 60 s while the other took more. There are cases where each of the two algorithm is catastrophically slower than the other. In some cases, a bug made the LinBox test program crash with a segmentation fault. We conclude there is no clear winner (even if GPLU is usually a bit faster. The hybrid algorithm described in Sect. 3 outperforms both.

A reviewer asked a comparison with GBLA [3]. This is difficult, because GBLA is tailored for matrices arising in Gröbner basis computations, and exploit their specific shape. For instance, they have (hopefully small) *dense* areas, which GBLA rightly store in dense data structures. This phenomenon does not occur in our benchmark collection, which is ill-suited to GBLA. GBLA is nevertheless undoubtedly more efficient on Gröbner basis matrices.

Direct Methods vs Iterative Methods. We now turn our attention to the remaining matrices of the collection, the "hard" ones. Some of these are yet unamenable to any form of elimination, because they cause too much fill-in. However, some matrices can be processed much faster by our hybrid algorithm than by any other existing method.

For instance, `relat9` is the third largest matrix of the collection; computing its rank takes a little more than two days using the Wiedemann algorithm. Using the "Tall and Narrow Schur Complement" technique described above, the hybrid algorithm computes its rank in 34 min. Most of this time is spent in forming a size-8937 dense matrix by computing random dense linear combination of the 9 million remaining sparse rows. The rank of the dense matrix is quickly computed using the `Rank` function of FFLAS-FFPACK [27]. A straightforward parallelization using OpenMP brings this down to 10 min using the 4 cores of our workstation (the dense rank computation is not parallel). The same goes for the `rel9` matrix, which is similar.

The rank of the `M0,6-D9` matrix, which is the 9-th largest of the collection, could not be computed by the right-looking algorithm within the limit of 8 GB of memory. It takes 42 h to the Wiedemann algorithm to find its rank. The standard version of the hybrid algorithm finds it in less than 30 s.

The `shar_te.b3` matrix is an interesting case. It is a very sparse matrix of size 200200 with only 4 non-zero entries per row. Its rank is 168310.

Table 2. Comparison of sparse elimination techniques. Times are in seconds.

Matrix	Right-looking	GPLU	Hybrid
Franz/47104x30144bis	39	488	1.7
G5/IG5-14	0.5	70	0.4
G5/IG5-15	1.6	288	1.1
G5/IG5-18	29	109	8
GL7d/GL7d13	11	806	0.3
GL7d/GL7d24	34	276	11.6
Margulies/cat_ears_4_4	3	184	0.1
Margulies/flower_7_4	7.5	667	2.5
Margulies/flower_8_4	37	9355	3.7
Mgn/MO,6.data/MO,6-D6	45	8755	0.1
Homology/ch7-8.b4	173	0.2	0.2
Homology/ch7-8.b5	611	45	10.7
Homology/ch7-9.b4	762	0.4	0.4
Homology/ch7-9.b5	3084	8.2	3.4
Homology/ch8-8.b4	1022	0.4	0.5
Homology/ch8-8.b5	5160	6	2.9
Homology/n4c6.b7	223	0.1	0.1
Homology/n4c6.b8	441	0.2	0.2
Homology/n4c6.b9	490	0.3	0.2
Homology/n4c6.b10	252	0.3	0.2
Homology/mk12.b4	72	9.2	1.5
Homology/shar_te2.b2	94	1	0.2
Kocay/Trec14	80	31	4
Margulies/wheel_601	7040	4	0.3
Mgn/MO,6.data/MO,6-D11	722	0.4	0.6
Smooshed/olivermatrix.2	75	0.6	0.1

The right-looking algorithm fails, and the Wiedemann algorithm takes 3650s. Both GPLU and the hybrid algorithm terminate in more than 5 h. However, performing *one* iteration of the hybrid algorithm computes a Schur complement of size 134645 with 7.4 non-zero entries per row on average. We see that the quantity $n|A|$, a rough indicator of the complexity of iterative methods, decreases a little. Indeed, computing the rank of the first Schur complement takes 2422 s using the Wiedemann algorithm. This results in a 1.5× speed-up.

All-in-all, the hybrid algorithm is capable of quickly computing the rank of the 3rd, 6th, 9th, 11th and 13th largest matrices of the collection, whereas previous elimination techniques could not. Previously, the only possible option

Table 3. Some harder matrices. Times are in seconds. M.T. stands for "Memory Thrashing"

Matrix	n	m	$|A|$	Right-looking	Wiedemann	Hybrid
kneser_10_4	349651	330751	992252	923	9449	0.1
mk13.b5	135135	270270	810810	M.T	3304	41
M0,6-D6	49800	291960	1066320	42	979	0.1
M0,6-D7	294480	861930	4325040	2257	20397	0.8
M0,6-D8	862290	1395840	8789040	20274	133681	7.7
M0,6-D9	1395480	1274688	9568080	M.T	154314	27.6
M0,6-D10	1270368	616320	5342400	22138	67336	42.7
M0,6-D11	587520	122880	1203840	722	4864	0.5
relat8	345688	12347	1334038		244	2
rel9	5921786	274667	23667185		127675	1204
relat9	9746232	274667	38955420		176694	2024

was the Wiedemann algorithm. The hybrid algorithm allows for large speedup of 100×, 1000× and 10000× in these cases (Table 3).

Acknowledgement. Claire Delaplace was supported by the french ANR under the BRUTUS project. We thank the anonymous reviewers for their comments.

References

1. Amestoy, P.R., Duff, I.S., Koster, J., L'Excellent, J.-Y.: A fully asynchronous multifrontal solver using distributed dynamic scheduling. SIAM J. Matrix Anal. Appl. **23**(1), 15–41 (2001)
2. Bosma, W., Cannon, J., Playoust, C.: The Magma algebra system. I. The user language. J. Symbolic Comput. 24(3–4), 235–265 (1997). http://dx.doi.org/10.1006/jsco.1996.0125. computational algebra and number theory, London (1993)
3. Boyer, B., Eder, C., Faugère, J., Lachartre, S., Martani, F.: GBLA - gröbner basis linear algebra package. CoRR abs/1602.06097 (2016). http://arxiv.org/abs/1602.06097
4. Cavallar, S.: Strategies in filtering in the number field sieve. In: Bosma, W. (ed.) ANTS 2000. LNCS, vol. 1838, pp. 209–232. Springer, Heidelberg (2000). http://dx.doi.org/10.1007/10722028_11
5. Chen, Y., Davis, T.A., Hager, W.W., Rajamanickam, S.: Algorithm 887: CHOLMOD, supernodal sparse cholesky factorization and update/downdate. ACM Trans. Math. Softw. **35**(3), 22:1–22:14 (2008). http://doi.acm.org/10.1145/1391989.1391995
6. Cleveland Ashcraft, C., Grimes, R.G., Lewis, J.G., Peyton, P.W., Simon, H.D., Bjørstad, P.E.: Progress in sparse matrix methods for large linear systems on vector supercomputers. Int. J. High Perform. Comput. Appl. **1**(4), 10–30 (1987). http://dx.doi.org/10.1177/109434208700100403

7. Coppersmith, D.: Solving homogeneous linear equations over $\mathbb{F}_2$ via block wiedemann algorithm. Math. Comput. **62**(205), 333–350 (1994)

8. Davis, T.A.: Algorithm 832: UMFPACK V4.3—an unsymmetric-pattern multifrontal method. ACM Trans. Math. Softw. **30**(2), 196–199 (2004). http://dx.doi.org/10.1145/992200.992206

9. Davis, T.A.: Direct Methods for Sparse Linear Systems (Fundamentals of Algorithms 2). Society for Industrial and Applied Mathematics, Philadelphia (2006)

10. Davis, T.A., Natarajan, E.P.: Algorithm 907: KLU, A direct sparse solver for circuit simulation problems. ACM Trans. Math. Softw. 37(3) (2010). http://doi.acm.org/10.1145/1824801.1824814

11. Demmel, J.W., Eisenstat, S.C., Gilbert, J.R., Li, X.S., Liu, J.W.H.: A supernodal approach to sparse partial pivoting. SIAM J. Matrix Anal. Appl. **20**(3), 720–755 (1999)

12. Duff, I.S., Erisman, A.M., Reid, J.K.: Direct Methods for Sparse Matrices. Numerical Mathematics and Scientific Computation, Oxford University Press, USA, first paperback edition edn. (1989)

13. Duff, I.S., Reid, J.K.: Some design features of a sparse matrix code. ACM Trans. Math. Softw. **5**(1), 18–35 (1979). http://doi.acm.org/10.1145/355815.355817

14. Duff, I.S., Reid, J.K.: The multifrontal solution of indefinite sparse symmetric linear. ACM Trans. Math. Softw. **9**(3), 302–325 (1983). http://doi.acm.org/10.1145/356044.356047

15. Dumas, J.G., Villard, G.: Computing the rank of sparse matrices over finite fields. In: Ganzha, V.G., Mayr, E.W., Vorozhtsov, E.V. (eds.) CASC 2002, Proceedings of the fifth International Workshop on Computer Algebra in Scientific Computing, Yalta, Ukraine, pp. 47–62. Technische Universität München, Germany, September 2002. http://ljk.imag.fr/membres/Jean-Guillaume.Dumas/Publications/sparseeliminationCASC2002.pdf

16. Dumas, J.-G.: Sparse integer matrices collection. http://hpac.imag.fr

17. Dumas, J.-G., Elbaz-Vincent, P., Giorgi, P., Urbanska, A.: Parallel computation of the rank of large sparse matrices from algebraic k-theory. In: Maza, M.M., Watt, S.M. (eds.) Parallel Symbolic Computation, PASCO 2007, International Workshop, 27–28 July 2007, University of Western Ontario, London, Ontario, Canada, pp. 43–52. ACM, New York (2007). http://doi.acm.org/10.1145/1278177.1278186

18. Faugére, J.-C., Lachartre, S.: Parallel gaussian elimination forgröbner bases computations in finite fields. In: Maza, M.M., Roch, J.-L. (eds.) PASCO, pp. 89–97. ACM, New York (2010)

19. Gilbert, J.R., Peierls, T.: Sparse partial pivoting in time proportional to arithmetic operations. SIAM J. Sci. Stat. Comput. **9**(5), 862–874 (1988). http://dx.doi.org/10.1137/0909058

20. Kleinjung, T., Aoki, K., Franke, J., Lenstra, A.K., Thomé, E., Bos, J.W., Gaudry, P., Kruppa, A., Montgomery, P.L., Osvik, D.A., te Riele, H., Timofeev, A., Zimmermann, P.: Factorization of a 768-Bit RSA modulus. In: Rabin, T. (ed.) CRYPTO 2010. LNCS, vol. 6223, pp. 333–350. Springer, Heidelberg (2010). http://dx.doi.org/10.1007/978-3-642-14623-7_18

21. LaMacchia, B.A., Odlyzko, A.M.: Solving large sparse linear systems over finite fields. In: Menezes, A., Vanstone, S.A. (eds.) CRYPTO 1990. LNCS, vol. 537, pp. 109–133. Springer, Heidelberg (1991). http://dx.doi.org/10.1007/3-540-38424-3_8

22. Markowitz, H.M.: The elimination form of the inverse and its application to linear programming. Manage. Sci. **3**(3), 255–269 (1957). http://dx.doi.org/10.1287/mnsc.3.3.255

23. May, J.P., Saunders, B.D., Wan, Z.: Efficient matrix rank computation with application to the study of strongly regular graphs. In: Wang, D. (ed.) Symbolic and Algebraic Computation, International Symposium, ISSAC 2007, Waterloo, Ontario, Canada, July 28–August 1, 2007, Proceedings, pp. 277–284. ACM (2007). http://doi.acm.org/10.1145/1277548.1277586

24. Saunders, B.D., Youse, B.S.: Large matrix, small rank. In: Proceedings of the 2009 International Symposium on Symbolic and Algebraic Computation, ISSAC 2009, pp. 317–324. ACM, New York (2009). http://doi.acm.org/10.1145/1576702.1576746

25. Saunders, D.: Matrices with two nonzero entries per row. In: Proceedings of the 2015 ACM on International Symposium on Symbolic and Algebraic Computation, ISSAC 2015, pp. 323–330. ACM, New York (2015). http://doi.acm.org/10.1145/2755996.2756679

26. The CADO-NFS Development Team: CADO-NFS, an implementation of the number field sieve algorithm (2015), release2.2.0. http://cado-nfs.gforge.inria.fr/

27. The FFLAS-FFPACK group: FFLAS-FFPACK: Finite Field Linear Algebra Subroutines/Package, v2.0.0 edn. (2014). http://linalg.org/projects/fflas-ffpack

28. The Sage Developers: Sage Mathematics Software (Version 5.7) (2013). http://www.sagemath.org

29. Wiedemann, D.H.: Solving sparse linear equations over finite fields. IEEE Trans. Inf. Theory **32**(1), 54–62 (1986). http://dx.doi.org/10.1109/TIT.1986.1057137

MATHCHECK2: A SAT+CAS Verifier for Combinatorial Conjectures

Curtis Bright[1], Vijay Ganesh[1], Albert Heinle[1(✉)], Ilias Kotsireas[2],
Saeed Nejati[1], and Krzysztof Czarnecki[1]

[1] University of Waterloo, Waterloo, Canada
{cbright,vganesh,a3heinle}@uwaterloo.ca
[2] Wilfred Laurier University, Waterloo, Canada
ikotsire@uwaterloo.ca

Abstract. In this paper we present MATHCHECK2, a tool which combines sophisticated search procedures of current SAT solvers with domain specific knowledge provided by algorithms implemented in computer algebra systems (CAS). MATHCHECK2 is aimed to finitely verify or to find counterexamples to mathematical conjectures, building on our previous work on the MATHCHECK system. Using MATHCHECK2 we validated the Hadamard conjecture from design theory for matrices up to rank 136 and a few additional ranks up to 156. Also, we provide an independent verification of the claim that Williamson matrices of order 35 do not exist, and demonstrate for the first time that 35 is the smallest number with this property. Finally, we provided more than 160 matrices to the MAGMA Hadamard database that are not equivalent to any matrices previously included in that database.

1 Introduction

"**Brute**-brute force has no hope. But clever, inspired brute force is the future." – Doron Zeilberger[1]

A recent important movement is the incorporation of modern solvers for satisfiability problems into suitable computations coming from the field of computer algebra. Projects like SC^2 [34] demonstrate that the interest is coming from both academia, as well as from industry.

The great strength of SAT solvers are their sophisticated search procedures. In recent years, conflict-driven clause-learning (CDCL) Boolean SAT solvers [3,20,21] have become very efficient general-purpose search procedures for a large variety of applications. Despite this remarkable progress these algorithms have worst-case exponential time complexity, and may not perform well by themselves for many search applications. However, as we will show in this paper, the run-time can be significantly reduced when adding domain knowledge to the

---

[1] From Doron Zeilberger's talk at the Fields institute in Toronto, December 2015 (http://www.fields.utoronto.ca/video-archive/static/2015/12/379-5401/mergedvid eo.ogv, minute 44).

© Springer International Publishing AG 2016
V.P. Gerdt et al. (Eds.): CASC 2016, LNCS 9890, pp. 117–133, 2016.
DOI: 10.1007/978-3-319-45641-6_9

search procedure. This is where modern computer algebra systems (CAS) come into play: MAPLE [6], MATHEMATICA [38], SAGE [35] and MAGMA [4] are often rich storehouses of algorithms to compute domain-specific knowledge. Hence, the complementary strengths of modern SAT solvers and CAS have a great potential when utilized in synergy. The domain-specific knowledge of a CAS can be crucially important in cutting down a search space , while at the same time the clever heuristics of SAT solvers, in conjunction with CAS, can efficiently search a wide variety of spaces.

The success of the SAT and CAS combination has been demonstrated in a previous paper on MATHCHECK [39]. There, Ganesh et al. explored one way of combining these two classes of systems wherein the CAS was used as a theory solver, à la DPLL(T), to add theory lemmas to the SAT solvers that was the primary driver of the search. They primarily used MATHCHECK to finitely verify (i.e., verify up to some finite bound) conjectures from graph theory.

Our main focus in this paper are conjectures in combinatorial mathematics, which are often simple to state but very hard to verify. For example, a conjecture like the Hadamard [7] might assert the existence of certain combinatorial objects in an infinite number of cases, which makes exhaustive search impossible. In such cases, mathematicians often resort to finite verification in the hopes of learning some meta property of the class of combinatorial structures they are investigating, or discover a counterexample to such conjectures. But even finite verification of combinatorial conjectures up to some finite bound is very difficult, because the search space for such conjectures is often exponential in the size of the structures they refer to. This makes straightforward brute-force search impractical, and also ansatz-driven methods (e.g., calculating Gröbner bases [8]) do not scale well enough in general.

We present a different way of combining SAT and CAS and use it to finitely verify the Hadamard conjecture. MATHCHECK2 can be viewed as a parallel systematic generator of combinatorial structures referred to by the conjecture-under-verification C. It uses the domain knowledge of a CAS to aid the parallelization and prune away structures that do not satisfy C, while the SAT solver is used to verify whether any of the remaining structures satisfy C. In addition, we use UNSAT cores [3] from the SAT solver to further prune the search in a CDCL-style learning feedback loop.

Hadamard Conjecture: We apply our system to the *Hadamard conjecture* which states that for any natural number n, there exists a $4n \times 4n$ matrix H with ± 1 entries for which HH^{T} is a diagonal matrix with each diagonal entry equal to $4n$. In particular, we specialize in Hadamard matrices generated by the so-called Williamson method. We verify that such Hadamard matrices do not exist in order $4 \cdot 35$, a result which was previously computed using a different methodology by D. Đoković [25]. However, due to the nature of the problem and the techniques used, no short certificate of the computations could be produced, making it difficult to check the work short of re-implementing the approach from scratch. In fact, the author specifically states that

In the case $n = 35$ our computer search did not produce any solutions [...] Although we are confident about the correctness of this claim, an independent verification of it is highly desirable since this is the first odd integer, found so far, with this property.

Because our system was written completely independently, uses different techniques internally, and makes use of well-tested SAT solvers and CAS functions, the results of our paper provide an independent verification solicited by Đoković. (The above notes equally apply to the verification in [14].)

Previously used techniques could search effectively for Williamson matrices for odd choices of n, leveraging additional symmetry in these cases. With the help of SAT solvers, we are additionally able to show $n = 35$ is the smallest natural number for which no Williamson matrices of order n exist, not merely the smallest odd number.

2   Architecture of MATHCHECK2

The architecture of our proposed MATHCHECK2 system is outlined in Fig. 1. At its heart is a generator of combinatorial structures, which uses data provided to it by CAS functions to prune the search space and interfaces with SAT solvers to verify the conjecture-in-question. The generator contains functions useful for translating combinatorial conditions into clauses which can be read by a SAT solver. It is possible to substitute the SAT solver with an SMT solver to simplify the encoding process, but this resulted in a too large overhead in our computations. The generator is currently optimized to deal with conjectures which concern Hadamard matrices from coding and combinatorial design theory.

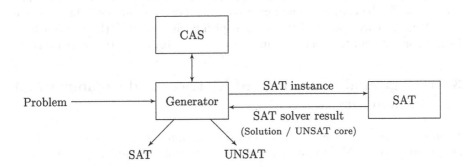

Fig. 1. Outline of the architecture of MATHCHECK2.

Once the class of combinatorial objects has been determined, the script accepts a parameter n which determines the size of the object to search for. For example, when searching for Hadamard matrices, the parameter n denotes the order (i.e., the number of rows) of the matrix. The generator then queries the CAS (we chose MAPLE in our calculations) it is interfaced with for properties that any order n instance of the combinatorial object in question must satisfy. However, since our generator is written in Python, many CAS functionality is provided by certain modules such as **numpy**; in order to avoid overhead in calling a CAS specifically, we tried to use these module functions whenever possible. The result returned by the CAS is read by the generator and then used to prune the space which will be searched by the SAT solver.

Once the generator determines the space to be searched it splits the space into distinct subspaces in a divide-and-conquer fashion. Once the partitioning of the search space has been completed, the script generates two types of files:

1. A single "master" file in DIMACS[2] format which contains the conditions specifying the combinatorial object being searched for. These are encoded as propositional formulae in conjunctive normal form. An assignment to the variables which makes all of them true would give a valid instance of the object being searched for (and a proof that no such assignment exists proves that no instance of the object in question exists).
2. A set of files which contain partial assignments of the variables in the master file. Each file corresponds to exactly one subspace of the search space produced by the generator.

The main advantage of splitting up the problem in such a way is that it easily facilitates parallelization. Using domain specific knowledge, we partition the search space into different classes of the same mathematical structure, and since these classes are independent of each other, a cluster of SAT solvers can search the space for each partition in parallel.

Furthermore, in cases that an instance is found to be unsatisfiable, some SAT solvers such as MAPLESAT [18,19], that support the generation of a so-called *UNSAT core*, can be used to further prune away other similar structures that do not satisfy the conjecture-under-verification. Given an unsatisfiable instance ϕ, its UNSAT core is a set of clauses that pithily characterizes the reason why ϕ is unsatisfiable and thus encodes an unsatisfying subspace of the search space.

3   Background on Hadamard Matrices and Combinatorial Mathematics

In this section we discuss the mathematical preliminaries necessary to understand our work on MATHCHECK2 and its application to Hadamard matrices.

3.1   Hadamard Matrices

First, we define the combinatorial objects known as Hadamard matrices and present some of their properties.

---

[2] For more information on this format, please refer to http://www.satcompetition. org/2009/format-benchmarks2009.html.

Definition 1. *A matrix $H \in \{\pm 1\}^{n \times n}$, $n \in \mathbb{N}$, is called a **Hadamard matrix**, if for all $i \neq j \in \{1, \ldots, n\}$, the dot product between row i and row j in H is equal to zero. We call n the **order** of the Hadamard matrix.*

First studied by Hadamard [10], he showed that if n is the order of a Hadamard matrix, then either $n = 1$, $n = 2$ or n is a multiple of 4. In other words, he gave a *necessary* condition on n for there to exist a Hadamard matrix of order n. The Hadamard conjecture is that this condition is also *sufficient*, so that there exists a Hadamard matrix of order n for all $n \in \mathbb{N}$ where n is a multiple of 4.

Hadamard matrices play an important role in many widespread branches of mathematics, for example in coding theory [23, 29, 36] and statistics [12]. Because of this, there is a high interest in the discovery of different Hadamard matrices up to equivalence. Two Hadamard matrices H_1 and H_2 are said to be *equivalent* if H_2 can be generated from H_1 by applying a sequence of negations/permutations to the rows/columns of H_1, i.e., if there exist signed permutation matrices U and V such that $U \cdot H_1 \cdot V = H_2$.

There are several known ways to construct sequences of Hadamard matrices. One of the simplest such constructions is by Sylvester [33]: given a known Hadamard matrix H of order n, $\left[\begin{smallmatrix} H & H \\ H & -H \end{smallmatrix} \right]$ is a Hadamard matrix of order $2n$. This process can of course be iterated, and hence one can construct Hadamard matrices of order $2^k n$ for all $k \in \mathbb{N}$ from H.

There are other methods which produce infinite classes of Hadamard matrices such as those by Paley [27]. However, no general method is known which can construct a Hadamard matrix of order n for arbitrary multiples of 4. The smallest unknown order is currently $n = 4 \cdot 167 = 668$ [7]. A database with many known matrices is included in the computer algebra system MAGMA [4]. Further collections are available online [31, 32].

Because there are $2^{(n^2)}$ matrices of order n with ± 1 entries, the search space of possible Hadamard matrices grows extremely quickly as n increases, and brute-force search is not feasible. Because of this, researchers have defined special types of Hadamard matrices which can be searched for more efficiently because they lie in a small subset of the entire space of Hadamard matrices.

3.2   Williamson Matrices

One prominent class of special Hadamard matrices are those generated by so-called Williamson matrices. These are described in this section.

Theorem 1 (cf. [37]). *Let $n \in \mathbb{N}$ and let A, B, C, $D \in \{\pm 1\}^{n \times n}$. Further, suppose that*

1. *A, B, C, and D are symmetric;*
2. *A, B, C, and D commute pairwise (i.e., $AB = BA$, $AC = CA$, etc.);*
3. *$A^2 + B^2 + C^2 + D^2 = 4nI_n$, where I_n is the identity matrix of order n.*

Then

$$\begin{bmatrix} A & B & C & D \\ -B & A & -D & C \\ -C & D & A & -B \\ -D & -C & B & A \end{bmatrix}$$

is a Hadamard matrix of order 4n.

For practical purposes, one considers A, B, C, and D in the Williamson construction to be *circulant* matrices, i.e., those matrices in which every row is the previous row shifted by one entry to the right (with wrap-around, so that the first entry of each row is the last entry of the previous row). Such matrices are completely defined by their first row $[x_0, \ldots, x_{n-1}]$ and always satisfy the commutativity property. If the matrix is also symmetric then we must further have $x_1 = x_{n-1}$, $x_2 = x_{n-2}$, and in general $x_i = x_{n-i}$ for $i = 1, \ldots, n-1$. Therefore, if a matrix is both symmetric and circulant its first row must be of the form

$$\begin{array}{ll} [x_0, x_1, x_2, \ldots, x_{(n-1)/2}, x_{(n-1)/2}, \ldots, x_2, x_1] & \text{if } n \text{ is odd} \\ [x_0, x_1, x_2, \ldots, x_{n/2-1}, x_{n/2}, x_{n/2-1}, \ldots, x_2, x_1] & \text{if } n \text{ is even.} \end{array} \tag{1}$$

Definition 2. *A **symmetric** sequence of length n is one of the form (1), i.e., one which satisfies $x_i = x_{n-i}$ for $i = 1, \ldots, n-1$.*

Williamson matrices are circulant matrices A, B, C, and D which satisfy the conditions of Theorem 1. Since they must be circulant, they are completely defined by their first row. (In light of this, we may simply refer to them as if they were sequences.) Furthermore, since they are symmetric the Hadamard matrix generated by these matrices is completely specified by the $4\lceil\frac{n+1}{2}\rceil$ variables

$$a_0, a_1, \ldots, a_{\lceil(n-1)/2\rceil}, b_0, \ldots, b_{\lceil(n-1)/2\rceil}, c_0, \ldots, c_{\lceil(n-1)/2\rceil}, d_0, \ldots, d_{\lceil(n-1)/2\rceil}.$$

Given an assignment of these variables, the rest of the entries of the matrices A, B, C, and D may be chosen in such a way that conditions 1 and 2 of Theorem 1 always hold. There is no trivial way of enforcing condition 3, but we will later derive consequences of this condition which will simplify the search for matrices which satisfy it.

There are three types of operations which, when applied to the Williamson matrices, produce different but essentially equivalent matrices. For our purposes, generating just one of the equivalent matrices will be sufficient, so we impose additional constraints on the search space to cut down on extraneous solutions and hence speed up the search.

1. Ordering: Note that the conditions on the Williamson matrices are symmetric with respect to A, B, C, and D. In other words, those four matrices can be permuted amongst themselves and they will still generate a valid Hadamard matrix. Given this, we enforce the constraint that

$$|\text{rowsum}(A)| \leq |\text{rowsum}(B)| \leq |\text{rowsum}(C)| \leq |\text{rowsum}(D)|,$$

where rowsum(X) denotes the sum of the entries of the first (or any) row of X. Any A, B, C, and D can be permuted so that this condition holds.

2. Negation: The entries in the sequences defining any of A, B, C, or D can be negated and the sequences will still generate a Hadamard matrix. Given this, we do not need to try both possibilities for the sign of the rowsum of A, B, C, and D. For example, we can choose to enforce that the rowsum of each of the generating matrices is nonnegative. Alternatively, when n is odd we can choose the signs so they satisfy rowsum(X) $\equiv n$ (mod 4) for $X \in \{A, B, C, D\}$. In this case, a result of Williamson [37] says that $a_i b_i c_i d_i = -1$ for all $1 \leq i \leq (n-1)/2$.

3. Permuting Entries: We can reorder the entries of the generating sequences with the rule $a_i \mapsto a_{ki \bmod n}$ where k is any number coprime with n, and similarly for b_i, c_i, d_i (the *same* reordering must be applied to each sequence for the result to still be equivalent). Such a rule effectively applies an automorphism of $\mathbb{Z}_n$ to the generating sequences.

3.3   Power Spectral Density

Because the search space for Hadamard matrices is so large, it is advantageous to focus on a specific construction method and describe properties which any Hadamard matrix generated by this specific method must satisfy; such properties can speed up a search by significantly reducing the size of the necessary space. One such set of properties for Williamson matrices is derived using the *discrete Fourier transform* from Fourier analysis, i.e., the periodic function $\mathrm{DFT}_A(s) := \sum_{k=0}^{n-1} a_k \omega^{ks}$ for a sequence $A = [a_0, a_1, \ldots, a_{n-1}]$, where $s \in \mathbb{Z}$ and $\omega := e^{2\pi i/n}$ is a primitive nth root of unity. Because $\omega^{ks} = \omega^{ks \bmod n}$ one has that $\mathrm{DFT}_A(s) = \mathrm{DFT}_A(s \bmod n)$, so that only n values of DFT_A need to be computed and the remaining values are determined through periodicity. In fact, when A consists of real entries, it is well-known that $\mathrm{DFT}_A(s)$ is equal to the complex conjugate of $\mathrm{DFT}_A(n - s)$. Hence only $\lfloor \frac{n+1}{2} \rfloor$ values of DFT_A need to be computed.

The *power spectral density* of the sequence A is given by

$$\mathrm{PSD}_A(s) := |\mathrm{DFT}_A(s)|^2 \qquad \text{for } s \in \mathbb{Z}.$$

3.4   Periodic Autocorrelation

As we will see, the defining properties of Williamson matrices (in particular, condition 3 of Theorem 1) can be re-cast using a function known as the periodic autocorrelation function (PAF). Re-casting the equations in this way is advantageous because many other combinatorial conjectures can also be stated in terms of the PAF. Hence, code which is used to counter-example or finitely verify one such conjecture can be re-applied to many other conjectures.

Definition 3. *The **periodic autocorrelation function** of the sequence A is the periodic function given by*

$$\mathrm{PAF}_A(s) := \sum_{k=0}^{n-1} a_k a_{(k+s) \bmod n} \qquad for \ s \in \mathbb{Z}.$$

Similar to the discrete Fourier transform, one has $\mathrm{PAF}_A(s) = \mathrm{PAF}_A(s \bmod n)$ and $\mathrm{PAF}_A(s) = \mathrm{PAF}_A(n - s)$ (see [17]), so that the PAF_A only needs to be computed for $s = 0, \ldots, \lfloor \frac{n-1}{2} \rfloor$; the other values can be computed through symmetry and periodicity.

Now we will see how to rewrite condition 3 of Theorem 1 using PAF values. Note that the sth entry in the first row of $A^2 + B^2 + C^2 + D^2$ is

$$\mathrm{PAF}_A(s) + \mathrm{PAF}_B(s) + \mathrm{PAF}_C(s) + \mathrm{PAF}_D(s).$$

Condition 3 requires that this entry should be $4n$ when $s = 0$ and it should be 0 when $s = 1, \ldots, n - 1$. The condition when s is 0 does not need to be explicitly checked because in that case the sum will always be $4n$, as $\mathrm{PAF}_A(0) = \sum_{k=0}^{n-1}(\pm 1)^2 = n$ and similarly for B, C, and D.

Additionally, the first row of $A^2 + B^2 + C^2 + D^2$ will be symmetric as each matrix in the sum has a symmetric first row. Thus ensuring that

$$\mathrm{PAF}_A(s) + \mathrm{PAF}_B(s) + \mathrm{PAF}_C(s) + \mathrm{PAF}_D(s) = 0 \quad \text{for } s = 1, \ldots, \lceil \tfrac{n-1}{2} \rceil \quad (2)$$

guarantees that every entry in the first row of $A^2 + B^2 + C^2 + D^2$ is 0 besides the first. Since $A^2 + B^2 + C^2 + D^2$ will also be circulant, ensuring that (2) holds will ensure condition 3 of Theorem 1.

3.5   Compression

Because the size of the space in which a combinatorial object lies is generally exponential in the size of the object, it is advantageous to instead search for *smaller* objects when possible. Recent theorems on so-called "compressed" sequences allow us to do that when searching for Williamson matrices.

Definition 4 (cf. [26]). *Let $A = [a_0, a_1, \ldots, a_{n-1}]$ be a sequence of length $n = dm$ and set*

$$a_j^{(d)} = a_j + a_{j+d} + \cdots + a_{j+(m-1)d}, \qquad j = 0, \ldots, d - 1.$$

*Then we say that the sequence $A^{(d)} = [a_0^{(d)}, a_1^{(d)}, \ldots, a_{d-1}^{(d)}]$ is the **m-compression** of A.*

Example 1. Consider the sequence $A = [1, 1, -1, -1, -1, 1, -1, 1, 1, -1, 1, -1, -1, -1, 1]$ of length 15. Since 15 factors uniquely as $15 = 3 \cdot 5$, there are two nontrivial choices for the tuple (d, m), namely $(d, m) = (3, 5)$ and $(d, m) = (5, 3)$. The sequence A then has the two compressions

$$A^{(3)} = [-3, 1, 1] \qquad \text{and} \qquad A^{(5)} = [3, -1, -1, -1, -1].$$

As we will see, the space of the compressed sequences that we are interested in will be much smaller than the space of the uncompressed sequences. What makes compressed sequences especially useful is that we can derive conditions that the compressed sequences must satisfy using our known conditions on the uncompressed sequences. To do this, we utilize the following theorem which is a special case of a result from [26].

Theorem 2. *Let A, B, C, and D be sequences of length $n = dm$ which satisfy*

$$\mathrm{PAF}_A(s) + \mathrm{PAF}_B(s) + \mathrm{PAF}_C(s) + \mathrm{PAF}_D(s) = \begin{cases} 4n, & s = 0 \\ 0, & 1 \le s < \mathrm{len}(A). \end{cases} \quad (3)$$

Then for all $s \in \mathbb{Z}$ we have

$$\mathrm{PSD}_A(s) + \mathrm{PSD}_B(s) + \mathrm{PSD}_C(s) + \mathrm{PSD}_D(s) = 4n. \quad (4)$$

Furthermore, both (3) and (4) hold if the sequences A, B, C, D are replaced with their compressions $A^{(d)}$, $B^{(d)}$, $C^{(d)}$, $D^{(d)}$.

Since $\mathrm{PSD}_X(s)$ is always nonnegative, Eq. (4) implies that $\mathrm{PSD}_{A^{(d)}}(s) \le 4n$ (and similarly for B, C, D). Therefore if a candidate compressed sequence $A^{(d)}$ satisfies $\mathrm{PSD}_{A^{(d)}}(s) > 4n$ for some $s \in \mathbb{Z}$ then we know that the uncompressed sequence A can never be one of the sequences which satisfies the preconditions of Theorem 2.

Useful Properties: Lastly, we derive some properties that the compressed sequences which arise in our context must satisfy. For a concrete example, note that the compressed sequences of Example 1 fulfill these properties.

Lemma 1. *If A is a sequence of length $n = dm$ with ± 1 entries, then the entries $a_i^{(d)}$, $i \in \{0, \ldots, d-1\}$, have absolute value at most m and $a_i^{(d)} \equiv m$ (mod 2).*

Proof. For all $0 \le j < d$ we have, using the triangle inequality, that

$$\left| a_j^{(d)} \right| = \left| \sum_{k=0}^{m-1} a_{j+kd} \right| \le \sum_{k=0}^{m-1} |a_{j+kd}| = m.$$

Additionally, $a_j^{(d)} \equiv \sum_{k=0}^{m-1} 1 \equiv m$ (mod 2) since $a_{j+kd} \equiv 1$ (mod 2).

In the course of our research we discovered the following useful property of compressed sequences which significantly reduces the number of SAT instances we need to generate.

Lemma 2. *The compression of a symmetric sequence is also symmetric.*

Proof. Suppose that A is a symmetric sequence of length $n = dm$. We want to show that $a_j^{(d)} = a_{d-j}^{(d)}$ for $j = 1, \ldots, d - 1$. By reversing the sum defining $a_j^{(d)}$ and then using the fact that $n = md$, we have

$$\sum_{k=0}^{m-1} a_{j+kd} = \sum_{k=0}^{m-1} a_{j+(m-1-k)d} = \sum_{k=0}^{m-1} a_{n+j-d(k+1)}.$$

By the symmetry of A, $a_{n+j-d(k+1)} = a_{d(k+1)-j}$, which equals a_{d-j+dk}. The sum in question is therefore equal to $\sum_{k=0}^{m-1} a_{d-j+dk} = a_{d-j}^{(d)}$, as required.

4   Encoding and Search Space Pruning Techniques

An attractive property of Hadamard matrices when encoding them in a SAT context is that each of their entries is one of two possible values, namely ± 1. We choose the encoding that 1 is represented by true and -1 is represented by false. We call this the *Boolean value* or *BV* encoding. Under this encoding, the multiplication function of two $x, y \in \{\pm 1\}$ becomes the XNOR function in the SAT setting, i.e., $BV(x \cdot y) = XNOR(BV(x), BV(y))$.

4.1   Encoding the Problem of Finding Hadamard Matrices as SAT Instances

For each multiplication of two entries in a given matrix, one can store one additional variable representing the result of the multiplication. The sum of variables (when thought of as ± 1 values) can be encoded using a network of binary adders. Both of these encodings add polynomially many extra variables to a given SAT instance.

In this way, it is easy to realize the naive encoding of the problem of finding a Hadamard matrix for a fixed order $n \in \mathbb{N}$, i.e., the problem of finding $h_{ij} \in \{\pm 1\}$ for $0 \leq i, j < n$ which satisfy $\sum_{k=0}^{n-1} h_{ik} \cdot h_{jk} = 0$ for all $i \neq j$. However, the naive encoding does not scale well, since the number of variables and conditions quickly becomes too large as n increases.

The encoding of a quadruple of circulant Williamson matrices relies on far less variables, which makes finding Hadamard matrices using the Williamson method scale better. In particular, if each Williamson matrix has order $n \in \mathbb{N}$, the entries of all four matrices are determined by the knowledge of $4\lceil \frac{n+1}{2} \rceil$ entries, namely

$$a_0, a_1, \ldots, a_{\lceil (n-1)/2 \rceil}, b_0, \ldots, b_{\lceil (n-1)/2 \rceil}, c_0, \ldots, c_{\lceil (n-1)/2 \rceil}, d_0, \ldots, d_{\lceil (n-1)/2 \rceil}.$$

The conditions can again be checked by introducing new variables for pairwise products and realizing binary adders.

In the subsequent subsections, we present techniques using the results from the previous section and allowing us to reach even higher orders for which we can find Hadamard matrices.

4.2 Technique 1: Sum-of-Squares Decomposition

As a special case of compression, consider what happens when $d = 1$ and $m = n$. In this case, the compression of A is a sequence with a single entry whose value is $\sum_{k=0}^{n-1} a_k = \text{rowsum}(A)$. If A, B, C, and D are $\{\pm1\}$-sequences which satisfy the conditions of Theorem 2, then the theorem applied to this m-compression says that

$$\text{PAF}_{A^{(1)}}(0) + \text{PAF}_{B^{(1)}}(0) + \text{PAF}_{C^{(1)}}(0) + \text{PAF}_{D^{(1)}}(0) = 4n$$

which simplifies to

$$\text{rowsum}(A)^2 + \text{rowsum}(B)^2 + \text{rowsum}(C)^2 + \text{rowsum}(D)^2 = 4n,$$

and by Lemma 1 each rowsum must have the same parity as n.

In other words, the rowsums of the sequences A, B, C, and D decompose $4n$ into the sum of four perfect squares whose parity matches the parity of n. Since there are usually only a few ways of writing $4n$ as a sum of four perfect squares this severely limits the number of sequences which could satisfy the hypotheses of Theorem 2. Furthermore, some computer algebra systems contain functions for explicitly computing what the possible decompositions are (e.g., PowersRepresentations in MATHEMATICA and nsoks by Joe Riel of Maplesoft [30]). We can query such CAS functions to determine all possible values that the rowsums of A, B, C, and D could possibly take. For example, when $n = 35$ we find that there are exactly three ways to write $4n$ as a sum of four positive odd squares in ascending order, namely,

$$1^2 + 3^2 + 3^2 + 11^2 = 1^2 + 3^2 + 7^2 + 9^2 = 3^2 + 5^2 + 5^2 + 9^2 = 4 \cdot 35.$$

As described in Sect. 3.2, any Williamson quadruple is equivalent to another quadruple whose rowsum sum-of-squares decomposition is of one of the above three types.

4.3 Technique 2: Divide-and-Conquer via Compression

Because each instance can take a significant amount of time to solve, it is beneficial to divide instances into multiple partitions, each instance encoding a subset of the search space. In our case, we found that an effective splitting method was to split by compressions, i.e., to have each instance contain one possibility of the compressions of A, B, C, and D. To do this, we first need to know all possible compressions of A, B, C, and D. These can be generated by applying Lemmas 1 and 2. For example, when $n = 35$ and $d = 5$ there are 28 possible compressions of A with rowsum$(A) = 1$. Of those, only 12 satisfy $\text{PSD}_A(s) \leq 4n$ for all $s \in \mathbb{Z}$. The calculation of $\text{PSD}_A(s)$ was possible to be performed within Python using numpy instead of directly querying a CAS. There are also 12 possible compressions for each of B, C, and D with rowsum$(B) = $ rowsum$(C) = 3$ and rowsum$(D) = 11$. Thus there are 12^4 total instances which would need to be generated for this selection of rowsums, however, only 41 of them satisfy the conditions given by Theorem 2.

Furthermore, if n has two nontrivial divisors m and d then we can find all possible m-compressions and d-compressions of A, B, C, and D. In this case, each instance can set *both* the m-compression and the d-compression of each of A, B, C, and D. Since there are more combinations to check when dealing with two types of compression this causes an increase in the number of instances generated, but each instance has more constraints and a smaller subspace to search through.

4.4   Technique 3: UNSAT Core

After using the divide-and-conquer technique one obtains a collection of instances which are almost identical. For example, the instances will contain variables which encode the rowsums of A, B, C, and D. Since there are multiple possibilities of the rowsums (as discussed in Sect. 4.2), not all instances will set those variables to the same values. However, since the instances are the same except for those variables, it is sometimes possible to use an *UNSAT core* result from one instance to learn that other instances are unsatisfiable.

MAPLESAT is one SAT solver which supports UNSAT core generation. Provided a master instance and a set of assumptions (variables which are set either true or false), the UNSAT core contains a subset of the assumptions which make the master instance unsatisfiable. Thus, any other instance which sets the variables in the UNSAT core in the same way must also be unsatisfiable.

For example, our instances for $n = 35$ contained 15,663 clauses which were identical among all instances. The instances contained 3376 variables of which only 168 were given as assumptions and assigned differently in each instance.

5   Verification of the Nonexistence of Williamson Matrices of Order 35

We searched for Williamson matrices of order 35 using the techniques described in Sect. 4 with both 5 and 7-compression. Despite the exponential growth of possible first rows of the matrices A, B, C, and D, the described pruning results in 21,674 SAT instances of three possible forms, as described in Fig. 2. Each instance has subsequently been checked with several SAT solvers, and each one has been discovered to be unsatisfiable. Using MAPLESAT with UNSAT core generation, 19,356 SAT solver calls were necessary to determine that all instances were unsatisfiable.

Our practice was to have people that were not involved in writing the respective code verify its correctness, and to have domain experts verify the application of the theorems used. Furthermore, our confidence of the correctness of our code was strengthened by the successful discovery of Williamson-generated Hadamard matrices for all the orders $4n$ with $n < 35$. These have been determined to be valid Hadamard matrices by the computer algebra system MAGMA.

rowsum(A)	rowsum(B)	rowsum(C)	rowsum(D)	Number of Instances
1	3	3	11	6960
1	3	7	9	8424
3	5	5	9	6290

Fig. 2. The number of instances of each type generated in the process of searching for Williamson matrices of order 35.

6   Experimental Results on Hadamard Matrices

We checked all of the Hadamard matrices we computed for equivalence against those in MAGMA's Hadamard matrix database. In total, our methods generated 160 pairwise inequivalent Hadamard matrices which were also not equivalent to any matrices in this database. We submitted these to the MAGMA team and one can download these on our project website (https://sites.google.com/site/uwmathcheck/hadamard-conjecture).

Experimental Setup and Methodology: The timings were run on the high-performance computing cluster SHARCNET@. Specifically, the cluster we used ran CentOS 5.4 and used 64-bit AMD Opteron processors running at 2.2 GHz. Each SAT instance was generated using MATHCHECK2 with the appropriate parameters and the instance was submitted to SHARCNET to solve by running MAPLESAT on a single core (with a timeout of 24 h).

Figure 3 contains a summary of the performance of our encoding and pruning techniques. The timings are for searching for Williamson matrices of order n with $25 \leq n \leq 35$ and for each of the techniques discussed in Sect. 4. We did not use Techniques 2 and 3 for orders 29 and 31 as they have no nontrivial divisors to perform compression with, but they were otherwise very effective at partitioning the search space in an efficient way. Technique 3 was effective at cutting down the number of instances generated in certain orders. Although the instances pruned tended to be those which would have been quickly solved, this technique would be especially valuable in a situation where few cores are available, as it would allow the overhead of many SAT solver calls to be avoided.

The timings given in Fig. 3 refer to the total amount of time used by MAPLE-SAT across all the jobs run on SHARCNET for each order and technique. The numbers in parentheses in Fig. 3 denote how many MAPLESAT calls returned a result and did not time out. The jobs using the base encoding and Technique 1 which did not time out all returned SAT. All of the jobs using Technique 2 completed without timing out and most the instances were found to be UNSAT. The Technique 3 results were the same as the Technique 2 results except with fewer calls to MAPLESAT as some instances could immediately be determined to be UNSAT.

Order	Base Encoding (Sec. 4.1)	Technique 1 (Sec. 4.2)	Technique 2 (Sec. 4.3)	Technique 3 (Sec. 4.4)
25	317s (1)	1702s (4)	408s (179)	408s (179)
26	865s (1)	3818s (3)	61s (3136)	34s (1592)
27	5340s (1)	8593s (3)	1518s (14994)	1439s (689)
28	7674s (1)	2104s (2)	234s (13360)	158s (439)
29	-	21304s (1)	N/A	N/A
30	1684s (1)	36804s (1)	139s (370)	139s (370)
31	-	83010s (1)	N/A	N/A
32	-	-	96011s (13824)	95891s (348)
33	-	-	693s (8724)	683s (7603)
34	-	-	854s (732)	854s (732)
35	-	-	31816s (21674)	31792s (19356)

Fig. 3. The numbers in parentheses denote how many MAPLESAT calls successfully returned a result for the given Williamson order. The timings refer to the total amount of time used during those calls. A hyphen denotes a timeout after 24 h.

7   Related Work

The idea of combining the capabilities of SAT/SMT solvers and computer algebra systems or domain-specific knowledge has been examined by various research groups. Junges et al. [15] studied an integration of Gröbner basis theory in the context of SMT solvers. Although they implemented their own version of Buchberger's algorithm, they describe that it is possible to "plug in an off-the-shelf GB procedure implementation such as the one in SINGULAR" as the core procedure. SINGULAR [9] is a computer algebra system with specialized algorithms for polynomial systems. Ábrahám later highlights the potentials of combining symbolic computation and SMT solving in [1]. The VERIT SMT solver [5] uses the computer algebra system REDUCE [11] to support non-linear arithmetic. The LEAN theorem prover [22] combines domain-specific knowledge with SMT solvers. Combining SAT and SMT with theorem proving has been done in the automated theorem prover COQ as well [2]. The idea of using equivalences in satisfiability problems to prune the search space has also been exploited by symmetry breaking [13, 28]. SAT-based results on the Erdős discrepancy conjecture [16] inspired the previous version of MATHCHECK [39]. This version also combined SAT with computer algebra systems but specialized in graph theory and used the CAS to uncover theory lemmas as the search progressed. Work related to finding Hadamard matrices has been referenced in Sect. 3.

8   Conclusions and Future Work

We have presented the advantages of utilizing the power of SAT solvers in combination with domain specific knowledge and algorithms provided by computer algebra systems. Our main mathematical problem was the verification of the Hadamard conjecture for some orders by using MATHCHECK2 to search for and discover Williamson matrices. We verified independently, as requested by D. Đoković, that there is no Hadamard Matrix of order $4 \cdot 35$ which is generated by Williamson matrices. Moreover, we discovered 160 Hadamard matrices that are not equivalent to any matrix in the MAGMA Hadamard database.

A future direction is to scale to Hadamard matrices of higher order. For this, we plan to refine the methods (e.g., by examining other construction types), and possibly implement certain search strategies directly into the SAT solver. We also want to analyze the UNSAT cores generated by Technique 3 to explain their effectiveness in certain cases, as well as exploring the usage of incremental SAT solvers [3,24]. Finally, we plan to use MATHCHECK2 and our newly acquired knowledge to consider other combinatorial problems. There are many problems which can be expressed as a search for objects which satisfy certain autocorrelation equations (as just one example, those involving complex Golay sequences). Since the ability to work with autocorrelation is already built-in to MATHCHECK2, we should be able to execute such searches with minor modifications.

References

1. Ábrahám, E.: Building bridges between symbolic computation and satisfiability checking. In: Proceedings of the 2015 ACM on International Symposium on Symbolic and Algebraic Computation, pp. 1–6. ACM, New York (2015)
2. Armand, M., Faure, G., Grégoire, B., Keller, C., Théry, L., Wener, B.: Verifying SAT and SMT in CoQ for a fully automated decision procedure. In: PSATTT 2011: International Workshop on Proof-Search in Axiomatic Theories and Type Theories, pp. 11–25 (2011)
3. Biere, A., Heule, M., van Maaren, H., Walsh, T. (eds.): Handbook of Satisfiability. Frontiers in Artificial Intelligence and Applications, vol. 185. IOS Press (2009)
4. Bosma, W., Cannon, J., Playoust, C.: The Magma algebra system I: the user language. J. Symbolic Comput. **24**(3), 235–265 (1997)
5. Bouton, T., Caminha B. de Oliveira, D., Déharbe, D., Fontaine, P.: VERIT: an open, trustable and efficient SMT-Solver. In: Schmidt, R.A. (ed.) CADE-22. LNCS, vol. 5663, pp. 151–156. Springer, Heidelberg (2009)
6. Char, B.W., Fee, G.J., Geddes, K.O., Gonnet, G.H., Monagan, M.B.: A tutorial introduction to Maple. J. Symbolic Comput. **2**(2), 179–200 (1986)
7. Colbourn, C.J., Dinitz, J.H. (eds.): Handbook of Combinatorial Designs. Discrete Mathematics and its Applications (Boca Raton), 2nd edn. Chapman & Hall/CRC, Boca Raton (2007)
8. Cox, D., Little, J., O'Shea, D.: Ideals, Varieties, and Algorithms, 2nd edn. Springer, New York (1992)
9. Decker, W., Greuel, G.M., Pfister, G., Schönemann, H.: SINGULAR 4-0-2 – A computer algebra system for polynomial computations (2015). http://www.singular.uni-kl.de
10. Hadamard, J.: Résolution d'une question relative aux déterminants. Bull. Sci. Math. **17**(1), 240–246 (1893)
11. Hearn, A.: REDUCE user's manual, version 3.8 (2004)
12. Hedayat, A., Wallis, W.: Hadamard matrices and their applications. Ann. Stat. **6**(6), 1184–1238 (1978)

13. Hnich, B., Prestwich, S.D., Selensky, E., Smith, B.M.: Constraint models for the covering test problem. Constraints **11**(2), 199–219 (2006)
14. Holzmann, W.H., Kharaghani, H., Tayfeh-Rezaie, B.: Williamson matrices up to order 59. Des. Codes Crypt. **46**(3), 343–352 (2008)
15. Junges, S., Loup, U., Corzilius, F., Ábrahám, E.: On Gröbner bases in the context of satisfiability-modulo-theories solving over the real numbers. In: Muntean, T., Poulakis, D., Rolland, R. (eds.) CAI 2013. LNCS, vol. 8080, pp. 186–198. Springer, Heidelberg (2013)
16. Konev, B., Lisitsa, A.: A SAT attack on the Erdős discrepancy conjecture. In: Sinz, C., Egly, U. (eds.) SAT 2014. LNCS, vol. 8561, pp. 219–226. Springer, Heidelberg (2014)
17. Kotsireas, I.S.: Algorithms and metaheuristics for combinatorial matrices. In: Handbook of Combinatorial Optimization, pp. 283–309. Springer, New York (2013)
18. Liang, J.H., Ganesh, V., Poupart, P., Czarnecki, K.: Learning rate based branching heuristic for SAT solvers. In: Creignou, N., Le Berre, D., Le Berre, D., Le Berre, D., Le Berre, D., Le Berre, D. (eds.) SAT 2016. LNCS, vol. 9710, pp. 123–140. Springer, Heidelberg (2016). doi:10.1007/978-3-319-40970-2_9
19. Liang, J.H., Ganesh, V., Poupart, P., Czarnecki, K.: Exponential recency weighted average branching heuristic for SAT solvers. In: Proceedings of AAAI 2016 (2016)
20. Marques-Silva, J.P., Sakallah, K., et al.: GRASP: a search algorithm for propositional satisfiability. IEEE Trans. Comput. **48**(5), 506–521 (1999)
21. Moskewicz, M.W., Madigan, C.F., Zhao, Y., Zhang, L., Malik, S.: Chaff: engineering an efficient SAT solver. In: Proceedings of the 38th Annual Design Automation Conference, pp. 530–535. ACM, New York (2001)
22. de Moura, L., Kong, S., Avigad, J., van Doorn, F., von Raumer, J.: The lean theorem prover (system description). In: Felty, P.A., Middeldorp, A. (eds.) CADE-25. LNCS, vol. 9195, pp. 378–388. Springer, Switzerland (2015)
23. Muller, D.E.: Application of Boolean Algebra to Switching Circuit Design and to Error Detection. Electron. Comput. Trans. IRE Prof. Group Electron. Comput. EC-3(3), 6–12 (1954)
24. Nadel, A., Ryvchin, V.: Efficient SAT solving under assumptions. In: Cimatti, A., Sebastiani, R. (eds.) SAT 2012. LNCS, vol. 7317, pp. 242–255. Springer, Heidelberg (2012)
25. Đoković, D.Ž.: Williamson matrices of order $4n$ for $n = 33, 35, 39$. Discrete Math. **115**(1), 267–271 (1993)
26. Đoković, D.Ž., Kotsireas, I.S.: Compression of periodic complementary sequences and applications. Des. Codes Crypt. **74**(2), 365–377 (2015)
27. Paley, R.E.: On orthogonal matrices. J. Math. Phys. **12**(1), 311–320 (1933)
28. Prestwich, S.D., Hnich, B., Simonis, H., Rossi, R., Tarim, S.A.: Partial symmetry breaking by local search in the group. Constraints **17**(2), 148–171 (2012)
29. Reed, I.: A class of multiple-error-correcting codes and the decoding scheme. Trans. IRE Prof. Group Inf. Theory **4**(4), 38–49 (1954)
30. Riel, J.: nsoks: A MAPLE script for writing n as a sum of k squares
31. Seberry, J.: Library of Williamson Matrices. http://www.uow.edu.au/~jennie/WILLIAMSON/williamson.html
32. Sloane, N.: Library of Hadamard Matrices. http://neilsloane.com/hadamard/
33. Sylvester, J.J.: Thoughts on inverse orthogonal matrices, simultaneous sign successions, and tessellated pavements in two or more colours, with applications to Newton's rule, ornamental tile-work, and the theory of numbers. London Edinb. Dublin Philos. Mag. J. Sci. **34**(232), 461–475 (1867)

34. SC^2: Satisfiability checking and symbolic computation. http://www.sc-square.org/
35. The Sage Developers: Sage Mathematics Software (Version 7.0) (2016). http://www.sagemath.org
36. Walsh, J.L.: A closed set of normal orthogonal functions. Am. J. Math. **45**(1), 5–24 (1923)
37. Williamson, J.: Hadamard's determinant theorem and the sum of four squares. Duke Math. J **11**(1), 65–81 (1944)
38. Wolfram, S.: The MATHEMATICA Book, version 4. Cambridge University Press (1999)
39. Zulkoski, E., Ganesh, V., Czarnecki, K.: MathCheck: a math assistant via a combination of computer algebra systems and SAT solvers. In: Felty, P.A., Middeldorp, A. (eds.) CADE-25. LNCS, vol. 9195, pp. 607–622. Springer, Switzerland (2015)

Incompleteness, Undecidability and Automated Proofs
(Invited Talk)

Cristian S. Calude[1]([✉]) and Declan Thompson[1,2]

[1] Department of Computer Science, University of Auckland, Auckland, New Zealand
cristian@cs.auckland.ac.nz
[2] Department of Philosophy, Stanford University, Stanford, USA
http://www.cs.auckland.ac.nz/~cristian
http://www.stanford.edu/~declan

Abstract. Incompleteness and undecidability have been used for many years as arguments against automatising the practice of mathematics. The advent of powerful computers and proof-assistants – programs that assist the development of formal proofs by human-machine collaboration – has revived the interest in formal proofs and diminished considerably the value of these arguments.

In this paper we discuss some challenges proof-assistants face in handling undecidable problems – the very results cited above – using for illustrations the generic proof-assistant Isabelle.

1 Introduction

Gödel's incompleteness theorem (1931) and Turing's undecidability of the halting problem (1936) form the basis of a largely accepted thesis that mathematics cannot be relegated to computers. However, the impetuous development of powerful computers and versatile software led to the creation of *proof-assistants* – programs that assist the development of formal proofs by human-machine collaboration. This trend has revived the interest in formal proofs and diminished considerably the value of the above thesis for the working mathematician.

An impressive list of deep mathematical theorems have been formally proved including Gödel's incompleteness theorem (1986), the fundamental theorem of calculus (1996), the fundamental theorem of algebra (2000), the four colour theorem (2004), Jordan's curve theorem (2005) and the prime number theorem (2008). In 2014 Hales's Flyspeck project team formally validated in [7] Hales's proof [21] of the Kepler conjecture. Why did Hales' proof by exhaustion of the conjecture – involving the checking of many individual cases using complex computer calculations – published in the prestigious *Annals of Mathematics*, need "a validation"? Because the referees of the original paper had not been "100 % certain" that the paper was correct.

Hilbert's standard of proof is becoming practicable due to proof-assistants.

V.P. Gerdt et al. (Eds.): CASC 2016, LNCS 9890, pp. 134–155, 2016.
DOI: 10.1007/978-3-319-45641-6_10

Are proof-assistants able to handle undecidable problems, the very results which have been used to argue the impossibility of doing mathematics with computers? In what follows we discuss some challenges proof-assistants face in handling two undecidable problems – termination and correctness – using for illustrations the generic proof-assistant Isabelle.

2   Truth and Provability

2.1   Incompleteness

In 1931 K. Gödel proved his celebrated incompleteness theorem which states that *no consistent system of axioms whose theorems can be effectively listed (e.g., by a computer program or algorithm) is capable of proving all true relations between natural numbers (arithmetic).* In such a system there are statements about the natural numbers that are true, but unprovable within the system – they are called *undecidable statements in the system.* For example, the *consistency* of such a system can be coded as a true property of natural numbers which the system cannot demonstrate, hence it is undecidable in the system. Furthermore, extending the system does not cure the problem. Examples of systems satisfying the properties of Gödel's incompleteness theorem are Peano arithmetic and Zermelo–Fraenkel set theory. One can interpret Gödel's theorem as saying that *no consistent system whose theorems can be effectively listed is capable of proving any mathematically true statement.*

2.2   Undecidability

Five years later A. Turing proved a computationally similar result, *the existence of (computationally) undecidable problems,* i.e. problems which have no algorithmic solution. To this goal Turing proposed a mathematical model of computability, based on what today we call a Turing machine (program), which is widely, but not unanimously, accepted as adequate (the Church-Turing thesis). The halting problem is the problem of determining in a finite time, from a description of an arbitrary Turing program and an input, whether the program will finish running or continue to run forever. Turing's theorem shows that *no Turing program can solve (correctly) every instance of the halting problem.* Our ubiquitous computers cannot do everything, a shock for some.

2.3   Incompleteness vs. Undecidability

Gödel's undecidable statements depend on the fixed formal system; Turing's undecidable problems depend on the adopted mathematical model of computation. They are both *relative.* There is a deep relation between these results coming from the impossibility of dealing algorithmically with some forms of infinity. In what follows we present a form of Gödel's incompleteness theorem by examining the halting problem. By $N(P, v)$ we mean that the Turing program P

will *never* halt on input v. For any particular program P and input v, $N(P,v)$ is a perfectly definite statement which is either true (in case P will never halt in the described situation) or false (in case P will eventually halt). When $N(P,v)$ is false, this fact can always be demonstrated by running P on v. No amount of computation will suffice to demonstrate the fact that $N(P,v)$ is true. We may still be able to prove that a particular $N(P,v)$ is true by a logical analysis of P's behaviour, but, because of the undecidability of the halting problem, no such automated method works correctly in *all* cases.

Suppose that certain strings of symbols (possibly paragraphs of a natural language (English, for example)) have been singled out – typically with the help of axioms and rules of inference – as proofs of particular statements of the form $N(P,v)$. Operationally, we assume that we have an *algorithmic syntactic test* that can determine whether an alleged proof Π that "$N(P,v)$ is true" is or is not correct. There are two natural requirements for the rules of proof:

Soundness: If there is a proof Π that $N(P,v)$ is true, then P will never halt on input v.

Completeness: If P will never halt on input v, then there is a proof Π that $N(P,v)$ is true.

Can we find a set of rules which is both sound and complete? The answer is *negative*. Suppose, by absurdity, we had found some "rules" of proof which are both sound and complete. Suppose that the proofs according to these "rules" are particular strings of symbols on some specific finite alphabet. Let $\Pi_1, \Pi_2, \Pi_3, \ldots$ be the quasi-lexicographic computable enumeration of all finite strings on this alphabet. This sequence includes all possible proofs, as well as a lot of other things (including a high percentage of total non-sense). But, hidden between the non-sense, we have all possible proofs. Next we show how we can use our "rules" to solve the halting problem – an impossibility. We are given a Turing program P and an input v and have to test whether or not P will eventually halt on v. To answer this question we run in parallel the following two computations:

(A) the computation of P on v,

(B) the computation consisting in generating the sequence $\Pi_1, \Pi_2, \Pi_3, \ldots$ of all possible proofs and using, as each Π_i is generated, the syntactic test to determine whether or not Π_i is a proof of $N(P,v)$.

The algorithm (A) and (B) stops when either (A) stops or in (B) a proof Π_i for $N(P,v)$ is validated as correct: in the first case the answer is "P stops on v" and in the second case the answer is "P does not stop on v".

First we prove that the algorithm cannot stop simultaneously on both (A) and (B). If the algorithm stops through (B) then a proof Π_i for $N(P,v)$ is found, so by soundness the computation (A) cannot stop; if, on the contrary, the computation (A) stops, then no valid proof for $N(P,v)$ can exist because, by soundness, then P will never halt on input v.

Second we prove that the algorithm stops either on (A) or on (B). Indeed, if the algorithm does not stop on (A), then P will never halt on input v, so by completeness the algorithm will stop on (B). If the algorithm does not stop on (B), then there is no valid proof for $N(P, v)$, so again by completeness, P will stop on v, hence the algorithm will stop through (A).

Finally we prove the correctness of the algorithm. If P will eventually halt on v, then the computation (A) will eventually stop and the algorithm will give the correct answer: "P stops on v". If P never halts on v, (A) will be of no use: however, because of completeness, there will be a valid proof Π_i of $N(P, v)$ which will be eventually *discovered* by the computation (B). Having obtained this Π_i we will be sure (because of soundness) that P will indeed never halt.

Thus, we have described an algorithm which would solve the halting problem, a contradiction! The conclusion is that *no rules of proof can be both sound and complete:* there is a true statement $N(P, v)$ which has no proof Π. This statement is *undecidable* in the system: it cannot be proved nor disproved (being true, this would contradict soundness).

The above proof does not indicate any particular pair (P, v) for which the "rules" cannot prove that $N(P, v)$ is true! We only know that there will be a pair (P, v) for which $N(P, v)$ is true, but not provable from the "rules". There always are other sound rules which decide the "undecidable" statement: for example, adding the "undecidable" (true) statement as an axiom we get a larger system which trivially proves the original "undecidable" statement. No matter how we try to avoid an "undecidable" statement the new and more powerful rules will in turn have their own undecidable statements.

2.4 Hilbert's Programme and Hilbert's Axiom

In the late 19th century mathematics – shaken by the discoveries of paradoxes (for example, Russell's paradox) – entered into a foundational crisis (in German, Grundlagenkrise der Mathematik) which prompted the search for proper foundations in the early 20th century. Three schools of philosophy of mathematics were opposing each other: formalism, intuitionism and logicism. D. Hilbert, the main exponent of formalism, held that *mathematics is only a language and a series of games, but not an arbitrary game with arbitrary rules.* Hilbert's programme proposed to ground all mathematics on a finite, complete set of axioms, and provide a proof that these axioms were consistent. Hilbert's programme included a formalisation of mathematics using a precise formal language with the following properties: *completeness*, a proof that all true mathematical statements can be proved in the adopted formal system, and *consistency*, a proof – preferably involving finite mathematical objects only – that no contradiction can be obtained in the formal system. The consistency of more complicated systems, such as complex analysis, could be proven in terms of simpler systems and, ultimately, the consistency of the whole of mathematics would be reduced to that of arithmetic.

In his famous lecture entitled "Mathematical problems", presented to the International Congress of Mathematicians held in Paris in 1900, Hilbert expressed his deep conviction in the solvability of all mathematical problems (cited from [18, p. 11]):

Is the axiom of solvability of every problem a peculiar characteristic of mathematical thought alone, or is it possibly a general law inherent in the nature of the mind, that all questions which it asks must be answerable?...
This conviction of the solvability of every mathematical problem is a powerful incentive to the worker. We hear within us the perpetual call: There is the problem. Seek its solution. You can find it by pure reason, for in mathematics there is no *ignorabimus.*

Thirty years later, on 8 September 1930, in response to *Ignoramus et ignorabimus* ("We do not know, we shall not know"), the Latin maxim used as motto by the German physiologist Emil du Bois-Reymond [10] to emphasise the limits of understanding of nature, Hilbert concluded his retirement address to the Society of German Scientists and Physicians[1] – the same meeting where Gödel presented his completeness and incompleteness theorems – with his now famous words:

We must not believe those, who today, with philosophical bearing and deliberative tone, prophesy the fall of culture and accept the *ignorabimus.* For us there is no *ignorabimus,* and in my opinion none whatever in natural science. In opposition to the foolish *ignorabimus* our slogan shall be: "We must know. We will know." (in German: *Wir müssen wissen. Wir werden wissen*[2]).

2.5   Objective vs. Subjective Mathematics

According to Gödel [19], see also [18], objective mathematics consists of *the body of those mathematical propositions which hold in an absolute sense, without any further hypothesis.*

A mathematical statement constitutes an *objective problem* if it is a candidate for objective mathematics, that is, if its truth or falsity is independent of any hypotheses and does not depend on where or how it can be demonstrated. Problems in arithmetic are objective problems in contrast with problems in axiomatic geometry, which depend on their provability in a specific axiomatic system.

Gödel's *subjective mathematics* is the body of all humanly demonstrable or knowable mathematically true statements, that is, the set of all propositions which the human mind can in principle prove in some well-defined system of axioms in which every axiom is recognised to belong to objective mathematics and every rule preserves objective mathematics.

Does objective mathematics coincide with subjective mathematics? Gödel's answer (1951, see [19]) based on his incompleteness theorem was: *Either ... the*

---

[1] Included in the short radio presentation, see [24].

[2] The words are engraved on Hilbert's tombstone in Göttingen. This is a triple irony: their use as an epitaph, the fact that the day before the talk, Hilbert's optimism was undermined by Gödel's presentation of the incompleteness theorem, whose exceptional significance was, with the exception of John von Neumann, completely missed by the audience.

human mind ... infinitely surpasses the powers of any finite machine, or else there exist absolutely unsolvable ... problems.

Working with a *constructive* interpretation of truth values "true", "false" and modal "can be known" Martin-Löf stated the following theorem [27]: *There are no propositions which can neither be known to be true nor be known to be false.*

For the non-constructive mathematician this means that no propositions can be effectively produced (i.e. by an algorithm) of which it can be shown that they can neither be proved constructively nor disproved constructively. There may be absolutely unsolvable problems, but one cannot effectively produce one for which one can show that it is unsolvable.

2.6   Hilbert's Programme After Incompleteness

In the standard interpretation, incompleteness shows that most of the goals of Hilbert's programme were impossible to achieve ... However, much of it can be and was salvaged by changing its goals slightly. With the following modifications some parts of Hilbert's programme have been successfully completed. Although it is not possible to formalise all mathematics, it is feasible to formalise essentially all the mathematics that "anyone uses". Zermelo–Fraenkel set theory combined with first-order logic gives a satisfactory and generally accepted formalism for essentially all current mathematics. Although it is not possible to prove completeness for systems at least as powerful as Peano arithmetic (if they have a computable set of axioms), it is feasible to prove completeness for many weaker but interesting systems, for example, first-order logic (Gödel's completeness theorem), Kleene algebras and the algebra of regular events and various logics used in computer science. Undecidability is a consequence of incomputability: there is no algorithm deciding the truth of statements in Peano arithmetic. Tarski's algorithm (see [32]) decides the truth of any statement in analytic geometry (more precisely, the theory of real closed fields is decidable). With the Cantor-Dedekind axiom, this algorithm can decide the truth of any statement in Euclidean geometry. Finally, Martin-Löf theorem cited in the previous section strongly limits the impact of absolutely unsolvable problems, if any exist, as one cannot effectively produce one for which one could show that it is unsolvable.

3   Can Computers Do Mathematics?

Mathematical proofs are essentially based on axiomatic-deductive reasoning. This view was repeatedly expressed by the most prominent mathematicians. For Bourbaki [11], *Depuis les Grecs, qui dit Mathématique, dit démonstration.*

A *formal proof*, written in a formal language consisting of certain strings of symbols from a fixed alphabet, satisfies Hilbert's criterion of mechanical testing:

> The rules should be so clear, that if somebody gives you what they claim is a proof, there is a mechanical procedure that will check whether the proof is correct or not, whether it obeys the rules or not.

By making sure that every step is correct, one can tell once and for all whether a proof is correct or not, i.e. whether a theorem has been proved. Hilbert's concept of formal proof is an ideal of rigour for mathematics which has important applications in mathematical logic (computability theory and proof theory), but for many years seemed to be irrelevant for the practice of mathematics which uses *informal* (pen-on-paper) *proofs*. Such a proof is a rigorous argument expressed in a mixture of natural language and formulae that is intended to convince a knowledgeable mathematician of the truth of a statement, the theorem. Routine logical inferences are omitted. "Folklore" results are used without proof. Depending on the area, arguments may rely on intuition. Informal proofs are the standard of presentation of mathematics in textbooks, journals, classrooms, and conferences. They are the product of a social process. In principle, an informal proof can be converted into a formal proof; however, this is rarely, almost never, done in practice. See more in [15].

In the last 30 years a new influence on the mathematical practice has started to become stronger and stronger: *the impact of software and technology*. Software tools – called interactive theorem provers or proof-assistants – aiding the development of formal proofs by human-machine collaboration have appeared and got better and better. They include an interactive proof editor with which a human can guide the search for, the checking of and the storing of formal proofs, using a computer.

As discussed in Sect. 1, an impressive list of deep mathematical theorems have been formally proved. The December 2008 issue of the *Notices of AMS* includes four important papers on formal proof. A formal proof in Isabelle for a sharper form of the Kraft-Chaitin theorem was given in [14]. In 2014 an automated proof of Gödel's ontological proof of God's existence was given in [9] and an automatic 13-gigabyte proof solved a special case of the Erdös discrepancy problem [26]; only a year later, Tao [31] gave a pen-and-paper general solution. The current longest automatic proof has almost 200-tb^3: it solves the Boolean Pythagorean triples problem [23], a long-standing open problem in Ramsey theory. A compressed 68-gb certificate allows anyone to reconstruct the proof for checking.

Hilbert's standard of proof is practicable, it's becoming reality. However, as noted in [17],

> [T]he majority of mathematicians remain hesitant to use software to help develop, organize, and verify their proofs. Yet concerns linger over usability and the reliability of computerized proofs, although some see technological assistance as being vital to avoid problems caused by human error.

There are three main obstacles to a wider use of automated proofs [17]: (a) the lack of trust in the "machine", (b) the necessity of repeatedly developing foundational material, (c) the apparent loss of understanding in favour of the syntactical correctness (see also [13]). Current solutions involve (a) using

---

^{3} The approximate equivalent of all the digitised texts held by the US Library of Congress.

a small "trusted" kernel on top of which employ a complicated software that parses the code, but ultimately calls the kernel to check the proof, or use an independent checker, (b) growing archives of formal proofs (see for example [2]) and developing more powerful automatic proof procedures, (c) developing environments in which users can write and check formal proofs as well as query them with reference to the symptoms of understanding [15,33] and write papers explaining formal proofs.

4   Formalised Computability Theory

Computability theory is an inherently interesting field for automated theorem proving. Since the limitations of computation being studied are true of the theorem provers themselves, formalised computability theory is like modifying the engine of a plane mid-flight.

Early work in formalised computability theory was completed by [30], who formalised the primitive recursive functions in the ALF (Another Logical Framework) proof-assistant [1], a predecessor of the contemporary Coq [3]. The purpose of this formalisation was to provide a computer-checked proof that the Ackermann function is not primitive recursive. As such, study into the relationship of the recursive functions to other forms of computation was not undertaken. A formalisation of Unlimited Register Machines (URM) was given in [36] (see also [35]), and it was shown that URMs can simulate partial recursive functions.[4] The converse was not shown however. The paper [36] also formalised partial recursive functions in Coq.

More recent work by Michael Norrish [28] has established a greater body of formal computability theory. Norrish formalises an implementation of the λ-calculus model of computation in the HOL4 system [4] and further defines the partial recursive functions and establishes the computational equivalence of these two models. A number of standard results are proven, including the existence of a universal machine (which is constructed), the identification of recursively enumerable sets with the ranges or domains of partially computable functions, the undecidability of the halting problem and Rice's theorem. For use in these results, Norrish implements the "dove-tailing" method, whereby a function is run on input $0, \ldots, n$ for n steps, and then $0, \ldots, n, n+1$ for $n+1$ steps, and so forth.

A formalisation of Turing machines is given in [8] in the Matita interactive theorem prover [6]. While the focus is on the use of Turing machines in complexity theory, a universal Turing machine is constructed, and its correctness proved.

An impressive formalisation of three models of computable functions can be found in [34]. Here, the authors define Turing machines, abacus machines

---

[4] Historically, the syntactic class of partial functions constructed recursively is called *partially recursive functions*, see [25,29]. This class coincides with the semantic partial functions implementable by standard models of computation (Turing machines, URMs, the λ-calculus etc.) – the *partially computable functions*.

and partial recursive functions in the Isabelle, and give a formal proof of the undecidability of the halting problem for Turing machines. Turing machines are shown to be able to model abacus machines, and abacus machines to model recursive functions. The bulk of [34]'s work is in creating a universal recursive function which takes encodings of Turing machines as inputs, and gives the same output as those machines would. This formally establishes not only the existence of universal functions, but also the equivalence of the three models of computation. While a universal Turing machine is not directly constructed, its existence can be inferred from the proofs that Turing machines can model abacus machines, which can in turn model recursive functions. This contrasts with the direct constructions of a universal Turing machine given in [8]. The establishment of the equivalence of recursive functions and the λ-calculus in [28] gives us formal proofs of the computational equivalence of the following four models: Turing machines, abacus machines, partial recursive functions and the λ-calculus.

4.1   Automated Proofs in Isabelle

Isabelle is a generic proof-assistant derived from the Higher Order Logic (HOL) theorem proving software, which in turn is a descendant of Logic for Computable Functions (LCF). LCF was developed in 1972, HOL became stable around 1988, and development of Isabelle started in the 1990s [20]. Isabelle is based on a small core set of logical principles from which theories can be built up. As such, the confidence with which we can claim any theorem proven in Isabelle to be true is the same confidence with which we can claim that the small core is true.

Isabelle provides a formal language to work in, and a set of proof methods, which allow it to prove statements using logical rules, definitions, and axioms, as well as already proved statements. Proofs in Isabelle are essentially natural deduction style. A structured proof language, Isar, is provided which aims to make proofs more human readable, and which serves to greatly reduce the learning curve required to use Isabelle. In addition to the standard proof methods, Isabelle has a feature called *sledgehammer*, which calls external automated theorem provers in an attempt to prove the current goal. Isabelle is developed jointly at the University of Cambridge, Technische Universität München and Université Paris-Sud [5].

An Isabelle proof proceeds in an interactive manner. The user makes a claim, and must then prove it. Isabelle's output (separate to the source code which the user writes) gives information like the goals currently needing to be proved and whether any redundancy in proofs has been detected. Using jEdit, the user can select a line of a completed proof, and check the output to see what was being done at that point – this "hook" into the proof allows for easier understanding of new proofs. Note that the Isabelle output is not included with the formal proof which the user ends up with. Since it is not exported to documentation, we have used an image of the output below.

4.2   Partial Recursive Functions in Isabelle

The formalisation of partial recursive functions in Isabelle given by Xu et al. in [34] makes use of a datatype *recf* of recursive functions. The constructors for this datatype follow standard conventions for partial recursive functions, [29]. Adapting the original Isabelle code definitions to more standard notation and using the notation $\boldsymbol{x}_{i;j}$ to mean $x_i, x_{i+1}, \ldots, x_j$ we have:

$$
\begin{aligned}
z(\boldsymbol{x}) &= 0, \\
s(\boldsymbol{x}) &= x_0 + 1, \\
id_n^m(\boldsymbol{x}) &= x_n, \\
Cn^n(f,g)(\boldsymbol{x}) &= f(g_0(\boldsymbol{x}), \ldots, g_m(\boldsymbol{x})), \\
Pr^n(f,g)(\boldsymbol{x}) &= \begin{cases} f(\boldsymbol{x}_{0;n-2}), & \text{if } x_{n-1} = 0, \\ g(\boldsymbol{x}_{0;n-2}, x_{n-1} - 1, Pr^n(f,g)(\boldsymbol{x}_{0;n-2}, x_{n-1} - 1)), & \text{otherwise}, \end{cases} \\
Mn^n(f)(\boldsymbol{x}) &= \mu\{y \mid f(\boldsymbol{x}_{0;n-1}, y) = 0\}.
\end{aligned}
\tag{1}
$$

Separately, the termination for partial recursive functions is defined in [34] as follows (notice that the clauses for z and s implicitly require a 1-ary list).

$$
termi\ z([n])
$$
$$
termi\ s([n])
$$
$$
(n < m \wedge |\boldsymbol{x}| = m) \rightarrow termi\ id_n^m(\boldsymbol{x})
$$
$$
termi\ f(g_0(\boldsymbol{x}), \ldots, g_m(\boldsymbol{x})) \wedge \forall i\ termi\ g_i(\boldsymbol{x}) \wedge |\boldsymbol{x}| = n \rightarrow termi\ Cn^n(f,g)(\boldsymbol{x})
$$
$$
\forall y < x_{n-1}\ termi\ g(\boldsymbol{x}_{0;n-2}, y, Pr^n(f,g)(\boldsymbol{x}_{0;n-2}, y)) \wedge
$$
$$
termi\ f(\boldsymbol{x}_{0;n-2}) \wedge |\boldsymbol{x}| = n + 1 \rightarrow termi\ Pr^n(f,g)(\boldsymbol{x})
$$
$$
|\boldsymbol{x}| = n \wedge termi\ f(\boldsymbol{x}, r) \wedge f(\boldsymbol{x}, r) = 0 \wedge
$$
$$
\forall i < r\ termi\ f(\boldsymbol{x}, i) \wedge f(\boldsymbol{x}, i) > 0 \rightarrow termi\ Mn^n(f)(\boldsymbol{x})
$$

To see how these definitions work, let us construct an implementation of the addition function $+$ within the framework [34]. We will take this opportunity to demonstrate how proofs proceed in the Isabelle system (the rest of this section has been generated from Isabelle code).

Addition is fairly easy to define using primitive recursion. We simply follow the Robinson Arithmetic approach of defining

$$
x + y = \begin{cases} x, & \text{if } y = 0, \\ (s(x)) + z, & \text{if } y = s(z). \end{cases}
$$

Using primitive recursion, we have $+ := Pr^1(id_0^1, Cn^3(s, [id_2^3]))$. We note that we use indices starting at 0. This is very similar to the Isabelle source code:

definition `"rec_add = (Pr 1 (id 1 0) (Cn 3 s [(id 3 2)]))"`

Within Isabelle, we can prove that `rec_add` is indeed the addition function +. The following lemma achieves this through an induction on the second argument.

lemma `[simp]` : `"rec_exec rec_add [m, n] = m + n"`
apply`(induction n)`
by`(simp_all add:rec_add_def)`

First we have stated the statement of the lemma. The command `rec_exec` tells Isabelle to evaluate the function `rec_add` (that is, we are using Definition (1) from above). Note that since they are unbound, there is an implicit universal quantification over the variables m, n. We have flagged this lemma as a simplification (`[simp]`), which tells Isabelle's *simp* proof method that whenever it sees `rec_exec rec_add [m, n]` it can be replaced by `m+n`. Our first step in the proof is to apply induction to argument n. Isabelle determines that n is a natural number (since the recursive functions are defined over them) and so adopts the appropriate inductive hypothesis. As such there are two goals to prove. The first is that adding 0 to m returns m, and the second is that if addition is correct for $m + n$ it is also correct for $m + (n + 1)$. At this point, the Isabelle output gives the information shown in Fig. 1. Both subgoals can be proved easily by unpacking the definition of `rec_add`. Our final command is to apply the *simp* proof method to all remaining subgoals, making use of the definition of `rec_add`. The *simp* method utilises a large number of built in simplification rules, as well as those rules added to it by `[simp]` flags in an attempt to prove the current goal(s). Here it succeeds. Both commands *by* and *apply* apply the proof methods indicated; the difference is that *by* tells Isabelle we have finished the proof – it is a streamlining of an *apply* command followed by the *qed* end-of-proof command.

```
proof (prove): step 1

goal (2 subgoals):
 1. rec_exec rec_add [m, 0] = m + 0
 2. ∧n. rec_exec rec_add [m, n] = m + n ⟹ rec_exec rec_add [m, Suc n] = m + Suc n
```

Fig. 1. The Isabelle output after applying the *induction* proof method.

Next, we will show that `rec_add` terminates on all inputs. This proceeds in a more complex fashion. First, we establish that the unpacked definition terminates (which requires a complete sub-proof), and then apply the definition to show that `rec_add` terminates.

```
lemma [simp] : "terminate rec_add [m, n]"
proof -
have "terminate (Pr 1 (id 1 0) (Cn 3 s [(id 3 2)])) ([m]@[n])"
  proof
  show "terminate (id 1 0) [m]" by (simp add: termi_id)
  show "length [m] = 1" by simp
  {fix y assume "y < n"

    have "terminate (Cn 3 s [id 3 2]) [m, y, rec_exec (Pr 1 (id 1 0) (Cn
3 s [id 3 2])) [m, y]]"
    proof
      show "length [m, y, rec_exec (Pr 1 (id 1 0) (Cn 3 s [id 3 2])) [m,
y]] = 3" by simp
      have "terminate (id 3 2) [m, y, rec_exec (Pr 1 (id 1 0) (Cn 3 s [id
3 2])) [m, y]]"
        by (simp add: termi_id)
      thus "∀g∈set [id 3 2]. terminate g [m, y, rec_exec (Pr 1 (id 1 0)
(Cn 3 s [id 3 2])) [m, y]]"
        by simp
      show "terminate s (map (λg. rec_exec g [m, y, rec_exec (Pr 1 (id 1
0) (Cn 3 s [id 3 2])) [m, y]]) [id 3 2])"
        by (simp add: termi_s)
    qed
  }
  hence "∀ y < n. terminate (Cn 3 s [(id 3 2)])
          ([m, y, rec_exec (Pr 1 (id 1 0) (Cn 3 s [(id 3 2)])) [m, y]])"
by blast
  thus "∀y<n. terminate (Cn 3 s [recf.id 3 2])
          ([m] @ [y, rec_exec (Pr 1 (recf.id 1 0) (Cn 3 s [recf.id 3
2])) ([m] @ [y])])" by simp
  qed
thus ?thesis by (simp add: rec_add_def)
qed
```

This proof uses a different style to the previous proof – the Isar mark up language. Isar is designed to reflect the style of informal proofs, and is intended to be fairly human readable. Commands such as *hence*, *show* and *thus* have strict meanings within the system, which are similar to their natural language meanings.

In the first step of the proof, we claim that $Pr^1(id_0^1, Cn^3(s, [id_2^3]))$ (our addition function) halts on the arbitrary inputs m, n. This is established inside the sub-proof. There we first show two simple facts – that the identity function terminates on $[m]$ and that $[m]$ is a list of length 1. Next, we must establish the following goal:

$$\forall y < n(terminate \ Cn^3(s, [id_2^3])([m, y, Pr^1(id_0^1, Cn^3(s, [id_2^3]))([m, y])))).$$

This establishes termination for the recursive cases of the addition function. We prove this statement by fixing an arbitrary $y < n$ and showing that it is true for that y. This requires another sub-proof, this time for the termination of

$Cn^3(s, [id_2^3])$. There are three goals to achieve: First that the correct number of arguments is supplied (i.e. the list is of length 3), second that id_2^3 terminates on a list of length 3 and third that the successor function terminates on a list of length 1. Each is a straightforward unpacking of definitions, achieved by *simp*.

Having established that $Pr^1(id_0^1, Cn^3(s, [id_2^3]))$ terminates on arbitrary inputs m, n, we show our *thesis* (namely, that `rec_add` terminates on arbitrary m, n) by applying the definition of `rec_add`. The command *qed* indicates the end of a (successful) proof.

5   Formalising the Halting Problem and Its Undecidability

In [34] a detailed mechanised proof of the undecidability of the halting problem, including proofs of correctness for all programs used, is given. The proof uses the Turing machine model of computation and follows, in broad strokes, the classical proof. The assumption is made of the existence of a Turing program H which can solve the halting problem. Specifically, given an encoding $\langle M, n \rangle$ of a Turing machine M and input n, H outputs 0 if M halts on n and 1 otherwise.

The following modification D of H is then constructed: $D\langle M \rangle = \infty$ if $M\langle M \rangle$ halts, and $D\langle M \rangle = 1$, otherwise, and a contradiction is reached by computing: $D\langle D \rangle = \infty$ iff $D\langle D \rangle$ halts.

Any formalisation requires a number of aspects in the proof to be made explicit. For example, the modification D must be constructed, and the changes to H shown to be correct. Furthermore, explicit notions of halting and correctness must be defined. Since proofs are computer checked, special care must be taken with the implementation of halting.

In what follows we give an overview of the formal proof provided in [34] of the undecidability of the halting problem. The formalisation of Turing machines uses a two-way infinite single tape Turing machine in which tape cells can be in one of two states – blank or occupied. The tape is represented by a pair (l, r) of lists, with l representing the cells to the left of the read/write head and r the cell being read and those to the right of it. Five actions are available; write blank, write occupied, move left, move right and do nothing. A Turing program is simply a list of pairs of actions and natural numbers representing states – the order of the pairs encodes which instruction maps to which state and input. An example program from [34] follows:

$$dither := [\overbrace{(W_{Bk}, 1)}^{\text{read } Bk}, \overbrace{(R, 2)}^{\text{read } Oc}, \underbrace{(L, 1), (L, 0)}_{\text{state 2}}]$$
$$\underbrace{\phantom{[(W_{Bk}, 1), (R, 2)}}_{\text{state 1 (start)}}$$

The program begins in state 1. If it reads a blank cell, it writes a blank cell and goes to state 1. If it reads an occupied cell, it moves right and goes to state 2. In state 2, it moves left, returning to state one if it saw a blank cell and going to state 0 (the halting state) if it saw an occupied cell. Hence this program halts on a tape containing two occupied cells, and loops indefinitely on any tape with fewer such cells.

In order to ease construction of programs, a sequential composition of Turing programs is introduced. Essentially, this modifies the programs by increasing the state numbers of any subsequent programs and changing the halting state to the start state of each next program. This composition is used to combine three Turing programs: a copy program (to copy a machine's code so it can read it), the supposed program to solve the halting problem (H above) and a program to loop infinitely in certain cases (the "dither" program). This result is the machine D from above. *Both the copying program and dither must be proved correct.* We will outline how this proceeds for dither.

Correctness of a Turing program is established through the use of Hoare triples. Essentially, the triple $\{P\}p\{Q\}$ indicates that program p run on a tape satisfying P will result in a tape satisfying Q. We can also write $\{P\}p \uparrow$ to indicate that p run on a tape satisfying P will never halt. The program dither should satisfy the triples

$$\{\lambda tp.\exists k.tp = (Bk^k, \langle 1\rangle)\} \; dither \; \{\lambda tp.\exists k.tp = (Bk^k, \langle 1\rangle)\}$$
$$\{\lambda tp.\exists k.tp = (Bk^k, \langle 0\rangle)\} \; dither \; \uparrow$$

if it is to match the description above. The first statement can be established in Isabelle by calculation; provided a tape matching the first condition, run the dither program and see what happens. Due to the design of the implementation, this is straightforward and very easily automated. A proof of the second statement clearly cannot proceed in the same manner – *running dither on such a tape should result in an infinite loop*, a phenomenon which will affect any Isabelle simulation of the machine. Instead, the second statement can be established by an induction on the number of steps performed, starting with the given input tape.

Having established the correctness of copy and dither, one can proceed to prove the undecidability of the halting problem, following the standard method. Here a definition of the halting problem must be introduced. The property of a Turing machine p halting on an input n is defined using Hoare triples as follows:[5]

$$halts \; p \; n := \{\lambda tp.tp = ([], \langle n\rangle)\} \; p \; \{\lambda tp.\exists k, m, l.tp = (Bk^k, \langle m\rangle @Bk^l)\}.$$

We then assume that a machine H exists which solves the halting problem. Formally within Isabelle this is captured by the following Hoare triples:

$$halts \; M \; n \rightarrow \{\lambda tp.tp = ([Bk], \langle(\langle M\rangle, n)\rangle)\} \; H \; \{\lambda tp.\exists k.tp = (Bk^k, \langle 0\rangle)\}$$
$$\neg halts \; M \; n \rightarrow \{\lambda tp.tp = ([Bk], \langle(\langle M\rangle, n)\rangle)\} \; H \; \{\lambda tp.\exists k.tp = (Bk^k, \langle 1\rangle)\}.$$

Then we define the diagonalising Turing machine *contra* by

$$contra := copy; H; dither$$

---

[5] In Isabelle, the @ symbol indicates concatenation of lists. Also note that this definition of *halts* assumes functions with some number of inputs and a single output.

where; indicates the sequential composition of the programs. The contradiction is now reached through reasoning established from the proofs of correctness for copy and dither, and the Hoare triple assumptions for H.

6   Correctness vs. Termination in Isabelle

In this section, we discuss some relations between the undecidable properties of correctness and termination of functions in Isabelle. As a formal proof-assistant, Isabelle is charged with being able to prove both these properties (or their negations) for arbitrary functions. A partially computable function is correct if it gives the expected (with respect to some specifications) output on every input. Correctness is a *relative notion* – a function may be syntactically fine, but if it was intended to do multiplication and actually does division, it is not correct. In contrast, the notion of termination of a function is *absolute*. The evaluation of a partially computable function f terminates on input x is equivalent to the mathematical property of f being defined on x. The two terminologies reflect the dual origins of computability theory: partially recursive functions are exactly the partially computable functions, i.e. the partial functions computed by Turing machines. Correctness is more undecidable than termination, see [16].

Proofs of correctness and termination form a critical part of computability theory and any formalisation of computability theory theorems cannot avoid such proofs. For this reason, any implementation of partial recursive functions within Isabelle needs to be able to handle correctness and termination. Clearly, as we have discussed above, this presents a challenge. All results discussed in this section have been generated from within the Isabelle system.

Suppose we have a unary partially computable function f (x and n are naturals) and define the following partial function:

$$g(x, n) = \begin{cases} f(x), & \text{if } n > 0, \\ 0, & \text{otherwise.} \end{cases}$$

Mathematically, for $n > 0$, $g(x, n)$ is defined if and only if $f(x)$ is defined. Since g is defined for all values of x when $n = 0$, we have $\text{dom}(g) = \{(x, 0) \mid x \in \mathbb{N}\} \cup \{(x, n + 1) \mid x \in \text{dom}(f) \wedge n \in \mathbb{N}\}$.

In practice, it is possible that n is the output of some other function which tests properties of x. For example, we might take $n = h(x)$. The resulting function $t(x) = g(x, h(x))$ has a domain which requires testing of h: $\text{dom}(t) = \{x \in \mathbb{N} \mid x \in \text{dom}(h) \wedge (h(x) > 0 \rightarrow x \in \text{dom}(f))\}$.

If h is total and has the property that $h(x) > 0 \rightarrow x \in \text{dom}(f)$ then this gives a computable restriction of f; in those cases where $f(x)$ is undefined (and possibly in some other cases), $g(x, h(x)) = 0$ since $h(x) = 0$. This can be useful for working with f without worrying about incomputability. Of course, it may be prudent to assume that $f(x) \neq 0$ for all x, so that we can identify when a potentially incomputable argument has been supplied.

Let us consider how g could be constructed in Isabelle, using the implementation of partial recursive functions from [34]. For any definition of g we should be able (in principle) to establish three facts. First, that it meets the definition of g (and thereby is *correct*). Second, specifically that it returns 0 when $n = 0$. Third, that in the case $n = 0$ it terminates. This third requirement may seem superfluous at first – if the function returns 0 then surely it terminates – but as we will see soon, the definitions for recursive functions in this interpretation do not always result in termination behaving as expected.

Arguably the most obvious implementation of g is primitive recursion on n. Indeed, if we define g this way we are able to establish all three requirements in Isabelle with no difficulty. Instead, let us consider a function which we can prove "correct" in some sense, but which does not terminate.

Take `rec_times` to be a computable function for multiplication, and `rec_sg` to be the *signature* function: $signature(0) = 0$, $signature(x) = 1$ if $x > 0$.

These functions are defined as recursive functions using the implementation from [34]. The following lemmata show that `rec_times` and are correctly defined.

```
lemma "rec_exec rec_times [x,y] = x*y" using rec_times_def by simp
lemma "rec_exec rec_sg [x] = (if x > 0 then 1 else 0)" using rec_sg_def
by simp
```

Here we notice an interesting interplay between object and meta languages. The left-hand side of each lemma references a formally defined recursive function, using [34]'s implementation in Isabelle. For example, `rec_exec rec_times [x,y] = x*y` evaluates the formal function `rec_times` on input `[x,y]`. The right-hand side utilises built-in Isabelle functions, such as multiplication `*`, `if then` statements and greater-than `>`. These can be seen as meta-language operations, with `rec_times` and `rec_sg` in the object language. An interesting observation is that the Isabelle language is meta *with respect to these functions*. However, as part of a formal system, it would generally be regarded as the object language. The two lemmata show correctness of the functions, which follows from their definitions (suppressed for clarity) using the `simp` proof method.

Now consider the following function intended to implement g.

```
definition "g2 F = Cn 2 rec_times [Cn 2 F [id 2 0], Cn 2 rec_sg [id 2 1]]"
```

In this definition, we simply multiply $F(x)$ by $signature(n)$. If $n > 0$ then the answer will be $F(x)$ and if $n = 0$ the answer will be $F(x) \times 0$. In mathematical notation, we have

$$g_2(F) := Cn^2(\times, [Cn^2(F, [id_0^2]), Cn^2(signature, [id_1^2])]).$$

Expanding out the compositions, we have simply $g_2(F)(x,n) = F(x) \times signature(n)$. Indeed, we can establish this in Isabelle by unpacking the definitions.

```
lemma "rec_exec (g2 F) [x, n] = (rec_exec F [x])*(rec_exec rec_sg [n])"
by(simp add:g2_def)
```

The reader familiar with partial recursive functions, however, should have noted an important problem with g_2, if it is to implement g. As defined, $g(F)(x, 0) = 0$, regardless of whether or not $F(x)$ is defined. However $F(x) \times 0 = 0$ only if $F(x)$ is defined. Specifically, for this construction we have $\mathrm{dom}(g_2) = \{(x, n) \mid x \in \mathrm{dom}(F)\}$, which differs from the domain for g. Hence g_2 does not implement g. Worryingly, we can still prove the following lemma in Isabelle.

lemma "rec_exec (g2 F) [x, n] = (if n>0 then rec_exec F [x] else 0)"
by (simp add: g2_def)

Once again this lemma follows by a simple unpacking of the definition. Notice that the else condition does not depend upon $F(x)$ being defined. According to this lemma, if $n = 0$ then $g_2(F)(x, 0) = 0$ for arbitrary F, x. Indeed, we can be more specific.

lemma "rec_exec (g2 F) [x, 0] = 0"
  by (simp add:g2_def)

Does this then mean that this implementation of partial recursive functions in Isabelle is flawed? Arguably, yes. We have been able to show an incorrect lemma, or at least a lemma which is incorrect given the natural understanding of the rec_exec command. However we have not yet shown all the three facts needed to be established. And it is with termination that we (as might be expected) encounter problems.

lemma "terminate (g2 F) [x, 0]"
apply (simp add:g2_def)
proof
show "length [x, 0] = 2" **by** simp
show "terminate rec_times (map (λg. rec_exec g [x, 0]) [Cn 2 F [id 2 0], Cn 2 rec_sg [id 2 (Suc 0)]])" **by** simp
show "∀g∈set [Cn 2 F [id 2 0], Cn 2 rec_sg [id 2 (Suc 0)]]. terminate g [x, 0]"
  proof –
  have "terminate (Cn 2 rec_sg [id 2 (Suc 0)]) [x, 0]" **using** termi_id termi_cn **by** simp
    moreover have "terminate (Cn 2 F [id 2 0]) [x, 0]"
      proof
      show "length [x, 0] = 2" **by** simp
      show "∀g∈set [id 2 0]. terminate g [x, 0]" **using** termi_id **by** simp
      show "terminate F (map (λg. rec_exec g [x, 0]) [id 2 0])" **sorry**
      qed
    ultimately show ?thesis **by** simp
  qed
oops

First we establish that the length of input is correct. We second show that rec_times terminates on the required inputs. Decoded, this second statement seems to be of the form $terminate\ F(x) \times signature(n)$. In fact it is slightly more subtle. We are asked to show that `rec_times` terminates on $F(x)$ and $signature(0)$, but an inherent assumption in the Isabelle system is that these are both defined natural numbers; that the inputs are in a correct format. Since `rec_times` is a total function, we are able to prove this statement.

Finally, we are required to establish that both $Cn^2(F, [id_0^2])(x, 0)$ and $Cn^2(signature, [id_1^2])(x, 0)$ terminate. The second claim is simple; $Cn^2(signature, [id_1^2])(x, 0)$ is a primitive recursive function and so will terminate – an unpacking of definitions will establish this. The first claim is impossible to establish however. We have $Cn^2(F, [id_0^2](x, 0) = F(x)$ and so to establish that $Cn^2(F, [id_0^2](x, 0)$ terminates we must establish first that $F(x)$ terminates (that is, $F(x)$ is defined).

Since F is arbitrary, we cannot establish termination for g_2 on $(x, 0)$. Hence g_2 does not implement g correctly. The problem of incorrectly establishing the correctness of g_2 can then be explained by requiring that correctness should include termination. The implementation of partial recursive functions by [34] has split evaluation of functions from their termination as a way to overcome termination issues within Isabelle. This has come at the cost of clarity in the implementation, and a departure from the standard definition of partial recursive functions, where termination and evaluation are inextricably linked.

6.1   Reuniting Evaluation and Termination

In standard definitions of partial recursive functions, evaluation of the function is explicitly linked to the termination of any functions involved. For example, in [12] the first mention of each function type specifies its domain. Functions are built up recursively, and, for example, a function θ obtained through composition $\theta(x_1, \ldots, x_n) = \psi(\phi_1(x_1, \ldots, x_n), \ldots, \phi_m(x_1, \ldots, x_m))$ has the domain defined as $\text{dom}(\theta) = \{(x_1, \ldots, x_n) \in \mathbb{N}^n \mid (x_1, \ldots, x_n) \in \bigcap_{i=1}^m \text{dom}(\phi_i) \text{ and } (\phi_1(x_1, \ldots, x_n), \ldots, \phi_m(x_1, \ldots, x_n)) \in \text{dom}(\psi)\}$.

This explicit mentioning of domain contrasts with [34]'s implementation. In their implementation, the function executions are defined as having $\mathbb{N}^* = \mathbb{N} \cup \mathbb{N}^2 \cup \mathbb{N}^3 \cup \ldots$ as their domain. That is, Isabelle will happily (attempt to) evaluate a function on any input, regardless of whether it is in the domain of that function. Restrictions to domain, and to correct arity of arguments, are implemented entirely within the termination definitions.[6]

Partial recursive functions have domains built into them directly. Divorcing domains from the function definitions – motivated by the wish to increase understandability – is not a correct solution. Creating an implementation of partial recursive functions in Isabelle in which termination and evaluation are presented at once would be difficult. Isabelle requires proofs of termination for certain functions, which is likely part of the reason [34] decided to split the definitions. Proofs involving combined definitions are likely to be much messier

---

[6] Of course it should be noted that if an input is not within that domain of a function, Isabelle's attempt to evaluate is likely not to terminate. However, consequential strange behaviours can be observed, such as in g_2.

than the current model, since domains must explicitly be dealt with. From a formal perspective, the separation allows for separate proofs, which are more easily digested. However, it would be enough for a combined model to establish equivalence with the [34] model. If we could implement partial recursive functions in Isabelle, using a model defined as closely to a standard pen-and-paper definition (such as that provided by [12]) as possible, we would have greater confidence that model adequately represents the mathematical notion of partial recursive function. Subsequently establishing the equivalence of this model with that in [34] would allow the "importing" of results proved by [34]. This solution would mean greater ease of formal proofs from the split model while maintaining connection to the original model of partial recursive functions.

A tempting diversion in implementing partial recursive functions is ensuring they can be evaluated by the proof system. This would mean the proof system acting as an interpreter, and actually running the programs specified by the functions. When we evaluate `rec_add`, actual recursive calls are made to find the result.

This would be a very interesting approach – enlisting a modern computer to simulate a decades old model of computation. However, from a standpoint of formal proof, it is unnecessary. Proofs involving partial recursive functions at most require unpacking general definitions – explicit evaluation of functions is rarely required. That is, while evaluating addition through recursive calls may be fun, it is highly unlikely that any proofs will require it; since proofs generally deal with the abstract, we are less concerned with what $1 + 2$ is and more with how $x + y$ works. Due to this, it would be acceptable for an implementation of partial recursive functions to combine "evaluation" and termination at the expense of the system actually being able to evaluate the functions.

7   'Symptoms' of Undecidability in Isabelle

The problems in [34]'s implementation give one example of how undecidability impacts Isabelle. The careful nature in which the model is constructed, and the split of evaluation from termination are direct consequences of the undecidability of the halting problem. The *sledgehammer* feature, which searches for proofs to given claims, has a time restriction built in, again to combat undecidability.

Isabelle is a programming language, so its programs may terminate or not. When dealing with models of computation, what happens when Isabelle attempts to simulate such programs?

Isabelle has the ability to evaluate functions within the system. The user can type `value "1+2"` and Isabelle's output will display the answer. For basic functions this acts as a calculator, and for functions defined in Isabelle it can be used to ensure they behave as expected. It is interesting to see how Isabelle handles non-terminating computations, since identifying them is undecidable.

Consider the partial recursive function defined by $Mn^1(+)(x) = \mu\{y \mid x+y = 0\}$. It is obvious that $Mn^1(+)(0) = 0$ and $Mn^1(+)(x) = \infty$ for $x > 0$.

How then will Isabelle evaluate `value "rec_exec (Mn 1 rec_add) [1]"`? In fact Isabelle *refuses to try*, throwing instead a well-sortedness error. This makes sense. The `Mn` function requires finding the least element of a possibly empty set. Since Isabelle has no guarantee the set is non-empty, it refuses to evaluate.

Isabelle's cautious nature comes at a cost, failing to evaluate *any* recursive function. The addition function is a recursive function and its definition does not use minimisation. This puts it into the class of computable functions. If Isabelle were to attempt to calculate `value "rec_exec rec_add [1, 1]"`, it would succeed. However, Isabelle again *refuses to try*, throwing the same well-sortedness error. Even a proof of general termination for `rec_add` does not help. Isabelle notices that the `rec_exec` definition incorporates a minimisation clause, and so refuses to have anything to do with evaluation. It even refuses to attempt evaluation of `value "rec_exec z [0]"`.

Yet all is not lost. We can still prove `rec_exec rec_add [1, 1] = 2`. In fact, this is almost trivial, since we have proved already the general statement that `rec_exec rec_add [m, n] = m + n`, and $m + n$ (the Isabelle function) can be calculated by Isabelle. This leads to an interesting contrast: *Isabelle will not attempt to evaluate partial recursive functions, but is happy to attempt to prove a claim made by the user.* While both operations involve skirting close to undefined functions, in proofs Isabelle can offload much responsibility to the user. For our original addition minimisation function $Mn^1(+)$, we can obviously not prove an output for any input other than 0, but `rec_exec (Mn 1 rec_add) [0] = 0` is provable in Isabelle.

8   Concluding Remarks

The nature of computability makes automated proofs a particularly interesting area. The landmark results of Gödel and Turing still loom large. Where they were discussing hypothetical computation models, we are using equivalent models to prove their own limitations. Yet we are still able to carve out larger sections of what can be achieved. To avoid the inherent complications, novel approaches need to be adopted, and the original proofs modified to achieve the required goals, given the abilities of modern proof-assistants (see, for example, [22]).

The formalisation of recursive functions we have considered is a good example of this. The model, though very similar, is not the traditional partial recursive functions as termination and evaluation have been separated. It would be nice if a combined model of partial recursive functions could be shown, within Isabelle, to be equivalent to the model provided in [34].

Great progress has been made in both formal proving and developing computability theory. Unexpectedly, the use of proof-assistants brings new connections between incompleteness and undecidability into sharp focus, so formal proving can contribute to semantics too. We expect that the role of proof-assistants for the working mathematician will steadily increase.

References

1. Alf homepage. http://homepages.inf.ed.ac.uk/wadler/realworld/alf.html. Accessed 25 Oct 2014
2. Archive of formal proofs. http://afp.sourceforge.net. Accessed 18 May 2016
3. Coq homepage. http://coq.inria.fr/. Accessed 25 Oct 2014
4. HOL4 homepage. http://hol.sourceforge.net/. Accessed 25 Oct 2014
5. Isabelle homepage. http://isabelle.in.tum.de/. Accessed 20 Oct 2014
6. Matita hompage. http://matita.cs.unibo.it/. Accessed 25 Oct 2014
7. Flyspeck project, September 2014. http://aperiodical.com/2014/09/the-flyspeck-project-is-complete-we-know-how-to-stack-balls
8. Asperti, A., Ricciotti, W.: Formalizing turing machines. In: Ong, L., de Queiroz, R. (eds.) WoLLIC 2012. LNCS, vol. 7456, pp. 1–25. Springer, Heidelberg (2012)
9. Benzmüller, C., Woltzenlogel Paleo, B.: Automating Gödel's ontological proof of God's existence with higher-order automated theorem provers. In: Schaub, T., Friedrich, G., O'Sullivan, B. (eds.) ECAI 2014, Frontiers in Artificial Intelligence and Applications, vol. 263, pp. 93–98. IOS Press (2014)
10. Du Bois-Reymond, E.H.: Über die Grenzen des Naturerkennens; Die sieben Welträthsel, zwei Vorträge. Von Veit, Leipzig (1898)
11. Bourbaki, N.: Theory of Sets. Elements of Mathematics. Springer, Heidelberg (1968)
12. Calude, C.: Theories of Computational Complexity, North-Holland, Amsterdam (1988)
13. Calude, C.S., Calude, E., Marcus, S.: Passages of proof. Bull. Eur. Assoc. Theor. Comput. Sci. **84**, 167–188 (2004)
14. Calude, C.S., Hay, N.J.: Every computably enumerable random real is provably computably enumerable random. Logic J. IGPL **17**, 325–350 (2009)
15. Calude, C.S., Müller, C.: Formal proof: reconciling correctness and understanding. In: Carette, J., Dixon, L., Coen, C.S., Watt, S.M. (eds.) MKM 2009, Held as Part of CICM 2009. LNCS, vol. 5625, pp. 217–232. Springer, Heidelberg (2009)
16. Cooper, S.B.: Computability Theory. Chapman Hall/CRC, London (2004)
17. Edwards, C.: Automated proofs. Math struggles with the usability of formal proofs. Commun. ACM **59**(4), 13–15 (2016)
18. Feferman, S.: Are there absolutely unsolvable problems? Gödel's dichotomy. Philosophia Math. **14**(2), 134–152 (2006)
19. Gödel, K.: Some basic theorems on the foundations of mathematics and their implications. In: Feferman, S., Dawson Jr., J.W., Goldfarb, W., Parsons, C., Solovay, R.M. (eds.) Collected Works. Unpublished Essays and Lectures. vol. III, pp. 304–323. Oxford University Press (1995)
20. Gordon, M.: From LCF to HOL: a short history. In: Proof, Language, and Interaction, pp. 169–186 (2000)
21. Hales, T.C.: A proof of the Kepler conjecture. Ann. Math. **162**(3), 1065–1185 (2005)
22. Hernández-Orozco, S., Hernández-Quiroz, F., Zenil, H., Sieg, W.: Rare speed-up in automatic theorem proving reveals tradeoff between computational time and information value (2015). http://arxiv.org/abs/1506.04349
23. Heule, M.J.H., Kullmann, O., Marek, V.W.: Solving and verifying the boolean Pythagorean triples problem via cube-and-conquer (2016). http://arxiv.org/abs/1605.00723v1 [cs.DM]

24. Hilbert, D.: Hilbert's 1930 radio speech. https://www.youtube.com/watch?v=EbgAu_X2mm4
25. Kleene, S.C.: Introduction to Metamathematics. North-Holland, Amsterdam (1952)
26. Konev, B., Lisitsa, A.: A SAT attack on the Erdös discrepancy conjecture (2014). http://arxiv.org/abs/1402.2184v2
27. Martin-Löf, P.: Verification then and now. In: De Pauli-Schimanovich, W., Koehler, E., Stadler, F. (eds.) The Foundational Debate, Complexity and Constructivity in Mathematics and Physics, pp. 187–196. Kluwer, Dordrecht (1995)
28. Norrish, M.: Mechanised computability theory. In: van Eekelen, M., Geuvers, H., Schmaltz, J., Wiedijk, F. (eds.) ITP 2011. LNCS, vol. 6898, pp. 297–311. Springer, Heidelberg (2011)
29. Soare, R.I.: Recursively Enumerable Sets and Degrees: A Study of Computable Functions and Computably Generated Sets. Springer, Heidelberg (1987)
30. Szasz, N.: A machine checked proof that Ackermann's function is not primitive recursive. In: Huet, G. (ed.) Logical Environments, pp. 31–7. University Press (1991)
31. Tao, T.: The Erdös discrepancy problem (2015). http://arxiv.org/abs/1509.05363v5
32. Tarski, A.: A Decision Method for Elementary Algebra and Geometry. University of California Press, Berkeley and Los Angeles (1951)
33. Thompson, D.: Formalisation vs. understanding. In: Calude, C.S., Dinneen, M.J. (eds.) UCNC 2015. LNCS, vol. 9252, pp. 290–300. Springer, Heidelberg (2015)
34. Xu, J., Zhang, X., Urban, C.: Mechanising turing machines and computability theory in Isabelle/HOL. In: Blazy, S., Paulin-Mohring, C., Pichardie, D. (eds.) ITP 2013. LNCS, vol. 7998, pp. 147–162. Springer, Heidelberg (2013)
35. Zammit, V.: A mechanisation of computability theory in HOL. In: von Wright, J., Harrison, J., Grundy, J. (eds.) TPHOLs 1996. LNCS, vol. 1125. Springer, Heidelberg (1996)
36. Zammit, V.: On the Readability of Machine Checkable Formal Proofs. Ph.D. Thesis, University of Kent, March 1999

A Numerical Method for Computing Border Curves of Bi-parametric Real Polynomial Systems and Applications

Changbo Chen and Wenyuan Wu[✉]

Chongqing Key Laboratory of Automated Reasoning and Cognition,
Chongqing Institute of Green and Intelligent Technology,
Chinese Academy of Sciences, Beijing, China
changbo.chen@hotmail.com, wuwenyuan@cigit.ac.cn

Abstract. For a bi-parametric real polynomial system with parameter values restricted to a finite rectangular region, under certain assumptions, we introduce the notion of border curve. We propose a numerical method to compute the border curve, and provide a numerical error estimation.

The border curve enables us to construct a so-called "solution map". For a given value u of the parameters inside the rectangle but not on the border, the solution map tells the subset that u belongs to together with a connected path from the corresponding sample point w to u. Consequently, all the real solutions of the system at u (which are isolated) can be obtained by tracking a real homotopy starting from all the real roots at w throughout the path. The effectiveness of the proposed method is illustrated by some examples.

1 Introduction

Parametric polynomial systems arise naturally in many applications, such as robotics [7], stability analysis of biological systems [25], model predictive control [8], etc. In these applications, often it is important to identify different regions in the real parametric space such that the system behaves the "same" in each region.

It is not a surprise that symbolic methods have been the dominant approaches for solving parametric systems due to their ability to describe exactly the structure of the solution sets. The symbolic methods can be classified into two categories, namely the approaches which are primarily interested in finding the solutions in an algebraically closed field (often the complex field) or a real closed field (often the real field). Methods belonging to the first category include the Gröbner basis method [2], the characteristic set or triangular decomposition method [29], the comprehensive Göbner bases method [26], the comprehensive triangular decomposition method [4,5], etc. Methods belonging to the second one include the quantifier elimination method [24], the cylindrical algebraic decomposition method [6], the Sturm-Habicht sequence method [9], the parametric

ⓒ Springer International Publishing AG 2016
V.P. Gerdt et al. (Eds.): CASC 2016, LNCS 9890, pp. 156–171, 2016.
DOI: 10.1007/978-3-319-45641-6_11

geometric resolution method [21], etc. Our method is directly motivated by the border polynomial method [30] and the discriminant variety method [12,18].

Another motivation of our work comes from the recent advances from the numeric community. The homotopy continuation method [13,22], which was initially used to compute all the complex solutions of zero-dimensional polynomial systems, has been developed to study the complex and real witness points of positive dimensional polynomial system [10,22,27], as well as describing the real algebraic curves [16] and surfaces [17]. The Cheater's homotopy [14,22] provides a way to compute the complex solutions of a parametric polynomial system at a particular parameter value u by first computing the solutions of the system at a generic random parametric value u_0 and then using these solutions as starting points for constructing a homotopy. In [15], a real homotopy continuation method for computing the real solutions of zero-dimensional polynomial systems is introduced. In [19], the authors provide a local approach to detect the singularities of a parametric polynomial system along a solution path when sweeping though the parametric space.

At last, we would like to mention that this work is also motivated by an invited talk given by Hoon Hong [11], where the speaker suggested some possible ideas for doing quantifier elimination by a symbolic-numeric approach.

In this paper, we propose a numeric method for computing all the real solutions of a bi-parametric polynomial system F with parameters values restricted in a rectangular region R. Under certain assumptions (see the beginning of Sect. 2), there exists a border curve B in R which divides the rectangle R into finitely many connected components (called cells) such that in each cell the real zero set of F defines finitely many smooth functions with disjoint graphs. We provide a numeric algorithm to compute such a curve and analyze its numerical error. The key idea is to reduce the computation of the border curve to tracing all its corresponding connected components in a higher dimensional space with the help of the critical point techniques [20,27].

To handle the numerical error, we define the notion of δ-connectedness and make use of the connected component of a graph G to represent the cells separated by the border curve. Moreover, in each cell, we choose a sample point far from the border curve and compute all the real solutions of F at this point. All in all, the border curve B, the connectivity graph G, the set of sample points W and the set Z of solutions at these sample points together form a so-called "solution map". Now if one wants to compute the real solutions of F at a given value u of parameters, instead of directly solving $F(u)$, one could easily make use of the solution map to choose the sample point w sharing the same cell with u and construct real homotopies from known results $V_{\mathbb{R}}(F(w))$ to unknown $V_{\mathbb{R}}(F(u))$.

Obviously, provided that the solution map has been computed during an offline phase, it will be very efficient to solve $V_R(F(u))$ online by the real homotopy approach when the number of the real solutions at u is much smaller than the number of complex ones. However, if one's interest is only on computing the real solutions of a zero-dimensional system, it is not recommended to go though the above costly offline computation.

The notion of border curve is inspired by the notions of border polynomial and discriminant variety. However, it applies directly to characterizing the real zero set of a parametric system while the later two also characterize the distinct complex solutions. We use the following example to illustrate the difference of border curve with them.

Example 1. *Consider a parametric system consisting of a single polynomial:*

$$f := (X_1 - U_1)(X_1 - U_2 - 1)(X_1^2 - 2U_1 + U_1^2 + 5)(X_1^2 - 2U_2 + U_2^2 + 5).$$

The border polynomial or discriminant variety of f is defined by the discriminant of f with respect to X_1, which has 9 irreducible factors: $U_1 - U_2, U_1 - 2 + U_2, U_1 - 1 - U_2, U_2^2 + 3, U_1^2 - 2U_1 + 5, 2U_1^2 - 2U_1 + 5, U_2^2 - 2U_2 + 5, U_1^2 + U_2^2 - 2U_2 + 5, U_1^2 + U_2^2 - 2U_1 + 2U_2 + 6$. Among them, only the zero sets of the first three factors are nonempty. They define the real counterpart of the border polynomial or discriminant variety, which characterizes when f has multiple complex roots. In contrast, the border curve is defined only by the polynomial $U_1 - 1 - U_2$, which characterizes when f has a multiple real root with respect to X_1. Indeed, the last two factors of f have no real points.

2 Border Curve

In this section, we introduce the concept of border curve for a bi-parametric polynomial system and present a numeric method to compute it.

Throughout this paper, let $F(X_1, \ldots, X_m, U_1, U_2) = 0$ be a bi-parametric polynomial system consisting of m polynomials with real coefficients. Let $\pi :$ $\mathbb{R}^{m+2} \to \mathbb{R}^2$ be the projection defined by $\pi(x_1, \ldots, x_m, u_1, u_2) = (u_1, u_2)$. Let R be a rectangle of the parametric space (U_1, U_2). We make the following assumptions for F and R:

(A_1) The set $V_\mathbb{R}(F) \cap \pi^{-1}(R)$ is compact.
(A_2) Let $F' := \{F, \det(\mathcal{J}_F)\}$, where $\det(\mathcal{J}_F)$ is the determinant of the Jacobian of F with respect to $(X_1, \ldots, X_m)$. We have $\dim(V_\mathbb{R}(F')) = 1$.
(A_3) At each regular point of $V_\mathbb{R}(F')$, the Jacobian of F' has full rank.

Definition 1. *Given a bi-parametric polynomial system $F(X_1, \ldots, X_m, U_1, U_2) = 0$. Assume that it satisfies the first two assumptions. Then the border curve B of F restricted to the rectangle R is defined as $\pi(V_\mathbb{R}(F, \det(\mathcal{J}_F))) \cap R$.*

Remark 1. *Assumption (A_3) is not needed for this definition. It is required by the subroutine RealWitnessPoint of Algorithm BorderCurve for numerically computing the border. Assumption (A_1) can be checked by symbolic methods [12].*

Proposition 1. *Let B be the border curve of F restricted to the rectangle R. Then $R \setminus B$ is divided into finitely many connected components, called cells, such that in each cell, the real zero set of F defines finitely many smooth functions, whose graphs are disjoint.*

Proof. By Assumption (A_2), the set $R \setminus B$ is non-empty. Moreover it is a semi-algebraic set and thus has finitely many connected components. Let $\mathcal{C}$ be any component and let u be any point of $\mathcal{C}$. Since $u \notin B$, by Definition 1, if the set $\pi^{-1}(u) \cap V_{\mathbb{R}}(F)$ is not empty, the Jacobian $\mathcal{J}_F$ is non-singular at each point of it. Thus the set $\pi^{-1}(u) \cap V_{\mathbb{R}}(F)$ must be finite. On the other hand, by the implicit function theorem, around each neighborhood of these points, $V_{\mathbb{R}}(F)$ uniquely defines a smooth function of (U_1, U_2). By the compactness assumption (A_1) and the connectivity of $\mathcal{C}$, the domain of these functions can be extended to the whole $\mathcal{C}$. For similar reasons, if the set $\pi^{-1}(u) \cap V_{\mathbb{R}}(F)$ is empty, the set $\pi^{-1}(\mathcal{C}) \cap V_{\mathbb{R}}(F)$ must also be empty. Thus the proposition holds.

Next we present a numeric algorithm for computing the border curve. Let RealWitnessPoint be the routine introduced in [27] for computing the witness points of a real variety $V_{\mathbb{R}}(F')$ satisfying Assumption (A_3). The basic idea of this routine is to introduce a random hyperplane H. Then "roughly speaking" the witness points of $V_{\mathbb{R}}(F')$ either belong to $V_{\mathbb{R}}(F') \cap H$ or are the critical points of the distance from the connected components of $V_{\mathbb{R}}(F')$ to H.

Algorithm BorderCurve
Input: a bi-parametric polynomial system $F = 0$; a rectangle R.
Output: an approximation of the border curve of F restricted to R.
Steps:
1. Let F' be the new polynomial system $F \cup \{\det(\mathcal{J}_F)\}$.
2. Set $W := \emptyset$.
3. Compute the intersection of $V_{\mathbb{R}}(F')$ with the fibers of the four edges of R by a homotopy continuation method and add the points into W.
4. Compute RealWitnessPoint(F') and add the points into W.
5. For each point p in W, starting from p, follow both directions of the tangent line of $V_{\mathbb{R}}(F')$ at p, trace the curve F' by a prediction-projection method (see Lemma 2), until a closed curve is found or the projections of the traced points onto (U_1, U_2) hit the boundary of R.
6. Return the projections of the traced points in R.

Remark 2. *In the above algorithm, it is possible that the determinant* $\det(\mathcal{J}_F)$ *is a polynomial of large degree or with large coefficients. For numerical stability, we can instead set* $F' := F \cup \mathcal{J}_F \cdot \mathbf{v} \cup \{\mathbf{n} \cdot \mathbf{v} - 1\}$ *with additional variables* $\mathbf{v} = \{v_1, \ldots, v_m\}$, *where* $\mathbf{n}$ *is a random vector of* $\mathbb{R}^m$ *and* $\mathbf{n} \cdot \mathbf{v} = 1$ *is a random real hyperplane.*

Example 2. *Consider a rectangle* $R := [-3, 3] \times [-3, 3]$ *and a parametric system*

```
F := [X_1^3+(11/6)*X_1^2-(8/15)*X_1*X_2-(8/3)*X_1*U_1-(4/3)*X_1*U_2
-(1/6)*U_2^2-77/30, X_2^3+(7/30)*X_1*U_1+(41/30)*X_1*U_2+(37/15)*X_2*U_2]
```

The discriminant variety of F *is the zero set of an irreducible polynomial of degree 24 with 301 terms.*

-32788673396080447979520000000000000000000*U_1^12*U_2^12
-...+158724170405225359792849346762880000 U_1
+580243205083580617232057253189120000 U_2
-10990793356014795647354316125568000000

Its zero set is the red curve plotted in Fig. 3. Algorithm BorderCurve *computes a space curve in $\mathbb{R}^4$, whose projection onto (U_1, U_2, X_1) is illustrated in Fig. 1. The projection of the space curve onto (U_1, U_2) (namely the border curve B), when drawn in the rectangle R, is "the same" (cannot tell the difference by eyes) as the red curve shown in Fig. 3.*

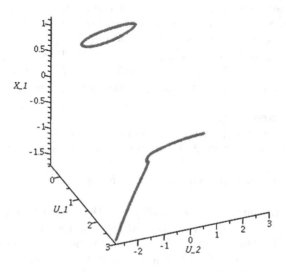

Fig. 1. The projection of the space curve onto (U_1, U_2, X_1).

3   Numerical Error Estimation

In last section, we introduced the concept of border curve and provided an algorithm to compute a numerical approximation of it. In this section, we estimate the numerical error of such an approximated border curve. We first recall several results from [28].

Lemma 1. *Let $P(X_1, \ldots, X_n) = \{P_1, \ldots, P_{n-1}\}$ be a set of $n - 1$ polynomials with n variables. Let $\mathcal{J}$ be the Jacobian of P. Let $\mathcal{J}_{ij} = \partial P_i / \partial X_j$, $i = 1, \ldots, n-1$, $j = 1, \ldots, n$. Let $K(P) := \max(\{\|\nabla \mathcal{J}_{ij}(z)\|_2 \mid z \in V_{\mathbb{R}}(\sum_{i=1}^{n} X_i^2 - 1)\})$. Let $\mu = \sqrt{(n-1)n}$. Assume that $K(P) \leq 1$ holds. Let z_0 and z_1 be two points of $V_{\mathbb{R}}(\sum_{i=1}^{n} X_i^2 - 1)$. Then we have*

$$\|\mathcal{J}(z_1) - \mathcal{J}(z_0)\|_2 \leq \mu \|z_1 - z_0\|_2.$$

Remark 3. *This lemma was proved in [28] as Eq. (21).*

Lemma 2 (Theorem 3.9 in [28]). *Let $P(X_1, \ldots, X_n) = \{P_1, \ldots, P_{n-1}\}$ be a set of $n-1$ polynomials with n variables such that $V_{\mathbb{R}}(P) \subset V_{\mathbb{R}}(\sum_{i=1}^{n} X_i^2 - 1)$ and $K(P) \leq 1$ hold. Let z_0 be a point of $V_{\mathbb{R}}(P)$. Let σ be the smallest singular value of $\mathcal{J}(z_0)$. Let $\mu = \sqrt{(n-1)n}$. Let $\omega = \sqrt{2(2\rho - 1)\left(2\rho - 2\sqrt{\rho(\rho - 1)} - 1\right)}$, for some $\rho \geq 1$. Assume that $2\rho > 3\omega$ holds (which is true for any $\rho \geq 1.6$). Let $s = \frac{\sigma}{2\mu\rho}$. Let L be a hyperplane which is perpendicular to the tangent line of $V_{\mathbb{R}}(P)$ at z_0 of distance s to z_0.*

We move z_0 in the tangent direction in distance s and apply the Newton Iteration to the zero dimensional system $P \cup \{L\}$. Assume that the Newton Iteration converges to z_1. Then z_1 is on the same component with z_0 if and only if

$$\|z_1 - z_0\| < \omega \cdot s. \tag{1}$$

Remark 4. *Note that one can always find a $\rho \geq 1.6$ (by increasing ρ and thus decreasing the step size s) such that $\|z_1 - z_0\| < \omega \cdot s$ holds.*

We have the following numerical error estimation of the border curve.

Theorem 1. *Let $F(X_1, ..., X_m, U_1, U_2) = 0$ be a bi-parametric polynomial system satisfying the assumptions A_1 and A_2. Let B be the border curve of F restricted to a rectangle R. Let $F' := F \cup \{\det(\mathcal{J}_F)\}$. We consider*

$$P = \{\bar{F}'_1, ..., \bar{F}'_{m+1}, \sum_{k=0}^{m} X_k^2/2 - 1/2 = 0\}, \tag{2}$$

where $\bar{F}'_i, 1 \leq i \leq m+1$ is homogenized from F'_i in variables $\{X_1, ..., X_m\}$ by an additional variable X_0.

Since we consider the solutions of P in a rectangular region, without loss of generality, we can assume $K(P) \leq 1$. Otherwise the polynomials can be rescaled by that upper bound of $K(P)$.

Let z_0 and z_1 be two points of $V_{\mathbb{R}}(P)$ satisfying Eq. (1). Let $C_{z_0 z_1}$ be the curve segment between z_0 and z_1 in $V_{\mathbb{R}}(P)$. Let $\tilde{z}_0$ and $\tilde{z}_1$ be computed points caused by numerical error within distance ϵ to z_0 and z_1 respectively.

Let $\mu = \sqrt{(m+2)(m+3)}$. Let $\tilde{\sigma}_0$ and $\tilde{\sigma}_1$ be respectively the smallest singular value of $\mathcal{J}_P(\tilde{z}_0)$ and $\mathcal{J}_P(\tilde{z}_1)$. Let $\tilde{\sigma} := \max(\tilde{\sigma}_0, \tilde{\sigma}_1)$. Let ρ and ω be as defined in Lemma 2. Let u_0 and u_1 be respectively the projection of $\tilde{z}_0$ and $\tilde{z}_1$. Let $B_{z_0 z_1} \subset B$ be the projection of $C_{z_0 z_1}$. Then the distance from $B_{z_0 z_1}$ to the segment $\overline{u_0 u_1}$ is at most

$$\tan\left(2\arccos(\frac{1}{\omega})\right) \frac{\omega}{4\mu\rho}(\mu\epsilon + \tilde{\sigma}) + \epsilon.$$

or $1.082\epsilon + 0.082\frac{\tilde{\sigma}}{\mu}$ if we choose $\rho = 1.6$.

Proof. Let σ_0 (resp. σ_1) be the smallest singular value of $\mathcal{J}_P(z_0)$ (resp. $\mathcal{J}_P(z_1)$). Let $s_i = \frac{\sigma_i}{2\mu\rho}, i = 1, 2$. Let L_0 (resp. L_1) be hyperplane which is perpendicular to the tangent line of $V_{\mathbb{R}}(P)$ at z_0 (resp. z_1) of distance s_0 (resp. s_1) to z_0 (resp. z_1). Let $h := \|z_0 - z_1\|$.

We define a cone with z_0 as the apex, the tangent line at z_0 as the axis, and the angle deviating from the axis being $\theta := \arccos(\frac{s_0}{\omega s_0})$. By Lemma 2, the curve from z_0 to z_1 must be in this cone when the step size is small. Similarly, we can construct another cone with z_1 as the apex, the tangent line at z_1 as the axis, and the angle deviating from the axis being $\theta := \arccos(\frac{s_1}{\omega s_1})$, such that it contains the curve from z_1 back to z_0. The projection of the intersection of these cones onto the parametric space is a triangle or quadrilateral. Figure 2 illustrates the two cones and the curve contained in them. From Fig. 2, we know that $|CE| < |AE|\tan(2\theta)$ and $|CE| < |EB|\tan(2\theta)$ hold. Since $|AE|+|EB| = h$, we deduce that $|CE| < \frac{h}{2}\tan(2\theta) = \frac{h}{2}\tan(2\arccos(1/\omega))$.

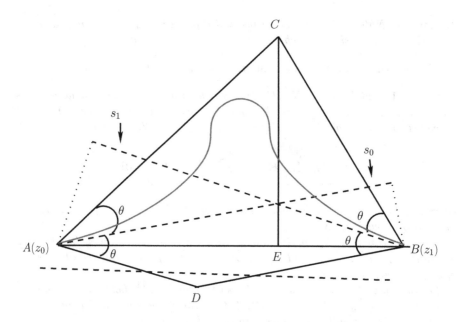

Fig. 2. A 2D image illustrating the intersection of two cones.

Next we estimate s_0 and s_1. By Weyl's theorem [23], $|\tilde{\sigma}_i - \sigma_i| \leq \|\mathcal{J}_P(\tilde{z}_i) - \mathcal{J}_P(z_i)\|_2$ holds. By Lemma 1, we have $\|\mathcal{J}_P(\tilde{z}_i) - \mathcal{J}_P(z_i)\|_2 \leq \mu\epsilon$. Thus $\sigma_i \leq \mu\epsilon + \tilde{\sigma}_i$, $i = 1, 2$, holds. Therefore we have $s_i \leq \frac{\mu\epsilon+\tilde{\sigma}}{2\mu\rho}$, $i = 1, 2$.

Since the distance from any point on the curve to the segment $\overline{z_0 z_1}$ is no greater than $|CD|$ and the distance from $\overline{z_0 z_1}$ to $\overline{\tilde{z}_0 \tilde{z}_1}$ is no greater than ϵ, we obtain the final estimation

$$\tan\left(2\arccos(\frac{1}{\omega})\right)\frac{\omega}{4\mu\rho}(\mu\epsilon + \tilde{\sigma}) + \epsilon.$$

Example 3. *For the polynomial system F, the rectangle R given in Example 2, the theoretical error estimation given by Theorem 1 is about 7.08×10^{-3} while the actual computed error is about 9.91×10^{-4}. Both errors can be reduced if a smaller step size is chosen.*

4   Constructing the Solution Map

As an extension of Proposition 1, in this section, we introduce a numerical version of the connected components of $R \setminus B$ and the notion of solution map to describe the real solutions of the parametric system F restricted to the rectangle R.

In Sect. 2, we provided an algorithm to compute a numerical approximation of the border curve. In Sect. 3, we estimated the distance between the approximated border curve and the true border curve. Since the border curve and its approximation usually do not overlap, two points are connected with respect to the approximated curve does not imply that they are connected with respect to the border curve. To handle this problem, we introduce the notion of "δ-connected", where δ should be no less than the error estimation provided by Theorem 1, to divide the interior of a rectangle into "numerically connected areas" with respect to an approximation of the border curve.

Definition 2 (δ-stripe). *Let $p \in \mathbb{R}^2$. Let $D_r(p)$ be the closed disk of center p and radius r. Let Γ be a path of $\mathbb{R}^2$. We define $\Gamma_\delta := \cup_{p \in \Gamma} D_{\delta/2}(p)$ as the associated δ-stripe of Γ.*

Definition 3 (δ-connected). *Let R be a rectangle of $\mathbb{R}^2$ and let S be a set of points in R. Let $\delta \geq 0$. Let p_1, p_2 be two points of $R \setminus S$. We say p_1 and p_2 are δ-connected with respect to S if there is a path Γ in R connecting p_1 with p_2 such that there are no points of S belonging to the associated δ-stripe of Γ connecting p_1 and p_2. A δ-connected set of R is a subset of R such that every two points of it is δ-connected.*

It is clear that the two notions connectedness and δ-connectedness coincide when $\delta = 0$ holds. The following two propositions establish the relations between connectedness and δ-connectedness when $\delta > 0$ holds.

Proposition 2. *Let B be the border curve of a bi-parametric polynomial system F restricted to a rectangle R. By Theorem 1, there exists a polyline $\tilde{B}$ and a $\delta \geq 0$ such that $B \subseteq \tilde{B}_\delta$. If two points p_1 and p_2 are δ-connected with respect to $\tilde{B}$, then they are connected with respect to B.*

Proof. Since p_1 and p_2 are δ-connected with respect to $\tilde{B}$, there exists a path Γ connecting p_1 and p_2 such that $\Gamma_\delta \cap \tilde{B} = \emptyset$ holds. To prove the proposition, it is enough to show that B has no intersection with Γ. Now assume that B intersects with Γ at a point p. Since Γ_δ is the associated δ-stripe of Γ, the distance between p and $\tilde{B}$ is greater than $\delta/2$, which is a contradiction to the fact that $B \subseteq \tilde{B}_\delta$.

Next we associate a rectangle with a grid graph.

Definition 4. *Let $\delta > 0$. Let R be a rectangle with width $m\delta$ and with length $n\delta$, where $m, n \in \mathbb{N}$. It can be naturally divided into a $m \times n$ grid of mn squares. The grid itself also defines an undirected graph, whose vertices and edges are exactly those of squares (overlapped vertices and edges are treated as one) in the grid. Such a grid (together with the graph) is called the* associated δ-grid *of R.*

Let G be a subgraph of the associated δ-grid of R. The set defined by G is the set of points on G and the set of points inside the squares of G. Let p be a point of R. Then p belongs to at least one square in the δ-grid of R. We pick one of them according to some fixed rule and call it the associated square *of p, denoted by A_p.*

Remark 5. *Here, for simplicity, we use a grid where each square has the same size. It is also possible to define a grid with different sizes of squares.*

We have the following key observation.

Proposition 3. *Let R be a rectangle and G be its associated δ-grid graph. Let S be a set of interior points of R. For every $p \in S$, we remove from G all the vertices of the associated square A_p of p and name the resulting graph still by G. Let $Z(G)$ be the set defined G. Then we have*

- *The distance between p and any point of $Z(G)$ at least δ.*
- *Let $q \in R \setminus Z(G)$. Then there exists a p, $1 \leq i \leq s$, such that the distance between p and q is at most $2\sqrt{2}\delta$.*

Proof. For a given A_p, let N_p be the set of all its neighboring squares (at most 8). When we delete the vertices (and the edges connected to them) of A_p, these vertices and edges are also removed from N_p. As a result, the distance between p and those undeleted vertices and edges of N_p is at least δ and at most $2\sqrt{2}\delta$. Thus the proposition holds.

Remark 6. *Informally speaking, this proposition tells us that the connected components in G is at least δ, but at most $2\sqrt{2}\delta$ far from the points in S.*

Definition 5 (Connectivity Graph). *Let $\delta > 0$. Let R be a rectangle of $\mathbb{R}^2$ of size $m\delta \times n\delta$. Let B be a set of sequences of points in R. The distance between two successive points in a sequence is at most δ. A* connectivity graph *of (R, B) is a subgraph G of the δ-grid of R such that*

- *Each connected component of G defines a δ-connected subset of R with respect to B.*
- *There exists $s \in \mathbb{N}$ such that every interior point of R, which is $s\delta$ far from points in S belongs to at least one of subsets defined by the connected components of G.*

The following algorithm computes a connectivity graph.

Algorithm ConnectivityGraph

Input: A rectangle R of size $m\delta \times n\delta$, a set B of sequence of points belonging to R. The distance between two successive points in a sequence is at most δ.

Output: A connectivity graph of (R, B).

Steps:

1. Let G be the δ-grid of R.
2. For each point p of B, delete the four vertices (and edges connected to them) of the associated square A_p from G.
3. Return G.

Proposition 4. *Algorithm* ConnectivityGraph *is correct.*

Proof. By Proposition 3, for any vertex v of G and any point b of B, the distance between them is at least δ. Thus, by Definition 3, the zero set defined by each connected component of G is δ-connected. By Proposition 3, if the minimal distance between an interior point p of R and points in B is greater than $2\sqrt{2}\delta$, then p must belong to the set defined by G. Thus, the algorithm is correct.

Finally we are able to define the solution map of a bi-parametric polynomial system restricted to a rectangle R.

Definition 6 (Solution map). *Given a rectangle R and a bi-parametric polynomial system $F(X_1, \ldots, X_m, U_1, U_2) = 0$. Assume that F satisfies the assumptions A_1, A_2. A solution map of F restricted to R, denoted by M, is a quadruple (B, G, W, Z) where*

- *B is a set of sequences of points approximating the border curve of F.*
- *G is a connectivity graph of (R, B).*
- *W is a set of points in R s.t. each point of W is a vertex of a connected component of G and each connected component has exactly one vertex in W.*
- *Z is a correction of sets of points in $\mathbb{R}^m$ such that each element of Z is a solution set of F at a point of W.*

Example 4. *Consider the system in Example 2. Its connectivity graph and solution map are shown in Fig. 3. The set of sample points is $W = \{(U_1 = \frac{1}{2}, U_2 = \frac{-6}{5}), (U_1 = \frac{1}{2}, U_2 = \frac{6}{5}), (U_1 = \frac{-12}{5}, U_2 = \frac{-3}{5})\}$. The corresponding set of solution sets Z is*

$\{\{(X_1 = 1.077556426, X_2 = 1.497479963), (X_1 = .9908314677, X_2 = .4355629194), (X_1 = .8136158335, X_2 = -1.930040765)\}, \{(X_1 = -.8199351413, X_2 = -.6397859006), (X_1 = -2.611117469, X_2 = .3233313936), (X_1 = 1.354310493, X_2 = -1.291254606)\}, \{X_1 = .2814014780, X_2 = -2.885892092\}\}.$

5 Real Homotopy Continuation

As an application of the solution map, in this section, we present how to construct a real homotopy to compute all the real zeros of a parametric polynomial system at a given value of parameters.

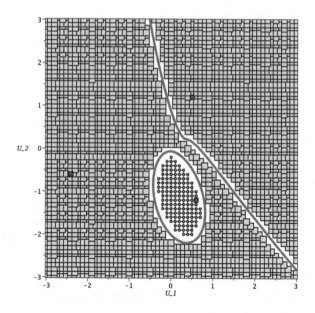

Fig. 3. The solution map (solution sets not shown).

Definition 7. *Let* $H(X,t) \subset \mathbb{R}[X_1,\ldots,X_n,t]$. *We call* $H(X,t)$ *a* real homotopy *if it satisfies the* smoothness property: *over the interval* $[0,1]$, *the real zero set of* $H(X,t)$ *defines a finite number of smooth functions of* t *with disjoint graphs.*

Given a real homotopy and the solutions of H at $t = 0$, one can use a standard prediction-projection method with the adaptive step size strategy in Lemma 2 to trace the solution curves of $H(X,t)$ to get the solutions of H at $t = 1$. We denote such an algorithm by RealHomotopy.

For a bi-parametric polynomial system F, in previous section, we have shown how to construct a solution map of F. To compute the real solutions of F at a given u of the parameters, RealHomotopy is called to trace the real homotopy starting from a known solution of F stored in M. More precisely, we have the following algorithm OnlineSolve.

Algorithm OnlineSolve
Input: a bi-parametric system $F \subset \mathbb{Q}[X,U]$, a rectangle R, an interior point u of R, a solution map M of F, and a point u of R.
Output: if u is not close to the border curve of F, return the real solutions of F at u, that is $V_{\mathbb{R}}(F(u))$; otherwise throw an exception.
Steps:
1. Let G be the connectivity graph in M.
2. Let A_u be the associated square of u. Let C_u be a connected component of G such that A_u is a subgraph of C_u. If C_u does not exist, throw an exception. If C_u exists, let v_u be one of the vertices of A_u.
3. Let w_u be the sample point of C_u in M.

4. Let $w_u \rightsquigarrow v_u$ be the shortest path between w_u and v_u computed for instance by Dijkstra's algorithm. Connecting v_u and u with a segment and denote the path $w_u \rightsquigarrow v_u \to u$ by Γ.
5. Let $p_0, p_1, \ldots, p_s$ be the sequence of successive vertices of Γ.
6. Let S be the solution set of $F(p_0)$ in M.
7. For i from 0 to $s - 1$ do
   (a) let $(\phi_i(t), \psi_i(t))$ be a parametrization of the segment $\overline{p_i p_{i+1}}$, $t \in [0,1]$;
   (b) let $H_i(X,t) := F(X, U_1 = \phi_i(t), U_2 = \psi_i(t))$;
   (c) let $S := \mathsf{RealHomotopy}(H_i, S)$;
8. Return S.

Proposition 5. *If an exception is not thrown, Algorithm* OnlineSolve *correctly computes the real solutions of F at u.*

Remark 7. *The correctness of* OnlineSolve *can be easily verified from its description. When the exception is thrown, it means that u is close to the border curve of F, which indicates a numerically difficult region with ill-conditioned Jacobians.*

A connected path between a sample point $(-5/2, -3/5)$ and a chosen point $(1, -1)$ is shown in Fig. 4. This path is a guide for building real homotopies.

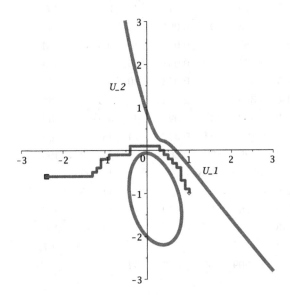

Fig. 4. A connected path for constructing real homotopies.

6 Experimentation

In this section, we illustrate the effectiveness of our method by some examples. Our method was implemented in MAPLE, which will make external calls to HOM4PS2 (by exchanging input and output files) when it needs to compute the solutions of a zero-dimensional polynomial system. The experimentation was conducted on a Ubuntu Laptop (Intel i7-4700MQ CPU @ 2.40 GHz, 8.0 GB memory). The memory usage was restricted to 60 % of the total memory. The timeout (represented by − in Table 1) was set to be 1800 s. The experimentation results are summarized in Table 1, where BP denotes the command RegularChains[ParametricSystemTools][BorderPolynomial] and DV denotes the command RootFinding[Parametric][DiscriminantVariety] in the computer algebra system MAPLE 18, and BC denotes BorderCurve.

Table 1. Experimentation results

| | Symbolic methods | | | | Numeric method | |
| | BP | | DV | | BC | |
Sys	time (s)	deg	time (s)	deg	time (s)	#(points)
1-2	1.340	1	0.562	1	3.798	171
1-3	0.575	4	0.019	4	2.147	307
1-4	0.433	9	0.021	9	1.097	252
1-5	0.385	16	0.024	16	0.668	153
1-6	0.575	25	0.031	25	1.940	313
1-7	0.396	30	0.053	30	1.579	127
1-8	0.389	48	2.724	48	3.399	668
2-2	0.552	4	0.028	4	3.641	839
2-3	0.800	24	0.372	24	14.748	2694
2-4	4.329	70	90.661	69	41.572	4084
2-5	68.930	173	-		10.695	266
2-6	-		-		110.657	4190
3-2	0.726	14	0.070	12	1.073	306
3-3	-		-		32.550	2301
3-4	-		-		286.638	9672
4-2	16.309	48	6.947	32	11.654	1073
4-3	-		-		188.058	5452
4-4	-		-		1415.822	11342
5-2	-		991.215	72	53.230	185
5-3	-		-		1054.712	11640
6-2	-		-		790.768	1486

The first column denotes the tested random sparse systems, each of which has a label i-j, where i denotes the number of equations (same as number of unknowns, or number of total variables minus 2) and j denotes the total degree. More precisely, the system i-j has the form $\{X_k^j + \text{low degree terms on X, U} \mid k = 1 \cdots i\}$. The third and fifth column denotes respectively the degree of the border polynomial and the degree of the polynomial representing the discriminant variety. The seventh column denotes the number of points obtained by the numeric method for representing the border curves. The rest columns denote the execution time for three methods. From the table, it is clear that the numeric method (BC) outperforms the symbolic counterparts (BP and DV) on computing the border curves of systems of larger size.

7   Conclusion and Future Work

In this paper, under some assumptions, we introduced the concept of border curve for a bi-parametric real polynomial system and proposed a numerical method to compute it. The border curve was applied to describing the real solutions of a parametric polynomial system through the construction of a solution map and computing the real solutions of the parametric system at a particular value of parameters through constructing a real homotopy.

We tested a preliminary implementation of our method in MAPLE on computing the border curves of a set of randomly chosen sparse polynomial systems and compared our implementation with similar symbolic solvers on these examples. The experimentation shows that the numerical one is much more efficient than the symbolic ones on examples with more than 2 unknowns (or 4 variables).

In a future work, we will consider how to relax the assumptions introduced in Sect. 2. To make this approach more practical, structures of the polynomial system F' in Remark 2 must be exploited, for example using ideas in [28]. An efficient (and parallel) implementation of the method in C like languages is also important for applications. Extending the method to the multi-parametric case with possibly the help of the roadmap method [1, 3] and the numerical cell decomposition method [17] will be investigated.

Acknowledgements. This work is partially supported by NSFC (11301524, 11471307, 61572024) and CSTC (cstc2015jcyjys40001).

References

1. Basu, S., Roy, M.F., Safey El Din, M., Schost, É.: A baby step-giant step roadmap algorithm for general algebraic sets. Found. Comput. Math. **14**(6), 1117–1172 (2014)
2. Buchberger, B.: An algorithm for finding a basis for the residue class ring of a zero-dimensional polynomial ideal. Ph.D. thesis, University of Innsbruck (1965)
3. Canny, J.: The Complexity of Robot Motion Planning. MIT Press, Cambridge (1987)

4. Chen, C., Golubitsky, O., Lemaire, F., Maza, M.M., Pan, W.: Comprehensive triangular decomposition. In: Ganzha, V.G., Mayr, E.W., Vorozhtsov, E.V. (eds.) CASC 2007. LNCS, vol. 4770, pp. 73–101. Springer, Heidelberg (2007)
5. Chou, S.C., Gao, X.S.: Computations with parametric equations. In: ISSAC 1991, pp. 122–127. ACM (1991)
6. Collins, G.E.: Quantifier elimination for real closed fields by cylindrical algebraic decompostion. In: Brakhage, H. (ed.) Automata Theory and Formal Languages 2nd GI Conference. LNCS, vol. 33, pp. 134–183. Springer, Heidelberg (1975)
7. Corvez, S., Rouillier, F.: Using computer algebra tools to classify serial manipulators. In: Winkler, F. (ed.) ADG 2002. LNCS (LNAI), vol. 2930, pp. 31–43. Springer, Heidelberg (2004)
8. Fotiou, I.A., Rostalski, P., Parrilo, P.A., Morari, M.: Parametric optimization and optimal control using algebraic geometry. Int. J. Control **79**(11), 1340–1358 (2006)
9. González Vega, L., Lombardi, H., Recio, T., Roy, M.F.: Sturm-Habicht sequence. In: ISSAC 1989, pp. 136–146. ACM (1989)
10. Hauenstein, J.D.: Numerically computing real points on algebraic sets. Acta Applicandae Mathematicae **125**(1), 105–119 (2012)
11. Hong, H.: Overview on real quantifier elimination. In: MACIS 2013, p. 1 (2013)
12. Lazard, D., Rouillier, F.: Solving parametric polynomial systems. J. Symb. Comput. **42**(6), 636–667 (2007)
13. Li, T.Y.: Numerical solution of multivariate polynomial systems by homotopy continuation methods. Acta Numerica **6**, 399–436 (1997)
14. Li, T.Y., Sauer, T., Yorke, J.A.: The Cheater's homotopy: an efficient procedure for solving systems of polynomial equations. SIAM J. Numer. Anal. **26**(5), 1241–1251 (1989)
15. Li, T.Y., Wang, X.S.: Solving real polynomial systems with real homotopies. Math. Comput. **60**, 669–680 (1993)
16. Lu, Y., Bates, D., Sommese, A., Wampler, C.: Finding all real points of a complex curve. Contemp. Math. **448**, 183–206 (2007)
17. Mario Besana, G., Di Rocco, S., Hauenstein, J.D., Sommese, A.J., Wampler, C.W.: Cell decomposition of almost smooth real algebraic surfaces. Numer. Algorithms **63**(4), 645–678 (2012)
18. Moroz, G.: Complexity of the resolution of parametric systems of polynomial equations and inequations. In: ISSAC 2006, pp. 246–253. ACM (2006)
19. Piret, K., Verschelde, J.: Sweeping algebraic curves for singular solutions. J. Comput. Appl. Math. **234**(4), 1228–1237 (2010)
20. Rouillier, F., Roy, M.F., Safey El Din, M.: Finding at least one point in each connected component of a real algebraic set defined by a single equation. J. Complex. **16**(4), 716–750 (2000)
21. Schost, E.: Computing parametric geometric resolutions. Appl. Algebra Eng. Commun. Comput. **13**(5), 349–393 (2003)
22. Sommese, A., Wampler, C.: The Numerical Solution of Systems of Polynomials Arising in Engineering and Science. World Scientific Press, Singapore (2005)
23. Stewart, G.W.: Perturbation theory for the singular value decomposition. In: SVD and Signal Processing, II: Algorithms, Analysis and Applications, pp. 99–109. Elsevier (1990)
24. Tarski, A.: A decision method for elementary algebra and geometry. Fund. Math. **17**, 210–239 (1931)
25. Wang, D.M., Xia, B.: Stability analysis of biological systems with real solution classification. In: Kauers, M. (ed.) ISSAC 2005, pp. 354–361. ACM (2005)

26. Weispfenning, V.: Comprehensive Gröbner bases. J. Symb. Comp. **14**, 1–29 (1992)
27. Wu, W., Reid, G.: Finding points on real solution components and applications to differential polynomial systems. In: ISSAC 2013, pp. 339–346. ACM (2013)
28. Wu, W., Reid, G., Feng, Y.: Computing real witness points of positive dimensional polynomial systems (2015). Accepted for Theoretical computer science. http://www.escience.cn/people/wenyuanwu
29. Wu, W.T.: Basic principles of mechanical theorem proving in elementary geometries. J. Sys. Sci. Math. Scis **4**(3), 207–235 (1984)
30. Yang, L., Xia, B.: Real solution classifications of a class of parametric semialgebraic systems. In: A3L 2005, pp. 281–289 (2005)

The Complexity of Cylindrical Algebraic Decomposition with Respect to Polynomial Degree

Matthew England[1](✉) and James H. Davenport[2]

[1] Faculty of Engineering, Environment and Computing, School of Computing, Electronics and Maths, Coventry University, Coventry CV1 5FB, UK
Matthew.England@coventry.ac.uk
[2] Department of Computer Science, University of Bath, Bath BA2 7AY, UK
J.H.Davenport@bath.ac.uk
http://computing.coventry.ac.uk/~mengland/
http://people.bath.ac.uk/masjhd/

Abstract. Cylindrical algebraic decomposition (CAD) is an important tool for working with polynomial systems, particularly quantifier elimination. However, it has complexity doubly exponential in the number of variables. The base algorithm can be improved by adapting to take advantage of any equational constraints (ECs): equations logically implied by the input. Intuitively, we expect the double exponent in the complexity to decrease by one for each EC. In ISSAC 2015 the present authors proved this for the factor in the complexity bound dependent on the number of polynomials in the input. However, the other term, that dependent on the degree of the input polynomials, remained unchanged.

In the present paper the authors investigate how CAD in the presence of ECs could be further refined using the technology of Gröbner Bases to move towards the intuitive bound for polynomial degree.

1 Introduction

A *cylindrical algebraic decomposition* (CAD) is a *decomposition* of $\mathbb{R}^n$ (under a given variable ordering, so that the projections considered are $(x_1, \ldots, x_k) \to (x_1, \ldots, x_j)$ for $j < k$) into cells. The cells are arranged *cylindrically*, meaning the projections of any pair with respect to the given ordering are either equal or disjoint. In this definition *algebraic* is short for semi-algebraic meaning each CAD cell can be described with a finite sequence of polynomial constraints. A CAD is produced to be invariant for input; originally *sign-invariant* for a set of input polynomials (so on each cell each polynomial is positive, zero or negative), and more recently *truth-invariant* for input Boolean-valued formulae built from the polynomials (so on each cell each formula is either true or false).

Introduced by Collins for real quantifier elimination (QE) [1], applications of CAD included parametric optimisation [26], epidemic modelling [10] and even motion planning [42]. Recent applications include theorem proving [41], deriving optimal numerical schemes [24] and reasoning with multi-valued functions [18].

V.P. Gerdt et al. (Eds.): CASC 2016, LNCS 9890, pp. 172–192, 2016.
DOI: 10.1007/978-3-319-45641-6_12

CAD has worst case complexity doubly exponential [9, 20], due to the nature of the information to be recorded rather than the algorithm used [9]. Let n be the number of variables, m the number of input polynomials, and d the maximum degree (in any one variable) of the input. Then a complexity analysis in Sect. 5 of [22] shows that the best known variant of Collins' algorithm to produce a sign-invariant CAD for the polynomials [37] has an upper bound on the size of the CAD (i.e. number of cells) with dominant term

$$(2d)^{2^n-1} m^{2^n-1} 2^{2^{n-1}-1}, \tag{1}$$

i.e. the CAD grows doubly exponentially with the number of variables n.

In fact, at the end of the projection stage, when we are considering $\mathbb{R}^1$, this analysis shows that we have M polynomials, each of degree D, where $D = d^{2^{O(n)}}$ and $M = m^{2^{O(n)}}$. Of course, by replacing $\{f, g\}$ with $\{fg\}$ we can reduce M at the cost of increasing D, but since it is much easier to find the roots of $\{f, g\}$ than $\{fg\}$, we do not want to. The lower bound in [20] shows that $D = d^{2^{\Omega(n)}}$, and in [9] that, without artificial combination, $M = m^{2^{\Omega(n)}}$. Both rely on the technique from [28], and the formulae demonstrating this growth are not straightforward: in particular needing $O(n)$ quantifier alternations. But the underlying polynomials *are* simple: all linear for [9] and all bar two linear for [20]. Furthermore, each polynomial only involves a bounded number of variables (generally two) independent of n, showing that the doubly-exponential difficulty of CAD resides in the complicated number of ways simple polynomials can interact.

To improve the CAD performance and this bound we now build CADs which are not sign-invariant for polynomials but truth-invariant for formulae. This can be achieved by identifying *equational constraints* (ECs): polynomial equations logically implied by formulae. The presence of an EC restricts the dimension of the solution space and if exploited properly by the algorithm we may expect a reduction in complexity accordingly. Intuitively, we expect the double exponent to decrease by 1 for each independent (to be made precise later) EC available.

In [22] the present authors described how to adapt CAD to make use of multiple (primitive) ECs. Suppose that our input formula consists of polynomials (as described above) and that ℓ suitable ECs can be identified. The algorithm in [22] was shown to have corresponding upper bound dominant term

$$(2d)^{2^n-1} m^{2^{n-\ell}-2} 2^{\ell 2^{n-\ell}-3\ell}. \tag{2}$$

So while the bound is still doubly exponential with respect to n, some of the double exponents have been reduced by ℓ. To be precise, the double exponent of m (and its corresponding constant factor) is reduced while the double exponent with respect to d (actually $2d$) has not. This is due to the focus of [22] being on reducing the number of polynomials created during the intermediate calculations with no attempt made to control degree growth.

Contribution and Plan. The present paper is concerned with how to gain the corresponding improvement to the factor dependent on d to achieve the intuitive complexity bound. The hypothesis is that this should be possible by making use of the theory of Gröbner bases in place of iterated resultants. We start in Sects. 2.1 and 2.2 by reviewing background material on CAD, and then focus on CAD in the presence of ECs in Sects. 2.3 and 2.4. In Sect. 3 we consider how the growth of degree in iterated resultants grows compared to that of the true multivariate resultant (which encodes what is needed for CAD with ECs). In Sect. 3.3 we propose controlling this using Gröbner Bases and in Sect. 4 we give a worked example of how these can precondition CAD. In Sect. 5 we sketch how this improves upon the bound (2) and then we finish in Sect. 6 by discussing some outstanding issues.

2   CAD with Respect to Equational Constraints

2.1   CAD Computation Scheme and Terminology

We describe the computation scheme and terminology that CAD algorithms derived from Collins share. We assume a set of input polynomials (possibly derived from input formulae) in ordered variables $x = x_1 \prec \ldots \prec x_n$. The *main variable* of a polynomial (mvar) is the highest ordered variable present.

The first phase of CAD, *projection*, applies projection operators recursively on the input polynomials, each time producing another set of polynomials with one less variable. Together these define the *projection polynomials* used in the second phase, *lifting*, to build CADs incrementally by dimension. First a CAD of the real line is built with cells (points and intervals) determined by the real roots of the univariate polynomials (those in x_1 only). Next, a CAD of $\mathbb{R}^2$ is built by repeating the process over each cell in $\mathbb{R}^1$ with the bivariate polynomials in (x_1, x_2) evaluated at a sample point of the cell in $\mathbb{R}^1$. This produces *sections* (where a polynomial vanishes) and *sectors* (the regions between) which together form the *stack* over the cell. Taking the union of these stacks gives the CAD of $\mathbb{R}^2$. The process is repeated until a CAD of $\mathbb{R}^n$ is produced.

All cells are represented by (at least) a sample point and an *index*. The latter is a list of integers, with the kth integer fixing variable x_k according to the ordered real roots of the projection polynomials in $(x_1, \ldots, x_k)$. If the integer is $2i$ the cell is over the ith root (counting low to high) and if $2i + 1$ over the interval between the ith and $(i+1)$th (or the unbounded intervals at either end).

In each lift we extrapolate the conclusions drawn from working at a sample point to the whole cell. The validity of this approach follows from the correct choice of projection operator. For sign-invariance to be maintained the operator must produce polynomials: *delineable* in a cell, meaning the portion of their zero set in the cell consists of disjoint sections; and, *delineable* as a set, meaning the sections of different polynomials are identical or disjoint. One of the projection operators used in this paper is

$$P(B) := \mathrm{coeff}(B) \cup \mathrm{disc}(B) \cup \mathrm{res}(B). \tag{3}$$

Here disc and coeff denote respectively the set of discriminants and coefficients of a set of polynomials; and res denotes either the resultant of a pair of polynomials or, when applied to a set, the set of polynomials

$$\mathrm{res}(A) = \{\mathrm{res}(f_i, f_j) \mid f_i \in A, f_j \in A, f_j \neq f_i\}.$$

We assume B is an irreducible basis for a set of polynomials in which every element has mvar x_n. For a general set of polynomials A we would proceed by letting B be an irreducible basis of the primitive part of A; apply the operators above; and take the union of the output with the content of A. The operator P was introduced in [37] along with proofs of related delineability results.

2.2   Brief Summary of Improvements to CAD

As discussed in the introduction, CAD has worst case complexity doubly exponential in the number of variables. For some problems there exist algorithms with better complexity [2], however, CAD implementations remain the best general purpose approach for many. This is due in large part to the numerous techniques developed to improve the efficiency of CAD since Collins' original work including: refinements to the projection operator [7,27,29,37]; the early termination of lifting, such as when sufficient for QE [17] or for building a sub-CAD [45]; and symbolic-numeric lifting schemes [32,43]. Some recent advances include further refinements to the projection operator when dealing with multiple formulae as input [4,5]; local projection approaches [8,44]; decompositions via complex space [3,14]; and the development of heuristics for CAD problem formulation [6,21,46] including machine learning approaches [31].

2.3   Equational Constraints

As discussed in the Introduction, identifying equational constraints can improve the performance of CAD.

Definition. A *QFF* is a quantifier free Tarski formula: a Boolean combination $(\wedge, \vee, \neg)$ of statements about the signs $(= 0, > 0, < 0)$ of integral polynomials.

An *equational constraint* (EC) is a polynomial equation logically implied by a QFF. An EC is *explicit* if an atom of the QFF, and *implicit* otherwise.

Collins first suggested that the projection phase of CAD could be simplified in the presence of an EC [16]. The insight is that a CAD sign-invariant for the defining polynomial of an EC, and sign-invariant for any others only on sections of that polynomial, would be sufficient. The intuitive restriction of (3) is to use only those coefficients, discriminants and resultants which are derived from the EC polynomial, as in (4) below where $F \subseteq B$ is a basis for the EC polynomial.

$$P_F(B) := P(F) \cup \{\mathrm{res}(f, g) \mid f \in F, g \in B \setminus F\} \tag{4}$$

The validity of using this operator for the first projection was verified in [39], with subsequent projections returning to (3). The operator could only be used for a single EC in the main variable of the system as the delineability result for (4) could not be applied recursively, excluding its use at a subsequent projection to take advantage of any EC with corresponding main variable. This led to the development of the operator (5) in [40] which suffered no such reduction at the cost of including the discriminants that had been removed from (3) by (4).

$$P_F^*(B) := P_F(B) \cup \operatorname{disc}(B \setminus F) \tag{5}$$

See Sect. 2 of [22] for examples demonstrating these operators.

A system to derive implicit ECs was also introduced by [40], based on the observation that the resultant of the polynomials defining two ECs itself defines an EC. This is essential for maximising the savings from ECs since the reduced operators (4), (5) are for use with a single EC; meaning the savings gained are dependent not on the number of ECs, but the number identified with different main variable. Note that such iterated resultants are already produced during CAD projection. So using them as ECs requires us only to identify them as such (rather than introducing new polynomials for consideration) and hence does not mean an increase in m. Also, while they may have higher degree than the initial input polynomials, their degree is no higher than the other polynomials at the stage they are used (rather than just passed as content) by a projection operator.

In [22] the present authors reviewed the theory of reduced projection operators and deduced how it could also yield savings in the lifting phase; reducing both the number of cells we must lift over with respect to polynomials; and the number of such polynomials we lift with. These approaches meant that the projection polynomials are no longer a fixed set (key to some CAD implementations) and that the invariance structure of the final CAD can no longer be expressed in terms of sign-invariance of polynomials. For the worked example in [22, Sect. 4] combining the advances in this subsection allowed a sign-invariant CAD with 1,118,205 cells to be replaced by a truth invariant CAD with 93 cells.

2.4   CAD with ECs

Algorithm 1 describes the CAD projection phase in the presence of multiple ECs described in the previous subsection. Note that (as with the previous theory of multiple ECs this is based on) we assume the ECs are primitive. Algorithm 1 applies the best possible (smallest validated) projection operator at each stage. The word *suitable* in the output declaration means a CAD lifting phase that makes well-orientedness checks in line with the theory of McCallum's projection operators (see [37,39,40] for details). Algorithm 2 is one such suitable lifting algorithm. It uses the F_i (knowledge of which projection steps made use of an EC) to tailor its lifts: lifting only with respect to EC polynomials (steps 7–10) and only over cells where an EC was satisfied (steps 11–15) (lifting trivially to the cylinder otherwise). The correctness of these algorithms was proven in [22].

Table 1 is recreated from [22] and shows the growth in the number and degree of the projection polynomials when following Algorithm 1 under the assumption that we have declared ECs for the first ℓ projections (so $0 < \ell \leq \min(m, n)$).

Algorithm 1. CAD Projection using multiple stated ECs

Input : A formula ϕ in variables $x_1, \ldots, x_n$, and a sequence of sets $\{E_k\}_{k=1}^n$; each either empty or containing a single primitive polynomial with mvar x_k which defines an EC for ϕ.

Output: A sequence of sets of polynomials ready for a suitable CAD lifting algorithm.

1 Extract from ϕ the set of defining polynomials A_n;
2 **for** $k = n, \ldots, 2$ **do**
3   Set B_k to the finest squarefree basis for $\mathrm{prim}(A_k)$;
4   Set C to $\mathrm{cont}(A_k)$;
5   Set F_k to the finest squarefree basis for E_k;
6   **if** F_k *is empty* **then**
7 | Set $A_{k-1} := C \cup P(B_k)$;
8   **else**
9 | Set $A_{k-1} := C \cup P_{F_i}^*(B_i)$;

10 **return** $A_1, \ldots, A_n; F_1, \ldots, F_n$.

Rather than the actual polynomials created the table keeps track of sets of polynomials known to have the *(M,D)-property*: the ability to be partitioned into M subsets, each with maximum combined degree D.

The *(M,D)*-property was introduced in McCallum's thesis and was used (along with tables like Table 1) to give a detailed comparison of the complexity of several different projection operators in [5, Sect. 2.3]. The key observation is that the number of real roots in a set with the *(M,D)*-property is at most MD (although in practice many will be in $\mathbb{C} \setminus \mathbb{R}$). Hence the number of cells in the CAD of $\mathbb{R}^1$ is bounded by twice the product of the final entries, plus 1.

Define d_i and m_i as the entries in the Number and Degree columns of Table 1 from the row with i Variables. Then the number of cells in the final CAD of $\mathbb{R}^n$ is bounded by

$$\prod_{i=1}^n \left[2m_i d_i + 1 \right]. \tag{6}$$

Omitting the +1s from each term will usually allow for a closed form expression of the dominant term in the bound.

The derivation of bound (2) from Table 1 was given in [22, Sect. 5]. It involved considering the two improvements to the lifting phase. The first was lifting only with respect to EC polynomials; meaning that for the purposes of the bound we could set m_i to 1 for $i = n, \ldots, n - \ell$. The second was to lift trivially (to a cylinder) over those cells where an EC was false.

Denote by (†) the bound on the CAD of $\mathbb{R}^{n-(\ell+1)}$ given by (6) but with the product terminating at $n - (\ell + 1)$, as there can be no reduced lifting until this point. The lift to $\mathbb{R}^{n-\ell}$ will involve stack generation over all cells, but only with respect to the EC which has at most $d_{n-\ell}$ real roots and thus the CAD of $\mathbb{R}^{n-\ell}$ at most $[2d_{n-\ell} + 1](\dagger)$ cells. The next lift, to $\mathbb{R}^{n-\ell-1}$, will lift the sections with respect to the EC, and the sectors only trivially. Hence the cell count bound is $[2d_{n-(\ell-1)} + 1]d_{n-\ell}(\dagger) + (d_{n-\ell} + 1)(\dagger)$ with dominant term $2d_{n-(\ell-1)}d_{n-\ell}(\dagger)$.

Algorithm 2. CAD Lifting using multiple stated ECs

Input : The output of Algorithm 1: two sequences of polynomials sets
$A_1, \ldots, A_n; F_1, \ldots, F_n$, the latter subsets of the former.

Output: Either: $\mathcal{D}$, a truth-invariant CAD of $\mathbb{R}^n$ for ϕ (described by lists I and
S of cell indices and sample points); or **FAIL**, if not well-oriented.

1 If F_1 is not empty then set p to be its element; otherwise set p to the product of
   polynomials in A_1;
2 Build $\mathcal{D}_1 := (I_1, S_1)$ according to the real roots of p;
3 **if** $n = 1$ **then**
4 ⌊ return $\mathcal{D}_1$;

5 **for** $k = 2, \ldots, n$ **do**
6 │ Initialise $\mathcal{D}_k = (I_k, S_k)$ with I_k and S_k empty sets;
7 │ **if** F_k *is empty* **then**
8 │ │ Set $L := B_k$;
9 │ **else**
10 │ ⌊ Set $L := F_k$;

11 │ **if** F_{k-1} *is empty* **then**
12 │ │ Set $\mathcal{C}_a := \mathcal{D}_{k-1}$ and $\mathcal{C}_b$ empty;
13 │ **else**
14 │ │ Set $\mathcal{C}_a$ to be cells in $\mathcal{D}_{k-1}$ with $I_{k-1}[-1]$ even;
15 │ ⌊ Set $\mathcal{C}_b := \mathcal{D}_{k-1} \setminus \mathcal{C}_a$;

16 │ **for** *each cell* $c \in \mathcal{C}_a$ **do**
17 │ │ **if** *An element of* L *is nullified over* c **then**
18 │ │ ⌊ **return FAIL**;
19 │ ⌊ Generate a stack over c with respect to the polynomials in L, adding
   │   cell indices and sample points to I_k and S_k;

20 │ **for** *each cell* $c \in \mathcal{C}_b$ **do**
21 │ ⌊ Extend to a single cell in $\mathbb{R}^k$ (cylinder over c), adding index and sample
   │   point to I_k and S_k;

22 **return** $\mathcal{D}_n = (I_n, S_n)$.

Subsequent lifts follow the same pattern and so the dominant term (omitting
the +1s) in the cell count bound for $\mathbb{R}^n$ is

$$2d_n d_{n-1} \ldots d_{n-(\ell-1)} d_{n-\ell} \prod_{i=1}^{n-(\ell+1)} \left[2m_i d_i + 1 \right]. \tag{7}$$

As shown in [22] using Table 1 (7) evaluates to (2).

3   Controlling Degree Growth

3.1   Iterated Resultant Calculations

As discussed in the Introduction, [22] showed that building truth-invariant CADs
by taking advantage of ECs reduced the CAD complexity bound from (1) to (2).

Table 1. Projection in CAD with projection operator (5) ℓ times and then (3).

Variables	Number	Degree
n	m	d
$n-1$	$2m$	$2d^2$
$n-2$	$4m$	$8d^4$
$\vdots$	$\vdots$	$\vdots$
$n-\ell$	$2^\ell m$	$2^{2^\ell-1}d^{2^\ell}$
$n-(\ell+1)$	$2^{2\ell}m^2$	$2^{2^{\ell+1}-1}d^{2^{\ell+1}}$
$n-(\ell+2)$	$2^{4\ell}m^4$	$2^{2^{\ell+2}-1}d^{2^{\ell+2}}$
$\vdots$	$\vdots$	$\vdots$
$n-(\ell+r)$	$2^{2^r\ell}m^{2^r}$	$2^{2^{\ell+r}-1}d^{2^{\ell+r}}$
$\vdots$	$\vdots$	$\vdots$
1	$2^{2^{(n-1-\ell)}\ell}m^{2^{n-1-\ell}}$	$2^{2^{n-1}-1}d^{2^{n-1}}$

Most notably, the double exponent of the term with base m (number of input polynomials) decreased by ℓ (the number of projections made with respect to an EC). However, the term with base d (degree of input polynomials) was unchanged. This term is doubly exponential due to the iterated resultant calculations during projection: the resultant of two degree d polynomials is the determinant of a $2d \times 2d$ matrix whose entries all have degree at most d, and thus a polynomial of degree at most $2d^2$. This increase in degree compounded by $(n-1)$ projections gives the first term of the bound (1).

When building CAD in the presence of ECs many of these iterated resultants are avoided (thus reducing the *number* of polynomials, but not their degree). Indeed, the derivation of ECs via propagation is itself an iterated resultant calculation. The purpose of the resultant in CAD construction is to ensure that the points in lower dimensional space where polynomials vanish together are identified, and thus that the behaviour over a sample point in a lower dimensional cell is indicative of the behaviour over the cell as a whole.

The iterated resultant (and discriminant) calculations involved in CAD have been studied previously, for example in [34,38]. We will follow the work of Busé and Mourrain in [13] who consider the iterative application of the univariate resultant to multivariate polynomials, demonstrating decompositions into irreducible factors involving the multivariate resultants (following the formalisation of Jouanolou [33]). They show that the approach will identify polynomials of higher degree than the true multivariate resultant and thus more than required for the purpose of identifying implicit equational constraints. For example, given 3 polynomials in 3 variables of degree d the true multivariate resultant has degree $\mathcal{O}(d^3)$ rather than $\mathcal{O}(d^4)$.

The key result of [13] for our purposes follows. Note that this, using the formalisation of resultants in [33] [13, Sect. 2], considers polynomials of a given *total degree*. However, the CAD complexity analysis discussed above and later is (following previous work on the topic) with regards to polynomials of *degree at most d* in a given variable. For clarity we use the Fraktur font when discussing total degree and Roman fonts when the maximum degree.

Corollary ([13, Corollary 3.4]). Given three polynomials $f_k(x, y, z)$ of the form

$$f_k(x, y, z) = \sum_{|\alpha| + i + j \leq \mathfrak{d}_k} a^{(k)}_{\alpha, i, j} x^\alpha y^i z^j \in S[x][y, z],$$

where S is any commutative ring, then the iterated univariate resultant

$$\mathrm{Res}_y\left(\mathrm{Res}_z(f_1, f_2), \mathrm{Res}_z(f_1, f_3)\right) \in S[x]$$

is of total degree at most $\mathfrak{d}_1^2 \mathfrak{d}_2 \mathfrak{d}_3$ in x, and we may express it in multivariate resultants (following the formalism of Jouanolou [33]) as

$$\mathrm{Res}_y\left(\mathrm{Res}_z(f_1, f_2), \mathrm{Res}_z(f_1, f_3)\right) = (-1)^{\mathfrak{d}_1 \mathfrak{d}_2 \mathfrak{d}_3} \mathrm{Res}_{y,z}(f_1, f_2, f_3) \\ \times \mathrm{Res}_{y,z,z'}\left(f_1(x, y, z), f_2(x, y, z), f_3(x, y, z'), \delta_{z,z'}(f_1)\right). \tag{8}$$

Moreover, if the polynomials f_1, f_2, f_3 are sufficiently generic and $n > 1$, then this iterated resultant has exactly total degree $\mathfrak{d}_1^2 \mathfrak{d}_2 \mathfrak{d}_3$ in x and both resultants on the right hand side of the above equality are distinct and irreducible.

[Although not stated as part of the result in [13], under these genericity assumptions, $\mathrm{Res}_{y,z}(f_1, f_2, f_3)$ has total degree $\mathfrak{d}_1 \mathfrak{d}_2 \mathfrak{d}_3$ and the second resultant on the right hand side of (8) has total degree $\mathfrak{d}_1(\mathfrak{d}_1 - 1)\mathfrak{d}_2 \mathfrak{d}_3$ (see [13, Proposition 3.3] and [38, Theorem 2.6]).]

In [13] the authors interpret this result as follows[1].

The resultant $R_{12} := \mathrm{Res}_z(f_1, f_2)$ defines the projection of the intersection curve between the two surfaces $\{f_1 = 0\}$ and $\{f_2 = 0\}$. Similarly, $R_{13} := \mathrm{Res}_z(f_1, f_3)$ defines the projection of the intersection curve between the two surfaces $\{f_1 = 0\}$ and $\{f_3 = 0\}$. Then the roots of $\mathrm{Res}_y(R_{12}, R_{13})$ can be decomposed into two distinct sets: the set of roots x_0 such that there exists y_0 and z_0 such that

$$f_1(x_0, y_0, z_0) = f_2(x_0, y_0, z_0) = f_3(x_0, y_0, z_0),$$

and the set of roots x_1 such that there exist two distinct points (x_1, y_1, z_1) and (x_1, y_1, z_1') such that

$$f_1(x_1, y_1, z_1) = f_2(x_1, y_1, z_1) \quad \text{and} \quad f_1(x_1, y_1, z_1') = f_3(x_1, y_1, z_1').$$

The first set gives rise to the term $\mathrm{Res}_{y,z}(f_1, f_2, f_3)$ in the factorization of the iterated resultant $\mathrm{Res}_y(\mathrm{Res}_{12}, \mathrm{Res}_{13})$, and the second set of roots corresponds to the second factor.

---

[1] We note that in this quote we made a small correction to the description of the second set of roots (removing a dash from y_1 in the second distinct point). We thank the anonymous referee of the present paper for identifying this correction.

Only the first set are of interest to us *if* the f_i are all ECs. However, for a general CAD construction, the second set of roots may also be necessary as they indicate points where the geometry of the sectors changes.

3.2 How Large Are These Resultants?

Suppose we are considering three ECs defined by f_1, f_2 and f_3; that we wish to eliminate two variables $z = x_n$ and $y = x_{n-1}$; and that the f_i have degree at most d in each variable *separately*. Then we may naïvely set each $\mathfrak{d}_i = nd$ to bound the total degree.

The following approach does better. Let $K = S[x_1, \ldots, x_{n-2}, y, z]$ and $L = S[\xi_1, \ldots, \xi_N, y, z]$. Only a finite number of monomials in $x_1, \ldots, x_{n-2}$ occur as coefficients of the powers of y, z in f_1, f_2 and f_3. Map each such monomial $x^\alpha = \prod_{i=1}^{n-2} x_i^{\alpha_i}$ to $\widetilde{m_j} := \xi_j^{\max \alpha_i}$ (using a different ξ_j for each monomial[2]) and let $\widetilde{f_i} \in L$ be the result of applying this map to the monomials in f_i. Note that the operation $\widetilde{}$ commutes with taking resultants in y and z (though not in the x_i of course).

The total degree in the ξ_j of $\widetilde{f_i}$ is the same as the maximum degree in all the $x_1, \ldots, x_{n-2}$ of f_i, i.e. bounded by d, and hence the total degree of the $\widetilde{f_i}$ in all variables is bounded by $3d$ (d for the ξ_i, d for y and d for z). If we apply (8) to the $\widetilde{f_i}$, we see that

$$\mathrm{Res}_y\big(\mathrm{Res}_z(\widetilde{f_1}, \widetilde{f_2}), \mathrm{Res}_z(\widetilde{f_1}, \widetilde{f_3})\big)$$

has a factor $\mathrm{Res}_{y,z}(\widetilde{f_1}, \widetilde{f_2}, \widetilde{f_3})$ of total degree (in the ξ_j) $(3d)^3$. Hence, by inverting $\widetilde{}$, we may conclude $\mathrm{Res}_{y,z}(f_1, f_2, f_3)$ has maximum degree, in each x_i, of $(3d)^3$.

The results of [13,33] apply to any number of eliminations. In particular, if we have eliminated not 2 but $\ell - 1$ variables we will have a polynomial $\mathrm{Res}_{x_{n-\ell+1}\ldots x_n}(f_{n-\ell}, \ldots, f_n)$ of maximum degree $\ell^\ell d^\ell$ in the remaining variables $x_1, \ldots, x_{n-\ell}$ as the last implicit EC.

These resultants $\mathrm{Res}_{x_{n-\ell+1}\ldots x_n}$ therefore only have singly-exponential growth, rather than the doubly-exponential growth of the iterated resultants: can we compute them?

3.3 Gröbner Bases in Place of Iterated Resultants

A *Gröbner Basis* G is a particular generating set of an ideal I defined with respect to a monomial ordering. One definition is that the ideal generated by the leading terms of I is generated by the leading terms of G. Gröbner Bases (GB) allow properties of the ideal to be deduced such as dimension and number of zeros and so are one of the main practical tools for working with polynomial systems. Their properties and an algorithm to derive a GB for any ideal were introduced by Buchberger in his PhD thesis of 1965 [11]. There has been much

---

[2] It would be possible to economise: if $x_1 x_2^2 \mapsto \xi_1^2$, then we could map $x_1^2 x_2^4$ to ξ_1^4 rather than a new ξ_2^4. Since this trick is used purely for the analysis and not in implementation, we ignore such possibilities.

research to improve and optimise GB calculation, with the F_5 algorithm [25] perhaps the most used approach currently.

Like CAD the calculation of GB is necessarily doubly exponential in the worst case [35] (when using a lexicographic order), although recent work in [36] showed that rather than being doubly exponential with respect to the number of variables present the dependency is in fact on the dimension of the ideal. Despite this worst case bound GB computation can often be done very quickly usually to the point of instantaneous for any problem tractable by CAD.

A reasonably common CAD technique is to precondition systems with multiple ECs by replacing the ECs by their GB. I.e. let $E = \{e_1, e_2, \dots\}$ be a set of polynomials; $G = \{g_1, g_2, \dots\}$ a GB for E; and B any Boolean combination of constraints, $f_i \sigma_i 0$, where $\sigma_i \in \{<, >, \leq, \geq, \neq, =\}$) and $F = \{f_1, f_2, \dots\}$ is another set of polynomials. Then

$$\Phi = (e_1 = 0 \land e_2 = 0 \land \dots) \land B$$
$$\Psi = (g_1 = 0 \land g_2 = 0 \land \dots) \land B$$

are equivalent and a CAD truth-invariant for either could be used to solve problems involving Φ.

As discussed, the cost of computing the GB itself is minimal so the question is whether it is beneficial to CAD. The first attempt to answer this question was given by Buchberger and Hong in 1991 [12] (using GB and CAD implementations in the SAC-2 system [15]). These experiments were carried out before the development of reduced projection operators and so the CADs computed were sign-invariant (and thus also truth-invariant for the formulae involved). Of the 10 problems studied: 6 were improved by the GB preconditioning, (speed-up from 2-fold to 1700-fold); 1 problem resulted in a 10-fold slow-down; 1 timed out after GB but completed without; and the other 2 were intractable both for CAD and GB. The problem was recently revisited by Wilson et al. [47] who studied the same problem set using QEPCAD-B for the CAD and MAPLE 16 for the GB. There had been a huge improvement to the time taken by GB computation but it was still the case that two of the problems were hindered by GB preconditioning. A recent machine learning experiment to decide when GB precondition should be applied [30] found that 75 % of a data set of 1200 randomly generated CAD problems benefited from GB preconditioning.

If we consider GB preconditioning of CAD in the knowledge of the improved projection schemes for ECs (Subsect. 2.4) then we see an additional benefit from the GB. It provides ECs which are not in the main variable of the system removing the need for iterated resultants to find implicit ECs to use in subsequent projections.

Since our aim is to produce one EC in each of the last ℓ variables, we need to choose an ordering on monomials which is lexicographic with respect to $x_n \succ x_{n-1} \succ \cdots \succ x_{n-\ell+1}$: it does not actually matter (from the point of view of the theory: general theory suggests that 'total degree reverse lexicographic in the rest' would be most efficient in practice) how we tie-break after that.

Let us suppose (in line with [22]) that we have ℓ ECs $f_1, \ldots, f_\ell$ (at least one of them, say f_1 must include x_n, and similarly we can assume f_2 includes x_{n-1} and so on), such that these imply (even over $\mathbb{C}$) that the last ℓ variables are determined (not necessarily uniquely) by the values of $x_1, \ldots, x_{n-\ell}$. Then the polynomials f_1, $\mathrm{Res}_{x_n}(f_1, f_2)$, $\mathrm{Res}_{x_n, x_{n-1}}(f_1, f_2, f_3)$ etc. are all implied by the f_i. Hence either they are in the GB, or they are reduced to 0 by the GB, which implies that smaller polynomials are in the GB. Hence our GB will contain polynomials (which are ECs) of degree (in each variable separately) at most

$$d, \ 4d^2, \ 27d^3, \ \ldots, \ ((\ell+1)d)^{\ell+1}.$$

Note that we are not making, and in the light of [36] cannot make, any similar claim about the polynomials in fewer variables. Note also that it is vital that the equations be in the last variables for this use of [13,33] to work. That is, our results do not directly extend from the case we study, of first applying ℓ reduced CAD projections in the presence of ECs (before reverting to the standard ones), to the more general case of having any ℓ of the projections be reduced.

4   Worked Example

We will work with the polynomials

$$f_1 := xy - z^2 - w^2 + 2z, \qquad f_2 := x^2 + y^2 + z^2 + w + z,$$
$$f_3 := -w^2 - y^2 - z^2 + x + z \qquad h := z + w,$$

and the semi-algebraic system

$$\phi := f_1 = 0 \wedge f_2 = 0 \wedge f_3 = 0 \wedge h > 0.$$

We assume a variable ordering $z \succ y \succ x \succ w$ (meaning we will first project with respect to z) and seek a CAD truth-invariant for ϕ.

In theory, we could analyse this system with a sign-invariant CAD for the four polynomials $\{f_1, f_2, f_3, h\}$. However in MAPLE neither our own CAD implementation [23] nor the CAD command within the REGULARCHAINS Library[3] detailed in [3,14] finished within 30 min.

Instead, we should take advantage of the ECs available. There are 3 explicit ECs within the input formula. However, they all have main variable z and so only one of them may be a designated EC for projection purposes (and trying to do this still results in a time-out after 30 min). The existing theory [22,40] would suggest propagating the ECs by calculating:

$$r_1 := \mathrm{res}(f_1, f_2, z) = y^4 + 2xy^3 + (3x^2 - 2w^2 + 2w + 6)y^2 + (2x^3 - 2w^2 x$$
$$+ 2wx - 3x)y + x^4 - 2w^2 x^2 + 2wx^2 + 6x^2 + w^4 - 2w^3 + 4w^2 + 6w,$$

$$r_2 := \mathrm{res}(f_1, f_3, z) = y^4 + 2xy^3 + (x^2 - 2x + 2)y^2 + (x - 2x^2)y + w^2 + x^2 - 2x,$$

$$r_3 := \mathrm{res}(f_2, f_3, z) = 4y^2 + x^4 + 2x^3 - 2w^2 x^2 + 2wx^2 + 3x^2 - 2w^2 x + 2wx - 2x$$
$$+ w^4 - 2w^3 + 3w^2 + 2w;$$

---

[3] As downloaded from www.regularchains.org on 11th March 2016.

three implicit ECs with main variable y. We may continue to calculate ECs with main variable x as:

$$R_1 := \text{res}(r_1, r_2, y), \qquad R_2 := \text{res}(r_1, r_3, y), \qquad R_3 := \text{res}(r_2, r_3, y);$$

which evaluate to three different degree 16 polynomials in x available in the Appendix. All possible resultants of these to eliminate x evaluate to 0 (and a numerical plot of the R_i shows them all to have overlapping real part).

We now have multiple choices for running Algorithm 1 since we can only declare one polynomial as an EC with a set main variable. There are hence $3 \times 3 \times 3 = 27$ possible configurations. We attempt to build the CAD for each choice (lifting using the improved procedure developed in [22]) and found that 6 configurations complete within 30 min. Of these there was an average of 152 cells calculated in 65 s. The optimal configuration gave 111 cells in 23 s using a designation of f_2, r_3 and R_2.

Now consider taking a GB of $\{f_1, f_2, f_3\}$. We use a plex monomial ordering on the same variable ordering as the CAD to achieve a basis defined by

$$g_1 = 2z + x^2 + x - w^2 + w,$$

$$\begin{aligned} g_2 = {} & 4y^2 + x^4 + 2x^3 + (-2w^2 + 2w + 3)x^2 + (2w^2 + 2w - 2)x \\ & + w^4 - 2w^3 + 3w^2 + 2w, \end{aligned}$$

$$\begin{aligned} g_3 = {} & 4yx - x^4 - 2x^3 + (2w^2 - 2w - 5)x^2 + (2w^2 - 2w - 4)x \\ & - w^4 + 2w^3 - w^2 - 4w, \end{aligned}$$

$$\begin{aligned} g_4 = {} & (4w^4 - 8w^3 + 4w^2 + 16w)y + x^7 + 4x^6 + (-4w^2 + 4w + 18)x^5 \\ & + (-12w^2 + 12w + 36)x^4 + (5w^4 - 10w^3 - 31w^2 + 40w + 53)x^3 \\ & + (10w^4 - 20w^3 - 34w^2 + 52w + 32)x^2 + (-2w^6 + 6w^5 + 7w^4 - 32w^3 \\ & + 13w^2 + 44w + 16)x - 2w^6 + 6w^5 - 2w^4 - 14w^3 + 12w^2 + 16w, \end{aligned}$$

$$\begin{aligned} g_5 = {} & x^8 + 4x^7 + (-4w^2 + 4w + 18)x^6 + (-12w^2 + 12w + 36)x^5 + (6w^4 - 12w^3 \\ & - 30w^2 + 44w + 53)x^4 + (12w^4 - 24w^3 - 32w^2 + 60w + 32)x^3 \\ & + (-4w^6 + 12w^5 + 6w^4 - 48w^3 + 26w^2 + 64w + 16)x^2 \\ & + (-4w^6 + 12w^5 - 4w^4 - 28w^3 + 24w^2 + 32w)x \\ & + w^8 - 4w^7 + 6w^6 + 4w^5 - 15w^4 + 8w^3 + 16w^2. \end{aligned}$$

This is an alternative generating set for the ideal defined by the explicit ECs and thus all the $g_i = 0$ are also ECs for ϕ. Hence we may consider using these as the designated ECs when building the CAD instead of the iterated resultants. Note that the degrees of the GB polynomials (with respect to any one variable) are on average lower (and never greater) than those of the (corresponding) iterated resultants.

There is no longer any choice regarding the EC with mvar z or x but there are 3 possibilities for the designation with mvar y. Designating g_2 yields 83 cells while either g_3 or g_4 result in 55 cells. All 3 configurations took less than 10 s to compute (with designating g_4 the quickest).

5   Sketch of the Effect on Complexity

Following Sect. 3 we see that when building a lexicographical basis the degree of the polynomials in the GB is restricted and thus this will be a better method for the identification of implicit ECs to use in subsequent projections than iterated resultants. Let us sketch how this will effect the complexity of CAD following the techniques set out in [5, 22] and summarised in Sect. 2.4.

The designated ECs will have lower degrees $d, 4d^2, 27d^3$ and in general $(sd)^s$ for the EC with mvar x_{n-s}. We use the word *sketch* in the section title partly because we will ignore the constant factors and focus on the exponents of d generated in what follows. This is both for simplicity in the analysis, and because we have not found a closed form solution for the product of the constant factors in the new analysis. But we do note that when using GB the constant factors grow exponentially in ℓ while with iterated resultants they grow doubly exponentially in ℓ (as in Table 1). Further, the constant term can be shown to be strictly lower for all but the first few projections, with the issue there a laxness of the analysis not the algorithm (as in Sect. 3.1 we saw that the multivariate resultants was itself a factor of the iterated resultant). The other issues which prompted us to use the word *sketch* are discussed in the next section.

We keep track of both the degree of the designated EC and the degree of the entire set of polynomials. The reduced projection operator $P_F(B)$ will still take discriminants and coefficients of these; and resultants of them with the other projection polynomials. Thus the highest degree polynomial produced grows with the exponent of d being the sum of the exponent from the designated EC and that from the other polynomials. This generates the top half of Table 2 and we see that the exponents form the so called *Lazy Caterer's sequence*[4] otherwise known as the Central Polygonal Numbers. The remaining projections use the sign-invariant projection operator and so the degree is squared each time, leading to the bottom half of Table 2.

We can now consider the generic bound (7) using the degrees from Table 2 as the d_i. The term with base d may be computed by

$$\prod_{s=0}^{\ell} \left(d^{s+1}\right) \prod_{r=1}^{n-\ell-1} \left(d^{2^{r-1}\ell(\ell+1)+2^r}\right).$$

The exponent of d evaluates to

$$2^{(n-\ell)}\tfrac{1}{2}(\ell^2 + \ell + 2) - \tfrac{1}{2}(\ell^2 + \ell) - 2. \tag{9}$$

Let us compare this with the term with base m from (2). As with the improvements in [22], the improvements here have allowed the reduction of the double exponent from by ℓ, the number of ECs used. However, the reduction is not quite as clean as the exponential term in the single exponent is multiplied by a quadratic in ℓ. This is to be expected as the singly exponential dependency on ℓ in the Number column of Table 1 was only in the term with constant base while for Table 2 the term with base d is itself single exponential in ℓ.

---

[4] The On-Line Encyclopedia of Integer Sequences, 2010, Sequence A000124, https://oeis.org/A000124.

Table 2. Maximum degree of projection polynomials produced for CAD when using projection operator (5) ℓ times and then (3).

Variables	Maximum degree	
	EC	Others
n	d	d
$n-1$	d^2	d^2
$n-2$	d^3	d^4
$n-3$	d^4	d^7
$\vdots$	$\vdots$	$\vdots$
$n-\ell$	$d^{\ell+1}$	$d^{\ell(\ell+1)(1/2)+1}$
$n-(\ell+1)$	$d^{\ell(\ell+1)+2}$	
$n-(\ell+2)$	$d^{2\ell(\ell+1)+2^2}$	
$n-(\ell+3)$	$d^{2^2\ell(\ell+1)+2^3}$	
$\vdots$	$\vdots$	
$n-(\ell+r)$	$d^{2^{r-1}\ell(\ell+1)+2^r}$	
$\vdots$	$\vdots$	
1	$d^{2^{n-\ell-2}\ell(\ell+1)+2^{n-\ell-1}}$	

6   Discussion

We have considered the issue of CAD in the presence of multiple ECs. We followed our recent work in [22] which reduced the complexity with respect to the number of polynomials m, and showed that similar improvements can be obtained with the respect to polynomial degree d by using Gröbner Bases in place of iterated resultants. We have sketched the complexity results but defer the full analysis until a number of issues can be cleared up. These include:

– Will using GB not risk increasing the base number of polynomials in m?

On one level this seems unlikely (since we are starting with a generating set all in the main variable and deriving another which would mostly not be) but we have yet to rule it out. Of course, the number of polynomials in the input can bear little relation to the number generated by projection.

   We note that there is an alternative way to use GB for CAD than that outlined in Sect. 3.3 (replacing a set of ECs by another). We could instead use the GB purely as an implicit EC generation tool; and just add selected polynomials from it to our input without replacing anything. For example, the GB in the worked example of Sect. 4 had 3 polynomials with main variable y only one of which can be the designated EC. Rather than replacing all the f_i by all the g_i we could instead just add 2 of the g_i (one in main variable y and one in x) to the input set to act as designated ECs at lower levels. This approach would cap the increase in m to the number of designated ECs we can identify.

- Will the GB always produce as many ECs with different main variables as the iterated resultant method?
- How to proceed in the case where we have non-primitive ECs?

As with most previous work on ECs, we have assumed primitive designated ECs. We refer the reader to: the final section of [22] where we sketch approaches that could be adapted to deal with this (including the theory of TTICAD [4,5]); and the final section of [19] where we demonstrate the importance of this issue by showing the examples from [9,20] to involve imprimitive ECs.

- How is the complexity affected when the projections using ECs are not in strict succession?
- Can we mix the orderings in the CAD and the GB?

Finally, we return to the fact acknowledged in Sect. 3.3 that previous work on using GB to precondition CAD [12,30,47] has found that it is not always beneficial and how that interacts with the claims of this paper. The simple answer is that the analysis offered here is of the worst case and makes no claim to the average complexity. However, we actually hypothesise that it was it was the fact that the CAD computations involved in those paper did not take advantage of the new multiple EC technology which will account for many of the cases were GB hindered CAD. We plan future experiments to test this hypothesis.

Acknowledgements. This work was originally supported by EPSRC grant: EP/J003247/1 and is now supported by EU H2020-FETOPEN-2016-2017-CSA project $\mathcal{SC}^2$ (712689). Thanks to the referees for their helpful comments, and Prof. Buchberger for reminding JHD that Gröbner bases were applicable here.

Appendix

A   The Iterated Resultants from Section 4

$$
\begin{aligned}
R_1 := \operatorname{res}(r_1, r_2, y) = {} & x^{16} + 8\,x^{15} + (-8\,w^2 + 8\,w + 64)x^{14} + (-56\,w^2 + 56\,w \\
& + 288)x^{13} + (28\,w^4 - 56\,w^3 - 332\,w^2 + 400\,w + 1138)x^{12} + (168\,w^4 \\
& - 336\,w^3 - 1144\,w^2 + 1552\,w + 2912)x^{11} + (-56\,w^6 + 168\,w^5 + 648\,w^4 \\
& - 1816\,w^3 - 2664\,w^2 + 5328\,w + 6336)x^{10} + (-280\,w^6 + 840\,w^5 \\
& + 1400\,w^4 - 5400\,w^3 - 2616\,w^2 + 11368\,w + 7808)x^9 + (70\,w^8 \\
& - 280\,w^7 - 500\,w^6 + 3080\,w^5 - 270\,w^4 - 11576\,w^3 + 4860\,w^2 \\
& + 20816\,w + 7381)x^8 + (280\,w^8 - 1120\,w^7 + 80\,w^6 + 6080\,w^5 - 8480\,w^4 \\
& - 11792\,w^3 + 22840\,w^2 + 20192\,w + 920)x^7 + (-56\,w^{10} + 280\,w^9 \\
& - 80\,w^8 - 2160\,w^7 + 4960\,w^6 + 3200\,w^5 - 22608\,w^4 + 2584\,w^3 \\
& + 40840\,w^2 + 16040\,w + 2024)x^6 + (-168\,w^{10} + 840\,w^9 - 1520\,w^8
\end{aligned}
$$

$$
\begin{aligned}
&- 1360\,w^7 + 12016\,w^6 - 11296\,w^5 - 23368\,w^4 + 30136\,w^3 + 22032\,w^2 \\
&+ 624\,w + 736)x^5 + (28\,w^{12} - 168\,w^{11} + 396\,w^{10} + 160\,w^9 - 3690\,w^8 \\
&+ 6576\,w^7 + 4520\,w^6 - 24712\,w^5 + 13154\,w^4 + 37456\,w^3 + 1464\,w^2 \\
&- 1568\,w + 5968)x^4 + (56\,w^{12} - 336\,w^{11} + 1192\,w^{10} - 1680\,w^9 \\
&- 2688\,w^8 + 12496\,w^7 - 13464\,w^6 - 16912\,w^5 + 37240\,w^4 + 13472\,w^3 \\
&- 16384\,w^2 + 1984\,w + 3072)x^3 + (-8\,w^{14} + 56\,w^{13} - 248\,w^{12} + 520\,w^{11} \\
&+ 72\,w^{10} - 3088\,w^9 + 7664\,w^8 - 2040\,w^7 - 16176\,w^6 + 20424\,w^5 \\
&+ 20056\,w^4 - 15360\,w^3 - 8544\,w^2 + 4608\,w + 2304)x^2 + (-8\,w^{14} \\
&+ 56\,w^{13} - 296\,w^{12} + 808\,w^{11} - 1144\,w^{10} - 776\,w^9 + 6184\,w^8 - 7048\,w^7 \\
&- 6944\,w^6 + 19696\,w^5 + 3872\,w^4 - 16832\,w^3 - 1152\,w^2 + 4608\,w)x + w^{16} \\
&- 8\,w^{15} + 52\,w^{14} - 184\,w^{13} + 454\,w^{12} - 440\,w^{11} - 772\,w^{10} + 3352\,w^9 \\
&- 2447\,w^8 - 4288\,w^7 + 8200\,w^6 + 2080\,w^5 - 7664\,w^4 - 384\,w^3 + 2304\,w^2
\end{aligned}
$$

$$
\begin{aligned}
R_2 := \mathrm{res}(r_1, r_3, y) = \; & x^{16} + 8\,x^{15} + (-8\,w^2 + 8\,w + 28)x^{14} + (-56\,w^2 + 56\,w \\
&+ 48)x^{13} + (28\,w^4 - 56\,w^3 - 116\,w^2 + 160\,w - 2)x^{12} + (168\,w^4 \\
&- 336\,w^3 + 80\,w^2 + 184\,w - 256)x^{11} + (-56\,w^6 + 168\,w^5 + 108\,w^4 \\
&- 592\,w^3 + 852\,w^2 - 240\,w - 12)x^{10} + (-280\,w^6 + 840\,w^5 - 1120\,w^4 \\
&+ 360\,w^3 + 1872\,w^2 - 1448\,w + 2000)x^9 + (70\,w^8 - 280\,w^7 + 220\,w^6 \\
&+ 560\,w^5 - 2742\,w^4 + 3232\,w^3 - 1428\,w^2 + 224\,w + 4537)x^8 + (280\,w^8 \\
&- 1120\,w^7 + 2720\,w^6 - 3280\,w^5 - 1280\,w^4 + 6016\,w^3 - 11696\,w^2 + 7496\,w \\
&+ 2552)x^7 + (-56\,w^{10} + 280\,w^9 - 620\,w^8 + 480\,w^7 + 2488\,w^6 - 6880\,w^5 \\
&+ 9384\,w^4 - 5744\,w^3 - 9404\,w^2 + 12008\,w - 4120)x^6 + (-168\,w^{10} \\
&+ 840\,w^9 - 2960\,w^8 + 5840\,w^7 - 4832\,w^6 - 3088\,w^5 + 21104\,w^4 \\
&- 27128\,w^3 + 12552\,w^2 + 3888\,w - 5888)x^5 + (28\,w^{12} - 168\,w^{11} + 612\,w^{10} \\
&- 1280\,w^9 + 498\,w^8 + 3648\,w^7 - 12424\,w^6 + 17360\,w^5 - 4546\,w^4 \\
&- 13928\,w^3 + 19032\,w^2 - 9344\,w - 176)x^4 + (56w^{12} - 336w^{11} + 1552w^{10} \\
&- 4200\,w^9 + 7296\,w^8 - 6080\,w^7 - 7440\,w^6 + 25880\,w^5 - 31352\,w^4 \\
&+ 13472\,w^3 + 1856\,w^2 - 10304\,w + 1536)x^3 + (-8\,w^{14} + 56\,w^{13} - 284\,w^{12} \\
&+ 880\,w^{11} - 1740\,w^{10} + 1616\,w^9 + 2468\,w^8 - 10704\,w^7 + 15828\,w^6 \\
&- 8040\,w^5 - 1064\,w^4 + 9792\,w^3 - 3168\,w^2 + 2304)x^2 + (-8\,w^{14} + 56\,w^{13} \\
&- 320\,w^{12} + 1096\,w^{11} - 2800\,w^{10} + 4600\,w^9 - 3968\,w^8 - 2152\,w^7 \\
&+ 9592\,w^6 - 10832\,w^5 + 5312\,w^4 + 4672\,w^3 - 5760\,w^2 + 4608\,w)x + w^{16} \\
&- 8\,w^{15} + 52\,w^{14} - 208\,w^{13} + 646\,w^{12} - 1376\,w^{11} + 2012\,w^{10} - 1136\,w^9 \\
&- 1295\,w^8 + 4328\,w^7 - 3992\,w^6 + 2368\,w^5 + 2320\,w^4 - 1920\,w^3 + 2304\,w^2
\end{aligned}
$$

$$R_3 := \operatorname{res}(r_3, r_3, y) = x^{16} + 8\,x^{15} + (-8\,w^2 + 8\,w + 44)x^{14} + (-56\,w^2 + 56\,w$$
$$+ 160)x^{13} + (28\,w^4 - 56\,w^3 - 228\,w^2 + 272\,w + 430)x^{12} + (168\,w^4$$
$$- 336\,w^3 - 592\,w^2 + 856\,w + 816)x^{11} + (-56\,w^6 + 168\,w^5 + 444\,w^4$$
$$- 1264\,w^3 - 812\,w^2 + 1952\,w + 1092)x^{10} + (-280\,w^6 + 840\,w^5 + 560\,w^4$$
$$- 3000\,w^3 + 32\,w^2 + 3032\,w + 736)x^9 + (70\,w^8 - 280\,w^7 - 340\,w^6$$
$$+ 2240\,w^5 - 902\,w^4 - 4208\,w^3 + 2716\,w^2 + 3120\,w - 183)x^8 + (280\,w^8$$
$$- 1120\,w^7 + 480\,w^6 + 3440\,w^5 - 4640\,w^4 - 2304\,w^3 + 5840\,w^2 + 1128\,w$$
$$- 1144)x^7 + (-56\,w^{10} + 280\,w^9 - 60\,w^8 - 1760\,w^7 + 3128\,w^6 + 960\,w^5$$
$$- 7352\,w^4 + 3216\,w^3 + 5860\,w^2 - 1320\,w - 824)x^6 + (-168\,w^{10} + 840\,w^9$$
$$- 1280\,w^8 - 880\,w^7 + 5568\,w^6 - 5008\,w^5 - 4464\,w^4 + 7848\,w^3 + 984\,w^2$$
$$- 2576\,w - 64)x^5 + (28\,w^{12} - 168\,w^{11} + 276\,w^{10} + 400\,w^9 - 2302\,w^8$$
$$+ 2848\,w^7 + 1880\,w^6 - 7440\,w^5 + 3582\,w^4 + 5704\,w^3 - 3208\,w^2 - 1216\,w$$
$$+ 720)x^4 + (56\,w^{12} - 336\,w^{11} + 880\,w^{10} - 840\,w^9 - 1424\,w^8 + 4800\,w^7$$
$$- 3856\,w^6 - 3464\,w^5 + 6968\,w^4 + 32\,w^3 - 3392\,w^2 + 448\,w + 512)x^3$$
$$+ (-8\,w^{14} + 56\,w^{13} - 172\,w^{12} + 208\,w^{11} + 308\,w^{10} - 1504\,w^9 + 1972\,w^8$$
$$+ 432\,w^7 - 3788\,w^6 + 2920\,w^5 + 2552\,w^4 - 3136\,w^3 - 864\,w^2 + 1024\,w$$
$$+ 256)x^2 + (-8\,w^{14} + 56\,w^{13} - 208\,w^{12} + 424\,w^{11} - 352\,w^{10} - 520\,w^9$$
$$+ 1744\,w^8 - 1416\,w^7 - 1176\,w^6 + 2928\,w^5 - 384\,w^4 - 1984\,w^3 + 384\,w^2$$
$$+ 512\,w)x + w^{16} - 8\,w^{15} + 36\,w^{14} - 96\,w^{13} + 150\,w^{12} - 48\,w^{11} - 308\,w^{10}$$
$$+ 672\,w^9 - 351\,w^8 - 648\,w^7 + 1096\,w^6 - 880\,w^4 + 128\,w^3 + 256\,w^2$$

References

1. Arnon, D., Collins, G.E., McCallum, S.: Cylindrical algebraic decomposition I: the basic algorithm. SIAM J. Comput. **13**, 865–877 (1984)
2. Basu, S., Pollack, R., Roy, M.F.: Algorithms in Real Algebraic Geometry. Algorithms and Computations in Mathematics, vol. 10. Springer, Heidelberg (2006)
3. Bradford, R., Chen, C., Davenport, J.H., England, M., Moreno Maza, M., Wilson, D.: Truth table invariant cylindrical algebraic decomposition by regular chains. In: Gerdt, V.P., Koepf, W., Seiler, W.M., Vorozhtsov, E.V. (eds.) CASC 2014. LNCS, vol. 8660, pp. 44–58. Springer, Heidelberg (2014)
4. Bradford, R., Davenport, J.H., England, M., McCallum, S., Wilson, D.: Cylindrical algebraic decompositions for boolean combinations. In: Proceedings of ISSAC 2013, pp. 125–132. ACM (2013)
5. Bradford, R., Davenport, J.H., England, M., McCallum, S., Wilson, D.: Truth table invariant cylindrical algebraic decomposition. J. Symbolic Comput. **76**, 1–35 (2016)
6. Bradford, R., Davenport, J.H., England, M., Wilson, D.: Optimising problem formulation for cylindrical algebraic decomposition. In: Carette, J., Aspinall, D., Lange, C., Sojka, P., Windsteiger, W. (eds.) CICM 2013. LNCS, vol. 7961, pp. 19–34. Springer, Heidelberg (2013)

7. Brown, C.W.: Improved projection for cylindrical algebraic decomposition. J. Symbolic Comput. **32**(5), 447–465 (2001)
8. Brown, C.W.: Constructing a single open cell in a cylindrical algebraic decomposition. In: Proceedings of ISSAC 2013, pp. 133–140. ACM (2013)
9. Brown, C.W., Davenport, J.H.: The complexity of quantifier elimination and cylindrical algebraic decomposition. In: Proceedings of ISSAC 2007, pp. 54–60. ACM (2007)
10. Brown, C.W., El Kahoui, M., Novotni, D., Weber, A.: Algorithmic methods for investigating equilibria in epidemic modelling. J. Symbolic Comput. **41**, 1157–1173 (2006)
11. Buchberger, B.: Bruno Buchberger's PhD thesis (1965): an algorithm for finding the basis elements of the residue class ring of a zero dimensional polynomial ideal. J. Symbolic Comput. **41**(3–4), 475–511 (2006)
12. Buchberger, B., Hong, H.: Speeding up quantifier elimination by Gröbner bases. Technical report, 91–06. RISC, Johannes Kepler University (1991)
13. Busé, L., Mourrain, B.: Explicit factors of some iterated resultants and discriminants. Math. Comput. **78**, 345–386 (2009)
14. Chen, C., Moreno, M.M., Xia, B., Yang, L.: Computing cylindrical algebraic decomposition via triangular decomposition. In: Proceedings of ISSAC 2009, pp. 95–102. ACM (2009)
15. Collins, G.E.: The SAC-2 computer algebra system. In: Caviness, B.F. (ed.) EUROCAL 1985. LNCS, vol. 204, pp. 34–35. Springer, Heidelberg (1985)
16. Collins, G.E.: Quantifier elimination by cylindrical algebraic decomposition - 20 years of progress. In: Caviness, B.F., Johnson, J.R. (eds.) Quantifier Elimination and Cylindrical Algebraic Decomposition. Texts & Monographs in Symbolic Computation, pp. 8–23. Springer, Heidelberg (1998)
17. Collins, G.E., Hong, H.: Partial cylindrical algebraic decomposition for quantifier elimination. J. Symbolic Comput. **12**, 299–328 (1991)
18. Davenport, J.H., Bradford, R., England, M., Wilson, D.: Program verification in the presence of complex numbers, functions with branch cuts etc. In: Proceedings of SYNASC 2012, pp. 83–88. IEEE (2012)
19. Davenport, J.H., England, M.: Need polynomial systems be doubly-exponential? In: Greuel, G.-M., Koch, T., Paule, P., Sommese, A. (eds.) ICMS 2016. LNCS, vol. 9725, pp. 157–164. Springer, Heidelberg (2016)
20. Davenport, J.H., Heintz, J.: Real quantifier elimination is doubly exponential. J. Symbolic Comput. **5**(1–2), 29–35 (1988)
21. England, M., Bradford, R., Chen, C., Davenport, J.H., Maza, M.M., Wilson, D.: Problem formulation for truth-table invariant cylindrical algebraic decomposition by incremental triangular decomposition. In: Watt, S.M., Davenport, J.H., Sexton, A.P., Sojka, P., Urban, J. (eds.) CICM 2014. LNCS, vol. 8543, pp. 45–60. Springer, Heidelberg (2014)
22. England, M., Bradford, R., Davenport, J.H.: Improving the use of equational constraints in cylindrical algebraic decomposition. In: Proceedings of ISSAC 2015, pp. 165–172. ACM (2015)
23. England, M., Wilson, D., Bradford, R., Davenport, J.H.: Using the regular chains library to build cylindrical algebraic decompositions by projecting and lifting. In: Hong, H., Yap, C. (eds.) ICMS 2014. LNCS, vol. 8592, pp. 458–465. Springer, Heidelberg (2014)
24. Erascu, M., Hong, H.: Synthesis of optimal numerical algorithms using real quantifier elimination (case study: square root computation). In: Proceedings of ISSAC 2014, pp. 162–169. ACM (2014)

25. Faugère, J.C.: A new efficient algorithm for computing groebner bases without reduction to zero (F5). In: Proceedings of ISSAC 2002, pp. 75–83. ACM (2002)
26. Fotiou, I.A., Parrilo, P.A., Morari, M.: Nonlinear parametric optimization using cylindrical algebraic decomposition. In: 2005 European Control Conference on Decision and Control, CDC-ECC 2005, pp. 3735–3740 (2005)
27. Han, J., Dai, L., Xia, B.: Constructing fewer open cells by gcd computation in CAD projection. In: Proceedings of ISSAC 2014, pp. 240–247. ACM (2014)
28. Heintz, J.: Definability and fast quantifier elimination in algebraically closed fields. Theor. Comput. Sci. **24**(3), 239–277 (1983)
29. Hong, H.: An improvement of the projection operator in cylindrical algebraic decomposition. In: Proceedings of ISSAC 1990, pp. 261–264. ACM (1990)
30. Huang, Z., England, M., Davenport, J.H., Paulson, L.: Using machine learning to decide when to precondition cylindrical algebraic decomposition with Groebner bases (2016, to appear)
31. Huang, Z., England, M., Wilson, D., Davenport, J.H., Paulson, L.C., Bridge, J.: Applying machine learning to the problem of choosing a heuristic to select the variable ordering for cylindrical algebraic decomposition. In: Watt, S.M., Davenport, J.H., Sexton, A.P., Sojka, P., Urban, J. (eds.) CICM 2014. LNCS, vol. 8543, pp. 92–107. Springer, Heidelberg (2014)
32. Iwane, H., Yanami, H., Anai, H., Yokoyama, K.: An effective implementation of a symbolic-numeric cylindrical algebraic decomposition for quantifier elimination. In: Proceedings of SNC 2009, pp. 55–64 (2009)
33. Jouanolou, J.P.: Le formalisme du résultant. Adv. Math. **90**(2), 117–263 (1991)
34. Lazard, D., McCallum, S.: Iterated discriminants. J. Symbolic Comput. **44**(9), 1176–1193 (2009)
35. Mayr, E.W., Meyer, A.R.: The complexity of the word problems for commutative semigroups and polynomial ideals. Adv. Math. **46**(3), 305–329 (1982)
36. Mayr, E.W., Ritscher, S.: Dimension-dependent bounds for gröbner bases of polynomial ideals. J. Symbolic Comput. **49**, 78–94 (2013)
37. McCallum, S.: An improved projection operation for cylindrical algebraic decomposition. In: Caviness, B.F., Johnson, J.R. (eds.) Quantifier Elimination and Cylindrical Algebraic Decomposition. Texts & Monograph in Symbolic Computation, pp. 242–268. Springer, Heidelberg (1998)
38. McCallum, S.: Factors of iterated resultants and discriminants. J. Symbolic Comput. **27**(4), 367–385 (1999)
39. McCallum, S.: On projection in CAD-based quantifier elimination with equational constraint. In: Proceedings of ISSAC 1999, pp. 145–149. ACM (1999)
40. McCallum, S.: On propagation of equational constraints in CAD-based quantifier elimination. In: Proceedings of ISSAC 2001, pp. 223–231. ACM (2001)
41. Paulson, L.C.: MetiTarski: past and future. In: Beringer, L., Felty, A. (eds.) ITP 2012. LNCS, vol. 7406, pp. 1–10. Springer, Heidelberg (2012)
42. Schwartz, J.T., Sharir, M.: On the "Piano-Movers" problem: I. the case of a two-dimensional rigid polygonal body moving amidst polygonal barriers. Commun. Pure Appl. Math. **36**(3), 345–398 (1983)
43. Strzeboński, A.: Cylindrical algebraic decomposition using validated numerics. J. Symbolic Comput. **41**(9), 1021–1038 (2006)
44. Strzeboński, A.: Cylindrical algebraic decomposition using local projections. In: Proceedings of ISSAC 2014, pp. 389–396. ACM (2014)
45. Wilson, D., Bradford, R., Davenport, J.H., England, M.: Cylindrical algebraic subdecompositions. Math. Comput. Sci. **8**, 263–288 (2014)

46. Wilson, D., England, M., Davenport, J.H., Bradford, R.: Using the distribution of cells by dimension in a cylindrical algebraic decomposition. In: Proceedings of SYNASC 2014, pp. 53–60. IEEE (2014)
47. Wilson, D.J., Bradford, R.J., Davenport, J.H.: Speeding up cylindrical algebraic decomposition by Gröbner bases. In: Jeuring, J., Campbell, J.A., Carette, J., Dos Reis, G., Sojka, P., Wenzel, M., Sorge, V. (eds.) CICM 2012. LNCS, vol. 7362, pp. 280–294. Springer, Heidelberg (2012)

Efficient Simplification Techniques for Special Real Quantifier Elimination with Applications to the Synthesis of Optimal Numerical Algorithms

Mădălina Eraşcu[1,2(✉)]

[1] Faculty of Mathematics and Computer Science, West University of Timişoara,
Timişoara, Romania
[2] Institute e-Austria Timişoara, Timişoara, Romania
madalina.erascu@e-uvt.ro

Abstract. This paper presents efficient simplification techniques tailored for sign semi-definite conditions (SsDCs). The SsDCs for a polynomial $f \in \mathbb{R}[y]$ with parametric coefficients are written as $\underset{\substack{y \\ L \leq y \leq U}}{\forall} f(y) \geq 0$ and $\underset{\substack{y \\ L \leq y \leq U}}{\forall} f(y) \leq 0$. We give sufficient conditions for the simplification techniques to be sound for linear and quadratic polynomials. We show their effectiveness compared to state of the art quantifier elimination tools for input formulae occurring in the optimal numerical algorithms synthesis problem by an implementation on top of **Reduce** command of Mathematica.

1 Introduction

Real quantifier elimination is a fundamental problem which arises in many challenging problems, in application areas like engineering, numerical analysis, program analysis, databases. Thus, there has been extensive research on developing efficient algorithms and software systems.

Due to Tarski [22], we have two remarkable results: (1) the theory of real closed fields (RCF) admits quantifier elimination and (2) a decision procedure for the theory was developed. Unfortunately, the decision procedure had non-elementary complexity so was totally impractical. It was Collins [7] who gave the first effective method of quantifier elimination (QE) using cylindrical algebraic decomposition (CAD). Since then, QE undergoes many improvements [6], however QE is inherently doubly exponential [9] which makes it difficult to be used to medium size problems. The improvements of the QE methods over RCF are performed in different phases of the CAD-based algorithm [3,5,6,21] or are taking into account the anatomy of the input formulae [6,15] or the structure of the quantifiers blocks [14], but also employ various heuristics [10]. State of the art tools implementing QE over RCF are QEPCAD-B [4], **Reduce** command of Mathematica [19], Redlog [11], SynRAC [2]. In the past years, one could notice

© Springer International Publishing AG 2016
V.P. Gerdt et al. (Eds.): CASC 2016, LNCS 9890, pp. 193–211, 2016.
DOI: 10.1007/978-3-319-45641-6_13

the efforts in adapting and integrating existing QE methods for RCF into the so-called satisfiability modulo theory (SMT) solvers in order to solve industrial-size problems [8,16]. In the SMT community, formula preprocessing is fundamental before actually feeding it to the SMT solver [1].

In this paper, we present preprocessing/simplification methods for input formulae of types:

$$\forall_{\substack{y \\ L \leq y \leq U}} f(y) \geq 0 \qquad\qquad \forall_{\substack{y \\ L \leq y \leq U}} f(y) \leq 0$$

where $f \in \mathbb{R}[y]$, $f(y) = y^n + a_{n-1}y^{n-1} + \ldots + a_1y + a_0$ where $a_{n-1}, \ldots, a_0$ are real parameters. More precisely, we give quantifier-free conditions equivalent to each of them. These conditions depend on the behavior (monotone, convex, concave) of f on the interval $[L, U]$, hence we give sufficient conditions for a certain behavior for linear and quadratic case.

These simplification methods were motivated by our previous work [12,13] and aim for the automation of the QE where state of the art techniques fail.

Motivating Case Study. Consider the fundamental operation of computing the square root of a given real number. For the problem, various numerical methods have been developed: [17,20]. We consider an interval version of the problem [18] : given a real number x and an error bound ε, find an interval such that it contains $\sqrt{x}$ and its width is less than or equal to ε. A typical interval method starts with an initial interval and repeatedly updates it by applying a *refinement map*, say R, on it until it becomes narrow enough (see below).

---

in: $x > 0$, $\varepsilon > 0$
out: I, an interval such that $\sqrt{x} \in I$ and width$(I) \leq \varepsilon$

$I \leftarrow [\min(1, x), \max(1, x)]$
while width$(I) > \varepsilon$
$\quad I \leftarrow R(I, x)$
return I

A well known hand-crafted refinement map (called *Secant-Newton*) is given by

$$R^*([L, U], x) \;\mapsto\; \left[L + \frac{x - L^2}{L + U}, U + \frac{x - U^2}{2U}\right]$$

The problem is to check if there is any refinement map which is better than Secant-Newton. At this aim, we fixed a search space, that is, a family of maps in which we search for a better map:

$$R_{p,q} : [L, U], x \mapsto [L', U']$$

$$L' = L + \frac{x + p_0L^2 + p_1LU + p_2U^2}{p_3L + p_4U} \qquad U' = U + \frac{x + q_0U^2 + q_1UL + q_2L^2}{q_3U + q_4L}$$

For solving the problem we need to find the values of p and q such that the resulting algorithm is optimal among the correct, terminating and quadratic convergent ones. It can be stated as the following constrained optimization problem:

Minimize
$$E(p,q) := \sup_{\substack{0 < L \le \sqrt{x} \le U \\ L \ne U}} \frac{U' - L'}{U - L}$$

subject to

$$C(p,q) \;:\Longleftrightarrow\; \underset{\substack{L,U,x \\ 0 < L \le \sqrt{x} \le U}}{\forall} \; 0 < L' \le \sqrt{x} \le U'$$

$$Q(p,q) \;:\Longleftrightarrow\; \underset{\substack{x \\ x > 0}}{\forall} \; \underset{\substack{c \\ c > 0}}{\exists} \; \underset{\substack{L,U \\ 0 < L \le \sqrt{x} \le U}}{\forall} \; U' - L' \le c\,(U - L)^2$$

The constraint C ensures the correctness of the algorithm. The constraint Q ensures the quadratic convergence of the algorithm. The objective function E is the Lipschitz constant and measures the complexity of the algorithm (the smaller, the faster).

For solving the constrained optimization problem, one cannot simply apply standard numerical optimization methods to the above optimization problem since: (1) the constraints C and Q are quantified formulae and (2) the objective function E is the result of parametric optimization (sup). Fortunately, one can easily translate it into a real quantifier elimination problem. Thus, the problem of synthesizing optimal algorithms can be reduced to a real quantifier elimination problem. In principle, the above problem can be carried out automatically using quantifier elimination algorithms. However, the computational requirement is so huge that the automatic synthesis is practically impossible with current real quantifier elimination software.

We overcame this difficulty by mainly following the same solution process as in our previous work [12,13]:

1. Reducing a complicated quantifier elimination formula into several simpler ones, by: (a) carefully dividing the formula, exploiting the logical structure of the formula, (b) designing and implementing some simplification techniques for automating the quantifier elimination process.
2. Eliminating quantifiers from the resulting several simpler formulas automatically, using the state of the art quantifier elimination software such as Mathematica [19] and QEPCAD-B [4].

As the result, we were able to synthesize semi-automatically an optimal quadratic convergent algorithm (see Sect. 3.1), which is significantly better than the well known hand-crafted Secant-Newton.

This paper extends/modifies our previous results [12,13] in the following aspects:

1. In [12,13], we optimized also among the terminating maps, that is among the maps satisfying the condition:

$$T(p,q) \;:\Longleftrightarrow\; \underset{\substack{x \\ x > 0}}{\forall} \; \underset{\substack{c \\ 1 > c > 0}}{\exists} \; \underset{\substack{L,U \\ 0 < L \le \sqrt{x} \le U}}{\forall} \; U' - L' \le c\,(U - L)$$

Since we know that we have at least one map for which $c \in (0,1)$ (Secant-Newton map has $c = \frac{1}{2}$) the condition T is left out.

2. In [12], we optimized among the correct and terminating maps which fulfilled a certain natural condition, obtaining an infinite family of optimal maps. In [13], we detected in the class of optimal maps *a single optimal quadratically convergent map*. In this paper, we removed the natural condition, enlarging the class of optimal maps of [12] while the main result of [13] is preserved. The removal of the additional constraint solves one of the open problems posed in [12,13].

3. In [12,13], in the quantifier elimination from C, high degree manual intervention was needed since the general QE tools failed to solve the QE problem. In this paper, by abstracting these hands-on techniques into systematic ideas, we devised simplification methods which assist the general QE by CAD algorithm from Mathematica (`Reduce` command) in delivering a quantifier-free formula. This partially solves one of the open problems posed in [12,13] and, therefore, further investigations are needed for polynomials of degree 3 and above.

The paper is structured as follows. In Sect. 2, we present simplification techniques tailored for the optimal algorithm synthesis problem whose main result, that is, a quadratically convergent optimal algorithm is presented in Sect. 3.1. In Sect. 3.2, we show how we derived the main result. Finally, in Sect. 4, we conclude and discuss future research directions.

2   Simplification Techniques for Sign Semi-definite Conditions

In this section we introduce the definitions of sign semi-definite conditions, simplification techniques tailored for them, as well as sufficient conditions for the simplification techniques to be sound.

Definition 1. Let $f(y) = y^n + a_{n-1}y^{n-1} + \ldots + a_1 y + a_0$ be a polynomial in y over $\mathbb{R}$. We call the following conditions sign semi-definite conditions (SsDCs)[1] for f:

$$\mathop{\forall}_{\substack{y \\ L \leq y \leq U}} f(y) \geq 0 \qquad\qquad \mathop{\forall}_{\substack{y \\ L \leq y \leq U}} f(y) \leq 0$$

Definition 2. f is monotone increasing on $[L, U]$ iff $\mathop{\forall}_{\substack{x,y \\ L \leq x \leq y \leq U}} f(L) \leq f(x) \leq f(y) \leq f(U)$.

The following lemmas (Lemmas 1–4) eliminate one universally quantified variable from an univariate polynomial expression providing equivalent necessary and sufficient conditions. Note, however, that they can be applied also to multivariate polynomials, as in lemmas from Sect. 3.2.1, by viewing the multivariate polynomials as univariate polynomials.

---

[1] We will not consider here the sign definite conditions but they can be treated in a similar fashion.

Lemma 1. *Let f be monotone increasing on $[L, U]$. Then we have*

(a) $\quad \underset{\substack{y \\ L \le y \le U}}{\forall} \; f(y) \ge 0 \quad \Longleftrightarrow \quad (L \le U \Rightarrow f(L) \ge 0)$

(b) $\quad \underset{\substack{y \\ L \le y \le U}}{\forall} \; f(y) \le 0 \quad \Longleftrightarrow \quad (L \le U \Rightarrow f(U) \le 0)$

Lemma 2. *Let f be monotone decreasing on $[L, U]$. Then we have*

(a) $\quad \underset{\substack{y \\ L \le y \le U}}{\forall} \; f(y) \ge 0 \quad \Longleftrightarrow \quad (L \le U \Rightarrow f(U) \ge 0)$

(b) $\quad \underset{\substack{y \\ L \le y \le U}}{\forall} \; f(y) \le 0 \quad \Longleftrightarrow \quad (L \le U \Rightarrow f(L) \le 0)$

Lemma 3. *Let f be convex on $[L, U]$. Then we have*

(a) $\quad \underset{\substack{y \\ L \le y \le U}}{\forall} \; f(y) \ge 0 \quad \Longleftrightarrow \quad (L \le c \le U \Rightarrow f(c) \ge 0)$, *where c is the critical point of f.*

(b) $\quad \underset{\substack{y \\ L \le y \le U}}{\forall} \; f(y) \le 0 \quad \Longleftrightarrow \quad (L \le U \Rightarrow f(L) \le 0 \wedge f(U) \le 0)$

Lemma 4. *Let f be concave on $[L, U]$. Then we have*

(a) $\quad \underset{\substack{y \\ L \le y \le U}}{\forall} \; f(y) \ge 0 \Longleftrightarrow (L \le U \Rightarrow f(L) \ge 0 \wedge f(U) \ge 0)$.

(b) $\quad \underset{\substack{y \\ L \le y \le U}}{\forall} \; f(y) \le 0 \Longleftrightarrow (L \le c \le U \Rightarrow f(c) \le 0)$, *where c is the critical point of f.*

Lemmas 1–4 are useful if one can determine algorithmically when a function is monotone increasing/decreasing or convex/concave on a certain interval $[L, U]$. Checking algorithmically these for arbitrary degree f is a challenging problem since we have to compute the real roots of f' (or to find isolating intervals for them), which are the critical points of f. However, for degree 1 or 2 these checks can be performed easily (Lemmas 5–8).

Lemma 5. *Let f be a polynomial function of degree 1 on y. If the leading coefficient of f on $[L, U]$ is positive then f is increasing on $[L, U]$.*

Lemma 6. *Let f be a polynomial function of degree 1 on y. If the leading coefficient of f on $[L, U]$ is negative then f is decreasing on $[L, U]$.*

Lemma 7. *Let f be a polynomial function of degree 2 on y and c its critical point.*

(a) *If $c < L \le U$ and the leading coefficient of f on $[L, U]$ is positive then f is monotone increasing on $[L, U]$.*

(b) *If $L \le c \le U$ and the leading coefficient of f on $[L, U]$ is positive then f is convex on $[L, U]$.*

(c) *If* $L \leq U < c$ *and the leading coefficient of* f *on* $[L, U]$ *is positive then* f *is monotone decreasing on* $[L, U]$.

Lemma 8. *Let* f *be a polynomial function of degree* 2 *on* y *and* c *its critical point.*

(a) *If* $c < L \leq U$ *and the leading coefficient of* f *on* $[L, U]$ *is negative then* f *is monotone decreasing on* $[L, U]$.
(b) *If* $L \leq c \leq U$ *and the leading coefficient of* f *on* $[L, U]$ *is negative then* f *is concave on* $[L, U]$.
(c) *If* $L \leq U < c$ *and the leading coefficient of* f *on* $[L, U]$ *is negative then* f *is monotone increasing on* $[L, U]$.

We implemented the results presented above in Mathematica (Simplifier routine http://www.risc.jku.at/projects/SPy/CASC2016/). It *(a)* uses the tactics for eliminating a quantifier from Lemmas 1–4 if their preconditions are fulfilled (Lemmas 5–8) *(b)* applies `Reduce` command of Mathematica to eliminate the rest of the quantifiers. It was successfully applied to all, but one, quantifier elimination problems appearing in the synthesis of optimal algorithms (Sect. 3), however *(a)* in some cases the quantifier-free formula obtained by Mathematica was further simplified by hand for esthetic reasons *(b)* in the case the quantifier-free formula could not be found by Mathematica, we manually eliminated one variable then applied QEPCAD-B, concluding that the formula simplification step of cylindrical algebraic decomposition plays a major role at delivering the final answer.

3 Application: Synthesis of Optimal Numerical Algorithms

In this section, we state the main result. We will use the notations and the results introduced in the previous sections.

3.1 Main Result

Theorem 1 (Main). *We have*

(A) $C \wedge Q \wedge 0 < E < 1 \implies E \geq \frac{1}{4}$
(B) $C \wedge Q \wedge E = \frac{1}{4} \iff p = (-1, 0, 0, 1, 1) \wedge q = \left(-\frac{3}{4}, -\frac{1}{2}, \frac{1}{4}, 1, 1\right)$

For the simplicity of the solution process, we will use the following notation.

Notation 1 [2]

$$y = \sqrt{x} \tag{1}$$
$$W = U - L \tag{2}$$
$$a = (a_0, a_1, a_2, a_3, a_4) = (p_0 + p_1 + p_2, p_1 + 2p_2, p_2, p_3 + p_4, p_4) \tag{3}$$
$$b = (b_0, b_1, b_2, b_3, b_4) = (q_0 + q_1 + q_2, 2q_0 + q_1, q_0, q_3 + q_4, q_3) \tag{4}$$

Recalling the formulation of the constrained optimization problem and the definitions of L' and U' we have:

Minimize
$$E(a, b) := \sup_{\substack{0 < L \le y \le L+W \\ W \ne 0}} \frac{U' - L'}{W}$$

subject to
$$C(a, b) \ : \Longleftrightarrow \ \mathop{\forall}_{\substack{L, U, y \\ 0 < L \le y \le L+W}} \quad 0 < L' \le y \le U'$$

$$Q(a, b) \ : \Longleftrightarrow \ \mathop{\forall}_{\substack{y \\ y > 0}} \mathop{\exists}_{\substack{c \\ c > 0}} \mathop{\forall}_{\substack{L, W \\ 0 < L \le y \le L+W}} U' - L' \le c \, (U - L)^2$$

where

$$L' = L + \frac{y^2 + a_0 L^2 + a_1 LW + a_2 W^2}{a_3 L + a_4 W} \qquad U' = L + W + \frac{y^2 + b_0 L^2 + b_1 LW + b_2 W^2}{b_3 L + b_4 W}$$

Note that the denominators of these maps must be non-zero, that is

$$(a_3 > 0 \ \wedge \ a_4 \ge 0) \ \vee \ (a_3 < 0 \ \wedge \ a_4 \le 0) \tag{5}$$
$$\wedge$$
$$(b_3 > 0 \ \wedge \ b_4 \ge 0) \ \vee \ (b_3 < 0 \ \wedge \ b_4 \le 0) \tag{6}$$

For the rest of the derivation process, we will use Notation 1. We will return to the initial notation at the end of Sect. 3.2.3.

3.2 Proof

In this section, we prove Theorem 1. The proof essentially consists in two quantifier elimination problems over real numbers, hence, in principle, could be carried out automatically. However, the computational requirement is huge making

---

[2] (2) was motivated by the desire for simplifying the bounds of $\sqrt{x}$. (3) and (4) are motivated by (1) and (2) and by the initial definitions of L' and U':

$$L' = L + \frac{y^2 + (p_0 + p_1 + p_2)L^2 + (p_1 + 2p_2)LW + p_2 W^2}{(p_3 + p_4)L + p_4 W}$$

$$U' = L + W + \frac{y^2 + (q_0 + q_1 + q_2)L^2 + (2q_0 + q_1)LW + q_0 W^2}{(q_3 + q_4)L + q_3 W}$$

the automatic proof practically impossible. To overcome this inconvenience, we divided the proof into three parts: (1) *simplify the constraint*, (2) *simplify the objective function*, and (3) *carry out constrained minimization*.

3.2.1   Simplify the Constraint

The aim of this section is to find a quantifier-free formula equivalent to C. Since this can not be achieved automatically, we perform simplifications on C (Lemmas 9–12).

It is natural to require that when $L = y = L + W$ we have $L' = y = U'$. Hence we have the following lemma:

Lemma 9. *We have*

$$\underset{\substack{L,W,y \\ 0<L=y=L+W}}{\forall} \quad 0 < L' = y = U' \implies a_0 = b_0 = -1$$

Proof. Assume the left hand side of the formula. Let K be the formula obtained from the left hand side formula by instantiating the universally quantified variables y and W with L and 0 respectively. Then K holds. By eliminating quantifiers from the formula K, using Mathematica, we have $a_0 = b_0 = -1$.

In the following, we continue deriving a quantifier-free necessary condition of the constraint C. First, we split C as follows:

$$C\,(a, b) \iff C_1\,(a) \wedge C_2\,(a) \wedge C_3\,(b)$$

where

$$C_1\,(a) : \iff \underset{\substack{L,W,y \\ 0<L\le y\le L+W}}{\forall} \quad 0 < L' \qquad C_2\,(a) : \iff \underset{\substack{L,W,y \\ 0<L\le y\le L+W}}{\forall} \quad L' \le y$$

$$C_3\,(b) : \iff \underset{\substack{L,W,y \\ 0<L\le y\le L+W}}{\forall} \quad y \le U'$$

This splitting is a natural thing to do because the new formulas are simpler. Moreover it is used in the subsequent lemmas.

Lemma 10. *We have*

$$C \implies a_3 > 0 \wedge a_4 \ge 0 \wedge a_1 \le 0 \wedge a_2 \le 0 \wedge a_1 - a_3 + 2 \le 0 \wedge a_2 - a_4 + 1 \le 0$$
$$\vee$$
$$a_3 < 0 \wedge a_4 \le 0 \wedge a_1 \ge 0 \wedge a_2 \ge 0$$

Proof. Assume C. From Lemma 9, we have $a_0 = -1$. From the C, we have C_2. By recalling the definition of L', we have

$$\underset{\substack{L,W,y \\ 0<L\le y\le L+W}}{\forall} \quad L + \frac{y^2 - L^2 + a_1 LW + a_2 W^2}{a_3 L + a_4 W} \le y$$

Motivated by (5), we consider two cases.

Case $a_3 > 0 \land a_4 \geq 0$. By factoring, collecting the terms and noting that the denominator of L' is positive, that is

$$a_3 > 0 \land a_4 \geq 0 \ \land \ 0 < L \leq L + W \implies a_3 L + a_4 W > 0$$

we obtain $a_3 > 0 \ \land \ a_4 \geq 0 \ \land \ \underset{\substack{L,W \\ 0 < L \leq L + W}}{\forall} \ \underset{\substack{y \\ L \leq y \leq L + W}}{\forall} \ f_1(y) \leq 0$ where

$f_1(y) := y^2 - y(a_3 L + a_4 W) - L^2 + a_1 LW + a_2 W^2 + L(a_3 L + a_4 W)$.

By using the Simplifier and further simplifying, we have

$$a_3 > 0 \ \land \ a_4 \geq 0 \ \land \ a_1 \leq 0 \ \land \ a_2 \leq 0 \ \land \ a_1 - a_3 + 2 \leq 0 \ \land \ a_2 - a_4 + 1 \leq 0$$

Case $a_3 < 0 \land a_4 \leq 0$. By factoring, collecting the terms and noting that the denominator of L' is negative, that is

$$a_3 < 0 \land a_4 \leq 0 \ \land \ 0 < L \leq L + W \implies a_3 L + a_4 W < 0$$

we obtain $a_3 < 0 \ \land \ a_4 \leq 0 \ \land \ \underset{\substack{L,W \\ 0 < L \leq L + W}}{\forall} \ \underset{\substack{y \\ L \leq y \leq L + W}}{\forall} \ f_1(y) \geq 0$

By using the Simplifier and further simplifying, we have

$$a_3 < 0 \ \land \ a_4 \leq 0 \ \land \ a_1 \geq 0 \ \land \ a_2 \geq 0$$

Lemma 11. *We have*

$$C \implies a_3 > 0 \land a_4 \geq 1 \land a_2 = 0 \land a_1 \leq 0 \land a_1 - a_3 + 2 \leq 0 \land a_1 + a_4 \geq 0$$

Proof. Assume C. From Lemma 9, we have $a_0 = -1$. From C, we have C_1. By recalling the definition of L', we have

$$\underset{\substack{L,W,y \\ 0 < L \leq y \leq L + W}}{\forall} \ L + \frac{y^2 - L^2 + a_1 LW + a_2 W^2}{a_3 L + a_4 W} > 0$$

Motivated by (5), we consider two cases.

Case $a_3 > 0 \land a_4 \geq 0$. By factoring, collecting the terms and noting that the denominator of L' is positive, that is,

$$a_3 > 0 \land a_4 \geq 0 \ \land \ 0 < L \leq L + W \implies a_3 L + a_4 W > 0$$

we obtain $a_3 > 0 \ \land \ a_4 \geq 0 \ \land \ \underset{\substack{L,W \\ 0 < L \leq L + W}}{\forall} \ \underset{\substack{y \\ L \leq y \leq L + W}}{\forall} \ f_2(y) > 0$ where

$f_2(y) := y^2 + (a_3 - 1)L^2 + (a_1 + a_4)LW + a_2 W^2$.

By combining with Lemma 10, using the Simplifier and further simplifying, we have

$$a_3 > 0 \ \land \ a_4 \geq 1 \ \land \ a_2 = 0 \ \land \ a_1 \leq 0 \ \land \ a_1 - a_3 + 2 \leq 0 \ \land \ a_1 + a_4 \geq 0$$

Case $a_3 < 0 \wedge a_4 \leq 0$. By factoring, collecting the terms and noting that the denominator of L' is negative, that is

$$a_3 < 0 \wedge a_4 \leq 0 \ \wedge \ 0 < L \leq L + W \implies a_3 L + a_4 W < 0$$

we obtain $a_3 < 0 \ \wedge \ a_4 \leq 0 \ \wedge \ \underset{\substack{L,W \\ 0 < L \leq L+W}}{\forall} \ \underset{\substack{y \\ L \leq y \leq L+W}}{\forall} \ f_2(y) < 0.$

By combining with Lemma 10 and using the Simplifier, we obtain *False*.

Lemma 12. *We have*

$$C \implies b_3 > 0 \wedge b_4 \geq 0 \wedge b_1 + 2 \geq 0 \wedge b_2 + 1 \geq 0 \wedge b_1 + b_3 \geq 0 \wedge (b_4 - 2 > 0 \vee -b_4^2 + 4b_4 + 4b_2 \geq 0)$$
$$\vee$$
$$b_3 < 0 \wedge b_4 \leq 0 \wedge b_1 + 2 \leq 0 \wedge b_2 + 1 \leq 0$$

Proof. Assume C. From Lemma 9, we have $b_0 = -1$. From C, we have C_3. By recalling the definition of U', we have

$$\underset{\substack{L,W,y \\ 0 < L \leq y \leq L+W}}{\forall} \quad y \leq L + W + \frac{y^2 - L^2 + b_1 LW + b_2 W^2}{b_3 L + b_4 W}$$

Motivated by (6), we consider two cases.
Case $b_3 > 0 \ \wedge \ b_4 \geq 0$. By factoring, collecting the terms and noting that the denominator of U' is positive, that is

$$b_3 > 0 \wedge b_4 \geq 0 \ \wedge \ 0 < L \leq L + W \implies b_3 L + b_4 W > 0$$

we obtain $b_3 > 0 \ \wedge \ b_4 \geq 0 \ \wedge \ \underset{\substack{L,W \\ 0 < L \leq L+W}}{\forall} \ \underset{\substack{y \\ L \leq y \leq L+W}}{\forall} \ f_3(y) \geq 0$ where

$f_3(y) := y^2 - y(b_3 L + b_4 W) - L^2 + b_1 LW + b_2 W^2 + (L+W)(b_3 L + b_4 W).$
Using the Simplifier does not succeed because **Reduce** command of Mathematica could not eliminate the quantifier from the formula $b_3 > 0 \wedge b_4 \geq 0 \wedge \underset{\substack{L,W \\ 0 < L \leq \frac{b_3 L + b_4 W}{2} \leq L+W}}{\forall} \ f_3\left(\frac{b_3 L + b_4 W}{2}\right) \geq 0.$ Hence we divide the problem and approach it using QEPCAD-B.
Since f_3 is convex and its critical point of is $\frac{b_3 L + b_4 W}{2}$, we have

$$b_3 > 0 \ \wedge \ b_4 \geq 0 \ \wedge \ \underset{\substack{L,W \\ 0 < L \leq L+W}}{\forall} \ f_3(L) \geq 0$$
$$\wedge \ \underset{\substack{L,W \\ 0 < L \leq \frac{b_3 L + b_4 W}{2} \leq L+W}}{\forall} \ f_3\left(\frac{b_3 L + b_4 W}{2}\right) \geq 0$$
$$\wedge \ \underset{\substack{L,W \\ 0 < L \leq L+W}}{\forall} \ f_3(L+W) \geq 0$$

By eliminating the quantifiers and combining using QEPCAD-B, we have

$$b_3 > 0 \wedge b_4 \geq 0 \wedge b_1 + 2 \geq 0 \wedge b_2 + 1 \geq 0 \wedge b_1 + b_3 \geq 0 \wedge (b_4 - 2 > 0 \vee -b_4^2 + 4b_4 + 4b_2 \geq 0)$$

Case $b_3 < 0 \ \wedge \ b_4 \leq 0$. By factoring, collecting the terms and noting that the denominator of U' is negative, that is

$$b_3 < 0 \ \wedge \ b_4 \leq 0 \ \wedge \ 0 < L \leq L + W \implies b_3 L + b_4 W < 0$$

we obtain $b_3 < 0 \ \wedge \ b_4 \leq 0 \ \wedge \ \underset{\substack{L,W \\ 0 < L \leq L+W}}{\forall} \ \underset{\substack{y \\ L \leq y \leq L+W}}{\forall} \ f_3(y) \leq 0.$

By using the Simplifier and further simplifying, we have

$$b_3 < 0 \ \wedge \ b_4 \leq 0 \ \wedge \ b_1 + 2 \leq 0 \ \wedge \ b_2 + 1 \leq 0$$

Lemma 13 (Simplified constraint). *We have* $C \implies F$ *where*

$F(a,b) : \iff$

$a_0 = -1 \wedge a_1 \leq 0 \wedge a_2 = 0 \wedge a_3 > 0 \wedge a_4 \geq 1 \wedge a_1 - a_3 + 2 \leq 0 \wedge a_1 + a_4 \geq 0$

$\wedge$

$$\begin{pmatrix} b_0 = -1 \wedge b_3 > 0 \wedge b_4 \geq 0 \wedge b_1 + 2 \geq 0 \wedge b_2 + 1 \geq 0 \wedge b_1 + b_3 \geq 0 \wedge (b_4 - 2 > 0 \vee -b_4^2 + 4b_4 + 4b_2 \geq 0) \\ \vee \\ b_0 = -1 \wedge b_3 < 0 \wedge b_4 \leq 0 \wedge b_1 + 2 \leq 0 \wedge b_2 + 1 \leq 0 \end{pmatrix}$$

Proof. Immediate from Lemmas 9, 11, and 12.

Lemma 14. *We have* $C \wedge 0 < E < 1 \implies F \wedge 0 < E < 1$

Proof. From Lemma 13, we have $C \implies F$ from which we trivially obtain the goal.

A careful reader might notice that quantifier elimination from the constrained Q should be performed. This could not be achieved using manual and automatic simplifications, so in the section *simplify the objective function* we eliminate the quantifier from E under $C \wedge 0 < E < 1$ and verify, in the section *carry out the constrained minimization*, if the optimal solution satisfies Q.

3.2.2   Simplify the Objective Function

In this subsection, we eliminate *max* and *sup* from the objective function E under the constraint $C \wedge 0 < E < 1$. Since this can not be achieved automatically using the routines implemented in e.g. Mathematica, we performed a series of strategic simplifications which eliminate the parameters a and b which do not lead to $0 < E < 1$.

Lemma 15. *Assume* $C \wedge 0 < E < 1$. *Then*

$$E(a,b) = \max\{E_1(a,b), E_2(a,b)\}$$

where

$$E_1(a,b) = \sup_{\substack{L,W \\ L>0 \\ W>0 \\ M\geq 0}} \left[\frac{M(L+W)^2 + N}{W} \right] \qquad E_2(a,b) = \sup_{\substack{L,W \\ L>0 \\ W>0 \\ M\leq 0}} \left[\frac{ML^2 + N}{W} \right]$$

where again

$$M = \frac{1}{b_3 L + b_4 W} - \frac{1}{a_3 L + a_4 W}; \qquad N = W + \frac{-L^2 + b_1 LW + b_2 W^2}{b_3 L + b_4 W} - \frac{-L^2 + a_1 LW}{a_3 L + a_4 W}$$

Proof. Assume $C \wedge 0 < E < 1$. From Lemma 14, we have $F \wedge 0 < E < 1$. By recalling the definitions of L' and U', we have

$$E\,(a,b) = \sup_{\substack{L,W,y \\ 0 < L \le y \le L+W \\ W > 0}} \left[\frac{L + W + \frac{y^2 - L^2 + b_1 LW + b_2 W^2}{b_3 L + b_4 W} - \left(L + \frac{y^2 - L^2 + a_1 LW}{a_3 L + a_4 W} \right)}{W} \right]$$

By combining the denominators, collecting the terms involving y^2, simplifying, and splitting the sup variables, we have

$$E\,(a,b) = \sup_{\substack{L,W \\ L > 0 \\ W > 0}} \sup_{\substack{y \\ L \le y \le L+W}} \left[\frac{My^2 + N}{W} \right]$$

By splitting the cases depending on the sign of M, we have

$$E\,(a,b) = \max\{E_1(a,b), E_2(a,b)\}$$

where

$$E_1\,(a,b) = \sup_{\substack{L,W \\ L > 0 \\ W > 0 \\ M \ge 0}} \sup_{\substack{y \\ L \le y \le L+W}} \left[\frac{My^2 + N}{W} \right] \qquad E_2\,(a,b) = \sup_{\substack{L,W \\ L > 0 \\ W > 0 \\ M \le 0}} \sup_{\substack{y \\ L \le y \le L+W}} \left[\frac{My^2 + N}{W} \right]$$

Note that $W > 0$. We consider two cases.

Case 1: $\frac{M}{W} \ge 0$. Since $f(y) = \frac{M}{W} y^2 + \frac{N}{W}$ is a convex function in y with the critical point $0 \notin [L, L + W]$, we have that it is also strictly positive on $[L, L + W]$ and

$$E_1\,(a,b) = \sup_{\substack{L,W \\ L > 0 \\ W > 0 \\ M \ge 0}} \left[\frac{M(L + W)^2 + N}{W} \right]$$

Case 2: $\frac{M}{W} \le 0$. Since f is a concave function in y with the critical point $0 \notin [L, L + W]$, we have that it is also strictly negative on $[L, L + W]$ and

$$E_2\,(a,b) = \sup_{\substack{L,W \\ L > 0 \\ W > 0 \\ M \le 0}} \left[\frac{ML^2 + N}{W} \right]$$

Note that $f(0) = \frac{N}{W}$, hence $E\,(a,b) = \max\{E_1(a,b), E_2(a,b)\} = E_1(a,b)$.

Notation 2.

$$h_1 = 1 - \frac{1}{a_4} + \frac{b_2 + 1}{b_4}$$
$$u_1 = -a_4^2 b_3(1 + b_2) + a_3 b_4^2 - a_4 b_4(-a_4(2 + b_1) + b_4(2 + a_1))$$
$$v_1 = a_3 b_3 b_4 - a_4 b_3 b_4(2 + a_1) - a_3 a_4(b_3(1 + b_2) - b_4(2 + b_1))$$

Lemma 16. *Assume* $C \wedge 0 < E < 1$.
(a) If $b_4 > 0 \wedge a_4 < b_4 \wedge a_3 > b_3$ *then*

$$E_1(a,b) = h_1 + \left(\frac{1}{a_4 b_4}\right)^2 \sup_{0 < V \le \frac{b_3 - a_3}{a_4 - b_4}} \left[\frac{u_1 V + v_1}{\left(\frac{a_3}{a_4} + V\right)\left(\frac{b_3}{b_4} + V\right)} \right]$$

(b) If $b_4 > 0 \wedge a_4 > b_4 \wedge a_3 < b_3$ *then*

$$E_1(a,b) := h_1 + \left(\frac{1}{a_4 b_4}\right)^2 \sup_{V \ge \frac{b_3 - a_3}{a_4 - b_4}} \left[\frac{u_1 V + v_1}{\left(\frac{a_3}{a_4} + V\right)\left(\frac{b_3}{b_4} + V\right)} \right]$$

(c) If $b_4 > 0 \wedge a_4 \ge b_4 \wedge a_3 \ge b_3$ *then*

$$E_1(a,b) = h_1 + \left(\frac{1}{a_4 b_4}\right)^2 \sup_{V > 0} \left[\frac{u_1 V + v_1}{\left(\frac{a_3}{a_4} + V\right)\left(\frac{b_3}{b_4} + V\right)} \right]$$

Proof. Assume $C \wedge 0 < E < 1$. From Lemma 14, we have $F \wedge 0 < E < 1$. Let $V = \frac{W}{L}$. Recalling the notations in Lemma 15, combining the denominators, and simplifying, we have

$$E_1(a,b) = 1 + \sup_{\substack{V \\ V > 0 \\ \frac{a_3 + a_4 V}{b_3 + b_4 V} \ge 1}} \left[\frac{(b_1 + 2 + (b_2 + 1)V)(a_3 + a_4 V) - (a_1 + 2 + V)(b_3 + b_4 V)}{(a_3 + a_4 V)(b_3 + b_4 V)} \right]$$

The expression of the denominator motivated us to split the analysis of $E_1(a,b)$ into two cases, based on the possible values of b_4: $b_4 > 0$, $b_4 \le 0$.
Case $b_4 > 0$. By carrying out polynomial division in V, simplifying and isolating the terms involving V

$$E_1(a,b) = h_1 + \left(\frac{1}{a_4 b_4}\right)^2 \sup_{\substack{V \\ V > 0 \\ (a_4 - b_4)V \ge b_3 - a_3}} \left[\frac{u_1 V + v_1}{\left(\frac{a_3}{a_4} + V\right)\left(\frac{b_3}{b_4} + V\right)} \right]$$

We find the domain of V. From Mathematica, we have

$$\exists_V (V > 0 \wedge (a_4 - b_4)V \ge b_3 - a_3) \iff (a_4 < b_4 \wedge a_3 > b_3) \vee (a_4 = b_4 \wedge a_3 \ge b_3) \vee a_4 > b_4$$

Hence, we proceed by case distinction.
Case $a_4 < b_4 \wedge a_3 > b_3$. We have

$$E_1(a,b) = h_1 + \left(\frac{1}{a_4 b_4}\right)^2 \sup_{0 < V \le \frac{b_3 - a_3}{a_4 - b_4}} \left[\frac{u_1 V + v_1}{\left(\frac{a_3}{a_4} + V\right)\left(\frac{b_3}{b_4} + V\right)} \right]$$

Hence (a) is proved.

Case $a_4 = b_4 \wedge a_3 \geq b_3$. We have

$$E_1(a,b) = h_1 + \left(\frac{1}{a_4 b_4}\right)^2 \sup_{\substack{V \\ V > 0}} \left[\frac{u_1 V + v_1}{\left(\frac{a_3}{a_4} + V\right)\left(\frac{b_3}{b_4} + V\right)}\right] \tag{7}$$

Case $a_4 > b_4$ necessitates further case distinction based on the sign of $b_3 - a_3$.

Case $a_4 > b_4 \wedge a_3 < b_3$. We have

$$E_1(a,b) = h_1 + \left(\frac{1}{a_4 b_4}\right)^2 \sup_{\substack{V \\ V \geq \frac{b_3 - a_3}{a_4 - b_4}}} \left[\frac{u_1 V + v_1}{\left(\frac{a_3}{a_4} + V\right)\left(\frac{b_3}{b_4} + V\right)}\right]$$

Hence, (b) is proved.

Case $a_4 > b_4 \wedge a_3 = b_3$. We have

$$E_1(a,b) = h_1 + \left(\frac{1}{a_4 b_4}\right)^2 \sup_{\substack{V \\ V > 0}} \left[\frac{u_1 V + v_1}{\left(\frac{a_3}{a_4} + V\right)\left(\frac{b_3}{b_4} + V\right)}\right] \tag{8}$$

Case $a_4 > b_4 \wedge a_3 > b_3$. We have

$$E_1(a,b) = h_1 + \left(\frac{1}{a_4 b_4}\right)^2 \sup_{\substack{V \\ V > 0}} \left[\frac{u_1 V + v_1}{\left(\frac{a_3}{a_4} + V\right)\left(\frac{b_3}{b_4} + V\right)}\right] \tag{9}$$

From (7), (8), and (9), we have:

$$E_1(a,b) = h_1 + \left(\frac{1}{a_4 b_4}\right)^2 \sup_{\substack{V \\ V > 0}} \left[\frac{u_1 V + v_1}{\left(\frac{a_3}{a_4} + V\right)\left(\frac{b_3}{b_4} + V\right)}\right]$$

and $(a_4 = b_4 \wedge a_3 \geq b_3) \vee (a_4 > b_4 \wedge a_3 = b_3) \vee (a_4 > b_4 \wedge a_3 > b_3) \iff a_4 \geq b_4 \wedge a_3 \geq b_3$. Hence, (c) is proved.

Note that, if $b_4 \leq 0$, we have

$$F \wedge V > 0 \wedge \frac{a_3 + a_4 V - (b_3 + b_4 V)}{b_3 + b_4 V} \geq 0 \iff \textit{False}$$

Hence, there are no a, b with $b_3 < 0 \wedge b_4 \leq 0$ such that $0 < E_1(a,b) < 1$.

Lemma 17. *Let b, w, t be strictly positive. Then we have*

(1) $\displaystyle\sup_{\substack{V \\ 0 < V \leq b}} \frac{uV + v}{(V+w)(V+t)} \geq \frac{ub + v}{(b+w)(b+t)}$

(2a) $\displaystyle\sup_{\substack{V \\ V > 0}} \frac{uV + v}{(V+w)(V+t)} \geq 0$

(2b) $\sup\limits_{\substack{V \\ V>0}} \frac{uV+v}{(V+w)(V+t)} = 0$ *iff* $v \le 0 \wedge u \le 0$

(3a) $\sup\limits_{\substack{V \\ V\ge b}} \frac{uV+v}{(V+w)(V+t)} \ge 0$

(3b) $\sup\limits_{\substack{V \\ V\ge b}} \frac{uV+v}{(V+w)(V+t)} = 0$ *iff* $(v \le 0 \wedge u \le 0) \vee (v > 0 \wedge u \le -\frac{v}{b})$

Proof. Proof of (1). We have two cases, depending whether the graph of the function is above or below 0 (since $y = 0$ is horizontal asymptote).
Case $v \ge 0 \wedge u \ge -\frac{v}{b}$. We have $\sup\limits_{\substack{V \\ 0<V\le b}} \frac{uV+v}{(V+w)(V+t)} \ge \frac{ub+v}{(b+w)(b+t)}$.

Case $v \le 0 \wedge u \le -\frac{v}{b}$. We have $\sup\limits_{\substack{V \\ 0<V\le b}} \frac{uV+v}{(V+w)(V+t)} = \frac{ub+v}{(b+w)(b+t)}$.

Hence $\sup\limits_{\substack{V \\ 0<V\le b}} \frac{uV+v}{(V+w)(V+t)} \ge \frac{ub+v}{(b+w)(b+t)}$.

Proof of (2a). We have

$$\sup\limits_{\substack{V \\ V>0}} \frac{uV+v}{(V+w)(V+t)} \ge \lim\limits_{V\to\infty} \frac{uV+v}{(V+w)(V+t)} = 0$$

Proof of (2b). We have $\sup\limits_{\substack{V \\ V>0}} \frac{uV+v}{(V+w)(V+t)} = 0 \iff \forall\limits_{\substack{V \\ V>0}} \frac{uV+v}{(V+w)(V+t)} \le 0 \iff$
$v \le 0 \wedge u \le 0$ (by *Mathematica*)

Proof of (3a). We have

$$\sup\limits_{\substack{V \\ V\ge b}} \frac{uV+v}{(V+w)(V+t)} \ge \lim\limits_{V\to\infty} \frac{uV+v}{(V+w)(V+t)} = 0$$

Proof of (3b). We have $\sup\limits_{\substack{V \\ V\ge b}} \frac{uV+v}{(V+w)(V+t)} = 0 \iff (v \le 0 \wedge u \le 0) \vee$
$(v > 0 \wedge u \le -\frac{v}{b})$ (by Mathematica)

Lemma 18. *We have*

$$C \wedge 0 < E < 1 \wedge b_4 > 0 \wedge a_4 \ge b_4 \wedge a_3 \ge b_3 \implies h_1 \ge \frac{1}{4} \wedge 0 < E < 1$$

Proof. Motivated by Lemma 16, we consider three cases. *Case $C \wedge 0 < E < 1 \wedge b_4 > 0 \wedge a_4 < b_4 \wedge a_3 > b_3$.* From Lemma 13 and Mathematica, we have

$$\min\limits_{\substack{a,b \\ C\wedge b_4>0\wedge a_4<b_4\wedge a_3>b_3}} h_1 \ge \min\limits_{\substack{a,b \\ F\wedge b_4>0\wedge a_4<b_4\wedge a_3>b_3}} h_1 = 0$$

Note that we have to compute

$$E = E_1(a,b) = h_1 + \left(\frac{1}{a_4 b_4}\right)^2 \sup\limits_{\substack{V \\ 0<V\le \frac{b_3-a_3}{a_4-b_4}}} \left[\frac{u_1 V + v_1}{\left(\frac{a_3}{a_4}+V\right)\left(\frac{b_3}{b_4}+V\right)}\right]$$

When the graph of $h(V) = \frac{u_1 V + v_1}{\left(\frac{a_3}{a_4} + V\right)\left(\frac{b_3}{b_4} + V\right)}$ is above 0, we have, by Lemma 17(1)

$$\sup_{\substack{V \\ 0 < V \leq \frac{b_3 - a_3}{a_4 - b_4}}} \frac{u_1 V + v_1}{(V + w)(V + t)} \geq \frac{u_1 \frac{b_3 - a_3}{a_4 - b_4} + \frac{b_3 - a_3}{a_4 - b_4}}{\left(\frac{b_3 - a_3}{a_4 - b_4} + w\right)\left(\frac{b_3 - a_3}{a_4 - b_4} + t\right)}$$

From Mathematica, we have

$$\underset{\substack{a_1, a_3, a_4 \\ b_1, b_2, b_3, b_4}}{\exists} \quad F \wedge b_4 > 0 \wedge a_4 < b_4 \wedge a_3 > b_3 \wedge \frac{a_3}{a_4} > 0 \wedge \frac{b_3}{b_4} > 0 \wedge b > 0 \wedge v_1 \geq 0 \wedge u_1 \geq -\frac{v_1}{b} \wedge$$
$$\frac{u_1 b + v_1}{(b + w)(b + t)} = 0 \iff True$$

and, therefore,

$$\sup_{\substack{V \\ 0 < V \leq \frac{b_3 - a_3}{a_4 - b_4}}} \frac{u_1 V + v_1}{(V + w)(V + t)} \geq \frac{u_1 \frac{b_3 - a_3}{a_4 - b_4} + \frac{b_3 - a_3}{a_4 - b_4}}{\left(\frac{b_3 - a_3}{a_4 - b_4} + w\right)\left(\frac{b_3 - a_3}{a_4 - b_4} + t\right)} \geq 0$$

When the graph of h is below 0, by a similar reasoning we have

$$\sup_{\substack{V \\ 0 < V \leq \frac{b_3 - a_3}{a_4 - b_4}}} \frac{u_1 V + v_1}{(V + w)(V + t)} = \frac{u_1 \frac{b_3 - a_3}{a_4 - b_4} + \frac{b_3 - a_3}{a_4 - b_4}}{\left(\frac{b_3 - a_3}{a_4 - b_4} + w\right)\left(\frac{b_3 - a_3}{a_4 - b_4} + t\right)} = 0$$

Summarizing

$$\sup_{\substack{V \\ 0 < V \leq \frac{b_3 - a_3}{a_4 - b_4}}} \frac{u_1 V + v_1}{(V + w)(V + t)} \geq \frac{u_1 \frac{b_3 - a_3}{a_4 - b_4} + \frac{b_3 - a_3}{a_4 - b_4}}{\left(\frac{b_3 - a_3}{a_4 - b_4} + w\right)\left(\frac{b_3 - a_3}{a_4 - b_4} + t\right)} \geq 0$$

and since $h_1 \geq 0$, we have that $E \geq 0$, which contradicts the assumption $0 < E < 1$.
Case $C \wedge 0 < E < 1 \wedge b_4 > 0 \wedge a_4 > b_4 \wedge a_3 < b_3$. From Lemma 13 and Mathematica we have

$$\min_{\substack{a,b \\ C \wedge b_4 > 0 \wedge a_4 > b_4 \wedge a_3 < b_3}} h_1 \geq \min_{\substack{a,b \\ F \wedge b_4 > 0 \wedge a_4 > b_4 \wedge a_3 < b_3}} h_1 = \frac{1}{4} \quad \text{but}$$

$$F \wedge b_4 > 0 \wedge a_4 > b_4 \wedge a_3 < b_3 \wedge h_1 = \frac{1}{4} \iff False$$

Case $C \wedge 0 < E < 1 \wedge b_4 > 0 \wedge a_4 \geq b_4 \wedge a_3 \geq b_3$. From Lemma 13 and Mathematica we have

$$\min_{\substack{a,b \\ C \wedge b_4 > 0 \wedge a_4 \geq b_4 \wedge a_3 \geq b_3}} h_2 \geq \min_{\substack{a,b \\ F \wedge b_4 > 0 \wedge a_4 \geq b_4 \wedge a_3 \geq b_3}} h_2 = \frac{1}{4}$$

Hence we have $C \wedge b_4 > 0 \wedge a_4 \geq b_4 \wedge a_3 \geq b_3 \implies h_1 \geq \frac{1}{4}$ from which we immediately obtain the goal.

Lemma 19 (Simplified objective function). *Let* $C \wedge 0 < E < 1 \wedge b_4 > 0 \wedge a_4 \geq b_4 \wedge a_3 \geq b_3$. *We have*
(a) $E \geq h_1$
(b) $E = h_1$ *iff* $u \leq 0 \wedge v \leq 0$

Proof. Immediate from Lemmas 15, 16, and 17.

3.2.3   Carry Out the Constrained Minimization

In this subsection, we finally derive the main result (Theorem 1) by carrying out the constrained minimization, using the results from the previous two subsections (Lemmas 14 and 19).

Proof. (Proof of (A)) We need to prove $C \wedge Q \wedge 0 < E < 1 \implies E \geq \frac{1}{4}$. Note

$$C \wedge 0 < E < 1$$

$$\implies C \wedge 0 < E < 1 \wedge b_4 > 0 \wedge\ a_4 \geq b_4\ \wedge\ a_3 \geq b_3 \wedge h_1 \geq \frac{1}{4}\ \text{(by Lemma 18)}$$

$$\implies C \wedge 0 < E < 1 \wedge b_4 > 0\ \wedge\ a_4 \geq b_4 \wedge a_3 \geq b_3 \wedge E \geq h_1 \wedge h_1 \geq \frac{1}{4} \text{(by Lemma 19(a))}$$

$$\implies E \geq \frac{1}{4}$$

Since $C \wedge Q \wedge 0 < E < 1 \implies C \wedge 0 < E < 1$, we obtain the goal.

Proof. (Proof of (B)) We need to prove

$$C \wedge Q \wedge E = \frac{1}{4} \quad \Longleftrightarrow \quad p = (-1, 0, 0, 1, 1) \wedge q = \left(-\frac{3}{4}, -\frac{1}{2}, \frac{1}{4}, 1, 1\right)$$

Direction $\implies$: For simplicity, we will use in the proof the a, b notation and return to the p, q notation in the end. Note

$$C \wedge E = \frac{1}{4}$$

$$\Longleftrightarrow C \wedge 0 < E < 1 \wedge E = \frac{1}{4}$$

$$\Longleftrightarrow C \wedge 0 < E < 1 \wedge b_4 > 0 \wedge\ a_4 \geq b_4\ \wedge\ a_3 \geq b_3 \wedge h_1 \geq \frac{1}{4} \wedge E = \frac{1}{4}$$
   (from Lemma 18)

$$\Longleftrightarrow C \wedge 0 < E < 1 \wedge b_4 > 0 \wedge\ a_4 \geq b_4\ \wedge\ a_3 \geq b_3 \wedge E \geq h_1 \wedge h_1 \geq \frac{1}{4} \wedge E = \frac{1}{4}$$
   (from Lemma 19(a))

$$\Longleftrightarrow C \wedge 0 < E < 1 \wedge b_4 > 0 \wedge a_4 \geq b_4 \wedge a_3 \geq b_3 \wedge E = h_1 = \frac{1}{4} \wedge u_1 \leq 0 \wedge v_1 \leq 0$$
   (from Lemma 19(b))

$$\Longleftrightarrow F \wedge 0 < E < 1 \wedge b_4 > 0 \wedge a_4 \geq b_4 \wedge a_3 \geq b_3 \wedge E = h_1 = \frac{1}{4} \wedge u_1 \leq 0 \wedge v_1 \leq 0$$
   (from Lemma 13)

$$\Longleftrightarrow a_4 = b_4 = 1\ \wedge\ b_2 = -\frac{3}{4}\ \wedge\ -\frac{1}{2} \leq a_1 \leq 0\ \wedge\ a_3 - a_1 - 2 \geq 0\ \wedge$$
$$b_1 + 2 \geq 0\ \wedge\ b_1 + b_3 \geq 0\ \wedge\ a_3 \geq b_3\ \wedge\ 4a_1 - 4a_3 - 4b_1 + b_3 \geq 0$$
   (by QEPCAD-B)

Recalling the Notation 1, we have

$$p_0 = -(1+p_1) \wedge p_2 = 0 \wedge p_4 = 1 \wedge -\tfrac{1}{2} \leq p_1 \leq 0 \wedge p_3 \geq 1 + p_1 \wedge$$
$$q_0 = -\tfrac{3}{4} \wedge q_2 = -\left(\tfrac{1}{4} + q_1\right) \wedge q_3 = 1 \wedge 1 + 2q_1 \geq 0 \wedge q_1 + q_4 \geq \tfrac{1}{2} \wedge$$
$$p_3 \geq q_4 \wedge 4(p_1 - p_3 - q_1) + q_4 + 3 \geq 0$$
$$\Longleftrightarrow Z(p,q)$$

We have shown that $C \wedge E = \tfrac{1}{4} \Longrightarrow Z(p,q)$. We have $C \wedge Q \wedge E = \tfrac{1}{4} \Longrightarrow Z \wedge Q$. Finding automatically that the quantifier-free equivalent condition to $Z \wedge Q$ is $p = (-1,0,0,1,1) \wedge q = \left(-\tfrac{3}{4}, -\tfrac{1}{2}, \tfrac{1}{4}, 1, 1\right)$ does not succeed. Hence we proceed manually.

We assume $Z \wedge Q \wedge \neg(p = (-1,0,0,1,1) \wedge q = \left(-\tfrac{3}{4}, -\tfrac{1}{2}, \tfrac{1}{4}, 1, 1\right))$ and derive a contradiction.

Case 1: $Z \wedge Q \wedge \quad p \neq (-1,0,0,1,1)$. Let $p = (-1, -\tfrac{1}{2}, 0, 1, 1)$. We immediately have, by Mathematica, Z and $\neg Q$.

Case 2: $Z \wedge Q \wedge \quad q \neq \left(-\tfrac{3}{4}, -\tfrac{1}{2}, \tfrac{1}{4}, 1, 1\right)$ Let $q = \left(-\tfrac{3}{4}, 0, \tfrac{1}{4}, 1, 1\right)$. We immediately have, by Mathematica, Z and $\neg Q$.

Direction $\Longleftarrow$: C and Q and $E = \tfrac{1}{4}$ are proved by Mathematica.

4 Conclusion and Future Work

In this paper, we presented efficient simplification techniques which proved to be useful for the semi-automated synthesis of an optimal quadratic convergent interval algorithm for square root computation, where state of the art techniques implemented in quantifier elimination tools fail.

As future work, we plan to further investigate in the direction of simplification techniques and to integrate them into state of the art quantifier elimination tools.

Acknowledgements. The author thanks Hoon Hong for providing feedback on an earlier draft of this paper and to anonymous referees.

References

1. Abraham, E.: Building bridges between symbolic computation and satisfiability checking. In: ISSAC 2015 Proceedings, pp. 1–6. ACM, New York (2015)
2. Anai, H., Yanami, H.: SyNRAC: a Maple-package for solving real algebraic constraints. In: Sloot, P.M.A., Abramson, D., Bogdanov, A.V., Gorbachev, Y.E., Dongarra, J., Zomaya, A.Y. (eds.) ICCS 2003, Part I. LNCS, vol. 2657, pp. 828–837. Springer, Heidelberg (2003)
3. Brown, C.W.: Improved projection for cylindrical algebraic decomposition. J. Symbolic Comput. **32**(5), 447–465 (2001)
4. Brown, C.W.: QEPCAD-B: a program for computing with semi-algebraic sets using CADs. SIGSAM Bull. **37**(4), 97–108 (2003)
5. Brown, C.W.: Constructing a single open cell in a cylindrical algebraic decomposition. In: ISSAC 2013 Proceedings, pp. 133–140. ACM, New York (2013)

6. Caviness, B., Johnson, J. (eds.): Quantifier Elimination and Cylindrical Algebraic Decomposition. Texts and Monographs in Symbolic Computation. Springer, Heidelberg (1998)
7. Collins, G.E.: Quantifier elimination for real closed fields by cylindrical algebraic decomposition. In: Brakhage, H. (ed.) Automata Theory and Formal Languages. LNCS, vol. 33, pp. 134–183. Springer, Heidelberg (1975)
8. Corzilius, F., Kremer, G., Junges, S., Schupp, S., Ábrahám, E.: SMT-RAT: an open source C++ toolbox for strategic and parallel SMT solving. In: Heule, M., Weaver, S. (eds.) SAT 2015. LNCS, vol. 9340, pp. 360–368. Springer, Heidelberg (2015). doi:10.1007/978-3-319-24318-4_26
9. Davenport, J., Heintz, J.: Real quantifier elimination is doubly exponential. J. Symbolic Comput. **5**(1–2), 29–35 (1988)
10. Dolzmann, A., Seidl, A., Sturm, T.: Efficient projection orders for CAD. In: ISSAC 2004 Proceedings, pp. 111–118. ACM, New York (2004)
11. Dolzmann, A., Sturm, T.: REDLOG: computer algebra meets computer logic. SIGSAM Bull. (ACM Spec. Interest Group Symbolic Algebraic Manipulation) **31**(2), 2–9 (1997)
12. Erascu, M., Hong, H.: Synthesis of optimal numerical algorithms using real quantifier elimination (case study: square root computation). In: ISSAC 2014 Proceedings, pp. 162–169. ACM, New York (2014)
13. Erascu, M., Hong, H.: Real quantifier elimination for the synthesis of optimal numerical algorithms (case study: square root computation). J. Symbolic Comput. **75**, 110–126 (2016)
14. Heintz, J., Roy, M.-F., Solern, P.: On the theoretical and practical complexity of the existential theory of reals. Comput. J. **36**(5), 427–431 (1993)
15. Iwane, H., Higuchi, H., Anai, H.: An effective implementation of a special quantifier elimination for a sign definite condition by logical formula simplification. In: Gerdt, V.P., Koepf, W., Mayr, E.W., Vorozhtsov, E.V. (eds.) CASC 2013. LNCS, vol. 8136, pp. 194–208. Springer, Heidelberg (2013)
16. Jovanović, D., de Moura, L.: Solving non-linear arithmetic. In: Gramlich, B., Miller, D., Sattler, U. (eds.) IJCAR 2012. LNCS, vol. 7364, pp. 339–354. Springer, Heidelberg (2012)
17. Meggitt, J.E.: Pseudo division and pseudo multiplication processes. IBM J. Res. Dev. **6**(2), 210–226 (1962)
18. Moore, R.E., Kearfott, R.B., Cloud, M.J.: Introduction to Interval Analysis. Society for Industrial and Applied Mathematics, Philadelphia (2009)
19. Wolfram Research. Mathematica Edition: Version 8.0. Wolfram Research (2010)
20. Revol, N.: Interval Newton iteration in multiple precision for the univariate case. Numer. Algorithms **34**(2–4), 417–426 (2003)
21. Strzebonski, A.: Cylindrical algebraic decomposition using validated numerics. J. Symbolic Comput. **41**(9), 1021–1038 (2006)
22. Tarski, A.: A decision method for elementary algebra and geometry. Bull. Am. Math. Soc. **59**, 91–93 (1953)

Symbolic-Numeric Algorithms for Solving BVPs for a System of ODEs of the Second Order: Multichannel Scattering and Eigenvalue Problems

A.A. Gusev[1(✉)], V.P. Gerdt[1], L.L. Hai[1], V.L. Derbov[2], S.I. Vinitsky[1,3], and O. Chuluunbaatar[1]

[1] Joint Institute for Nuclear Research, Dubna, Russia
gooseff@jinr.ru
[2] Saratov State University, Saratov, Russia
[3] RUDN University, 6 Miklukho-Maklaya Street, Moscow 117198, Russia

Abstract. Symbolic-numeric algorithms for solving multichannel scattering and eigenvalue problems of the waveguide or tunneling type for systems of ODEs of the second order with continuous and piecewise continuous coefficients on an axis are presented. The boundary-value problems are formulated and discretized using the FEM on a finite interval with interpolating Hermite polynomials that provide the required continuity of the derivatives of the approximated solutions. The accuracy of the approximate solutions of the boundary-value problems, reduced to a finite interval, is checked by comparing them with the solutions of the original boundary-value problems on the entire axis, which are calculated by matching the fundamental solutions of the ODE system. The efficiency of the algorithms implemented in the computer algebra system Maple is demonstrated by calculating the resonance states of a multichannel scattering problem on the axis for clusters of a few identical particles tunneling through Gaussian barriers.

Keywords: Eigenvalue problem · Multichannel scattering problem · System of ODEs · Finite element method

1 Introduction

At present, the physical processes of electromagnetic wave propagation in multilayered optical waveguide structures and metamaterials [8], near-surface quantum diffusion of molecules and clusters [5,7], and transport of charge carriers in quantum semiconductor structures [6] are a subject of growing interest and intense studies. The mathematical formulation of these physical problems leads to the boundary-value problems (BVPs) for partial differential equations, which are reduced by the Kantorovich method to a system of ordinary differential equations (ODEs) of the second order with continuous or piecewise continuous

V.P. Gerdt et al. (Eds.): CASC 2016, LNCS 9890, pp. 212–227, 2016.
DOI: 10.1007/978-3-319-45641-6_14

potentials in an infinite region (on axis or semiaxis). The asymptotic boundary conditions depend upon the kind of the considered physical problem, e.g., multi-channel scattering, eigenvalue problem, or calculation of metastable states.

There is a number of unresolved problems in constructing calculation schemes and implementing them algorithmically. For example, the conventional calculation scheme for solving the scattering problem on axis was constructed only for the same number of open channels in the left-hand and right-hand asymptotic regions [1]. Generally, the lack of symmetry in the coefficient functions entering the ODE system with respect to the sign of the independent variable makes it necessary to construct more general calculation schemes. In the eigenvalue problem for bound or metastable states of the BVPs with piecewise constant potentials, the desired set of real or complex eigenvalues is conventionally calculated from the dispersion equation using the method of matching the general solutions with the unknown coefficients calculated from a system of algebraic equations. This method is quite a challenge, when the number of equations and/or the number of discontinuities of the potentials is large [8]. The aim of this paper is to present the construction of algorithms and programs implemented in the computer algebra systems Maple that allow progress in solving these problems and developing high-efficiency symbolic-numeric software.

In earlier papers [2,3], we developed symbolic-numeric algorithms of the finite element method (FEM) with Hermite interpolation polynomials (IHP) to calculate high-accuracy approximate solutions for a single ODE with piecewise continuous potentials and reduced boundary conditions on a finite interval. Here this algorithm is generalized to a set of ODEs and implemented as KANTBP 4M in the computer algebra system Maple [4]. For the multichannel scattering problem with piecewise constant potentials on the axis, the numerical estimates of the accuracy of the approximate solution of the BVP reduced to finite interval are presented using an auxiliary algorithm of matching the fundamental solutions at each boundary between the adjacent axis subintervals. The efficiency of the algorithms is demonstrated by the example of calculating the resonance and metastable states of the multichannel scattering problem on the axis for clusters formed by a few identical particles tunneling through Gaussian barriers.

The paper has the following structure. Section 2 formulates the eigenvalue problem and the multichannel scattering problem of the waveguide type for a system of ODEs with continuous and piecewise continuous coefficients on an axis. Sections 3 and 4 present the algorithms for solving the multichannel scattering problem and the eigenvalue problem. The comparative analysis of the solutions of the ODE system with piecewise constant potentials is given. In Sect. 5 the quantum transmittance induced by metastable states of clusters is analysed. Finally, the summary is given, and the possible use of algorithms and programs is outlined.

2   Formulation of the Boundary Value Problems

The symbolic-numeric algorithm realized in Maple is intended for solving the BVP and the eigenvalue problem for the system of second-order ODEs with respect to the unknown functions $\boldsymbol{\Phi}(z) = (\Phi_1(z),\ldots,\Phi_N(z))^T$ of the independent variable $z \in (z^{\min}, z^{\max})$ numerically using the Finite Element Method:

$$(\mathbf{D} - E\,\mathbf{I})\,\boldsymbol{\Phi}^{(i)}(z) \equiv \left(-\frac{1}{f_B(z)}\mathbf{I}\frac{d}{dz}f_A(z)\frac{d}{dz} + \mathbf{V}(z) \right.$$

$$\left. +\frac{f_A(z)}{f_B(z)}\mathbf{Q}(z)\frac{d}{dz} + \frac{1}{f_B(z)}\frac{d\,f_A(z)\mathbf{Q}(z)}{dz} - E\,\mathbf{I} \right)\boldsymbol{\Phi}(z) = 0. \qquad (1)$$

Here $f_B(z) > 0$ and $f_A(z) > 0$ are continuous or piecewise continuous positive functions, $\mathbf{I}$ is the identity matrix, $\mathbf{V}(z)$ is a symmetric matrix, $V_{ij}(z) = V_{ji}(z)$, and $\mathbf{Q}(z)$ is an antisymmetric matrix, $Q_{ij}(z) = -Q_{ji}(z)$, of the effective potentials having the dimension $N \times N$. The elements of these matrices are continuous or piecewise continuous real or complex-valued coefficients from the Sobolev space $\mathcal{H}_2^{s\geq1}(\Omega)$, providing the existence of nontrivial solutions subjected to homogeneous mixed boundary conditions: Dirichlet and/or Neumann, and/or third-kind at the boundary points of the interval $z \in \{z^{\min}, z^{\max}\}$ at given values of the elements of the real or complex-valued matrix $\mathcal{R}(z^t)$ of the dimension $N \times N$

$$(\mathrm{I}): \quad \boldsymbol{\Phi}(z^t) = 0, \quad z^t = z^{\min} \text{ and/or } z^{\max}, \qquad (2)$$

$$(\mathrm{II}): \quad \lim_{z \to z^t} f_A(z)\left(\mathbf{I}\frac{d}{dz} - \mathbf{Q}(z) \right) = 0, \quad z^t = z^{\min} \text{ and/or } z^{\max}, \qquad (3)$$

$$(\mathrm{III}): \quad \left(\mathbf{I}\frac{d}{dz} - \mathbf{Q}(z) \right)\bigg|_{z=z^t} = \mathcal{R}(z^t)\boldsymbol{\Phi}(z^t), \quad z^t = z^{\min} \text{ and/or } z^{\max}. \qquad (4)$$

One needs to note that the boundary conditions (2)–(4) can be applied to both ends of the domain independently, e.g. the boundary condition (2) to $z^{\min}$ and, at the same time, the boundary condition (4) to $z^{\max}$. The solution $\boldsymbol{\Phi}(z)\in\mathcal{H}_2^{s\geq1}(\bar{\Omega})$ of the BPVs (1)–(4) is determined using the Finite Element Method(FEM) by numerical calculation of stationary points for the symmetric quadratic functional

$$\Xi(\boldsymbol{\Phi}, E, z^{\min}, z^{\max}) \equiv \int_{z^{\min}}^{z^{\max}} \boldsymbol{\Phi}^\bullet(z)\,(\mathbf{D}-E\,\mathbf{I})\,\boldsymbol{\Phi}(z)dz = \Pi(\boldsymbol{\Phi}, E, z^{\min}, z^{\max}) + C,$$

$$C = -f^A(z^{\max})\boldsymbol{\Phi}^\bullet(z^{\max})\mathbf{G}(z^{\max})\boldsymbol{\Phi}(z^{\max}) + f^A(z^{\min})\boldsymbol{\Phi}^\bullet(z^{\min})\mathbf{G}(z^{\min})\boldsymbol{\Phi}(z^{\min}),$$

$$\Pi(\boldsymbol{\Phi}, E, z^{\min}, z^{\max}) = \int_{z^{\min}}^{z^{\max}} \left[f^A(z)\frac{d\boldsymbol{\Phi}^\bullet(z)}{dz}\frac{d\boldsymbol{\Phi}(z)}{dz} + f^B(z)\boldsymbol{\Phi}^\bullet(z)\mathbf{V}(z)\boldsymbol{\Phi}(z) \right. \qquad (5)$$

$$\left. +f^A(z)\boldsymbol{\Phi}^\bullet(z)\mathbf{Q}(z)\frac{d\boldsymbol{\Phi}(z)}{dz} - f^A(z)\frac{d\boldsymbol{\Phi}(z)^\bullet}{dz}\mathbf{Q}(z)\boldsymbol{\Phi}(z) - f^B(z)E\boldsymbol{\Phi}^\bullet(z)\boldsymbol{\Phi}(z) \right] dz,$$

where $\mathbf{G}(z) = \mathcal{R}(z) - \mathbf{Q}(z)$ is a symmetric matrix of the dimension $N \times N$, and the symbol $\bullet$ denotes either the transposition ^{T}, or the Hermitian conjugation †.

Problem 1. For the multichannel scattering problem on the axis $z \in (-\infty, +\infty)$ at fixed energy $E \equiv \Re E$, the desired matrix solutions $\boldsymbol{\Phi}(z) \equiv \{\boldsymbol{\Phi}_v^{(i)}(z)\}_{i=1}^N$,

$\Phi_v^{(i)}(z) = (\Phi_{1v}^{(i)}(z), \dots, \Phi_{Nv}^{(i)}(z))^T$ (the subscript v takes the values $\to$ or $\leftarrow$ and indicates the initial direction of the incident wave) of the BVP for the system of N ordinary differential equations of the second order (1) in the interval $z \in (z^{\min}, z^{\max})$ are calculated by the code. These matrix solutions are to obey the homogeneous third-kind boundary conditions (4) at the boundary points of the interval $z \in \{z^{\min}, z^{\max}\}$ with the asymptotes of the "incident wave + outgoing waves" type in the open channels $i = 1, \dots, N_o$:

$$\Phi_v(z \to \pm\infty) = \begin{cases} \begin{cases} \mathbf{X}^{(+)}(z)\mathbf{T}_v, & z \in [z^{\max}, +\infty), \\ \mathbf{X}^{(+)}(z) + \mathbf{X}^{(-)}(z)\mathbf{R}_v, & z \in (-\infty, z^{\min}], \end{cases} & v = \to, \\ \begin{cases} \mathbf{X}^{(-)}(z) + \mathbf{X}^{(+)}(z)\mathbf{R}_v, & z \in [z^{\max}, +\infty), \\ \mathbf{X}^{(-)}(z)\mathbf{T}_v, & z \in (-\infty, z^{\min}], \end{cases} & v = \leftarrow, \end{cases} \quad (6)$$

where $\mathbf{T}_v$ and $\mathbf{R}_v$ are unknown rectangular and square matrices of transmission and reflection amplitudes, respectively, used to construct the scattering matrix $\mathbf{S}$ of the dimension $N_o \times N_o$:

$$\mathbf{S} = \begin{pmatrix} \mathbf{R}_\to & \mathbf{T}_\leftarrow \\ \mathbf{T}_\to & \mathbf{R}_\leftarrow \end{pmatrix}, \quad (7)$$

which is symmetric and unitary in the case of real-valued potentials.

For the multichannel scattering problem on a semiaxis $z \in (z^{\min}, +\infty)$ or $z \in (-\infty, z^{\max})$, the desired matrix solution $\Phi(z)$ of the BVP for the system of N ordinary differential equations of the second order (1) is calculated in the interval $z \in (z^{\min}, z^{\max})$. This matrix solution is to obey the homogeneous third-kind boundary conditions (4) at the boundary point $z^{\max}$ or $z^{\min}$ of the interval, with the asymptotes of the "incident wave + outgoing waves" type in the open channels $i = 1, \dots, N_o$:

$$\Phi_\leftarrow(z \to +\infty) = \mathbf{X}^{(-)}(z) + \mathbf{X}^{(+)}(z)\mathbf{R}_\leftarrow, \quad z \in [z^{\max}, +\infty) \quad (8)$$
$$\text{or} \quad \Phi_\to(z \to -\infty) = \mathbf{X}^{(+)}(z) + \mathbf{X}^{(-)}(z)\mathbf{R}_\to, \quad z \in (-\infty, z^{\min}],$$

and obeying the homogeneous boundary conditions (Dirichlet and/or Neumann, and/or third-kind (see (2)–(4))) at the boundary point $z^{\min}$ or $z^{\max}$ to construct the scattering matrix $\mathbf{S} = \mathbf{R}_\leftarrow$ or $\mathbf{S} = \mathbf{R}_\to$, which is symmetric and unitary in the case of real-valued potentials.

In the solution of a multichannel scattering problem, the closed channels are taken into account. In this case, the asymptotic conditions (6), (8) have the form

$$LR: \Phi_\to(z \to \pm\infty) = \begin{cases} \mathbf{X}_{\max}^{(\to)}(z)\mathbf{T}_\to + \mathbf{X}_{\max}^{(c)}(z)\mathbf{T}_\to^c, & z \to +\infty, \\ \mathbf{X}_{\min}^{(\to)}(z) + \mathbf{X}_{\min}^{(\leftarrow)}(z)\mathbf{R}_\to + \mathbf{X}_{\min}^{(c)}(z)\mathbf{R}_\to^c, & z \to -\infty \end{cases} \quad (9)$$

$$RL: \Phi_\leftarrow(z \to \pm\infty) = \begin{cases} \mathbf{X}_{\max}^{(\leftarrow)}(z) + \mathbf{X}_{\max}^{(\to)}(z)\mathbf{R}_\leftarrow + \mathbf{X}_{\max}^{(c)}(z)\mathbf{R}_\leftarrow^c, & z \to +\infty, \\ \mathbf{X}_{\min}^{(\leftarrow)}(z)\mathbf{T}_\leftarrow + \mathbf{X}_{\min}^{(c)}(z)\mathbf{T}_\leftarrow^c, & z \to -\infty. \end{cases} \quad (10)$$

where $\mathbf{X}_{\max}^{(\to)}(z) = \mathbf{X}^{(+)}(z), z \geq z^{\max}$, $\mathbf{X}_{\min}^{(\to)}(z) = \mathbf{X}^{(+)}(z), z \leq z^{\min}$, $\mathbf{X}_{\min}^{(\leftarrow)}(z) = \mathbf{X}^{(-)}(z), z \leq z^{\min}$ in Eq. (9) and $\mathbf{X}_{\max}^{(\leftarrow)}(z) = \mathbf{X}^{(-)}(z), z \geq z^{\max}$ $\mathbf{X}_{\max}^{(\to)}(z) =$

$\mathbf{X}^{(+)}(z), z \geq z^{\max}, \mathbf{X}^{(\leftarrow)}_{\min}(z) = \mathbf{X}^{(-)}(z), z \leq z^{\min}$ in Eq. (10). It is assumed that the leading terms of the asymptotic solutions $\mathbf{X}^{(\pm)}(z)$ of the BVP at $z \leq z^{\min}$ and/or $z \geq z^{\max}$ have the following form:

in the open channels $V^t_{i_o i_o} < E$ are oscillating solutions $j = 1, \ldots, N$, $i_o = 1, \ldots, N_o$:

$$X^{(\pm)}_{i_o j}(z) \rightarrow \frac{\exp\left(\pm \imath p^t_{i_o} z\right)}{\sqrt{f_A(z) p^t_i}} \delta_{i_o j}, \quad p^t_{i_o} = \sqrt{\frac{f_B(z^t)}{f_A(z^t)}} \sqrt{E - V^t_{i_o i_o}} \quad (11)$$

in the closed channels $V^t_{i_c i_c} \geq E$ are exponentially decreasing solutions $j = 1, \ldots, N$, $i_c = N_o + 1, \ldots, N$

$$X^{(c)}_{i_c j}(z) \rightarrow \frac{1}{\sqrt{f_A(z)}} \exp\left(-p^t_{i_c} |z|\right) \delta_{i_c j}, \quad p^t_{i_c} = \sqrt{\frac{f_B(z^t)}{f_A(z^t)}} \sqrt{V^t_{i_c i_c} - E}. \quad (12)$$

These relations are valid if the coefficients of the equations with $z \leq z^{\min}$ and/or $z \geq z^{\max}$ satisfy the following conditions $t = \min, \max$:

$$\frac{f_A(z)}{f_B(z)} = \frac{f_A(z^t)}{f_B(z^t)} + o(1), \quad V_{ij}(z) = V^t_{ii} \delta_{ij} + o(1), \quad Q^t_{ij}(z) = o(1). \quad (13)$$

In the procedure of solving the BVP (1)–(4), the corresponding symmetric quadratic functional (5) is used, where the symbol $\bullet$ denotes the transposition and the complex conjugation $\dagger$ for real-valued potentials and the transposition T for complex-valued potentials required for discretisation of the problem using the FEM.

Problem 2. For the eigenvalue problem the code calculates a set of M energy eigenvalues E: $\Re E_1 \leq \Re E_2 \leq \ldots \leq \Re E_M$ and the corresponding set of eigenfunctions $\boldsymbol{\Phi}(z) \equiv \{\boldsymbol{\Phi}^{(m)}(z)\}^M_{m=1}$, $\boldsymbol{\Phi}^{(m)}(z) = (\Phi^{(m)}_1(z), \ldots, \Phi^{(m)}_N(z))^T$ from the space $\mathcal{H}^2_2$ for the system of N ordinary differential equations of the second order (1) subjected to the homogeneous boundary conditions of the first and/or second, and/or third kind (see (2)–(4)) at the boundary points of the interval $z \in (z^{\min}, z^{\max})$. In the case of real-valued potentials, the solutions are subjected to the normalisation and orthogonality conditions

$$\langle \boldsymbol{\Phi}^{(m)} | \boldsymbol{\Phi}^{(m')} \rangle = \int_{z^{\min}}^{z^{\max}} f_B(z)(\boldsymbol{\Phi}^{(m)}(z))^\bullet \boldsymbol{\Phi}^{(m')}(z) dz = \delta_{mm'}, \quad (14)$$

and the corresponding symmetric quadratic functional (5) is used, in which $\bullet$ denotes the Hermitian conjugation $\dagger$ needed for discretisation of the problem by the FEM. In the case of complex valued potentials, the solutions are to obey the normalisation and orthogonality conditions (14), and the corresponding symmetric quadratic functional (5) is used, in which $\bullet$ denotes the transposition T.

To solve the bound-state problem on the axis or on the semiaxis, the original problem is approximated by the BVP (1)–(4) on a finite interval $z \in (z^{\min}, z^{\max})$ under the boundary conditions of the third kind (4) with the given matrices $\mathcal{R}(z^t)$, which are independent of the unknown eigenvalue E, and the set of approximate eigenvalues and eigenfunctions is calculated. If the matrices $\mathcal{R}(z^t)$ depend on the unknown eigenvalue E, then $\mathcal{R}(z^t, E)$ is determined by the known asymptotic expansion of the desired solution. In this case, the Newtonian iteration scheme is implemented to calculate the approximate eigenfunctions and eigenvalues. The appropriate initial approximations are chosen from the solutions calculated previously with the boundary conditions independent of E.

Problem 3. For the calculation of metastable states with unknown complex eigenvalues E, the program solves the BVP for the set of equations (1) on a finite interval with the homogeneous conditions of the third kind (4), depending on the unknown eigenvalue E, using the appropriate symmetric quadratic functional (5). In this case, the symbol $\bullet$ denotes the transposition T, which is necessary for the discretisation of the problem in the FEM. In contrast to the scattering problem, the asymptotic solutions for metastable states contain only outgoing waves, considered in the sufficiently large, but finite interval of the spatial variable. For the metastable states on the axis $z \in (-\infty, +\infty)$, the eigenfunctions obey the boundary conditions of the third kind (4), where the matrix $\mathcal{R}(z^t) = \mathrm{diag}(\mathcal{R}(z^t))$ depends on the desired complex energy eigenvalue $E \equiv E_m = \Re E_m + \imath\Im E_m$, $\Im E_m < 0$ and is given by [9]

$$\mathcal{R}_{i_o i_o}(z^t, E_m) = \pm\sqrt{f_B(z^t)/f_A(z^t)}\sqrt{V^t_{i_o i_o} - E_m}, \quad t = \min, \max, \quad (15)$$

where $+$ or $-$ corresponds to $t = \max$ or $t = \min$, respectively, because the asymptotic solution of this problem contains only outgoing waves in the open channels $V^t_{i_o i_o} < \Re E$, $i_o = 1, \ldots, N_o$, while in the closed channels, there are only decay waves $V^t_{i_c i_c} > \Re E$, $i_c = N_o + 1, \ldots, N$

$$\mathcal{R}_{i_c i_c}(z^t, E_m) = \mp\sqrt{f_B(z^t)/f_A(z^t)}\sqrt{E_m - V^t_{i_c i_c}}, \quad t = \min, \max, \quad (16)$$

where $+$ or $-$ corresponds to $t = \min$ or $t = \max$, respectively.

For the metastable states on the semiaxis $z \in (z^{\min}, +\infty)$ or $z \in (-\infty, z^{\max})$, the solution is to obey the boundary condition (4) at the boundary point $z^{\max}$ or $z^{\min}$ and the boundary condition of the first, second, or third kind (see (2), (3) or (4), respectively) at the boundary point $z^{\min}$ or $z^{\max}$.

In this case, the eigenfunctions obey the orthogonality and normalisation conditions

$$(\boldsymbol{\Phi}^{(m')}|\boldsymbol{\Phi}^{(m)}) = (E_m - E_{m'})\left[\int_{z^{\min}}^{z^{\max}}(\boldsymbol{\Phi}^{(m')}(z))^T\boldsymbol{\Phi}^{(m)}(z)f_B(z)dz - \delta_{m'm}\right] + C_{m'm} = 0, \quad (17)$$

$$C_{m'm} = \sum_{t=\min,\max}\mp f_A(z^t)(\boldsymbol{\Phi}^{(m')}(z^t))^T[\mathcal{R}_{i_o i_o}(z^t, E_m) - \mathcal{R}_{i_o i_o}(z^t, E_{m'}) - 2\mathbf{Q}(z^t)]\boldsymbol{\Phi}^{(m)}(z^t),$$

where $+$ or $-$ corresponds to $t = \min$ or $t = \max$, respectively. Note that the orthogonality condition is derived by calculating the difference of two functionals (5) with the substitution of eigenvalues E_m, $E_{m'}$, eigenfunctions $\boldsymbol{\Phi}^{(m)}(z)$,

$\boldsymbol{\Phi}^{(m')}(z)$, and elements of matrices $\mathcal{R}(z^{\max}, E_m)$, $\mathcal{R}(z^{\min}, E_{m'})$ from Eq. (16). The calculation of the complex eigenvalues and eigenfunctions of metastable states is performed using the Newton iteration method. The appropriate initial approximations are chosen from the solutions calculated previously with the boundary conditions at fixed E.

3 The Algorithm for Solving the Scattering Problem

We consider a discrete representation of the solutions $\boldsymbol{\Phi}(z)$ of the problem (1)–(4) reduced by means of the FEM to the variational functional (5), on the finite-element grid, $\Omega^p_{h_j(z)}[z^{\min}, z^{\max}] = [z_0 = z^{\min}, z_l, l = 1, \ldots, np - 1, z_{np} = z^{\max}]$, with the mesh points $z_l = z_{jp} = z_j^{\max} \equiv z_{j+1}^{\min}$ of the grid $\Omega^{h_j(z)}[z^{\min}, z^{\max}]$ and the nodal points $z_l = z_{(j-1)p+r}$, $r = 0, \ldots, p$ of the sub-grids $\Omega_j^{h_j(z)}[z_j^{\min}, z_j^{\max}]$, $j = 1, \ldots, n$.

The solution $\boldsymbol{\Phi}^h(z) \approx \boldsymbol{\Phi}(z)$ is sought in the form of a finite sum over the basis of local functions $N^g_\mu(z)$ at each nodal point $z = z_l$ of the grid $\Omega^p_{h_j(z)}[z^{\min}, z^{\max}]$ of the interval $z \in \Delta = [z^{\min}, z^{\max}]$ (see [2]):

$$\boldsymbol{\Phi}^h(z) = \sum_{\mu=0}^{L-1} \boldsymbol{\Phi}^h_\mu N^g_\mu(z), \quad \boldsymbol{\Phi}^h(z_l) = \boldsymbol{\Phi}^h_{l\kappa^{\max}}, \quad \left.\frac{d^\kappa \boldsymbol{\Phi}^h(z)}{dz^\kappa}\right|_{z=z_l} = \boldsymbol{\Phi}^h_{l\kappa^{\max}+\kappa}, \quad (18)$$

where $L = (pn+1)\kappa^{\max}$ is the number of basis functions and $\boldsymbol{\Phi}^h_\mu$ (matrices of the dimension $N \times 1$) at $\mu = l\kappa^{\max} + \kappa$ are the nodal values of the κ-th derivatives of the function $\boldsymbol{\Phi}^h(z)$ (including the function $\boldsymbol{\Phi}^h(z)$ itself for $\kappa=0$) at the points z_l.

The substitution of the expansion (18) into the variational functional (5) reduces the solution of the problem (1)–(4) to the solution of the algebraic problem with respect to the matrix functions, $\boldsymbol{\Phi}^h \equiv ((\chi^{(1)})^h, \ldots, (\chi^{(N_o)})^h)$ at $E = E^h$,

$$\mathbf{G}^p \boldsymbol{\Phi}^h \equiv (\mathbf{A}^p - E^h \mathbf{B}^p)\boldsymbol{\Phi}^h = \mathbf{M}\boldsymbol{\Phi}^h, \quad \mathbf{M} = \mathbf{M}^{\max} - \mathbf{M}^{\min}, \quad (19)$$

with the matrices $\mathbf{A}^p$ and $\mathbf{B}^p$ of the dimension $NL \times NL$ obtained by integration in the variational functional (5) (see, e.g., [2]). The matrices $\mathbf{M}^{\max}$ and $\mathbf{M}^{\min}$ arise due to the approximation of the boundary conditions of the third kind at the left-hand and right-hand boundaries of the interval $z \in (z^{\min}, z^{\max})$

$$\frac{d\boldsymbol{\Phi}^h(z)}{dz} = (\mathbf{G}(z) + \mathbf{Q}(z))\boldsymbol{\Phi}^h(z), \quad z = z^{\min}, \quad z = z^{\max}. \quad (20)$$

The elements of the matrix $\mathbf{M} = \{M_{l'_1, l'_2}\}^{NL}_{l'_1, l'_2=1}$ equal zero except those, for which both indexes $l'_1 = (l_1 - 1)N + \nu_1$, $l'_2 = (l_2 - 1)N + \nu_2$ belong to the interval $1, \ldots, N$ or to the interval $(L - \kappa_{\max})N + 1, \ldots, (L - \kappa_{\max})N + N$, where N is the number of equations (1) and L is the number of basis functions $N^g_\mu(z)$ in the expansion of the desired solutions (18) in the interval $z \in \Delta = [z^{\min}, z^{\max}]$.

Input. We present the matrix $\boldsymbol{\Phi}^h$ of the dimension $NL\times 1$ in the form of three submatrices: matrix $\boldsymbol{\Phi}_a$ of the dimension $N\times 1$, such that $(\boldsymbol{\Phi}_a)_{i1} = (\boldsymbol{\Phi}^h)_{i1}$, matrix $\boldsymbol{\Phi}_c$ of the dimension $N\times 1$, such that $(\boldsymbol{\Phi}_c)_{i1} = (\boldsymbol{\Phi}^h)_{(L-\kappa_{\max})N+i,1}$, and the matrix $\boldsymbol{\Phi}_b$ of the dimension $(L-2)N\times 1$ is derived by omitting the submatrices $\boldsymbol{\Phi}_a$ and $\boldsymbol{\Phi}_c$ from the solution matrix. Then the matrices in l.h.s. and r.h.s. of Eq. (19) take the form

$$(\mathbf{A}^p - E\,\mathbf{B}^p) = \begin{pmatrix} \mathbf{G}^p_{aa} & \mathbf{G}^p_{ab} & 0 \\ \mathbf{G}^p_{ba} & \mathbf{G}^p_{bb} & \mathbf{G}^p_{bc} \\ 0 & \mathbf{G}^p_{cb} & \mathbf{G}^p_{cc} \end{pmatrix}, \quad \mathbf{M} = \begin{pmatrix} -\mathbf{G}^p_{\min} & 0 & 0 \\ 0 & 0 & 0 \\ 0 & 0 & \mathbf{G}^p_{\max} \end{pmatrix}. \quad (21)$$

The matrices $\mathbf{G}^p_{bb}$ of the dimension $(L-2)N\times(L-2)N$, $\mathbf{G}^p_{ba}$ and $\mathbf{G}^p_{bc}$ of the dimension $(L-2)N\times N$, $\mathbf{G}^p_{ab}$ and $\mathbf{G}^p_{cb}$ of the dimension $N\times(L-2)N$, $\mathbf{G}^p_{aa}$, $\mathbf{G}^p_{cc}$, of the dimension $N\times N$ are determined from the finite element approximation and considered as known. The existence of zero submatrices is related to the band structure of the matrix $\mathbf{G}^p$ from Eq. (19). The matrices $\mathbf{G}_{\min}$ and $\mathbf{G}_{\max}$ of the dimension $N\times N$ correspond to nonzero blocks of the matrix $\mathbf{M}$, and the matrices $\boldsymbol{\Phi}_a$ and $\boldsymbol{\Phi}_c$ of the dimension $N\times 1$, are given by the asymptotic values (9), (10) and will be considered below, the matrix $\boldsymbol{\Phi}_b$ of the dimension $(L-2)N\times 1$ is derived by omitting the submatrices $\boldsymbol{\Phi}_a$ and $\boldsymbol{\Phi}_c$ from the solution matrix. We rewrite problem (19) in the following form

$$\begin{aligned} \mathbf{G}^p_{aa}\boldsymbol{\Phi}_a + \mathbf{G}^p_{ab}\boldsymbol{\Phi}_b &= -\mathbf{G}^p_{\min}\boldsymbol{\Phi}_a, \\ \mathbf{G}^p_{ba}\boldsymbol{\Phi}_a + \mathbf{G}^p_{bb}\boldsymbol{\Phi}_b + \mathbf{G}^p_{bc}\boldsymbol{\Phi}_c &= 0, \\ \mathbf{G}^p_{cb}\boldsymbol{\Phi}_b + \mathbf{G}^p_{cc}\boldsymbol{\Phi}_c &= \mathbf{G}^p_{\max}\boldsymbol{\Phi}_c. \end{aligned} \quad (22)$$

Step 1. Let us eliminate $\boldsymbol{\Phi}_b$ from the problem. From the second equation, the explicit expression follows

$$\boldsymbol{\Phi}_b = -(\mathbf{G}^p_{bb})^{-1}\mathbf{G}^p_{ba}\boldsymbol{\Phi}_a - (\mathbf{G}^p_{bb})^{-1}\mathbf{G}^p_{bc}\boldsymbol{\Phi}_c, \quad (23)$$

however, it requires the inversion of a large-dimension matrix. To avoid it, we consider the auxiliary problems

$$\mathbf{G}^p_{bb}F_{ba} = \mathbf{G}^p_{ba}, \qquad \mathbf{G}^p_{bb}F_{bc} = \mathbf{G}^p_{bc}. \quad (24)$$

Since $\mathbf{G}^p_{bb}$ is a non-degenerate matrix, each of the matrix equations (24) has a unique solution

$$F_{ba} = (\mathbf{G}^p_{bb})^{-1}\mathbf{G}^p_{ba}, \qquad F_{bc} = (\mathbf{G}^p_{bb})^{-1}\mathbf{G}^p_{bc}. \quad (25)$$

Step 2. Then for the function $\boldsymbol{\Phi}_b$ we have the expression

$$\boldsymbol{\Phi}_b = -F_{ba}\boldsymbol{\Phi}_a - F_{bc}\boldsymbol{\Phi}_c, \quad (26)$$

and the problem (19) with the matrix of the dimension $NL\times NL$ is reduced to two algebraic problems with the matrices of the dimension $N\times N$

$$\begin{aligned} \mathbf{Y}^p_{aa}\boldsymbol{\Phi}_a + \mathbf{Y}^p_{ac}\boldsymbol{\Phi}_c &= -\mathbf{G}^p_{\min}\boldsymbol{\Phi}_a, \\ \mathbf{Y}^p_{ca}\boldsymbol{\Phi}_a + \mathbf{Y}^p_{cc}\boldsymbol{\Phi}_c &= \mathbf{G}^p_{\max}\boldsymbol{\Phi}_c, \end{aligned} \quad (27)$$

where $\mathbf{Y}^p_{**}$ is expressed in terms of the solutions $\mathbf{F}_{ba}$ and $\mathbf{F}_{bc}$ of the problems (24)

$$\mathbf{Y}^p_{aa} = \mathbf{G}^p_{aa} - \mathbf{G}^p_{ab}\mathbf{F}_{ba}, \quad \mathbf{Y}^p_{ac} = -\mathbf{G}^p_{ab}\mathbf{F}_{bc}, \tag{28}$$
$$\mathbf{Y}^p_{ca} = -\mathbf{G}^p_{cb}\mathbf{F}_{ba}, \, \mathbf{Y}^p_{cc} = \mathbf{G}^p_{cc} - \mathbf{G}^p_{cb}\mathbf{F}_{bc}.$$

Note that the system of equations (28) is solved at **step 4** for each of $N_o^L + N_o^R$ incident waves.

Step 3. Consider the solution (9) for the incident wave travelling from left to right (LR) and the solution (10) for the incident wave travelling from right to left (RL). $\mathbf{\Phi}_\rightarrow(z \rightarrow \pm\infty)$ and $\mathbf{\Phi}_\leftarrow(z \rightarrow \pm\infty)$ are matrix solutions of the dimension $1 \times N_o^L$ and $1 \times N_o^R$. In other words, there are N_o^L linearly independent solutions, describing the incident wave traveling from left to right and N_o^R linearly independent solution, describing the incident wave traveling from right to left, respectively. The matrices $\mathbf{X}^{(\rightarrow)}_{\min}(z)$, $\mathbf{X}^{(\leftarrow)}_{\min}(z)$ of the dimension $1 \times N_o^L$ and the matrices $\mathbf{X}^{(\rightarrow)}_{\max}(z)$, $\mathbf{X}^{(\leftarrow)}_{\max}(z)$ of the dimension $1 \times N_o^R$ represent the fundamental asymptotic solution at the left and right boundaries of the interval, describing the motion of the wave in the arrow direction. The matrices $\mathbf{X}^{(c)}_{\min}(z)$ of the dimension $1 \times (N - N_o^L)$ and $\mathbf{X}^{(c)}_{\max}(z)$ of the dimension $1 \times (N - N_o^R)$ are fundamental asymptotically decreasing solutions at the left and right boundaries of the interval. The elements of these matrices are column matrices of the dimension $N \times 1$. It follows that the matrices of reflection amplitudes $\mathbf{R}_\rightarrow$ and $\mathbf{R}_\leftarrow$ are square matrices of the dimension $N_o^L \times N_o^L$ and $N_o^R \times N_o^R$, while the matrices of transmission amplitudes $\mathbf{T}_\rightarrow$, $\mathbf{T}_\leftarrow$ are rectangular matrices of the dimension $N_o^R \times N_o^L$ and $N_o^L \times N_o^R$. The auxiliary matrices $\mathbf{R}^c_\rightarrow$, $\mathbf{T}^c_\rightarrow$, $\mathbf{R}^c_\leftarrow$ and $\mathbf{T}^c_\leftarrow$ are rectangular matrices of the dimension $(N-N_o^L) \times N_o^L$, $(N-N_o^R) \times N_o^L$, $(N-N_o^R) \times N_o^R$ and $(N-N_o^L) \times N_o^R$. Then the components of the wave functions (9) and (10) take the form for LR and RL waves:

$$(\mathbf{\Phi}_a)_{i_o i_o^L} = X^{(\rightarrow)}_{i_o i_o^L}(z^{\min}) + \sum_{i_o'=1}^{N_o^L} X^{(\leftarrow)}_{i_o i_o'}(z^{\min})R^{(\rightarrow)}_{i_o' i_o^L} + \sum_{i_c'=1}^{N-N_o^L} X^{(c)}_{i_o i_c'}(z^{\min})R^{(c\rightarrow)}_{i_c' i_o^L},$$

$$(\mathbf{\Phi}_c)_{i_o i_o^L} = \sum_{i_o'=1}^{N_o^R} X^{(\leftarrow)}_{i_o i_o'}(z^{\max})T^{(\rightarrow)}_{i_o' i_o^L} + \sum_{i_c'=1}^{N-N_o^R} X^{(c)}_{i_o i_c'}(z^{\max})T^{(c\rightarrow)}_{i_c' i_o^L},$$

$$(\mathbf{\Phi}_a)_{i_o i_o^R} = \sum_{i_o'=1}^{N_o^L} X^{(\rightarrow)}_{i_o i_o'}(z^{\min})T^{(\leftarrow)}_{i_o' i_o^R} + \sum_{i_c'=1}^{N-N_o^L} X^{(c)}_{i_o i_c'}(z^{\min})T^{(c\leftarrow)}_{i_c' i_o^R}, \tag{29}$$

$$(\mathbf{\Phi}_c)_{i_o i_o^R} = X^{(\leftarrow)}_{i_o i_o^R}(z^{\max}) + \sum_{i_o'=1}^{N_o^R} X^{(\rightarrow)}_{i_o i_o'}(z^{\max})R^{(\leftarrow)}_{i_o' i_o^R} + \sum_{i_c'=1}^{N-N_o^R} X^{(c)}_{i_o i_c'}(z^{\max})R^{(c\leftarrow)}_{i_c' i_o^R},$$

where the asymptotic solutions $\mathbf{X}^{(\rightarrow)}(z) \equiv \mathbf{X}^{(+)}(z)$, $\mathbf{X}^{(\leftarrow)}(z) \equiv \mathbf{X}^{(-)}(z)$ of the BVP at $z \leq z^{\min}$ and/or $z \geq z^{\max}$ are given by Eqs. (11)–(12). RL: The products in the r.h.s. of Eq. (27) in accordance with (4) and (20) are calculated via the first derivatives of the asymptotic solutions $X'^{(*)}_{**}(z^t) = \left.\frac{dX^{(*)}_{**}(z)}{dz}\right|_{z=z^t}$ for LR: $(\mathbf{G}^p_{\min}\boldsymbol{\Phi}_a)_{i_o i_o^L}$, $(\mathbf{G}^p_{\max}\boldsymbol{\Phi}_c)_{i_o i_o^L}$ and RL: $(\mathbf{G}^p_{\min}\boldsymbol{\Phi}_a)_{i_o i_o^R}$, $(\mathbf{G}^p_{\max}\boldsymbol{\Phi}_c)_{i_o i_o^R}$.

Step 4. Substituting the expressions (29) and their derivatives into Eq. (27), we form and solve the system of inhomogeneous equations for LR at $i_o^L = 1, ..., N_o^L$

$$\sum_{i_o'=1}^{N_o^L} \left(X'^{(\leftarrow)}_{i_o i_o'}(z^{\min}) + \sum_{j_o=1}^{N} (\mathbf{Y}^p_{aa})_{i_o j_o} X^{(\leftarrow)}_{j_o i_o'}(z^{\min}) \right) R^{(\rightarrow)}_{i_o' i_o^L}$$

$$+ \sum_{i_c'=1}^{N-N_o^L} \left(X'^{(c)}_{i_o i_c'}(z^{\min}) + \sum_{j_o=1}^{N} (\mathbf{Y}^p_{aa})_{i_o j_o} X^{(c)}_{j_o i_c'}(z^{\min}) \right) R^{(c\rightarrow)}_{i_c' i_o^L}$$

$$+ \sum_{i_o'=1}^{N_o^R} \sum_{j_o=1}^{N} (\mathbf{Y}^p_{ac})_{i_o j_o} X^{(\leftarrow)}_{j_o i_o'}(z^{\max}) T^{(\rightarrow)}_{i_o' i_o^L} + \sum_{i_c'=1}^{N-N_o^R} \sum_{j_o=1}^{N} (\mathbf{Y}^p_{ac})_{i_o j_o} X^{(c)}_{j_o i_c'}(z^{\max}) T^{(c\rightarrow)}_{i_c' i_o^L}$$

$$= -X'^{(\rightarrow)}_{i_o i_o^L}(z^{\min}) - \sum_{j_o=1}^{N} (\mathbf{Y}^p_{aa})_{i_o j_o} X^{(\rightarrow)}_{j_o i_o^L}(z^{\min}),$$

$$+ \sum_{i_o'=1}^{N_o^L} \sum_{j_o=1}^{N} (\mathbf{Y}^p_{ca})_{i_o j_o} X^{(\leftarrow)}_{j_o i_o'}(z^{\min}) R^{(\rightarrow)}_{i_o' i_o^L} + \sum_{i_c'=1}^{N-N_o^L} \sum_{j_o=1}^{N} (\mathbf{Y}^p_{ca})_{i_o j_o} X^{(c)}_{j_o i_c'}(z^{\min}) R^{(c\rightarrow)}_{i_c' i_o^L}$$

$$+ \sum_{i_o'=1}^{N_o^R} \left(-X'^{(\leftarrow)}_{i_o i_o'}(z^{\max}) + \sum_{j_o=1}^{N} (\mathbf{Y}^p_{cc})_{i_o j_o} X^{(\leftarrow)}_{j_o i_o'}(z^{\max}) \right) T^{(\rightarrow)}_{i_o' i_o^L}$$

$$+ \sum_{i_c'=1}^{N-N_o^R} \left(-X'^{(c)}_{i_o i_c'}(z^{\max}) + \sum_{j_o=1}^{N} (\mathbf{Y}^p_{cc})_{i_o j_o} X^{(c)}_{j_o i_c'}(z^{\max}) \right) T^{(c\rightarrow)}_{i_c' i_o^L})$$

$$= - \sum_{j_o=1}^{N} (\mathbf{Y}^p_{ca})_{i_o j_o} X^{(\rightarrow)}_{j_o i_o^L}(z^{\min}),$$

or a similar one for RL at $i_o^R = 1, ..., N_o^R$ that has a unique solution.

Remark. When solving the problem on a semiaxis with the Neumann or the third-kind boundary conditions at the boundary $z^{\min}$ or $z^{\max}$ of the semiaxis, the role of unknowns is played by the elements of the matrices $\boldsymbol{\Phi}_a$ or $\boldsymbol{\Phi}_c$, instead of $\mathbf{R}$ and $\mathbf{T}$, while for the Dirichlet boundary conditions, we have $\boldsymbol{\Phi}_a = 0$ or $\boldsymbol{\Phi}_c = 0$, so that in this case the corresponding equation is not taken into account.

4   The BVP with Piecewise Constant Potentials

The accuracy of the approximate solutions of the reduced BVPs on the finite interval calculated by FEM is checked by comparison with the solutions of the

BVPs for the system of Eq. (1) at $f_A(z) = f_A$, $f_B(z) = f_B$, $Q_{ij}(z) = 0$ on the entire axis with the matrix of piecewise constant potentials

$$V_{ij}(z) = V_{ji}(z) = \{V_{ij;1}, z \le z_1; \dots; V_{ij;k-1}, z \le z_{k-1}; V_{ij;k}, z > z_{k-1}\}. \quad (30)$$

Algorithm for solving the BVP by matching the fundamental solutions. In the algorithm, the following series of steps are implemented in two cycles $i_o = i_o^L = 1, \dots, N_o^L$ and $i_o = i_o^R = 1, \dots, N_o^R$:

Step 1. In the intervals $z \in (-\infty, z_1)$, $z \in (z_{k-1}, +\infty)$, one of the asymptotic states of the multichannel scattering problem is constructed, $\boldsymbol{\Phi}_0 \equiv \boldsymbol{\Phi}_a = \{(\boldsymbol{\Phi}_a)_i \equiv (\boldsymbol{\Phi}_a)_{ii_o^L} \text{ or } (\boldsymbol{\Phi}_a)_{ii_o^R}\}$ and $\boldsymbol{\Phi}_k \equiv \boldsymbol{\Phi}_c = \{(\boldsymbol{\Phi}_c)_i \equiv (\boldsymbol{\Phi}_c)_{ii_o^L} \text{ or } (\boldsymbol{\Phi}_c)_{ii_o^R}\}$, corresponding to Eq. (9) or (10), its explicit form given in Eq. (29).

Step 2. In the cycle by l for each of the internal subintervals $z \in [z_{l-1}, z_l]$, $l = 2, \dots, k-1$, the general solution is calculated that depends on $2N$ parameters $C_{2N(l-2)+1}, \dots, C_{2N(l-1)}$, $\boldsymbol{\Phi}_l = \mathbf{X}_{l;1}C_{2N(l-2)+1} + \dots + \mathbf{X}_{l;2N}C_{2N(l-1)}$, of the ODE system (1) with constant coefficients $V_{ij;l}$ from (30), the spectral parameter E being fixed, and the first derivative of the obtained solution is calculated.

Step 3: In the cycle by l, the differences $\boldsymbol{\Phi}_l(z_l) - \boldsymbol{\Phi}_{l-1}(z_l)$ and $(d/dz)(\boldsymbol{\Phi}_l(z) - \boldsymbol{\Phi}_{l-1}(z))|_{z_l}$, $l = 1, \dots, k$ are calculated and set equal to zero. As a result, the system of $2N(k-1)$ inhomogeneous equations with respect to $2N(k-1)$ unknown expansion coefficients $C_1, \dots, C_{2N(k-2)}$, as well as the corresponding elements of the matrices $\mathbf{T}_*$, $\mathbf{R}_*$ listed in Eq. (29) are obtained and solved.

Remark. For solving the bound state problem or calculating metastable states, the algorithm is modified as follows.

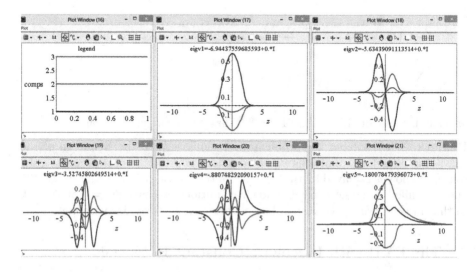

Fig. 1. A screenshot of the FEM algorithm run showing the components of five solutions $\boldsymbol{\Phi}_m^h(z)$, $m = 1, \dots, 5$, of the bound state problem.

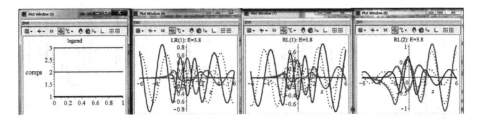

Fig. 2. The screenshot of the FEM algorithm run, showing the real (solid lines) and the imaginary (dotted lines) components of the solution of the scattering problem for the wave incident from the left, LR(1), and the waves incident from the right from the first, RL(1), and the second RL(2) open channels.

1. The sequence of **steps 1–3** is performed only once.
2. At **Step 1**, instead of the asymptotic expressions (9) and (10), one uses

$$\boldsymbol{\Phi}(z \to \pm\infty) = \left\{ \mathbf{X}^{(c)}_{\max}(z)\mathbf{C}_+, \, z \to +\infty, \, \mathbf{X}^{(c)}_{\min}(z)\mathbf{C}_-, \, z \to -\infty, \right\} \quad (31)$$

where $\mathbf{C}_\pm$ is a column matrix with the dimension $1 \times N$, and $\mathbf{X}^{(c)}_*(z)$ is the specially selected fundamental solution that for bound states should decrease exponentially at $z \to \pm\infty$, while for metastable states must describe diverging waves in open channels and decrease exponentially in closed channels.
3. In **step 3**, a system of $2N(k-1)$ linear homogeneous algebraic equations for $2N(k-1)+1$ unknown coefficients $C_1, \ldots, C_{2N(k-2)}$ and the corresponding elements of the matrices $\mathbf{C}_\pm$, which is nonlinear and transcendent with respect to the spectral parameter E, is obtained and solved.

Benchmark Calculations. We solved the BVP for the system of equations (1) with the effective potentials (30) and the third-kind boundary conditions (4) on a finite interval, which is determined from the asymptotic solutions (9), (10), (11), (12) of the multichannel scattering problem on the axis

$$\mathbf{V}(z) = \left\{ \begin{pmatrix} 0 & 0 & 0 \\ 0 & 5 & 0 \\ 0 & 0 & 10 \end{pmatrix}, z < -2; \begin{pmatrix} -5 & 4 & 4 \\ 4 & 0 & 4 \\ 4 & 4 & 10 \end{pmatrix}, -2 \leq z \leq 2; \begin{pmatrix} 0 & 0 & 0 \\ 0 & 0 & 0 \\ 0 & 0 & 10 \end{pmatrix}, z > 2 \right\}.$$

For solving the BVP the uniform finite-element grid $z^{\min} = -6$, $h_{j=1,\ldots,30} = 0.4$, $z^{\max} = 6$ with seventh-order Hermitian elements $(\kappa^{\max}, p) = (2, 3)$, $p' = 7$ preserving the continuity of the first derivative in the approximate solutions was chosen. The calculations were performed with 16 significant digits. Given $E = 3.8$, for the wave incident from the left there is one open channel, $N^L_o = 1$, and for the wave incident from the right, there are two open channels, $N^R_o = 2$. The comparison of FEM results with those of solving the system of algebraic equations yields the error estimate $accuracy = S_{an} - S_{matr} \sim 10^{-13}$. for the computation of the square matrices of reflection amplitudes $\mathbf{R}_\rightarrow$ and $\mathbf{R}_\leftarrow$, having the dimension 1×1 and 2×2, and the rectangular matrices of transmission amplitudes $\mathbf{T}_\rightarrow$ and

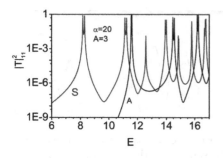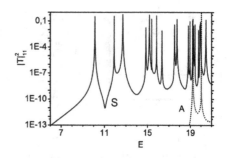

Fig. 3. The total probability $|\mathbf{T}|^2_{11}$ of transmission through the repulsive Gaussian barrier versus the energy E (in oscillator units) at $\sigma=1/10$, $\alpha=20$ for the cluster of three ($n=3$, left panel) and four ($n=4$, right panel) identical particles initially being in the in the ground symmetric (solid lines) and antisymmetric (dashed lines) state.

$\mathbf{T}_{\leftarrow}$ having the dimension 2×1 and 1×2. With the error of the same order, the conditions of symmetry, $\mathbf{S} - \mathbf{S}^T$ and S-matrix unitarity $\mathbf{SS}^{dag} - \mathbf{I}$ are satisfied. For five eigenvalues, the differences $\delta E_m = |E^h_m - E^{ex}_m|$ between the results of two above methods appeared to be of the order of 10^{-9} in the calculations performed with 12 significant figures. The components $\mathbf{\Phi}_m$ of the bound state solutions and the solutions $\mathbf{\Phi}_v$ of the scattering problem on a finite-element grid are shown in Figs. 1 and 2. The running time for this example using KANTBP 4M implemented in Maple 16 is 232 s for the PC Intel Pentium CPU 1.50 GHz 4 GB 64 bit Windows 8.

5   Quantum Transmittance Induced by Metastable States

In Ref. [5], the problem of tunneling of a cluster of n identical particles, coupled by pair harmonic oscillator potentials, through the Gaussian barriers $V(x_i) = \alpha/(2\pi\sigma^2)^{1/2} \exp(-x_i^2/\sigma^2)$, $i = 1,...,n$, with averaging over the basis of the cluster eigenfunctions was formulated as a multichannel scattering problem for the system of ODEs (1) with the center-of-mass independent variable $z = (x_1 + ... + x_n)/\sqrt{n}$ and the boundary conditions (4) that follow from the asymptotic conditions (6) at $f_A(z) = 1$, $f_B(z) = 1$, $Q_{ij}(z) = 0$. The elements $V_{ij}(z)$ of the effective potentials matrix were calculated analytically and plotted in [5].

Let us apply the technique developed in the present paper and implemented as KANTBP 4M to the tunneling problem for the cluster comprising three and four identical particles in symmetric (S) and antisymmetric (A) states.

At first we solve the scattering problem with fixed energy $E = \Re E$. The solutions of the BVP were discretised on the finite-element grid $\Omega_h = (-11(11)11)$ for $n = 3$ and $\Omega_h = (-13(13)13)$ for $n = 4$, with the number of Lagrange elements of the twelfth order $p' = 12$ shown in brackets. The boundary points of the interval z^t were chosen in accordance with the required accuracy of the approximate solution $\max\{|V_{ij}(z^t)/\alpha|; i, j = 1,...,j_{\max}\} < 10^{-8}$. The number N of the cluster basis functions in the expansion of solutions of the original problem

[5] and, correspondingly, the number of equations for S-states for $n = 3, 4$ was chosen equal to $N = 21, 39$ and for A-states $N = 16, 15$. The results of the calculations for three and four particles are presented in Fig. 3. The resonance values of energy $E = E_l^{S(A)}$ and the corresponding maximal values of the transmission coefficient $|\mathbf{T}|^2_{11}$ clearly visible in Fig. 3 are presented in Table 1.

Table 1. The first resonance energy values $E_l^{S(A)}$, at which the maximum of the transmission coefficient $|\mathbf{T}|^2_{11}$ is achieved, and the complex energy eigenvalues $E_m^M = \Re E_m^M + \imath \Im E_m^M$ of the metastable states for symmetric S (antisymmetric A) states of $n = 3$ and $n = 4$ particles at $\sigma = 1/10$, $\alpha = 20$.

l	E_l^S	$|T|^2_{11}$	m	E_m^M		l	E_l^S	$|T|^2_{11}$	m	E_m^M
1	8.175	0.775	1	$8.175 - \imath 5.1(-3)$		1	10.121	0.321	1	$10.119 - \imath 4.0(-3)$
	8.306	0.737	2	$8.306 - \imath 5.0(-3)$					2	$10.123 - \imath 4.0(-3)$
2	11.111	0.495	3	$11.110 - \imath 5.6(-3)$		2	11.896	0.349	3	$11.896 - \imath 6.3(-5)$
	11.229	0.476	4	$11.229 - \imath 5.5(-3)$		3	12.713	0.538	4	$12.710 - \imath 4.5(-3)$
3	12.598	0.013	5	$12.598 - \imath 6.4(-3)$			12.717	0.538	5	$12.720 - \imath 4.5(-3)$
			6	$12.599 - \imath 6.3(-3)$		4	14.858	0.017	6	$14.857 - \imath 4.3(-3)$
4	13.929	0.331	7	$13.929 - \imath 4.5(-3)$					7	$14.859 - \imath 4.3(-3)$
	14.003	0.328	8	$14.004 - \imath 4.6(-3)$		5	15.188	0.476	8	$15.185 - \imath 3.9(-3)$
5	14.841	0.014	9	$14.841 - \imath 3.5(-3)$					9	$15.191 - \imath 3.9(-3)$
	14.877	0.008	10	$14.878 - \imath 3.5(-3)$		6	15.405	0.160	10	$15.405 - \imath 1.4(-5)$
						7	15.863	0.389	11	$15.863 - \imath 5.3(-5)$

l	E_l^A	$|T|^2_{11}$	m	E_m^M		l	E_l^A	$|T|^2_{11}$	m	E_m^M
1	11.551	1.000	1	$11.551 - \imath 1.8(-3)$		1	19.224	0.177	1	$19.224 - \imath 4.0(-4)$
	11.610	1.000	2	$11.610 - \imath 2.0(-3)$					2	$19.224 - \imath 4.0(-4)$
2	14.459	0.553	3	$14.459 - \imath 2.9(-3)$		2	20.029	0.970	3	$20.029 - \imath 3.3(-7)$
	14.564	0.480	4	$14.565 - \imath 2.7(-3)$						

For metastable states, the eigenfunctions obey the boundary conditions of the third kind (4), where the matrices $\mathcal{R}(z^t) = \mathrm{diag}(\mathcal{R}(z^t))$ depend on the desired complex energy eigenvalue, $E \equiv E_m^M = \Re E_m^M + \imath \Im E_m^M$, $\Im E_m^M < 0$, are given by (15), (16), since the asymptotic solutions of this problem contain only outgoing waves in the open channels. In this case, the eigenfunctions obey the orthogonality and normalisation conditions (17). The discretisation of the solutions of the BVP was implemented on the above finite-element grid. The algebraic eigenvalue problem was solved using the Newton method with the optimal choice of the iteration step [3] using the additional condition $\Xi_h(\boldsymbol{\Phi}^{(m)}, E_m, z^{\min}, z^{\max}) = 0$ obtained as a result of the discretisation of the functional (5) and providing the upper estimates for the approximate eigenvalue. As the initial approximation we used the real eigenvalues and the eigenfunctions orthonormalised by the condition that the expression in square brackets in Eq. (17) is zero. They were found as a result of solving the bound-state problem with the functional (5) at $\mathcal{R}(z^t) = 0$ on the grid $\Omega_h = (-5(5)5)$ for $n = 3$ and $n = 4$. The results of the calculations performed with the variational functional (5), (17), defined in the interval $[z^{\min}, z^{\max}]$, for the complex values of energy of the metastable states

$E_m^M \equiv E_m = \Re E_m^M + i\Im E_m^M$ for $n = 3$ and $n = 4$ are presented in Table 1. The resonance values of energy corresponding to these metastable states are responsible for the peaks of the transmission coefficient, i.e., the quantum transparency of the barriers. The position of peaks presented in Fig. 3 is seen to be in quantitative agreement with the real part $\Re E_m^M$, and the half-width of the $|\mathbf{T}|_{11}^2(E_l)$ peaks agrees with the imaginary part $\Gamma = -2\Im E_m^M$ of the complex energy eigenvalues $E_m^M = \Re E_m^M + i\Im E_m^M$ of the metastable states by the order of magnitude.

6   Summary and Perspectives

The developed approach, algorithms, and programs can be adapted and applied to study the waveguide modes in a planar optical waveguide, the quantum diffusion of molecules and micro-clusters through surfaces, and the fragmentation mechanism in producing very neutron-rich light nuclei.

The work was partially supported by the Russian Foundation for Basic Research, grant No. 14-01-00420, and the Bogoliubov-Infeld JINR-Poland program.

References

1. Gusev, A.A., Chuluunbaatar, O., Vinitsky, S.I., Abrashkevich, A.G.: KANTBP 3.0: new version of a program for computing energy levels, reflection and transmission matrices, and corresponding wave functions in the coupled-channel adiabatic approach. Comput. Phys. Commun. **185**, 3341–3343 (2014)
2. Gusev, A.A., Chuluunbaatar, O., Vinitsky, S.I., Derbov, V.L., Góźdź, A., Hai, L.L., Rostovtsev, V.A.: Symbolic-numerical solution of boundary-value problems with self-adjoint second-order differential equation using the finite element method with interpolation hermite polynomials. In: Gerdt, V.P., Koepf, W., Seiler, W.M., Vorozhtsov, E.V. (eds.) CASC 2014. LNCS, vol. 8660, pp. 138–154. Springer, Heidelberg (2014)
3. Gusev, A.A., Hai, L.L., Chuluunbaatar, O., Ulziibayar, V., Vinitsky, S.I., Derbov, V.L., Gozdz, A., Rostovtsev, V.A.: Symbolic-numeric solution of boundary-value problems for the Schrodinger equation using the finite element method: scattering problem and resonance states. In: Gerdt, V.P., Koepf, W., Seiler, W.M., Vorozhtsov, E.V. (eds.) CASC 2015. LNCS, vol. 9301, pp. 182–197. Springer, Heidelberg (2015)
4. Gusev, A.A., Hai, L.L., Chuluunbaatar, O., Vinitsky, S.I.: Program KANTBP 4M for solving boundary-value problems for systems of ordinary differential equations of the second order (2015). http://wwwinfo.jinr.ru/programs/jinrlib/kantbp4m/
5. Gusev, A.A., Vinitsky, S.I., Chuluunbaatar, O., Rostovtsev, V., Hai, L.L., Derbov, V., Krassovitskiy, P.: Symbolic-numerical algorithm for generating cluster eigenfunctions: tunneling of clusters through repulsive barriers. In: Gerdt, V.P., Koepf, W., Mayr, E.W., Vorozhtsov, E.V. (eds.) CASC 2013. LNCS, vol. 8136, pp. 427–442. Springer, Heidelberg (2013)
6. Harrison, P.: Quantum Well, Wires and Dots. Theoretical and Computational Physics of Semiconductor Nanostructures. Wiley, New York (2005)

7. Krassovitskiy, P.M., Pen'kov, F.M.: Contribution of resonance tunneling of molecule to physical observables. J. Phys. B: At. Mol. Opt. Phys. **47**, 225210 (2014)

8. Sevastyanov, L.A., Sevastyanov, A.L., Tyutyunnik, A.A.: Analytical calculations in maple to implement the method of adiabatic modes for modelling smoothly irregular integrated optical waveguide structures. In: Gerdt, V.P., Koepf, W., Seiler, W.M., Vorozhtsov, E.V. (eds.) CASC 2014. LNCS, vol. 8660, pp. 419–431. Springer, Heidelberg (2014)

9. Siegert, A.J.F.: On the derivation of the dispersion formula for nuclear reactions. Phys. Rev. **56**, 750–752 (1939)

Symbolic Algorithm for Generating Irreducible Rotational-Vibrational Bases of Point Groups

A.A. Gusev[1], V.P. Gerdt[1], S.I. Vinitsky[1(✉)], V.L. Derbov[2], A. Góźdź[3], A. Pędrak[3], A. Szulerecka[3], and A. Dobrowolski[3]

[1] Joint Institute for Nuclear Research, Dubna, Russia
gooseff@jinr.ru, vinitsky@theor.jinr.ru
[2] Saratov State University, Saratov, Russia
[3] Institute of Physics, Maria Curie-Skłodowska University, Lublin, Poland

Abstract. Symbolic algorithm implemented in computer algebra system for generating irreducible representations of the point symmetry groups in the rotor + shape vibrational space of a nuclear collective model in the intrinsic frame is presented. The method of generalized projection operators is used. The generalized projection operators for the intrinsic group acting in the space $L^2(SO(3))$ and in the space spanned by the eigenfunctions of a multidimensional harmonic oscillator are constructed. The efficiency of the scheme is investigated by calculating the bases of irreducible representations subgroup $\overline{D}_{4y}$ of octahedral group in the intrinsic frame of a quadrupole-octupole nuclear collective model.

Keywords: Generalized projection operators · Dihedral group · Irreducible representations · Quadrupole-octupole nuclear collective model

1 Introduction

The motivation of the present work comes from searching for higher point symmetries in nuclei, in particular, tetrahedral nuclei [5,7]. To reveal such symmetries, e.g., the inter-band and intra-band E2 electromagnetic reduced transition probabilities can be investigated [6].

For a more realistic description of oscillation modes it is necessary to construct the rotational-vibration bases using the generalized projection operators for the intrinsic rotation groups in the framework of nuclear collective models [15]. In particular, general collective orthogonal bases of irreducible representations (Irrs) of the intrinsic point group $\overline{D}_{4y}$, i.e., the subgroup of the octahedral point group, combining the zero-phonon and one-phonon excitations in the quadrupole and octupole modes together with the rotational motion has been constructed. This offers the possibility to diagonalize the quadrupole-octupole-rotational collective Hamiltonians in future. Such a collective approach would enable searching for fingerprints of the high-rank symmetries (e.g., octahedral, tetrahedral, etc.) in the nuclear bands. This task could be performed by considering, e.g., the inter-band

© Springer International Publishing AG 2016
V.P. Gerdt et al. (Eds.): CASC 2016, LNCS 9890, pp. 228–242, 2016.
DOI: 10.1007/978-3-319-45641-6_15

Table 1. Character table for the dihedral group D_{4y}

Irrs\Classes	C_1	C_2	C_3	C_4	C_5	Cartesian bases
$\Gamma_1 = A_1$	1	1	1	1	1	$v_1 = R^2 = x^2 + y^2 + z^2$
$\Gamma_2 = B_1$	1	1	-1	1	-1	$v_1 = (x^2 - z^2)$
$\Gamma_3 = A_2$	1	1	1	-1	-1	$v_1 = y$
$\Gamma_4 = B_2$	1	1	-1	-1	1	$v_1 = -xz$
$\Gamma_5 = E$	2	-2	0	0	0	$v_1 = x, \quad v_2 = -z$

$$C_1 = \{E\}, \ C_2 = \{C_{2y}\}, \ C_3 = \{C_{4y}, C_{4y}^{-1}\}, C_4 = \{C_{2x}, C_{2z}\}, \ C_5 = \{C_{2c}, C_{2d}\}$$

and intra-band electromagnetic reduced transition probabilities $B(E\lambda)$ within the set of eigensolutions of these Hamiltonians [10].

In this paper, we present a symbolic algorithm implemented in computer algebra system for generating irreducible representations of point symmetry groups in the rotor+shape vibrational space of a nuclear collective model in the intrinsic frame. The algorithm was implemented in Reduce while Maple or Mathematica can be used also. The algorithm for calculating the rotor bases has been elaborated in our previous paper [11] using the method of generalized projection operators (GPOs) [2]. Here we formulate the GPO approach for the intrinsic point symmetry group acting in the space $L^2(SO(3))$ and in the space spanned by eigenfunctions of the multidimensional harmonic oscillator. The efficiency of this scheme is investigated by deriving the bases of irreducible representations of the subgroup $\overline{D}_{4y}$ of the octahedral group for the rotor space and for the (2+7) or $(2 + 4)$ dimensional oscillators in the intrinsic frame of the quadrupole-octupole nuclear collective model.

The structure of the paper is the following. In Sect. 2, we describe the multipole collective variables and the quadrupole-octupole collective model. In Sects. 3 and 4, we describe the algorithms and the results of calculations of the rotation and shape vibrational bases, respectively. In the conclusion, we discuss the perspectives of further studies.

2   Quadrupole-Octupole Collective Model

Let $X^{(lab)} = \{q^{(lab)} : q^{(lab)} = (q_1^{(lab)}, q_2^{(lab)}, \ldots, q_f^{(lab)}))\}$ be the configuration space of a nucleus (single-particle+collective variables+ ...) in the laboratory frame. And, let the set $\{e_n(q^{(lab)})\}$ represent an orthonormal basis in the space of states $L^2(X)$ having the required physical meaning. These functions can be determined, e.g., by a set of commuting physical observables $\hat{A}_l$, where $l = 1, 2, \ldots, r$, and the functions $\{e_n(q^{(lab)})\}$ are the common eigenfunctions of $\hat{A}_l$.

Let the functions $Z_\nu(q^{(lab)})$, where $\nu = 1, 2, \ldots, \nu_M$, represent a set of "global" properties of this nucleus (shape, density, multipole momenta, etc.), then the expansion coefficients $\alpha_{\nu n}$, defined as

$$Z_\nu(q^{(lab)}) = \sum_n \alpha_{\nu n}^{(lab)} \, e_n(q^{(lab)}) \tag{1}$$

can be considered as collective variables.

Table 2. The operators in representations Γ_* the dihedral group D_{4y}

$\Gamma_1(E)=\begin{pmatrix}1\end{pmatrix}$	$\Gamma_1(C_{2x})=\begin{pmatrix}1\end{pmatrix}$	$\Gamma_1(C_{2y})=\begin{pmatrix}1\end{pmatrix}$	$\Gamma_1(C_{2z})=\begin{pmatrix}1\end{pmatrix}$
$\Gamma_1(C_{4y})=\begin{pmatrix}1\end{pmatrix}$	$\Gamma_1(C_{4y}^{-1})=\begin{pmatrix}1\end{pmatrix}$	$\Gamma_1(C_{2c})=\begin{pmatrix}1\end{pmatrix}$	$\Gamma_1(C_{2d})=\begin{pmatrix}1\end{pmatrix}$
$\Gamma_2(E)=\begin{pmatrix}1\end{pmatrix}$	$\Gamma_2(C_{2x})=\begin{pmatrix}1\end{pmatrix}$	$\Gamma_2(C_{2y})=\begin{pmatrix}1\end{pmatrix}$	$\Gamma_2(C_{2z})=\begin{pmatrix}1\end{pmatrix}$
$\Gamma_2(C_{4y})=\begin{pmatrix}-1\end{pmatrix}$	$\Gamma_2(C_{4y}^{-1})=\begin{pmatrix}-1\end{pmatrix}$	$\Gamma_2(C_{2c})=\begin{pmatrix}-1\end{pmatrix}$	$\Gamma_2(C_{2d})=\begin{pmatrix}-1\end{pmatrix}$
$\Gamma_3(E)=\begin{pmatrix}1\end{pmatrix}$	$\Gamma_3(C_{2x})=\begin{pmatrix}-1\end{pmatrix}$	$\Gamma_3(C_{2y})=\begin{pmatrix}1\end{pmatrix}$	$\Gamma_3(C_{2z})=\begin{pmatrix}-1\end{pmatrix}$
$\Gamma_3(C_{4y})=\begin{pmatrix}1\end{pmatrix}$	$\Gamma_3(C_{4y}^{-1})=\begin{pmatrix}1\end{pmatrix}$	$\Gamma_3(C_{2c})=\begin{pmatrix}-1\end{pmatrix}$	$\Gamma_3(C_{2d})=\begin{pmatrix}-1\end{pmatrix}$
$\Gamma_4(E)=\begin{pmatrix}1\end{pmatrix}$	$\Gamma_4(C_{2x})=\begin{pmatrix}-1\end{pmatrix}$	$\Gamma_4(C_{2y})=\begin{pmatrix}1\end{pmatrix}$	$\Gamma_4(C_{2z})=\begin{pmatrix}-1\end{pmatrix}$
$\Gamma_4(C_{4y})=\begin{pmatrix}-1\end{pmatrix}$	$\Gamma_4(C_{4y}^{-1})=\begin{pmatrix}-1\end{pmatrix}$	$\Gamma_4(C_{2c})=\begin{pmatrix}1\end{pmatrix}$	$\Gamma_4(C_{2d})=\begin{pmatrix}1\end{pmatrix}$
$\Gamma_5(E)=\begin{pmatrix}1&0\\0&1\end{pmatrix}$	$\Gamma_5(C_{2x})=\begin{pmatrix}1&0\\0&-1\end{pmatrix}$	$\Gamma_5(C_{2y})=\begin{pmatrix}-1&0\\0&-1\end{pmatrix}$	$\Gamma_5(C_{2z})=\begin{pmatrix}-1&0\\0&1\end{pmatrix}$
$\Gamma_5(C_{4y})=\begin{pmatrix}0&1\\-1&0\end{pmatrix}$	$\Gamma_5(C_{4y}^{-1})=\begin{pmatrix}0&-1\\1&0\end{pmatrix}$	$\Gamma_5(C_{2c})=\begin{pmatrix}0&-1\\-1&0\end{pmatrix}$	$\Gamma_5(C_{2d})=\begin{pmatrix}0&1\\1&0\end{pmatrix}$

One of the most frequently used parametrizations of the nuclear surface is the Rayleigh expansion. The two-dimensional manifold (nuclear surface) is described by the spherical angles $\{q_1^{(lab)} = \theta, q_2^{(lab)} = \phi\}$. The angular momentum observables $\hat{A}_1 = \hat{J}^2$, $\hat{A}_2 = \hat{J}_z$ generate the physical eigenbasis $\{e_n(q^{(lab)}) = Y_{lm}(\theta, \phi)\}$. The required property is defined as the shape of the nucleus written in spherical coordinates: $Z(\theta, \phi) \equiv r = R(\theta, \phi) \in L^2(S_2)$.

This results in the expansion of this surface in terms of the chosen basis:

$$R(\theta, \phi) = R_0(1 + \sum_{l,m} \alpha_{lm}^{(lab)\star} Y_{lm}(\theta, \phi)), \qquad (2)$$

where $\alpha_{lm}^{(lab)\star} = (-1)^m \alpha_{l,-m}^{(lab)}$ are multipole collective surface variables [9].

One can notice that a more natural description of the collective nuclear motions can be performed in the intrinsic frame. The intrinsic frame in respect to the motion defined by a group of motions like rotations, translations, scaling modes, etc., can be defined using the group theoretical tools.

Let the symbol G denote a dim(G)-dimensional Lie group of transformations in the configuration space $X^{(lab)}$. The operators $T(g)$ represent a realization of the action of $g \in G$ on $X^{(lab)}$ [3]:

$$q = T(g)\, q^{(lab)}, \quad F_k(q, g) = 0, \qquad (3)$$

where $k = 1, 2, \ldots, \dim(G)$. The functions $F_k(q, g) = 0$ describe the appropriate constraints required to eliminate the set redundant variables within the set of all intrinsic variables $\{q\}$. In this way, the intrinsic variables consist of independent variables $\{q\}$ and $\{g\}$. Usually, the group elements g are parametrized by a set of real parameters which are considered as intrinsic collective coordinates describing the motion generated by the group G. The minimal and, at the same time, the most important group of motions G is the group describing the rotational nuclear motion.

The standard choice of collective variables in the laboratory frame is given by the deformation parameters $\{\alpha_\lambda^{(lab)}\}$, were $\lambda=1,2,3,...,\lambda_{max}$, for details see Ref. [13]. In our case, we consider $\lambda_{max}=3$. The classical collective Hamiltonian in the laboratory frame [8] can be written as

$$\mathcal{H}_{cl} = \mathcal{T}_{cl} + \mathcal{V}_{cl}(\alpha^{(lab)}), \tag{4}$$

where the kinetic energy with the time derivatives denoted by a dot above the variable:

$$\mathcal{T}_{cl} = \frac{1}{2}\sum_{\lambda=1}^{3} B_\lambda \sum_\mu |\dot{\alpha}_{\lambda\mu}^{(lab)}|^2 = \frac{1}{2}\sum_{\lambda=1}^{3} B_\lambda \sum_\mu (-1)^\mu \dot{\alpha}_{\lambda\mu}^{(lab)}\dot{\alpha}_{\lambda,-\mu}^{(lab)} \tag{5}$$

and the potential energy (for the harmonic oscillator) is expressed as

$$\mathcal{V}_{cl}(\alpha^{(lab)}) = \frac{1}{2}\sum_{\lambda=1}^{3} B_\lambda\omega_\lambda^2 \sum_\mu |\alpha_{\lambda\mu}^{(lab)}|^2. \tag{6}$$

In this model, the most part of interesting quantum observables are expressed in terms of spherical tensors. The scalar product of such SO(3)–tensors is defined as

$$\xi_\lambda \cdot \eta_\lambda = \sum_{(\lambda\mu),(\lambda'\nu)} h^{(\lambda\mu)(\lambda'\nu)}\xi_{\lambda\mu}\eta_{\lambda'\nu}, \tag{7}$$

where the basic geometric metric tensor is:

$$h^{(\lambda\mu)(\lambda'\nu)} = \sqrt{2\lambda+1}\,(\lambda\mu\lambda'\nu|00) = (-1)^\mu\delta_{(\lambda,\mu)}^{(\lambda',-\nu)} \tag{8}$$

Note that the Clebsch–Gordan coupling gives the only scalar product in the manifold of spherical tensors. Then the metric form in the α manifold, which is probably the most general one, is given by:

$$ds^2 = \sum_{\lambda\mu,\lambda'\nu} S_\lambda(\alpha)\sqrt{2\lambda+1}\,(\lambda\mu\lambda'\nu|00)\,d\alpha_{\lambda\mu}d\alpha_{\lambda'\nu}, \tag{9}$$

where $S_\lambda(\alpha)$ are SO(3)–invariants. The conventional transformation of the collective variables $\alpha_{\lambda\mu'}^{(lab)}$ and the spherical harmonics $Y_{\lambda\mu'}(\theta,\varphi)$ from the laboratory frame to $\alpha_{\lambda\mu}$ and $Y_{\lambda\mu}(\theta',\varphi')$ of the rotating (intrinsic, body-fixed) frame defined by the Euler angles $\Omega = (\Omega_1,\Omega_2,\Omega_3)$ reads as [13]:

$$\alpha_{\lambda\mu} = \sum_{\mu'} D_{\mu'\mu}^\lambda(\Omega)\alpha_{\lambda\mu'}^{(lab)}, \quad Y_{\lambda\mu}(\theta',\varphi') = \sum_{\mu'} D_{\mu'\mu}^\lambda(\Omega)Y_{\lambda\mu'}(\theta,\varphi), \tag{10}$$

where $D_{\mu'\mu}^\lambda(\Omega) \in L_2(SO(3))$ are Wigner functions [16]. The constraints F_k imposed on these transformations proposed by Bohr for the case of $\lambda = 2$ are

$$F_{1,2}(\alpha,\Omega) = \alpha_{2,\pm1} = 0, \quad F_3(\alpha,\Omega) = \alpha_{22} - \alpha_{2,-2} = 0. \tag{11}$$

Below we use these conditions as the standard form of constraints defining the intrinsic frame.

Now we need the quantum collective Hamiltonian. The quantization procedure (Pauli prescription) of the intrinsic classic Hamiltonian leads to the following quantum form [8]:

$$T_{cl} \rightarrow \hat{T} = -\frac{\hbar^2}{2} \sum_{\mu\mu'} \frac{1}{\sqrt{|\eta|}} \frac{\partial}{\partial q^\mu} \sqrt{|\eta|} (\eta^{-1})^{\mu\mu'} \frac{\partial}{\partial q^{\mu'}}. \tag{12}$$

Here the metric tensor and the scalar product in the intrinsic collective space is obtained from (8) as:

$$\eta_{k'k'} = \sum_{\lambda\mu\lambda'\mu'} \frac{\partial \alpha_{\lambda\mu}^{(lab)*}}{\partial q^k} \frac{\partial \alpha_{\lambda'\mu'}^{(lab)*}}{\partial q^{k'}} h_{(\lambda\mu)(\lambda'\mu')}, \tag{13}$$

In the following we restrict the intrinsic collective space $q = (q_1, q_2, \ldots, q_{f'})$, Ω, where $f' = f - \dim(SO(3))$ ($\dim(SO(3)) = 3$), of multipole shape vibrations to the intrinsic deformation parameters $\{\alpha_{1\mu}\}$, $\{\alpha_{20}, \alpha_{22}\}$, $\{\alpha_{3\mu}\}$, and to 3D rotations: $\Omega = (\Omega_1, \Omega_2, \Omega_3)$. It is usually supposed that the dipole deformation parameters $\{\alpha_{1\mu}\}$ are responsible for the center-of-mass motion. So, to obtain the collective basis functions $\Psi(q, \Omega) \in L^2(\mathbf{R}^{2+7} \otimes SO(3))$ of Irrs Γ_* of intrinsic group $\bar{O}$ and its subgroups, for example $\bar{D}_{4y}$, in the center-of-mass frame we apply the idea of the adiabatic separation of vibrational $q = \{q_2, q_3\}$, $q_2 = \{\alpha_{20}, \alpha_{22}\}$, $q_3 = \{\alpha_{3\mu}\}$ and rotational $\Omega = (\Omega_1, \Omega_2, \Omega_3)$ motion [15]

$$\Psi_{a_2b_{02}t_2;a_3b_{03}t_3;t_1pq}^{\Gamma_{n_1}\Gamma_{n_2}\Gamma_{n_3}JMK_2K_3}(q, \Omega) = v_{a_2b_{02}t_2}^{n_2K_2}(q_2) v_{a_3b_{03}t_3}^{n_3K_3}(q_3) v_{n_1t_1pq}^{JM}(\Omega). \tag{14}$$

In fact, we can always choose the basis as a product of the vibrational quadrupole $v_{a_2b_{02}t_2}^{n_2K_2}(q_2)$, octupole $v_{a_3b_{03}t_3}^{n_3K_3}(q_3)$ and rotational $v_{n_1t_1pq}^{JM}(\Omega)$ states corresponding to uncoupled harmonic oscillator Hamiltonians.

3   Generalized Projection Operators and Rotor Basis

Generalized projection operators (GPOs) for the intrisic group $\bar{G}$ act in the Hilbert space $\mathcal{K} = L_2(G)$ of functions on the laboratory group G. As a result, they create the subspace of this space of states $\mathcal{K}_b^\Gamma = \hat{P}_{bb}^\Gamma \mathcal{K}$. These operators are defined as

$$\hat{P}_{ab}^\Gamma = \frac{\dim(\Gamma)}{\mathrm{card}(\overline{G})} \sum_{g \in G} \left(\Delta_{ab}^\Gamma(g) \right)^* \hat{\bar{g}}, \tag{15}$$

where $\dim(\Gamma)$ denotes the dimension of Irrs Γ of the group G, $\mathrm{card}(G)$ is the order of the group G, and $\Delta_{ab}^\Gamma(g)$ denotes the matrix elements of the appropriate Irr Γ for the element g. The symbol $\hat{\bar{g}}$ denotes the unitary operator of the intrinsic group $\bar{G}$ acting in the space $\mathcal{K}$. The octahedral group $\bar{O}$ has been presented

Table 3. Output for $v_{n_1tpq_0}^{JM}(\Omega)$: $\Gamma = \Gamma_{n_1}$, $J = 0, 1, 2, 3$; t distinguishes equiv Irrs of $\bar{D}_{4y}$ intrinsic point group, $r_{MK}^{J\pm}(\Omega) = (r_{MK}^J(\Omega) \pm r_{M-K}^J(\Omega))/(\sqrt{2})$, $r_{M0}^J(\Omega) = r_{M0}^J(\Omega)$.

J	Γ	p	q_0	t	$v_{ntpq_0}^{JM}(\Omega)$
0	$\Gamma_1 = A_1$	1	1	1	$v_{1111}^{0M}(\Omega) = r_{M0}^0(\Omega)$
1	$\Gamma_3 = A_2$	1	1	1	$v_{3111}^{1M}(\Omega) = r_{M1}^{1+}(\Omega)$
1	$\Gamma_5 = E$	1	1	1	$v_{5111}^{1M}(\Omega) = -r_{M1}^{1-}(\Omega)$
1	$\Gamma_5 = E$	1	2	1	$v_{5112}^{1M}(\Omega) = r_{M1}^{1-}(\Omega)$
1	$\Gamma_5 = E$	2	1	1	$v_{5121}^{1M}(\Omega) = -r_{M0}^1(\Omega)$
2	$\Gamma_1 = A_1$	1	1	1	$v_{1111}^{2M}(\Omega) = (\sqrt{3}r_{M0}^2(\Omega) + 3r_{M2}^{2+}(\Omega))/(2\sqrt{3})$
2	$\Gamma_2 = B_1$	1	1	1	$v_{2111}^{2M}(\Omega) = (-\sqrt{3}r_{M0}^2(\Omega) + r_{M2}^{2+}(\Omega))/2$
2	$\Gamma_4 = B_2$	1	1	1	$v_{4111}^{2M}(\Omega) = -r_{M1}^{2-}(\Omega)$
2	$\Gamma_5 = E$	1	1	1	$v_{5111}^{2M}(\Omega) = r_{M1}^{2+}(\Omega)$
2	$\Gamma_5 = E$	2	2	1	$v_{5122}^{2M}(\Omega) = -r_{M2}^{2-}(\Omega)$
3	$\Gamma_2 = B_1$	1	1	1	$v_{2111}^{3M}(\Omega) = -r_{M2}^{3-}(\Omega)$
3	$\Gamma_3 = A_2$	1	1	1	$v_{3111}^{3M}(\Omega) = (\sqrt{15}r_{M1}^{3+}(\Omega) + 5r_{M3}^{3+}(\Omega))/(2\sqrt{10})$
3	$\Gamma_4 = B_2$	1	1	1	$v_{4111}^{3M}(\Omega) = (-\sqrt{15}r_{M1}^{3+}(\Omega) + 3r_{M3}^{3+}(\Omega))/(2\sqrt{6})$
3	$\Gamma_5 = E$	1	1	1	$v_{5111}^{3M}(\Omega) = -r_{M3}^{3-}(\Omega)$
3	$\Gamma_5 = E$	1	2	1	$v_{5112}^{3M}(\Omega) = (\sqrt{15}r_{M1}^{3-}(\Omega) + 3r_{M3}^{3-}(\Omega))/(2\sqrt{6})$
3	$\Gamma_5 = E$	2	1	1	$v_{5121}^{3M}(\Omega) = (-\sqrt{5}r_{M0}^3(\Omega) - \sqrt{3}r_{M2}^{3+}(\Omega))/(2\sqrt{2})$
3	$\Gamma_5 = E$	2	2	1	$v_{5122}^{3M}(\Omega) = r_{M2}^{3+}(\Omega)$

in [11] while its subgroup $\overline{D}_{4y}$ is described in Tables 1 and 2. The operators of representations Γ_* differ by transposition from Cornwell's ones [4]. The rotational basis $v_{n_1t_1pq}^{JM}(\Omega)$ of the intrinsic octahedral group $\overline{O}$ in the space $L^2(SO(3))$ spanned by the orthonormalized (in respect to the scalar product $\langle \psi_1 | \psi_2 \rangle = \frac{1}{8\pi^2} \int_0^{2\pi} d\Omega_1 \int_0^\pi d\Omega_2 \sin(\Omega_2) \int_0^{2\pi} d\Omega_3 \psi_1(\Omega)^* \psi_2(\Omega))$ Wigner functions $r_{MK}^J(\Omega) = \sqrt{2J+1}D_{MK}^J(\Omega)^*$ was calculated using the symbolic algorithm GPO presented in [11]. The algorithm was implemented in the computer algebra system Reduce. The typical running time of calculating the rotational basis $v_{n_1t_1pq}^{JM}(\Omega)$ for the Irrs $\Gamma_1,...,\Gamma_5$ of the required group $\bar{D}_{4y}$ for $J \leq 10$ is 380 s using the PC Intel Pentium CPU 1.50 GHz 4 GB 64 bit Windows 8. The results for $J \leq 3$ are shown in Table 3.

4   Shape Vibrational Basis

In this section, we present the calculation scheme aimed to generate irreducible bases of intrinsic point groups in the shape vibration space of the nuclear collective model state space using the generalized projection operator method. We start from the precise definition of the rotation group action in both the geometrical space of multipole collective surface variables in the intrinsic frame and the functional spaces spanned by the vibrational states of the quadrupole-octupole harmonic oscillator. To define the rotations of intrinsic spherical tensors with respect to intrinsic rotation group, one can write the expansion of the nuclear shape (2) in terms of spherical harmonics $Y_{lm}(\theta', \phi')$ in the intrinsic frame:

$$R(\theta', \phi') = R_0 \left(1 + \sum_{lm} \alpha^*_{lm} Y_{lm}(\theta', \phi')\right) \tag{16}$$

$$= R_0 \left(1 + \sum_l a_{l0} Y_{l0}(\theta', \phi') + \sum_{lm} \left(a_{lm} Y^{(+)}_{lm}(\theta', \phi') + b_{lm} Y^{(-)}_{lm}(\theta', \phi')\right)\right).$$

Here $\alpha^*_{l,m} = (-1)^m \alpha_{l,-m}$ are the intrinsic tensors describing nuclear deformation with respect to the intrinsic frame. They can be expressed in terms of real-valued components a and b:

$$\alpha_{\lambda 0} = a_{\lambda 0}, \quad \alpha_{\lambda \mu} = (a_{\lambda \mu} - \imath b_{\lambda \mu})/\sqrt{2}, \quad \alpha_{\lambda - \mu} = (-1)^\mu (a_{\lambda \mu} + \imath b_{\lambda \mu})/\sqrt{2}, \tag{17}$$

by making use of the special combinations of spherical harmonics

$$Y^{(+)}_{lm}(\theta', \phi') = (Y_{lm}(\theta', \phi') + (-1)^m Y_{l-m}(\theta', \phi'))/\sqrt{2},$$
$$Y^{(-)}_{lm}(\theta', \phi') = \imath(Y_{lm}(\theta', \phi') - (-1)^m Y_{l-m}(\theta', \phi'))/\sqrt{2}.$$

Note that the spherical harmonics are tensors in respect to intrinsic rotations. We rewrite the transformation $\hat{\bar{g}} \alpha_{\lambda \mu} = \bar{\alpha}_{\lambda \mu}$, $\hat{\bar{g}} \in \bar{G}$,

$$\bar{\alpha}_{\lambda \mu} = \sum_{\mu'=-\lambda}^{\lambda} D^\lambda_{\mu \mu'}(g)^* \alpha_{\lambda \mu'}, \quad Y_{\lambda \mu}(\bar{\theta}', \bar{\varphi}') = \sum_{\mu'=-\lambda}^{\lambda} D^\lambda_{\mu \mu'}(g)^* Y_{\lambda \mu'}(\theta', \varphi') \tag{18}$$

between two intrinsic coordinate frames. We use the relations between the complex-valued coefficients $\bar{\alpha}_{\lambda \mu}$, $\alpha_{\lambda \mu}$ and real-valued ones $\bar{a}_{\lambda \mu}$, $\bar{b}_{\lambda \mu}$, $a_{\lambda \mu}$, $b_{\lambda \mu}$

$$\bar{\alpha}_{\lambda 0} = \bar{a}_{\lambda 0}, \quad \bar{\alpha}_{\lambda \mu} = (\bar{a}_{\lambda \mu} - \imath \bar{b}_{\lambda \mu})/\sqrt{2}, \quad \bar{\alpha}_{\lambda - \mu} = (-1)^\mu (\bar{a}_{\lambda \mu} - \imath \bar{b}_{\lambda \mu})/\sqrt{2}. \tag{19}$$

Using the relations (17), (19), and the definition of Wigner functions [16]

$$D^\lambda_{\mu \mu'}(g)^* = \exp(\imath \mu \alpha) d^\lambda_{\mu \mu'}(\beta) \exp(\imath \mu' \gamma), \quad d^\lambda_{-\mu-\mu'}(\beta) = (-1)^{\mu+\mu'} d^\lambda_{\mu \mu'}(\beta), \tag{20}$$

$$\exp(\pm \imath \mu \alpha) = \cos(\imath \mu \alpha) \pm \imath \sin(\imath \mu \alpha), \exp(\pm \imath \mu' \gamma) = \cos(\imath \mu' \gamma) \pm \imath \sin(\imath \mu' \gamma),$$

we find the orthogonal transformation $\bar{\mathbf{q}} = \bar{\mathbf{M}}(g) \mathbf{q}$ of the intrinsic variables $\mathbf{q} = \{a_{\lambda \mu}, b_{\lambda \mu}\}$ to new intrinsic variables $\bar{\mathbf{q}} = \{\bar{a}_{\lambda \mu}, \bar{b}_{\lambda \mu}\}$:

$$\bar{a}_{\lambda \mu} = \sqrt{2} C^\lambda_{\mu 0} a_{\lambda 0} + \sum_{\mu'=1}^{\lambda} (C^\lambda_{\mu \mu'} + (-1)^{\mu'} C^\lambda_{\mu - \mu'}) a_{\lambda \mu'} + (S^\lambda_{\mu \mu'} - (-1)^{\mu'} S^\lambda_{\mu - \mu'}) b_{\lambda \mu'},$$

$$\bar{a}_{\lambda 0} = C^\lambda_{00} a_{\lambda 0} + \sum_{\mu'=1}^{\lambda} \sqrt{2} C^\lambda_{0 \mu'} a_{\lambda \mu'} + \sqrt{2} S^\lambda_{0 \mu'} b_{\lambda \mu'}, \tag{21}$$

$$\bar{b}_{\lambda \mu} = -\sqrt{2} S^\lambda_{\mu 0} a_{\lambda 0} + \sum_{\mu'=1}^{\lambda} (-S^\lambda_{\mu \mu'} - (-1)^{\mu'} S^\lambda_{\mu - \mu'}) a_{\lambda \mu'} + (C^\lambda_{\mu \mu'} - (-1)^{\mu'} C^\lambda_{\mu - \mu'}) b_{\lambda \mu'},$$

where $C^\lambda_{\mu\mu'} = \cos(\mu\alpha+\mu'\gamma)d^\lambda_{\mu\mu'}(\beta)$, $S^\lambda_{\mu\mu'} = \sin(\mu\alpha+\mu'\gamma)d^\lambda_{\mu\mu'}(\beta)$. The transformation between both intrinsic frames $\bar{\mathbf{q}} = \bar{\mathbf{M}}(g)\mathbf{q}$ is represented by a block diagonal matrix $\lambda = 2,3,...,\lambda^{\max}$, with block matrices of the dimension $(2\lambda+1)\times(2\lambda+1)$:

$$\bar{\mathbf{q}}=\bar{\mathbf{M}}(g)\mathbf{q}, \quad \bar{\mathbf{q}}=\begin{pmatrix}\bar{\mathbf{q}}_2(g)\\ \bar{\mathbf{q}}_3(g)\\ \vdots\end{pmatrix}, \quad \mathbf{q}=\begin{pmatrix}\mathbf{q}_2(g)\\ \mathbf{q}_3(g)\\ \vdots\end{pmatrix}, \quad \bar{\mathbf{M}}(g)=\begin{pmatrix}\bar{\mathbf{M}}_2(g) & 0 & \cdots\\ 0 & \bar{\mathbf{M}}_3(g) & \cdots\\ \vdots & \vdots & \ddots\end{pmatrix}, \quad (22)$$

The transformation of pairs of components $\bar{\mathbf{q}}_\lambda = \begin{pmatrix}\bar{a}_{\lambda\lambda} & \cdots & \bar{a}_{\lambda1} & \bar{a}_{\lambda0} & \bar{b}_{\lambda1} & \cdots & \bar{b}_{\lambda\lambda}\end{pmatrix}^T$ and $\mathbf{q}_\lambda = \begin{pmatrix}a_{\lambda\lambda}, & \cdots & a_{\lambda1}, & a_{\lambda0} & b_{\lambda1} & \cdots & b_{\lambda\lambda}\end{pmatrix}^T$ are connected by the transformations $\bar{\mathbf{q}}_\lambda = \bar{\mathbf{M}}_\lambda(g)\mathbf{q}_\lambda$, with the blocks matrices $\bar{\mathbf{M}}_\lambda \equiv \bar{\mathbf{M}}_\lambda(g)$ of the following form

$$\bar{\mathbf{M}}_\lambda = \begin{pmatrix} \vdots & \vdots & \vdots & \vdots & \vdots \\ \cdots & C^\lambda_{22}+C^\lambda_{2-2} & C^\lambda_{21}-C^\lambda_{2-1} & \sqrt{2}C^\lambda_{20} & S^\lambda_{21}+S^\lambda_{2-1} & S^\lambda_{22}-S^\lambda_{2-2} & \cdots \\ \cdots & C^\lambda_{12}+C^\lambda_{1-2} & C^\lambda_{11}-C^\lambda_{1-1} & \sqrt{2}C^\lambda_{10} & S^\lambda_{11}+S^\lambda_{1-1} & S^\lambda_{12}-S^\lambda_{1-2} & \cdots \\ \cdots & \sqrt{2}C^\lambda_{02} & \sqrt{2}C^\lambda_{01} & C^\lambda_{00} & \sqrt{2}S^\lambda_{01} & \sqrt{2}S^\lambda_{02} & \cdots \\ \cdots & -S^\lambda_{12}-S^\lambda_{1-2} & -S^\lambda_{11}+S^\lambda_{1-1} & -\sqrt{2}S^\lambda_{10} & C^\lambda_{11}+C^\lambda_{1-1} & C^\lambda_{12}-C^\lambda_{1-2} & \cdots \\ \cdots & -S^\lambda_{22}-S^\lambda_{2-2} & -S^\lambda_{21}+S^\lambda_{2-1} & -\sqrt{2}S^\lambda_{20} & C^\lambda_{21}+C^\lambda_{2-1} & C^\lambda_{22}-C^\lambda_{2-2} & \cdots \\ \vdots & \vdots & \vdots & \vdots & \vdots \end{pmatrix}. \quad (23)$$

The inverse transformation is implemented using the transposed matrix $\bar{\mathbf{M}}^{-1} = \bar{\mathbf{M}}^T$ because $\bar{\mathbf{M}}$ is the orthogonal matrix, i.e., $\bar{\mathbf{M}}\bar{\mathbf{M}}^T = \bar{\mathbf{M}}^T\bar{\mathbf{M}} = \mathbf{I}$. The matrix $\bar{\mathbf{M}}$ transforming the intrinsic variables between the intrinsic frames is the transposed matrix to the matrix $\mathbf{M}$ transforming the 'laboratory' variables to the intrinsic ones, i.e., $\bar{\mathbf{M}} = \mathbf{M}^T$ corresponding to the transformation (10).

The quadrupole-octupole vibrational Hamiltonian $\hat{H}_v$ having the form of that of a harmonic oscillator can be written in the intrinsic frame as

$$\hat{H}_v=\left[-\frac{\hbar^2}{2B_2}\left(\frac{\partial^2}{\partial(a_{20})^2}+\frac{\partial^2}{\partial(a_{22})^2}\right)+\frac{B_2\omega_2^2}{2}\left((a_{20})^2+(a_{22})^2\right)\right] \quad (24)$$

$$+\left[-\frac{\hbar^2}{2B_3}\left(\sum_{\mu=0}^3\frac{\partial^2}{\partial(a_{3\mu})^2}+\sum_{\mu=1}^3\frac{\partial^2}{\partial(b_{3\mu})^2}\right)+\frac{B_3\omega_3^2}{2}\left(\sum_{\mu=0}^3(a_{3\mu})^2+\sum_{\mu=1}^3(b_{3\mu})^2\right)\right].$$

Since we do not introduce any interaction between the quadrupole and the octupole vibrational modes, the eigenvalues of $\hat{H}_v$ (24) are sums of 2D and 7D harmonic oscillator eigenvalues with the appropriate frequencies ω_2 and ω_3:

$$E_v^{K_{23}}=\hbar\omega_2(k_1+k_2+1) + \hbar\omega_3(k_3+...+k_9+7/2), \quad K_{23}=K_2+K_3=k_1+...+k_9.$$

The corresponding eigenfunctions characterized by a set of quantum numbers $\nu = \{k_1,...,k_9\}$ of 9D harmonic oscillator can be expressed as the products of 1D harmonic oscillator eigenfunctions [12,13]:

$$\psi_\nu^{\check{K}}(\mathbf{q}) = \psi_{k_1,k_2}^{K_2}(\mathbf{q}_2)\psi_{k_3,\dots,k_9}^{K_3}(\mathbf{q}_3)\delta_{K_{23},k_1+\dots+k_9}, \tag{25}$$

$$\psi_{k_1,k_2}^{K_2}(\mathbf{q}_2) = \psi_{k_1}(\eta_2 a_{20})\psi_{k_2}(\eta_2 a_{22}), \quad \psi_{k_3,\dots,k_9}^{K_3}(\mathbf{q}_3) = \psi_{k_3 k_4 k_5 k_6}^{K_3}(\mathbf{q}_3^a)\psi_{k_7 k_8 k_9}^{K_3}(\mathbf{q}_3^b),$$

$$\psi_{k_3 k_4 k_5 k_6}^{K_3}(\mathbf{q}_3^a) = \psi_{k_3}(\eta_3 a_{33})\psi_{k_4}(\eta_3 a_{32})\psi_{k_5}(\eta_3 a_{31})\psi_{k_6}(\eta_3 a_{30}),$$

$$\psi_{k_7 k_8 k_9}^{K_3}(\mathbf{q}_3^b) = \psi_{k_7}(\eta_3 b_{31})\psi_{k_8}(\eta_3 b_{32})\psi_{k_9}(\eta_3 b_{33}),$$

where $\check{K} = (K_2, K_3, K_{23})$. They are to satisfy the orthonormalization conditions

$$\int \psi_\nu^{\check{K}}(\mathbf{q})\psi_{\nu'}^{\check{K}'}(\mathbf{q})d\mathbf{q} = \delta_{\check{K},\check{K}'}\delta_{k_1 k_1'}\dots\delta_{k_9 k_9'}. \tag{26}$$

The functions $\psi_{k_*}(\eta_* a_{**})$ in the above expressions are 1D harmonic oscillator eigenfunctions with the parameter η_λ defined as $\eta_\lambda = \sqrt{B_\lambda \omega_\lambda / \hbar}$:

$$\psi_{k_*}(\eta_* a_{**}) = \sqrt{\frac{\sqrt{\eta_*}}{2^k k! \sqrt{\pi}}} H_{k_*}(\sqrt{\eta_*} a_{**})\exp(-\eta_* a_{**}^2/2), \tag{27}$$

where $H_{k_*}(x) = (2x)^{k_*} + \dots$ are the Hermite polynomials [1].

4.1   GPOs Action – Intrinsic Groups

To construct the appropriate bases in the space of vibrational functions (25) for the intrinsic point groups $\overline{G}$ we have to use the action of the operators (15) corresponding to the *intrinsic* group:

$$\Psi_{n a b_0 \nu}^{\check{K}}(\mathbf{q}) = \frac{\dim(\Gamma)}{\mathrm{card}(G)}\sum_{g\in G}\Delta^\Gamma(g)_{ab_0}{}^* \hat{\bar{g}}\psi_\nu^{\check{K}}(\mathbf{q})$$

$$= \frac{\dim(\Gamma)}{\mathrm{card}(G)}\sum_{g\in G}\Delta^\Gamma(g)_{ab_0}{}^* \psi_\nu^{\check{K}}(\bar{\mathbf{M}}(g)\mathbf{q})$$

$$= \frac{\dim(\Gamma)}{\mathrm{card}(G)}\sum_{g\in G}\Delta^\Gamma(g)_{ab_0}{}^* \psi_{k_1,k_2}^{K_2}(\bar{\mathbf{M}}_2(g)\mathbf{q}_2)\psi_{k_3,\dots,k_9}^{K_3=K_{23}-K_2}(\bar{\mathbf{M}}_3(g)\mathbf{q}_3), \tag{28}$$

where the matrices of the representations $\Delta^\Gamma(g)_{ab_0}$ obtained from the action of $g \in G$ in the space of basis functions of the Cartesian variables are used. In particular, for $\overline{D}_{4y}$, see Tables 1 and 2. In the above expressions, the labels $K_{23} = k_1 + \dots + k_9$, or $K_{23} = K_2 + K_3$, $K_2 = k_1 + k_2$, $K_3 = k_3 + \dots + k_9$, $k_1, k_2 = K_2 - k_1$ run over the full range $k_1, k_2 = 0, \dots, K_2$, and $k_3, k_4, \dots, k_9$ run over the full range $k_3, \dots, k_9 = 0, \dots, K_3$ that provides the decomposition of the vibration space for a given K_2 and K_3 into the irreducible representations Γ.

The matrices $\bar{\mathbf{M}}_2(g)$ and $\bar{\mathbf{M}}_3(g)$ for $g = (C_{4y})$ have the following form:

$$\bar{\mathbf{M}}_2(C_{4y}) = \begin{bmatrix} \frac{1}{2} & 0 & \frac{\sqrt{3}}{2} & 0 & 0 \\ 0 & -1 & 0 & 0 & 0 \\ \frac{\sqrt{3}}{2} & 0 & -\frac{1}{2} & 0 & 0 \\ 0 & 0 & 0 & 0 & 1 \\ 0 & 0 & 0 & -1 & 0 \end{bmatrix}, \quad \bar{\mathbf{M}}_3(C_{4y}) = \begin{bmatrix} 0 & -\frac{\sqrt{6}}{4} & 0 & -\frac{\sqrt{10}}{4} & 0 & 0 & 0 \\ \frac{\sqrt{6}}{4} & 0 & \frac{\sqrt{10}}{4} & 0 & 0 & 0 & 0 \\ 0 & -\frac{\sqrt{10}}{4} & 0 & \frac{\sqrt{6}}{4} & 0 & 0 & 0 \\ \frac{\sqrt{10}}{4} & 0 & -\frac{\sqrt{6}}{4} & 0 & 0 & 0 & 0 \\ 0 & 0 & 0 & 0 & -\frac{1}{4} & 0 & \frac{\sqrt{15}}{4} \\ 0 & 0 & 0 & 0 & 0 & -1 & 0 \\ 0 & 0 & 0 & 0 & \frac{\sqrt{15}}{4} & 0 & \frac{1}{4} \end{bmatrix}. \tag{29}$$

Note that the matrices $\bar{\mathbf{M}}_2(g)$ act on the vector $\mathbf{q}_2 = (a_{22}, a_{21}, a_{20}, b_{21}, b_{22})^T$ under the condition $a_{21} = b_{21} = b_{22} = 0$, that follows from Eq. (11). They transform pairs of components a_{22} and a_{20} into combination of $\bar{a}_{22}$ and $\bar{a}_{20}$. In this case, instead of Eq. (29) we have

$$\begin{pmatrix} \bar{a}_{22} \\ \bar{a}_{20} \end{pmatrix} = \bar{\mathbf{M}}_2(g) \begin{pmatrix} a_{22} \\ a_{20} \end{pmatrix}, \; \bar{\mathbf{M}}_2(C_{4y}) = \begin{bmatrix} \frac{1}{2} & \frac{\sqrt{3}}{2} \\ \frac{\sqrt{3}}{2} & -\frac{1}{2} \end{bmatrix}. \tag{30}$$

In turn, the matrices $\bar{\mathbf{M}}_3(g)$ act on the vector $\mathbf{q}_3 = (a_{33}, a_{32}, a_{31}, a_{30}, b_{31}, b_{32}, b_{33})^T$ and transform the pairs a_{32}, a_{30} into a combination of $\bar{a}_{32}$ and $\bar{a}_{30}$, a_{33}, a_{31} into a combination of $\bar{a}_{33}$ and $\bar{a}_{31}$ and b_{33}, b_{31} into a combination of $\bar{b}_{33}$ and $\bar{b}_{31}$. An exception is the variable b_{32} which transforms into itself because it is an invariant of the tetrahedral symmetry. In this case, instead of Eq. (29) we have

$$\begin{pmatrix} \bar{a}_{33} \\ \bar{a}_{32} \\ \bar{a}_{31} \\ \bar{a}_{30} \end{pmatrix} = \bar{\mathbf{M}}_3(g) \begin{pmatrix} a_{33} \\ a_{32} \\ a_{31} \\ a_{30} \end{pmatrix}, \; \bar{\mathbf{M}}_3(C_{4y}) = \begin{bmatrix} 0 & -\frac{\sqrt{6}}{4} & 0 & -\frac{\sqrt{10}}{4} \\ \frac{\sqrt{6}}{4} & 0 & \frac{\sqrt{10}}{4} & 0 \\ 0 & -\frac{\sqrt{10}}{4} & 0 & \frac{\sqrt{6}}{4} \\ \frac{\sqrt{10}}{4} & 0 & -\frac{\sqrt{6}}{4} & 0 \end{bmatrix}. \tag{31}$$

Note that in the decomposition (25) for the group $\overline{D}_{4y}$, instead of $\psi^{K_3}_{k_3,\ldots,k_9}(\mathbf{q}_3)$ one can use the vectors $\psi^{K_a}_{k_3,k_4,k_5,k_6}(a_{33}, a_{32}, a_{31}, a_{30}) \psi^{K_b}_{k_7,k_9}(b_{33}, b_{31}) \psi^{K_c}_{k_8}(b_{32})$. This allows one to construct the vibrational basis $\psi^{K_a}_{k_3,k_4,k_5,k_6}(a_{33}, a_{32}, a_{31}, a_{30})$ of Irrs of $\overline{D}_{4y}$ separately, in accordance with the conditions $b_{33} = b_{32} = b_{31} = 0$ [15].

The eigenfunctions $\Psi^N_{l,k}(\bar{q}_0, \bar{q}_2) \sim H_l(\bar{q}_0(g)) H_k(\bar{q}_2(g))$ transformed by the orthogonal transformation $\bar{\mathbf{M}}(g)$, where each pair of variables $(\bar{q}_0, \bar{q}_2)$ is the transformed pair of variables (q_0, q_2), are calculated using the formula [17]:

$$H_l(\sin\theta q_0 + \cos\theta q_2) H_k(\cos\theta q_0 - \sin\theta q_2) = \sum_{j=0}^{l+k=N} \alpha_{j,lk}(g) H_{l+k-j}(q_0) H_j(q_2). \tag{32}$$

The required coefficients $\alpha_{j,lk}(g)$ are determined by the following expression:

$$\alpha_{j,lk}(g) = \sum_{i=\max(0,j-k)}^{\min(j,l)} \frac{l! k! (-1)^{i-j}}{i!(j-i)!(l-i)!(k-j+i)!} (\cos\theta)^{k-j+2i} (\sin\theta)^{j+l-2i}, \tag{33}$$

where the values $\cos\theta$ and $\sin\theta$ are given by the direct substitutions of the corresponding pairs of elements of the matrices $\bar{M}_2(g)$ and $\bar{M}_3(g)$.

The expressions (32) and (33) make it possible to express the action of the group $\overline{G}$ onto the 2D oscillator functions for $K_2 = k_1 + k_2$ as follows:

$$\hat{g}\Psi^{K_2}_{k_1 k_2}(q_0, q_2) = \sum_{j=0}^{K_2} \alpha_{j,k_1 k_2}(g) \Psi^{K_2}_{k_1+k_2-j,j}(q_0, q_2). \tag{34}$$

Table 4. The vibrational functions $v_{abot}^{nK_{23}}$ of Irrs $\Gamma_1,...,\Gamma_5$, $K_{23}=0,1$

$\Gamma_1, K_{23}=0$	$v_{111}^{10}=\psi_{[0]}^0(\mathbf{q})\equiv\psi_{[000000]}^0(\mathbf{q})$

$K_{23}=1$	
$\Gamma_1,$	$v_{111}^{11}=[\frac{\sqrt{3}}{2}\frac{\psi_1(a_{22})}{\psi_0(a_{22})}+\frac{1}{2}\frac{\psi_1(a_{20})}{\psi_0(a_{20})}]\psi_{[0]}^0(\mathbf{q})$
$\Gamma_2,$	$v_{111}^{21}=[-\frac{\sqrt{3}}{2}\frac{\psi_1(a_{20})}{\psi_0(a_{20})}+\frac{1}{2}\frac{\psi_1(a_{22})}{\psi_0(a_{22})}]\psi_{[0]}^0(\mathbf{q})$
$\Gamma_5,$	$v_{111}^{51}=v_{211}^{51}=\frac{\psi_1(a_{33})}{\psi_0(a_{33})}\psi_{[0]}^0(\mathbf{q})$
$\Gamma_5,$	$v_{112}^{51}=v_{212}^{51}=\frac{\psi_1(a_{31})}{\psi_0(a_{31})}\psi_{[0]}^0(\mathbf{q})$
$\Gamma_5,$	$v_{121}^{51}=v_{221}^{51}=\frac{\psi_1(a_{32})}{\psi_0(a_{32})}\psi_{[0]}^0(\mathbf{q})$
$\Gamma_5,$	$v_{122}^{51}=v_{222}^{51}=\frac{\psi_1(a_{30})}{\psi_0(a_{30})}\psi_{[0]}^0(\mathbf{q})$

Table 5. The vibrational functions $v_{abot}^{nK_{23}}$ of Irrs Γ_1,Γ_2, $K_{23}=2$

$\Gamma_1, K_{23}=2$

$$v_{111}^{12}=[\frac{\sqrt{15}}{10}\frac{\psi_1(a_{20})}{\psi_0(a_{20})}\frac{\psi_1(a_{22})}{\psi_0(a_{22})}+\frac{3\sqrt{10}}{20}\frac{\psi_2(a_{20})}{\psi_0(a_{20})}+\frac{\sqrt{10}}{4}\frac{\psi_2(a_{22})}{\psi_0(a_{22})}]\psi_{[0]}^0(\mathbf{q})$$
$$v_{112}^{12}=[\frac{\sqrt{15}}{5}\frac{\psi_1(a_{20})}{\psi_0(a_{20})}\frac{\psi_1(a_{22})}{\psi_0(a_{22})}-\frac{\sqrt{10}}{5}\frac{\psi_2(a_{20})}{\psi_0(a_{20})}]\psi_{[0]}^0(\mathbf{q})]$$
$$v_{113}^{12}=[\frac{\sqrt{15}}{8}\frac{\psi_1(a_{30})}{\psi_0(a_{30})}\frac{\psi_1(a_{32})}{\psi_0(a_{32})}+\frac{5\sqrt{2}}{16}\frac{\psi_2(a_{30})}{\psi_0(a_{30})}+\frac{3\sqrt{2}}{16}\frac{\psi_2(a_{32})}{\psi_0(a_{32})}+\frac{\sqrt{2}}{2}\frac{\psi_2(a_{33})}{\psi_0(a_{33})}]\psi_{[0]}^0(\mathbf{q})$$
$$v_{114}^{12}=[-\frac{3\sqrt{110}}{176}\frac{\psi_2(a_{30})}{\psi_0(a_{30})}+\frac{\sqrt{110}}{22}\frac{\psi_2(a_{31})}{\psi_0(a_{31})}+\frac{\sqrt{110}}{16}\frac{\psi_2(a_{32})}{\psi_0(a_{32})}$$
$$-\frac{3\sqrt{33}}{88}\frac{\psi_1(a_{30})}{\psi_0(a_{30})}\frac{\psi_1(a_{32})}{\psi_0(a_{32})}+\frac{\sqrt{33}}{11}\frac{\psi_1(a_{31})}{\psi_0(a_{31})}\frac{\psi_1(a_{33})}{\psi_0(a_{33})}]\psi_{[0]}^0(\mathbf{q})$$
$$v_{115}^{12}=[\frac{\sqrt{110}}{22}\frac{\psi_1(a_{30})}{\psi_0(a_{30})}\frac{\psi_1(a_{32})}{\psi_0(a_{32})}+\frac{\sqrt{110}}{22}\frac{\psi_1(a_{31})}{\psi_0(a_{31})}\frac{\psi_1(a_{33})}{\psi_0(a_{33})}-\frac{\sqrt{33}}{11}\frac{\psi_2(a_{30})}{\psi_0(a_{30})}$$

$\Gamma_2, K_{23}=2$

$$v_{111}^{22}=[-\frac{\sqrt{6}}{4}\frac{\psi_2(a_{20})}{\psi_0(a_{20})}+\frac{\sqrt{6}}{4}\frac{\psi_2(a_{22})}{\psi_0(a_{22})}-\frac{1}{2}\frac{\psi_1(a_{20})}{\psi_0(a_{20})}\frac{\psi_1(a_{22})}{\psi_0(a_{22})}]\psi_{[0]}^0(\mathbf{q})$$
$$v_{112}^{22}=[-\frac{\sqrt{15}}{8}\frac{\psi_1(a_{30})}{\psi_0(a_{30})}\frac{\psi_1(a_{32})}{\psi_0(a_{32})}-\frac{5\sqrt{2}}{16}\frac{\psi_2(a_{30})}{\psi_0(a_{30})}-\frac{3\sqrt{2}}{16}\frac{\psi_2(a_{32})}{\psi_0(a_{32})}+\frac{\sqrt{2}}{2}\frac{\psi_2(a_{33})}{\psi_0(a_{33})}]\psi_{[0]}^0(\mathbf{q})$$
$$v_{113}^{22}=[-\frac{3\sqrt{110}}{176}\frac{\psi_2(a_{30})}{\psi_0(a_{30})}-\frac{\sqrt{110}}{22}\frac{\psi_2(a_{31})}{\psi_0(a_{31})}+\frac{\sqrt{110}}{16}\frac{\psi_2(a_{32})}{\psi_0(a_{32})}$$
$$-\frac{3\sqrt{33}}{88}\frac{\psi_1(a_{30})}{\psi_0(a_{30})}\frac{\psi_1(a_{32})}{\psi_0(a_{32})}-\frac{\sqrt{33}}{11}\frac{\psi_1(a_{31})}{\psi_0(a_{31})}\frac{\psi_1(a_{33})}{\psi_0(a_{33})}]\psi_{[0]}^0(\mathbf{q})$$
$$v_{114}^{22}=[-\frac{\sqrt{110}}{22}\frac{\psi_1(a_{30})}{\psi_0(a_{30})}\frac{\psi_1(a_{32})}{\psi_0(a_{32})}+\frac{\sqrt{110}}{22}\frac{\psi_1(a_{31})}{\psi_0(a_{31})}\frac{\psi_1(a_{33})}{\psi_0(a_{33})}$$
$$+\frac{\sqrt{33}}{11}\frac{\psi_2(a_{30})}{\psi_0(a_{30})}-\frac{\sqrt{33}}{11}\frac{\psi_2(a_{31})}{\psi_0(a_{31})}]\psi_{[0]}^0(\mathbf{q})$$

A more effective way for calculating the coefficients $\alpha_{j,lk}(g)$ of the expansion (32) is to use only the leading terms $H_{k_*}(x)=(2x)^{k_*}+...$ of the Hermite polynomials [12,14]. It is important that this method can be also applied in a general case of products of a finite number of Hermite polynomials. This means that for the harmonic oscillator functions with $K_2+K_3=K_{23}$, used in Eq. (28), we can apply the following transformation:

$$\psi_{k_1,k_2}^{K_2}(\bar{\mathbf{M}}_2(g)\mathbf{q}_2)=\sum_{k_1'+k_2'=K_2}\alpha[2]_{k_1,k_2;k_1',k_2'}(g)\psi_{k_1',k_2'}^{K_2}(\mathbf{q}_2),$$

$$\psi_{k_3,...,k_9}^{K_3}(\bar{\mathbf{M}}_3(g)\mathbf{q}_3)=\sum_{k_3'+...+k_9'=K_3}\alpha[3]_{k_3,...,k_9;k_3',...,k_9'}(g)\psi_{k_3',...,k_9'}^{K_3}(\mathbf{q}_3). \quad (35)$$

For convenience, these formulae can be written in the matrix form. In the considered case, expression (28) can be presented as follows:

$$\Psi_{\Gamma ab_0 \nu}^{\check{K}}(\mathbf{q}_2, \mathbf{q}_3) = \sum_{[k']} A_{\Gamma;\nu\nu';ab_0}^{\check{K}} \psi_{k_1',k_2'}^{K_2}(\mathbf{q}_2) \psi_{k_3',\dots,k_9'}^{K_3}(\mathbf{q}_3)$$

$$= \sum_{[k']} A_{\Gamma;\nu\nu';ab_0}^{\check{K}} \psi_{\nu'}^{\check{K}}(\mathbf{q}_2, \mathbf{q}_3), \qquad (36)$$

$$A_{\Gamma;\nu\nu';ab_0} = \frac{\dim(\Gamma)}{\operatorname{card}(G)} \sum_{g \in G} \Delta^{\Gamma}(g)_{ab_0}^{*} \, \alpha[2]_{k_1,k_2;k_1',k_2'}(g) \alpha[3]_{k_3,\dots,k_9;k_3',\dots,k_9'}(g). \qquad (37)$$

where $[k'] = \{(k_1', \dots, k_9'): k_1' + k_2' = K_2, k_3' + \dots + k_9' = K_3\}$ and $\check{K} = K_2 + K_3$.

4.2   Algorithm for Construction of Shape-Vibrational Basis

Let us consider the algorithm for the explicit construction of the required bases in the case of the intrinsic octahedral group $\bar{O}$ [11] and its subgroup $\bar{D}_{4y}$ from Tables 1 and 2 acting in the space of eigenfunctions of the quadrupole-octupole harmonic oscillator (25)–(27) in accordance with (28) and (22), using as **INPUT**.

Step 1. First one needs to organize two loops: the first one is over the main oscillator quantum number $K_{23} = 0, 1, \dots, K_{23\,\mathrm{max}}$ and the second one is running over all irreducible representations $\Gamma \in \{\Gamma_1, \Gamma_2, \dots, \Gamma_{n_\Gamma}\}$ of the group G, where n_Γ denotes the number of irreducible representations.

For given K_{23} and Γ, in Eq. (36) choose a single (though we are doing the loop over all $b_0 = 1, \dots, \dim(\Gamma)$) fixed pair of indices $(a = b_0, b_0)$ such that there

Table 6. The vibrational functions $v_{ab_0 t}^{n K_{23}}$ of Irrs $\Gamma_3, \dots, \Gamma_5$, $K_{23} = 2$

Γ_3, $K_{23} = 2$
$v_{111}^{32} = [-\dfrac{\sqrt{5}}{4}\dfrac{\psi_1(a_{30})\,\psi_1(a_{31})}{\psi_0(a_{30})\,\psi_0(a_{31})} + \dfrac{\sqrt{5}}{4}\dfrac{\psi_1(a_{32})\,\psi_1(a_{33})}{\psi_0(a_{32})\,\psi_0(a_{33})} - \dfrac{\sqrt{3}}{4}\dfrac{\psi_1(a_{30})\,\psi_1(a_{33})}{\psi_0(a_{30})\,\psi_0(a_{33})}$
$\qquad -\dfrac{\sqrt{3}}{4}\dfrac{\psi_1(a_{31})\,\psi_1(a_{32})}{\psi_0(a_{31})\,\psi_0(a_{32})}]\psi_{[0]}^0(\mathbf{q})$

Γ_4, $K_{23} = 2$
$v_{111}^{42} = [\dfrac{\sqrt{165}}{44}\dfrac{\psi_1(a_{30})\,\psi_1(a_{33})}{\psi_0(a_{30})\,\psi_0(a_{33})} + \dfrac{\sqrt{165}}{44}\dfrac{\psi_1(a_{31})\,\psi_1(a_{32})}{\psi_0(a_{31})\,\psi_0(a_{32})}$
$\qquad +\dfrac{5\sqrt{11}}{44}\dfrac{\psi_1(a_{30})\,\psi_1(a_{31})}{\psi_0(a_{30})\,\psi_0(a_{31})} + \dfrac{\sqrt{11}}{4}\dfrac{\psi_1(a_{32})\,\psi_1(a_{33})}{\psi_0(a_{32})\,\psi_0(a_{33})}]\psi_{[0]}^0(\mathbf{q})$
$v_{112}^{42} = [-\dfrac{\sqrt{330}}{44}\dfrac{\psi_1(a_{30})\,\psi_1(a_{31})}{\psi_0(a_{30})\,\psi_0(a_{31})} - \dfrac{3\sqrt{22}}{44}\dfrac{\psi_1(a_{30})\,\psi_1(a_{33})}{\psi_0(a_{30})\,\psi_0(a_{33})} + \dfrac{2\sqrt{22}}{11}\dfrac{\psi_1(a_{31})\,\psi_1(a_{32})}{\psi_0(a_{31})\,\psi_0(a_{32})}]\psi_{[0]}^0(\mathbf{q})$
$v_{113}^{42} = [\dfrac{\sqrt{10}}{4}\dfrac{\psi_1(a_{30})\,\psi_1(a_{33})}{\psi_0(a_{30})\,\psi_0(a_{33})} - \dfrac{\sqrt{6}}{4}\dfrac{\psi_1(a_{30})\,\psi_1(a_{31})}{\psi_0(a_{30})\,\psi_0(a_{31})}]\psi_{[0]}^0(\mathbf{q})$

Γ_5, $K_{23} = 2$, $b_0 = 1$	Γ_5, $K = 2$, $b_0 = 2$
$v_{111}^{52} = v_{211}^{52} = \dfrac{\psi_1(a_{22})\,\psi_1(a_{33})}{\psi_0(a_{22})\,\psi_0(a_{33})}\psi_{[0]}^0(\mathbf{q})$	$v_{121}^{52} = v_{221}^{52} = \dfrac{\psi_1(a_{22})\,\psi_1(a_{32})}{\psi_0(a_{22})\,\psi_0(a_{32})}\psi_{[0]}^0(\mathbf{q})$
$v_{112}^{52} = v_{212}^{52} = \dfrac{\psi_1(a_{20})\,\psi_1(a_{33})}{\psi_0(a_{20})\,\psi_0(a_{33})}\psi_{[0]}^0(\mathbf{q})$	$v_{122}^{52} = v_{222}^{52} = \dfrac{\psi_1(a_{20})\,\psi_1(a_{32})}{\psi_0(a_{20})\,\psi_0(a_{32})}\psi_{[0]}^0(\mathbf{q})$
$v_{113}^{52} = v_{213}^{52} = \dfrac{\psi_1(a_{22})\,\psi_1(a_{31})}{\psi_0(a_{22})\,\psi_0(a_{31})}\psi_{[0]}^0(\mathbf{q})$	$v_{123}^{52} = v_{223}^{52} = \dfrac{\psi_1(a_{22})\,\psi_1(a_{30})}{\psi_0(a_{22})\,\psi_0(a_{30})}\psi_{[0]}^0(\mathbf{q})$
$v_{114}^{52} = v_{214}^{52} = \dfrac{\psi_1(a_{20})\,\psi_1(a_{31})}{\psi_0(a_{20})\,\psi_0(a_{31})}\psi_{[0]}^0(\mathbf{q})$	$v_{124}^{52} = v_{224}^{52} = \dfrac{\psi_1(a_{20})\,\psi_1(a_{30})}{\psi_0(a_{20})\,\psi_0(a_{30})}\psi_{[0]}^0(\mathbf{q})$

exists a set of the oscillator quantum numbers ν for which $\Psi^{\check{K}}_{\Gamma b_0 b_0 \nu}(\mathbf{q}_2, \mathbf{q}_3)$ are not identically equal to zero. Note that, it can happen, so that all the vectors $\Psi^{\check{K}}_{\Gamma b_0 b_0 \nu}(\mathbf{q}_2, \mathbf{q}_3) = 0$ for the given b_0 or the representation Γ.

Step 2. Given the subset of the vectors $\tilde{u}_{\check{K},\Gamma,\nu,b_0}(\mathbf{q}_2, \mathbf{q}_3) = \Psi^{\check{K}}_{\Gamma b_0 b_0 \nu}(\mathbf{q}_2, \mathbf{q}_3)$, where $\nu = \{k_1, ..., k_9\}$, run over all solutions of the equations $k_1 + k_2 = K_2$, $k_3 + ... + k_9 = K_3$ for $K_{23} = K_2 + K_3$.

Now we have to choose the maximal subset of linearly independent vectors from the set of the vectors $\tilde{u}_{\check{K},\Gamma,\nu,b_0}(\mathbf{q}_2, \mathbf{q}_3)$.

To solve this problem we use the symmetric orthonormalization procedure. The required overlaps of the vectors $\tilde{u}_{\check{K},\Gamma,\nu,b_0}(\mathbf{q}_2, \mathbf{q}_3)$ are expressed in terms of the coefficients $A^{\check{K}}_{\Gamma n; \nu'; a b_0}$ of Eq. (36):

$$\mathcal{N}_{\nu\nu'} = \langle \tilde{u}_{\check{K},\Gamma,\nu,b_0} | \tilde{u}_{\check{K},\Gamma,\nu',b_0} \rangle = \sum_{[k'']} A^{\check{K}}_{\Gamma; \nu\nu''; b_0 b_0} A^{\check{K}}_{\Gamma; \nu'\nu''; b_0 b_0}. \tag{38}$$

For given $K_{23} = K_2 + K_3$, Γ and b_0, these overlaps form a finite dimensional Hermitian matrix referred to as the Gram matrix. The indices ν represent the sets of oscillatory quantum numbers, i.e., $\nu = \{k_1, ..., k_9\}$.

Solving the eigenvalue problem for the matrix $\mathcal{N}$ allows one to find the orthonormal basis, in which this matrix is diagonal:

$$\mathcal{N} w_t = \lambda_t w_t, \tag{39}$$

where the first s values of t, i.e., $t = 1, 2, ..., s$ correspond to the eigenvalues $\lambda_t \neq 0$, and the indices $t = s+1, s+2, ...$ are related to $\lambda_t = 0$.

The coefficients w_t allow the construction of states, which in the Generator Coordinate Method [13] are named the "natural states". They furnish the required basis in our space spanned with the vectors $|\tilde{u}_{\check{K},\Gamma,\nu,b_0}\rangle$ corresponding to the non-zero eigenvalues $\lambda_t \neq 0$:

$$|u_{\check{K},\Gamma,t,b_0}\rangle = A_t \sum_\nu w_t(\nu) |\tilde{u}_{\check{K},\Gamma,\nu,b_0}\rangle = A_t \sum_{[k]} w_t(\nu) \sum_{[k']} A^{\check{K}}_{\Gamma m; \nu; b_0 b_0; \nu'} |\Psi^{\check{K}}_{\nu'}\rangle, \tag{40}$$

where $\check{K}$, Γ, and b_0 are fixed and $A_t = \{\lambda_t \sum_{[k']} |w_{t(\nu')}|^2\}^{-1/2}$ is the normalization coefficient. Thus, we obtain the required basis in the projected subspace $\mathcal{K}^\Gamma_{K_{23} b_0}$.

Next, the vectors $u_{\check{K},\Gamma,t,b_0}(\mathbf{q}_2, \mathbf{q}_3)$ are expanded using the harmonic oscillator basis (25). We denote the expansion coefficients by $B^{\check{K}}_{\Gamma n, t, b_0; \nu'} = \langle \nu' | u_{\check{K},\Gamma n,t,b_0} \rangle$. They can be calculated from the formula (40) with $t = 1, ..., s$

$$u_{\check{K},\Gamma,t,b_0}(\mathbf{q}_2, \mathbf{q}_3) = \sum_{[k']} B^{\check{K}}_{\Gamma,t,b_0;\nu'} \Psi^{\check{K}}_{\nu'}(\mathbf{q}_2, \mathbf{q}_3)$$

$$B^{\check{K}}_{\Gamma,t,b_0\nu'} = \delta_{k_1'+...+k_9', k_1+...+k_9} A_t \sum_{[k]} w_t(\nu) A^{\check{K}}_{\Gamma; \nu; a b_0; \nu'} \tag{41}$$

Step 3. Now we are able to apply the projection operators (28) for the irreducible representation Γ of the group $\overline{G}$ to every vector $u_{\check{K},\Gamma,t,b_0}(\mathbf{q}_2, \mathbf{q}_3)$, $t = 1, ..., s$ and for every $a = 1, 2, ..., \dim(\Gamma)$, with fixed b_0:

$$\bar{v}_{\check{K},\Gamma,t,a b_0}(\mathbf{q}_2, \mathbf{q}_3) = \hat{P}^\Gamma_{a b_0} u_{\check{K},\Gamma,t,b_0}(\mathbf{q}_2, \mathbf{q}_3) = \sum_{[k']} \bar{C}^{\check{K},\Gamma,t,a,b_0}_{\nu,\nu'} \Psi^{\check{K}}_{\nu'}(\mathbf{q}_2, \mathbf{q}_3). \tag{42}$$

Using (36) and (42), the coefficients $\bar{C}^{\check{K},\Gamma,t,a,b_0}_{\nu,\nu'}$ can be written as

$$\bar{C}^{\check{K},\Gamma,t,a,b_0}_{\nu,\nu'} = B^{\check{K}}_{\Gamma,t,b_0;\nu'} A^{\check{K}}_{\Gamma;\nu;ab_0;\nu'}. \tag{43}$$

Using the orthonormalization conditions (26) for $\bar{v}_{\check{K},\Gamma,t,ab_0}(\mathbf{q}_2, \mathbf{q}_3)$, we calculate the normalization factor

$$N_2^2(\check{K}, \Gamma, t, ab_0) = \sum_{[k']} \sum_{[k]} |\bar{C}^{\check{K},\Gamma,t,a,b_0}_{\nu,\nu'}|^2 \tag{44}$$

and the required set of orthonormalized vectors $v_{\check{K},\Gamma,t,ab_0}(\mathbf{q}_2, \mathbf{q}_3)$

$$v^{nK_{23}}_{ab_0t} \equiv v_{\check{K},\Gamma,t,ab_0}(\mathbf{q}_2, \mathbf{q}_3) = \sum_{[k']} C^{\check{K},\Gamma,t,a,b_0}_{\nu,\nu'} \Psi^{\check{K}}_{\nu'}(\mathbf{q}_2, \mathbf{q}_3), \tag{45}$$

$$C^{\check{K},\Gamma,t,a,b_0}_{\nu,\nu'} = \bar{C}^{\check{K},\Gamma,t,a,b_0}_{\nu,\nu'}/N_2(\check{K}, \Gamma, t, ab_0).$$

The above algorithm (**Steps 1–3**) with **OUTPUT**, containing a set of the vectors $v^{nK_{23}}_{ab_0t}$ from (45) for all irreducible representations of the intrinsic octahedral group $\bar{O}$ or its subgroup from **INPUT**, was implemented in the computer algebra system Reduce. The typical running time of calculating the irreducible representations $\Gamma_1,...,\Gamma_5$ for the octahedral group for $K_{23} \leq 5$ is 3080 s using the PC Intel Core i5 CPU 3.33 GHz 4 GB 64bit Windows 7.

Example of **OUTPUT** for the group $\bar{D}_{4y}$, for all irreducible representations $\Gamma_1, ..., \Gamma_5$, $a=1,2$, $b_0=1,2$, $K_{23}=k_1 + ... + k_9 \leq 2$ the vectors $\bar{v}_{\check{K},\Gamma_n,t,ab_0}(\mathbf{q}_2, \mathbf{q}_3)$, denoted for practical reason as $v^{nK_{23}}_{ab_0t}$, are shown in Tables 4, 5, 6. The index $t = 1, ..., s$ distinguishes between the equivalent representations. In the tables, one should upload the basis functions $\psi^0_{[0]}(\mathbf{q}) \equiv \psi^0_{[000000]}(\mathbf{q}) = \psi^0_{[00]}(\mathbf{q}_2)\psi^0_{[0000]}(\mathbf{q}_3^a)$, because we put $b_{33}=b_{32}=b_{31} =0$ to compare our results with Ref. [15].

5   Conclusion and Perspectives

The symbolic algorithms for evaluating the bases of irreducible representations for the laboratory and intrinsic point groups in both the rotor space $L^2(SO(3))$ and the shape vibrational space $L^2(\mathbf{R}^{2+7})$ or $L^2(\mathbf{R}^{2+4})$ of the octupole-quadrupole harmonic oscillator are presented.

These algorithms can be implemented in any computer algebra system dealing with non-commuting operations calculus and used for generating a FORTRAN code for more efficient numerical calculations.

These special realizations of bases are useful for calculating spectra and electromagnetic transitions in molecular and nuclear physics.

These algorithms can be adopted to any subgroup of the octahedral group and applied for constructing the rotational–vibrational bases (14), especially for collective models of molecules and nuclei.

The work was partially supported by the Russian Foundation for Basic Research, grant No. 14-01-00420 and the Bogoliubov-Infeld JINR-Poland program.

References

1. Abramowitz, M., Stegun, I.A.: Handbook of Mathematical Functions. Dover, New York (1965)
2. Barut, A., Rączka, R.: Theory of Group Representations and Applications. PWN, Warszawa (1977)
3. Chen, J.Q., Ping, J., Wang, F.: Group Representation Theory for Physicists. World Sci., Singapore (2002)
4. Cornwell, J.F.: Group Theory in Physics. Academic Press, New York (1984)
5. Doan, Q.T., et al.: Spectroscopic information about a hypothetical tetrahedral configuration in ^{156}Gd. Phys. Rev. C **82**, 067306 (2010)
6. Dobrowolski, A., Góźdź, A., Szulerecka, A.: Electric transitions within the symmetrized tetrahedral and octahedral states. Phys. Scr. **T154**, 014024 (2013)
7. Dudek, J., Góźdź, A., Schunck, N., Miśkiewicz, M.: Nuclear tetrahedral symmetry: possibly present throughout the Periodic Table. Phys. Rev. Lett. **88**, 252502 (2002)
8. Góźdź, A., Dobrowolski, A., Pȩdrak, A., Szulerecka, A., Gusev, A.A., Vinitsky, S.I.: Structure of Bohr type nuclear collective spaces - a few symmetry related problems. Nucl. Theory **32**, 108–122 (2013)
9. Góźdź, A., Pȩdrak, A., Dobrowolski, A., Szulerecka, A., Gusev, A.A., Vinitsky, S.I.: Shapes and symmetries of nuclei. Bulg. J. Phys. **42**, 494–502 (2015)
10. Góźdź, A., Szulerecka, A., Dobrowolski, A., Dudek, J.: Nuclear collective models and partial symmetries. Acta Phys. Pol. B **42**, 459–463 (2011)
11. Gusev, A.A., Gerdt, V.P., Vinitsky, S.I., Derbov, V.L., Góźdź, A.: Symbolic algorithm for irreducible bases of point groups in the space of SO(3) group. In: Gerdt, V.P., Koepf, W., Seiler, W.M., Vorozhtsov, E.V. (eds.) CASC 2015. LNCS, pp. 166–181. Springer, Heidelberg (2015)
12. Pogosyan, G.S., Smorodinsky, A.Y., Ter-Antonyan, V.M.: Oscillator Wigner functions. J. Phys. A **14**, 769–776 (1981)
13. Ring, P., Schuck, P.: The Nuclear Many-Body Problem. Springer, New York (1980)
14. Rojansky, V.: On the theory of the Stark effect in hydrogenic atoms. Phys. Rev. **33**, 1–15 (1929)
15. Szulerecka, A., Dobrowolski, A., Góźdź, A.: Generalized projection operators for intrinsic rotation group and nuclear collective model. Phys. Scr. **89**, 054033 (2014)
16. Varshalovich, D.A., Moskalev, A.N., Khersonskii, V.K.: Quantum Theory of Angular Momentum. World Sci., Singapore (1989)
17. Vilenkin, J.A., Klimyk, A.U.: Representation of Lie Group and Special Functions, vol. 2. Kluwer Academic Publ., Dordrecht (1993)

A Symbolic Investigation of the Influence of Aerodynamic Forces on Satellite Equilibria

Sergey A. Gutnik[1]([envelope]) and Vasily A. Sarychev[2]

[1] Moscow State Institute of International Relations (University),
76, Prospekt Vernadskogo, Moscow 119454, Russia
s.gutnik@inno.mgimo.ru

[2] Keldysh Institute of Applied Mathematics, Russian Academy of Sciences,
4, Miusskaya Square, Moscow 125047, Russia
vas31@rambler.ru

Abstract. Computer algebra methods are used to study the properties of a nonlinear algebraic system that determines equilibrium orientations of a satellite moving along a circular orbit under the action of gravitational and aerodynamic torques. An algorithm for the construction of a Gröbner basis is proposed for determining the equilibrium orientations of a satellite with given principal central moments of inertia and given aerodynamic torque in special cases, when the center of pressure of aerodynamic forces is located in one of the principal central planes of inertia of the satellite. The conditions of the equilibria existence are obtained, depending on three dimensionless parameters of the problem. The number of equilibria depending on the parameters is found by the analysis of real roots of algebraic equations from the constructed Gröbner basis. The evolution of domains with fixed number of equilibria from 24 to 8 is investigated in detail. All bifurcation values of the system parameters corresponding to the qualitative change of these domains are determined.

1 Introduction

In this work, a symbolic investigation of a satellite dynamics under the influence of gravitational and aerodynamic torques is presented. The gravity orientation systems are based on the result that a satellite with different moments of inertia in the central Newtonian force field in a circular orbit has 24 equilibrium orientations [1]. However, at altitudes from 250 up to 500 km, rotational motion of a satellite is subjected to aerodynamic torque too. Therefore, it is necessary to study the joint action of gravitational and aerodynamic torques and, in particular, to analyse all possible satellite equilibria in a circular orbit. Such solutions are used in practical space technology in the design of attitude control systems of satellites.

The basic problems of satellite dynamics with an aerodynamic attitude control system have been presented in [1]. The problem of determining the classes of equilibrium orientations for general values of aerodynamic torque was considered in [2,3]. In [4–6], some equilibrium orientations were found in special cases when

© Springer International Publishing AG 2016
V.P. Gerdt et al. (Eds.): CASC 2016, LNCS 9890, pp. 243–254, 2016.
DOI: 10.1007/978-3-319-45641-6_16

the center of pressure is located on a satellite principal central axis of inertia and on a satellite principal central plane of inertia. In [7], all equilibrium orientations were found in the case of axisymmetric satellite.

The present work continues the study started in [6]. In this paper, all cases when the center of pressure is located in the satellite principal central plane of inertia are considered. All possible equilibrium orientations are investigated, and their existence conditions are obtained. The equilibrium orientations are determined by real roots of the system of nonlinear algebraic equations. The investigation of equilibria was possible due to application of Computer Algebra Gröbner basis method. The evolution of domains with a fixed number of equilibria is investigated in dependence of three dimensionless system parameters.

2   Equations of Motion

Consider the motion of a satellite subjected to gravitational and aerodynamic torques in a circular orbit. We assume that (1) the gravity field of the Earth is central and Newtonian, (2) the satellite is a triaxial rigid body, (3) the effect of atmosphere on a satellite is reduced to the drag force applied at the center of pressure and directed against the velocity of the satellite center of mass relative to the air, and the center of pressure is fixed in the satellite body. To write the equations of motion we introduce two right-handed Cartesian coordinate systems with origin at the satellite center of mass O. $OXYZ$ is the orbital reference frame. The axis OZ is directed along the radius vector from the Earth center of mass to the satellite center of mass, the axis OX is in the direction of a satellite orbital motion. $Oxyz$ is the satellite body reference frame; Ox, Oy, and Oz are the principal central axes of inertia of the satellite. The orientation of the satellite body coordinate system $Oxyz$ with respect to the orbital coordinate system is determined by means of the aircraft angles of pitch α, yaw β, and roll γ, and the direction cosines in the transformation matrix between the orbital coordinate system $OXYZ$ and $Oxyz$ are represented by the following expressions:

$$
\begin{aligned}
a_{11} &= \cos(x, X) = \cos\alpha\cos\beta, \\
a_{12} &= \cos(y, X) = \sin\alpha\sin\gamma - \cos\alpha\sin\beta\cos\gamma, \\
a_{13} &= \cos(z, X) = \sin\alpha\cos\gamma + \cos\alpha\sin\beta\sin\gamma, \\
a_{21} &= \cos(x, Y) = \sin\beta, \\
a_{22} &= \cos(y, Y) = \cos\beta\cos\gamma, \\
a_{23} &= \cos(z, Y) = -\cos\beta\sin\gamma, \\
a_{31} &= \cos(x, Z) = -\sin\alpha\cos\beta, \\
a_{32} &= \cos(y, Z) = \cos\alpha\sin\gamma + \sin\alpha\sin\beta\cos\gamma, \\
a_{33} &= \cos(z, Z) = \cos\alpha\cos\beta - \sin\alpha\sin\beta\sin\gamma.
\end{aligned}
\tag{1}
$$

Then equations of the satellite attitude motion can be written in the Euler form [1,4]:

$$A\dot{p} + (C - B)qr - 3\omega_0^2(C - B)a_{32}a_{33} = \omega_0^2(H_2a_{13} - H_3a_{12}),$$
$$B\dot{q} + (A - C)rp - 3\omega_0^2(A - C)a_{31}a_{33} = \omega_0^2(H_3a_{11} - H_1a_{13}),$$
$$C\dot{r} + (B - A)pq - 3\omega_0^2(B - A)a_{31}a_{32} = \omega_0^2(H_1a_{13} - H_3a_{11}), \tag{2}$$
$$p = (\dot{\alpha} + \omega_0)a_{21} + \dot{\gamma}, \quad q = (\dot{\alpha} + \omega_0)a_{22} + \dot{\beta}\sin\gamma,$$
$$r = (\dot{\alpha} + \omega_0)a_{23} + \dot{\beta}\cos\gamma.$$

Here p, q, and r are the projections of the satellite angular velocity onto the axes Ox, Oy, and Oz; A, B, and C are the principal central moments of inertia of the satellite (without loss of generality, we assume that $B > A > C$); ω_0 is the angular velocity of the orbital motion of the satellite center of mass. $H_1 = -aQ/\omega_0^2$, $H_2 = -bQ/\omega_0^2$, $H_3 = -cQ/\omega_0^2$, Q is the atmospheric drug force acting on a satellite; a, b, and c are the coordinates of the center of pressure of a satellite in the reference frame $Oxyz$. The dot designates differentiation with respect to time t.

3   Equilibrium Orientations of a Satellite

Setting in (2) $\alpha = \alpha_0 = $ const, $\beta = \beta_0 = $ const, $\gamma = \gamma_0 = $ const, we obtain at $A \neq B \neq C$ the equations

$$(C - B)(a_{22}a_{23} - 3a_{32}a_{33}) = H_2a_{13} - H_3a_{12},$$
$$(A - C)(a_{21}a_{23} - 3a_{31}a_{33}) = H_3a_{11} - H_1a_{13}, \tag{3}$$
$$(B - A)(a_{21}a_{22} - 3a_{31}a_{32}) = H_1a_{12} - H_2a_{11},$$

which allow us to determine the satellite equilibria in the orbital reference frame. Substituting the expressions for the direction cosines from (1) in terms of the aircraft angles and into Eq. (3), we obtain three equations with three unknowns α, β, and γ. The second procedure for closing Eq. (3) is to add the following six orthogonality conditions for the direction cosines

$$a_{i1}a_{j1} + a_{i2}a_{j2} + a_{i3}a_{j3} - \delta_{ij} = 0, \tag{4}$$

where δ_{ij} is the Kronecker delta and $i, j = 1, 2, 3$. Equations (3) and (4) form a closed system with respect to the direction cosines, which also specifies the equilibrium solutions of the satellite.

The system (3) and (4) has been solved for general case of the problem when $H_1 \neq 0$, $H_2 \neq 0$, $H_3 \neq 0$ [2,3]. With the help of computer algebra method it was shown that equilibrium orientations are determined by real solutions of algebraic equation of the twelfth degree. The equilibrium orientations and their stability were analysed numerically. The problem has been solved analytically only for some specific cases when the center of pressure is located on a satellite principal central axis of inertia Ox, when $H_1 \neq 0$, $H_2 = H_3 = 0$ [4,5] and for the case of axisymmetric satellite when $A \neq B = C$ [7]. When the pressure center locates in the satellite principal central plane of inertia Oxz of the frame $Oxyz$, in the case when $H_1 \neq 0$, $H_2 = 0$, and $H_3 \neq 0$, very complex analytical study of the system (3) and (4) was conducted, and the equilibria were analysed numerically [6]. In the present work, the problem of determination of the classes of equilibrium orientations for all cases when the pressure center locates in one of satellite principal central planes of inertia of the frame $Oxyz$, when (1) $H_1 \neq 0$, $H_2 \neq 0$, and $H_3 = 0$, (2) $H_1 \neq 0$, $H_2 = 0$, and $H_3 \neq 0$ and (3) $H_1 = 0$, $H_2 \neq 0$, and $H_3 \neq 0$, with the help of Computer Algebra methods is investigated. The existence of flat solutions in these cases is specified.

4   Investigation of Equilibria

4.1   Equilibria in the Case $H_3 = 0$ ($H_1 \neq 0$, $H_2 \neq 0$)

We begin by considering the first case $H_1 \neq 0$, $H_2 \neq 0$, and $H_3 = 0$ when the pressure center is located in the plane Oxy. Introducing dimensionless parameters $h_i = H_i/(B - C)$, $\nu = (B - A)/(B - C)$, $(0 < \nu < 1)$, system (3) takes the form

$$a_{22}a_{23} - 3a_{32}a_{33} + h_2a_{13} = 0,$$
$$(1 - \nu)(a_{23}a_{21} - 3a_{33}a_{31}) + h_1a_{13} = 0, \qquad (5)$$
$$\nu(a_{21}a_{22} - 3a_{31}a_{32}) - h_1a_{12} + h_2a_{11} = 0.$$

To solve the algebraic system (4), (5) we applied the algorithm of constructing the Gröbner bases [8]. The method of constructing the Gröbner bases is an algorithmic procedure for complete reduction of the problem in the case of the system of polynomials in several variables to the polynomial of one variable. Using the Gröbner[gbasis] Maple 15 package [9] for constructing Gröbner bases, the lexicographic monomial order was chosen. We constructed the Gröbner basis for the system of nine polynomials (4), (5) with nine variables direction cosines a_{ij} ($i, j = 1, 2, 3$), and in the list of polynomials, we include the polynomials from the left-hand sides f_i ($i = 1, 2, ...9$) of the algebraic equations (4), (5):

G:=map(factor,Gröbner[gbasis]([f1, ... f9], plex(a11, ... a33))).

Here we write down the polynomial in the Gröbner basis that depends only on one variable $x = a_{33}$. This polynomial has the form

$$P(x) = P_1(x)P_2(x) = 0, \qquad (6)$$

where

$$P_1(x) = x(x^2 - 1),$$
$$P_2(x) = p_0x^4 + p_1x^2 + p_2,$$
$$p_0 = 9(1 - \nu)^2p_3^2,$$
$$p_1 = p_3(h_1^6 - (1 - \nu)(6\nu - (3 - \nu)h_2^2)h_1^4$$
$$\quad - (1 - \nu)^2((2\nu - 3)h_2^4 + 9\nu^2(2h_2^2 - 1))h_1^2 + (1 - \nu)^4(h_2^2 + 3\nu)^2),$$

$$p_2 = -\nu^2 h_1^2 h_2^2 (h_1^6 - (1-\nu)(5\nu+1-(3-\nu)h_2^2)h_1^4$$
$$- (1-\nu)^2((2\nu-3)h_2^4 + 2(5\nu^2 - \nu + 1)h_2^2 - 3\nu(2+\nu))h_1^2$$
$$+ (1-\nu)^4(h_2^2 - 1)(h_2^2 + 3\nu)^2),$$
$$p_3 = h_1^4 + 2(1-\nu)(h_2^2 - 3\nu)h_1^2 + (1-\nu)^2(h_2^2 + 3\nu)^2.$$

It is necessary to consider three cases $a_{33} = 0$, $a_{33}^2 = 1$, and $P_2(a_{33}^2) = 0$ to investigate system (4), (5).

In the first case when $a_{33} = 0$, system (4), (5) takes the form

$$\begin{aligned} 3\nu a_{31} a_{32} + (h_1 a_{31} + h_2 a_{32})a_{23} &= 0, \\ a_{31}^2 + a_{32}^2 &= 1, \\ a_{23}^2 &= 1, \\ a_{33} = a_{13} = a_{21} = a_{22} &= 0, \\ a_{11} = -a_{23}a_{32}, \quad a_{12} &= a_{23}a_{21}. \end{aligned} \tag{7}$$

The first two equations of system (7) can be written in a simpler form

$$9\nu^2 a_{31}^4 + 6\nu h_2 a_{31}^3 + (h_1^2 + h_2^2 - 9\nu^2)a_{31}^2 - 6\nu h_2 a_{31} - h_1^2 = 0, \tag{8}$$

$$a_{32} = \mp \frac{h_1 a_{31}}{(3\nu a_{31} \pm h_2)}.$$

Having solved system (8) one can determine the remaining direction cosines from the equations of system (7). The first equation of system (7) represents the equations of four hyperbolas. Their two branches pass through the origin of coordinate system ($a_{31} = 0$ and $a_{32} = 0$) in the plane of variables a_{31} and a_{32}, while the second equation determines the unit circle in this plane. The number of real solutions to system (7) (and, hence, to system (8)) depends on the character of intersections of the hyperbolas with the circle. It is clear that two branches of the hyperbolas that pass through the origin of coordinates always intersect the circle at four points. If two other branches of the hyperbolas also intersect the circle, we have four more solutions. In the case when hyperbola branches touch the circle four solutions merge into two (there are two pairs of multiple roots) [7]. Thus, system (7), and hence system (8) too, has either eight or four solutions. It follows from the reasoning presented above that bifurcation points are those points of plane a_{31}, a_{32}, through which the branches of hyperbolas and the circle pass simultaneously, and where tangents to these curves coincide. The condition of coincidence of the tangents to two hyperbola branches and the circle has the following form

$$\frac{d(a_{31})}{d(a_{32})} = -\frac{3\nu a_{32} \pm h_1}{3\nu a_{31} \pm h_2} = -\frac{2a_{31}}{2a_{32}}$$

or

$$3\nu(a_{32}^2 - a_{31}^2) \pm h_1 a_{32} \mp h_2 a_{31} = 0. \tag{9}$$

Substituting the expression for a_{32} from (8) into the second equation of (7) and equation (9), we get the following system of equations

$$\frac{h_1^2 a_{31}^2}{(3\nu a_{31} + h_2)^2} = 1 - a_{31}^2, \qquad \frac{h_1^2 h_2}{(3\nu a_{31} + h_2)^2} = -(3\nu a_{31} + h_2). \qquad (10)$$

Excluding h_1^2 from system of equations (10), after some simple transformations we get the relationship $a_{31} = -(3\nu)^{-1/3} h_2^{1/3}$. Finally, substituting the expression for a_{31} into the second equation of (10) we arrive at the equation of astroid

$$h_1^{2/3} + h_2^{2/3} = (3\nu)^{2/3}. \qquad (11)$$

There are eight solutions inside the region $h_1^{2/3} + h_2^{2/3} < (3\nu)^{2/3}$; when passing through curve (11) (which is a bifurcation curve), the number of solutions changes to four; there exist four solutions in the region $h_1^{2/3} + h_2^{2/3} > (3\nu)^{2/3}$.

Now let us consider the second case $a_{33}^2 = 1$. In this case, system (4), (5) takes the form

$$\begin{aligned}
\nu a_{21} a_{22} + (h_1 a_{21} + h_2 a_{22}) a_{33} &= 0, \\
a_{21}^2 + a_{22}^2 &= 1, \\
a_{33}^2 &= 1, \\
a_{31} = a_{32} = a_{13} = a_{23} &= 0, \\
a_{11} = a_{22} a_{33}, \quad a_{12} = -a_{21} a_{33}.
\end{aligned} \qquad (12)$$

The first two equations of system (12) can be written in the following form:

$$\nu^2 a_{21}^4 + 2\nu h_2 a_{21}^3 + (h_1^2 + h_2^2 - \nu^2) a_{21}^2 - 2\nu h_2 a_{31} - h_1^2 = 0, \qquad (13)$$

$$a_{22} = \mp \frac{h_1 a_{21}}{(\nu a_{21} \pm h_2)}.$$

Applying the approach suggested above for investigating system (7) one can demonstrate that for system (12), the bifurcation curve separating the region of existence of eight solutions from the region of existence of four solutions is also the astroid

$$h_1^{2/3} + h_2^{2/3} = (\nu)^{2/3}. \qquad (14)$$

In Figs. 1, 2 and 3, astroids (11) and (14) for the ν values equal to 0.2, 0.5, and 0.8 are presented, that separate in plane (h_1, h_2) three regions with different numbers of equilibrium orientations of the satellite under the action of gravitational and aerodynamic torques. There exist 8, 6, and 4 real solutions (16, 12, and 8 equilibria) of both Eqs. (8) and (13) for the first and second cases in regions $h_1^{2/3} + h_2^{2/3} < (\nu)^{2/3}$; $(\nu)^{2/3} < h_1^{2/3} + h_2^{2/3} < (3\nu)^{2/3}$, and $h_1^{2/3} + h_2^{2/3} > (3\nu)^{2/3}$, respectively.

Let us consider the third case for which the satellite equilibria are determined by the real roots of the biquadratic equation $P_2(x) = 0$. The number of real roots of the biquadratic equation (6) is even and not greater than 4. For each solution, one can find from the second polynomial from of the constructed Gröbner base two values of a_{32} and, then, their respective values a_{31}. For each set of values a_{31}, a_{32}, and a_{33}, one can unambiguously determine from original system (4) and (5) the respective values of the direction cosines a_{11}, a_{12}, a_{13} a_{21}, a_{22}, and a_{23}. Thus, each real root of the biquadratic equation (6) is matched with two sets of values a_{ij} (two equilibrium orientations). Since the number of real roots of biquadratic equation (6) does not exceed 4, the satellite in the third case can have no more than 8 equilibrium orientations.

For the variable $t = x^2 = a_{33}^2$, we get the quadratic equation

$$P_2(t) = p_0 t^2 + p_1 t + p_2 = 0. \tag{15}$$

Equation (15) has two solutions

$$t_1 = \frac{-p_1 - \sqrt{p_1^2 - 4p_0 p_2}}{2p_0}; \quad t_2 = \frac{-p_1 + \sqrt{p_1^2 - 4p_0 p_2}}{2p_0}. \tag{16}$$

It is possible to show that the discriminant $p_1^2 - 4p_0 p_2 \geq 0$ at any values of the system parameters. Thus, in case of the inequality $t_1 = a_{33}^2 > 0$ satisfaction, Eq. (15) has two real roots t_1 and t_2 which correspond to four a_{33} values, and system (4), (5) (at $a_{33} \neq 0, a_{33} \neq \pm 1$) has 8 solutions, and these solutions correspond to 8 satellite equilibrium orientations. These equilibria exist in the domain bounded by the curve $t_1(h_1, h_2, \nu) = 0$. In Figs. 1, 2 and 3, these curves are marked as t_1.

In the domain bounded by the curves $t_1(h_1, h_2, \nu) = 0$, $t_2(h_1, h_2, \nu) = 1$ for which inequalities $t_1(h_1, h_2, \nu) < 0$ and $0 < t_2(h_1, h_2, \nu) < 1$ take place, only four equilibria exist, which correspond only to one root t_2. Outside the boundary $t_2(h_1, h_2, \nu) = 1$, there are no solutions of the third case. In Figs. 1, 2 and 3, these curves are denoted as t_2.

The results of the analysis of the equilibria total number in the third case can be summarized as follows. The curves $t_1(h_1, h_2, \nu) = 0$, $t_2(h_1, h_2, \nu) = 1$ decompose the plane (h_1, h_2) into three domains where 8 equilibria, 4 equilibria, and no equilibria exist.

The final decomposition of the plane (h_1, h_2) for all three cases is presented in Figs. 1, 2 and 3 for $\nu = 0.2$, $\nu = 0.5$, and $\nu = 0.8$. Curves (11), (14) and $t_1(h_1, h_2, \nu) = 0$, $t_2(h_1, h_2, \nu) = 1$ separate the plane into domains with the fixed number of equilibria equal to 24, 20, 16, 12, and 8.

4.2 Equilibria in the Case $H_1 = 0$ ($H_2 \neq 0$, $H_3 \neq 0$)

Let us consider the next case $H_1 = 0$, $H_2 \neq 0$, and $H_3 \neq 0$ when the pressure center locates in the plane Oyz. System (3) in that case takes the form

$$a_{22}a_{23} - 3a_{32}a_{33} + h_2 a_{13} - h_3 a_{12} = 0,$$
$$(1 - \nu)(a_{23}a_{21} - 3a_{33}a_{31}) - h_3 a_{11} = 0, \tag{17}$$
$$\nu(a_{21}a_{22} - 3a_{31}a_{32}) + h_2 a_{11} = 0.$$

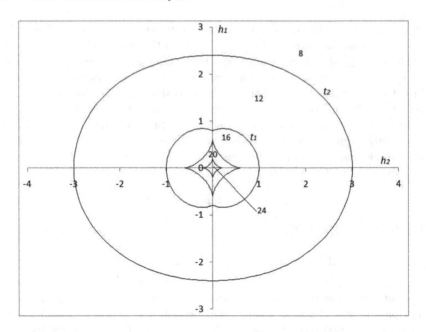

Fig. 1. The regions with the fixed number of equilibria for $\nu = 0.2$

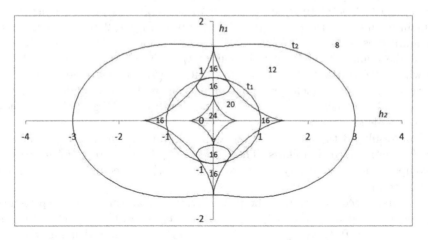

Fig. 2. The regions with the fixed number of equilibria for $\nu = 0.5$

Applying the approach suggested above for investigating system (4), (5), we used the algorithm of constructing the Gröbner bases for the polynomials on the left-hand sides of the system (4), (17). The polynomial in the Gröbner basis that depends only on one variable in that case a_{31} has the form

$$P_{h1}(a_{31}) = P_3(a_{31})P_4(a_{31}) = 0, \tag{18}$$

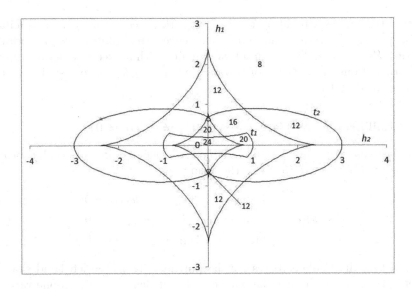

Fig. 3. Regions with the fixed number of equilibria for $\nu = 0.8$

where

$$P_3(a_{31}) = a_{31}(a_{31}^2 - 1),$$
$$P_4(a_{31}) = p_{40}a_{31}^4 + p_{41}a_{31}^2 + p_{42},$$
$$p_{40} = 9\nu^2(1-\nu)^2 p_{43}^2,$$
$$p_{41} = p_{43}(\nu^4 h_3^6 + \nu^2(1-\nu)(6\nu^2 - (3\nu - 1)h_2^2)h_3^4$$
$$- \nu(1-\nu)^2((2\nu - 3)h_2^4 + 9\nu(2h_2^2 - \nu^2))h_3^2 + (1-\nu)^4 h_2^2(h_2^2 + 3\nu)^2),$$
$$p_{42} = -h_2^2 h_3^2(\nu^4 h_3^6 + \nu^2(1-\nu)(\nu^2(5+\nu) + (1-3\nu)h_2^2)h_3^4$$
$$+ \nu(1-\nu)^2((3\nu - 2)h_2^4 + 2\nu(\nu(1-\nu) - 5)h_2^2 + 3\nu^3(1+2\nu))h_3^2$$
$$+ (1-\nu)^4(h_2^2 - \nu^2)(h_2^2 + 3\nu)^2),$$
$$p_{43} = \nu^2 h_3^4 + 2\nu(1-\nu)(h_2^2 - 3\nu)h_3^2 + (1-\nu)^2(h_2^2 + 3\nu)^2.$$

It is also necessary to consider three cases, $a_{31} = 0$, $a_{31}^2 = 1$, and $P_{42}(a_{31}^2) = 0$ to investigate system (4), (17).

In the first case when $a_{31} = 0$, using the approach described above in Subsect. 4.1, it is possible to obtain the bifurcation curve

$$h_2^{2/3} + h_3^{2/3} = 3^{2/3},$$

which separates the plane (h_2, h_3) into two regions with eight and four equilibrium orientations of the satellite. In the second case when $a_{31}^2 = 1$, applying the above approach one can demonstrate that for system (17), the bifurcation curve separating the region of existence of eight solutions from the region of existence of four solutions is also the astroid

$$h_2^{2/3} + h_3^{2/3} = 1.$$

Another two curves separating the regions with an equal number of equilibria can be obtained from the conditions of existence of real roots of the biquadratic equation $P_4(a_{31}) = 0$. The evolution of domains with a fixed number of equilibrium orientations in the plane of two parameters (h_2, h_3) is very similar to the case described in 4.1.

4.3 Equilibria in the Case $H_2 = 0$ ($H_1 \neq 0$, $H_3 \neq 0$)

In the last case $H_1 \neq 0$, $H_2 = 0$, and $H_3 \neq 0$ when the pressure center locates in the plane Oxz system (3) takes the form

$$a_{22}a_{23} - 3a_{32}a_{33} - h_3 a_{12} = 0,$$
$$(1 - \nu)(a_{23}a_{21} - 3a_{33}a_{31}) - h_3 a_{11} + h_1 a_{13} = 0, \qquad (19)$$
$$\nu(a_{21}a_{22} - 3a_{31}a_{32}) - h_1 a_{12} = 0.$$

Constructing the Gröbner bases for the polynomials on the left-hand sides of system (4), (19), we will get the polynomial that depends only on one variable a_{32} in the form

$$P_{h2}(a_{32}) = P_5(a_{32})P_6(a_{32}) = 0, \qquad (20)$$

where

$$
\begin{aligned}
P_5(a_{32}) &= a_{32}(a_{32}^2 - 1), \\
P_6(a_{32}) &= p_{60}a_{32}^4 + p_{61}a_{32}^2 + p_{62}, \\
p_{60} &= 9\nu^2 p_{63}^2, \\
p_{61} &= p_{63}(h_1^6 + \nu((2 + \nu)h_3^2 - 6(1 - \nu))h_1^4 \\
&\quad + \nu^2((2\nu + 1)h_3^4 - 18(1 - \nu)^2 h_3^2 + 9(1 - \nu)^2)h_1^2 + \nu^4 h_3^2(h_3^2 + 3(1 - \nu))^2), \\
p_{62} &= -h_1^2 h_3^2(h_1^6 + \nu((2 + \nu))h_3^2 + 5\nu - 6)h_1^4 \\
&\quad + \nu^2((2\nu + 1)h_3^4 + 2(9\nu - 5\nu^2 - 5)h_3^2 + 3(\nu^2 - 4\nu + 3))h_1^2 \\
&\quad + \nu^4(h_3^2 - 1)(h_3^2 + 3(1 - \nu))^2), \\
p_{63} &= h_1^4 + 2\nu(h_3^2 - 3(1 - \nu))h_1^2 + \nu^2(h_3^2 + 3(1 - \nu))^2.
\end{aligned}
$$

It is necessary to consider three cases $a_{32} = 0$, $a_{32}^2 = 1$, $P_6(a_{32}) = 0$, to investigate the system (4), (19). In the first case when $a_{32} = 0$, using the approach described in Subsect. 4.1, it is possible to obtain the bifurcation curve

$$h_1^{2/3} + h_3^{2/3} = (3(1 - \nu))^{2/3},$$

which separates the plane (h_1, h_3) into two regions with eight and four equilibrium orientations of the satellite. In the second case when $a_{32}^2 = 1$, the bifurcation curve separating the region of existence of eight solutions from the region of existence of four solutions is also the astroid

$$h_1^{2/3} + h_3^{2/3} = (1 - \nu)^{2/3}.$$

Another two curves separating the regions with an equal number of equilibria can be obtained from the conditions of existence of real roots of the biquadratic equation $P_6(a_{32}) = 0$.

The evolution of domains with a fixed number of equilibrium orientations in the plane of two parameters (h_1, h_3) is also very similar to the first case. In [6], the sufficient conditions for stability of the equilibrium orientations for the last case $h_1 \neq 0$, $h_2 = 0$, and $h_3 \neq 0$ are obtained using the Lyapunov theorem.

Conditions $a_{31} = 0$, $a_{31}^2 = 1$; $a_{32} = 0$, $a_{32}^2 = 1$ and $a_{33} = 0$, $a_{33}^2 = 1$ define all flat solutions of the problem.

5 Conclusion

In this work, the attitude motion of the satellite under the action of gravitational and aerodynamic torques in a circular orbit has been investigated. The main attention was given to determination of the satellite equilibrium orientation in the orbital reference frame and to the analysis of their evolutions in three cases when the center of pressure of aerodynamic forces is located in one of the principal central planes of inertia of the satellite Oxy, Oxz, and Oyz.

The symbolic method of determination of all the satellite equilibria is suggested in the cases when $h_1 = 0$, or $h_2 = 0$, or $h_3 = 0$. The Computer algebra system Maple is applied to reduce the satellite stationary motion system of nine algebraic equations with nine variables to a single algebraic equation with one variable, using the algorithm for the construction of Gröbner basis.

These results permit us to describe the change of the number of equilibrium orientations of the satellite as a function of the parameters h_i and ν. When the aerodynamic torque is small enough, there exist 24 equilibria; when it is large enough, there are 8. The evolution of domains with a fixed number of equilibrium orientations was investigated both analytically and numerically in the plane of two parameters (h_1, h_2) at different values of parameter ν. All bifurcation values of the system parameters corresponding to the qualitative change of domains with fixed number of equilibria were determined. The existence of flat solutions of the problem was specified. The results of the study can be used at the stage of preliminary design of the satellite with aerodynamic control system.

References

1. Sarychev, V.A.: Problems of orientation of satellites, Itogi Nauki i Tekhniki. Ser. "Space Research", vol. 11. VINITI, Moscow (1978)
2. Gutnik, S.A.: Symbolic-numeric investigation of the aerodynamic forces influence on satellite dynamics. In: Gerdt, V.P., Koepf, W., Mayr, E.W., Vorozhtsov, E.V. (eds.) CASC 2011. LNCS, vol. 6885, pp. 192–199. Springer, Heidelberg (2011)
3. Sarychev, V.A., Gutnik, S.A.: Dynamics of a satellite subject to gravitational and aerodynamic torques. Investigation of equilibrium positions. Cosm. Res. **53**, 449–457 (2015)
4. Sarychev, V.A., Mirer, S.A.: Relative equilibria of a satellite subjected to gravitational and aerodynamic torques. Celest. Mech. Dyn. Astron. **76**(1), 55–68 (2000)

5. Sarychev, V.A., Mirer, S.A., Degtyarev, A.A., Duarte, E.K.: Investigation of equilibria of a satellite subjected to gravitational and aerodynamic torques. Celest. Mech. Dyn. Astron. **97**, 267–287 (2007)
6. Sarychev, V.A., Mirer, S.A., Degtyarev, A.A.: Equilibria of a satellite subjected to gravitational and aerodynamic torques with pressure center in a principal plane of inertia. Celest. Mech. Dyn. Astron. **100**, 301–318 (2008)
7. Sarychev, V.A., Gutnik, S.A.: Dynamics of an axisymmetric satellite under the action of gravitational and aerodynamic torques. Cosm. Res. **50**, 367–375 (2012)
8. Buchberger, B.: A theoretical basis for the reduction of polynomials to canonical forms. SIGSAM Bull. **10**(3), 19–29 (1976)
9. Char, B.W., Geddes, K.O., Gonnet, G.H., Monagan, M.B., Watt, S.M.: Maple Reference Manual. Watcom Publications Limited, Waterloo (1992)

Computer Algebra in High-Energy Physics (*Invited Talk*)

Thomas Hahn[✉]

Max Planck Institute for Physics, Föhringer Ring 6, 80805 Munich, Germany
hahn@feynarts.de

Abstract. Paper and pencil are no longer sufficient to obtain the predictions mandated by modern colliders. This is due to the required precision, the number of final-state particles, and also the number of particles in the model. More than any other collider, the Large Hadron Collider (LHC) at CERN has to rely on precise theoretical predictions to even look in the right place, let alone test measurements at a quantitative level. The methods of perturbative quantum field theory, Feynman diagrams, have not changed much over time, and their application remains a formidable, though fully algorithmic, calculational problem. This contribution focusses on how Computer Algebra plays an essential role in this programme and shows by a few examples how the methods are actually implemented in a Computer Algebra system.

1 Introduction

Elementary particles and their interactions are mathematically described by quantum-field-theoretical models given in the form of a Lagrangian density $\mathcal{L}$, usually made up of monomials in the quantum fields where bilinear terms describe propagation and multilinear terms describe interactions. The quantum fields are evaluated at the same point in space-time, hence the interactions are pointlike and automatically satisfy the postulates of special relativity, i.e. in the field-theoretical picture two matter particles interact by exchanging a force particle, which propagates with at most light speed like all the other particles.

What is measured in collider experiments, on the other hand, are quantities like the cross-section σ of a scattering reaction, which is the (suitably normalized) probability to find a given particle in a certain part of the detector. The connection between the Lagrangian and the cross-section is made through the scattering operator S, where the differential cross-section for the reaction IN → OUT is proportional to the squared matrix element $|\mathcal{M}|^2$, $\mathcal{M} = \langle \text{OUT}| S |\text{IN}\rangle$.

The S-matrix element $\mathcal{M}$ is typically computed in perturbation theory with the help of Feynman diagrams. For example, the following diagram contributes to the cross-section of $e^+e^- \to t\bar{t}$ (top-pair production at an e^+–e^- collider):

V.P. Gerdt et al. (Eds.): CASC 2016, LNCS 9890, pp. 255–275, 2016.
DOI: 10.1007/978-3-319-45641-6_17

The diagram is read from left to right: an electron–positron pair annihilates into a photon which subsequently decays into the top–anti-top pair. But Feynman diagrams not only display the physics intuitively, they can be translated into formulas, the Feynman amplitudes, according to rules determined by the Lagrangian $\mathcal{L}$. Each of the dots (•) represents a coupling between the fermions and the photon. This coupling is given by the fermion's charge times $\sqrt{\alpha}$, where $\alpha \simeq 1/137$ is the fine-structure constant. The diagram above is hence of order α, which is the lowest order.

In the next higher order many more diagrams contribute (314 in the Standard Model, to be precise) of which only three are shown here:

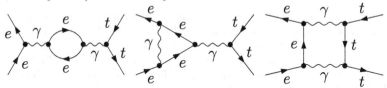

Each diagram contains four dots now and is hence of order α^2. At the same time each diagram contains a closed loop, and this is no coincidence: the expansion in α is at the same time an expansion in the number of loops. Each loop order enhances the accuracy of the result but also contributes significantly to the difficulty of the computation.

2   Particle-Physics Models

The Lagrangians considered in current high-energy physics are constructed on symmetry principles. One posits that under special 'field rotations', the gauge transformations, the Lagrangian is invariant. This restricts the terms allowed in the Lagrangian.

The simplest quantum field theory realized in Nature, Quantum Electrodynamics (QED), is symmetric under U(1) rotations and describes the electromagnetic force, while the weak force transforms under SU(2) and the strong force (Quantum Chromodynamics, QCD) under SU(3). These three theories are combined in the Standard Model (SM) with product gauge group SU(3) × SU(2) × U(1). Despite its deceptively simple formula, which even fits on a T-shirt, it embodies the results of more than 50 years of work in experimental and theoretical high-energy physics. Its particle content is summarized in Fig. 1.

Nearly all models in particle physics today are based on the SM, which means they either contain the SM as a subgroup or admit a low-energy limit which reproduces the SM.

Whereas in the 1950s and 1960s many discoveries were made experimentally (unexpected signals showing up in the detector) and only later explained by theory, these days (almost) no discovery is made without theory. This is certainly true for the top quark, discovered 1994 at Fermilab, and the Higgs boson, discovered 2012 at CERN, where extensive theoretical analysis preceded the experiment and pointed experimenters to the right place to look.

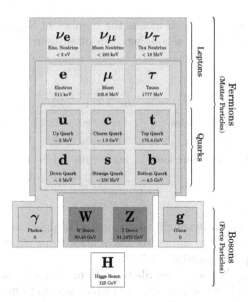

Fig. 1. The Standard Model and its particle content.

For instance, Otto Nachtmann's textbook on particle physics from 1986 [33] contained a footnote stating that the top quark "likely" has a mass of around 40 GeV. This is clear indication that the experiments did not really know where to look. The mass measurements honed in only after LEP had collected enough data on electroweak precision observables, in particular the W-boson mass, which was the most important input parameter for the prediction of the top-quark mass. Such indirect constraints proceed through loop calculations and make heavy use of computer algebra.

The SM is only considered an effective theory, however, as it contains many parameters it does not predict and also it does not unify the fourth force of Nature, gravity. With all SM particles established, we are currently posed for the next big discovery of a particle 'Beyond the Standard Model' (BSM). Needless to say, there is an overabundance of BSM models, and again, phenomenologists are already excluding many of these models or restricting their parameter space e.g. by considering their effects on precision observables through calculations that rely on computer algebra.

3   The Challenges

Calculating Feynman diagrams is difficult for a number of reasons.

Firstly, combinatorics leads to a massive growth of terms both with number of loops and number of external legs. To illustrate this effect the following table lists the number of $2 \to 2$ topologies, i.e. the number of ways of connecting two incoming with two outgoing legs for different numbers of loops. At this stage no physics is involved yet, which means the number of Feynman diagrams will typically be even higher since a topology can often be realized multiply within a model's particles.

# loops	0	1	2	3
# $2 \to 2$ topologies	4	99	2214	50051
Typical accuracy	30%	10%	2%	1%
General procedure known	yes	yes	$1 \to 1$	no
Limits		$2 \to 8$	$2 \to 5$	$1 \to 2$	$1 \to 1$

Researchers have constantly sought to devise better algorithms and there is certainly some progress. For example, recursive algorithms effectively re-use parts of diagrams that appear multiply [1,7,30].

Secondly, solving the loop integrals is far from trivial beyond one-loop order and has become a field by itself, which also heavily profits from computer algebra.

Thirdly there are also difficult numerical issues such as efficient phase-space integration, treatment of unstable particles, or significant numerical cancellations between diagrams, which shall not be discussed in depth here.

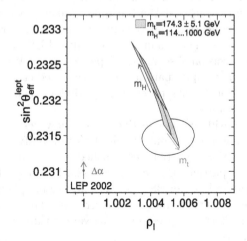

Fig. 2. The LEP measurement of two precision observables (the rho-parameter and effective weak mixing angle) with its one-sigma ellipse (68% confidence level) vs. the theoretical prediction, varied within the then-current bounds on m_t and m_H. Shown in green is the prediction according to pure QED (no weak interaction). (colour figure in online)

If the calculation is so formidable, do we really need higher orders after all? The answer is Yes, for three reasons.

1. **Higher orders are seen experimentally.**

   Figure 2 is a plot from the end of the LEP era which shows the plane in two precision observables. The ellipse indicates the one-sigma range around the measured value and the theoretical prediction in yellow takes into account the then-current ranges of the top-quark and Higgs-boson masses. What makes this plot special is the small green arrow in the lower left corner, however: it indicates the prediction from pure QED alone. That is, a loop calculation can completely rule out the theory without weak interactions.

   In precision observables even higher-loop effects are seen, here for example are prediction and measurement from the anomalous magnetic moment of the muon, a_μ:

$$
\begin{array}{rll}
10^{10}a_\mu = 11614097.29 & \text{QED} & \text{1-loop} \\
41321.76 & & \text{2-loop} \\
3014.19 & & \text{3-loop} \\
36.70 & & \text{4-loop} \\
.63 & & \text{5-loop} \\
690.6 & \text{Had.} & \\
19.5 & \text{EW} & \text{1-loop} \\
-4.3 & & \text{2-loop} \\
\hline
11659176 & \text{theory, total} & \\
11659204 & \text{experiment (BNL)} &
\end{array}
$$

   This calculation is actually a very special case where loop integrals far beyond the known results were computed numerically with brute force [5].

2. **Indirect access to particles beyond the direct kinematic reach.**

   Models may contain particles too heavy to be produced directly at available collider energies. They nevertheless contribute to available observables through loop effects. While this usually precludes a discovery, limits obtained from such indirect sources can be used to limit the model's parameter space and focus experimental searches.

   Both the top quark and the Higgs boson were discovered only after indirect analyses based on loop calculations had narrowed down the search range.

3. **Loop-mediated effects.**

   For a lot of interesting physics there exists no tree level at all, i.e. the lowest-order contribution is from a one-loop diagram.

   This happens to be the case for the main Higgs production mode at the LHC, $gg \to H$, which is mediated by a top loop since the (color-neutral) Higgs boson does not directly couple to the gluon. In a particular approximation, the infinite top-mass limit, an effective theory can be constructed where the top loop is shrunk to a single vertex, and in this theory there is then a tree-level result.

   Also most reactions considered in flavor physics start at loop level. This is intentionally so, for the idea is precisely to eliminate a 'large' SM contribution that would screen subtle BSM effects in the flavor sector.

The present state of the art of perturbative calculations in particle physics is visualized in Fig. 3. The progress in this chart is unfortunately both slow and not uniform.

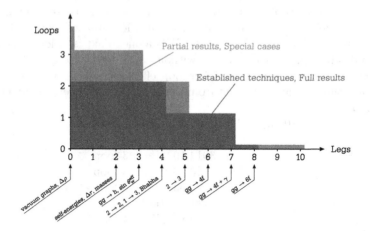

Fig. 3. The present reach of perturbative calculations in high-energy physics. A third axis, for the number of mass scales in the calculation (which likewise affects computational difficulty significantly), has been left out for visual clarity but would show similar behavior as the other two axes.

4 Loop Integrals

Feynman diagrams are usually evaluated in momentum space. Momentum conservation fixes all momenta flowing through the 'tree parts' of a Feynman diagram but for each loop one momentum is unconstrained which, in the quantum-mechanical sense of summing over all possibilities, has to be integrated over:

$$= \int \frac{\mathrm{d}^4 q}{(q^2 - m_1^2)\left((p+q)^2 - m_2^2\right)}. \tag{1}$$

Loop integrals can be divergent both in the ultraviolet (UV), which means they diverge due to the upper integration bound, for example

$$= \int \frac{\mathrm{d}^4 q}{q^2 \, q^2} = \int \mathrm{d}\Omega \int^\Lambda \frac{q^3 \mathrm{d}q}{q^4} \propto \log \Lambda, \tag{2}$$

and in the infrared (IR), which means they diverge due to the lower integration bound, for example

$$
\int \frac{\mathrm{d}^4 q}{q^2 \left((q+k_1)^2 - m_1^2\right)\left((q-k_2)^2 - m_2^2\right)}
$$

on-shell:
$$k_i^2 = m_i^2$$

$$
= \int_\lambda \frac{q^3 \mathrm{d}q\, \mathrm{d}\Omega}{q^2 \left(2q \cdot k_1\right)\left(-2q \cdot k_2\right)} \propto \log \lambda.
$$
(3)

How can finite and sensible results be obtained in the presence of divergent Feynman integrals?

- In the first place the integrals have to be regularized, i.e. a regularization parameter is introduced which renders the integral finite and reproduces the divergence in a certain limit. Various regularization schemes are on the market, but by far the most popular one is dimensional regularization [10] – it introduces the fewest extra terms and is compatible with the most symmetries. In dimensional regularization the integrals are evaluated formally in D instead of the original 4 dimensions, and the divergence then shows up as a pole in $D - 4$.
- The UV divergences can be absorbed by renormalizing the theory, i.e. through redefinitions of the model parameters. The parameters appearing in the Lagrangian, masses m_0 and couplings g_0, have no physical significance beyond tree level. Renormalization means to express observables computed in terms of m_0 and g_0 by physical ones, m and g. In a renormalizable model the UV divergences can (always!) be absorbed in the relations between m_0, g_0 and m, g. Formally one substitutes $m_0 \to m + \delta m$, $g_0 \to g + \delta g$ in the Lagrangian which leads to new vertices containing δm and δg, and hence to additional Feynman diagrams, the so-called counter-term diagrams. There are graphical correspondences, for example the UV divergences arising from the left diagram (a correction to the mass) are cancelled by the right diagram, where the loop has been shrunk to a counter-term vertex:

- IR divergences have a different origin: since the photon (or gluon) is massless, it can carry away an arbitrarily small amount of energy, hence an N-particle reaction with charged external lines (electric or color) is indistinguishable experimentally from the corresponding $(N+1)$-particle reaction with an additional soft photon (or gluon) emitted. Also here some combinatorics is at work, for example the IR divergences cancel between the diagram on the left and the two on the right, obtained by cutting the photon line:

For practical calculations the following points are relevant:

- The (scalar) one-loop integrals are known and available in several public implementations [12,16,27,28,34].
- Beyond one-loop only special cases are exactly known (e.g. [6,32]), or expansions are made [29]. There exist various general strategies to attack the integration [38], and many of them make heavy use of computer algebra, yet there exists no algorithmic procedure.
- Loop integrals can be evaluated numerically, too, after the divergences have been extracted e.g. by sector decomposition [11]. Computation times are easily a factor 1000 higher then, however.
- The numerator of a loop integral can also contain powers of the loop momentum, such integrals are known as tensor integrals. Algorithms exist to reduce the tensor integrals, i.e. write them as linear combinations of scalar integrals [31]. The one-loop procedure is known for a long time [36] but suffers from a significant increase in number of terms with increasing number of external legs. Recent improvements have mostly targeted this increase [35].
- The "D-dimensional" computation of the amplitude differs from the 4-dim. one only by few explicit substitutions, for example the self-contracted metric tensor becomes $g_\mu^\mu = D$. Since the regulator will ultimately be removed again ($D \to 4$), most of these D's are harmless, except if they multiply a divergent loop integral which contains a $1/(D-4)$ pole. Schematically:

$$D \cdot (\text{div. integral}) = 4 \cdot (\text{div. integral}) + \text{extra term}. \tag{4}$$

This substitution essentially required all loop calculations to be done analytically, i.e. the automated ones by computer algebra. Only rather recently have the extra terms generated in this way been re-cast, at one-loop order, in the form of counter-terms, so that they can be added by computing extra Feynman diagrams [18], and hence it is now possible to also do one-loop calculations (semi)numerically.

5   The Recipe

We shall now give a methodological survey of the steps one would manually perform to compute a cross-section from Feynman diagrams for the example of the $e^+e^- \to \bar{t}t$ reaction at tree level considered before. It shall serve as a template for the computer-based implementation later.

1. Draw all topologically inequivalent diagrams with the desired number of external legs and loops:

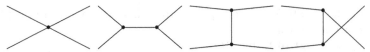

This is a **graph-theoretical step** with no physics input yet.

2. Figure out what particles can run on each topology:

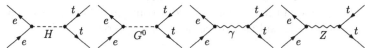

This is a **combinatorial step** and requires physics input, i.e. one needs to know which particles exist in the model and which interactions between them are possible.

Observe that of the four topologies above only one could be realized due to (electric and color) charge conservation, but this remaining one gives rise to four Feynman diagrams.

Also certain approximations are usually carried out in this step. For example, the coupling of fermions to scalars (which can easily be identified graphically) is suppressed by a factor m_f/M_W which, for the electron, works out to 10^{-7}. It is thus a good approximation to leave out the left two diagrams already at this relatively early stage of the computation.

3. Translate the diagrams into formulas by applying the Feynman rules:

$$
= \underbrace{\langle v_1|\, i e \gamma^\mu\, |u_2\rangle}_{\text{left vertex}} \underbrace{\frac{g_{\mu\nu}}{(k_1 + k_2)^2}}_{\text{propagator}} \underbrace{\langle u_4|\left(-\tfrac{2}{3} i e \gamma^\nu\right)|v_3\rangle}_{\text{right vertex}}.
$$

(5)

This is an (arguably unusual) database look-up.

4. Contract the indices, take the traces, introduce invariants, etc.:

$$
= \frac{8\pi\alpha}{3s} F_1\,, \qquad F_1 = \langle v_1|\, \gamma_\mu\, |u_2\rangle \langle u_4|\, \gamma^\mu\, |u_3\rangle\,.
$$

(6)

There are several ways of computing fermionic matrix elements like the F_1 appearing here. The textbook approach is to square the amplitude and then take the trace over Dirac matrices:

$$
|F_1|^2 = \mathrm{Tr}\left\{(\slashed{k}_1 - m_e)\gamma_\mu(\slashed{k}_2 + m_e)\gamma_\nu\right\} \mathrm{Tr}\left\{(\slashed{k}_4 + m_t)\gamma^\mu(\slashed{k}_3 - m_t)\gamma^\nu\right\}
$$
$$
= \tfrac{1}{2}s^2 + st + (m_e^2 + m_t^2 - t)^2.
$$

(7)

This step of **algebraic simplification** is not really necessary at tree-level, and many programs indeed implement it numerically (e.g. [4,21]) to avoid having to use a computer algebra system. With the help of extra Feynman diagrams one can meanwhile even work (semi)numerically at one loop [18], see Sect. 4. Beyond one-loop there is presently no purely numerical alternative, however.

5. Write the results up as a computer program, debug and run that program to produce numerical values.
This is a **programming step**.

The Nobel Prize for physics 1999 to Gerard 't Hooft and Martinus Veltman was awarded "for having shown how the theory may be used for precise calculations of physical quantities." The procedure outlined above may be complicated, but it is algorithmic and can hence be given to a computer.

6   Implementation

As the Recipe in Sect. 5 has already demonstrated, the computation of Feynman diagrams involves very different tasks. In particular it involves steps for which computer algebra is naturally suited, such as treatment of D-dimensional and tensorial objects, or Dirac traces. An analytical calculation can further take better care of certain cancellations, e.g. due to gauge symmetry which are typical for processes with external weak gauge bosons in the high-energy limit [17].

Computing the Feynman diagrams results in an expression for the squared S-matrix element which finally needs to be integrated over the phase space spanned by the outgoing particles. This integration is multidimensional (e.g. 4-, 7-dimensional) and is done numerically. The integrand is far from a smooth function due to the propagators, $1/(p^2 - m^2)$, which lead to significant 'ridge structures' in the vicinity of $p^2 \approx m^2$. Even with dedicated Monte Carlo algorithms, many points need to be sampled (10^7, say), hence for this purpose one needs a fast integrand function, implemented in some high-level language, e.g. Fortran or C++.

The computational model will thus be: Symbolic manipulation for the structural and algebraic operations. Compiled high-level language for the numerical evaluation.

Automated computation of radiative corrections, at least at one-loop order, has become an industry today due to the precision demands of the LHC. Many packages have become available only in the last few years [3,8,9,13,15,19], most of which are geared explicitly towards LHC calculations.

In the following we will have a closer look at two (by this standard) 'old' packages, FeynArts [22] and FormCalc [27], together with the loop-integral library LoopTools, which date from the mid-1990s. They implement the 'traditional' Feynman-diagrammatic method (no recursions etc.) and perform an analytic calculation as far as possible, for "any" model. The user can either inspect/modify the analytic results in Mathematica or proceed to generate a Fortran or C code for the numerical evaluation of the squared matrix element. Figure 4 gives a 'flow-chart' of the evaluation.

7   Generating Feynman Diagrams with FeynArts

FeynArts is a Mathematica package for the generation and visualization of Feynman diagrams and amplitudes. The generation of amplitudes is a three-step process which corresponds closely to Steps 1, 2, and 3 of the Recipe in Sect. 5.

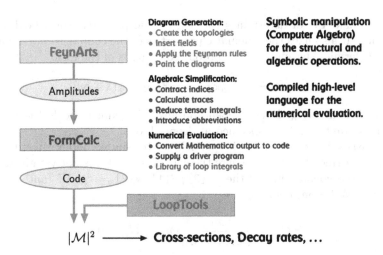

Fig. 4. The evaluation of Feynman diagrams with FeynArts, FormCalc, and LoopTools.

In the first step, the distinct topologies for a given number of loops and external legs are produced, e.g.

```
top = CreateTopologies[1, 1 → 2]
```

The internal algorithm starts from predefined zero-leg 'starting' topologies of the requested loop order and successively adds legs:

This is not an entirely self-sufficient procedure, but the starting topologies also carry some meta-information which would be difficult to obtain from scratch and are currently available up to three loops.

In the second step, the physics information is read from a Model File which contains the particle content and the coupling list, and the fields are distributed over the topologies in all admissible ways, e.g.

```
ins = InsertFields[top, F[4,{3}] → {F[4,{2}], V[1]}]
```

The field labelling here is the one of the default model, SM.mod and corresponds to the decay $b \to s\gamma$, where b and s are the third and second members of the down-type quark class F[4], and V[1] is the photon. This notation is part of the more general concept of levels of fields:

– The **Generic Level** determines the space–time properties of a field, e.g. a fermion F. It also fixes the kinematic properties of the couplings. For example, the FFV (fermion–fermion–vector boson) coupling is of the form

$$G_L \gamma_\mu \mathbb{P}_L + G_R \gamma_\mu \mathbb{P}_R, \qquad \mathbb{P}_{L,R} = \tfrac{1}{2}(\mathbb{1} \mp \gamma_5). \tag{8}$$

The coefficients $G_{L,R}$ do not depend on any other kinematical objects (only model constants) and are specified at deeper levels.

– The **Classes Level** specifies the particle up to 'simple' index substitutions, e.g. the neutrino class F[1] (where the generation index is not yet given). For example, the coefficients for the -F[2], F[1], V[3] $= \bar{\ell}_i \nu_j W$ (anti-lepton$_i$–neutrino$_j$–W-boson) coupling are

$$G_L = \frac{iem_{\ell,i}}{\sqrt{2}\sin\theta_w M_W}\delta_{ij}, \quad G_R = 0. \tag{9}$$

This particular coupling was chosen for demonstration because, if neutrino masses are neglected, the neutrino does not couple to right-handed fermions, hence $G_R = 0$ provides an easy sanity check. Observe that i and j, the generation indices of the two fermions, are not specified yet.

– The **Particles Level** spells out any indices left unspecified, e.g. the electron F[2,{1}], being the first member of the lepton class F[2].

The reason for this splitting is mainly economy: kinematic simplifications can be performed at Generic Level, where there are typically much fewer diagrams than at lower levels. Likewise, 'trivial' sums e.g. over fermion generations need not be written out explicitly in terms of Particles-Level Feynman diagrams.

The diagrams returned by **CreateTopologies** and **InsertFields** can be drawn with **Paint**, with output as Mathematica Graphics object, PostScript, or LATEX. LATEX code produced by **Paint** can be post-processed (e.g. 'touched up' for publication) with the FeynEdit editor [25].

Finally, the Feynman rules are applied with

```
amp = CreateFeynAmp[ins]
```

The output contains Feynman diagrams represented by the Mathematica function **FeynAmp**. For illustration, consider the diagram

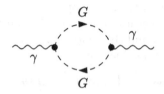

This diagram is given by **FeynAmp**[*id, loopmom, genamp, ins*], where

– *id* is an identifier for bookkeeping, e.g. GraphID[Topology == 1, Generic == 1],

- *loopmom* identifies the loop momenta in the form `Integral[q1]`,
- *genamp* is the generic amplitude,

$\dfrac{\text{I}}{\text{32 Pi}^4}$ `RelativeCF` ①

`FeynAmpDenominator[`$\dfrac{1}{\text{q1}^2 - \text{Mass[S[Gen3]]}^2}$, $\dfrac{1}{(-\text{p1} + \text{q1})^2 - \text{Mass[S[Gen4]]}^2}$`]` ②

`(p1 − 2 q1)[Lor1]  (−p1 + 2 q1)[Lor2]` ③

`ep[V[1],p1,Lor1]  ep`*`[V[1],k1,Lor2]` ④

$\text{G}_{\text{SSV}}^{(0)}$`[(Mom[1] − Mom[2])[KI1[3]]]` $\text{G}_{\text{SSV}}^{(0)}$`[(Mom[1] − Mom[2])[KI1[3]]]` ⑤

where individual items can easily be identified: prefactor ①, loop denominators ②, kinematic coupling structure ③, polarization vectors ④, coupling constants ⑤.

- *ins* is a list of rules substituting the unspecified items in the generic amplitude,

`{ Mass[S[Gen3]], Mass[S[Gen4]],`
  $\text{G}_{\text{SSV}}^{(0)}$`[(Mom[1] − Mom[2])[KI1[3]]],`
  $\text{G}_{\text{SSV}}^{(0)}$`[(Mom[1] − Mom[2])[KI1[3]]], RelativeCF }` →
`Insertions[Classes][{MW, MW, I EL, − I EL, 2}]`

7.1 Model Files

Model Files are the 'software within the software' and provide FeynArts with the necessary physics input. Technically they are ordinary Mathematica text files loaded by FeynArts during model initialization which provide two main objects: the list of particles, `M$ClassesDescription`, and the list of couplings, `M$CouplingMatrices`.

The Model Files come in two kinds corresponding to the Generic and Classes level of fields. For example, the FFV coupling of Eq. (8) can be written in the form

$$C(F, F, V) = \begin{pmatrix} G_L \\ G_R \end{pmatrix} \cdot \begin{pmatrix} \gamma_\mu \mathbb{P}_L \\ \gamma_\mu \mathbb{P}_R \end{pmatrix} \equiv \boldsymbol{G} \cdot \boldsymbol{K}. \tag{10}$$

The kinematic vector $\boldsymbol{K}$ is stored in the **Generic Model File** (`.gen`) and the coupling vector $\boldsymbol{G}$ in the **Classes Model File** (`.mod`). The Generic Model File changes only if the user wants to work in a different representation of the Poincaré group (i.e. rarely), whereas the Classes Model File is different for every model considered.

The generic FFV coupling is stored as

```
AnalyticalCoupling[s1 F[j1,p1], s2 F[j2,p2], s3 V[j3,p3,{li3}]]
== G[1][s1 F[j1], s2 F[j2], s3 S[j3]] .
      { NonCommutative[DiracMatrix[li3], ChiralityProjector[-1]],
        NonCommutative[DiracMatrix[li3], ChiralityProjector[+1]] }
```

where the dot product is explicit and the coupling vector for the $\bar{\ell}_i \nu_j W$ coupling of Eq. (9) is stored as

```
C[ -F[2,{i}], F[1,{j}], V[3] ]
== { {I EL Mass[F[2,{i}]]/(Sqrt[2] SW MW) IndexDelta[i, j]},
     {0} }
```

Only `Mass` and `IndexDelta` are FeynArts functions here, all other symbols have been chosen by the Model File's creator. Observe that the r.h.s. is a list of lists, hence the name `M$CouplingMatrices` is indeed appropriate: the inner lists may also contain higher-order (i.e. counter-term) vertices.

Model Files for FeynArts can currently be generated by FeynRules [2] and LanHEP [37]. The SARAH package [39] is useful for the high-level derivation of SUSY models. FeynArts itself includes the ModelMaker tool which turns a suitably defined Lagrangian into a Model File.

8   FormCalc

The output of FeynArts (see p. 261) is not in a good shape for immediate numerical evaluation. It contains uncontracted indices, unregularized loop integrals, fermion traces, SU(N) generators, etc. The symbolic expressions for the diagrams are thus first simplified algebraically with FormCalc, which performs the following steps: indices are contracted, fermion traces evaluated, open fermion chains simplified, color structures standardized, tensor integrals reduced, abbreviations introduced.

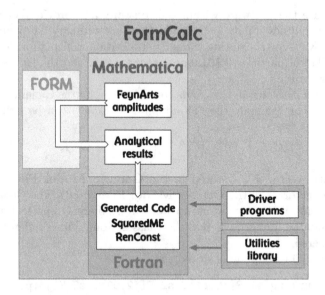

Fig. 5. The control flow in FormCalc.

Most of these steps are internally executed in FORM [40], a computer-algebra system whose instruction set has many adaptations especially useful in high-energy physics, see Sect. 9. The interfacing with FORM is transparent to the user, i.e. the user does not have to work with the FORM code. FormCalc thus combines the speed of FORM with the powerful instruction set of Mathematica and the latter greatly facilitates further processing of the results. Figure 5 shows the control flow in FormCalc.

The main function is `CalcFeynAmp` which is applied to a FeynArts amplitude (the output of `CreateFeynAmp`) and combines the steps outlined above. Its output is in general a linear combination of loop integrals with prefactors that contain model parameters, kinematic variables, and abbreviations introduced by FormCalc, e.g.

```
C0i[cc0, MW2, MW2, S, MW2, MZ2, MW2] *
  ( -4 Alfa^2 CW2 MW2/SW2 S AbbSum16 +
    32 Alfa^2 CW2/SW2 S² AbbSum28 +
    4 Alfa^2 CW2/SW2 S² AbbSum30 -
    8 Alfa^2 CW2/SW2 S² AbbSum7 +
    Alfa^2 CW2/SW2 S (T-U) Abb1 +
    8 Alfa^2 CW2/SW2 S (T-U) AbbSum29 )
```

The first line represents the one-loop integral $C_0(M_W^2, M_W^2, s, M_W^2, M_Z^2, M_W^2)$, multiplied with a linear combination of abbreviations like `Abb1` or `AbbSum29` with coefficients containing kinematical invariants like the Mandelstam variables S, T, and U, and model parameters such as the fine-structure constant `Alfa` = α.

8.1 Abbreviations

The automated introduction of abbreviations is a key concept in FormCalc. It is crucial in rendering an amplitude as compact as possible. The main effect comes from three layers of recursively defined abbreviations, introduced when the amplitude is read back from FORM, i.e. during `CalcFeynAmp`. For example:

$$\text{AbbSum29} = \text{Abb2} + \boxed{\text{Abb22}} + \text{Abb23} + \text{Abb3}$$

$$\boxed{\text{Abb22} = \text{Pair1}\,\text{Pair3}\,\text{Pair6}}$$

$$\boxed{\text{Pair3} = \text{Pair[e[3], k[1]]}}$$

Written out, this abbreviation is equivalent to

```
Pair[e[1],e[2]] Pair[e[3],k[1]] Pair[e[4],k[1]] +
Pair[e[1],e[2]] Pair[e[3],k[2]] Pair[e[4],k[1]] +
Pair[e[1],e[2]] Pair[e[3],k[1]] Pair[e[4],k[2]] +
Pair[e[1],e[2]] Pair[e[3],k[2]] Pair[e[4],k[2]]
```

Singled out for abbreviation are objects of a particular kind, e.g. all kinematic objects such as the dot products (`Pair`) here, and also subexpressions occurring multiply, which amounts to a sort of common subexpression elimination. The abbreviationing is also a beautiful example of an extremely efficient handshake between FORM and Mathematica: FORM singles out the subexpressions and Mathematica introduces abbreviations for them.

For numerical evaluation, the abbreviations are grouped into categories:

1. Abbreviations that depend on the helicities.
2. Abbreviations that depend on angular variables.
3. Abbreviations that depend only on $\sqrt{s}$.

Correct execution of the different categories guarantees that almost no redundant evaluations are made, e.g. in a $2 \to 2$ process with external unpolarized fermions, statements in the innermost loop over the helicities are executed 2^4 times as often as those in the loop over the angle. This technique of moving invariant expressions out of the loop is known as 'hoisting' in computer science and makes the generated code execute essentially as fast as hand-tuned code.

8.2 Code Generation

Numerical evaluation of the FormCalc results is done in Fortran or C99, firstly for speed, and secondly for ease of inclusion into other programs.

Code generation for the squared amplitude is a highly automated procedure shown in Fig. 6. FormCalc has two fairly advanced functions for code generation, `WriteSquaredME` and `WriteRenConst`, and also offers low-level code generation functions with which it is very easy to turn an arbitrary Mathematica expression into code. The philosophy is that the user should not have to modify the generated code. This means that the code has to be encapsulated (i.e. no loose ends the user has to bother with), and that all necessary subsidiary files (include files, makefile) have to be produced, too.

One can also 'come full circle' and, with only minor changes in the setup, turn the FormCalc-generated program into a MathLink executable [23] so that the cross-section becomes available as a Mathematica function of its model parameters. This is particularly useful for performing e.g. contour plots over parameter space.

9 FORM

The open-source computer-algebra system FORM [40] is one of the major workhorses for symbolic manipulation in high-energy physics. Since it is maybe not so well-known outside of the field, a short introduction is in order here.

FORM is a non-interactive program (edit–run cycle) with a strongly typed language, i.e. variables must be declared as symbols, vectors, indices, tensors, functions, etc. Its main strength is on polynomial expressions, where it can handle huge expressions, many times the amount of RAM available. The following trivial program shall suffice for a look and feel:

```
Symbols a, b, c, d;
Local expr = (a + b)^2;
```

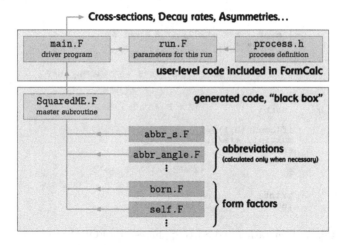

Fig. 6. Numerical evaluation of a Feynman amplitude in FormCalc: the `SquaredME` subroutine is fully generated, for the user-level code templates are inserted, to be modified as required for the application.

```
id b = c - d;
print;
.end
```

A FORM program is organized in modules, each terminated by a "dot" statement (`.sort`, `.store`, `.end`, etc.). Figure 7 shows how FORM works through a program:

– In the Generation Phase ("normal statements") terms are only generated. All operations in this phase are local, i.e. operate only on one term at a time.
– In the Sorting Phase ("dot statements") the generated terms are inspected and similar terms collected. This is the only 'global' operation which requires FORM to look at all terms 'simultaneously.'

This distinction gives the user strategic control over potentially time-consuming global operations. It also allows FORM to concurrently execute the Generation Phase over the terms on the available cores, which results in a degree of parallelization not matched by other computer-algebra systems.

10 Validation

Whenever a procedure is automated, i.e. humans become 'decoupled' from the actual computation, methods are needed to validate the results.

For Feynman amplitudes there are various things one can test:

– **Cancellation of divergences**
The divergences from loop integrals must cancel in the final result. For the UV

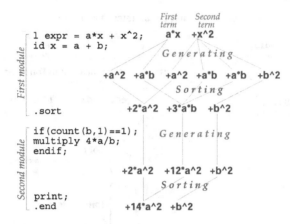

Fig. 7. Generating and Sorting phases in a FORM program.

divergences this is relatively straightforward and can even be done analytically. For IR divergences it is more complicated due to different phase spaces and can usually only be achieved numerically.

– **Gauge invariance**

Gauge theories like the SM must be invariant under gauge transformations. A pretty strong check is to compute the amplitude in an arbitrary gauge and check that the gauge-parameter dependence cancels. The standard R_ξ gauges unfortunately lead to higher-rank tensor integrals and complicate the computation a lot but one can use non-linear gauges which have been specially engineered to avoid this effect [20]. The background-field method [17] amounts to a different gauge choice, too, and constitutes a useful check for processes involving external weak gauge bosons.

– **Special limits**

Certain limits of the calculation may be known from theoretical considerations. For example, the high-energy behavior of reactions involving external weak gauge bosons is known from the Equivalence Theorem [14] to be identical to the one where the vector bosons have been substituted by their Goldstone partners.

– **Comparison with independent calculations**

Comparing with other calculations is of course the "gold-plated mode" but requires a lot of work. Standard software packages gain in reputation from passing such tests and can in turn be used as reliable cross-checks. The availability of several independent packages for essentially the same task is not a luxury but testifies to the difficulty of these calculations with plenty of error sources.

11   Extensions

FeynArts and FormCalc have originally been designed, as many other software packages, too, to do a 'complete' job. That is, all the steps from the generation of

Feynman diagrams to the numerical computation of a cross-section are executed from a single control program (e.g. a single Mathematica session) – that at least is how the demo programs insinuate usage. Such a 'monolithic' approach becomes problematic when wanting to use the programs beyond their designed scope, for example beyond one loop.

In a recent proof-of-concept implementation for a particular class of two-loop Higgs-mass corrections [26], a suite of scripts was devised to reorganize this so-far monolithic procedure: The calculation was compartmentalized into several steps, each implemented as an independent shell script and invoked from the command line. The Mathematica Kernel is run through the shell's 'here documents' [24], hence the script can be sent in the background or to some compute cluster. The suite is coordinated through a makefile, so that only the out-of-date parts of the calculation are redone. In lieu of 'in vivo' debugging (e.g. setting breakpoints) detailed logs are kept. For the parts of the calculation not available in FormCalc, e.g. the two-loop tensor reduction, external packages were used. This first attempt already looks very promising to make flexible use of existing programs, even for very specific tasks, while retaining core functionality of the established packages.

12 Conclusions

With the present experimental situation in high-energy physics, where precision is becoming more and more important to distinguish signal from background and model candidates are having many particles and parameters, software engineering has constantly been gaining ground. Theoretical calculations performed with tools developed in recent years have increased the discrimination power of experimental searches at LHC and other colliders and made discoveries such as that of the Higgs boson 2012 possible.

The nature of perturbative calculations with Feynman diagrams make hybrid programming techniques necessary. Computer algebra is an indispensible tool because many calculations must be done symbolically. On the other hand, fast number crunching can only be achieved in a compiled language.

Further development of the tools is important to extend the theoretical reach and match the expected experimental precision of future colliders and discover what lies beyond the Standard Model.

References

1. Actis, S., Denner, A., Hofer, L., Lang, J.-N., Scharf, A., Uccirati, S.: RECOLA: REcursive Computation of One-Loop Amplitudes. arXiv:1605.01090
2. Alloul, A., Christensen, N., Degrande, C., Duhr, C., Fuks, B.: FeynRules 2.0 - a complete toolbox for tree-level phenomenology. Comp. Phys. Comm. **185**, 2250 (2014). arXiv:1310.1921
3. Alwall, J., et al.: The automated computation of tree-level and next-to-leading order differential cross sections, and their matching to parton shower simulations. JHEP **1407**, 079 (2014). arXiv:1405.0301

4. Alwall, J., Herquet, M., Maltoni, F., Mattelaer, O., Stelzer, T.: MadGraph 5: going beyond. JHEP **1106**, 128 (2011). arXiv:1106.0522

5. Aoyama, T., Hayakawa, M., Kinoshita, T., Nio, M.: Complete tenth-order QED contribution to the muon g-2. Phys. Rev. Lett. **109**, 111808 (2012). arXiv:1205.5370

6. Bauberger, S., Böhm, M.: Simple one-dimensional integral representations for two loop selfenergies: the Master diagram. Nucl. Phys. B 445, 25 (1995). arXiv:hep-ph/9501201

7. Berends, F., Giele, W.: On the construction of scattering amplitudes for spinning massless particles. Nucl. Phys. B 507, 157 (1997). arXiv:hep-th/9704008

8. Bern, Z., Dixon, L., Febres Cordero, F., Höche, S., Ita, H., Kosower, D., Maître, D., Ozeren, K.: The BlackHat library for one-loop amplitudes. J. Phys. Conf. Ser. **523**, 012051 (2014). arXiv:1310.2808

9. Bevilacqua, G., Czakon, M., Garzelli, M., van Hameren, A., Kardos, A., Papadopoulos, C., Pittau, R., Worek, M.: Helac-NLO. Comp. Phys. Comm. **184**, 986 (2013). arXiv:1110.1499

10. Bollini, C., Giambiagi, J.J.: Dimensional Renormalization. Il Nuovo Cimento B **12**, 20–26 (1972)

11. Borowka, S., Heinrich, G., Jones, S., Kerner, M., Schlenk, J., Zirke, T.: SecDec-3.0: numerical evaluation of multi-scale integrals beyond one loop. Comp. Phys. Comm. 196, 470 (2015). arXiv:1502.06595

12. Carrazza, S., Ellis, R.K., Zanderighi, G.: QCDLoop: a comprehensive framework for one-loop scalar integrals. arXiv:1605.03181

13. Cascioli, F., Höche, S., Krauss, F., Moretti, N., Pozzorini, S., Schönherr, M., Siegert, F., Maierhöfer, P.: Next-to-leading order simulations with Sherpa+OpenLoops. PoS LL **2014**, 022 (2014)

14. Cornwall, J., Levin, D., Tiktopoulos, G.: Derivation of gauge invariance from high-energy unitarity bounds on the s matrix. Phys. Rev. D **10**, 1145 (1974)

15. Cullen, G., et al.: GoSam-2.0: a tool for automated one-loop calculations within the Standard Model and beyond. Eur. Phys. J. C **74**(8), 3001 (2014). arXiv:1404.7096

16. Denner, A., Dittmaier, S., Hofer, L.: Collier: a Fortran-based Complex One-Loop LIbrary in Extended Regularizations. arXiv:1604.06792

17. Denner, A., Weiglein, G., Dittmaier, S.: Application of the background field method to the electroweak standard model. Nucl. Phys. B 440, 95 (1995). arXiv:hep-ph/9410338

18. Draggiotis, P., Garzelli, M., Papadopoulos, C., Pittau, R.: Feynman rules for the rational part of the QCD 1-loop amplitudes, JHEP 0904, 072 (2010). arXiv:0903.0356. Garzelli, M., Malamos, I., Pittau, R.: Feynman rules for the rational part of the Electroweak 1-loop amplitudes. JHEP 1001, 040 (2009). arXiv:0910.3130

19. Ellis, R.K., Giele, W., Kunszt, Z., Melnikov, K., Zanderighi, G.: One-loop amplitudes for W^+ 3 jet production in hadron collisions. JHEP **0901**, 012 (2009). arXiv:0810.2762

20. Gajdosik, T., Pasukonis, J.: Non Linear Gauge Fixing for FeynArts. arXiv:0710.1999

21. Gleisberg, T., Hoeche, S., Krauss, F., Schönherr, M., Schumann, S., Siegert, F., Winter, J.: Event generation with SHERPA 1.1. JHEP **0902**, 007 (2009). arXiv:0811.4622

22. Hahn, T.: Generating Feynman diagrams and amplitudes with FeynArts 3. Comp. Phys. Comm. 140, 418 (2001). arXiv:hep-ph/0012260

23. Hahn, T.: A Mathematica interface for FormCalc-generated code. Comp. Phys. Comm. **178**, 217 (2008). arXiv:hep-ph/0611273

24. Hahn, T., Illana, J.I.: Excursions into FeynArts and FormCalc. Nucl. Phys. Proc. Suppl. **160**, 101 (2006). arXiv:hep-ph/0607049

25. Hahn, T., Lang, P.: FeynEdit: a tool for drawing Feynman diagrams. Comp. Phys. Comm. 179, 931 (2008). arXiv:0711.1345

26. Hahn, T., Paßehr, S.: Implementation of the $\mathcal{O}(\alpha_t^\xi)$ MSSM Higgs-mass corrections in FeynHiggs. arXiv:1508.00562

27. Hahn, T., Pérez-Victoria, M.: Automated one-loop calculations in four and D dimensions. Comp. Phys. Comm. **118**, 153 (1999). arXiv:hep-ph/9807565

28. van Hameren, A.: OneLOop: for the evaluation of one-loop scalar functions. Comp. Phys. Comm. 182, 2427 (2011). arXiv:1007.4716

29. Harlander, R., Seidensticker, T., Steinhauser, M.: Complete corrections of $\mathcal{O}(\alpha\alpha_f)$ to the decay of the Z boson into bottom quarks. Phys. Lett. B **426**, 125 (1998). arXiv:hep-ph/9712228

30. Kanaki, A., Papadopoulos, C.: HELAC: a Package to compute electroweak helicity amplitudes. Comp. Phys. Comm. **132**, 306 (2000). arXiv:hep-ph/0002082

31. Laporta, S.: High precision calculation of multiloop Feynman integrals by difference equations. Int. J. Mod. Phys. A **15**, 5087 (2000). arXiv:hep-ph/0102033

32. Martin, S., Robertson, D.: TSIL: a program for the calculation of two-loop self-energy integrals. Comp. Phys. Comm. **174**, 133 (2006). arXiv:hep-ph/0501132

33. Nachtmann, O.: Phänomene und Konzepte der Elementarteilchenphysik, Vieweg (1986)

34. van Oldenborgh, G.J., Vermaseren, J.A.M.: New algorithms for one-loop integrals. Z. Phys. C **46**, 425 (1990)

35. Ossola, G., Papadopoulos, C., Pittau, R.: Reducing full one-loop amplitudes to scalar integrals at the integrand level. Nucl. Phys. B **763**, 147 (2007). arXiv:hep-ph/0609007

36. Passarino, G., Veltman, M.: One Loop Corrections for e+ e− Annihilation Into mu+ mu− in the Weinberg Model. Nucl. Phys. B **160**, 151 (1979)

37. Semenov, A.: LanHEP - a package for automatic generation of Feynman rules from the Lagrangian. Version 3.2. Comp. Phys. Comm. 201, 167 (2016). arXiv:1412.5016

38. Smirnov, V.: Feynman Integral Calculus. Springer, Berlin (2006)

39. Staub, F.: SARAH 3.2: Dirac Gauginos, UFO output, and more. Comp. Phys. Comm. 184, 1792 (2013). arXiv:1207.0906

40. Ueda, T., Vermaseren, J.: Recent developments on FORM. J. Phys. Conf. Ser. **523**, 012047 (2014)

A Note on Dynamic Gröbner Bases Computation

Amir Hashemi[1,2]([⊠]) and Delaram Talaashrafi[3]

[1] Department of Mathematical Sciences, Isfahan University of Technology,
84156-83111 Isfahan, Iran
Amir.Hashemi@cc.iut.ac.ir
[2] School of Mathematics, Institute for Research in Fundamental Sciences (IPM),
19395-5746 Tehran, Iran
[3] Department of Electrical and Computer Engineering,
Isfahan University of Technology, 84156-83111 Isfahan, Iran
d.talaashrafi@ec.iut.ac.ir

Abstract. For most applications of Gröbner bases, one needs only a *nice* Gröbner basis of a given ideal and does not need to specify the monomial ordering. From a nice basis, we mean a basis with small size. For this purpose, Gritzmann and Sturmfels [14] introduced the method of *dynamic Gröbner bases computation* and also a variant of Buchberger's algorithm to compute a nice Gröbner basis. Caboara and Perry [6] improved this approach by reducing the size and number of intermediate linear programs. In this paper, we improve the latter approach by proposing an algorithm to compute nicer Gröbner bases. The proposed algorithm has been implemented in SAGE and its efficiency is discussed via a set of benchmark polynomials.

1 Introduction

Gröbner bases are a fundamental tool in computer algebra which provide efficient algorithmic computation in polynomial ideal theory. A Gröbner basis is a (generally not minimal) generating set for an ideal which allows to determine easily many important properties of the ideal. The notion of Gröbner bases together with the basic algorithm to compute them were originally introduced in 1965 by Buchberger in his Ph.D. thesis [3,4]. Later on, he proposed two criteria for removing superfluous reductions to improve his algorithm [2]. Lazard [15] in 1983 established the link between Gröbner bases and linear algebra. In 1988, Gebauer and Möller [11] installed in an efficient way Buchberger's criteria on Buchberger's algorithm. In 1999, Faugère [8] presented his F_4 algorithm to compute Gröbner bases which relies on performing fast linear algebra techniques on sparse matrices (this algorithm has been efficiently implemented in MAPLE and MAGMA). In 2002, Faugère presented the famous F_5 algorithm (a signature-based algorithm benefits from an incremental structure to apply two new criteria) for computing Gröbner bases [9].

Although Robbiano [17] proved that there are infinitely many monomial orderings on a multivariate polynomial ring, there are finitely many distinct

© Springer International Publishing AG 2016
V.P. Gerdt et al. (Eds.): CASC 2016, LNCS 9890, pp. 276–288, 2016.
DOI: 10.1007/978-3-319-45641-6_18

reduced Gröbner bases for a given ideal (which leads to the definition of universal Gröbner bases), see [16]. On the other hand, for many applications of Gröbner bases, we need only a Gröbner basis of a given ideal and do not need to specify the monomial ordering. For such applications, we can point out the applications of Gröbner bases in ideal membership test, computation of intersection of ideals, computing a generating set for the syzygy module of a sequence of polynomials, radical membership test and computation in residue class rings. Therefore, a natural question that may arise is how to find the *nicest* Gröbner basis for an ideal. In this article, from a *nice* Gröbner basis, we mean a basis having small size (one may also consider a basis whose polynomials have small coefficients). In this direction, Gritzmann and Sturmfels [14] presented the dynamic version of Buchberger's algorithm (by using the Hilbert function) for polynomial ideals to obtain a nice Gröbner basis. Caboara [5] implemented this algorithm by using the Gebauer-Möller algorithm [11]. The main tool in the dynamic Gröbner bases computation consists in solving linear programs. Caboara and Perry [6] improved this computation by reducing the size and the number of the intermediate linear programs which have to be solved during the computation.

In this paper, we improve the approach proposed by Caboara and Perry by performing exact computations to find vertices of a polyhedron instead of finding the boundary vectors with approximate methods. Remark that these vertices are essential to finding the compatible terms (Theorem 1). Furthermore, their approach is based on choosing initial heuristic monomial orderings. However the efficiency of the computation may be affected by the choice of these monomial orderings. In our approach we do not need to choose any monomial ordering and this may enhance the efficiencyof our algorithm. In particular, we have implemented in SAGE the dynamic version of the Gebauer-Möller algorithm based on these techniques and have been able to find nicer Gröbner bases (for several examples) than the Caboara-Perry algorithm.

The rest of the paper is organized as follows. In the next section, we will review the basic definitions and notations which will be used throughout this paper. Section 3 is devoted to the description of the Caboara-Perry algorithm. In Sect. 4, we present our new techniques and apply it on the dynamic variant of Gebauer-Möller algorithm. We analyze the performance of the proposed algorithm in the last section. Finally, in Sect. 6 we conclude the paper by highlighting the advantages of this work.

2   Preliminaries

In this section, we review the basic definitions and notations from the theory of Gröbner bases which will be used in the paper. Throughout this paper we assume that $\mathcal{P} = \Bbbk[x_1, \ldots, x_n]$ is the polynomial ring and $\Bbbk$ is a field. We shall consider a sequence of polynomials $f_1, \ldots, f_m \in \mathcal{P}$ and the ideal $\mathcal{I} = \langle f_1, \ldots, f_m \rangle$ generated by the f_i's. For a polynomial $f \in \mathcal{P}$, we let $\mathrm{Supp}(f)$ to be the set of all monomials contained in f. For a monomial x^α, with $\alpha \in \mathbb{N}^n$ we write $\mathrm{Exp}(x^\alpha) = \alpha$. If $\prec$ is a monomial ordering on $\mathcal{P}$, the leading monomial of a polynomial $f \in \mathcal{P}$ w.r.t. $\prec$ is denoted by $\mathrm{LM}(f)$. A monomial in $\mathrm{Supp}(f)$ is called a *potential leading monomial* if there exists a monomial ordering $\prec$ on $\mathcal{P}$ such that it is the leading monomial of f w.r.t. $\prec$. If $F \subset \mathcal{P}$ is a finite set of polynomials, we denote by $\mathrm{LM}(F)$ the set $\{\mathrm{LM}(f) \mid f \in F\}$. The leading coefficient of f is the coefficient of $\mathrm{LM}(f)$ and is denoted by $\mathrm{LC}(f)$. The leading term of f is defined to be $\mathrm{LT}(f) = \mathrm{LM}(f)\,\mathrm{LC}(f)$. A finite set $G = \{g_1, \ldots, g_k\} \subset \mathcal{P}$ is called a *Gröbner basis* of $\mathcal{I}$ w.r.t $\prec$ if $\mathrm{LT}(\mathcal{I}) = \langle \mathrm{LT}(g_1), \ldots, \mathrm{LT}(g_k) \rangle$ where $\mathrm{LT}(\mathcal{I}) = \langle \mathrm{LT}(f) \mid f \in \mathcal{I} \rangle$. We refer e.g. to [7] for further details on Gröbner bases. To avoid confusion, we may denote the leading monomial of f by $\mathrm{LM}_\prec(f)$ (of course the same holds also for other presented notions).

Let us recall the definition of Hilbert function and Hilbert series of a *homogeneous* ideal. Let $X \subset \mathcal{P}$ and s a positive integer. We define the degree s part X_s of X to be the set of all homogeneous elements of X of degree s.

Definition 1. *The* Hilbert function *of a homogeneous ideal $\mathcal{I}$ is* $\mathrm{HF}_\mathcal{I}(s) = \dim\,(\mathcal{P}_s/\mathcal{I}_s)$ *where the right-hand side is the dimension of $\mathcal{P}_s/\mathcal{I}_s$ as a $\Bbbk$-vector space.*

Recall that the *Hilbert series* of a homogeneous ideal $\mathcal{I} \subset \mathcal{P}$ is the following power series $\mathrm{HS}_\mathcal{I}(t) = \sum_{s=0}^\infty \mathrm{HF}_\mathcal{I}(s)t^s$. It is well-known that the Hilbert series of a homogeneous ideal may be expressed as the quotient of two polynomials.

Proposition 1. *There exists a univariate polynomial $p(t)$ so that $\mathrm{HS}_\mathcal{I}(t) = p(t)/(1-t)^D$ with $p(1) \neq 0$ and $D = \dim(\mathcal{I})$.*

For the proof we refer to [10, Theorem 7, p. 130]. From [7, Proposition 4, p. 458] we deduce that the Hilbert function of $\mathcal{I}$ is the same as that of $\mathrm{LT}(\mathcal{I})$ and this provides an effective method to compute the Hilbert series of an ideal using Gröbner bases, see e.g. [13]. This observation shows that the set of monomials not contained in $\mathrm{LT}(\mathcal{I})$ forms a basis for $\mathcal{P}_s/\mathcal{I}_s$ as a $\Bbbk$-vector space (Macaulay's theorem). Further, if we add a new polynomial f into a set of polynomials F with $\mathrm{LT}(f) \notin \langle \mathrm{LM}(F) \rangle$ then we have $\mathrm{HF}_{\langle F \rangle + \langle f \rangle}(s) \leq \mathrm{HF}_{\langle F \rangle}(s)$ for each s. Furthermore, we will need an ordering to compare the Hilbert functions of two ideals. In doing so, we use the Hilbert series of the ideals. Let $\mathcal{I}, \mathcal{J} \subset \mathcal{P}$ be two ideals. Then we write $\mathrm{HS}_\mathcal{I}(t) < \mathrm{HS}_\mathcal{J}(t)$ (or equivalently we say that $\mathrm{HF}_\mathcal{I} < \mathrm{HF}_\mathcal{J}$) if there exists a positive integer s so that $\mathrm{HF}_\mathcal{I}(i) = \mathrm{HF}_\mathcal{J}(i)$ for each $i < s$ and $\mathrm{HF}_\mathcal{I}(s) < \mathrm{HF}_\mathcal{J}(s)$. To compare effectively the Hilbert series of two ideals, one may use Proposition 1 and read the sign of the first coefficient of $\mathrm{HS}_\mathcal{J}(t) - \mathrm{HS}_\mathcal{I}(t)$ in its series expansion.

Remark 1. The defined Hilbert function (and therefore Hilbert series) is valid only for homogeneous ideals, however, we consider not necessarily homogeneous ideals. For this purpose, if F is a set of not necessarily homogeneous polynomials we consider $\mathrm{HF}_{\langle \mathrm{LM}(F) \rangle}(s)$ for each s and $\mathrm{HS}_{\langle \mathrm{LM}(F) \rangle}(s)$.

3   Dynamic Gröbner Bases Computation

Finding a suitable order is a crucial ingredient to obtain nice Gröbner bases with less polynomials and less terms. For this purpose, one solution is to change the monomial ordering each time a new polynomial is constructed during the computation. It is proven in [14] that such an algorithm terminates with a Gröbner basis. Keeping in mind that changing monomial orderings is done in a way that previous computations remain valid. We call this method the *Dynamic Gröbner Basis Computation*. In this section, we recall briefly the method proposed by Caboara and Perry [6] which reduces the size and number of intermediate linear programs (which one requires to solve) in dynamic Gröbner bases computation.

For each nonzero $w \in \mathbb{R}^n$ one can define an ordering on $\mathcal{P}$ using the inner product of vectors. More precisely, for $\alpha, \beta \in \mathbb{N}^n$, we write $\alpha \prec_w \beta$ if $w.\alpha < w.\beta$ where $\alpha.w$ denotes the inner product of two vectors $\alpha, w \in \mathbb{R}^n$. If the elements of w are linearly independent over $\mathbb{Q}$ then $\prec_w$ is a monomial ordering. However, for a finite set F of polynomials, we can find a vector $w \in \mathbb{R}^n$ with independent components so that $\prec_w$ can be used to identify $\mathrm{LM}(F)$. More precisely, assume that a monomial ordering $\prec$ has been fixed on $\mathcal{P}$. Then, there exists $w \in \mathbb{R}^n$ so that for each $f \in F$ we have $\mathrm{LM}_\prec(f) = \mathrm{LM}_{\prec_w}(f)$.

Lemma 1. *For any monomial ordering $\prec$ and a positive integer a, there is a vector $v \in \mathbb{R}^d_+$ such that for all monomials x^α and x^β of total degree less than or equal to a we have $x^\alpha \prec x^\beta$ if and only if $v.\alpha < v.\beta$.*

This observation leads to an integer programming model to study this relationship which is reviewed in the following. A *polyhedron* in $\mathbb{R}^n$ is an intersection of finite half-spaces in $\mathbb{R}^n$. A polyhedron of the form $\{\mathbf{x} \in \mathbb{R}^n \mid A\mathbf{x} \leq 0\}$ where A is a matrix with n columns is called a *cone*. The following lemma [14, Lemma 1.3.1], relates monomial orderings and inner product of vectors.

Suppose that $G \subset \mathcal{P}$ is a finite set. Two monomial orderings $\prec_1$ and $\prec_2$ on $\mathcal{P}$ are called *equivalent* for G if $\mathrm{LM}_{\prec_1}(G) = \mathrm{LM}_{\prec_2}(G)$. For every $G \subset \mathcal{P}$ there are finitely many equivalence classes of monomial orderings. These equivalence classes are called *cone associated classes to* G and each class is denoted by $C(\prec, G)$ where $\prec$ is a monomial ordering in that cone. Let $G = \{g_1, \ldots, g_m\}$. The associated cone $C(\prec, G)$ can be described as a system of linear inequalities, say S, as follows: For each $i = 1, \ldots, m$, we let $T_i = \{t \mid t \in \mathrm{Supp}(g_i) \setminus \{\mathrm{LM}_\prec(g_i)\}\}$. Then, we set

$$S = \{(y_1, \ldots, y_n).(\mathrm{Exp}(\mathrm{LM}_\prec(g_i)) - \mathrm{Exp}(t)) > 0 \mid \forall\, t \in T_i,\ 0 \leq i \leq m\} \quad (1)$$
$$\cup\, \{y_i > 0 \mid 0 \leq i \leq n\}.$$

This system has a crucial role in the dynamic Gröbner basis computation.

Definition 2. *Boundary vectors of a cone are defined to be extreme points of the intersection of the closure of the system (1) and $y_1 + \cdots + y_n = d$ for some $d \in \mathbb{R}_+$.*

For $G \subset \mathcal{P}$ and a cone $C(\prec, G)$ associated to G, a monomial t of a polynomial f is said to be a *compatible term* w.r.t. $C(\prec, G)$ if there exists an order $\prec' \in C(\prec, G)$ such that $\mathrm{LM}_{\prec'}(f) = t$.

Theorem 1 ([6, *Theorem 2*]). *Let $G \subset \mathcal{P}$, $f \in \mathcal{P}$ and Ω be a set of boundary vectors of the cone $C(\prec, G)$ for somse monomial ordering. Consider $\mathrm{LM}_{\prec}(f) = t$. If there exists a term order $\prec' \in C(\prec, G)$ such that $\mathrm{LM}_{\prec'}(f) = u$, then there exists $\omega \in \Omega$ so that $\omega.(\mathrm{Exp}(u) - \mathrm{Exp}(t)) > 0$.*

It is important to know that converse of above theorem does not hold, but it has the following consequence, see [6, Corollary 1].

Corollary 1. *If we know a set of boundary vectors Ω of $C(\prec, G)$, then any monomial u in f is incompatible if $\omega.(\mathrm{Exp}(u) - \mathrm{Exp}(t)) < 0$ for each $\omega \in \Omega$ where $t = \mathrm{LM}(f)$.*

The dynamic Buchberger algorithm (only boundary vector technique part) proposed in [6] is as follows. The required functions are introduced briefly after the algorithm. Let $\mathbf{e}_1, \ldots, \mathbf{e}_n$ be the standard basis for $\mathbb{R}^n$. By convention, the Spolynomial of a critical pair $(f, 0)$ is defined to be f.

Algorithm 1. DYNAMICBUCHBERGER

Input: $F \subset \mathcal{P}$ a finite set of polynomials
Output: $G \subset \mathcal{P}$ and a monomial ordering $\prec$ such that G is a Gröbner basis for $\langle F \rangle$
select a monomial ordering $\prec$
$G := \{\}$
$P := \{(f, 0) \mid \forall\, f \in F\}$
$\Omega := \{\mathbf{e}_1, \ldots, \mathbf{e}_n\}$
while $P \neq \emptyset$ **do**
    select a pair $(p, q) \in P$ and remove it from P
    $r := $ A remainder of Spolynomial of (p, q) by G
    if $r \neq 0$ **then**
        Add (g, r) to P for each $g \in G$ and add r to G
        $t := \mathrm{LM}_{\prec}(r)$
        $U := \mathrm{Supp}(r) \setminus \{t\}$
        find a new ordering $\prec'$, using $identify\_clts(\prec, t, U, \Omega)$ to eliminate incompatible
        terms and *Hilbert function* to rank compatible terms.
        $\prec := monitor(G, \prec, \prec', lp(\prec', G))$
        remove useless pairs from P
    end if
    $\Omega := compute\_boundary\_vectors(lp(G, \prec))$
end while
return $(G, \prec)$

Below, we explain all the used functions.

- The function $lp(G, \prec)$ returns the inequality system corresponding to $C(\prec, G)$.

- The function *compute_boundary_vectors*$(lp(G, \prec))$ computes approximate boundary vectors of the input cone using linear programming techniques.
- The function *identify_clts*$(\prec, t, U, \Omega)$ uses Theorem 1 to eliminate incompatible terms of a polynomial with respect to the current cone and monomial ordering. The Hilbert function helps to select a term among the compatible terms that minimizes the basis of the ideal generated by the leading terms.
- To minimize number of constraints, in the algorithm we add only certain inequalities to the system. These inequalities will only be for the terms that are compatible. Taking into account that boundary vectors are approximated, this can lead to inconsistency in the system. Finally, to avoid this inconsistency, the algorithm calls *monitor*$(G, \prec, \prec', lp(\prec', G))$ function at each step. In more detail, this function verifies whether or not the previously-determined leading terms remain invariant and if some leading terms would change, it finds a compromise ordering if any exist.

4   Improved Dynamic Gröbner Bases Computation

In this section, we propose an improved version of the previous algorithm which allows us to compute Gröbner bases with less polynomials and less terms. Let us denote by $\text{LM}(f)$ a potential leading monomial which is selected as the leading monomial of f. Let $F = \{(f_1, t_1), \ldots, (f_m, t_m)\}$ such that $f_i \in \mathcal{P}$ and $\text{LM}(f_i) = t_i$ for each i. The linear inequality system associated to F is presented as follows:

$$S = \{ \bigcup_{i=1}^{m} (y_1, \ldots, y_n).(\text{Exp}(t_i) - \text{Exp}(t)) > 0 \mid t \in \text{Supp}(f_i) \setminus \{t_i\} \} \qquad (2)$$
$$\cup \{ y_i > 0 \mid 1 \le i \le n \}.$$

Definition 3. *If the closure of the system (2) joined with $d = y_1 + \cdots + y_n$ for some $d \in \mathbb{R}_+$ is consistent, we refer to this intersection as the* polyhedron *of F.*

Having a system of linear inequalities, there are algorithms for computing a polyhedron defined by such a system and its vertices. We show that the exact computation of vertices (see the end of this section) leads to nicer Gröbner bases. In this direction, we modify the dynamic Buchberger's algorithm to select a leading term for each polynomial (and even for the input polynomials) at each step of the algorithm. To this end, after computing a new polynomial, we find its compatible terms with respect to already computed polynomials and leading terms. Then, we select the best term among compatible terms as the leading term of the new polynomial. In order to find compatible terms of a polynomial w.r.t. a set of polynomials, we use the next trivial theorem which is in fact an improved version of Theorem 1.

Theorem 2. *Let $F = \{(f_1, t_1), \ldots, (f_m, t_m)\}$ and P the polyhedron corresponding to F. Further, let $V = \{v_1, \ldots, v_k\}$ be the set of vertices of P. Then, for a polynomial $f \in \mathcal{P}$, a monomial $t \in \text{Supp}(f)$ is a compatible term if and only if for each $p \in \text{Supp}(f) \setminus \{t\}$ there exists $v \in V$ with $v.(\text{Exp}(t) - \text{Exp}(p)) > 0$.*

Proof. That is a consequence of the well-known Corner Point Theorem which implies that a maximum of any objective function occurs at an extreme point, see [6, Theorem 2] for further details.   □

Now, assume that we have a list of exact vertices of the polyhedron of a set F. The next procedure finds compatible terms of a polynomial w.r.t F by Theorem 2.

Algorithm 2. FINDCOMPATIBLETERMS

Input: $f \in \mathcal{P}$ and V set of vertices of its polyhedron
Output: All compatible terms of f
$M := \text{Supp}(f)$
$ComTerms := \{\}$
for $p \in M$ **do**
  $flag := true$
  for $t \in M \setminus \{p\}$ **do**
    if $\nexists v \in V$ such that $v.(\text{Exp}(p) - \text{Exp}(t)) > 0$ **then**
      $flag := false$
    end if
  end for
  if $flag = true$ **then**
    Add p into $ComTerms$
  end if
end for
return ($ComTerms$)

In the following, for the seek of simplicity, we consider V, G, B, L, P as global variables; i.e. each of these variables can be read and changed anywhere by each of the functions. In order to describe the main algorithm of this section (the improved dynamic Gröbner bases algorithm), we establish the next auxiliary algorithm to update the used monomial ordering.

Algorithm 3. UPDATEORDERING

Input: $f \in \mathcal{P}$
Output: All compatible terms of f
$ComTerms := FindCompatibleTerms(f, V)$
$t := HilbertHeuristic(ComTerms, \text{LT}(G))$
$G, B := Update(G, B, (f, t))$
if $|ComTerms| > 1$ **then**
  $L := UpdateSystem(L, (f, t))$
  $P :=$ The polyhedron of L
  $V :=$ The set of exact vertices of P
  $VecOrder := center(P)$
  $L := eliminate(P, L)$
end if
return ($VecOrder$)

Let us briefly explain the main idea of the next algorithm. At each step, we maintain a set L of linear equations which defines a polyhedron. Any vector in the polyhedron describes an ordering which is compatible with past computations. When a new and nonzero polynomial f is considered to be added to the set that becomes a Gröbner basis, we choose a vector in the polyhedron (representing a compatible monomial ordering) such that the best possible monomial appearing in f moves to the leading position.

Algorithm 4. IMPDYNGB

Input: $F = \{f_1, \ldots, f_m\} \subset \mathcal{P}$
Output: A Gröbner basis for $\langle F \rangle$
$B := \{\}$
$G := \{\}$
$L := \{y_1 \geq 0, \ldots, y_n \geq 0 \, , \, y_1 + \cdots + y_n = d\}$ for some $d \in \mathbb{R}_+$
$P :=$ The polyhedron of L
$V :=$ The set of exact vertices of P
$VecOrder := center(P)$
for i **from** 1 **to** m **do**
   $VecOrder :=$ UPDATEORDERING(f_i)
end for
while $B \neq \emptyset$ **do**
   select and remove a pair $\{f, g\}$ from B
   $s := Spolynomial(f, g, VecOrder)$
   $r :=$ A remainder of the division of s by G and w.r.t. $VecOrder$
   $VecOrder :=$ UPDATEORDERING(r)
end while
return (G)

We explain below the used functions.

- The function *center(P)* finds a vector in the polyhedron P.
- We use *HilbertHeuristic* procedure to select one of the compatible terms as the leading term of the new polynomial. Indeed, this function receives as input a set of compatible terms $T = \{t_1, \ldots, t_k\}$ and a set of monomials M and returns t_i where the ideal generated by $M \cup \{t_i\}$ has the minimum Hilbert function (according to the ordering defined in Sect. 2) among all possible terms in T. In the case that Hilbert function can not decide between two terms, we break the tie using *degree reverse lexicographical ordering*.
- In addition, we use *Update* algorithm from [1] to refine the set of critical pairs by applying Buchberger's criteria when a new polynomial is constructed.
- *UpdateSystem* is a function to add inequalities related to the new polynomial to the existing system.
- We should note that not all inequalities are needed. By eliminating them using the *eliminate* function, we keep the sizes of systems of linear inequalities of small as possible, which in the end makes the computation faster. One of

the approaches proposed by Caboara and Perry in [6] to reduce the size of inequalities was to add only those constraints that correspond to terms that the boundary vectors identify as compatible leading terms. In our SAGE implementation, we call the function $eliminate(P, L)$ where L is a set of inequalities and P is the corresponding polyhedron and using the SAGE function `inequalities_list` we compute the set of inequalities corresponding to P and this would be a minimal representation for L.

Finally, in this algorithm, we use sugar strategy [12] for selecting pairs.

Remark 2. Algorithm 4 proceeds by computing Spolynomial and reduction computations with respect to $VecOrder$ (which can be any arbitrary vector in the polyhedron, and we choose the center point of the polyhedron). As a consequence, the new algorithm does not need any monomial ordering.

Remark 3. Another interesting fact about the new algorithm is that the computations which are involved in changing of monomial orderings are only performed when the new added polynomial has more than one compatible term. In particular, when the added polynomial has one compatible term, the new inequalities won't affect the polyhedron, and therefore we do not have to update the polyhedron. It is worthwhile noting that the only heuristic part of Algorithm 4 is the *HilbertHeuristic* procedure and in adition, the algorithm is deterministic.

Remark 4. We shall note that since all the cones pass through the origin then we can assume d to be any positive integer to be sure that the plane $y_1 + \cdots + y_n = d$ intersects all the cones.

Theorem 3. IMPDYNGB *algorithm terminates in finitely many steps and returns a Gröbner basis for the input ideal.*

Proof. The termination and correctness of the algorithm are inherited from those of GRÖBNERNEW2 (see [1, Theorem 5.73]). Indeed, we shall remark that when we turn to a new monomial ordering then the already computed leading terms remain invariant and this completes the proof. □

Example 1. In this example, we present SAGE [18] commands to construct a linear inequality system lp and add new inequality to it. Further, we show how one can compute the exact vertices corresponding to a polyhedron.

$$lp = MixedIntegerLinearProgram()$$
$$y = lp.new\_variable(real = True, nonnegative = True)$$
$$lp.set\_objective(None)$$
$$for\ i\ in\ range\ (1, 5):$$
$$\quad lp.add\_constraint(y[i] >= 0)$$
$$\quad lp.add\_constraint(-y[1] + 2y[3] + y[5] >= 0)$$
$$\quad lp.add\_constraint(2y[3] + 3y[4] + y[1] >= 0)$$
$$\quad lp.add\_constraint(y[1] + y[2] + y[3] + y[4] == 50)$$

We can compute the polyhedron corresponding to lp using the next command.

$$P = lp.polyhedron()$$

The list of vertices of the polyhedron P, can be found using the following command

$$P.vertices\_list()$$

which is equal to

$$[[0, 50, 0, 0, 0], [0, 0, 0, 50, 0], [100/3, 0, 50/3, 0, 0], [50, 0, 0, 0, 50], [0, 0, 50, 0, 0]].$$

Conversely, we can find inequalities of a polyhedron P with the following command (this function is used in *eliminate* function)

$$P.inequalities\_list().$$

5   Experimental Results

We have implemented IMPDYNGB in SAGE[1] and we have compared the behavior of this algorithm with the Caboara-Perry algorithm [6]. Further, we compare the results with the Gröbner basis w.r.t. reverse lexicographical (DRL) ordering. For this purpose, we used some examples (see below) from [6] (except Example 3). All computations were done over $\mathbb{Q}$. The results are shown in the following tables where the columns # Poly and # Monomials indicate, respectively, the number of polynomials and the size of the set of all monomials appearing in the output Gröbner basis (Table 1).

Test 1:

$$[u_0 + 2u_1 + 2u_2 + 2u_3 + 2u_4 + 2u_5 - 1, 2u_0u_1 + 2u_1u_2 + 2u_2u_3 + 2u_3u_4 + 2u_4u_5 - u_1,$$

$$2u_0u_2 + u_1^2 + 2u_1u_3 + 2u_2u_4 + 2u_3u_5 - u_2, 2u_0u_3 + 2u_1u_2 + 2u_1u_4 + 2u_2u_5 - u_3,$$

$$2u_0u_4 + 2u_1u_3 + 2u_1u_5 + u_2^2 - u_4, u_0^2 - u_0 + 2u_1^2 + 2u_2^2 + 2u_3^2 + 2u_4^2 + 2u_5^2]$$

Test 2:

$$[t^4zb + x^3ya, tx^8yz - ab^4cde, xy^2z^2d + zc^2e^2, tx^2y^3z^4 + ab^2c^3e^2]$$

Test 3:

$$[(z_1 - 6)^2 + z_2^2 + z_3^2 - 104, z_4^2 + (z_5 - 6)^2 + z_6^2 - 104, z_7^2 + (z_8 - 12)^2 + (z_9 - 6)^2 - 80,$$

$$z_1(z_4 - 6) + z_5(z_2 - 6) + z_3z_6 - 52, z_1(z_7 - 6) + z_8(z_2 - 12) + z_9(z_3 - 6) + 64,$$

$$z_4z_7 + z_8(z_5 - 12) + z_9(z_6 - 6) - 6z_5 + 32, 2z_2 + 2z_3 - 2z_6 - z_4 - z_5 - z_7 - z_9 + 18,$$

$$z_1 + z_2 + 2z_3 + 2z_4 + 2z_6 - 2z_7 + z_8 - z_9 - 38, z_1 + z_3 + z_5 - z_6 + 2z_7 - 2z_8 - 2z_4 + 8]$$

Test 4:

$$[x^{33}z^{23} - y^{82}a, x^{45} - y^{13}z^{21}b, x^{41}c - y^{33}z^{12}, x^{22} - y^{33}z^{12}d, x^5y^{17}z^{22}e - 1, xyzt - 1]$$

---

[1] The SAGE code of the implementations of our algorithms and examples are available at http://amirhashemi.iut.ac.ir/softwares.

Test 5:

$$[xb - ya, (x - l)d - y(c - l), b^2 - a^2 - r^2, (c - l)^2 + d^2 - s^2, (a - c)^2 + (b - d)^2 - t^2]$$

Test 6:

$$[45p + 35s - 165b - 36, 35p + 40z + 25t - 27s, 15w + 25ps + 30z - 18t - 165b^2,$$
$$- 9w + 15pt + 20zs, wp + 2zt - 11b^3, 99w - 11sb + 3b^2]$$

Test 7:

$$[-x^2yz^4 + t, -x^5y^7 + uz^2, vx^3z - y^2, z^5 - xy^3]$$

Test 8:

$$[abcde - 1, abcd + bcde + acde + deab + eabc, abc + bcd + cde + dea + eab,$$
$$ab + ed + bc + cd + ea, a + b + c + d + e]$$

Test 9:

$$[abcdef - 1, abcde + bcdef + cdefa + defab + efabc + fabcd$$
$$abcd + bcde + cdef + defa + efab + fabc, abc + bcd + cde + def + efa + fab,$$
$$ab + bc + cd + de + ef + fa, a + b + c + d + e + f]$$

Test 10:

$$[u + v + y - 1, t + 2u + z - 3, t + 2v + y - 1, -t - u - v + x - y - z,$$
$$tux^2 - 1569/31250yz^3, 587/15625ty + vz]$$

As one observes the new described algorithm outputs nicer Gröbner bases than the Caboara-Perry algorithm (and also that the Gröbner bases w.r.t. reverse lexicographical ordering).

Table 1. Comparison of IMPDYNGB algorithm with Caboara-Perry algorithm.

| System | Caboara-Perry | | DRL ordering | | IMPDYNGB | |
	# Poly	# Terms	# Poly	# Terms	# Poly	# Terms
Test 1	22	54	22	55	6	38
Test 2	35	137	239	479	83	166
Test 3	–	–	39	1337	9	369
Test 4	21	42	529	1058	6	12
Test 5	12	67	25	637	8	34
Test 6	9	20	13	23	6	16
Test 7	7	14	38	76	5	10
Test 8	11	68	21	88	11	52
Test 9	20	129	45	199	17	114
Test 10	7	15	7	15	6	14

6   Conclusion and Perspective

In this paper, a modification of Caboara-Perry algorithm [6] which is a variant of Buchberger's algorithm to compute Gröbner bases dynamically is suggested. The main idea is as the Buchberger algorithm progresses, it may change the monomial ordering (without starting with a fixed monomial ordering) so that the new monomial ordering compromises the past computations and reduces the volume below the Gröbner staircase using an ordering on Hilbert functions. We shall note that our aim in this paper is not to speed-up the computation of a Gröbner basis and is only to find a small Gröbner basis. With the improvements presented in this paper one could expect that locally one finds the best monomial ordering, where from *locally best* we mean that for small variations of the monomial ordering, no smaller Gröbner bases could be found. However, the global optimum requires an exhaustive enumeration of all Gröbner bases and can be obtained using the concept of *Gröbner fan* [16]. The proposed algorithm has been implemented in SAGE and it was shown that over several examples the described algorithm can find nicer Gröbner bases.

Acknowledgments. The research of the first author was in part supported by a grant from IPM (No. 94550420). The authors are grateful to anonymous referees for their useful and helpful comments on preliminary version of this paper.

References

1. Becker, T., Weispfenning, V.: Gröbner Bases: A Computational Approach to Commutative Algebra. Springer, New York (1993). In cooperation with Heinz Kredel
2. Buchberger, B.: A criterion for detecting unnecessary reductions in the construction of Gröbner-bases. In: Ng, E.W. (ed.) EUROSAM 1979. LNCS, vol. 72, pp. 3–21. Springer, Heidelberg (1979)
3. Buchberger, B.: Ein Algorithmus zum Auffinden der Basiselemente des Restklassenringes nach einem nulldimensionalen Polynomideal. Innsbruck: Univ. Innsbruck, Mathematisches Institut (Diss.) (1965)
4. Buchberger, B.: Bruno Buchberger's Ph.d. thesis 1965: an algorithm for finding the basis elements of the residue class ring of a zero dimensional polynomial ideal. Translation from the German. J. Symb. Comput. **41**(3–4), 475–511 (2006)
5. Caboara, M.: A dynamic algorithm for Gröbner basis computation. In: Proceedings of International Symposium on Symbolic and Algebraic Computation, ISSAC 1993, pp. 275–283 (1993)
6. Caboara, M., Perry, J.: Reducing the size and number of linear programs in a dynamic Gröbner basis algorithm. Appl. Algebra Eng. Commun. Comput. **25**(1–2), 99–117 (2014)
7. Cox, D., Little, J., O'Shea, D.: Ideals, Varieties, and Algorithms. An Introduction to Computational Algebraic Geometry and Commutative Algebra, 3rd edn. Springer, New York (2007)
8. Faugère, J.C.: A new efficient algorithm for computing Gröbner bases (F_4). J. Pure Appl. Algebra **139**(1–3), 61–88 (1999)

9. Faugère, J.C.: A new efficient algorithm for computing Gröbner bases without reduction to zero (F_5). In: Proceedings of International Symposium on Symbolic and Algebraic Computation, ISSAC 2002, pp. 75–83 (2002)

10. Fröberg, R.: An Introduction to Gröbner Bases. John Wiley & Sons, Chichester (1997)

11. Gebauer, R., Möller, H.: On an installation of Buchberger's algorithm. J. Symb. Comput. **6**(2–3), 275–286 (1988)

12. Giovini, A., Mora, T., Niesi, G., Robbiano, L., Traverso, C.: "One sugar cube, please" or selection strategies in the Buchberger algorithm. In: Proceedings of International Symposium on Symbolic and Algebraic Computation, ISSAC 1991, pp. 49–54 (1991)

13. Greuel, G.M., Pfister, G.: A singular introduction to commutative algebra. With contributions by Olaf Bachmann, Christoph Lossen and Hans Schönemann, 2nd extended edn. Springer, Berlin (2007)

14. Gritzmann, P., Sturmfels, B.: Minkowski addition of polytopes: computational complexity and applications to Gröbner bases. SIAM J. Discrete Math. **6**(2), 246–269 (1993)

15. Lazard, D.: Gröbner bases, Gaussian elimination and resolution of systems of algebraic equations. In: van Hulzen, J.A. (ed.) EUROCAL 1983. LNCS, vol. 162, pp. 146–156. Springer, Heidelberg (1983)

16. Mora, T., Robbiano, L.: The Gröbner fan of an ideal. J. Symb. Comput. **6**(2–3), 183–208 (1988)

17. Robbiano, L.: Term orderings on the polynomial ring. In: Caviness, B.F. (ed.) EUROCAL 1985. LNCS, vol. 204, pp. 513–517. Springer, Heidelberg (1985)

18. Stein, W.: Sage: Open Source Mathematical Software (Version 7.0). The Sage Group (2016). http://www.sagemath.org

Qualitative Analysis of the Reyman – Semenov–Tian–Shansky Integrable Case of the Generalized Kowalewski Top

Valentin Irtegov and Tatiana Titorenko[✉]

Institute for System Dynamics and Control Theory SB RAS,
134, Lermontov Street, Irkutsk 664033, Russia
{irteg,titor}@icc.ru

Abstract. Qualitative analysis for the general integrable case of the problem of motion of a rigid body in double force field is conducted. We seek invariant manifolds and the families of invariant manifolds of various dimension and levels, which possess some extremum property, and investigate their stability in the Lyapunov sense. It is shown that in the case of parallel force fields, there exist the families of permanent rotations of the body, the questions of their stability are considered. We also find some classic analogues for the solutions of the original problem and study their properties.

1 Introduction

In this paper, we consider the problem of motion of a rigid body with a fixed point under the influence of two force fields. These fields can be gravitational and magnetic (electric), and others. Such problems arise in many applications, e.g., space dynamics [1]. There are known several integrable cases of such systems when the inertia moments of the body are related as follows: $A = B = 2C$, and the force centers lie in the equatorial plane of inertia ellipsoid [2–4]. A series of works are devoted to the study of these cases (see., e.g., [5–7]). Their main topic is topological analysis of the system's phase space. The questions of stability of found solutions are not discussed. Our approach to the study of the system's phase space is based on solving the extremum problem for the elements of the algebra of the problem's first integrals. It allows us to apply the 2nd Lyapunov method for the analysis of stability and other properties of found solutions. In [8], we have conducted qualitative analysis of the Bogoyavlenskii particular integrable case [2] via this approach. In the given paper, the general integrable case revealed by Reyman A.G., Semenov–Tian–Shansky M.A. et al. [4] is studied. They have found an additional first integral in the problem under study, which can be considered as a generalization of classical area integral.

The approach used by us reduces the problem of qualitative analysis of motion equations to an algebraic one that gives a possibility to apply computer algebra tools in our study. We applied "Mathematica" computer algebra system (CAS).

© Springer International Publishing AG 2016
V.P. Gerdt et al. (Eds.): CASC 2016, LNCS 9890, pp. 289–304, 2016.
DOI: 10.1007/978-3-319-45641-6_19

2   Formulation of the Problem

The rotation of a rigid body around a fixed point in double force field in the moving coordinate axes is described by the following differential equations:

$$
\begin{aligned}
2\dot{p} &= qr + b\delta_3, & \dot{\gamma}_1 &= \gamma_2 r - \gamma_3 q, & \dot{\delta}_1 &= \delta_2 r - \delta_3 q, \\
2\dot{q} &= x_0\gamma_3 - pr, & \dot{\gamma}_2 &= \gamma_3 p - \gamma_1 r, & \dot{\delta}_2 &= \delta_3 p - \delta_1 r, \\
\dot{r} &= -b\delta_1 - x_0\gamma_2, & \dot{\gamma}_3 &= \gamma_1 q - \gamma_2 p, & \dot{\delta}_3 &= \delta_1 q - \delta_2 p,
\end{aligned}
\tag{1}
$$

where p, q, and r are the projections of the vector of angular velocity onto the axes related to the body; γ_i $(i = 1, 2, 3)$ are the components of the direction vector of the 1st force field; δ_i $(i = 1, 2, 3)$ are the components of the direction vector of the 2nd force field; x_0 and b are the components of the radius-vectors of force centers, respectively.

Equations (1) admit the polynomial first integrals:

$$
\begin{aligned}
2H &= 2(p^2 + q^2) + r^2 + 2(x_0\gamma_1 - b\delta_2) = 2h, \\
V_1 &= (p^2 - q^2 - x_0\gamma_1 - b\delta_2)^2 + (2pq - x_0\gamma_2 + b\delta_1)^2 = c_1, \\
V_2 &= \gamma_1^2 + \gamma_2^2 + \gamma_3^2 = 1, \quad V_3 = \delta_1^2 + \delta_2^2 + \delta_3^2 = 1, \\
V_4 &= \gamma_1\delta_1 + \gamma_2\delta_2 + \gamma_3\delta_3 = c_2, \\
V_5 &= x_0^2(p\gamma_1 + q\gamma_2 + \tfrac{r}{2}\gamma_3)^2 + b^2(p\delta_1 + q\delta_2 + \tfrac{r}{2}\delta_3)^2 \\
&\quad - x_0 b\, r[(\gamma_2\delta_3 - \gamma_3\delta_2)p + (\gamma_3\delta_1 - \gamma_1\delta_3)q + \tfrac{r}{2}(\gamma_1\delta_2 - \gamma_2\delta_1)] \\
&\quad + x_0 b^2\gamma_1(\delta_1^2 + \delta_2^2 + \delta_3^2) - x_0^2 b\,\delta_2(\gamma_1^2 + \gamma_2^2 + \gamma_3^2) \\
&\quad - b x_0(b\delta_1 - \gamma_2 x_0)(\delta_1\gamma_1 + \delta_2\gamma_2 + \delta_3\gamma_3) = c_3.
\end{aligned}
\tag{2}
$$

Here V_5 is the additional first integral found in [4].

Our problem is to find the stationary sets of equations (1) and to investigate their qualitative properties. By stationary sets, we mean sets of any finite dimension on which the necessary extremum conditions for the elements of the algebra of first integrals in the problem under study are satisfied. Zero-dimension sets having this property are known as stationary solutions, while nonzero-dimension sets are called stationary invariant manifolds (IMs).

3   Obtaining Invariant Manifolds

For finding the desired solutions on the basis of the approach used, a combination of the first integrals of the original problem is constructed. Here we restrict ourselves to the linear one:

$$
2K = 2\lambda_0 H - \lambda_1 V_1 - \lambda_2 V_2 - \lambda_3 V_3 - 2\lambda_4 V_4 - 4\lambda_5 V_5.
\tag{3}
$$

For a complete analysis of the problem, nonlinear combinations of the integrals have to be considered as well.

Next, the necessary conditions for the integral K to have an extremum with respect to the phase variables $p, q, r, \gamma_i, \delta_i$ are written down:

$$\partial K/\partial p = \lambda_0 p + \lambda_1 [b\,(\delta_2 p - \delta_1 q) + x_0\,(\gamma_1 p + \gamma_2 q) - p\,(p^2 + q^2)] + \lambda_5 [b\,x_0 r$$
$$(\delta_3\gamma_2 - \delta_2\gamma_3) - b^2\delta_1\varrho_1 - x_0^2\gamma_1\varrho_2] = 0,$$

$$\partial K/\partial q = -\lambda_0 q + \lambda_1 [b\,(\delta_1 p + \delta_2 q) + q(p^2 + q^2) + x_0\,(\gamma_1 q - \gamma_2 p)]$$
$$+ \lambda_5 [b\,x_0 r\,(\delta_3\gamma_1 - \delta_1\gamma_3) + b^2\delta_2\varrho_1 + x_0^2\gamma_2\varrho_2] = 0,$$

$$\partial K/\partial r = \lambda_0 r + \lambda_5 [2b\,x_0[p\,(\delta_3\gamma_2 - \delta_2\gamma_3) + q(\delta_1\gamma_3 - \delta_3\gamma_1) + r(\delta_2\gamma_1 - \delta_1\gamma_2)]$$
$$- b^2\delta_3\varrho_1 - x_0^2\gamma_3\varrho_2] = 0,$$

$$\partial K/\partial\gamma_1 = \lambda_0 x_0 - \lambda_2\gamma_1 - \lambda_4\delta_1 - \lambda_1 x_0\varrho_3 + \lambda_5 x_0\,[b\,[2x_0(2\delta_2\gamma_1 - \delta_1\gamma_2)$$
$$+ r\,(\delta_2 r - 2\delta_3 q)] - 2b^2(\delta_2^2 + \delta_3^2) - 2x_0 p\varrho_2] = 0,$$

$$\partial K/\partial\gamma_2 = \lambda_1 x_0\varrho_4 - \lambda_2\gamma_2 - \lambda_4\delta_2 + \lambda_5 x_0[b x_0\,(2b\delta_1\delta_2 + r\,(2\delta_3 p - \delta_1 r))$$
$$- 2(b\,(\delta_1\gamma_1 + \delta_3\gamma_3) + q\varrho_2)] = 0,$$

$$\partial K/\partial\gamma_3 = -\lambda_2\gamma_3 - \lambda_4\delta_3 + \lambda_5 x_0[2b x_0\,(b\delta_1\delta_3 + r(\delta_1 q - \delta_2 p)) \qquad (4)$$
$$- (2b\,(\delta_3\gamma_2 - 2\delta_2\gamma_3) + r\varrho_2)] = 0,$$

$$\partial K/\partial\delta_1 = -\lambda_1 b\varrho_4 - \lambda_3\delta_1 - \lambda_4\gamma_1 + \lambda_5 b\,[2b\,(x_0(\delta_2\gamma_2 + \delta_3\gamma_3) - p\varrho_1)$$
$$- x_0(\gamma_2(r^2 + 2x_0\gamma_1) - 2\gamma_3 q\,r)] = 0,$$

$$\partial K/\partial\delta_2 = -\lambda_0 b - \lambda_1 b\,\varrho_3 - \lambda_3\delta_2 - \lambda_4\gamma_2 + \lambda_5 b\,[2x_0^2\,(\gamma_1^2 + \gamma_3^2) + x_0\,(2b\,(\delta_1\gamma_2$$
$$- 2\delta_2\gamma_1) + r\,(\gamma_1 r - 2\gamma_3 p)) - 2bq\varrho_1] = 0,$$

$$\partial K/\partial\delta_3 = -\lambda_3\delta_3 - \lambda_4\gamma_3 - \lambda_5 b\,[b\,[2r(\delta_1 p + \delta_2 q + \delta_3 r) + 2x_0(2\delta_3\gamma_1 - \delta_1\gamma_3)]$$
$$+ 2x_0(r(\gamma_1 q - \gamma_2 p) + x_0\gamma_2\gamma_3)] = 0.$$

Here λ_i are the parameters of the family of the integrals K, $\varrho_1 = 2\delta_1 p + 2\delta_2 q + \delta_3 r$, $\varrho_2 = 2\gamma_1 p + 2\gamma_2 q + \gamma_3 r$, $\varrho_3 = b\delta_2 - p^2 + q^2 + x_0\gamma_1$, $\varrho_4 = b\delta_1 + 2pq - x_0\gamma_2$.

The above stationary conditions for the family of the integrals K represent a system of nonhomogeneous cubic equations with the parameters λ_i, b, x_0. The solutions of these equations when they are dependent allow one to define IMs and the IMs families of differential equations (1) which correspond to the integral K. We call the IMs of equations (1) the first-level IMs.

3.1 First-Level Invariant Manifolds

Firstly, we find maximal dimension IMs and investigate their qualitative properties. As any first integral defines a family of IMs of codimension 1, we start with the IMs of codimension 2.

In [2,6], three IMs of codimension 2 in the problem under study are presented. The first corresponds to the zero level of integral V_1 (2), the others have been obtained by chains of differential consequences of motion equations. In the given work, we find the IMs of such dimension from equations (4) by solving them with respect to part of the phase variables and parameters λ_i. The latter, as was shown in [8], gives a possibility to obtain the desired IMs as well as the first integrals of vector fields on these IMs.

We take as unknowns, e.g., the following combination $\delta_1, \gamma_2, \lambda_0, \lambda_1, \lambda_2, \lambda_3, \lambda_4$ from the phase variables and the parameters λ_i, and construct a Gröobner basis for the polynomials of system (4) with the "Mathematica" procedure *Groebner-Basis*, where the option "EliminationOrder" is used for monomial ordering. As a result, we have a system of equations equivalent to the initial one, but it is decomposed into two subsystems. Below, one of them is analysed.

$$
\begin{cases}
2\rho_2\lambda_0 - \lambda_1\Big[p\,(q\,r + b\,\delta_3)[2(p^2+q^2)+r^2-2b\,\delta_2] - x_0\gamma_3\,[2q\,(p^2+q^2)\\
+r(q\,r+b\,\delta_3)] + 2x_0q\,\gamma_1\,(p\,r-x_0\gamma_3)\Big] = 0,\ \rho_2\lambda_1 + bx_0\,\rho_1\lambda_5 = 0,\\
\rho_1\rho_2\lambda_2 + b\,x_0\lambda_5\Big[(q\,r+b\,\delta_3)^2[4p^2\,(\delta_2^2+\delta_3^2)+(2q\,\delta_2+r\,\delta_3)^2]\\
+4x_0\,q\,(q\,r+b\,\delta_3)[2(q\,\delta_3\gamma_1-p\,\delta_2\gamma_3)\,\delta_2+(r\gamma_1-2p\gamma_3)\,\delta_3^2]+x_0^2\,(4(2q\,\delta_2\gamma_3\\
+(2p\,\gamma_1+r\,\gamma_3)\,\delta_3)\,q\delta_2\gamma_3 + [4q^2\,(\gamma_1^2+\gamma_3^2)+(2p\,\gamma_1+r\gamma_3)^2]\,\delta_3^2)\Big] = 0,\\
\rho_1\rho_2\lambda_3 + bx_0\lambda_5\Big[(b^2((2p\,\gamma_1+r\,\gamma_3)^2\,\delta_3^2+4\,[p^2\delta_3^2(\gamma_1^2+\gamma_3^2)+(p^2+q^2)\,\delta_2^2\gamma_3^2\\
+(2p\,\gamma_1+r\,\gamma_3)\,q\,\delta_2\delta_3\gamma_3]) + [4q^2\,(\gamma_1^2+\gamma_3^2)+(2p\,\gamma_1+r\,\gamma_3)^2](p\,r-\gamma_3x_0)^2\\
+4b\,p\,[\delta_2\gamma_3(2p\,\gamma_1+r\,\gamma_3)-2(\gamma_1^2+\gamma_3^2)\,q\,\delta_3](x_0\gamma_3-p\,r))\Big] = 0,\\
\rho_1\lambda_4 + 2bx_0\lambda_5\Big[(b\,[2\delta_2\,(q\,\delta_3\gamma_1-p\,\delta_2\gamma_3)+(r\,\gamma_1-2p\,\gamma_3)\,\delta_3^2]+r\,[2(p^2+q^2)\,\delta_2\gamma_1\\
+p\,r\delta_2\gamma_3+(r\gamma_1-2p\gamma_3)\,q\delta_3]+x_0[2(\gamma_1^2+\gamma_3^2)\,q\delta_3-\delta_2\gamma_3(2p\gamma_1+r\gamma_3)])\Big] = 0.
\end{cases}
\tag{5}
$$

$$
\begin{cases}
2bp\,(\delta_2\gamma_3 - \delta_3\gamma_2) - (2p\gamma_1+2q\gamma_2+r\gamma_3)(p\,r-x_0\gamma_3) = 0,\\
2x_0q\,(\delta_1\gamma_3 - \delta_3\gamma_1) - (2p\delta_1+2q\delta_2+r\delta_3)(q\,r+b\delta_3) = 0.
\end{cases}
\tag{6}
$$

Here $\rho_1 = 2p\delta_3\gamma_1 + 2q\delta_2\gamma_3 + r\delta_3\gamma_3$, $\rho_2 = x_0q\gamma_3 - p\,(q\,r+b\delta_3)$.

As can easily be verified by IM definition, equations (6) define the IM of codimension 2 of differential equations (1): the derivative of expressions (6) calculated by virtue of equations (1) must vanish on the given expressions. Indeed, the derivative of expressions (6) calculated by virtue of equations (1) can be written as

$$
[(2x_0q^2\gamma_1 - (q\,r+b\delta_3)^2)\,y_1 + b\,(2p^2\gamma_1 - x_0\gamma_3^2)\,y_2]/(2\rho_2) = 0,
$$
$$
[x_0(b\delta_3^2+2q^2\delta_2)\,y_1 + (2bp^2\delta_2+(p\,r-x_0\gamma_3)^2)\,y_2]/(2\rho_2) = 0.
\tag{7}
$$

with the aid of "Mathematica" procedure *PolynomialReduce*. Here y_1, y_2 are the left-hand sides of equations (6), respectively.

As equalities (7) become identities when $y_1 = y_2 = 0$, hence the latter proves invariance of solution (6).

The equations of vector field on IM (6) are given by:

$$
2\dot{p} = q\,r + b\,\delta_3, \quad 2\dot{r} = -\frac{1}{\rho_2}[b\,((2q\delta_2+r\delta_3)(q\,r+b\,\delta_3) + 2x_0(q\delta_3\gamma_1 - p\delta_2\gamma_3))
$$
$$
2\dot{q} = x_0\gamma_3 - p\,r, \qquad\quad + x_0(2p\gamma_1+r\gamma_3)(p\,r-x_0\gamma_3)],
$$
$$
\dot{\gamma}_1 = -q\gamma_3 + \frac{r}{2\rho_2}[(2p\gamma_1+r\gamma_3)(p\,r-x_0\gamma_3) - 2b\,p\,\delta_2\gamma_3],
$$
$$
\dot{\gamma}_3 = q\gamma_1 - \frac{p}{2\rho_2}[(2p\gamma_1+r\gamma_3)(p\,r-x_0\gamma_3) - 2b\,p\delta_2\gamma_3],
$$
$$
\dot{\delta}_2 = p\delta_3 - \frac{r}{2\rho_2}[(2q\delta_2+r\delta_3)(q\,r+b\,\delta_3) + 2x_0q\,\delta_3\gamma_1],
\tag{8}
$$
$$
\dot{\delta}_3 = -p\delta_2 + \frac{q}{2\rho_2}[(2q\delta_2+r\delta_3)(q\,r+b\,\delta_3) + 2x_0q\delta_3\gamma_1].
$$

They are derived from the original differential equations by elimination of the variables δ_1, γ_2 from them with the aid of (6).

From (5), we find $\lambda_0, \lambda_1, \lambda_2, \lambda_3, \lambda_4$:

$$
\begin{aligned}
\lambda_0 &= -b\,x_0\,\lambda_5\,\rho_1\Big[p\,(q\,r+b\,\delta_3)[2(p^2+q^2)+r^2-2b\,\delta_2]-x_0\gamma_3\,[2q(p^2+q^2)\\
&\quad +r(q\,r+b\,\delta_3)]+2x_0q\gamma_1\,(p\,r-x_0\gamma_3)\Big]/(2\rho_2^2),\quad \lambda_1 = -\lambda_5\,b\,x_0\,\rho_1/\rho_2,\\
\lambda_2 &= -b\,x_0\lambda_5\Big[(q\,r+b\delta_3)^2[4p^2(\delta_2^2+\delta_3^2)+(2q\delta_2+r\delta_3)^2]+4x_0\,q\,(q\,r+b\delta_3)\\
&\quad \times[2(q\delta_3\gamma_1-p\delta_2\gamma_3)\,\delta_2+(r\gamma_1-2p\gamma_3)\,\delta_3^2]+x_0^2\,(4\,(2q\delta_2\gamma_3+(2p\gamma_1+r\gamma_3)\,\delta_3)\\
&\quad \times q\delta_2\gamma_3+[4q^2(\gamma_1^2+\gamma_3^2)+(2p\gamma_1+r\gamma_3)^2]\,\delta_3^2)\Big]/(\rho_1\rho_2),\\
\lambda_3 &= -b x_0\lambda_5\Big[(b^2((2\gamma_1p+\gamma_3r)^2\,\delta_3^2+4\,[p^2\delta_3^2(\gamma_1^2+\gamma_3^2)+(p^2+q^2)\,\delta_2^2\gamma_3^2\\
&\quad +(2\gamma_1p+\gamma_3r)\,q\delta_2\delta_3\gamma_3])+[4q^2(\gamma_1^2+\gamma_3^2)+(2p\gamma_1+r\gamma_3)^2](p\,r-\gamma_3x_0)^2\\
&\quad +4\,bp\,[\delta_2\gamma_3(2p\gamma_1+r\gamma_3)-2(\gamma_1^2+\gamma_3^2)\,q\delta_3](x_0\gamma_3-p\,r))\Big]/(\rho_1\rho_2),\\
\lambda_4 &= -2b x_0\lambda_5\Big[(b\,[2\delta_2(q\delta_3\gamma_1-p\delta_2\gamma_3)+(r\gamma_1-2p\gamma_3)\,\delta_3^2]+r\,[2(p^2+q^2)\,\delta_2\gamma_1\\
&\quad +p\,r\delta_2\gamma_3+(r\gamma_1-2p\gamma_3)\,q\delta_3]+x_0[2(\gamma_1^2+\gamma_3^2)\,q\delta_3-\delta_2\gamma_3(2p\gamma_1+r\gamma_3)])\Big]/\rho_1.
\end{aligned}
\tag{9}
$$

The derivatives of right-hand sides of (9), which are computed by virtue of equations (8), are identically equal to zero. The latter means that relations (9) are the first integrals of equations (8). Note that the right-hand sides of the above expressions as well as differential equations (8) are rational.

The structure of the 2nd subsystem is similar to the first one: its latter two equations (from 7) define the IM of codimension 2 of motion equations (1), and the top five equations allow one to obtain the first integrals of vector field on the given IM. In this work, for space reasons, we represent the IM equations only:

$$
\sigma_2\gamma_2^2 + \sigma_1\gamma_2 + \sigma_0 = 0, \quad \hat\sigma_2\delta_1 + \hat\sigma_1\gamma_2 + \hat\sigma_0 = 0.
\tag{10}
$$

Here σ_i and $\hat\sigma_i$ are the polynomials of $p, q, r, \delta_2, \delta_3, \gamma_1, \gamma_3$. These polynomials are bulky, and their full form is given in Appendix. The verification of invariance for this manifold is performed likewise as above.

It seems likely that there exist other IMs of codimension 2 of the original differential equations which correspond to the integral K. In order to reveal them by the above technique, it is necessary to construct Gröobner bases for the polynomials of system (4) with respect to all the possible combinations of two phase variables and five parameters λ_i. Altogether, the number of such different combinations will be $C_9^2 C_6^5 = 216$. We could not construct the bases for all these combinations so far because the latter demands significant computer resources, in particular, processor time. The bases which we could obtain have not given new results.

It is not difficult to show that the intersection of IM (6) and IM (10) is a non-empty set. To this end, let us construct a Gröobner basis for the polynomials of systems (6), (10) with respect to, e.g., the variables $\delta_1, \gamma_1, \gamma_2$:

$$x_0^2 \left[b^2 \delta_3^2 + (pr - x_0\gamma_3)^2\right]\left[4q^2\left(\gamma_1^2 + \gamma_3^2\right) + (2p\gamma_1 + r\gamma_3)^2\right] + b^2\left[(qr + b\delta_3)^2\right.$$

$$\left.+x_0^2\gamma_3^2\right]\left[4p^2\left(\delta_2^2 + \delta_3^2\right) + (2q\delta_2 + r\delta_3)^2\right] + 2b\,x_0\left[x_0\gamma_3 q\left(2(p^2 + q^2) + r^2\right)\right.$$

$$\left.-p\left(qr + b\delta_3\right)[2(p^2 + q^2) + r^2 - 2b\,\delta_2]\right]\left(2p\gamma_1\delta_3 + 2q\gamma_3\delta_2 + r\gamma_3\delta_3\right)$$

$$+4b\,x_0^2\,\gamma_3(x_0\gamma_3 - pr)\left[2(p^2 + q^2)\gamma_1\delta_2 + pr\gamma_3\delta_2 + q\,\delta_3(r\gamma_1 - 2p\gamma_3)\right]$$

$$+4b^2x_0q\,(qr + b\,\delta_3)\left[2\delta_2(q\gamma_1\,\delta_3 - p\gamma_3\delta_2) + (r\gamma_1 - 2p\gamma_3)\,\delta_3^2\right] = 0, \tag{11}$$

$$2bp\,(\delta_2\gamma_3 - \delta_3\gamma_2) - (2p\gamma_1 + 2q\gamma_2 + r\gamma_3)(pr - x_0\gamma_3) = 0,$$

$$2x_0q\,(\delta_1\gamma_3 - \delta_3\gamma_1) - (2p\delta_1 + 2q\delta_2 + r\delta_3)(qr + b\delta_3) = 0.$$

As can easily be verified by IM definition, equations (11) define the IM of codimension 3 of motion equations (1).

On substituting expressions (11) resolved with respect to $\delta_1, \gamma_1, \gamma_2$ into equations (6) and (10), these become identities. It means that IM (11) is a submanifold of both IM (6) and IM (10), i.e., it belongs to their intersection.

3.2   The Invariant Manifolds of 2nd Level and Higher

Let us consider the problem of seeking IMs for differential equations (8) on first-level IM (6). We shall call such IMs the 2nd-level IMs.

As mentioned above, any first integral defines the IMs family of codimension 1. Hence, first integrals (9) of differential equations (8) and their combinations can be considered as the IMs families of these equations. The integrals can also be used for obtaining other IMs, including the stationary ones.

The zero level of one of rational integrals (9)

$$\hat{K} = \frac{2(p\delta_3\gamma_1 + q\delta_2\gamma_3) + r\delta_3\gamma_3}{x_0q\gamma_3 - p\,(qr + b\delta_3)} = c \quad (c = \text{const}) \tag{12}$$

defines the IM of equations (8) which is given by

$$2(p\delta_3\gamma_1 + q\delta_2\gamma_3) + r\delta_3\gamma_3 = 0. \tag{13}$$

Obviously, this IM is stable as any first integral.

The equations of vector field on IM (13)

$$2\dot{p} = qr + b\delta_3, \qquad\qquad \dot{\gamma_1} = -\gamma_3 q - \frac{(2\gamma_1 p + \gamma_3 r)r}{2q},$$

$$2\dot{q} = x_0\gamma_3 - pr, \qquad\qquad 2\dot{\gamma_3} = \frac{2\gamma_1(p^2 + q^2) + \gamma_3 pr}{q},$$

$$\dot{r} = -\frac{b\delta_3\gamma_1}{\gamma_3} + \frac{x_0(2\gamma_1 p + \gamma_3 r)}{2q}, \quad 2\dot{\delta_3} = \frac{\delta_3(2\gamma_1(p^2 + q^2) + \gamma_3 pr)}{\gamma_3 q} \tag{14}$$

admit the first integral

$$\bar{K} = \frac{\gamma_3}{\delta_3} = c_0 \quad (c_0 = \text{const}) \tag{15}$$

which is found directly from the above equations.

We consider (15) as the equation of IMs family, where c_0 is the family parameter. It will be the family of 3rd-level IMs on 2nd-level IM (13). As any first integral is stable, the elements of this IMs family are stable.

The vector field on the elements of the given family can be written as

$$2\dot{p} = q\,r + b\delta_3, \quad 2\dot{q} = c_0 x_0 \delta_3 - p\,r, \quad 2\dot{\gamma}_1 = -\frac{2\gamma_1 p\,r + c_0 \delta_3(2q^2 + r^2)}{q},$$

$$\dot{r} = -\frac{b\gamma_1}{c_0} + \frac{x_0(2\gamma_1 p + c_0 \delta_3 r)}{2q}, \quad 2\dot{\delta}_3 = \frac{2\gamma_1(p^2 + q^2) + c_0 \delta_3 p\,r}{c_0 q}. \quad (16)$$

Equations (16) possess the first integral

$$\tilde{V}_1 = \left(p^2 - q^2 + \frac{b\,(2\gamma_1 p + c_0 \delta_3 r)}{2 c_0 q} - x_0 \gamma_1\right)^2$$

$$+ \left(\frac{x_0\,(2\gamma_1 p + c_0 \delta_3 r)}{2q} + \frac{b\,\gamma_1}{c_0} + 2p\,q\right)^2 = \tilde{c}_1 \quad (\tilde{c}_1 = \text{const}). \quad (17)$$

It is derived from integral V_1 (2) by eliminating the variables $\delta_1, \delta_2, \gamma_2, \gamma_3$ from it with the aid of (6), (13), (15).

The zero level of integral (17) defines the family of IMs of equations (16) on the elements of IMs family (15):

$$y_1 = p^2 - q^2 + \frac{b\,(2\gamma_1 p + c_0 \delta_3 r)}{2 c_0 q} - x_0 \gamma_1 = 0,$$

$$y_2 = \frac{x_0\,(2\gamma_1 p + c_0 \delta_3 r)}{2q} + \frac{b\,\gamma_1}{c_0} + 2p\,q = 0. \quad (18)$$

It will be the family of the 4th-level stationary IMs. Integral (17) takes a stationary value on the elements of this family.

Since for the equations of perturbed motion, there exists the sign-definite integral $\Delta \tilde{V}_1 = y_1^2 + y_2^2$ obtained in the neighbourhood of the elements of IMs family (18), one can conclude that the elements of the family under consideration are stable.

The equations of vector field on the elements of IMs family (18) are given by:

$$2\dot{p} = \frac{2b\,(p^2 + q^2)\,\sigma + q\,r}{r\varrho}, \quad 2\dot{q} = \frac{2 c_0 x_0 (p^2 + q^2)\,\sigma - p\,r}{r\varrho}, \quad \dot{r} = \frac{2(b\,p + c_0 x_0 q)\,\sigma}{\varrho}.$$

These have the following first integrals:

$$\tilde{H} = r^2 + \frac{4\sigma^2}{\varrho}, \quad \tilde{V}_2 = \frac{(p^2 + q^2)^2[b^2(4q^2 + r^2) + c_0 x_0(c_0 x_0(4p^2 + r^2) - 8bp\,q)]}{r^2 \varrho^2}.$$

Here $\varrho = b^2 + c_0^2 x_0^2$, $\sigma = b\,q - c_0 x_0 p$.

Using the above first integrals, we can find the 5th-level IMs and so on. Thus, with the aid of first integrals on the IM of next level, it is possible to obtain the IMs of various levels and to study their properties.

Further, we consider the problem of "lifting up" the above found IMs of 2nd-level and higher into the original phase space and the study of their properties there.

3.3 The "Lifted Up" Invariant Manifolds

The "lifting up" problem is resolved immediately. In order to obtain in the phase space of system (1) the equations of IM corresponding to IM (13) on IM (6) it is sufficient to add equations (6) to equation (13):

$$2bp\,(\delta_2\gamma_3 - \delta_3\gamma_2) - (2p\gamma_1 + 2q\gamma_2 + r\gamma_3)(p\,r - x_0\gamma_3) = 0,$$
$$2x_0q\,(\delta_1\gamma_3 - \delta_3\gamma_1) - (2p\delta_1 + 2q\delta_2 + r\delta_3)(q\,r + b\delta_3) = 0, \qquad (19)$$
$$2(p\delta_3\gamma_1 + q\delta_2\gamma_3) + r\delta_3\gamma_3 = 0.$$

By IM definition, it is verified that equations (19) define the IM of codimension 3 of motion equations (1). This IM is stationary one: the integral $2K_0 = V_2V_3 - V_4^2$ takes a stationary value on it.

Analogously, IMs families (15) and (18) are "lifted up" into the original phase space.

Having added equations (13), (6) to equation (15), we have the equations of the IMs family of codimension 4 which correspond to IM (15) in the phase space of system (1):

$$2bp\,(\delta_2\gamma_3 - \delta_3\gamma_2) - (2p\gamma_1 + 2q\gamma_2 + r\gamma_3)(p\,r - x_0\gamma_3) = 0,$$
$$2x_0q\,(\delta_1\gamma_3 - \delta_3\gamma_1) - (2p\delta_1 + 2q\delta_2 + r\delta_3)(q\,r + b\delta_3) = 0, \qquad (20)$$
$$2(p\delta_3\gamma_1 + q\delta_2\gamma_3) + r\delta_3\gamma_3 = 0, \quad \gamma_3 - c_0\delta_3 = 0.$$

These represent the family of stationary IMs. The integral $2K_1 = V_2/c_0 + c_0V_3 - 2V_4$ takes a stationary value on the elements of the family.

The family of stationary IMs of codimension 6 corresponds to IMs family (18) in the original phase space. Its equations can be written as

$$2bp\,(\delta_2\gamma_3 - \delta_3\gamma_2) - (2p\gamma_1 + 2q\gamma_2 + r\gamma_3)(p\,r - x_0\gamma_3) = 0,$$
$$2x_0q\,(\delta_1\gamma_3 - \delta_3\gamma_1) - (2p\delta_1 + 2q\delta_2 + r\delta_3)(q\,r + b\delta_3) = 0,$$
$$2(p\delta_3\gamma_1 + q\delta_2\gamma_3) + r\delta_3\gamma_3 = 0, \quad \gamma_3 - c_0\delta_3 = 0, \qquad (21)$$
$$p^2 - q^2 + \frac{b\,(2\gamma_1 p + c_0\delta_3 r)}{2c_0 q} - x_0\gamma_1 = 0, \quad \frac{x_0\,(2\gamma_1 p + c_0\delta_3 r)}{2q} + \frac{b\,\gamma_1}{c_0} + 2p\,q = 0.$$

The integral $2K_2 = -\lambda_1 V_1 - \lambda_2(V_2 + c_0^2 V_3 - 2c_0 V_4)$ takes a stationary value on the elements of this family. Note that both the integral V_1 and the combination of integrals $V_2 + c_0^2 V_3 - 2c_0 V_4$ take stationary values on the elements of this IMs family.

By combining the rest of first integrals (9) with IMs equations (6) or the above presented, we can derive the equations of IM (or IMs families) which differ from found already. Add, e.g., equations (20) to the first expression of (9). The resulting equations define the family of stationary IMs of codimension 5 for differential equations (1):

$$2bp\,(\delta_2\gamma_3 - \delta_3\gamma_2) - (2p\gamma_1 + 2q\gamma_2 + r\gamma_3)(p\,r - x_0\gamma_3) = 0,$$
$$2x_0q\,(\delta_1\gamma_3 - \delta_3\gamma_1) - (2p\delta_1 + 2q\delta_2 + r\delta_3)(q\,r + b\delta_3) = 0,$$
$$2(p\delta_3\gamma_1 + q\delta_2\gamma_3) + r\delta_3\gamma_3 = 0, \quad \gamma_3 - c_0\delta_3 = 0,$$

$$\frac{1}{2(x_0 q \gamma_3 - p(q r + b \delta_3))} \Big[p(q r + b \delta_3) [2(p^2 + q^2) + r^2 - 2b \delta_2]$$

$$- x_0 \gamma_3 [2q(p^2 + q^2) + r(q r + b \delta_3)] + 2x_0 q \gamma_1 (p r - x_0 \gamma_3) \Big] = \tilde{c} \quad (\tilde{c} = \text{const}). (22)$$

Here $\tilde{c}$ is the family parameter. The integral $2K_3 = V_2 + c_0^2 V_3 - 2c_0 V_4$ takes a stationary value on the elements of this family.

IMs and IMs families (19)–(22) are related: the IMs of lesser dimension are submanifolds of IMs of greater dimension, and all these IMs belong to IM (6). The latter is revealed by direct calculations.

Indeed, substitute expressions (19) resolved with respect to $\delta_1, \delta_2, \gamma_2$ into equations (6). They become identities. Hence, one can conclude that IM (19) is a submanifold of IM (6). Analogously, relations between IMs family (20) and IM (19) as well as the rest of IMs are established.

Let us investigate the stability of IMs and IMs families "lifted up" into the original phase space. The integrals K_i are used for constructing the corresponding Lyapunov functions.

Consider the family of IMs (21). For the equations of perturbed motion, the variation of the integral K_2 in the neighbourhood of the elements of IMs family (21) can be written as

$$2\Delta K_2 = [\zeta_1^2 + \zeta_2^2] + [\xi_1^2 + \xi_2^2 + \xi_3^2],$$

where $y_1, y_2, y_3, y_4, y_5, y_6$ are the deviations from the elements of the IMs family under study, $\zeta_1 = by_2 + x_0 y_4$, $\zeta_2 = by_1 - x_0 y_5$, $\xi_1 = y_4 - c_0 y_1$, $\xi_2 = y_5 - c_0 y_2$, $\xi_3 = y_6 - c_0 y_3$.

The above quadratic form is sign-definite for the variables $\zeta_1, \zeta_2, \xi_1, \xi_2, \xi_3$. The latter is sufficient for the stability of the elements of IMs family (21) with respect to the variables $\zeta_1 = b\delta_2 + x_0 \gamma_1 + q^2 - p^2, \zeta_2 = b\delta_1 - x_0 \gamma_2 + 2qp, \xi_1 = \gamma_1 - c_0 \delta_1, \xi_2 = \gamma_2 - c_0 \delta_2, \xi_3 = \gamma_3 - c_0 \delta_3$.

Analogously, the stability of the elements of IMs families (20) and (22) is studied. We have proved the stability of the elements of both families with respect to part of the phase variables.

4   The Case of Parallel Force Fields

In this section, we study the solutions of equations (1) which can be obtained directly from them.

First, we equate the right-handed parts of the original differential equations to zero:

$$\begin{array}{lll} q r + b \delta_3 = 0, & \gamma_2 r - \gamma_3 q = 0, & \delta_2 r - \delta_3 q = 0, \\ x_0 \gamma_3 - p r = 0, & \gamma_3 p - \gamma_1 r = 0, & \delta_3 p - \delta_1 r = 0, \\ -b\delta_1 - x_0 \gamma_2 = 0, & \gamma_1 q - \gamma_2 p = 0, & \delta_1 q - \delta_2 p = 0 \end{array} \quad (23)$$

Equations (23) can be used for finding IMs in the problem under study. Following the technique chosen for obtaining IMs, we construct a lexicographical basis for the polynomials of system (23) with respect to some part of the phase variables, e.g., $\gamma_1, \gamma_2, \gamma_3, \delta_1, \delta_2, \delta_3, r$. The rest of the variables are considered as parameters. The resulting system is equivalent to the initial one, but it can be decomposed into two subsystems.

The subsystem 1:

$$q\,r + b\,\delta_3 = 0,\; x_0\gamma_3 - p\,r = 0,\; q^2 + b\,\delta_2 = 0,$$
$$p^2 - x_0\gamma_1 = 0,\; p\,q - x_0\gamma_2 = 0,\; p\,q + b\,\delta_1 = 0 \qquad (24)$$

The subsystem 2:

$$r = 0,\; \delta_3 = 0,\; \gamma_3 = 0,\, b p^2\,\delta_2 + x_0\gamma_1 q^2 = 0,$$
$$-b p\,\delta_2 - x_0\gamma_2 q = 0,\; q\,\delta_1 - p\,\delta_2 = 0. \qquad (25)$$

As can easily be verified by IM definition, equations (24), taking into account the constraints imposed on the constants of cosines integrals V_2, V_3 (2), define the one-dimensional IM of motion equations (1). The vector field on the given IM is described by the equation $\dot{p} = 0$. So, geometrically, in space R^9, this IM corresponds to a curve, over each point of which the family of the solutions $p = p_0 = \text{const}$ of the latter equation is defined. Let us show that all the solutions belonging to the IM under consideration are stationary.

Equations (24) together with V_2, V_3 (2), $p = p_0$ define in the original phase space the family of solutions of system (1):

$$p = p_0,\; q = -\frac{bp_0}{x_0},\; r = -\frac{z_1}{p_0 x_0},\; \gamma_1 = \frac{p_0^2}{x_0},\; \gamma_2 = -\frac{bp_0^2}{x_0^2},$$
$$\gamma_3 = -\frac{z_1}{x_0^2},\; \delta_1 = \frac{p_0^2}{x_0},\; \delta_2 = -\frac{bp_0^2}{x_0^2},\; \delta_3 = -\frac{z_1}{x_0^2}, \qquad (26)$$

where p_0 is the family parameter, $z_1 = \sqrt{x_0^4 - (b^2 + x_0^2)\,p_0^4}\;(p_0 < |x_0/(b^2 + x_0^2)^{1/4}|$ is the condition for the above solutions to be real).

Substitute (26) into the equations of stationarity (4) and find $\lambda_2, \lambda_3, \lambda_5$ from the resulting equations as some functions of $\lambda_0, \lambda_4, p_0, b, x_0$. Next, having substituted these functions into (3), we have the family of integrals

$$2\check{K}_1 = \lambda_0\left[2H + \frac{x_0^2}{(b^2 + x_0^2)p_0^2}\left(x_0^2 V_2 + b^2 V_3 - \frac{4x_0^2 p_0^2}{(b^2 + x_0^2)p_0^4 + x_0^4}\right)V_5\right] - \lambda_1 V_1$$
$$+ \lambda_4(V_2 + V_3 - 2V_4).$$

The above family of the integrals is split up into three subfamilies which are the coefficients of $\lambda_0, \lambda_1, \lambda_4$, respectively. Each of the subfamilies assumes a stationary value on the elements of family (26). Hence, the solutions under study are stationary. Note that the above one-dimensional IM itself is stationary: the integral V_1 takes a stationary value on this IM.

On substituting expressions (26) into V_4 (2), the latter becomes identically equal to 1. So, the above solutions correspond to the case when the vectors γ_i

and δ_i ($i = 1, 2, 3$) are parallel. From a mechanical viewpoint, the elements of the family of stationary solutions (26) correspond to the permanent rotations of the body around the direction of the coinciding force fields with angular velocity $\omega^2 = x_0^2 p_0^{-2}$. The axis position of the latter in the body depends on the parameter p_0 and does not coincide with the principal axes of the body, besides the case $\omega^2 = \sqrt{x_0^2 + b^2}$.

Next, substitute expressions (24), V_2, V_3 (2) resolved with respect to the variables $\gamma_1, \gamma_2, \ \gamma_3, \delta_1, \delta_2, \delta_3, q, r$ into (6). The latter equations turn into identities. Whence, one concludes that the one-dimensional IM under consideration is a submanifold of IM (6).

Finally, let us investigate the stability of the IM defined by equations (24), V_2, V_3 (2), using the integral V_1 for constructing a Lyapunov function.

The variation of the integral V_1 in the neighbourhood of a solution belonging to this IM in the deviations

$$y_1 = \delta_1 - \frac{p_0^2}{x_0}, \ y_2 = \delta_2 + \frac{bp_0^2}{x_0^2}, \ y_3 = \delta_3 + \frac{z_1}{x_0^2}, \ y_4 = \frac{p_0^2}{x_0} - \gamma_1, \ y_5 = \gamma_2 + \frac{bp_0^2}{x_0^2},$$

$$y_6 = \gamma_3 + \frac{z_1}{x_0^2}, \ y_7 = q + \frac{bp_0}{x_0}, \ y_8 = r + \frac{z_1}{p_0 x_0}, \ y_9 = p - p_0$$

can be written as $\Delta V_1 = \zeta_1^2 + \zeta_2^2$, where $\zeta_1 = [b\,(x_0 y_1 + p_0^2) - (x_0^2 y_5 - bp_0^2) + 2(x_0 y_7 - bp_0)(y_9 + p_0)]x_0^{-1}$, $\zeta_2 = -[(b\,(x_0^2 y_2 - bp_0^2) + (x_0 y_7 - bp_0)^2]x_0^{-2} + (y_9 + p_0)^2 - (x_0 y_4 + p_0^2)$. Since the latter quadratic form is sign-definite for the variables appearing in it, the IM under study is stable with respect to the variables $\zeta_1 = 2p\,q - x_0 \gamma_2 + b\,\delta_1$, $\zeta_2 = p^2 - q^2 - x_0 \gamma_1 - b\,\delta_2$.

Similarly, equations (25), taking into account the constraints imposed on the constants of the cosines integrals, define the one-dimensional IM of motion equations (1), the vector field on which is described by the equation $\dot{p} = 0$ ($p = \tilde{p}_0 = $ const), and everything aforesaid for the IM defined by equations (24), V_2, V_3 (2) is true for the given IM.

Having substituted equations (25), V_2, V_3 (2) resolved with respect to $\gamma_1, \gamma_2, \gamma_3, \delta_1, \delta_2, \delta_3, q, r$ into (11), we find that these turn into identities. Hence, the IM under consideration is a submanifold of IM (11).

4.1   On Some Classic Analogues of the Solutions of the Original Problem

When $b = 0$, equations (1) correspond to the Kowalewski integrable case of motion of the body under the influence of one force field only [9]. This section of the paper studies the solutions of the Kowalewski differential equations which are some analogues for a number of the solutions of equations (1).

A. On the stationary IM $p^2 - q^2 - x_0 \gamma_1 - b\,\delta_2 = 0$, $2pq - x_0 \gamma_2 + b\,\delta_1 = 0$ of equations (1) (the case $V_1 = 0$) there exists the Bogoyavlenskii integral [2]:

$$V_6 = (p^2 + q^2)r - 2px_0 \gamma_3 + 2b\,q\delta_3.$$

When $b = 0$, it turns into the integral $\bar{V}_6 = (p^2 + q^2)r - 2px_0\gamma_3$ of the Kowalewski equations on the IM:

$$p^2 - q^2 - x_0\gamma_1 = 0, \quad 2pq - x_0\gamma_2 = 0. \tag{27}$$

The following duality condition

$$\Omega_1 = 2\bar{H}\bar{V}_2 - \bar{V}_1^2 = \bar{V}_6^2, \quad \Omega_2 = 2\bar{H}\bar{V}_2 - \bar{V}_6^2 = \bar{V}_1^2. \tag{28}$$

holds for the integrals $\bar{V}_1$ and $\bar{V}_6$, where $\bar{H}, \bar{V}_1$ and $\bar{V}_2$ are the energy, area, and cosines integrals of the Kowalewski equations on IM (27), respectively.

Expressions (28) are the envelopes for the following combinations of the integrals:

$$2K_1 = 2\bar{H} - 2\mu\bar{V}_6 + \mu^2\bar{V}_2, \quad 2K_2 = 2\bar{H} - 2\mu\bar{V}_1 + \mu^2\bar{V}_2.$$

As follows from (28), the zero levels of the integrals $\bar{V}_1$ and $\bar{V}_6$ define the IMs on which the envelopes Ω_1, Ω_2 take stationary values, respectively. Obviously, these IMs are stable as any first integrals.

B. When $b = 0$, equations (6) are

$$\begin{aligned}
(2p\gamma_1 + 2q\gamma_2 + r\gamma_3)(pr - x_0\gamma_3) &= 0, \\
q\left[2x_0(\delta_1\gamma_3 - \delta_3\gamma_1) - (2p\delta_1 + 2q\delta_2 + r\delta_3)r\right] &= 0.
\end{aligned} \tag{29}$$

Note that these are not the IM equations for the problem of the Kowalewski top, since the variables $\delta_i(i = 1, 2, 3)$ appear in the 2nd equation.

The first equation is split up as follows:

$$2p\gamma_1 + 2q\gamma_2 + r\gamma_3 = 0, \quad pr - x_0\gamma_3 = 0. \tag{30}$$

Here the first equation is the area integral (its constant is zero) in the Kowalewski problem. It defines the stationary IM of the Kowalewski top. This IM is stable as any first integral.

Second equation (30) together with the solution $q = 0$ for the 2nd equation of (29) define the family of stationary IMs of the Kowalewski top:

$$pr - x_0\gamma_3 = 0, \quad q = 0, \quad \lambda_1 = x_0/p, \quad \lambda_2 = 1/p^2, \quad \lambda_3 = -x_0^2/p^2. \tag{31}$$

This solution can be obtained from the equations of stationarity for the following family of integrals $2K = 2\tilde{H} - 2\lambda_1\tilde{V}_1 - \lambda_2\tilde{V}_2 - \lambda_3\tilde{V}_3$ by solving these equations with respect to the unknowns $q, r, \lambda_1, \lambda_2, \lambda_3$. In this case, the projection of angular velocity $p = p_0$ is constant, and it is the family parameter. Here $\tilde{H}, \tilde{V}_1, \tilde{V}_2, \tilde{V}_3$ are the energy, area, cosines and Kowalewski integrals of the Kowalewski top in the original phase space of this problem.

The vector field on the elements of IMs family (31) is given by:

$$\dot{r} = -x_0\gamma_2, \quad \dot{\gamma}_1 = \gamma_2 r, \quad \dot{\gamma}_2 = \gamma_3 p_0 - \gamma_1 r, \quad \dot{\gamma}_3 = -\gamma_2 p_0.$$

So, the motions on the elements of this family are the rotations of the body with variable angular velocity around an axis lying in the principal xOz plane of the body.

Having introduced the deviations $\xi_1 = q$, $\xi_2 = pr - x_0\gamma_3$ from the elements of the IMs family under consideration, we have the positive-definite variation of the integral K when $x_0\gamma_1 < 0$:

$$2\Delta K = p_0^{-2}(-2x_0\gamma_1\xi_1^2 + \xi_2^2 - \xi_1^4).$$

Hence, the elements of the IMs family under study are stable. The above sufficient condition of stability is equivalent to the restriction $h - p_0^2 < 0$.

C. When $b = 0$, equations (10) are decomposed into several subsystems. One of them is represented below.

$$2\gamma_3(pr - x_0\gamma_3)((p^2 + q^2)r - x_0p\gamma_3) - r\gamma_1[(p^2 + q^2)r^2 - 2prx_0\gamma_3 + x_0^2\gamma_3^2] = 0,$$
$$x_0\gamma_3[x_0\gamma_3\gamma_2 + 2p(r\gamma_2 - q\gamma_3)] - r(p^2 + q^2)(r\gamma_2 - 2q\gamma_3) = 0. \tag{32}$$

As can easily be verified, equations (32) define the IM of the Kowalewski top. The vector field on this IM can be written as:

$$2\dot{p} = qr, \quad 2\dot{q} = x_0\gamma_3 - pr, \quad \dot{r} = -\frac{2x_0q\,\gamma_3\,\sigma_1}{\sigma_2}, \quad \dot{\gamma}_3 = -\frac{2x_0q\,\gamma_3\,\sigma_1}{r\sigma_2}. \tag{33}$$

These are obtained by eliminating γ_1 and γ_2 from the Kowalewski equations with the aid of (32). Here $\sigma_1 = (p^2 + q^2)r - px_0\gamma_3$, $\sigma_2 = (p^2 + q^2)r^2 - x_0\gamma_3(2pr - x_0\gamma_3)$.

The above equations admit the first integrals

$$2\hat{H} = 2p^2 + 2q^2 + r^2 + 4x_0\gamma_3\left(\frac{p}{r} - \frac{x_0q^2\gamma_3}{\sigma_2}\right), \quad \hat{V}_1 = 4p\gamma_3\left(\frac{p}{r} - \frac{x_0q^2\gamma_3}{\sigma_2}\right)$$
$$+ \frac{4q^2\gamma_3\sigma_1}{\sigma_2} + r\gamma_3, \quad \hat{V}_2 = 4\gamma_3^2\left(\frac{p}{r} - \frac{x_0q^2\gamma_3}{\sigma_2}\right)^2 + \left(\frac{2q\gamma_3\sigma_1}{\sigma_2}\right)^2 + \gamma_3^2,$$
$$\hat{V}_3 = \left(p^2 - q^2 - 2x_0\gamma_3\left(\frac{p}{r} - \frac{x_0q^2\gamma_3}{\sigma_2}\right)\right)^2 + \left(2pq - \frac{2x_0q\gamma_3\sigma_1}{\sigma_2}\right)^2, \quad \frac{r}{\gamma_3} = c_0 = \text{const.}$$

The top four of them are derived by eliminating γ_1 and γ_2 from the integrals $\hat{H}, \hat{V}_1, \hat{V}_2, \hat{V}_3$, using expressions (32). The latter is found directly from equations (33). Likely, these integrals are dependent, and they can be used for the further analysis of the problem.

Note that the IM given by the equations $p = r = \gamma_2 = 0$ belongs to IM (32). The motions on this IM are the pendulum oscillations of the body around the horizontal y axis. The IM defined by equations $p = r = \gamma_2 = \delta_1 = \delta_3 = 0$ corresponds to the above IM in the phase space of system (1).

Equations (1) also have the IM given by the equations $q = r = \gamma_2 = \gamma_3 = \delta_1 = 0$. These describe the oscillations of the body around the x axis coinciding with the direction of the 1st force field.

The latter two IMs of equations (1) are found from the conditions of stationarity for the nonlinear combination of the original problem integrals $2K_r = 2H^2 - \lambda_1V_1 - \lambda_2V_2 - \lambda_3V_3$ by resolving these equations with respect to some part of the phase variables and the parameters λ_i.

5 Conclusion

In this paper, the procedures based on computer algebra methods, in particular, Gröbner bases, have been proposed for obtaining the IMs of various dimension and levels, and their analysis. These allow one to find IMs together with the first integrals of vector fields on them. The integrals can also be used in the problem of qualitative analysis. The procedures proposed can be applied for the study of conservative systems, including complete integrable ones.

With the aid of the above technique, qualitative analysis of the problem of motion of a rigid body in double force field for the general integrable case has been conducted. The IMs and the families of IMs of various dimension and levels have been found. Their relationships have been established. For a series of the stationary IMs, the sufficient conditions of their stability with respect to part of the phase variables have been obtained. It was shown that in the case of parallel force fields, there exist the families of permanent rotations of the body. The sufficient conditions of their stability with respect to part of the phase variables have also been derived. For some IMs of the original problem, the solutions corresponding to them in the problem for one force field, have been obtained.

This research was supported financially by the Russian Foundation for Basic Research (16-07-00201). This work was also supported in part by the Council for Grants of the President of Russian Foundation for state support of the leading scientific schools, project NSh-8081.2016.9.

Appendix

The coefficients of equations (10):

$$\sigma_2 = 4x_0^2 r u \left[(b\delta_2 + x_0\gamma_1)z - 4(p\,(q\,r + b\delta_3) - x_0\gamma_3\,q)(b\delta_3\,p + x_0\gamma_3 q) \right],$$

$$\begin{aligned}
\sigma_1 = 4x_0 \Big[&2b^4\,\delta_3^3\,p\,(\delta_2\,(r^2 + 2u) - \delta_3 q\,r) + b^3\,\delta_3\,p\,[2\delta_2 r\,(\delta_3\,q\,(r^2 + u) - \delta_2\,r\,u) \\
&- \delta_3^2\,(r^2\,(3u + 2q^2) + 4p^2 u)] + b^2\delta_3\,p\,r\,u\,[\delta_2\,r\,(r^2 + 2v) - \delta_3 q\,(4p^2 + 3r^2)] \\
&+ 2b^3\,x_0\,\delta_3^2\,[\gamma_3\,(\delta_2\,q\,(r^2 - 2u) + \delta_3\,r\,u) + \delta_3\,\gamma_1\,p\,(r^2 + 2u)] \\
&+ b^2\,x_0\,[\delta_3^2\,q\,(2\gamma_1\,p\,r\,(r^2 + 3u) + \gamma_3\,(r^2\,(2p^2 + u) + 4p^2 u)) - 2\delta_2\gamma_3\,r\,(\delta_3\,(u^2 + r^2 v) \\
&+ \delta_2\,q\,r\,u)] + bx_0^2\,r^2\,u\,[\delta_3\,\gamma_1\,p\,(4q^2 + r^2) + \gamma_3\,(\delta_2\,q\,(r^2 + 2v) - \delta_3 r u)] \\
&+ x_0^2\,\gamma_3\,q\,r\,u\,[\gamma_3\,p\,(4q^2 - 3r^2) - \gamma_1 r\,(4p^2 - r^2)] + 2b^2\,x_0^2\,\delta_3\,\gamma_3\,[\delta_3\,q\,(\gamma_1\,(r^2 - 2u) \\
&- 2\gamma_3\,p\,r) + \delta_2\,\gamma_3 p\,(r^2 + 2u)] + bx_0^2\,[\gamma_3^2\,p\,(2\delta_2\,q\,r\,(u - r^2) + \delta_3\,(4q^2 u - r^2(u + 2q^2))) \\
&+ 2\delta_3\gamma_1 r(\gamma_1\,p\,r\,u - \gamma_3(3u^2 + r^2 v))] + 2b\,x_0^3\,\gamma_3^2\,[\delta_2\gamma_3\,q\,(r^2 - 2u) + \delta_3\,(\gamma_3\,r\,u \\
&+ \gamma_1 p(r^2 + 2u))] + x_0^3\gamma_3[2\gamma_1\,q\,r\,(\gamma_1\,r\,u + \gamma_3\,p\,(3u - r^2)) + \gamma_3^2\,(q\,r^2\,(2p^2 + 3u) \\
&- 4q^3 u)] + 2x_0^4\,\gamma_3^3 q\,(\gamma_1\,(r^2 - 2u) - \gamma_3 p r) \Big],
\end{aligned}$$

$$\begin{aligned}
\sigma_0 = b^5\delta_3^2\,&[4\delta_2^2\,\delta_3\,q\,(r^2 - 2u) - 2\delta_3^3\,q\,(4p^2 + r^2) + 4\delta_2^3\,r\,u + \delta_2\,\delta_3^2\,r\,(r^2 + 4(v - q^2))] \\
&+ 2b^4\delta_3\,[\delta_2\,\delta_3^2\,q\,r^2\,(r^2 - 8q^2) - 4\delta_2^2\,\delta_3\,r\,(u^2 + q^2\,(u - r^2)) + 4\delta_3^3 q r^2\,u - \delta_3^3 r\,(4p^2 + r^2) \\
&\times (q^2 + u)] + b^3 r^2\,u\,[4\delta_2^2\delta_3 q\,(r^2 - 2u) - 2\delta_3^3 q\,(4p^2 + r^2) + 4\delta_2^3 r\,u + \delta_2\delta_3^2 r\,(r^2 + 4(v
\end{aligned}$$

$$- q^2))] + b^2 x_0 r\, u\, [\, 4\delta_2\delta_3\gamma_1 q\, r\, (r^2 - 4p^2) - 2\delta_3^2\gamma_3 p\, r\, (r^2 - 4q^2) + \delta_3^2\gamma_1 (16p^2 q^2 + r^4)$$

$$+ 4\delta_2\, u\, (4\delta_3\gamma_3 p\, q + \delta_2\gamma_1 r^2) - 4\delta_2^2\gamma_3 p\, r\, (r^2 + 2u)] + 2b^3 x_0\, [\, 4\delta_3\gamma_3 p\, q\, r\, (3\delta_3^2 u - 2\delta_2^2\, r^2)$$

$$+ 4\delta_2^2 r^2\, u\, (2\delta_3\gamma_1 q - \delta_2\gamma_3 p) + \delta_3^3\gamma_1 q\, (r^2(r^4 - 4q^2) + 8p^2 u) + 2\delta_2\delta_3^2\gamma_1 r\, (2q^2(r^2 - u)$$

$$+ (r^2 - 2u)u) - \delta_2\delta_3^2\gamma_3 p\, (r^2(r^2 + 4v) - 8u^2)] + b^4 x_0\delta_3^2\, [\, 4\delta_2\gamma_1(2\delta_3 q\, (r^2 - 2u)$$

$$+ 3\delta_2 r\, u) - 4\delta_2^2\gamma_3 p\, (r^2 + 2u) + 2\delta_3^2\gamma_3 p\, (4(q^2 + u) + r^2) + \delta_3^2\gamma_1 r\, (r^2 + 4(v - q^2))]$$

$$+ b x_0^2\gamma_3 r^2 u\, [\, \delta_2\gamma_3 r\, (4(p^2 + v) + r^2) - 2\delta_3\gamma_3 q\, (4p^2 + r^2) - 4\delta_2\gamma_1 p\, (4q^2 + r^2)]$$

$$+ 2b x_0^3\gamma_3\, [\, 4p\, r\, u\, (3\delta_3\gamma_3^2 q - \delta_2\gamma_1^2 r) - \delta_2\gamma_3^2 p\, r^2\, (8p^2 + r^2) + \delta_3\gamma_1\gamma_3 q(r^2(r^2 - 4q^2)$$

$$+ 8p^2 u) + 2\delta_2\gamma_1\gamma_3 r(2p^2(r^2 + u) + u(r^2 + 2u))] + b x_0^4\gamma_3^2\, [\, 2\delta_3\gamma_3^2 q(r^2 - 4(u + p^2))$$

$$+ 4\gamma_1^2(\delta_3 q\, (r^2 - 2u) + 3\delta_2 r\, u) - 8\delta_2\gamma_1\gamma_3 p\, (r^2 + 2u) + \delta_2\gamma_3^2 r\, (4(p^2 + v) + r^2)]$$

$$+ b^2 x_0^2\, [\, 2\delta_3^2\gamma_3^2 r\, (r^2 v - 4(2p^2 q^2 + u^2)) - 16\delta_2\delta_3\gamma_1\gamma_3 p\, q\, r^3 - 2\delta_3^2\gamma_1\gamma_3 p\, (8q^4 + r^4$$

$$+ 4p^2(2q^2 + r^2)) + 8\delta_2^2\gamma_3^2 r\, (2p^4 + q^4 + p^2(3q^2 + r^2)) + 4\gamma_1 r^2 u\, (2\delta_2\delta_3\gamma_1 q - 4\delta_2^2\gamma_3 p$$

$$+ \delta_3^2\gamma_1 r) + 2\delta_2\delta_3\gamma_3^2 q\, (r^4 - 8u^2 + 4r^2 v)] + 2b^3 x_0^2\, [\, 4\delta_2\delta_3\gamma_1\gamma_3^2 q\, (r^2 - 2u)$$

$$+ 2\gamma_1 r\, u\, (\delta_3^2\gamma_1^2 + 3\delta_2^2\gamma_3^2) + 4\delta_3^3\gamma_3^3 p\, (2q^2 + u) - 2\gamma_3 p\, (r^2 + 2u)(\delta_3^2\gamma_1^2 + \delta_2^2\gamma_3^2)$$

$$+ \delta_3^2\gamma_1\gamma_3^2 r\, (r^2 + 6v)] + 2b^3 x_0^2\, [\, 2q\, (r^2 - 2u)(\delta_3^3\gamma_1^2 + \delta_2^2\delta_3\gamma_3^2) - 4\delta_3^3\gamma_3^2 q\, (3p^2 + q^2)$$

$$+ 2\delta_2 r\, u\, (3\delta_3^2\gamma_1^2 + \delta_2^2\gamma_3^2) - 4\delta_2\delta_3^2\gamma_1\gamma_3 p\, (r^2 + 2u) + \delta_2\delta_3^2\gamma_3^2 r\, (r^2 + 6v)]$$

$$+ x_0^3\gamma_3^2 r\, u\, [\, 2\gamma_3 p\, r\, (4q^2 - r^2) + \gamma_1\, (16p^2 q^2 + r^4)] + 2x_0^4\gamma_3^2\, [\, 2\gamma_1^2 r^3\, u + \gamma_3^2 r\, (r^2 - 4q^2)$$

$$\times (p^2 + u) - \gamma_1\gamma_3 p\, (8q^4 + r^4 + 4p^2\, (2q^2 + r^2))] + x_0^5\gamma_3^3\, [\, 2\gamma_3^3 p\, (4q^2 - r^2) + 4\gamma_1^3 r\, u$$

$$- 4\gamma_1^2\gamma_3 p\, (r^2 + 2u) + \gamma_1\gamma_3^2 r\, (4(p^2 + v) + r^2)],$$

$$\hat{\sigma}_2 = 2b^4\delta_3^2\, (2\delta_2 u + \delta_3 q\, r) + b^2 r\, u\, [2\delta_2\, r\, u - \delta_3 q(4p^2 - r^2)] + b^3\delta_3\, [\, 6\delta_2 q\, r\, u$$

$$+ \delta_3\, ((p^2 + 3q^2)r^2 - 4p^2 u)] - b x_0\gamma_3 p\, r\, u\, (4q^2 + r^2) + 2b^3 x_0\delta_3^2\, (2\gamma_1 u - \gamma_3 p\, r)$$

$$+ 2b^2 x_0\, r[\delta_3 q\, (\gamma_1 u - 2\gamma_3 p\, r) - 3\delta_2\gamma_3 p\, u] + 2b^2 x_0^2\gamma_3^2\, (\delta_3 q\, r + 2\delta_2 u)$$

$$+ b x_0^2\gamma_3\, [\gamma_3((3p^2 + q^2)r^2 + 4q^2 u) - 2\gamma_1 p\, r\, u] + 2b x_0^3\gamma_3^2\, (2\gamma_1 u - \gamma_3\, p\, r),$$

$$\hat{\sigma}_1 = 2x_0^4\gamma_3^2\, (2\gamma_1 u + \gamma_3\, p\, r) + x_0^2 r\, u\, [2\gamma_1 r\, u - \gamma_3 p\, (4q^2 - r^2)] - x_0^3\gamma_3\, [\, 6\gamma_1 p\, r\, u$$

$$- \gamma_3(4q^2 u - (3p^2 + q^2)r^2)] - b x_0\delta_3 q\, r\, u\, (4p^2 + r^2) + 2b^3 x_0\delta_3^2\, (2\delta_2 u - \delta_3\, q\, r)$$

$$+ 2b x_0^2 r\, (\gamma_3 p\, (2\delta_3 q\, r - \delta_2 u) + 3\delta_3\gamma_1 q\, u) + 2b^2 x_0^2\delta_3^2\, (\gamma_3 p\, r + 2\gamma_1\, u)$$

$$- b^2 x_0\delta_3\, [\delta_3\, ((p^2 + 3q^2)r^2 + 4p^2 u) - 2\delta_2 q\, r\, u] + 2b x_0^3\gamma_3^2\, (2\delta_2 u - \delta_3 q\, r),$$

$$\hat{\sigma}_0 = 2b^4\delta_3^3\, (\delta_2 p\, r - 2\delta_3 p\, q) + 2b^3\delta_3\, [\delta_3 p\, (\delta_2 q\, (r^2 - 2u) - 2\delta_3\, (u + q^2)r) - \delta_2^2 p\, r\, u]$$

$$- b^2\delta_3 p\, r\, u\, [4\delta_3 q\, r + \delta_2\, (4q^2 - r^2)] + b x_0 r\, [\delta_2\gamma_3 q\, (4p^2 + r^2)u + \delta_3\gamma_1\, p\, (4q^2 + r^2)u]$$

$$+ 2b^3 x_0\delta_3^2\, [\gamma_3(\delta_2 q r + \delta_3 u) + \delta_3\, (\gamma_1 p\, r + \gamma_3 u)] + b^2 x_0\, [2\delta_3^2(2\gamma_3 q\, r\, (2p^2 + q^2)$$

$$+ \gamma_1\, p\, q\, (r^2 + 2u)) + \delta_2\gamma_3\, (\delta_3(4u^2 - 2r^2 v) - 2\delta_2 q\, r u)]$$

$$+ 2b^2 x_0^2\delta_3\gamma_3\, [\delta_3\gamma_1 q\, r + \gamma_3 p\, (\delta_2 r - 4\delta_3 q)] - x_0^2\gamma_3 q\, r\, u\, [4\gamma_3 p\, r + \gamma_1\, (4p^2 - r^2)]$$

$$+ b x_0^2\, [2\delta_3 p\, r\, (\gamma_1^2 u - 2\gamma_3^2(p^2 + 2q^2)) - \gamma_3\, (\delta_3\gamma_1\, (4u^2 + 2r^2 v)$$

$$+ 2\delta_2\gamma_3 p\, q\, (r^2 + 2u))] + 2b x_0^3\gamma_3^2\, [r\, (\delta_3\gamma_1 p + \delta_2\gamma_3 q) + 2\delta_3\gamma_3 u]$$

$$+ x_0^3\gamma_3 q\, [2\gamma_1^2 r u + \gamma_3(4\gamma_3 r(2p^2 + q^2) + \gamma_1 p\, (4u - 2r^2))] + 2x_0^4\gamma_3^3 q\, (\gamma_1 r - 2\gamma_3 p),$$

$$z = 4[b\delta_3\, (q\, r + b\delta_3) - x_0\gamma_3\, (p\, r - x_0\gamma_3)] + r^2 u, \quad u = p^2 + q^2, \quad v = p^2 - q^2.$$

References

1. Sarychev, V.A., Mirer, S.A., Degtyarev, A.A., Duarte, E.K.: Investigation of equilibria of a satellite subjected to gravitational and aerodynamic torques. Celest. Mech. Dyn. Astron. **97**(4), 267–287 (2007)
2. Bogoyavlenskii, O.I.: Two integrable cases of a rigid body dynamics in a force field. USSR Acad. Sci. Doklady **275**(6), 1359–1363 (1984)
3. Yehia, H.: New integrable cases in the dynamics of rigid bodies. Mech. Res. Comm. **13**(3), 169–172 (1986)
4. Bobenko, A.I., Reyman, A.G., Semenov-Tian-Shansky, M.A.: The Kowalewski Top 99 years later: a Lax pair, generalizations and explicit solutions. Commun. Math. Phys. **122**, 321–354 (1989)
5. Zotev, D.V.: Fomenko-Zieschang invariant in the Bogoyavlenskyi case. Regul. Chaotic Dyn. **5**(4), 437–458 (2000)
6. Ryabov, P.E., Kharlamov, M.P.: Classification of singularities in the problem of motion of the Kovalevskaya top in a double force field. Sb. Math. **203**(2), 257–287 (2012)
7. Ryabov, P.E.: Phase topology of one irreducible integrable problem in the dynamics of a rigid body. Theoret. Math. Phys. **176**(2), 1000–1015 (2013)
8. Irtegov, V., Titorenko, T.: On Invariant Manifolds and Their Stability in the Problem of Motion of a Rigid Body under the Influence of Two Force Fields. In: Gerdt, V.P., Koepf, W., Seiler, W.M., Vorozhtsov, E.V. (eds.) CASC 2015. LNCS, vol. 9301, pp. 220–232. Springer, Heidelberg (2015). doi:10.1007/978-3-319-24021-3_17
9. Kowalewski, S.: Sur le problème de la rotation d'un corps solide autour d'un point fixe. Acta Math. **12**, 177–232 (1889)

On Multiple Eigenvalues of a Matrix Dependent on a Parameter

Elizabeth A. Kalinina[(✉)]

Saint-Petersburg State University, Saint-Petersburg, Russia
ekalinina69@gmail.com

Abstract. In this paper, a square matrix with elements linearly dependent on a parameter is considered. We propose an algorithm to find all the values of the parameter such that the matrix has a multiple eigenvalue. We construct a polynomial whose roots are these values of the parameter. A numerical example shows how the algorithm works.

1 Introduction

For two square $k \times k$ complex matrices A and B without common eigenvalues, we would like to find all values of λ (real or complex) such that matrix $A + \lambda B$ has a multiple eigenvalue, i.e., an eigenvalue of multiplicity greater than 1.

It is well known that a generic matrix has only simple eigenvalues. Nevertheless, multiple eigenvalues typically appear in matrix families. Earlier the consideration of such singular matrices was, on the whole, only theoretically significant. However, now it has different practical applications in quantum mechanics, nuclear physics, optics and kinetics of material systems [14,18].

Similar problems are studied in perturbation theory. Many papers deal with condition numbers of matrices with multiple eigenvalues (see, e.g., the recent works [2,9,10]). But in these papers, only the values of λ close to zero are considered. Here we assume that λ can take on arbitrary values.

A detailed description of this problem, its properties and applications together with the solution based on the Newton method are presented in [6]. Also it is noted that there appears to be no globally convergent numerical method for the problem of finding all such values of λ. So the problem remains topical. In the recent paper [15], a method that allows us to find all required values of λ using only standard matrix computations is proposed. Unfortunately, as was mentioned by the authors themselves, this method is very sensitive and can be applied only for small matrices.

In this paper, an algorithm to construct a polynomial whose zeroes are the required values of λ is presented. Zeroes of the obtained polynomial can be found with arbitrary accuracy by any suitable method. The algorithm is based on properties of the Kronecker product of matrices and the Leverrier method for the computation of the characteristic polynomial of a matrix [11,21]. Also we present a numerical example that shows how the algorithm works. The results are compared with the ones obtained by the method described in [15].

© Springer International Publishing AG 2016
V.P. Gerdt et al. (Eds.): CASC 2016, LNCS 9890, pp. 305–314, 2016.
DOI: 10.1007/978-3-319-45641-6_20

It is shown how the suggested algorithm could be generalized to the case of matrix with elements polynomially dependent on a parameter.

2   Preliminary Results

Consider a polynomial

$$f(x) = a_0 x^n + a_1 x^{n-1} + a_2 x^{n-2} + \ldots + a_n, \qquad a_0 \neq 0, a_j \in \mathbb{C}, j = 0, 1, \ldots, n.$$

Let $\lambda_1, \lambda_2, \ldots, \lambda_n \in \mathbb{C}$ be the zeros of f counted with their multiplicity.

Definition 1. *For a polynomial $f(x)$, the Newton sums $s_0, s_1, s_2, \ldots$ are defined as*

$$s_0 = n;$$
$$s_p = \lambda_1^p + \lambda_2^p + \ldots + \lambda_n^p, \qquad p = 1, 2, \ldots$$

The Newton identities [19] give relations between the coefficients of a polynomial f and the Newton sums:

$$s_p = \begin{cases} -(a_1 s_{p-1} + a_2 s_{p-2} + \ldots + a_{p-1} s_1 + p a_p)/a_0, & \text{if } p \leq n, \\ -(a_1 s_{p-1} + a_2 s_{p-2} + \ldots + a_n s_{p-n})/a_0, & \text{if } p > n. \end{cases} \tag{1}$$

Suppose, for a polynomial $f(x)$ of degree n, we know the Newton sums $s_1, s_2, \ldots, s_n$. Assuming $a_0 = 1$, we can find the coefficients of $f(x)$ by the following formulae (the Newton formulae):

$$a_1 = -s_1; a_2 = -(s_2 + a_1 s_1)/2;$$
$$a_p = -(s_p + a_1 s_{p-1} + a_2 s_{p-2} + \ldots + a_{p-1} s_1)/p, \text{ if } p \leq n. \tag{2}$$

These formulae can be rewritten as follows [12]:

$$a_p = \frac{(-1)^p}{p!} \begin{vmatrix} s_1 & 1 & 0 & 0 & \ldots & 0 \\ s_2 & s_1 & 2 & 0 & \ldots & 0 \\ s_3 & s_2 & s_1 & 3 & \ldots & 0 \\ \multicolumn{6}{c}{\ldots} \\ s_p & s_{p-1} & s_{p-2} & s_{p-3} & \ldots & s_1 \end{vmatrix}_{p \times p}. \tag{3}$$

For any matrix $D_{k \times k}$, the Newton sums of its characteristic polynomial can be expressed through the traces of the powers of D. The following formulae give these expressions:

$$s_p = \operatorname{Sp} D^p, \qquad p = 1, 2, \ldots, \tag{4}$$

where $\operatorname{Sp} D$ stands for the trace of matrix D.

Definition 2. *If A is a $k \times k$ matrix and B is an $\ell \times \ell$ matrix, then the Kronecker product $A \otimes B$ is the $k\ell \times k\ell$ block matrix*

$$[A \otimes B]_{k\ell \times k\ell} = \begin{bmatrix} a_{11} B & a_{12} B & \ldots & a_{1k} B \\ a_{21} B & a_{22} B & \ldots & a_{2k} B \\ \multicolumn{4}{c}{\ldots} \\ a_{k1} B & a_{k2} B & \ldots & a_{kk} B \end{bmatrix}.$$

By $\alpha_1, \ldots, \alpha_k$ and $\beta_1, \ldots, \beta_\ell$ denote the eigenvalues of matrices A and B, respectively.

Given A and B, construct the matrix

$$C = A \otimes I_{\ell \times \ell} - I_{k \times k} \otimes B.$$

Here $I_{k \times k}$ stands for the $k \times k$ identity matrix.

Suppose we know the eigenvalues of matrices A and B, then to find the eigenvalues of C we can use the following theorem [13].

Theorem 1. *The eigenvalues of the matrix C are $\alpha_i - \beta_j$, where $i = 1, 2, \ldots, k$ and $j = 1, 2, \ldots, \ell$.*

Hence, we obtain the necessary and sufficient condition for two matrices A and B to have a common eigenvalue.

Corollary 1. *Matrices $A_{k \times k}$ and $B_{\ell \times \ell}$ have a common eigenvalue iff $\det C = 0$.*

We will use the following properties of the Kronecker product of matrices and the trace [8]:

$$(A \otimes B)(C \otimes D) = (AC) \otimes (BD), \tag{5}$$
$$\mathrm{Sp}\,(A \otimes B) = \mathrm{Sp}\,A \cdot \mathrm{Sp}\,B. \tag{6}$$

For a $k \times k$-matrix D, by $\delta_1, \delta_2, \ldots, \delta_k$ denote its eigenvalues counted with their multiplicity. Consider the matrix

$$\mathcal{C}_D = D \otimes I - I \otimes D. \tag{7}$$

According to Theorem 1, the eigenvalues of $\mathcal{C}_D$ are equal to $\delta_i - \delta_j$, where $i, j \in \{1, 2, \ldots, k\}$. Therefore, all the eigenvalues of D are simple iff $\mathrm{rank}\,\mathcal{C}_D = k^2 - k$.

3   Multiple Eigenvalues of a Matrix Dependent on a Parameter

Consider two square $k \times k$ complex matrices A and B. We would like to find all the values of λ such that the matrix $D(\lambda) = A + \lambda B$ has a multiple eigenvalue.

Remark 1. Suppose that A and B have no eigenvalues in common. This condition can be verified by Corollary 1 to theorem 1.

By s_p and S_p $(p = 0, 1, 2, \ldots)$ denote the Newton sums of characteristic polynomials of matrices D and $\mathcal{C}_D$, respectively. The following theorem establishes the relation between s_p and S_p (recall that by (4) $s_p = \mathrm{Sp}\,D^p$, $S_p = \mathrm{Sp}\,\mathcal{C}_D^p$):

Theorem 2. *The trace of the matrix* C_D^p ($p = 1, 2, \ldots$) *can be found by the formulae*

$$S_{2p} = 2ks_{2p} - 2C_{2p}^1 s_{2p-1}s_1 + 2C_{2p}^2 s_{2p-2}s_2 - \ldots + (-1)^p C_{2p}^p s_p^2,$$
$$S_{2p-1} = 0. \tag{8}$$

Here $C_n^p = \dfrac{n!}{p!(n-p)!}$, $p = 0, 1, 2, \ldots$.

Proof. By formula (5), we get:

$$C_D^p = D^p \otimes I - C_p^1 D^{p-1} \otimes D + C_p^2 D^{p-2} \otimes D^2 - \ldots + (-1)^p I \otimes D^p.$$

To conclude the proof, it remains to use property (6). □

Now we can prove the necessary and sufficient condition for a matrix to have a multiple eigenvalue.

Theorem 3. *Matrix D has a multiple eigenvalue iff*

$$\begin{vmatrix} S_2 & 2 & 0 & \ldots & 0 \\ S_4 & S_2 & 4 & \ldots & 0 \\ \ldots & & & & \\ S_{k^2-k} & S_{k^2-k-2} & S_{k^2-k-4} & \ldots & S_2 \end{vmatrix}_{(k^2-k)/2 \times (k^2-k)/2} = 0. \tag{9}$$

Proof. By formulae (3), the coefficient A_{k^2-k} of μ^k in the characteristic polynomial of matrix C_D

$$\det(C_D - \mu I_{k^2 \times k^2}) = A_0 \mu^{k^2} + A_1 \mu^{k^2-1} + A_2 \mu^{k^2-2} + \ldots + A_{k^2}$$

can be expressed through the Newton sums $S_p = \operatorname{Sp} C_D^p$:

$$A_{k^2-k} = \frac{1}{(k^2-k)!} \begin{vmatrix} 0 & 1 & 0 & 0 & \ldots & 0 & 0 \\ S_2 & 0 & 2 & 0 & \ldots & 0 & 0 \\ 0 & S_2 & 0 & 3 & \ldots & 0 & 0 \\ \ldots & & & & & & \\ 0 & S_{k^2-k-2} & 0 & S_{k^2-k-4} & \ldots & 0 & k^2-k-1 \\ S_{k^2-k} & 0 & S_{k^2-k-2} & 0 & \ldots & S_2 & 0 \end{vmatrix}.$$

Using the Laplace expansion formula for the determinant [7] along the columns with even indices, we get:

$$A_{k^2-k} = \frac{1}{(k^2-k)!} \begin{vmatrix} 1 & 0 & 0 & \ldots & 0 \\ S_2 & 3 & 0 & \ldots & 0 \\ S_4 & S_2 & 5 & \ldots & 0 \\ \ldots & & & & \\ S_{k^2-k-2} & S_{k^2-k-4} & S_{k^2-k-6} & \ldots & k^2-k-1 \end{vmatrix}.$$

$$\times \begin{vmatrix} S_2 & 2 & 0 & \ldots & 0 \\ S_4 & S_2 & 4 & \ldots & 0 \\ & \ldots & & & \\ S_{k^2-k} & S_{k^2-k-2} & S_{k^2-k-4} & \ldots & S_2 \end{vmatrix}$$

$$= \frac{1}{(k^2-k)!!} \begin{vmatrix} S_2 & 2 & 0 & \ldots & 0 \\ S_4 & S_2 & 4 & \ldots & 0 \\ & \ldots & & & \\ S_{k^2-k} & S_{k^2-k-2} & S_{k^2-k-4} & \ldots & S_2 \end{vmatrix},$$

where $(k^2 - k)!!$ stands for the product of all even natural numbers from 1 to $k^2 - k$. The last equality concludes the proof. □

The following formulae similar to the Newton formulae allow us to find A_{k^2-k} without using determinant.

Corollary 2.

$$A_2 = -S_2; A_4 = -(S_4 + A_2 S_2)/2;$$
$$A_{2p} = -(S_{2p} + A_2 S_{2p-2} + A_4 S_{2p-4} + \ldots + A_{2p-2} S_2)/2p,$$
$$\text{if } p \leq (k^2 - k)/2. \tag{10}$$

Theorems 1, 2, and 3 together with formulae (10) allow us to provide an algorithm to find all values of λ such that matrix $D = A + \lambda B$ has a multiple eigenvalue. In order to do this, first we must find the traces of the powers of D. We will use the following properties of matrix trace [19]:

Lemma 1. $\operatorname{Sp}(AB) = \operatorname{Sp}(BA)$, $\operatorname{Sp}(A + B) = \operatorname{Sp} A + \operatorname{Sp} B$.

Hence, we have the following equality:

$$s_p = \operatorname{Sp} D^p = \operatorname{Sp} A^p + \lambda C_p^1(\operatorname{Sp} A^{p-1}B) + \lambda^2 C_p^2 \operatorname{Sp}(A^{p-2}B^2) + \ldots + \lambda^p \operatorname{Sp} B^p$$
$$(p = 1, 2, 3, \ldots) \tag{11}$$

Thus, it is necessary to find the first 1 to $k^2 - k$ powers of matrices A and B.

Remark 2. We can omit the last operation of matrix multiplication. It is sufficient to find only the elements situated on the main diagonal of the resulting matrix.

Remark 3. The calculation of matrix powers is rather expensive problem. For large matrices, the Strassen algorithm [17] for fast matrix multiplication can be useful.

Remark 4. For large k, we can compute the first 1 to $k - 1$ powers of matrices A and B. Then we can find the traces of obtained matrices and characteristic polynomials of matrices A and B, using the Newton formulae (2). After that, we can use the Cayley – Hamilton theorem [19] and Lemma 1 for calculating the traces of matrices A^p and B^p for $p = k, k + 1, \ldots, k^2 - k$.

3.1   The Algorithm

Let A and B be two square $k \times k$-matrices. It is necessary to find all values of λ (real or complex) such that the matrix $A + \lambda B$ has a multiple eigenvalue.

1. Calculate the powers of matrices A and B: A^p, B^p, $(p = 1, 2, 3, \ldots, k^2 - k - 1)$.

2. Calculate the traces of matrices $A^p B^q$ $p, q \in \{0, 1, 2, \ldots, k^2 - k\}$, $p + q \leq k^2 - k$.

3. By (11), calculate the Newton sums s_p of the characteristic polynomial of matrix $D = A + \lambda B$ $(p = 1, 2, \ldots, k^2 - k)$.

4. By (8), find the Newton sums $S_{2p} = \mathrm{Sp}\, D^{2p}$ of the characteristic polynomial of matrix $\mathcal{C}_D$ $(p = 1, 2, \ldots, (k^2 - k)/2)$.

5. Calculate the zeroes of the polynomial (9).

The obtained zeroes are the required values of λ.

Remark 5. The method can be generalized for a matrix $D(\lambda)$ with elements polynomially dependent on λ, i.e., for the matrix polynomial

$$D(\lambda) = A_0 \lambda^m + A_1 \lambda^{m-1} + \ldots + A_m,$$

where A_j $j = 0, 1, \ldots, m$ is a square $k \times k$ complex matrix. In this case, it is more difficult to calculate the Newton sums of the characteristic polynomial of matrix $D(\lambda)$. All the other steps do not change. (In [20], a survey of different problems concerning the eigenvalues of a matrix and known methods of their solutions is presented.)

3.2   Asymptotic Complexity of the Algorithm and Accuracy of Computations

In the algorithm, matrix multiplication is the most expensive operation. The square matrix multiplication has an asymptotic complexity of $O(n^3)$, if carried out naively. The current $O(n^p)$ algorithm with the lowest known exponent p is a generalization of the Coppersmith – Winograd algorithm [3] that has an asymptotic complexity of $O(n^{2.3728639})$. The Strassen algorithm has an asymptotic complexity of $O(n^{\log_2 7}) \approx O(n^{2.807})$. To compute A^p, B^p $(p = 1, 2, \ldots, k^2 - k - 1)$ we have to do $2(k^2 - k - 2)$ matrix multiplications (we have $O(k^5)$ operations), then we find traces of matrices $A^p B^q$, $p, q \in \{0, 1, 2, \ldots, k^2 - k\}, p + q \leq k^2 - k$ — approximately $O(k^6)$ operations in all. Hence, there are $O(k^6)$ operations, if we do not take into account finding of zeroes of polynomial (9).

We can evaluate all the complex zeros of an kth degree univariate polynomial with relative errors $\leq \epsilon$ using $O(k^2 \log k(\log k + \log(1/\epsilon)))$ arithmetic operations [16]. For $\epsilon = 1/(10^{k^3})$, we have $O(k^6)$ operations. Therefore, the suggested algorithm has an asymptotic complexity of $O(k^6)$.

It is difficult to estimate how many times the singular value decomposition (SVD) must be computed during the work of algorithm described in [15]. However, because the SVD computation complexity is $O(n^2)$ [5], the asymptotic complexity of algorithm described in [15] is not less than $O(k^6)$. Therefore, the asymptotic computational complexities of two algorithms mentioned above may be considered equal.

Matrix multiplication is the base operation in the algorithm suggested in the present paper. Hence, we would like to reduce the numerical errors that occur in this operation. For matrix multiplication in floating point arithmetic, the accuracy of computations can be improved without much extra cost using result presented in [1]. For every inner product, computations are made in double precision, and the result is rounded to a single precision number. For $k \times 1$ vectors U, V, suppose that $|U^T||V| \leq v|U^T V|$, where $|U|, |V|$ are vectors whose components equal the absolute values of components of the vectors U, V, respectively. Then for an inner product $U^T V$, the computed inner product is almost as accurate as the correctly rounded exact product. The bound for an inner product $U^T V$ equals

$$|fl(fl_e(U^T V)) - U^T V| < v|U^T V| + \frac{k v_e}{1 - k v_e/2}(1 + v)|U^T||V|.$$

Here v ("unit roundoff") is the maximum value of relative error, $v = \varepsilon/2$ (for example, for *float* (4 bytes) $\varepsilon \approx 1.19 \cdot 10^{-7}$, for *double* (8 bytes) $\varepsilon \approx 2.22 \cdot 10^{-16}$, for *long double* (10 or 12 bytes depending on the system) $\varepsilon \approx 1.08 \cdot 10^{-19}$). The values of ε can be found in the standard included file *float.h* for **C**-compiler for the architecture x86. By $fl(a)$ denote computation with ordinary precision (with t binary digits), by $fl_e(a)$ denote computation with extended precision (with $2t$ binary digits), v_e is the corresponding unit roundoff.

Remark 6. In single precision, the condition $|U^T||V| \leq v|U^T V|$ holds for the inner product with absolute value of more than $0.1862645142 \cdot 10^{-8} k$. For example, if $k = 4000$, then the absolute value of the inner product must be more than $0.7450580568 \cdot 10^{-5}$.

Remark 7. There exists a direct way to solve the problem under consideration. We can find the characteristic polynomial of matrix $D = A + \lambda B$ and then compute the values of parameter corresponding to its multiple eigenvalues. But this approach seems to be quite expensive. Coefficients of the obtained polynomial are polynomially dependent on λ. The maximum possible degree of such a coefficient equals k. To find the required values of λ we must compute the discriminant of the characteristic polynomial, i.e., the determinant of a $(k^2 - k) \times (k^2 - k)$ polynomial matrix. According to results presented in [4], the computational complexity of this approach is not less than $O(k^7)$.

3.3   A Numerical Example

The computations in the following example were made using Maple 13.0.

Example 1. Consider the problem from [15]. For matrices

$$A = \begin{pmatrix} 1 & -2 & 3 \\ -1 & 1 & 2 \\ 1 & 1 & -1 \end{pmatrix}, \qquad B = \begin{pmatrix} 1 & -1 & 1 \\ 1 & 1 & 3 \\ -1 & 1 & 1 \end{pmatrix},$$

it is necessary to find all values of the parameter λ such that matrix $D = A + \lambda B$ has a multiple eigenvalue.

Compute the Newton sums of the characteristic polynomial of matrix $A + \lambda B$:

$$s_1 = 3\lambda + 1;$$
$$s_2 = 5\lambda^2 + 6\lambda + 17;$$
$$s_3 = 21\lambda^3 + 33\lambda^2 + 30\lambda - 8;$$
$$s_4 = 65\lambda^4 + 124\lambda^3 + 82\lambda^2 + 4\lambda + 117;$$
$$s_5 = 173\lambda^5 + 415\lambda^4 + 485\lambda^3 + 285\lambda^2 + 240\lambda - 134;$$
$$s_6 = 473\lambda^6 + 1386\lambda^5 + 2121\lambda^4 + 1620\lambda^3 + 522\lambda^2 - 324\lambda + 890.$$

The Newton sums of the characteristic polynomial of matrix $\mathcal{C}_D$ are:

$$S_2 = 12\lambda^2 + 24\lambda + 100;$$
$$S_4 = 186\lambda^4 + 504\lambda^3 + 1980\lambda^2 + 2424\lambda + 4234;$$
$$S_6 = -11280\lambda^6 - 33840\lambda^5 - 35904\lambda^4 + 6480\lambda^3 + 36282\lambda^2$$
$$+42300\lambda + 64058.$$

The coefficients are:

$$A_2 = -S_2/2 = -6\lambda^2 - 12\lambda - 50;$$
$$A_4 = -(S_4 + A_2 S_2)/4 = -57/2\lambda^4 - 54\lambda^3 - 123\lambda^2 - 6\lambda + 383/2;$$
$$A_6 = -(S_6 + a_2 S_4 + A_4 S_2)/6$$
$$= 2123\lambda^6 + 6738\lambda^5 + 11459\lambda^4 + 10908\lambda^3 + 21226\lambda^2 + 20952\lambda + 64246/3.$$

The zeroes of equation $A_6 = 0$ are:

$$-2.333069484;\ -1.401818975 \pm 0.6190045476i,$$

$$0.2836993683 \pm 0.1543575855i,\ 1.933794680.$$

The obtained values of λ differ from the ones presented in [15] in the first decimal places:
$$1.5628;\ -2.2078;\ -1.1690 \pm 0.8436i;\ 0.2735 \pm 0.0988i.$$

Compare the eigenvalues of matrix $A + \lambda B$ at the point $\lambda = -2.333069484$ and at the point $\lambda = -2.2078$.

For $\lambda = -2.333069484$, the eigenvalues are:

$$-0.257071186441648280,\ -0.257116318877424144,\ -5.48502094668092610.$$

As we can see, the first two eigenvalues coincide up to three first decimal places.

For $\lambda = -2.2078$, the eigenvalues are:

$$0.567197754532772214;\ -1.01664058066736440;\ -5.17395717386540620.$$

As we can see, all the eigenvalues are distinct.

The authors of the paper [15] note that their method has an error that is quite large. They propose to use it only for matrices of small sizes. But we see that in this case, the obtained error can be large, too.

4 Conclusions

For the problem considered in this paper, we propose an algorithm based on the Kronecker product and the Newton sums. The roots of the constructed univariate polynomial could be found by any suitable method with desirable accuracy.

Certainly, the suggested algorithm can be generalized for matrices with elements polynomially dependent on a parameter. For further investigation, there remains to develop such an algorithm in detail.

Acknowledgments. The author is grateful to the anonymous referees for valuable suggestions that helped to improve the paper.

References

1. Björk, Å., Dahlquist, G.: Numerical Mathematics and Scientific Computations, vol. 1. SIAM, Philadelphia (2008)
2. Burke, J.V., Lewis, A.S., Overton, M.: Optimization and pseudospectra, with applications to robust stability. SIAM J. Matrix Anal. Appl. **25**(1), 80–104 (2003)
3. Coppersmith, D., Winograd, Sh.: Matrix multiplication via arithmetic progressions. J. Symbolic Comput. **9**(3), 251–280 (1990)
4. Giorgi, P., Jeannerod, C.-P., Villard, G.: On the complexity of polynomial matrix computations. In: ISSAC 2003, pp. 135–142. ACM Press, New York (2003)
5. Golub, G.H., Van Loan, Ch.F: Matrix Computations. The Johns Hopkins University Press, Baltimore and London (1996)
6. Jarlebring, E., Kvaal, S., Michiels, W.: Computing all pairs $(\lambda; \mu)$ such that λ is a double eigenvalue of $A + \mu B$. SIAM J. Matrix Anal. Appl. **32**, 902–927 (2011)
7. Horn, R.A., Johnson, Ch.R.: Matrix Analysis. Cambridge University Press, New York (2013)
8. Horn, R.A., Johnson, Ch.R.: Topics in Matrix Analysis. Cambridge University Press, New York (1991)
9. Karow, M.: Eigenvalue condition numbers and a formula of Burke, Lewis and Overton. Electron. J. Linear Algebra **15**, 143–153 (2006)
10. Kressner, D., Peláez, M.J., Moro, J.: Structured Hölder condition numbers for multiple eigenvalues. SIAM J. Matrix Anal. Appl. **31**(1), 175–201 (2009)
11. Leverrier, U.J.J.: Sur les variations séculaires des élements des orbites pour les sept planètes principales. J. de Math. **1**(5), 220–254 (1840)
12. Littlewood, D.E.: The Theory of Group Characters and Matrix Representations of Groups. Oxford University Press, Oxford (1950)
13. MacDuffee, C.C.: The Theory of Matrices. Chelsea Publishing Company, New York (1956)
14. Mailybaev, A.A.: Computation of multiple eigenvalues and generalized eigenvectors for matrices dependent on parameters. Numer. Linear Algebra Appl. **13**, 419–436 (2006)

15. Muhič, A., Plestenjak, B.: A method for computing all values λ such that $A + \lambda B$ has a multiple eigenvalue. Linear Algebra Appl. **440**, 345–359 (2014)
16. Pan, V.: Algebraic complexity of computing polynomial zeros. Comput. Math. Appl. **14**(4), 285–304 (1987)
17. Strassen, V.: Gaussian elimination is not optimal. Num. Math. **13**, 354–356 (1969)
18. Schucan, T.H., Weidenmüller, H.A.: Perturbation theory for the effective interaction in nuclei. Ann. Phys. **76**, 483–501 (1973)
19. Gantmacher, F.R.: Theory of Matrices, vol. 2. AMS Chelsea Publishing Company, Providence (1960)
20. Wilkinson, J.H.: The Algebraic Eigenvalue Problem. Clarendon Press, Oxford (1965)
21. Wayland, H.: Expansion of determinantal equations into polynomial form. Quart. Appl. Math. **2**(4), 277–305 (1945)

A Generalised Branch-and-Bound Approach and Its Application in SAT Modulo Nonlinear Integer Arithmetic

Gereon Kremer$^{(\boxtimes)}$, Florian Corzilius, and Erika Ábrahám

RWTH Aachen University, Aachen, Germany
`gereon.kremer@cs.rwth-aachen.de`

Abstract. The branch-and-bound framework has already been successfully applied in SAT-modulo-theories (SMT) solvers to check the satisfiability of linear integer arithmetic formulas. In this paper we study how it can be used in SMT solvers for non-linear integer arithmetic on top of two real-algebraic decision procedures: the virtual substitution and the cylindrical algebraic decomposition methods. We implemented this approach in our SMT solver `SMT-RAT` and compared it with the currently best performing SMT solvers for this logic, which are mostly based on bit-blasting. Furthermore, we implemented a combination of our approach with bit-blasting that outperforms the state-of-the-art SMT solvers for most instances.

1 Introduction

Satisfiability checking [7] aims to develop algorithms and tools to check the satisfiability of existentially quantified logical formulas. Driven by the success of SAT solving for propositional logic, fruitful initiatives were started to enrich propositional SAT solving with solver modules for different theories. *SAT-modulo-theories (SMT) solvers* [6] make use of an efficient SAT solver to check the logical (Boolean) structure of formulas, and use different theory solver modules for checking the consistency of theory constraint sets in the underlying theory.

Besides theories, such as equality logic, uninterpreted functions, bit-vectors and arrays, SMT solvers also support *arithmetic* theories. Apart from interval constraint propagation (ICP), SMT solving for quantifier-free *linear real* arithmetic (*QF_LRA*) often makes use of linear programming techniques such as the *simplex* method [15]. For quantifier-free *non-linear real* arithmetic (*QF_NRA*), algebraic decision procedures, *e.g.* the *virtual substitution* (*VS*) method [28] or the *cylindrical algebraic decomposition* method (*CAD*) [12] can be used.

There are several powerful SMT solvers, *e.g.*, `CVC4` [4], `MathSAT5` [11], `Sateen` [23], `veriT` [9], `Yices2` [18] and `Z3` [25] which offer solutions for quantifier-free *linear integer* arithmetic (*QF_LIA*) problems that mostly build on the ideas in [17]. Closely related to our work is the SMT-adaptation of the *branch-and-bound* (*BB*) framework [26] in `MathSAT5`.

© Springer International Publishing AG 2016
V.P. Gerdt et al. (Eds.): CASC 2016, LNCS 9890, pp. 315–335, 2016.
DOI: 10.1007/978-3-319-45641-6_21

When moving from the real to the integer domain, non-linear arithmetic becomes undecidable. Despite this fact, there are a few SMT solvers that support the satisfiability check of quantifier-free *non-linear integer* arithmetic (*QF_NIA*) formulas, either for restricted domains (in which case the domain becomes finite and the problem becomes decidable) or in an incomplete manner. Some of these SMT solvers apply linearisation [8], whereas other tools, such as iSAT3 [19] and raSAT [27], use interval constraint propagation adapted to the integer domain. To the best of our knowledge, all other prominent solvers, such as APROVE [20], CVC4 or Z3, apply mainly bit-blasting, which exploits SAT solvers by the use of a binary encoding of bounded integer domains.

Although the satisfiability problem for QF_NIA is undecidable, we see in the employment and adaptation of algebraic decision procedures a promising alternative to achieve incomplete but practically efficient satisfiability checking solutions for QF_NIA problems. Such solutions are urgently needed in several research areas to open new possibilities and enable novel approaches. A typical example is the field of program verification where, for instance, deductive proof systems often generate QF_NIA formulas as verification conditions (see *e.g.* [3]). Just to mention a second example, for the termination analysis of programs often QF_NIA termination conditions need to be checked for satisfiability [8]. Currently, non-linear integer arithmetic problems appearing in these areas are often solved using, *e.g.*, theorem proving, linearisation or bit-blasting.

Today's SMT solvers neither exploit adaptations of algebraic QF_NRA decision procedures for finding QF_NIA solutions[1] nor use the BB framework to check QF_NIA formulas for satisfiability. In this paper we investigate these issues. The main contributions of this paper are:

- We show on the example of the CAD method that some algebraic decision procedures for QF_NRA can be adapted to drive their search towards integer solutions. Experimental results show that this approach works surprisingly well on the satisfiable problem instances which we considered.
- We propose improvements for the general integration of BB in SMT solving.
- We show on the examples of the VS and the CAD methods, how algebraic QF_NRA decision procedures can be embedded into the BB framework to solve QF_NIA problems.
- Finally, we provide experimental results to illustrate how different decision procedures can be strategically adapted, combined and embedded in the BB framework to tackle the challenge of QF_NIA satisfiability checking.

The rest of the paper is structured as follows. We start in Sect. 2 with preliminaries on QF_NIA, SMT solving, and the VS and the CAD methods. In Sect. 3 we present a general framework for BB. Sections 4 and 5 are devoted to the

---

[1] Z3 has a command-line option to solve, instead of QF_NIA problems, their QF_NRA relaxations. This way Z3 can detect unsatisfiability (no real solution is found) or sometimes even satisfiability (the found real solution happens to be integer), but otherwise it returns "unknown".

integration of the VS and the CAD methods, respectively, into the BB framework. After an experimental evaluation of the presented approach in Sect. 7, we conclude the paper in Sect. 8.

2 Preliminaries

(Quantifier-free non-linear arithmetic) formulas φ are Boolean combinations of *constraints* c which compare *polynomials* p to 0. A polynomial p can be a *constant*, a *variable* x, or a sum, difference or product of polynomials:

$$
\begin{aligned}
p &::= \quad 0 \quad | \quad 1 \quad | \quad x \quad | \quad (p+p) \quad | \quad (p-p) \quad | \quad (p \cdot p) \\
c &::= p = 0 \quad | \quad p < 0 \\
\varphi &::= \quad c \quad | \quad (\neg\varphi) \quad | \quad (\varphi \wedge \varphi)
\end{aligned}
$$

We use further operators such as disjunction $\vee$, implication $\Rightarrow$ and comparisons $>, \leq, \geq, \neq$, which are defined as syntactic sugar the standard way, and standard syntactic simplifications (e.g., we omit parentheses based on the standard operator binding order, write $p_1 p_2$ instead of $p_1 \cdot p_2$, $-p$ instead of $0 - p$).

We use $\mathbb{Z}[x_1, \ldots, x_n]$ to denote the set of all polynomials (with integer coefficients) over the variables $x_1, \ldots, x_n$ for some $n \geq 1$. A polynomial $p \in \mathbb{Z}[x_1, \ldots, x_n]$ is called *univariate* if $n = 1$, and *multivariate* otherwise. By $Var(\varphi)$, $Pol(\varphi)$ and $Con(\varphi)$ we refer to the set of all variables, polynomials and constraints occurring in the formula φ, respectively; especially, $Pol(p \sim 0)$ denotes the polynomial p of a given constraint $p \sim 0$ with $\sim \in \{<, \leq, =, \neq, \geq, >\}$.

Each polynomial $p \in \mathbb{Z}[x_1, \ldots, x_n]$ can be equivalently transformed to the form $a_k x_1^{e_{1,k}} \ldots x_n^{e_{n,k}} + \ldots + a_1 x_1^{e_{1,1}} \ldots x_n^{e_{n,1}} + a_0$ with *coefficients* $a_j \in \mathbb{Z}$ for $0 \leq i \leq n$ and *exponents* $e_{i,j} \in \mathbb{N}_0$ for $1 \leq i \leq n$ and $1 \leq j \leq k$. We call $m_j := x_1^{e_{1,j}} \ldots x_n^{e_{n,j}}$ a *monomial*, $t_j := a_j m_j$ a *term*, and a_0 the *constant part* of p. In the following, we assume that polynomials are in the above form with pairwise different monomials; note that this form is unique up to the ordering of the terms. By $p_1 \equiv p_2$ ($\varphi_1 \equiv \varphi_2$) we denote that the polynomials p_1 and p_2 (the formulas φ_1 and φ_2) can be transformed to the same form. By $\deg(t_j) := \sum_{i=1}^{n} e_{i,j}$ we denote the *degree* of the term t_j. By $\deg(p := \max_{1 \leq j \leq k} \deg(t_j)$ we denote the *degree of* p and by $\deg(x_i, p) := \max_{1 \leq j \leq k} e_{i,j}$ the *degree of* x_i in p. A polynomial p is *linear*, if $\deg(p) \leq 1$, and *non-linear* otherwise. A formula φ is *linear*, if all polynomials $p \in Pol(\varphi)$ are linear, and *non-linear* otherwise.

We use the standard semantics of arithmetic formulas. In the theory of *quantifier-free non-linear integer arithmetic (QF_NIA)*, all variables x_i are integer-valued (the *domain* of x_i, denoted by $Dom(x_i)$, is $\mathbb{Z}$); in *quantifier-free non-linear real arithmetic (QF_NRA)* all variables x_i are real-valued ($Dom(x_i) = \mathbb{R}$); in the theory of *non-linear mixed integer-real arithmetic (QF_NIRA)*, variables can have either domain. We denote by $\varphi_{\mathbb{Z}}$ ($\varphi_{\mathbb{R}}$) that φ is interpreted as a QF_NIA (QF_NRA) formula, and call $\varphi_{\mathbb{R}}$ the *real relaxation* of $\varphi_{\mathbb{Z}}$.

As a preprocessing step for QF_NIA formulas, we replace inequalities $p \neq 0$ by $p < 0 \vee p > 0$. Furthermore, based on the integrality of all variables, we simplify

Table 1. Simplification of a QF_NIA constraint $(\sum_{i=1}^{k} a_i m_i) + a_0 \sim 0$, where g is the greatest common divisor of $a_1, \ldots, a_k$, $a_0' := \frac{a_0}{g}$ and $r := \sum_{i=1}^{k} \frac{a_i}{g} m_i$

	$<$	$\leq$	$=$	$\geq$	$>$
$a_0' \in \mathbb{Z}$	$r + a_0' + 1 \leq 0$	$r + a_0' \leq 0$	$r + a_0' = 0$	$r + a_0' \geq 0$	$r + a_0' - 1 \geq 0$
$a_0' \notin \mathbb{Z}$	$r + \lceil a_0' \rceil \leq 0$	$r + \lceil a_0' \rceil \leq 0$	false	$r + \lfloor a_0' \rfloor \geq 0$	$r + \lfloor a_0' \rfloor \geq 0$

constraints in the formula according to Table 1. After these simplifications, only the relations $=$, $\geq$ and $\leq$ (but no $<$, $>$ nor $\neq$) appear in the formulas.

The *substitution* of a variable x by a value $v \in Dom(x)$ in a formula φ is denoted by $\varphi[v/x]$. A value $(v_1, \ldots, v_n) \in \mathbb{R}^n$ is a *real root* or *zero* of a polynomial $p \in \mathbb{Z}[x_1, \ldots, x_n]$ if $p(v_1, \ldots, v_n) := p[v_1/x_1] \ldots [v_n/x_n] \equiv 0$. A value $(v_1, \ldots, v_n) \in \mathbb{R}^n$ is a *solution* of a formula φ with $Var(\varphi) = \{x_1, \ldots, x_n\}$ if $\varphi(v_1, \ldots, v_n) := \varphi[v_1/x_1] \ldots [v_n/x_n] \equiv \texttt{true}$.

The *satisfiability checking problem* is the problem to decide whether there exists a solution for a given formula φ. Note that checking the satisfiability of a quantifier-free formula φ with $Var(\varphi) = \{x_1, \ldots, x_n\}$ and checking the validity of the existentially quantified formula $\exists x_1 \ldots \exists x_n . \varphi$ define the same problem.

2.1 SAT-modulo-theories Solving

For the satisfiability check of logical formulas over some theories, *SAT-modulo-theories (SMT) solvers* combine a SAT solver with one or more *theory solvers*. The SAT solver is used for the efficient exploration of the logical structure of the input formula, whereas the theory solver(s), implementing some decision procedures for the underlying theory, are used to check the consistency of sets (conjunctions) of certain theory constraints. The SMT-solving framework is illustrated in Fig. 1.

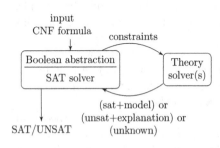

Fig. 1. The SMT solving framework

In this work we consider a less-lazy SMT-solving approach based on DPLL [16] SAT solving. The input formula is brought to negation normal form, negation is applied to atomic constraints, and finally conjunctive normal form (CNF) is built (using Tseitin's transformation), resulting in a conjunction of disjunctions of atomic (not negated) constraints. The SAT solver tries to find a satisfying solution for the *Boolean skeleton* of the input formula, which is the propositional logic formula obtained by replacing each constraint c by a fresh proposition h_c. In other words, the SAT solver tries to determine a set of constraints such that the input formula is satisfiable if the determined constraints have a common solution. During its search, the SAT solver works "less lazily", *i.e.*, it asks the theory solver(s) regularly whether the set of those constraints, whose abstraction variables are `true` in the current partial assignment, are together consistent. Note that we do *not* need to pass the negation of constraints with `false` abstraction variables to the theory solver, because all abstraction variables appear in the transformed formula without negation. Therefore, all clauses are still satisfiable even if we wrongfully assign a variable to `false`.

The DPLL algorithm, used in most state-of-the-art SAT solvers, executes the following loop: It repeatedly makes a *decision*, assuming a certain value for some proposition. Then it applies *Boolean constraint propagation (BCP)*, thereby identifying variable assignments that are implied by the last decision. If the propagation succeeds without conflicts, a new decision is made. However, the propagation might also lead to a *conflict*, which means that the current partial assignment cannot be extended to a full satisfying solution. In the latter case *conflict resolution* is applied to determine which decisions led to the conflict. After *backtracking*, *i.e.*, undoing some of the previous decisions, the SAT solver learns a new clause to exclude the found conflict (and other similar ones) from the future search and continues the search in other parts of the search space.

In the less-lazy SMT-solving context, theory solvers need to be able to work *incrementally*, *i.e.*, to extend a previously received set of constraints with new constraints and to check the extended set for consistency while reusing collected information from the last check in order to improve the performance; similarly, they need to be able to backtrack, *i.e.*, to remove constraints from their current constraint sets. Furthermore, to enable the SAT solver to resolve theory-rooted conflicts, the theory solver has to return an *explanation* if it detects inconsistency, usually in form of an inconsistent subset of its received constraints.

We use our SMT-solving toolbox `SMT-RAT` [14] to implement the approaches described in this work, and to compare the results to other approaches. `SMT-RAT` provides a rich set of SMT-compliant implementations of QF_NRA/ QF_NIA procedures. These procedures are encapsulated in *modules*, which are executed by our SMT solver according to a user-defined *strategy*. In this work, we will use only *sequential* strategies, where a preprocessing module, a SAT solver module, and one or more theory solver modules are arranged sequentially. The SAT solver sends theory constraints in an incremental fashion to the first theory solver module, which tries to determine whether its received constraints have a common satisfying solution. The first theory solver might also pass on sub-problems[2] to be checked for satisfiability to the next theory solver module and so on. Each theory solver module returns to its caller module (i) either satisfiability along with a model if requested, (ii) or unsatisfiability and an explanation in form of an infeasible subset of its received formulas, (iii) or unknown. Besides modules for parsing the input problem and transforming it to conjunctive normal form as requested by SAT solving, we use a `MiniSat`-based SAT solver module and theory solver modules implementing the simplex, the VS and the CAD methods, and a theory solver module for bit-blasting.

---

[2] These sub-problems are not necessarily constraints or conjunctions of constraints, but in general formulas with arbitrary Boolean structure.

2.2   Virtual Substitution

The *virtual substitution* (*VS*) method [28] is an incomplete decision procedure for non-linear real arithmetic. As we aim at satisfiability checking, we restrict ourselves to the quantifier-free fragment QF_NRA (where we understand all free variables as existentially quantified). Given a QF_NRA formula $\varphi_{\mathbb{R}}$ in variables $Var(\varphi_{\mathbb{R}}) = \{x_1, \dots, x_n\}$, the VS iteratively eliminates variables that appear at most quadratic in $\varphi_{\mathbb{R}}$. Assume *w.l.o.g.* that we want to eliminate a variable x_n, such that $\deg(x_n, p) \leq 2$ for all $p \in Pol(\varphi_{\mathbb{R}})$.

In the univariate case, *i.e.* $Pol(\varphi_{\mathbb{R}}) \subseteq \mathbb{Z}[x_n]$, we can use the solution equation for quadratic polynomials to determine the real roots of all $p \in Pol(\varphi_{\mathbb{R}})$. They separate regions in which each $p \in Pol(\varphi_{\mathbb{R}})$ is sign-invariant. We determine for each of those sign-invariant regions a representative element, which we collect in the set of *test candidates* $T(x_n, \varphi_{\mathbb{R}})$. Due to the sign-invariance of those regions, $\varphi_{\mathbb{R}}$ is satisfiable if and only if one of the test candidates satisfies the formula.

Even if the formula contains multivariate polynomials, we can apply the solution equation to determine real roots t_i and the side conditions $SC(t_i)$ for their existence (radicand non-negative, denominator is not zero), if we interpret multivariate polynomials $p \in \mathbb{Z}[x_1, \dots, x_n]$ as univariate polynomials with polynomial coefficients $p \in \mathbb{Z}[x_1, \dots, x_{n-1}][x_n]$.

However, now the results are parametric in $x_1, \dots, x_{n-1}$, therefore we know neither the existence nor the order of the roots; here, the VS uses $-\infty$ as a test candidate from the left-most sign-invariant region, and infinitesimals ϵ in order to represent with $t + \epsilon$ the sign-invariant region on the right of t. As the test candidates might contain fractions, radicals, $-\infty$ and ϵ, standard substitution $[t_i/x_n]$ can lead to improper expressions which are not arithmetic formulas. Instead we use the *virtual* substitution $[t_i /\!/ x_n]$ which

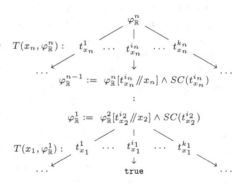

Fig. 2. Possible VS depth-first search

resolves these improper expressions with special rules [28] resulting in a QF_NRA formula $\varphi_{\mathbb{R}}[t_i /\!/ x_n]$ that is satisfiability-equivalent to $\varphi_{\mathbb{R}}[t_i/x_n]$.

In summary, the VS specifies a finite set $T(x_n, \varphi_{\mathbb{R}})$ of (symbolic) *test candidates* (TCs) for x_n in $\varphi_{\mathbb{R}}$, and for each TC $t \in T(x_n, \varphi_{\mathbb{R}})$ some *side conditions* $SC(t)$, such that

$$\varphi_{\mathbb{R}} \text{ is satisfiable} \quad \Leftrightarrow \quad \bigvee_{t \in T(x_n, \varphi_{\mathbb{R}})} (\varphi_{\mathbb{R}}[t /\!/ x_n] \wedge SC(t)) \quad \text{is satisfiable.} \quad (1)$$

In [13] we presented an implementation of the VS for satisfiability checking, which executes a depth-first search for a **true** leaf as illustrated in Fig. 2; a solution can be read off the *solution path* from the root to the **true** leaf.

2.3 Cylindrical Algebraic Decomposition

The *cylindrical algebraic decomposition (CAD)* [1,2,12] is a complete decision procedure for non-linear real arithmetic, which we will use in the SMT solving context for checking the satisfiability of QF_NRA formulas. Due to space restriction, we give only a high-level description here.

Given a formula $\varphi_{\mathbb{R}}$ in variables $x_1, \ldots, x_n$, the CAD method partitions $\mathbb{R}^n$ into a finite number of disjoint n-dimensional *cells*. Each of them is a connected semi-algebraic set over which all polynomials in $Pol(\varphi_{\mathbb{R}})$ are sign-invariant and thus either all or none of the points in a cell satisfy $\varphi_{\mathbb{R}}$. The cells are constructed to be *cylindrical*: the projections of any two n-dimensional cells onto the k-dimensional space ($1 \leq k < n$) are either identical or disjoint. The n-dimensional cells with identical k-dimensional projections S form *cylinders* $S \times \mathbb{R}^{n-k}$.

The CAD method uses a two-phase approach to compute such a partitioning. In the *projection* phase, a *projection operator* is applied to the input set $P_n = Pol(\varphi_{\mathbb{R}})$ of n-dimensional polynomials with respect to the variable x_n, yielding a set $P_{n-1} \subset \mathbb{Z}[x_1, \ldots, x_{n-1}]$ of $(n-1)$-dimensional polynomials, whose real roots constitute the boundaries of the $(n-1)$-dimensional projections of the n-dimensional CAD cells. The projection operator is applied recursively until a set $P_1 \subset \mathbb{Z}[x_1]$ of univariate polynomials is obtained. Note that this imposes a fixed variable ordering that cannot easily be changed without recomputing the projection. Much work has been done on providing efficient methods for computing preferably small projections, for example in [10,22,24].

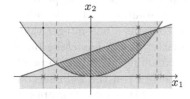

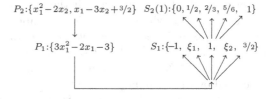

(a) Plot of the solution space and the sample points

(b) Schematic overview of the CAD method

Fig. 3. CAD example for $c_1 : x_1^2 - 2x_2 \leq 0$ and $c_2 : x_1 - 3x_2 + 3/2 \geq 0$

In the second phase called *lifting* or *construction*, the CAD method constructs a *sample point* for each of the cells. It first isolates the real roots $\xi_1, \ldots, \xi_k$ of the polynomials in P_1 which constitute the boundaries of the 1-dimensional projections of the cells which are the intervals $I = \{(-\infty, \xi_1), [\xi_1, \xi_1], (\xi_1, \xi_2), \ldots, (\xi_k, \infty)\}$. We choose a sample point from each of these intervals I_i, resulting in a sample set $S_1 = \{s_1, \ldots, s_{2k+1}\}$. Each sample point $s \in S_1$ from some interval I_i is now *lifted* using the polynomials from P_2: each polynomial from P_2 is partially evaluated on s which results in a set of univariate polynomials whose real roots $\{\xi_1^s, \ldots \xi_{l_s}^s\}$ are again isolated. For each real root $\{\xi_1^s, \ldots \xi_{l_s}^s\}$ there exists a surface $I_i \times \xi_j^s$ and together these surfaces separate the cylinder $I_i \times \mathbb{R}$ into $2 \cdot l_s + 1$ individual cells. Assuming that $\xi_1^s < \ldots < \xi_{l_s}^s$, these cells are the surfaces $I_i \times \xi_j^s$ (for $j = 1, \ldots, l_s$), the regions between two surfaces $I_i \times (\xi_j^s, \xi_{j+1}^s)$ (for $j = 1, \ldots, l_s - 1$), and the regions below and above all separating surfaces $I_i \times (-\infty, \xi_1^s)$ and $I_i \times (\xi_{l_s}^s, \infty)$. We can again take sample points from these cells and repeat the lifting procedure until we obtain n-dimensional sample points that are representatives for the n-dimensional cells that are sign-invariant with respect to the polynomials in P_n. We evaluate $\varphi_{\mathbb{R}}$ for each n-dimensional sample point to check whether one of them satisfies the formula, in which case we obtain a satisfying solution, otherwise the formula is not satisfiable.

The CAD method is illustrated on a 2-dimensional example in Fig. 3. Figure 3a depicts the solution space, while Fig. 3b visualises the CAD computation. The sample point $(1, 2/3)$ satisfies the formula $x_1^2 - 2x \leq 0 \wedge x_1 - 3x_2 + 3/2 \geq 0$. Note that we can *choose* a sample point from intervals between real roots. This choice is important when searching for integer solutions: while there is no integer solution for the sample point 1 from the interval (ξ_1, ξ_2), selecting 0 instead of 1 would have resulted in the integer solution $(0, 0)$.

3   A General Branch-and-Bound Framework

A popular approach to check QF_LIA formulas $\varphi_{\mathbb{Z}}$ for satisfiability is the *branch-and-bound (BB)* framework [26]. It first considers the relaxed problem $\varphi_{\mathbb{R}}$ in the real domain. If the relaxed problem is unsatisfiable then the integer problem is unsatisfiable, too. Otherwise, if there exists a real solution then it is either integer-valued, in which case $\varphi_{\mathbb{Z}}$ is satisfiable, or it contains a non-integer value $r \in \mathbb{R} \setminus \mathbb{Z}$ for an integer-valued variable x. In the latter case a *branching* takes place: BB reduces the relaxed solution space by excluding all values between $\lfloor r \rfloor = \max\{r' \in \mathbb{Z} \mid r' \leq r\}$ and $\lceil r \rceil = \min\{r' \in \mathbb{Z} \mid r' \geq r\}$ in the x-dimension, described by the formula $\varphi' = \varphi \wedge (x \leq \lfloor r \rfloor \vee x \geq \lceil r \rceil)$. This procedure is applied iteratively, *i.e.*, BB will now search for real-valued solutions for φ'. BB terminates if either an integer solution is found or the relaxation is unsatisfiable. Note that BB is in general incomplete even for the decidable logic QF_LIA.

The most well-known application combines BB with the simplex method. As branching introduces disjunctions and thus in general non-convexity, branching is implemented by case splitting: in one search branch we assume $x \leq \lfloor r \rfloor$, and in a second search branch we assume $x \geq \lceil r \rceil$. Depending on the heuristics, the search can be depth-first (full check of one of the branches, before the other branch is considered), breadth-first (check real relaxations in all current open branches before further branching is applied), or it can follow a more sophisticated strategy.

The combination of BB with the simplex method was also explored in the SMT-solving context [17]. The advantage in this setting is that we have more possibilities to design the branching.

- We can integrate a theory solver module based on the simplex method as described above, implementing BB internally in the theory solver by case splitting. It comes with the advantage that case splitting is always *local* to the current problem of the theory solver and does not affect later problems, and with the disadvantage that we cannot exploit the advantages of *learning*, *i.e.*, to remember reasons of unsatisfiability in certain branches and use this information to speed up the search in other branches.
- Alternatively, given a non-integer solution r for a variable x found by the theory solver on a relaxed problem, we can lift the branching to the SAT solver by extending the current formula with a new clause $(x \leq \lfloor r \rfloor \vee x \geq \lceil r \rceil)$ [5]. The newly added clause must be satisfied in order to satisfy the extended formula. Therefore, the SAT solver assigns (the Boolean abstraction variable of) either $x \leq \lfloor r \rfloor$ or $x \geq \lceil r \rceil$ to true, *i.e.*, the branching takes place. On the positive side, lifting branching information and branching decisions to the SAT solver allows us to learn from information collected in one branch, and to use this learnt information to speed up the search in other branches. On the negative side, the branching is not local anymore as it is remembered in a learnt clause. Therefore, it might cause unwanted splittings in later searches.

To unify advantages, MathSAT5 [21] implements a combined approach with theory-internal splitting up to a given depth and splitting at the logical level beyond this threshold.

Following the BB approach in combination with the simplex method, we can transfer the idea also to non-linear integer arithmetic: We can use QF_NRA decision procedures to find solutions for the relaxed problem and branch at non-integer solutions of integer-valued variables. However, there are some important differences. Most notably, the computational effort for checking the satisfiability of non-linear real-arithmetic problems is *much* higher than in the linear case. If we have found a real-valued solution and apply branching to find integer solutions, the branching will *refine* the search in the VS and CAD methods: it will create additional test candidates for the VS and new sample points for the CAD method, which will serve as roots for new sub-trees in the search tree. However, the search trees in both branches have a lot in common, that means, a lot of the same work has to be done for both sides of the branches. To prevent the solvers from doing much unnecessary work, we have to carefully design the BB procedure.

- Branching has to be *lifted* to the SAT solver level to enable *learning*, both in the form of *branching lemmas* as well as *explanations* for unsatisfiability in different branches.
- Learning explanations will allow us to speed up the search by transferring useful information between different branches. However, we need to handle branching lemmas thoughtfully and assure that learnt branching lemmas will not lead to branching for all future sub-problems, but only for "*similar*" ones where the branching will probably be useful.
- As branching refines the search, it has to work in an *incremental* fashion without resetting solver states.

- If possible, the search strategies of the underlying QF_NRA decision procedures have to be tuned to *prefer integer solutions* (and if they can choose between different integer values, they must choose the most "promising" one).
- Last but not least, as the performance of solving QF_NRA formulas for satisfiability highly improves if different theory solvers implementing different decision procedures are used in combination, a practically relevant BB approach for QF_NIA should support this option.

3.1   Processing Branching Lemmas in DPLL(T) SAT Solving

As mentioned before, an SMT solver combines a SAT solver and one or more theory solver modules. First we discuss our general approach of adapting these modules to implement BB for non-linear arithmetic, as described in Algorithm 1.

If we remove the Lines 14, 18–19, 22–27 from Algorithm 1 (printed in italic font) then we achieve the standard SMT framework without BB embedding. This basic algorithm first applies Boolean constraint propagation (BCP, Line 2) to detect implications of current decisions. If BCP does not lead to a conflict, the consistency in the theory domain is checked (Lines 3–4). If the theory constraints, which have to hold according to the current Boolean assignment, are inconsistent (Line 5) then the theory solver provides an explanation (an infeasible subset of its input constraints). We negate this explanation (resulting in a tautology) and add its Boolean abstraction as a new clause to the clause set to exclude this theory conflict from future search (Line 6); If either the BCP led to a Boolean conflict or a theory conflict occurred, then the solver tries to resolve the conflict (Line 10). If the conflict can be resolved (Line 11), then backtracking removes some of the decisions that led to the conflict; note that also the corresponding theory constraints will be removed from the input constraint list of the underlying theory solver (Line 12). As a result of a successful conflict resolution, a new clause will be learnt that will cause new implications in the next BCP iteration. Otherwise, if the conflict cannot be resolved, the input formula is unsatisfiable (Line 15).

Otherwise, if no conflict occurred, either all variables are assigned, in which case we have found a full solution (Line 28), or we choose one unassigned variable to which we assign a certain value (Lines 20–21) and propagate this when executing the next iteration of the main loop.

We modify this algorithm as follows. Firstly, additionally to sat and unsat, we allow theory solvers to return unknown. We do so because it is possible that the underlying theory solving procedure cannot determine the consistency of the set of constraints at hand. For instance, since QF_NIA is undecidable, a theory solver for QF_NIA can only be incomplete. This means either that the theory solver has non-terminating cases (which none of our theory solvers do), or that the theory solver relaxes each problem to a version that is decidable (*e.g.* QF_NRA). The latter case is only possible if we allow inconclusive answers, in which case the theory solver will return unknown. If the Boolean assignment is partial and the theory solver returns unknown, the SAT solver continues its search. If a full satisfying Boolean assignment was found by the SAT solver, but the theory solver cannot determine whether the solution is consistent in the

---

Algorithm 1. Extended SAT solving algorithm for BB in SMT

---

extended SAT algorithm()
begin
1 :   while **true do**
2 :   if BCP returns no conflict **then**
3 :   send newly assigned theory constraints to theory solver
4 :   check theory consistency
5 :   if theory solver returned **unsat then**
6 :   learn theory conflict
7 :   **end if**
8 :   **end if**
9 :   if Boolean or theory conflict occurred **then**
10 :   try to resolve conflict
11 :   if conflict can be resolved **then**
12 :   backtrack in SAT and theory solving
13 :   **else**
14 :   if *Line 26 was visited* **then return unknown**
15 :   **else return unsat**
16 :   **end if**
17 :   **else**
18 :   if *theory solver returned urgent branching lemmas* **then**
19 :   *learn them and branch*
20 :   **else if** not all propositional variables are assigned **then**
21 :   assign a value to an unassigned propositional variable
22 :   **else if** *theory solver returned* **unknown** **then**
23 :   if *theory solver returned final branching lemmas* **then**
24 :   *learn them and branch*
25 :   **else**
26 :   *exclude current Boolean assignment from further search*
27 :   **end if**
28 :   **else return sat**   // *theory solver returned sat*
29 :   **end if**
30 :   **end while**
end

(integer) theory, then the SAT solver excludes the current Boolean assignment from further search (by learning a clause in Line 26) and continues its search; if the following search detects a satisfying solution then the SMT solver returns **sat**, otherwise it returns **unknown** (Line 14).

Additionally to returning **unknown**, a theory solver can also return a so-called lemma that explains the inconclusive answer in more detail and that helps the SMT solver in finding a conclusive answer. This can happen for example if the theory solver has found a solution for the real relaxation of its input problem φ, but it is not integer-valued. In this case, the theory solver might return a *branching lemma* of the form

$$(c_1 \wedge \ldots \wedge c_k) \Rightarrow (x \leq \lfloor r \rfloor \vee x \geq \lceil r \rceil), \tag{2}$$

demanding to split the domain of the integer-valued variable x at the non-integer value $r \in \mathbb{R} \setminus \mathbb{Z}$, under the condition that the *branching premise* $c_1 \wedge \ldots \wedge c_k$ with $\{c_1, \ldots, c_k\} \subseteq Con(\varphi)$ holds. Additionally, the theory solver can specify which of the two branches it prefers to start with. We call the Boolean abstraction $(\neg h_{c_1} \vee \ldots \vee \neg h_{c_k} \vee h_{x \leq \lfloor r \rfloor} \vee h_{x \geq \lceil r \rceil})$ of the branching lemma in Eq. 2 a *branching clause* and its last two (possibly fresh) literals *branching literals*.

Branching lemmas can be either *urgent* or *final*[3]. Urgent branching lemmas are immediately abstracted, added to the SAT solver's clause set and used for branching (Lines 18–19). Final branching lemmas are relevant only if the SAT solver has a full satisfying Boolean assignment (Lines 23–24). When a branching clause is added, one of its branching literals (the one that was *not* preferred by the theory solver) will be assigned `false` (thus, if the branching premise is `true`, BCP will assign `true` to the preferred branching literal; this way we prevent that both branching literals become `true`, what would result in a theory conflict). Afterwards, we handle the branching clause just as any learnt clause and benefit from the usual reasoning and learning[4] process, which yields the best performance according to our experience.

To prevent unnecessary branchings, we assign always the value `false` to branching literals as decision variables in Line 29. Remember that only constraints with `true` abstraction variables will be passed to the theory solver. *I.e.*, only branching clauses whose premise is `true` play a role in the theory, and for those clauses only one of the branching literals.

4   Branch-and-Bound with Virtual Substitution

In this section we present how the VS method as introduced in Sect. 2.2 can be embedded into the BB framework to check the satisfiability of a given QF_NIA formula $\varphi_{\mathbb{Z}}^n$ over (theory) variables $x_1, \ldots, x_n$. First we apply VS on the real relaxation $\varphi_{\mathbb{R}}^n$ of $\varphi_{\mathbb{Z}}^n$. If we determine unsatisfiability, we know that $\varphi_{\mathbb{Z}}^n$ is also unsatisfiable. Otherwise, if we have found a solution S with the VS for $\varphi_{\mathbb{R}}^n$, as illustrated in Fig. 2, then S maps the variables $Var(\varphi_{\mathbb{R}}^n) = \{x_1, \ldots, x_n\}$ to TCs $S(x_j) = t_{x_j}^{ij}$ $(1 \leq j \leq n)$. For QF_NIA formulas we can omit to consider strict inequalities as described in Table 1. This saves us from considering TCs with infinitesimals as introduced in [28] and the comparably higher complexity they entail. Therefore, $S(x_j)$ is either $-\infty$ or of the form $\frac{q_{j,1} + q_{j,2}\sqrt{q_{j,3}}}{q_{j,4}}$ with $q_{j,1}, \ldots, q_{j,4} \in \mathbb{Z}[x_1, \ldots, x_{j-1}]$ (roots parametrised in some polynomials).

If a solution S for the relaxation $\varphi_{\mathbb{R}}^n$ is found then there is a `true` leaf in the search tree, as illustrated in Fig. 2. We now try to construct an *integer* solution S^* from the parametrised solution S, as illustrated in Fig. 4, traversing the solution path from the `true` leaf backwards. If the TC $t_{x_1}^{i_1}$ for x_1 is not $-\infty$, it does not contain any variables, thus we can determine whether its value is integer and set $S^*(x_1)$ to this value. If $t_{x_1}^{i_1} = -\infty$, we can take any integer which is

---

[3] In the experimental results we use final lemmas only.

[4] Note that modern SAT solvers also allow to forget learnt clauses that did not contribute to conflicts recently. This applies also to branching clauses.

strictly smaller than all the other TCs in $T(x_1, \varphi_{\mathbb{R}}^1)$. Now we iterate backwards: for each test candidate $t_{x_j}^{ij}$ on the solution path, which is not $-\infty$, we substitute the values $S^*(x_1), \ldots, S^*(x_{j-1})$ for the variables $x_1, \ldots, x_{j-1}$, resulting in $S^*(x_j) := S(x_j)[S^*(x_1)/x_1] \ldots [S^*(x_{j-1})/x_{j-1}]$, which again does not contain any variables and we can evaluate whether its value is integer. If $t_{x_j}^{ij} = -\infty$ then we evaluate all test candidates from $T(x_j, \varphi_{\mathbb{R}}^j)$ whose side conditions hold by substituting $S^*(x_1), \ldots, S^*(x_{j-1})$ for $x_1, \ldots, x_{j-1}$ in the TC expressions, and we set $S^*(x_j)$ to an integer value that is strictly smaller than all those TC values. We repeat this procedure until either a full integer solution is found or the resulting value is not integer in one dimension.

If all TC values are integer then VS returns $\mathtt{sat}$. Otherwise, if we determine that $S^*(x_j)$ for some j is not integer-valued, then there is some $z \in \mathbb{Z}$ such that $S^*(x_j) \in (z-1, z)$. In this case we return the branching lemma $(\bigwedge_{\psi \in Orig_{x_j}(S(x_j))} \psi) \Rightarrow (x_j \leq z-1 \lor x_j \geq z)$, where $Orig_{x_j}(S(x_j))$ denotes the VS module's received constraints being responsible for the creation of the TC $S(x_j)$. We can determine this set recursively with $Orig_{x_j}(S(x_j)) := Orig_{x_j}(c)$ if we used constraint $c \in Con(\varphi_{\mathbb{R}}^j)$ for generating the TC $S(x_j)$, and where

$$Orig_{x_j}(c) := \begin{cases} c & \text{, if } j = n \\ Orig_{x_{j+1}}(c) & \text{, if } x_j \notin Var(c) \\ Orig_{x_{j+1}}(S(x_{j+1})) \cup Orig_{x_{j+1}}(c') & \text{, if } c' \in Con(\varphi_{\mathbb{R}}^{j+1}) \text{ such that} \\ & \quad c \in Con(c'[S(x_{j+1})/\!/x_{j+1}]) \\ Orig_{x_{j+1}}(S(x_{j+1})) & \text{, otherwise.} \end{cases}$$

Note that the last case occurs if the given constraint is introduced through a TC's side condition.

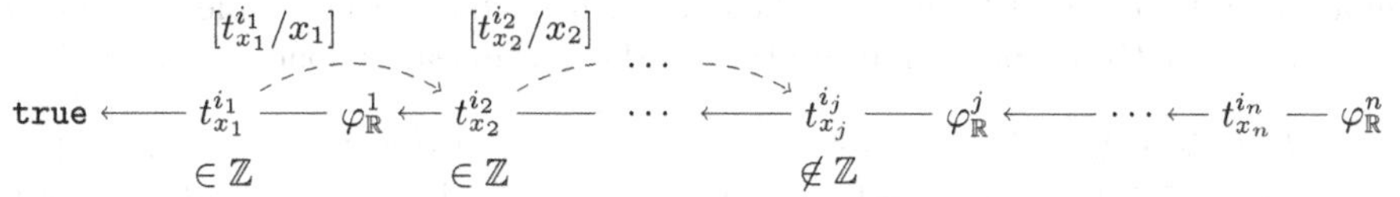

Fig. 4. Solution path from Fig. 2 traversed backwards from the leaf to the root

Basically, if we have found a non-integer valued TC $S^*(x_j) \notin \mathbb{Z}$, we can still continue the procedure to determine all other non-integer-valued TCs, but the gain (enabling a heuristics to select on which variable value we want to branch) comes at high computational costs, as we need to compute with nested fractions and square roots. Therefore, we do not consider other heuristics but branch always on the first detected non-integer value. In contrast, as we will see in the next section, the CAD methods offers more freedom to design other heuristics.

This procedure is sound, as we do not prune any integer solutions. It is not complete, as it might branch infinitely often for the same variable at an always increasing or always decreasing value. This procedure can also be used to check a QF_NIRA formula for satisfiability, if we eliminate real-valued variables first.

5   Branch-and-Bound with the CAD Method

Also the CAD method can be embedded into the BB approach the usual way, however, it offers more flexibility to tune its search towards integer solutions.

Sample point selection. The computation time of the lifting phase heavily depends on the representation size of the numbers involved. The representation of an integer is inherently smaller than that of a fractional number of a similar value due to the lack of a denominator and a smaller numerator. Therefore, when selecting a sample point from a given interval, we always choose an integer value whenever one exists. As a side effect, this is not only faster due to the smaller representation but also generates integer solutions automatically which generally helps to avoid unnecessary branches.

When several integer values are available in a given interval as possible sample points, some of them might lead to a full integer solution whereas some others not. Unfortunately, there is no generally valid rule to determine more "promising" samples. In our implementation we choose integer samples around the interval middle.

Note that we could even stop lifting for the given partial sample if a complete integer extension of the current sample becomes impossible during the lifting. However, if at an earlier level we had the choice between different integer values for a cell sample then we cannot conclude unsatisfiability in the current sub-tree. Therefore, even if we cannot choose integer samples, in the current implementation we continue lifting and search for a satisfying real extension of the partial integer sample. If the search leads to a (non-integer) solution, we request branching at the SAT level. Otherwise, if the current branch is unsatisfiable in the real domain, we continue the search in other parts of the search space.

Example 1. Consider $P_2 = \{x_2^2 + x_1 = 0, x_1 < -1\}$. Projecting x_2 yields $P_1 = \{x_1, x_1 + 1\}$ with real roots $\{-1, 0\}$. To satisfy $x_1 < -1$, we only need to choose a sample for x_1 from $(-\infty, -1)$. Assume we choose -2. Lifting -2 yields the polynomials $\{x_2^2 - 2, -1\}$ with real roots $\{-\sqrt{2}, \sqrt{2}\}$, both being non-integer. All other cells around these roots violate the sign condition. However, we cannot infer unsatisfiability in the integer domain: selecting -4 instead of -2 would have produced the polynomials $\{x_2^2 + 4, -3\}$ with integer roots $\{-2, 2\}$.

Remark 1. We would like to mention an idea, which is not yet implemented but might lead to further improvements. Assume as above an integer sample $s = (s_1, \ldots, s_j) \in \mathbb{Z}^j$ for which lifting yields no integer extension. Assume now additionally that in each dimension $i = 1, \ldots, j$ the sample s_i is the only integer point in the respective interval; we say that s is *unique*. In this special case the current sub-tree cannot contain any integer solutions; we can safely stop lifting for s and continue in other parts of the search space.

If we find a solution elsewhere, we can return sat. However, if the input formula has no integer solution, CAD needs to return an *explanation* for unsatisfiability. In the real domain, we generate such an explanation by specifying for each full-dimensional sample s (*i.e.*, for each leaf in the lifting tree) the set E_s of all original constraints that are violated by the leaf, and computing a possibly small covering set E which contains at least one constraint from each leaf's set E_s.

Now, if we do not complete the lifting for some sub-trees because we determined unsatisfiability at an earlier level, we cannot use the same approach to generate explanations. Instead, we can proceed as follows: Remember that $s = (s_1, \ldots, s_j) \in \mathbb{Z}^j$ is the sample for which lifting was stopped because s is unique and it has no integer extension. Each s_i samples an interval, whose endpoints are zeros of some polynomials from P_i at $(s_1, \ldots, s_{i-1})$; let $P_i^s \subseteq P_i$ be the set of those polynomials for $i = 1, \ldots, j$ and let $P_i^s = \emptyset$ for $i = j+1, \ldots, n$. Now we follow back the projection tree, and for $i = 1, \ldots, n-1$ we iteratively add to P_{i+1}^s all "projection parents"[5] of all polynomials in P_i^s, *i.e.*, all those polynomials that were used in the projection to generate P_i^s. As a result we achieve a set $P_n^s \subseteq P_n$ of original constraints, which serve as an explanation for the unsatisfiability of the sub-tree rooted at s. We compute this set P_n^s for each unique non-completed sample, build their union, and further extend it with additional constraints from P_n to cover all sets $E_{s'}$ of full-dimensional sample leafs s'. The resulting set is an infeasible subset of the input constraint set.

Remark 2. As the selection of sample points might be crucial for discovering integer solutions, we also experimented with choosing (if possible) multiple sample points for an interval instead of a single one. However, the overhead due to these redundant sample points greatly outweighs any gain, even if only two samples for a single interval are chosen. This is because the redundancy increases with every dimension and often leads to an additional exponential growth.

Sample point lifting order. The order in which sample points are lifted is crucial for fast solution finding. As we want to find integer solutions, we first lift integer sample points before considering non-integer ones. Furthermore, if we already have a partial lifting tree due to a previous incremental call to the CAD method, for further lifting we choose partial integer sample points of high dimension first.

Constructing branching lemmas. If CAD finds a solution $s = (s_1, \ldots, s_n) \in \mathbb{R}^n \setminus \mathbb{Z}^n$, it returns unknown and requests branching at the SAT level. We have tried three alternative strategies to generate branching lemmas. The first strategy branches on the value of x_i with $i = \min\{i = 1, \ldots, n \mid s_i \notin \mathbb{Z}\}$. The second strategy is similar but takes the highest index. In both cases, the branching premise is the set of all received constraints; in the future we will also experiment with the set P_n^s (see Remark 1).

---

[5] A projected polynomial can have several "parents"; in this case we can choose any of them. In practice, one could store the chronologically "oldest" parents, as backtracking removes input polynomials in chronologically reverse order.

The third strategy makes use of the sampling heuristics of the CAD that strongly prefers integers. That means that the longest integer prefix $(s_1, \ldots, s_j) \in \mathbb{Z}^j$ of s cannot be further extended with an integer sample component. This strategy generates the branching lemma[6]

$$C \rightarrow \left(\bigvee_{i=1}^{j} x_i \leq s_i - 1 \vee x_i \geq s_i + 1 \right). \tag{3}$$

Currently, the branching premise C is again the set of all received constraints. In the future we will also investigate collecting all constraints that reject integer sample points in the vicinity of the first non-integer component s_{j+1}. Let $s_{j+1\downarrow}$ $(s_{j+1\uparrow})$ be s where s_{j+1} is replaced by $\lfloor s_{j+1} \rfloor$ ($\lceil s_{j+1} \rceil$). We define the branching premise by $\{c \in C \mid c(s_{j+1\downarrow}) \equiv \mathtt{false} \vee c(s_{j+1\uparrow}) \equiv \mathtt{false}\}$.

6   Combination of Procedures

We can often improve the performance for solving QF_NRA formulas if we have different decision procedures at hand and use them in combination such that theory modules can pass on sub-problems for satisfiability check to other modules. In SMT-RAT, such combinations of decision procedures were already available for QF_NRA problems. In this section we discuss how to extend the approach for QF_NIA and the BB framework, on the examples of theory modules implementing the simplex, VS and CAD methods.

Given a set C of non-linear integer arithmetic constraints, the simplex method can be used to check the consistency of the relaxed linear constraints in C, first neglecting the non-linear ones. If simplex determines that the real relaxation of the linear part of the problem is unsatisfiable, it returns unsat. If it finds an integer solution that also satisfies the non-linear constraints, it returns sat. If it finds a solution that is not completely integer, but satisfies the real relaxation of the non-linear constraints, it creates a branching lemma and returns unknown. Otherwise, it forwards the whole input C to another theory solving module, and passes back the result and, if constructed, the branching lemma to its caller.

For VS, assume that we eliminate the variable x_j ($1 \leq j \leq n$) from the formula $\varphi_{\mathbb{R}}^j$ as illustrated in Fig. 2. In general, we can also use the virtual substitution if in some of the polynomials in $\varphi_{\mathbb{R}}^j$ the degree of x_i is higher than 2: We generate all test candidates for x_j from all constraints in which x_j appears at most quadratic. If any of those test candidates leads to a satisfying solution, we can conclude the satisfiability of $\varphi_{\mathbb{R}}^j$. Otherwise, we can pass the sub-problem $\varphi_{\mathbb{R}}^j$ to another theory solving module for satisfiability check. If it returns unsat, we have to consider another path in the search tree of Fig. 2. If it returns sat, we can use the integer assignment of the variables in the passed sub-problem to construct an integer solution for the remaining variables as explained in Sect. 4. Finally, if the sub-call returns unknown and constructs a branching lemma, then

---

[6] This form of multiple-branch lemmas are handled analogously to the 2-branch-case.

VS returns unknown and passes the branching lemma through. If there is no other theory solving module to be called, the sub-call also returns unknown.

The CAD theory solving module implements a complete decision procedure for QF_NRA. For this logic, the CAD module does not pass on any sub-problems to other solver modules.

7   Experimental Results

We consider a *sequential strategy* a sequence of modules that call each other sequentially as described before and we denote it by $M_1 \rightarrow \ldots \rightarrow M_k$ where M_1 may issue sub-calls to M_2 and so on. We evaluated different sequential strategies for solving QF_NIA formulas, using the following modules M_i:

- The SAT solver module M_{SAT} behaves as explained in Sect. 3.
- $M_{SAT_{Stop}}$ works similarly except that it returns unknown if an invoked theory solver module returns unknown, instead of continuing the search for further Boolean assignments. The module $M_{SAT_{Stop}}$ provides us a reference: if this module is able to solve a problem then the problem can be considered irrelevant for BB (as BB was not involved).
- The module M_{LRA} implements the simplex method with branching lemma generation, as explained in Sect. 6.
- The theory solver modules M_{VS} (implementing VS) and M_{CAD} (implementing CAD) check the real relaxation of a QF_NIA input formula. If the relaxation is unsatisfiable they return unsat, if they coincidentally find an integer solution they return sat, otherwise they return unknown (without applying BB).
- The VS module $M_{VS_{\mathbb{Z}}}$ constructs branching lemmas as explained in Sect. 4.
- The CAD modules $M_{CAD_{\mathbb{Z}}^{First}}$ and $M_{CAD_{\mathbb{Z}}^{Last}}$ construct branching lemmas (Sect. 5) based on the first - resp. last-lifted variable with a non-integer assignment.
- The CAD module $M_{CAD_{\mathbb{Z}}^{Path}}$ constructs branching lemmas which exclude the longest integer prefix of the found non-integer solutions (Eq. 3).
- Bit-blasting is implemented in the module $M_{IntBlast}$. In our strategies it will be combined with a preceding incremental variable bound widening module $M_{IncWidth}$, which constrains, for instance, first that all variables are in $[-1, 2]$, if no solution can be found, it requires all variables to be in $[-3, 4]$ etc.

All experiments were carried out on AMD Opteron 6172 processors. Every solver was allowed to use up to 4 GB of memory and 200 s of wall clock time.

For our experiments we used the largest benchmark sets for QF_NIA from the last SMT-COMP: APROVE, LEIPZIG (both generated by automated termination analysis) and CALYPTO (generated by sequential equivalence checking). Additionally, we crafted a new benchmark set CALYPTO$_\infty$ by removing all variable bound constraints from CALYPTO and thereby obtaining unbounded problems (together 8572 problem instances, see headline in Fig. 5e for the size of each set).

Selection of a VS heuristic. The SMT-RAT strategy $M_{SAT_{Stop}} \rightarrow M_{VS}$ could solve 7215 sat and 84 unsat instances, ran out of time or memory for 1146 instances, and returned **unknown** for 127 instances. Applying the SMT-RAT strategy $M_{SAT} \to M_{VS}$ to those 127 instances, we can solve an additional 30 **sat** instances. If we replace the module M_{VS} by the M_{VS_Z} module, which applies branching lemmas, we can solve further 63 **sat** and 10 **unsat** instances (see Fig. 5b).

RAT_Z: $M_{SAT} \to M_{LRA} \to M_{VS_Z} \to M_{CAD_Z^{First}}$

RAT_{blast}: $M_{IncWidth} \longrightarrow M_{IntBlast}$

$RAT_{blast.Z}$: $M_{IncWidth} \longrightarrow M_{IntBlast} \longrightarrow RAT_Z$

(a) The **SMT-RAT** strategies [14] used for the experimental results

	VS$_\mathbb{R}$		VS$_Z$	
	#	time	#	time
sat	30	714.2	93	487.3
unsat	0	0.0	10	9.2

(b) Comparison of 2 VS heuristics on 126 (101 **sat**, 25 **unsat**) for BB relevant instances

	CAD$_\mathbb{R}$		CAD$_Z^{First}$		CAD$_Z^{Last}$		CAD$_Z^{Path}$	
	#	time	#	time	#	time	#	time
sat	12	137.7	12	183.5	11	182.7	7	58.4
unsat	0	0.0	13	150.8	13	151.0	2	131.9

(c) Comparison of 4 CAD heuristics on 55 (27 **sat**, 28 **unsat**) for BB relevant instances

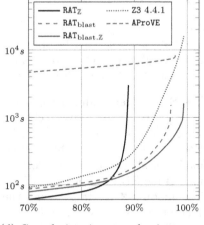

(d) Cumulative time to solve instances from all benchmark sets

| Benchmark→ | | AProve | (8129) | Calypto | (138) | Leipzig | (167) | Calypto | ∞ (138) | all | (8572) |
Solver↓		#	time	#	time	#	time	#	time	#	time
RAT$_Z$	sat	7283	2294.8	67	71.2	9	260.4	133	298.9	7492	2925.3
	unsat	73	14.3	52	40.7	0	0.0	3	< 0.1	128	55.1
RAT$_{blast}$	sat	8025	866.3	21	35.6	156	603.3	87	16.0	8289	1521.2
	unsat	12	0.4	5	0.1	0	0.0	0	0.0	17	0.5
RAT$_{blast.Z}$	sat	8025	780.7	79	122.3	156	511.5	134	21.8	8394	1436.3
	unsat	71	42.6	46	127.5	0	0.0	3	0.1	120	170.2
Z3	sat	7992	14695.5	78	19.1	158	427.6	126	57.3	8354	15199.5
	unsat	102	595.9	57	117.6	0	0.0	3	2.3	162	715.8
AProVE	sat	8025	7052.2	74	559.1	159	696.5	127	685.2	8385	8993.0
	unsat	0	0.0	0	0.0	0	0.0	0	0.0	0	0.0

(e) Comparison of 3 **SMT-RAT** strategies to currently fastest SMT solvers for QF_NIA

Fig. 5. The column # contains the number of solved instances and the column *time* contains the amount of seconds needed for solving these instances

Selection of a CAD heuristic. The **SMT-RAT** strategy $M_{SAT_{Stop}} \to M_{CAD}$ could solve 6656 **sat** and 26 **unsat** instances. The main reason why this approach can already solve more than 77 % of the examples lies in the nature of the CAD to choose preferably integer sample points and, of course, in the structure of the benchmark instances. The strategy ran out of time or memory for 1835 instances, and returned **unknown** for 55 instances. On these 55 examples, we compared the **SMT-RAT** strategy $M_{SAT} \to M_{CAD}$ and 3 other strategies replacing the M_{CAD} module by $M_{CAD_Z^{First}}$, $M_{CAD_Z^{Last}}$ and $M_{CAD_Z^{Path}}$. As shown in Fig. 5c, we find 12 additional **sat** instances with the M_{CAD} module. The BB modules $M_{CAD_Z^{First}}$ and $M_{CAD_Z^{Last}}$ perform very similar and find 13 additional **unsat** instances. This is due to the fact that almost always the assignment of only one variable was not yet integer. With the heuristic in the module $M_{CAD_Z^{Path}}$ we could solve less instances.

Combined strategies. We crafted three strategies, depicted in Fig. 5a, to combine different theory solver modules[7]. The strategy $RAT_{blast.Z}$ combines RAT_{blast} and RAT_Z by first using bit-blasting up to a width of 4 bits. If this does not yield a solution, it continues to use RAT_Z.

We compared these three strategies with the two fastest SMT solvers from the 2015 SMT-COMP for QF_NIA: Z3 and APrOVE. Though CVC4 performed worse than these two solvers, its experimental version solved slightly more instances than APrOVE in about half of the time; we did not include it here but expect it to perform between Z3 and APrOVE. Figure 5e shows that RAT_Z and RAT_{blast} complement each other well, especially for satisfiable instances. Compared to Z3 and APrOVE, $RAT_{blast.Z}$ solves more satisfiable instances and does this even faster by a factor of more than 10 and 6, respectively. The strategy RAT_Z solves less instances, but, as shown in Fig. 5d, this strategy solves the first 85 percent of the examples faster than any other SMT-RAT strategy or SMT solver. On unsatisfiable instances, however, Z3 is still better than SMT-RAT while APrOVE is not able to deduce unsatisfiability due to its pure bit-blasting approach.

We also tested all SMT-RAT strategies which use BB, once with and once without using a branching premise. Here we could not detect any notable difference, which we mainly relate to the fact that those problem instances, for which BB comes to application, are almost always pure conjunctions of constraints and involve only a small number of branching lemma liftings. For a more reliable evaluation a larger set of QF_NIA benchmarks would be needed.

8 Conclusion and Future Work

The efficiency of solving QF_NIA formulas highly depends on a good strategic combination of different procedures. In this paper we comprised two algebraic procedures, the virtual substitution and the cylindrical algebraic decomposition methods, in a combination with the branch-and-bound approach, which has already been applied effectively in combination with the simplex method. We showed by experimental evaluation that this combination highly complements bit-blasting, the currently most efficient approach for QF_NIA.

The next steps to enhance the strategy for solving QF_NIA formulas could involve interval constraint propagation in order to infer better bounds for the variables. We also plan to further optimise the generation of branching lemmas and of the explanations of unsatisfiability in the theory solving modules.

---

[7] Additionally, all of these strategies employ a common preprocessing.

References

1. Arnon, D.S., Collins, G.E., McCallum, S.: Cylindrical algebraic decomposition I: the basic algorithm. SIAM J. Comput. **13**(4), 865–877 (1984)
2. Arnon, D.S., Collins, G.E., McCallum, S.: Cylindrical algebraic decomposition II: an adjacency algorithm for the plane. SIAM J. Comput. **13**(4), 878–889 (1984)
3. Barnett, M., Chang, B.-Y.E., DeLine, R., Jacobs, B., M. Leino, K.R.: Boogie: a modular reusable verifier for object-oriented programs. In: de Boer, F.S., Bonsangue, M.M., Graf, S., de Roever, W.-P. (eds.) FMCO 2005. LNCS, vol. 4111, pp. 364–387. Springer, Heidelberg (2006)
4. Barrett, C., Conway, C.L., Deters, M., Hadarean, L., Jovanović, D., King, T., Reynolds, A., Tinelli, C.: CVC4. In: Gopalakrishnan, G., Qadeer, S. (eds.) CAV 2011. LNCS, vol. 6806, pp. 171–177. Springer, Heidelberg (2011)
5. Barrett, C.W., Nieuwenhuis, R., Oliveras, A., Tinelli, C.: Splitting on demand in SAT modulo theories. In: Hermann, M., Voronkov, A. (eds.) LPAR 2006. LNCS (LNAI), vol. 4246, pp. 512–526. Springer, Heidelberg (2006)
6. Barrett, C., Sebastiani, R., Seshia, S.A., Tinelli, C.: Satisfiability modulo theories. In: Handbook of Satisfiability. Frontiers in Artificial Intelligence and Applications, vol. 185, Chap. 26, pp. 825–885. IOS Press, Amsterdam (2009)
7. Biere, A., Heule, M.J.H., van Maaren, H., Walsh, T. (eds.): Handbook of Satisfiability. Frontiers in Artificial Intelligence and Applications, vol. 185. IOS Press, Amsterdam (2009)
8. Borralleras, C., Lucas, S., Navarro-Marset, R., Rodríguez-Carbonell, E., Rubio, A.: Solving non-linear polynomial arithmetic via SAT modulo linear arithmetic. In: Schmidt, R.A. (ed.) CADE-22. LNCS, vol. 5663, pp. 294–305. Springer, Heidelberg (2009)
9. Bouton, T., Caminha B. de Oliveira, D., Déharbe, D., Fontaine, P.: veriT: an open, trustable and efficient SMT-solver. In: Schmidt, R.A. (ed.) CADE-22. LNCS, vol. 5663, pp. 151–156. Springer, Heidelberg (2009)
10. Brown, C.W.: Improved projection for cylindrical algebraic decomposition. J. Symbolic Comput. **32**(5), 447–465 (2001)
11. Cimatti, A., Griggio, A., Schaafsma, B.J., Sebastiani, R.: The MathSAT5 SMT solver. In: Piterman, N., Smolka, S.A. (eds.) TACAS 2013 (ETAPS 2013). LNCS, vol. 7795, pp. 93–107. Springer, Heidelberg (2013)
12. Collins, G.E.: Quantifier elimination for real closed fields by cylindrical algebraic decomposition. In: Brakhage, H. (ed.) Automata Theory and Formal Languages, vol. 33, pp. 134–183. Springer, Berlin (1975)
13. Corzilius, F., Ábrahám, E.: Virtual substitution for SMT-solving. In: Owe, O., Steffen, M., Telle, J.A. (eds.) FCT 2011. LNCS, vol. 6914, pp. 360–371. Springer, Heidelberg (2011)
14. Corzilius, F., Kremer, G., Junges, S., Schupp, S., Ábrahám, E.: SMT-RAT: an open source C++ toolbox for strategic and parallel SMT solving. In: Heule, M., et al. (eds.) SAT 2015. LNCS, vol. 9340, pp. 360–368. Springer, Heidelberg (2015)
15. Dantzig, G.B.: Linear Programming and Extensions. Princeton University Press, Princeton (1963)
16. Davis, M., Logemann, G., Loveland, D.: A machine program for theorem-proving. Commun. ACM **5**(7), 394–397 (1962)
17. Dutertre, B., de Moura, L.: A fast linear-arithmetic solver for DPLL(T). In: Ball, T., Jones, R.B. (eds.) CAV 2006. LNCS, vol. 4144, pp. 81–94. Springer, Heidelberg (2006)

18. Dutertre, B.: Yices 2.2. In: Biere, A., Bloem, R. (eds.) CAV 2014. LNCS, vol. 8559, pp. 737–744. Springer, Heidelberg (2014)
19. Fränzle, M., Herde, C., Teige, T., Ratschan, S., Schubert, T.: Efficient solving of large non-linear arithmetik constraint systems with complex Boolean structure. J. Satisfiability Boolean Model. Comput. **1**, 209–236 (2007)
20. Fuhs, C., Giesl, J., Middeldorp, A., Schneider-Kamp, P., Thiemann, R., Zankl, H.: SAT solving for termination analysis with polynomial interpretations. In: Marques-Silva, J., Sakallah, K.A. (eds.) SAT 2007. LNCS, vol. 4501, pp. 340–354. Springer, Heidelberg (2007)
21. Griggio, A.: A practical approach to satisfiability modulo linear integer arithmetic. J. Satisfiability Boolean Model. Comput. **8**, 1–27 (2012)
22. Hong, H.: An improvement of the projection operator in cylindrical algebraic decomposition. In: Watanabe, S., Nagata, M. (eds.) Proceedings of the ISSAC 1990, pp. 261–264. ACM, New York (1990)
23. Kim, H., Somenzi, F., Jin, H.S.: Efficient term-ITE conversion for satisfiability modulo theories. In: Kullmann, O. (ed.) SAT 2009. LNCS, vol. 5584, pp. 195–208. Springer, Heidelberg (2009)
24. McCallum, S.: An improved projection operation for cylindrical algebraic decomposition of three-dimensional space. J. Symbolic Comput. **5**(1), 141–161 (1988)
25. de Moura, L., Bjørner, N.S.: Z3: an efficient SMT solver. In: Ramakrishnan, C.R., Rehof, J. (eds.) TACAS 2008. LNCS, vol. 4963, pp. 337–340. Springer, Heidelberg (2008)
26. Schrijver, A.: Theory of Linear and Integer Programming. John Wiley & Sons, Inc., New York (1986)
27. Tung, V.X., Van Khanh, T., Ogawa, M.: raSAT: an SMT solver for polynomial constraints. In: Olivetti, N., Tiwari, A. (eds.) IJCAR 2016. LNCS, vol. 9706, pp. 228–237. Springer, Heidelberg (2016)
28. Weispfenning, V.: Quantifier elimination for real algebra - the quadratic case and beyond. Appl. Algebra Eng. Commun. Comput. **8**(2), 85–101 (1997)

Computing Characteristic Polynomials of Matrices of Structured Polynomials

Marshall Law[(⊠)] and Michael Monagan[(⊠)]

Department of Mathematics, Simon Fraser University, Burnaby, BC, Canada
mylaw@sfu.ca, mmonagan@cecm.sfu.ca

Abstract. We present a parallel modular algorithm for finding characteristic polynomials of matrices with integer coefficient bivariate polynomials. For each prime, evaluation and interpolation gives us the bridge between polynomial matrices and matrices over a finite field so that the Hessenberg algorithm can be used.

1 Introduction

We are interested in specific structured matrices obtained from [9] which arise from combinatorial problems. The goal is to compute their respective characteristic polynomials. Let $A(x, y)$ represent the matrix of interest with dimension $n \times n$. The entries of A are polynomials in x and y of the form

$$A_{ij}(x, y) = c_{ij} x^a y^b$$

with $a, b \in \mathbb{N} \cup \{0\}, c_{ij} \in \mathbb{Z}, 1 \leq i, j \leq n$. Please see Appendix A1 for the 16 by 16 example. Let $C(\lambda, x, y) \in \mathbb{Z}[\lambda, x, y]$ be the characteristic polynomial, which is

$$C(\lambda, x, y) = \det (A - \lambda I_n)$$

by definition, where I_n is the $n \times n$ identity matrix.

The matrix sizes range from 16 to 256, so using the general purpose routine in a computer algebra system like Maple will work only for the small cases. Finding the characteristic polynomial using Maple takes over one day for the 128 by 128 case. Fortunately, there is much structure in the coefficients of the characteristic polynomial. We are able to automatically find this structure and take advantage of it. This paper presents the optimizations to computing the characteristic polynomial. On multi-core machines, we can compute the characteristic polynomial of the largest size 256 matrix in less than 24 h.

Paper Outline. Section 2 discusses some background along with core routines in our method. Section 3 summarizes a naive first approach. Section 4 is the query phase of our algorithm, which determines the structure to be taken advantage of. Section 5 presents how the optimizations work, based on the structure discovered from Sect. 4. Section 6 is on the parallel algorithm. Section 7 and onwards include timings, appendix and references.

© Springer International Publishing AG 2016
V.P. Gerdt et al. (Eds.): CASC 2016, LNCS 9890, pp. 336–348, 2016.
DOI: 10.1007/978-3-319-45641-6_22

2   Background

One method for computing the characteristic polynomial $C(\lambda, x, y)$ for a polynomial matrix is to construct the characteristic matrix $A - \lambda I_n$ and use the Bareiss fraction-free algorithm [2] to compute its determinant. Magma uses this method. The Bariess algorithm is a modification of Gaussian elimination based on Sylvester's identity. Since the core routine is similar to that of Gaussian elimination, it does $O(n^3)$ arithmetic operations in the ring $\mathbb{Z}[\lambda, x, y]$ which include polynomial subtraction, multiplication and exact division.

Maple uses the Berkowitz method [3] when the "CharacteristicPolynomial" routine in the "LinearAlgebra" package is called. This algorithm does $O(n^4)$ arithmetic operations in the ring $\mathbb{Z}[x, y]$ with no divisions. For our matrices, it is much faster than the $O(n^3)$ fraction-free method, over 10 times faster for the 16 by 16 case. The reason is that the intermediate polynomials in the Bareiss fraction free method grow in size and the multiplications and divisions are more expensive than the multiplications in the Berkowitz algorithm.

2.1   Hessenberg

For matrix entries over a field F, the Hessenberg algorithm [4] does $O(n^3)$ arithmetic operations in F to find $C(\lambda)$ in $F[\lambda]$. This algorithm builds up the characteristic polynomial from submatrices, and is the core of our modular algorithm. The first stage transforms the matrix A into a matrix H in Hessenberg form while preserving the characteristic polynomial. The Hessenberg form of H is given below.

$$H = \begin{bmatrix} h_{1,1} & h_{1,2} & h_{1,3} & \ldots & h_{1,n} \\ k_2 & h_{2,2} & h_{2,3} & \ldots & h_{2,n} \\ 0 & k_3 & h_{3,3} & \ldots & h_{3,n} \\ \vdots & \ddots & \ddots & \ddots & \vdots \\ 0 & \ldots & 0 & k_n & h_{n,n} \end{bmatrix}$$

The second stage computes the characteristic polynomial of H using a recurrence. Let $C_m(\lambda)$ represent the characteristic polynomial of the top left submatrix formed by the first m rows and columns. We have the following recurrence relation starting with $C_0(\lambda) = 1$,

$$C_m(\lambda) = (\lambda - h_{m,m})C_{m-1}(\lambda) - \sum_{i=1}^{m-1} \left(h_{i,m} C_{i-1}(\lambda) \prod_{j=i+1}^{m} k_j \right).$$

2.2   The Modular Algorithm

We compute the image of the characteristic polynomial for a sequence of primes $p_1, p_2, \ldots p_m$. Then to recover the solution over the integers we simply use Chinese remainder algorithm. Below is the outline, followed by the homomorphism diagram in Fig. 1 for the modular algorithm.

1. For each prime p in $p_1, p_2, \ldots p_m$, do the following:
   (a) Evaluate the matrix entries at $x = \alpha_i$ and $y = \beta_j$ modulo p.
   (b) Apply the Hessenberg algorithm to compute $C(\lambda, \alpha_i, \beta_j)$ the characteristic polynomial of the evaluated matrix $A(\alpha_i, \beta_j)$ modulo p.
   (c) Interpolate the coefficients of λ in y and x for $C(\lambda, x, y)$ modulo p.
2. Recover the integer coefficients of $C(\lambda, x, y)$ using the Chinese remainder algorithm (CRA).

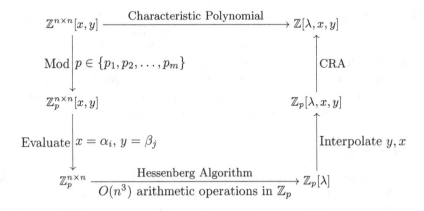

Fig. 1. Modular algorithm homomorphism diagram

2.3 Degree Bounds

To interpolate x and y in $C(\lambda, x, y)$ the modular algorithm needs $1 + \deg_x C$ points for x and $1 + \deg_y C$ for y. The degrees in x, y of the characteristic polynomial are bounded by the following:

$$\deg_x C(\lambda, x, y) \le D_x = \min \left(\sum_{i=1}^{n} \max_{j=1}^{n} \deg_x A_{ij}, \sum_{j=1}^{n} \max_{i=1}^{n} \deg_x A_{ij} \right)$$

$$\deg_y C(\lambda, x, y) \le D_y = \min \left(\sum_{i=1}^{n} \max_{j=1}^{n} \deg_y A_{ij}, \sum_{j=1}^{n} \max_{i=1}^{n} \deg_y A_{ij} \right)$$

Each sum within the two equations above adds the largest degree in each row and column respectively. Then we take the minimum of the two to obtain the best degree bound in that variable.

Kronecker Substitution. There is no doubt that the problem will be simpler if the matrices of interest only consist of one variable. This can be achieved by a Kronecker substitution. To ensure a reversible substitution, let $b > \deg_x C(\lambda, x, y)$, and apply it to $A(x, y = x^b)$. Then $C(\lambda, x, y)$ can be recovered from the characteristic polynomial of $A(x, y = x^b)$.

We use $b = D_x + 1$, as this is the smallest possible value for an invertible mapping. As expected, the degrees of the polynomial entries in $A(x, y = x^b)$ become quite large. Now the problem has effectively become solving for

$$\det(A(x, x^b) - \lambda I_n) = C(\lambda, x, x^b).$$

Note that in our benchmarks section, no Kronecker substitution was involved.

Coefficient Bound. The classical Hadamard inequality for n by n integer matrix $M = (m_{ij})$ asserts

$$|\det(M)| \le H(M) = \left(\prod_{i=1}^{n} \sum_{j=1}^{n} |m_{ij}|^2 \right)^{1/2}.$$

A similar bound [11] exists for matrices with polynomial entries. Let $M(x) = (m_{ij})$ with m_{ij} a polynomial in $\mathbb{Z}[x]$. Let $s_0, s_1, \ldots$ be the coefficients of the polynomial representation of $\det(M(x))$. Let $T = (t_{ij})$ be the n by n matrix obtained from M as follows. Let t_{ij} be the sum of the absolute values of the coefficients of $m_{ij}(x)$. Then the equivalent bound is given by

$$\| \det(M) \|_2 = \left(\sum |s_i|^2 \right)^{1/2} \le H(T) = \left(\prod_{i=1}^{n} \sum_{j=1}^{n} |t_{ij}|^2 \right)^{1/2}.$$

This bound generalizes to matrices of multivariate polynomials with integer coefficients by using a Kronecker substitution. The entries of our matrix $A(x, y)$ are unit monomials. Thus for $M = A - \lambda I_n$, $t_{ii} = 2$ and $t_{ij} = 1$ for $i \ne j$. The height of $C(\lambda, x, y)$, denoted $\|C(\lambda, x, y)\|_\infty$, is bounded by

$$\|C(\lambda, x, y)\|_\infty \le \|C(\lambda, x, y)\|_2 \le \left(\prod^{n}(n+3) \right)^{1/2} = (n+3)^{n/2}.$$

This bound tells us how many primes are needed at most to recover the integer coefficients of $C(\lambda, x, y)$. Namely, we need $\prod_{i=1}^{m} p_i > 2(n+3)^{n/2}$. Note that, with optimizations to be mentioned, the integer coefficients can be recovered with fewer primes.

Let $C(\lambda) = \sum_{i=0}^{n} c_i \lambda^i$ be the characteristic polynomial of A. Because $c_0 = C(0) = \det(A)$ we suspected that $\|C(\lambda)\|_\infty \le H(T)$. Thus for our matrices, we thought we could replace the above bound $(n+3)^{n/2}$ with the Hadamard bound $n^{n/2}$ on $\det(T)$. But in [5], an example of integer matrix with entries ± 1 is given where $|c_1| > n^{n/2} > |c_0|$.

3   Naive First Approach

For starters, we will proceed with a Kronecker substitution. With that, the degrees of the matrix entries are much larger, so we decided to use fast evaluation and interpolation. The fast algorithms used are based on the FFT (Fast Fourier Transform), which have been optimized. Since the transform itself is not the main concern, we will take advantage of the transforms decimated in time and frequency, as the authors did in [10].

3.1   Structures Found

Here we identify the foundation and justification for the following sections/phases, as there is much room for optimization. Let $C(\lambda, x, y) = \sum_{i=0}^{n} c_i(x, y)\lambda^i$. Then the coefficients of λ^i can be written in the form

$$c_i(x, y) = f_i(x, y)x^{g_i}y^{h_i}(x^2 - 1)^{k_i}$$

where the f_i are bivariate polynomials with even degrees in x. See Appendix A3 for $c_0(x, y)$ and $c_1(x, y)$ for 16 by 16 matrix. The exponent values $g_i, h_i, k_i \in \mathbb{N} \cup \{0\}$, for $0 \leq i < n$, represent factors. Table 1 below contains the values for these parameters for the $n = 16$ matrix (see Appendix A1) and also other information about $f_i(x, y)$. The columns $\deg_x$ are the degrees of f_i in x and columns $\deg_y$ are the degrees of f_i in y. Notice that the largest integers in $c_i(x, y)$ in magnitude (see columns $\|c_i\|_\infty$) decrease as i increases but those in f_i (see columns $\|f_i\|_\infty$) increase and decrease.

Table 1. Data for the coefficients of $C(\lambda, x, y)$ for $n = 16$.

i	g_i	h_i	k_i	$\deg_x$	$\deg_y$	$\|f_i\|_\infty$	$\|c_i\|_\infty$	i	g_i	h_i	k_i	$\deg_x$	$\deg_y$	$\|f_i\|_\infty$	$\|c_i\|_\infty$
0	32	32	32	0	0	1	601080390	8	12	10	10	20	12	4730	35264
1	32	28	28	4	4	4	160466400	9	14	8	8	16	12	3740	10876
2	24	25	25	14	6	31	28428920	10	8	6	6	20	12	2116	3242
3	26	22	22	12	8	128	16535016	11	10	4	4	16	12	806	1556
4	20	19	19	18	10	382	3868248	12	4	3	3	18	10	454	322
5	22	16	16	16	12	684	946816	13	6	2	2	12	8	142	108
6	16	14	14	20	12	1948	183648	14	0	1	1	14	6	31	22
7	18	12	12	16	12	3738	82492	15	4	0	0	4	4	4	4

3.2   Method of Approach

As the structure suggests, the most complicated factors to recover are the bivariate polynomials $f_i(x, y)$. In the next two sections, we present two phases to recover $C(\lambda, x, y)$. The first phase is to find g_i, h_i, k_i for $0 \leq i < n$. The second phase is to compute the "cofactors" $f_i(x, y)$.

4 Phase 1 - Query

The factors that need to be found for each λ coefficient are namely

$$x^{g_i} y^{h_i} (x^2 - 1)^{k_i}.$$

We can find the lowest degrees g_i and h_i by making two queries mod a prime, one for each variable. In each query, we evaluate one variable of the matrix A to obtain an image of the characteristic polynomial in the other variable. Let p be the prime and γ be chosen at random from $\mathbb{Z}_p$. We evaluate the matrix A to obtain the two univariate matrices with integer coefficients modulo p

$$A(x, \gamma) \mod p \text{ and } A(\gamma, y) \mod p.$$

Their respective characteristic polynomials are

$$C(\lambda, x, y = \gamma) \text{ and } C(\lambda, x = \gamma, y).$$

This is a much simpler problem, as the entries have much smaller degree and no Kronecker substitution is necessary. Even for our large matrices, the characteristic polynomial of univariate matrices can solved within minutes. The query phase concludes by finding the necessary parameters for phase 2, which are the factor degrees g_i, h_i, k_i for $0 \leq i < n$, and hence the minimal number of evaluation points e_x and e_y.

Due to two random evaluations, there is a possibility of failure in this phase. But we will show that the failure probability is small.

4.1 Lowest Degree Factors

After making the two queries, we have images $C(\lambda, x, y = \gamma)$ and $C(\lambda, x = \gamma, y)$. For each λ coefficient, simply search for first and last non-zero coefficient in x or y. The lowest degrees correspond to g_i, h_i for $0 \leq i < n$. Now also let $\bar{g}_i, \bar{h}_i$ be the largest degrees with non-zero coefficient in x, y respectively. To see it in perspective, the coefficient of λ^i in $C(\lambda, x, y = \gamma)$ has the form

$$\bullet x^{\bar{g}_i} + \cdots + \bullet x^{g_i}$$

where the symbol $\bullet$ represents integers modulo p. Similarly for $C(\lambda, x = \gamma, y)$,

$$\bullet y^{\bar{h}_i} + \cdots + \bullet y^{h_i}$$

and the key values $\bar{g}_i, g_i, \bar{h}_i, h_i$ can found easily by searching for first non-zeroes.

Non-zero Factors. Most coefficients of λ have a factor of $(x^2 - 1)^{k_i}$ with k_i large. Removing this reduces the number of evaluation points needed and the integer coefficient size of $c_i(x, y)/(x^2 - 1)^{k_i}$, hence also the number of primes needed. To determine k_i, we pick $0 < \gamma < p$ at random and divide $c_i(x, \gamma)$ by $(x^2 - 1)$ modulo p repeatedly. For our matrices it happens that $k_i = h_i$.

In general, to determine if $(ax + b)$ is a factor of a coefficient of $c_i(x, y)$, for small integers $a > 0, b$, we could compute the roots of $c_i(x, \gamma)$, a polynomial in $\mathbb{Z}_p[x]$, using Rabin's algorithm [12]. From each root, we try to reconstruct a small fraction $-\frac{b}{a}$ using rational number reconstruction (Sect. 5.10 of [6]). We have not implemented this.

4.2 Required Points

Now to find the minimal points required to recover all the $f_i(x, y)$, for $0 \le i < n$. Since we have already computed the largest, smallest and factor degrees, we can know the maximal degrees of $f_i(x, y)$ in both variables. The minimal number of evaluation points in x, y needed to interpolate x and y are given respectively by

$$e_x := \max \frac{1}{2}\{\bar{g}_i - g_i - 2k_i\} + 1, \text{ for } 0 \le i < n$$

$$e_y := \max\{\bar{h}_i - h_i\} + 1, \text{ for } 0 \le i < n$$

The scalar of half in e_x is due an optimization to be mentioned later (see Sect. 5.3.). The subtraction of $2k_i$ corresponds to the factor $(x^2 - 1)^{k_i}$.

4.3 Unlucky Evaluations

As mentioned earlier, random evaluations may cause the algorithm to return an incorrect answer, as we will explain here. Without loss of generality, consider the query of randomly evaluating at $y = \gamma$. Let $d_i = \deg_x f_i(x, y)$, then

$$f_i(x, y) = \sum_{j=0}^{d_i} f_{ij}(y)x^j, \text{ where } f_{ij} \in \mathbb{Z}[y].$$

When $f_{i0}(\gamma) = 0$, then the lowest degree is strictly greater than g_i, which is incorrect. If the algorithm continues, the target for interpolation is compromised, and the final answer is incorrect.

If $f_{id_i}(\gamma) = 0$, then the largest degree becomes less than $\bar{g}_i$. This may affect e_x, the required number of evaluation points, which takes the maximum of a set (see Sect. 4.2). The final answer will still be correct as long as e_x is correct.

Definition 1. *Let p be a prime, and $0 \le \gamma < p$. Then γ is an unlucky evaluation if for any $0 \le i < n$,*

$$f_{i0}(\gamma) = 0 \ (mod \ p) \ or \ f_{id_i}(\gamma) = 0 \ (mod \ p)$$

where $D_y \ge \deg_y C(\lambda, x, y)$ (from Sect. 2.3).

Theorem 1. *The probability that γ is unlucky is at most $\frac{2nD_y}{p}$.*

Proof. Since $D_y \ge \deg_y C(\lambda, x, y)$, so $\deg_y f_{ij} \le D_y$ for all i, j. For each $0 \le i < n$, there are at most $2D_y$ points where

$$f_{i0}(\gamma) = 0 \ (mod \ p) \ or \ f_{id_i}(\gamma) = 0 \ (mod \ p).$$

There are n of these cases, giving a total of $2nD_y$ unlucky evaluations. Therefore the probability of an unlucky evaluation (for $0 \le i < n$) is given by

$$\Pr\left[f_{i0}(\gamma) = 0 \ (mod \ p) \ or \ f_{id_i}(\gamma) = 0 \ (mod \ p)\right] \le \frac{2nD_y}{p}.$$

We use a 31 bit prime $p = 2^{27} \times 15 + 1$ in the query phase. For our largest matrix, the parameters are $n = 256$, $D_x = 3072$ and $D_y = 1024$. Our algorithm makes two queries, one for each variable, so the probability of an unlucky evaluation is less than 0.105%.

5   Phase 2 - Optimizations

The structure for each λ coefficient is already known, so in this section we will use a specific matrix and its characteristic polynomial. We have implemented each of the following optimizations with Newton interpolation. We note that these optimizations apply for fast interpolation (FFT) too.

We will illustrate the optimizations on the λ^6 coefficient from the 16 by 16 matrix. The coefficient of λ^6 is

$$c_6(x,y) = f_6(x,y)x^{16}y^{14}\left(x^2-1\right)^{14}$$

where $f_6(x,y)$ (see Appendix A2) has 91 terms and is irreducible over $\mathbb{Z}$. The parameters for the 16 by 16 matrix include $g_6 = 16, h_6 = 14 = k_6$, and $\bar{g}_6 = 64, \bar{h}_6 = 26$.

If the Kronecker substitution were to be used, it will use the substitution $y = x^{97}$, implying Fourier transform size of $s = 4096 > 64 + 26(97) = 2586$. If we work on one variable at a time, it will require $(64+1)(26+1) = 1755$ points. Keep in mind that the total number of evaluation points is the same as the number of calls to the Hessenberg algorithm, which is the bottleneck of the whole algorithm. Our implementation of the FFT involves the staircase increments as well, but can be eliminated if the truncated FFT (see [1,8]) were to be used.

Consider the step of interpolating x after y is interpolated. Let $E = \{\alpha_1, \alpha_2, \dots\}$ be the evaluation points, and $V_i = \{c_i(\alpha_1, y), c_i(\alpha_2, y), \dots\}$ be the values. By the end of this section, we will only require $(10+1)(12+1) = 143$ points, which gives a gain of more than a factor of 12. Note that the gain is greater for larger matrices.

5.1   Lowest Degree

Since the lowest degree is known, that is $g_6 = 16$, we need to interpolate $c_6(x,y)/x^{16}$. For each $\alpha_j \in E$, divide V_i by $\alpha_j^{g_6}$ point wise. Then regular interpolation will give

$$c_6(x,y)/x^{16} = f_6(x,y)y^{14}(x^2-1)^{14}.$$

In this example there is a saving of $g_6 = 16$ points. This optimization also applies to the other variable y, as $h_i = 14$. When both variables are taken into account, the total number of evaluation points is reduced from 1775 to $(64-16+1)(26-14+1) = 637$.

5.2   Even Degree

All the terms in $f_i(x,y)$ have even degrees in x. So if we interpolate $f_i(x^{1/2},y)$ instead of $f_i(x,y)$, the degree of the target is halved, and the number of evaluation points is also (approximately) halved. To do so, simply square each value in

E, and proceed with interpolation as usual. The polynomial recovered will have half of the true degree, then double each exponent to recover $f_i(x, y)$.

The even degrees structure for our matrices only applies to variable x. With this optimization the number of evaluation points decreases from $(64 + 1)$ to $(32 + 1)$.

5.3   Non-zero Factors

Here we have multiplicity of $k_6 = 14$. This optimization is similar to that of the lowest degree. For each $\alpha_j \in E$, we divide each V_i value by $(\alpha_j^2 - 1)^{h_i}$. Then regular interpolation will return

$$c_i(x, y)/(x^2 - 1)^{14} = f_i(x, y)x^{16}y^{14}.$$

E cannot contain ± 1, because there will be divisions by zero. This optimization is only applicable to variable x, and it alone decreases the number of evaluation points from $(64 + 1)$ to $(36 + 1)$.

Since k_i is large, $||f_i(x, y)||_\infty$ will be much smaller than $||c_i(x, y)||_\infty$. Therefore the algorithm needs fewer primes to recover $C(\lambda, x, y)$. The largest coefficient of $(x^2 - 1)^{k_i}$ is $\binom{k_i}{\lceil k_i/2 \rceil}$. For the 16 by 16 case, the coefficient bound for $C(\lambda, x, y)$ is $(16 + 3)^8$ a 34 bit integer. The actual height $||C(\lambda, x, y)||_\infty$ is a 30 bit integer (see Table 1) and $\max ||f_i(x, y)||_\infty = 4730 = ||f_8(x, y)||_\infty$ is a 13 bit integer. For the 64 by 64 case, $||C(\lambda, x, y)||_\infty$ is 188 bits and $\max ||f_i(x, y)||_\infty$ is 72 bits.

Due to this loose bound, the problem has effectively become much smaller in terms of integer coefficient size. The target is $\max_{0 \le i < n} ||f_i(x, y)||_\infty$, instead of the much larger $\max_{0 \le i < n} ||c_i(x, y)||_\infty$. So $C(\lambda, x, y)$ can be recovered with fewer primes. We give more details in Sect. 5.5.

5.4   Combined

If all three improvements in Sects. 5.1, 5.2 and 5.3 are combined together, the number of evaluation points required for x is $e_x = (64 - 16 - 2 \times 14)/2 + 1 = 11$. Likewise for y, $e_y = (26 - 14) + 1 = 13$. So the total the number of evaluations and hence Hessenberg calls decreases from 1755 to $11 \times 13 = 143$. Since the degrees in y are dense with the lowest degree optimization, we evaluate x first then y and interpolate in reverse order.

5.5   Chinese Remainder Algorithm

The last step of the modular algorithm is to recover the solution over the integers. For m primes and each non-zero coefficient $0 \le c_i < p_i$, we have to solve the system of congruences

$$u \equiv c_i \pmod{p_i} \quad \text{for } 1 \le i \le m.$$

To build up the final coefficients u over $\mathbb{Z}$, we use the mixed radix representation for the integer u, namely, we compute integers $v_1, v_2, \ldots, v_m$ such that

$$u = v_1 + v_2 p_1 + v_3 p_1 p_2 + \cdots + v_m p_1 p_2 \cdots p_{m-1}.$$

See p. 206 of [7]. To account for negative coefficients in $C(\lambda)$ we solve for v_i in the symmetric range $-\frac{p_i}{2} < v_i < \frac{p_i}{2}$ to obtain u in the range $-\frac{M}{2} < u < \frac{M}{2}$ where $M = \prod_{i=1}^{m} p_i$. From the non-zero factors optimization, the coefficient size bound on $||f_i(x, y)||_\infty$ becomes a very loose one. As stated previously for the 64 by 64 matrix, the bound is is 188 bits, but we can recover $C(\lambda, x, y)$ with only 72 bits. So after computing each image of $C(\lambda, x, y) \bmod p_1, p_2, \ldots$, we build the solution in mixed radix form until it stabilizes, that is when $v_k = 0$.

6   Parallelization

The modular algorithm was originally chosen since the computation for each prime can be done in parallel. Each evaluation, Hessenberg call and interpolation may also be computed in parallel. We chose to run each prime sequentially and look to parallelize within each prime for two reasons. First, we don't know how many primes are necessary because of the loose bound from Sect. 5.3. Second, memory may become an issue for computers with low RAM.

Our implementation in Cilk C does the x evaluations in parallel, and it suits the incremental method more. With each prime, the algorithm starts by evaluating x, and this evaluation is trivial since the matrix has unit monomials. In the previous section we have seen that $e_x = 11$ is required for the smallest case (16 by 16). But this matrix is too small to see any significant speed up. The 64 by 64 matrix on a computer with 4 cores shows a good speed up close to the theoretical maximum. Please see the benchmarks section for more details on timings.

7   Benchmarks

Table 2 consists of timings of our modular algorithm. Column min is the minimum number of 30 bit primes needed to recover the integer coefficients in $C(\lambda, x, y)$. Column bnd is the number of primes needed using the bound for $||C(\lambda, x, y)||_\infty$. The number of calls to the Hessenberg algorithm is $(xy)(\min + k)$ assuming we require k check primes. Our implementation uses $k = 1$ check prime. Table 3 includes data for Maple 2016 and Magma V2.22-2. The names and details of machines we ran our software on are given below and they all run Fedora 22.

- sarah: Intel Core i5-4590 quad core at 3.3 GHz, 8 GB RAM
- mark: Intel Core i5-4670 quad core at 3.4 GHz, 16 GB RAM
- luke: AMD FX8350 eight core at 4.2 GHz, 32 GB RAM
- ant: Intel Core i7-3930K six core at 3.2 GHz, 64 GB RAM

Table 2. Modular algorithm timings in seconds (s), minutes (m) or hours (h)

Size	#Points	#Primes	Sarah		Mark		Luke		Ant	
n	x, y	min,bnd	1 Core	4 Cores	1 Core	4 Cores	1 Core	8 Cores	1 Core	6 Cores
16	11,13	1, 2	0.04 s	0.01 s	0.04 s	0.01 s	0.04 s	0.01 s	0.04 s	0.01 s
32	28,31	1, 3	0.48 s	0.10 s	0.58 s	0.10 s	0.60 s	0.10 s	0.64 s	0.14 s
64	67,61	3, 7	19.12 s	5.19 s	18.64 s	5.08 s	32.93 s	4.84 s	21.52 s	4.16 s
128	131,141	6, 16	14.72 m	3.92 m	14.30 m	4.41 m	30.36 m	4.41 m	16.85 m	2.87 m
256	261,281	12, 35	14.42 h	3.75 h	14.23 h	3.77 h	31.29 h	4.28 h	16.53 h	2.74 h

Table 3. Maple and Magma timings in seconds (s), minutes (m) or hours (h)

Size	Maple								Magma
	Sarah		Mark		Luke		Ant		Ant
n	real	cpu	real	cpu	real	cpu	real	cpu	cpu
16	0.28 s	0.30 s	0.21 s	0.23 s	0.46 s	0.53 s	0.32 s	0.36 s	0.32 s
32	34.1 s	45.2 s	30.2 s	41.1 s	50.3 s	83.7 s	32.7 s	46.3 s	99.7 s
64	19.9 h	32.6 h	12.1 h	23.2 h	3.63 h	5.42 h	2.86 h	3.91 h	15.1 h
128,256	Not attempted								

A   Appendix

A1: 16 x 16 matrix

$$
\begin{bmatrix}
x^8 & x^5y & x^5y & x^4y^2 & x^5y & x^2y^2 & x^4y^2 & x^3y^3 & x^5y & x^4y^2 & x^2y^2 & x^3y^3 & x^4y^2 & x^3y^3 & x^3y^3 & x^4y^4 \\
x^7 & x^6y & x^4y & x^5y^2 & x^4y & x^3y^2 & x^3y^2 & x^4y^3 & x^4y & x^5y^2 & xy^2 & x^4y^3 & x^3y^2 & x^4y^3 & x^2y^3 & x^5y^4 \\
x^7 & x^4y & x^6y & x^5y^2 & x^4y & xy^2 & x^5y^2 & x^4y^3 & x^4y & x^3y^2 & x^3y^2 & x^4y^3 & x^3y^2 & x^2y^3 & x^4y^3 & x^5y^4 \\
x^6 & x^5y & x^5y & x^6y^2 & x^3y & x^2y^2 & x^4y^2 & x^5y^3 & x^3y & x^4y^2 & x^2y^2 & x^5y^3 & x^2y^2 & x^3y^3 & x^3y^3 & x^6y^4 \\
x^7 & x^4y & x^4y & x^3y^2 & x^6y & x^3y^2 & x^5y^2 & x^4y^3 & x^4y & x^3y^2 & xy^2 & x^2y^3 & x^5y^2 & x^4y^3 & x^4y^3 & x^5y^4 \\
x^6 & x^5y & x^3y & x^4y^2 & x^5y & x^4y^2 & x^4y^2 & x^5y^3 & x^3y & x^4y^2 & y^2 & x^3y^3 & x^4y^2 & x^5y^3 & x^3y^3 & x^6y^4 \\
x^6 & x^3y & x^5y & x^4y^2 & x^5y & x^2y^2 & x^6y^2 & x^5y^3 & x^3y & x^2y^2 & x^2y^2 & x^2y^3 & x^3y^3 & x^4y^2 & x^3y^3 & x^5y^3 & x^6y^4 \\
x^5 & x^4y & x^4y & x^5y^2 & x^4y & x^3y^2 & x^5y^2 & x^6y^3 & x^2y & x^3y^2 & xy^2 & x^4y^3 & x^3y^2 & x^4y^3 & x^4y^3 & x^7y^4 \\
x^7 & x^4y & x^4y & x^3y^2 & x^4y & xy^2 & x^3y^2 & x^2y^3 & x^6y & x^5y^2 & x^3y^2 & x^4y^3 & x^5y^2 & x^4y^3 & x^4y^3 & x^5y^4 \\
x^6 & x^5y & x^3y & x^4y^2 & x^3y & x^2y^2 & x^2y^2 & x^3y^3 & x^5y & x^6y^2 & x^2y^2 & x^5y^3 & x^4y^2 & x^5y^3 & x^3y^3 & x^6y^4 \\
x^6 & x^3y & x^5y & x^4y^2 & x^3y & y^2 & x^4y^2 & x^3y^3 & x^5y & x^4y^2 & x^4y^2 & x^5y^3 & x^4y^2 & x^3y^3 & x^5y^3 & x^6y^4 \\
x^5 & x^4y & x^4y & x^5y^2 & x^2y & xy^2 & x^3y^2 & x^4y^3 & x^4y & x^5y^2 & x^3y^2 & x^6y^3 & x^3y^2 & x^4y^3 & x^4y^3 & x^7y^4 \\
x^6 & x^3y & x^3y & x^2y^2 & x^5y & x^2y^2 & x^4y^2 & x^3y^3 & x^5y & x^4y^2 & x^2y^2 & x^3y^3 & x^6y^3 & x^5y^3 & x^5y^3 & x^6y^4 \\
x^5 & x^4y & x^2y & x^3y^2 & x^4y & x^3y^2 & x^3y^2 & x^4y^3 & x^4y & x^5y^2 & xy^2 & x^4y^3 & x^5y^2 & x^6y^3 & x^4y^3 & x^7y^4 \\
x^5 & x^2y & x^4y & x^3y^2 & x^4y & xy^2 & x^5y^2 & x^4y^3 & x^4y & x^3y^2 & x^3y^2 & x^4y^3 & x^5y^2 & x^4y^3 & x^6y^3 & x^7y^4 \\
x^4 & x^3y & x^3y & x^4y^2 & x^3y & x^2y^2 & x^4y^2 & x^5y^3 & x^3y & x^4y^2 & x^2y^2 & x^5y^3 & x^4y^2 & x^5y^3 & x^5y^3 & x^8y^4
\end{bmatrix}
$$

A2: $f_6(x, y)$ of 16 x 16 matrix

$$(2x^{16} + 4x^{14})y^{12} + (4x^{18} + 32x^{16} + 28x^{14} + 8x^{12})y^{11} +$$

$$(x^{20} + 34x^{18} + 149x^{16} + 188x^{14} + 43x^{12} - 22x^{10} + 3x^8)y^{10} +$$

$$(16x^{20} + 128x^{18} + 452x^{16} + 568x^{14} + 268x^{12} - 32x^{10} - 72x^8 - 8x^6)y^9 +$$

$$(52x^{20} + 348x^{18} + 910x^{16} + 1172x^{14} + 704x^{12} + 68x^{10} - 136x^8 - 120x^6 - 28x^4)y^8 +$$

$$(112x^{20} + 596x^{18} + 1344x^{16} + 1788x^{14} + 1224x^{12} + 216x^{10} - 220x^8 - 184x^6 - 92x^4 - 32x^2)y^7 +$$

$$(133x^{20} + 734x^{18} + 1551x^{16} + 1948x^{14} + 1476x^{12} + 428x^{10} - 320x^8 - 276x^6 - 81x^4 - 34x^2 - 15)y^6 +$$

$$(112x^{20} + 596x^{18} + 1344x^{16} + 1788x^{14} + 1224x^{12} + 216x^{10} - 220x^8 - 184x^6 - 92x^4 - 32x^2)y^5 +$$

$$(52x^{20} + 348x^{18} + 910x^{16} + 1172x^{14} + 704x^{12} + 68x^{10} - 136x^8 - 120x^6 - 28x^4)y^4 +$$

$$(16x^{20} + 128x^{18} + 452x^{16} + 568x^{14} + 268x^{12} - 32x^{10} - 72x^8 - 8x^6)y^3 +$$

$$(x^{20} + 34x^{18} + 149x^{16} + 188x^{14} + 43x^{12} - 22x^{10} + 3x^8)y^2 +$$

$$(4x^{18} + 32x^{16} + 28x^{14} + 8x^{12})y + (2x^{16} + 4x^{14})y^0$$

A3: First two coefficients of $C(\lambda, x, y)$ for 16 by 16 matrix

$$c_0(x, y) = x^{32}y^{32} \left(x^2 - 1\right)^{32}$$

$$c_1(x, y) = -x^{32}y^{28} \left(x^2 - 1\right)^{28} \left(2\,x^4y^2 + 4\,x^2y^3 + 4\,x^2y^2 + y^4 + 4\,x^2y + 1\right)$$

B1: Time 16 by 16 on Maple

```
A := Matrix(16, 16, [ [x^8, x^5*y, ...], ... ]);
with(LinearAlgebra):
C := CodeTools[Usage]( CharacteristicPolynomial(A, lambda) ):
```

B2: Time 16 by 16 on Magma

```
P<x,y> := PolynomialRing( IntegerRing(), 2);
A := Matrix(P, 16, 16, [ [x^8, x^5*y, ...], ... ]);
time C := CharacteristicPolynomial(A);
```

References

1. Arnold, A.: A new truncated fourier transform algorithm. In: Proceedings of ISSAC 2013, pp. 15–22. ACM Press (2013)
2. Bareiss, E.H.: Sylvester's identity and multistep integer-preserving Gaussian elimination. Math. Comput. **22**(103), 565–578 (1968)
3. Berkowitz, S.J.: On computing the determinant in small parallel time using a small number of processors. Inf. Process. Lett. **18**(3), 147–150 (1984)

4. Cohen, H.: A Course in Computational Algebraic Number Theory, p. 138. Springer, Heidelberg (1995)
5. Dumas, J.G.: Bounds on the coefficients of the characteristic and minimal polynomials. J. Inequalities Pure Appl. Math. **8**(2), 1–6 (2007). Article ID 31
6. von zur Gathen, J., Gerhard, J.: Modern Computer Algebra. Cambridge University Press, New York (2003)
7. Geddes, K.O., Czapor, S.R., Labahn, G.: Algorithms for Computer Algebra. Kluwer Academic Publishers, Boston (1992)
8. van der Hoeven, J.: The truncated fourier transform and applications. In: Proceedings of ISSAC 2004, pp. 290–296. ACM Press (2004)
9. Kauers, M.: Personal Communication
10. Law, M., Monagan, M.: A parallel implementation for polynomial multiplication modulo a prime. In: Proceedings of PASCO 2015, pp. 78–86. ACM Press (2015)
11. Lossers, O.P.: A Hadamard-type bound on the coefficients of a determinant of polynomials. SIAM Rev. **16**(3), 394–395 (2006). Solution to an exercise by Goldstein, A.J., Graham, R.L. (2006)
12. Rabin, M.: Probabilistic algorithms in finite fields. SIAM J. Comput. **9**(2), 273–280 (1979)

Computing Sparse Representations of Systems of Rational Fractions

François Lemaire$^{(\boxtimes)}$ and Alexandre Temperville

University of Lille, CNRS, Centrale Lille, UMR 9189 - CRIStAL - Centre de
Recherche en Informatique Signal et Automatique de Lille, 59000 Lille, France
`francois.lemaire@univ-lille1.fr`, `a.temperville@ed.univ-lille1.fr`

Abstract. We present new algorithms for computing sparse representations of systems of parametric rational fractions by means of change of coordinates. Our algorithms are based on computing sparse matrices encoding the degrees of the parameters occurring in the fractions. Our methods facilitate further qualitative analysis of systems involving rational fractions. Contrary to symmetry based approaches which reduce the number of parameters, our methods only increase the sparsity, and are thus complementary. Previously hand made computations can now be fully automated by our methods.

Keywords: Parametric systems · Simplification of rational fractions · Sparse basis of vector spaces

1 Introduction

This article presents new algorithms for computing sparse representations of systems of parametric rational fractions by means of change of variables. The goal of these algorithms is to help the analysis of parametric systems of rational fractions by producing sparser and equivalent formulations. Simplifying parametric systems is a central task, since many qualitative analyses (such as steady point analysis, bifurcation analysis, . . .) rely on quite costly computations in real algebraic geometry (see [1] and references therein).

Symmetry based approaches [2–6] reduce the number of parameters and as a consequence usually help the analysis of parametric systems. On the contrary, our approach keeps the number of parameters (in the same spirit as [2, Algorithm SemiRectifySteadyPoints]) and makes the systems sparsest (in the sense of Algorithms getSparsestFraction and getSparsestSumOfFractions given later).

This work was motivated by the differential system (see Example 14):

$$\begin{cases} G' = \theta(1 - G) - \alpha k_1 k_2 k_3 P^4 G \\ M' = \rho_b(1 - G) + \rho_f G - \delta_M M \\ P' = \dfrac{4\theta(1 - G) - 4\alpha k_1 k_2 k_3 P^4 G - \delta_P P + \beta M}{16 k_1 k_2 k_3 P^3 + 9 k_1 k_2 P^2 + 4 k_1 P + 1} \end{cases} \tag{1}$$

© Springer International Publishing AG 2016
V.P. Gerdt et al. (Eds.): CASC 2016, LNCS 9890, pp. 349–366, 2016.
DOI: 10.1007/978-3-319-45641-6_23

where the unknown functions are $G = G(t)$, $M = M(t)$ and $P = P(t)$, and where the parameters are θ, α, k_1, k_2, k_3, ρ_b, ρ_f, δ_M, δ_P and β.

Equation (1) was considered in [7], and rewritten with the guessed change of variables $\bar{k}_3 = k_1 k_2 k_3$, $\bar{k}_2 = k_1 k_2$, $\bar{k}_1 = k_1$, in order to obtain [7, Equation (3.4)] (in the case $n = 4$ and $\gamma_0 = 1$). Our new algorithm getSparsestSumOfFractions (see Sect. 4) was designed to automatically compute this change of variables, which yields the following simpler system:

$$
\begin{cases}
G' = \theta(1 - G) - \alpha\bar{k}_3 P^4 G \\
M' = \rho_b(1 - G) + \rho_f G - \delta_M M \\
P' = \dfrac{4\theta(1 - G) - 4\alpha\bar{k}_3 P^4 G - \delta_P P + \beta M}{16\bar{k}_3 P^3 + 9\bar{k}_2 P^2 + 4\bar{k}_1 P + 1}.
\end{cases}
\tag{2}
$$

The relatively small improvement between (1) and (2) in terms of degrees proves useful while searching for a Hopf Bifurcation. Indeed, the Routh-Hurwitz criterion applied on systems (1) and (2) leads to semi-algebraic systems of the form $h_1, h_2, h_3, h_4 = 0$, $h_5, h_6, h_7 > 0$, with respective degrees 9, 2, 9, 42, 12, 20, 23 in the case of (1), and smaller degrees 7, 2, 7, 32, 8, 14, 19 in the case of (2).

Consider a system of parametric rational fractions, involving parameters of a set U. To this system, we associate the so-called *matrix representation* encoding the degrees of the monomials in U. We then consider an invertible monomial map ϕ (i.e. an application which sends each parameter of U to a monomial in the elements of a set $\bar{U}$). The monomial map ϕ acts linearly on the matrix representation above, which allows us to look for a sparsest matrix representation using only linear algebra techniques. However, the use of fractions brings some difficulty since fractions are invariant when both numerators and denominators are multiplied by the same value. Example 10 shows how to automatically rewrite the fraction $\frac{x}{1+x+ix\tau_1 w}$ into the sparser fraction $\frac{1}{y+1+i\tau_1 w}$ by first dividing both numerators and denominators by x and then introducing $y = 1/x$.

Section 2 introduces the basic concepts. Section 3 introduces two new algorithms. Roughly speaking, the first one called CSBmodulo computes a sparsest representation of a vector space modulo another vector space, where both vector spaces are given by their basis. The second one called getSparsestFraction is a direct application of CSBmodulo for computing a sparsest representation of a fraction. Section 4 describes Algorithm getSparsestSumOfFractions which is a generalization of getSparsestFraction for systems or sums of rational fractions. Section 5 details the (technical) proof of CSBmodulo, relying on Corollary 4 which clarifies the structure of the sparsest bases.

2   Preliminaries

Consider $\mathbb{K}$ a commutative field and $U = \{u_1, \ldots, u_n\}$ a set of variables. We consider monomials in U of the form $u_1^{\alpha_1} \cdots u_n^{\alpha_n}$ where the α_i are in $\mathbb{Z}$. Since negative integer exponents are allowed, $u_1 u_2^{-1} u_3$ is considered as a monomial. The row vector $(1, \ldots, 1)$ of length ℓ is denoted by $\mathbb{1}_\ell$ (or simply $\mathbb{1}$ when the context is clear).

2.1   Matrix Representation of a Fraction

Definition 1. *Consider a matrix $N = (\alpha_{i,j})$ in $\mathbb{Z}^{n \times \ell}$, a set $T = \{t_1, \ldots, t_\ell\}$ of elements of $\mathbb{K}$, and an integer g with $1 \leq g < \ell$. By definition, the triple (N, T, g) written in the form*

$$
\begin{pmatrix}
\alpha_{1,1} & \cdots & \alpha_{1,g} & \alpha_{1,g+1} & \cdots & \alpha_{1,\ell} & u_1 \\
\vdots & \ddots & \vdots & \vdots & \ddots & \vdots & \vdots \\
\alpha_{n,1} & \cdots & \alpha_{n,g} & \alpha_{n,g+1} & \cdots & \alpha_{n,\ell} & u_n \\
t_1 & \cdots & t_g & t_{g+1} & \cdots & t_\ell
\end{pmatrix}, \tag{3}
$$

represents the fraction $q = \dfrac{\sum_{i=1}^{g} t_i m_i}{\sum_{i=g+1}^{\ell} t_i m_i}$ where $m_i = u_1^{\alpha_{1,i}} u_2^{\alpha_{2,i}} \cdots u_n^{\alpha_{n,i}}$. The triple (N, T, g) (or simply N) is called a matrix representation *of q.*

Example 1. Taking $g = 2$, $U = \{a, b\}$, and $\mathbb{K} = \mathbb{Q}(x, y)$, the triple

$$
N = \begin{pmatrix}
1 & 0 & 1 & 0 & a \\
0 & 2 & 0 & 1 & b \\
t_1 & t_2 & t_3 & t_4
\end{pmatrix} \tag{4}
$$

with $t_1 = 2, t_2 = 3xy, t_3 = y^2, t_4 = 5x$ represents the fraction

$$
q_1 = \frac{2a + 3b^2 xy}{ay^2 + 5bx} \in \mathbb{K}(U). \tag{5}
$$

Proposition 1. *Consider a triple (N, T, g) representing a fraction q. Then, for any column vector $v \in \mathbb{Z}^n$, the triple $(N + v\mathbb{1}, T, g)$ also represents the fraction q.*

Proof. For any integer $\delta \in \mathbb{Z}$, adding $\delta\mathbb{1}$ to the k-th row of N amounts to multiply both numerator and denominator of q by the same monomial u_k^δ.

Example 2. By Proposition 1 with $v = (-1, 2)$, the two triples

$$
\begin{pmatrix}
1 & 2 & 1 & 1 & \bar{a} \\
-2 & -2 & -2 & -1 & \bar{b} \\
t_1 & t_2 & t_3 & t_4
\end{pmatrix} \tag{6}
$$

$$
\begin{pmatrix}
0 & 1 & 0 & 0 & \bar{a} \\
0 & 0 & 0 & 1 & \bar{b} \\
t_1 & t_2 & t_3 & t_4
\end{pmatrix} \tag{7}
$$

represent the same fraction $\bar{q}_1$ of $\mathbb{K}(\bar{a}, \bar{b})$ written in two different ways:

$$
\bar{q}_1 = \frac{2\frac{\bar{a}}{\bar{b}^2} + 3\frac{\bar{a}^2}{\bar{b}^2} xy}{\frac{\bar{a}}{\bar{b}^2} y^2 + 5\frac{\bar{a}}{\bar{b}} x} \tag{8}
$$

$$
\bar{q}_1 = \frac{2 + 3\bar{a}xy}{y^2 + 5\bar{b}x}. \tag{9}
$$

2.2   Monomial Map

In the following definition, the ring $\mathbb{K}(\bar{U}^{1/p})$, where $\bar{U} = \{\bar{u}_1, \ldots, \bar{u}_n\}$, denotes the ring of fractions $\mathbb{K}\left(\bar{u}_1^{1/p}, \ldots, \bar{u}_n^{1/p}\right)$.

Definition 2. *Consider an invertible matrix $C \in \mathbb{Q}^{n \times n}$, and two sets of variables $U = \{u_1, u_2, \ldots, u_n\}$ and $\bar{U} = \{\bar{u}_1, \bar{u}_2, \ldots, \bar{u}_n\}$. The matrix C defines the ring homomorphism ϕ^C from $\mathbb{K}(U)$ to $\mathbb{K}(\bar{U}^{1/p})$, where p is the lcm of all denominators of the elements of C, in the following way: $\phi^C(u_k) = \prod_{i=1}^{n} \bar{u}_i^{c_{i,k}}$ for $1 \leq k \leq n$. The map ϕ^C is called a monomial map. One simply denotes ϕ^C by ϕ when no confusion is possible.*

In this article, we will consider special monomials maps ϕ and fractions q such that $\phi(q)$ is also a rational fraction of $\mathbb{K}(\bar{U})$.

2.3   Action of a Monomial Map

Proposition 2. *Consider a triple (N, T, g) representing a fraction q in $\mathbb{K}(U)$, and an invertible matrix C in $\mathbb{Q}^{n \times n}$. If CN only contains elements of $\mathbb{Z}$, then the triple (CN, T, g) is a matrix representation of the fraction $\phi^C(q)$ of $\mathbb{K}(\bar{U})$.*

Proof. Immediate.

Corollary 1. *Consider a triple (N, T, g) representing a fraction q, an invertible matrix C in $\mathbb{Q}^{n \times n}$, and a column vector v in $\mathbb{Q}^n$. If $\bar{N} = CN + v\mathbb{1}$ only contains elements of $\mathbb{Z}$, then the triple $(\bar{N}, T, g)$ is a matrix representation of $\phi^C(q)$.*

Proof. Direct consequence of Propositions 1 and 2.

Example 3. Let us respectively denote by N and $\bar{N}$ the matrices from Eqs. (4) and (7), and consider the invertible matrix C

$$\begin{pmatrix} 1 & 1 \\ -2 & -1 \end{pmatrix} \begin{matrix} \bar{a} \\ \bar{b} \end{matrix} \atop \begin{matrix} a & b \end{matrix} . \tag{10}$$

Recall the fractions q_1 and $\bar{q}_1$ from Examples 1 and 2. One easily checks that $\bar{N} = CN + v\mathbb{1}$, where $v = (-1, 2)$. Consequently Corollary 1 implies that $\bar{q}_1$ is indeed equal to $\phi^C(q_1)$.

3   Sparsifying a Fraction

Consider a triple (N, T, g) representing a fraction $q \in \mathbb{K}(U)$ with the notations of Definition 1. A sparse matrix N means that many monomials m_i involve a few u_j, implying that the fraction q is sparse w.r.t. to the u_j.

In this article, we have chosen to make the matrix N sparsest in order to simplify the corresponding fraction q. To do so, we allow ourselves monomial changes of variables on U. More precisely, by relying on Corollary 1, we look for an invertible matrix C in $\mathbb{Q}^{n \times n}$ and a column vector v in $\mathbb{Q}^n$, such that the matrix $\bar{N} = CN + v\mathbb{1}$ only has integer values and is sparsest. Anticipating on the algorithms, the matrix C and the vector v given in Example 3 yield a sparsest possible matrix $\bar{N}$.

[8, Algorithm CSB (Compute Sparsest Basis)] solves the problem above in the particular case where v is the zero vector. Indeed, [8, Algorithm CSB] takes as input a full row rank matrix M and returns a sparsest (i.e. with the least number of nonzeros) matrix $\bar{M}$ with entries in $\mathbb{Z}$, which is *row-equivalent* to M (recall two matrices A and B of the same dimension are called row-equivalent if $A = PB$ for some invertible matrix P).

As a consequence, Algorithm CSB needs to be generalized to compute a sparsest basis of the rows of N "modulo" some other basis; this is what Algorithm CSBmodulo does.

3.1   Algorithm CSBmodulo

Algorithm CSBmodulo below takes as input two matrices N and P such that $\begin{pmatrix} N \\ P \end{pmatrix}$ has a full row rank. It returns a matrix $\bar{N}$ and an invertible matrix $C \in \mathbb{Q}^{n \times n}$ such that $\bar{N} = CN + VP$ for some matrix V in $\mathbb{Q}^{n \times s}$. Moreover, $\bar{N}$ is sparsest and only has entries in $\mathbb{Z}$. The proof of Algorithm CSBmodulo is quite technical, especially proving that the computed $\bar{N}$ is sparsest. Since the proof is not necessary to understand the rest of the article, it has been placed in the appendix.

The idea of Algorithm CSBmodulo is the following. One first considers N and P in a symmetric way by building a sparsest basis of $M = \begin{pmatrix} N \\ P \end{pmatrix}$ at Line 2, thus obtaining a matrix $\bar{M}$, whose rows are then sorted by increasing number of nonzeros at Line 4. As a consequence, $\bar{M} = DM$ for some invertible matrix D. Then the matrix $\bar{N}$ is obtained by choosing n rows $\bar{M}_{r_1}, \ldots, \bar{M}_{r_n}$ of $\bar{M}$. Denoting by E the matrix composed of the n first columns of D, for any choice of rows in $\bar{M}$, one gets $\bar{N} = CN + VP$ for some matrix V, and where C is composed by the rows $E_{r_1}, \ldots, E_{r_n}$. The choice of rows has to be done carefully. First, the matrix C should be invertible. This condition is ensured by Line 9. Second, the matrix $\bar{N}$ should be sparsest. This condition will be ensured by first selecting the sparsest rows (i.e. the first rows of $\bar{M}$ since its rows are sorted).

Example 4. Take the following full row rank matrices

$$N = \begin{pmatrix} 1\,1\,1\,0\,1\,0\,1\,1\,1 \\ 0\,1\,0\,1\,1\,0\,1\,0\,0 \\ 0\,0\,0\,0\,1\,0\,1\,0\,0 \end{pmatrix} \text{ and } P = \begin{pmatrix} 1\,1\,1\,1\,0\,0\,0\,0\,0 \\ 0\,0\,0\,0\,1\,1\,0\,0\,0 \\ 0\,0\,0\,0\,0\,0\,1\,1\,1 \end{pmatrix}$$

Algorithm 1. CSBmodulo(N, P)

Input: Two matrices $N \in \mathbb{Z}^{n \times \ell}$ and $P \in \mathbb{Z}^{s \times \ell}$ such that $\begin{pmatrix} N \\ P \end{pmatrix}$ has full row rank.

Output: A matrix $\bar{N} \in \mathbb{Z}^{n \times \ell}$ and an invertible matrix $C \in \mathbb{Q}^{n \times n}$ such that
$\bar{N} = CN + VP$ for some matrix V in $\mathbb{Q}^{n \times s}$. Moreover, $\bar{N}$ is sparsest.

1 **begin**

2 $\quad M \leftarrow \begin{pmatrix} N \\ P \end{pmatrix}$;

3 $\quad \bar{M} \leftarrow \mathsf{CSB}(M)$;

4 $\quad$ Sort the rows of $\bar{M}$ by increasing number of nonzeros (from top to bottom) ;

5 $\quad$ Compute the invertible matrix $D \in \mathbb{Q}^{(n+s) \times (n+s)}$ such that $\bar{M} = DM$;

6 $\quad$ Denote by $E \in \mathbb{Q}^{(n+s) \times n}$ the n first columns of D ;

7 $\quad$ Consider empty matrices $C \in \mathbb{Q}^{0 \times n}$ and $\bar{N} \in \mathbb{Z}^{0 \times \ell}$;

8 $\quad$ **for** i *from* 1 *to* $n + s$ **do**

9 $\quad\quad$ **if** $\mathrm{Rank} \begin{pmatrix} C \\ E_i \end{pmatrix} > \mathrm{Rank}(C)$ **then**

10 $\quad\quad\quad C \leftarrow \begin{pmatrix} C \\ E_i \end{pmatrix}$; $\bar{N} \leftarrow \begin{pmatrix} \bar{N} \\ \bar{M}_i \end{pmatrix}$;

11 $\quad$ **return** $\bar{N}, C$;

---

considered later in Example 12. The matrices D and $\bar{M}$ computed by Lines 2–5 of Algorithm CSBmodulo are (using our implementation of CSB)

$$\bar{M} = \begin{pmatrix} 0\,1\,0\,0\,1\,0\,0\,0\,0 \\ 0\,0\,0\,0\,1\,1\,0\,0\,0 \\ 1\,0\,1\,0\,0\,0\,0\,0\,0 \\ 0\,1\,0\,1\,0\,0\,0\,0\,0 \\ 0\,0\,0\,0\,1\,0\,1\,0\,0 \\ 0\,0\,0\,0\,0\,0\,1\,1\,1 \end{pmatrix} \text{ and } D = \begin{pmatrix} 1 & 1 & -1 & -1 & 0 & -1 \\ 0 & 0 & 0 & 0 & 1 & 0 \\ 0 & -1 & 1 & 1 & 0 & 0 \\ 0 & 1 & -1 & 0 & 0 & 0 \\ 0 & 0 & 1 & 0 & 0 & 0 \\ 0 & 0 & 0 & 0 & 0 & 1 \end{pmatrix}.$$

The extraction from $\bar{M}$ of the matrix $\bar{N}$ by Lines 6–10 will consider the three first columns of D and ignore rows 2, 4 and 6 of $\bar{M}$, yielding

$$\bar{N} = \begin{pmatrix} 0\,1\,0\,0\,1\,0\,0\,0\,0 \\ 1\,0\,1\,0\,0\,0\,0\,0\,0 \\ 0\,0\,0\,0\,1\,0\,1\,0\,0 \end{pmatrix}.$$

Remark 1. Algorithm CSBmodulo extracts the matrix C from the n first columns of D. This extraction could be made by computing the row rank profile [9]: the *row rank profile* of an $m \times n$ matrix A of rank r is the lexicographically smallest sequence of r indices of linearly independent rows of A. An $PLUQ$ decomposition algorithm using Gaussian elimination is proposed in [9] to compute such a set.

3.2   Algorithm getSparsestFraction

This section presents Algorithm getSparsestFraction which computes a sparsest representation of a fraction q. It relies on Algorithm CSBmodulo. We first present Propositions 3 and 4 which are needed when the number of variables in U can be decreased by a monomial map. Propositions 3 and 4 are in fact a particular case of a more general treatment based on scaling type symmetries.

Proposition 3. *Consider a triple (N, T, g) representing a fraction $q \in \mathbb{K}(U)$. If N has not full row rank, then there exist a monomial map ϕ^C and a full row rank matrix N' such that (N', T, g) represents the fraction $\bar{q} = \phi^C(q)$.*

Proof. If $N \in \mathbb{Z}^{n \times \ell}$ with $\mathrm{Rank}(N) = p < n$, then there exists a full row rank matrix $N' \in \mathbb{Z}^{p \times \ell}$ and an invertible matrix $C \in \mathbb{Q}^{n \times n}$ such that $CN = \begin{pmatrix} N' \\ 0 \end{pmatrix}$. By Proposition 2, the fraction $\bar{q}$ represented by the triple (CN, T, g) is equal to $\phi^C(q)$. Because the matrix CN has $n - p$ zero rows, one can discard the $n - p$ last variables and the triple (N', T, g) still represents the fraction $\bar{q}$.

Example 5. Take $q_2 = \frac{2+3abxy}{y^2+5abx} \in \mathbb{K}(U)$ with $\mathbb{K} = \mathbb{Q}(x, y)$ and $U = \{a, b\}$. A matrix representation of q_2 is $N = \begin{pmatrix} 0 & 1 & | & 0 & 1 \\ 0 & 1 & | & 0 & 1 \end{pmatrix} \begin{matrix} a \\ b \end{matrix}$ with $t_1 = 2, t_2 = 3xy, t_3 = y^2, t_4 = 5x$. By taking $C = \begin{pmatrix} 1 & 0 \\ -1 & 1 \end{pmatrix}$ and $N' = (0\ 1\ 0\ 1)$, one has $CN = \begin{pmatrix} N' \\ 0 \end{pmatrix}$. Consequently, the monomial map ϕ^C satisfies $\phi^C(a) = \bar{a}/\bar{b}$ and $\phi^C(b) = \bar{b}$, thus $\phi^C(q_2) = \frac{2+3\bar{a}xy}{y^2+5\bar{a}x} \in \mathbb{K}(\bar{U})$.

Proposition 4. *Consider a triple (N, T, g) representing a fraction $q \in \mathbb{K}(U)$, such that N has full row rank. If $\begin{pmatrix} N \\ \mathbb{1} \end{pmatrix}$ has not full row rank, then there exist a monomial map ϕ^C and a matrix N' such that (N', T, g) represents the fraction $\bar{q} = \phi^C(q)$, where $\begin{pmatrix} N' \\ \mathbb{1} \end{pmatrix}$ has full row rank.*

Proof. Since N has full row rank and $\begin{pmatrix} N \\ \mathbb{1} \end{pmatrix}$ has not full row rank, one has $\mathbb{1} = \sum_{i=1}^{n} \beta_i N_i$ for some $\beta_i \in \mathbb{Q}$. Without loss of generality, one assumes that $\beta_n \neq 0$ (by exchanging some variables in the set U). Using Corollary 1 and the relation

$$CN - v\mathbb{1} = \begin{pmatrix} N' \\ 0 \end{pmatrix} \text{ where } C = \begin{pmatrix} & & 0 \\ & I & \vdots \\ & & 0 \\ \beta_1 & \cdots & \beta_n \end{pmatrix}, v = \begin{pmatrix} 0 \\ \vdots \\ 0 \\ 1 \end{pmatrix} \text{ and } N' = \begin{pmatrix} N_1 \\ \vdots \\ N_{n-1} \end{pmatrix}, \text{ the}$$

fraction $\bar{q}$ represented by the (CN, T, g) is equal to $\phi^C(q)$. Because the last row of CN is zero, one can discard the last variable of U, and the triple (N', T, g) still represents the fraction $\bar{q}$. Moreover, the matrix $\begin{pmatrix} N' \\ \mathbb{1} \end{pmatrix}$ has full row rank.

Example 6. Take $q_3 = \frac{2a+3bxy}{ay^2+5bx} \in \mathbb{K}(U)$ with $\mathbb{K} = \mathbb{Q}(x,y)$ and $U = \{a,b\}$. A matrix representation of q_3 is $N = \begin{pmatrix} 1\ 0\,|\,1\ 0 \\ 0\ 1\,|\,0\ 1 \end{pmatrix}\begin{matrix} a \\ b \end{matrix}$ with $t_1 = 2, t_2 = 3xy, t_3 = y^2, t_4 = 5x$. One has $CN - v\mathbb{1} = \begin{pmatrix} N' \\ 0 \end{pmatrix}$ where $C = \begin{pmatrix} 1\ 0 \\ 1\ 1 \end{pmatrix}$, $v = (0,1)$ and $N' = (1\ 0\ 1\ 0)$. Consequently, $\bar{q}_3$ equals $\phi^C(q_3) = \frac{2\bar{a}\bar{b}+3\bar{b}xy}{\bar{a}by^2+5bx} = \frac{2\bar{a}+3xy}{\bar{a}y^2+5x} \in \mathbb{K}(\bar{a})$. Furthermore, $\bar{q}_3$ can be represented by $N' = (1\ 0\,|\,1\ 0)\ \bar{a}$, where $\begin{pmatrix} N' \\ 1 \end{pmatrix}$ has full row rank.

Algorithm 2. getSparsestFraction(N, T, g)

Input: A triple (N, T, g) representing a fraction $q \in \mathbb{K}(U)$

Output: A triple $(\bar{N}, T, g)$ and an invertible matrix C in $\mathbb{Q}^{n\times n}$, such that $(\bar{N}, T, g)$ represents $\phi^C(q)$. Moreover, $\bar{N}$ is sparsest and only has entries in $\mathbb{Z}$.

1 **begin**

2   Apply Propositions 3 and 4 for ensuring that $\begin{pmatrix} N \\ 1 \end{pmatrix}$ has full row rank ;

3     $\bar{N}, C \leftarrow$ CSBmodulo($N, \mathbb{1}$) ;

4   **return** $(\bar{N}, T, g), C$;

Example 7. Let us apply Algorithm getSparsestFraction on the triple (N, T, g) of (4), representing the rational fraction q_1 of (5). The matrix $\begin{pmatrix} N \\ 1 \end{pmatrix}$ has full row rank (which is 3) so getSparsestFraction(N, T, g) calls CSBmodulo($N, \mathbb{1}$). During the CSBmodulo($N, \mathbb{1}$) call, the matrix $\bar{M}$ computed at Line 4 and the invertible matrix D computed at Line 5 are:

$$\bar{M} = \begin{pmatrix} 0\ 1\,|\,0\ 0 \\ 0\ 0\,|\,0\ 1 \\ 1\ 0\,|\,1\ 0 \end{pmatrix} \text{ and } D = \begin{pmatrix} 1\ \ 1\ -1 \\ -2\ -1\ \ 2 \\ 1\ \ 0\ \ 0 \end{pmatrix}.$$

Moreover, the matrix C computed by Lines 6–10 is simply the upper-left two by two submatrix of D, which is exactly the matrix (10). Finally the matrix $\bar{N}$ computed by Lines 6–10 is obtained by selecting the two first rows of $\bar{M}$, yielding the matrix (7) and its corresponding fraction (9).

Remark that if one simply computes CN, one gets the matrix representation (6) corresponding to the fraction (8), which is not as nice as the fraction (9) represented by $\bar{N}$.

The following example shows that the monomial map ϕ^C computed by Algorithm getSparsestFraction can involve fractional exponents.

Example 8. Take the matrix representation

$$N = \begin{pmatrix} 1\,0 & 0\,0 \\ 0\,3 & 1\,0 \\ 0\,0 & 0\,1 \\ x\,y & x\,y \end{pmatrix} \begin{matrix} a \\ b \\ c \\ \ \end{matrix}$$

representing the fraction $q = \frac{ax+b^3y}{bx+cy} \in \mathbb{K}(a,b,c)$ where $\mathbb{K} = \mathbb{Q}(x,y,z)$. Then getSparsestFraction(N,T,g) returns

$$\bar{N} = \begin{pmatrix} 1\,0 & 0\,0 \\ 0\,1 & 0\,0 \\ 0\,0 & 1\,0 \\ x\,y & x\,y \end{pmatrix} \begin{matrix} \bar{a} \\ \bar{b} \\ \bar{c} \\ \ \end{matrix} \quad \text{and} \quad C = \begin{pmatrix} 1 & 0 & 0 \\ 1/2 & 1/2 & 1/2 \\ -3/2 & -1/2 & -3/2 \end{pmatrix}.$$

The following example shows that negative exponents for the parameters are sometimes needed.

Example 9. Take the matrix representation

$$N = \begin{pmatrix} 0\ 1\ -1 & 0 \\ 2\,x\ \ y & z+1 \end{pmatrix} \begin{matrix} a \\ \ \end{matrix}$$

representing the fraction $q = \frac{2+ax+\frac{y}{a}}{z+1} \in \mathbb{K}(a)$ where $\mathbb{K} = \mathbb{Q}(x,y,z)$. Then getSparsestFraction(N,T,g) returns N, showing that the representation above, which contains a negative exponent, is already sparsest. Moreover, N is the unique sparsest representation (in the sense of this article), since any addition of a multiple of $\mathbb{1}$ to N will produce at least three nonzeros.

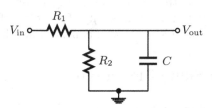

Fig. 1. An electric circuit with two resistors R_1 and R_2 and a capacitor C.

Example 10. The transfer function $H = V_{out}/V_{in}$ of the circuit given in Fig. 1 is

$$H = \frac{R_2}{R_1 + R_2 + iR_1R_2Cw},$$ where $i^2 = -1$ and w is the frequency. We consider $H \in \mathbb{K}(R_1, R_2, C)$ with $\mathbb{K} = \mathbb{C}(w)$ i.e. we consider R_1, R_2 and C as parameters. It can be shown that H admits a scaling symmetry acting on R_1, R_2, and C. As a consequence, using symmetry based techniques, one can discard one parameter

by a suitable (nonunique) change of variables. For example, H can be rewritten as

$$H = \frac{x}{1 + x + ix\tau_1 w} \tag{11}$$

by taking $x = R_2/R_1$ and $\tau_1 = R_1 C$. Also, H can be rewritten as

$$H = \frac{1}{y + 1 + iy\tau_2 w} \tag{12}$$

by taking $y = R_1/R_2$ and $\tau_2 = R_2 C$. Other changes of variables would also be possible but are not given here.

The relations (11) and (12) are not sparsest. Relation (12) can easily be made sparsest using the change of variables $a = y\tau_2$ yielding $H = \frac{1}{y+1+iaw}$. However, Relation (11) requires a slightly more subtle treatment, by first dividing both numerators and denominators by x thus writing $H = \frac{1}{\frac{1}{x}+1+i\tau_1 w}$ and then taking $y = 1/x$ (please note that $a = \tau_1$ since $a = y\tau_2 = (R_1/R_2)R_2 C = R_1 C = \tau_1$). The division by x will be automatically discovered by getSparsestFraction applied to the triple

$$N = \begin{pmatrix} 1 & \vline & 0\ 1\ 1 \\ 0 & \vline & 0\ 0\ 1 \\ 1 & & 1\ 1\ iw \end{pmatrix} \begin{matrix} x \\ \tau_1 \\ \; \end{matrix}$$

thanks to the line $\mathbb{1}$ added to N during the call of CSB, allowing to replace the first line N_1 of N by $N_1 - \mathbb{1} = \begin{pmatrix} 0 & \vline & -1\ 0\ 0 \end{pmatrix}$. Moreover, our implementation of CSB will negate the row $\begin{pmatrix} 0 & \vline & -1\ 0\ 0 \end{pmatrix}$ since it only contains nonpositive entries.

3.3   Complexity of getSparsestFraction

Except the CSBmodulo call, the instructions of Algorithm getSparsestFraction are done at most in $O(n^3)$. In fact, the complexity of getSparsestFraction is dominated by the one of CSBmodulo, which can be exponential in n in the worst case [8].

4   Sparsifying a Sum of Fractions

We now consider a sum of rational fractions in $\mathbb{K}(U)$. It could be rewritten as a single fraction by reducing the fractions to the same denominator. However, one will avoid such a manipulation for two reasons. First, this can increase the sizes of the numerator and denominator. Second, a practitioner might want to keep an expression as a sum of fractions. This last point occurs for example in the Biology context with expressions of the shape $p + \sum \frac{V_i x_i}{x_i + k_i}$ where the x_i are concentrations, p is a polynomial in the x_i and some other parameters, and the V_i and k_i are the constants of some Michaelis-Menten terms [10,11].

4.1   Matrix Representation of a Sum of Fractions

Definition 3. *Consider s fractions $q_i \in \mathbb{K}(U)$, where each fraction q_i is represented by a triple (N^i, T^i, g^i). Then, the set of the triples $H = \{(N^i, T^i, g^i)_{1 \leq i \leq s}\}$ is called the* matrix representation *of the sum $S = \sum_{i=1}^{s} q_i$. The set H of triples can be written as the following matrix, where the double bar separates two different fractions:*

$$N = (N^1 \| N^2 \| \ldots \| N^s). \tag{13}$$

Example 11. Consider the sum of rational fractions S_1 defined by

$$S_1 = \frac{2a + abxy}{ay^2 + 7bx} + abcy + \frac{abcy^2 + 3ay}{ax}. \tag{14}$$

A possible matrix representation of S_1 is

$$\begin{pmatrix} 1 & 1 & 1 & 0 & 1 & 0 & 1 & 1 & 1 \\ 0 & 1 & 0 & 1 & 1 & 0 & 1 & 0 & 0 \\ 0 & 0 & 0 & 0 & 1 & 0 & 1 & 0 & 0 \end{pmatrix} \begin{matrix} a \\ b \\ c \end{matrix} \tag{15}$$
$$t_1 \; t_2 \; t_3 \; t_4 \; t_5 \; t_6 \; t_7 \; t_8 \; t_9$$

with $t_1 = 2, t_2 = xy, t_3 = y^2, t_4 = 7x, t_5 = y, t_6 = 1, t_7 = y^2, t_8 = 3y, t_9 = x$.

Consider a matrix representation $(N^i, T^i, g^i)_{1 \leq i \leq s}$ of a sum $S = \sum q_i$. Because Proposition 1 can be applied independently on each fraction q_i, one introduces Proposition 5 which generalizes Proposition 1, after introducing the matrix $\mathbb{1}^S$ of size $s \times (\sum_{i=1}^{s} \mathrm{card}(T^i))$ defined as

$$\mathbb{1}^S = \begin{pmatrix} \mathbb{1}_{\mathrm{card}(T^1)} & & & \\ & \mathbb{1}_{\mathrm{card}(T^2)} & & \\ & & \ddots & \\ & & & \mathbb{1}_{\mathrm{card}(T^s)} \end{pmatrix}. \tag{16}$$

Proposition 5. *Consider a set of triples (N^i, T^i, g^i) representing a sum of rational fractions $S = \sum_{i=1}^{s} q_i$ as written in Definition 3. For any $V \in \mathbb{Z}^{n \times s}$, $\bar{N} = N + V\mathbb{1}^S$ is a matrix representation of S.*

Proof. Apply Proposition 1 on each fraction q_i.

4.2   Action of a Monomial Map

The following Proposition 6 and Corollary 2 are respectively the generalizations of Proposition 2 and Corollary 1 for sum of fractions.

Proposition 6. *Consider a set of triples (N^i, T^i, g^i) representing a sum of rational fractions $S = \sum_{i=1}^{s} q_i$ as written in Definition 3. Consider an invertible matrix C in $\mathbb{Q}^{n \times n}$. If CN only have entries in $\mathbb{Z}$, then CN is a matrix representation of $\bar{S} = \phi^C(S)$.*

Corollary 2. *Consider a set of triples (N^i, T^i, g^i) representing a sum of rational fractions $S = \sum_{i=1}^{s} q_i$ as written in Definition 3. Consider an invertible matrix C in $\mathbb{Q}^{n \times n}$ and a vector V in $\mathbb{Q}^{n \times s}$. If $CN + V\mathbb{1}^S$ only have integer entries, then $CN + V\mathbb{1}^S$ is a matrix representation of $\bar{S} = \phi^C(S)$.*

Example 12. Take the map $\phi(a) = \bar{a}$, $\phi(b) = \frac{\bar{a}}{\bar{b}}$, $\phi(c) = \frac{\bar{b}\bar{c}}{\bar{a}}$ defined by the following invertible matrix C:

$$\begin{pmatrix} 1 & 1 & -1 \\ 0 & -1 & 1 \\ 0 & 0 & 1 \end{pmatrix} \begin{matrix} \bar{a} \\ \bar{b} \\ \bar{c} \end{matrix} \\ \begin{matrix} a & b & c \end{matrix} \qquad (17)$$

With this change of variables, the fraction S_1 of Example 11 becomes

$$\bar{S}_1 = \frac{2\bar{a} + \frac{\bar{a}^2}{\bar{b}}xy}{\bar{a}y^2 + 7\frac{\bar{a}}{\bar{b}}x} + \bar{a}\bar{c}y + \frac{\bar{a}\bar{c}y^2 + 3\bar{a}y}{\bar{a}x}. \qquad (18)$$

Taking the matrix representation N of S_1 as written in (15), CN is a matrix representation of $\bar{S}_1 = \phi^C(S_1)$. The matrix CN can however be made sparser by considering

$$V = \begin{pmatrix} -1 & 0 & -1 \\ 1 & 0 & 0 \\ 0 & 0 & 0 \end{pmatrix}, \qquad (19)$$

and by applying Corollary 2. Indeed the following matrix $\bar{N} = CN + V\mathbb{1}^{S_1}$ is also a matrix representation of $\bar{S}_1$

$$\begin{pmatrix} 0 & 1 & 0 & 0 & 1 & 0 & 0 & 0 & 0 \\ 1 & 0 & 1 & 0 & 0 & 0 & 0 & 0 & 0 \\ 0 & 0 & 0 & 0 & 1 & 0 & 1 & 0 & 0 \end{pmatrix} \begin{matrix} \bar{a} \\ \bar{b} \\ \bar{c} \end{matrix}, \qquad (20)$$

and represents

$$\bar{S}_1 = \frac{2\bar{b} + \bar{a}xy}{\bar{b}y^2 + 7x} + \bar{a}\bar{c}y + \frac{\bar{c}y^2 + 3y}{x}. \qquad (21)$$

Remark that adding $V\mathbb{1}^{S_1}$ to CN corresponds to multiplying by $\frac{\bar{b}}{\bar{a}}$ the numerator and the denominator of the first fraction, and dividing by $\bar{a}$ the numerator and the denominator of the third fraction in (18).

4.3 Algorithm getSparsestSumOfFractions

Algorithm getSparsestSumOfFractions relies on the same ideas as Sect. 3: given a set of triples (N^i, T^i, g^i) representing a sum of rational fractions $S = \sum_{i=1}^{s} q_i$ as written in Definition 3, one looks for an invertible matrix C and a matrix V such that $CN + V\mathbb{1}^S$ is sparsest.

We first present Proposition 7, which is a slight generalization of Proposition 4. As in Sect. 3, Proposition 7 is needed when the sum of fractions admits scaling type symmetries in the U variables.

Proposition 7. *Consider a set of triples (N^i, T^i, g^i) representing a sum of rational fractions $S = \sum_{i=1}^{s} q_i$ as written in Definition 3. Assume that N has full row rank. If $\begin{pmatrix} N \\ \mathbb{1}^S \end{pmatrix}$ has not full row rank, then there exists a monomial map ϕ^C and matrices N'^i such that the set of triples (N'^i, T^i, g^i) represents the fraction $\bar{S} = \phi^C(S)$, where $\begin{pmatrix} N' \\ \mathbb{1}^S \end{pmatrix}$ has full row rank.*

Proof. The matrix $M = \begin{pmatrix} N \\ \mathbb{1}^S \end{pmatrix}$ has not full row rank, so there exists a non trivial linear dependency between the rows of M. Since both matrices N and $\mathbb{1}^S$ have full row rank, the linear dependency necessarily involves a row of N with a nonzero coefficient. The end of the proof is similar to the one of Proposition 4.

Algorithm 3. getSparsestSumOfFractions(H)

Input: A set $H = \{(N^i, T^i, g^i)_{1 \leq i \leq s}\}$ representing a sum of rational fractions
$\quad\quad S = \sum_{i=1}^{s} q_i \in \mathbb{K}(U)$ as in Definition 3.

Output: A set $\bar{H} = \{(\bar{N}^i, T^i, g^i)_{1 \leq i \leq s}\}$ and an invertible matrix C in $\mathbb{Q}^{n \times n}$,
$\quad\quad$ such that $\bar{H}$ represents $\phi^C(q)$. Moreover $\bar{N} = (\bar{N}^1 || \cdots || \bar{N}^s)$ is
$\quad\quad$ sparsest and only has entries in $\mathbb{Z}$.

1 **begin**

2 $\quad$ Apply Propositions 3 and 7 for ensuring that $\begin{pmatrix} N \\ \mathbb{1}^S \end{pmatrix}$ has full row rank ;

3 $\quad$ $\bar{N}, C \leftarrow$ CSBmodulo($N, \mathbb{1}^S$) ;

4 $\quad$ Write $\bar{N}$ as $(\bar{N}^1 || \cdots || \bar{N}^s)$;

5 $\quad$ **return** $\{(\bar{N}^i, T^i, g^i)_{1 \leq i \leq s}\}, C$;

Example 13. Recall the fraction S_1 from Example 11 and its matrix representation N of Eq. (15). One checks that the matrix $\begin{pmatrix} N \\ \mathbb{1}^S \end{pmatrix}$ has full row rank. Consequently getSparsestSumOfFractions(N) calls CSBmodulo($N, \mathbb{1}^S$). Our implementation of CSBmodulo returns the matrix $\bar{N}$ of Eq. (20) and the matrix C of Eq. (17). Consequently, getSparsestSumOfFractions computes the sparsest sum of fractions $\bar{S}_1 = \phi^C(S_1)$ of Eq. (21).

4.4   Application to Systems of ODEs

The techniques presented for the sum of fractions can be adapted for systems of differential equations of the form $X'(t) = F(\Theta, X(t))$ (where $X(t)$ is a vector of functions, $F(\Theta, X(t))$ is a vector of fractions and the Θ are parameters), such as Eq. (1). Indeed, one can consider the sum of fractions $\sum_{i=1}^{s} F_i(\Theta, X(t))$, where the $F_i(\Theta, X(t))$ denotes the components of the vector $F(\Theta, X(t))$, and apply Algorithm getSparsestSumOfFractions.

Example 14. Consider the sum S of the three right-hand sides of Eq. (1) seen as a fraction of $\mathbb{K}(k_1, k_2, k_3)$ with $\mathbb{K} = \mathbb{Q}(\theta, \alpha, \rho_b, \rho_f, \delta_M, \delta_P, \beta)$. It can represented by

$$N = \left(\begin{array}{cc|c||c|c||cc|cccc} 0 & 1 & 0 & 0 & 0 & 0 & 1 & 1 & 1 & 1 & 0 \\ 0 & 1 & 0 & 0 & 0 & 0 & 1 & 1 & 1 & 0 & 0 \\ 0 & 1 & 0 & 0 & 0 & 0 & 1 & 1 & 0 & 0 & 0 \end{array} \right) \begin{array}{c} k_1 \\ k_2 \\ k_3 \end{array}.$$
$$\quad\; t_1\, t_2 \quad t_3 \quad t_4 \quad t_5 \quad t_6\, t_7 \quad t_8\, t_9\, t_{10}\, t_{11}$$

with $t_1 = \theta(1 - G)$, $t_2 = -\alpha P^4 G$, $t_3 = 1$, $t_4 = \rho_b(1 - G) + \rho_f G - \delta_M M$, $t_5 = 1$, $t_6 = 4\theta(1 - G) - \delta_P P + \beta M$, $t_7 = -4\alpha P^4 G$, $t_8 = 16P^3$, $t_9 = 9P^2$, $t_{10} = 4P$, $t_{11} = 1$.

Algorithm getSparsestSumOfFractions(S) yields the sparsest Eq. (2) and the monomial map $\phi(k_1) = \bar{k}_1, \phi(k_2) = \bar{k}_2/\bar{k}_1, \phi(k_3) = \bar{k}_3/\bar{k}_2$ encoded by the matrix

$$C = \begin{pmatrix} 1 & -1 & 0 \\ 0 & 1 & -1 \\ 0 & 0 & 1 \end{pmatrix}.$$ Please note that the matrix $\mathbb{1}^S$ was not useful in the computations, since $\bar{N} = CN$.

5 Proof of Algorithm **CSBmodulo**

This last section is the most technical part of the article. We first present some definitions and intermediate results. We then present a new corollary showing that all the sparsest row bases of the same matrix (see Definition 4) share some common structure (see Corollary 4). Finally the proof of Algorithm CSBmodulo is presented.

Let M (resp. v) be a matrix (resp. vector). We denote by $\mathcal{N}(M)$ (resp. $\mathcal{N}(v)$) the number of nonzero coefficients of M (resp. v).

Definition 4. *Let $M \in \mathbb{Q}^{n \times \ell}$ a matrix with full row rank. One calls* sparsest row basis *of M any matrix $\bar{M}$ which is sparsest and row-equivalent to M.*

Definition 5. *Let $N \in \mathbb{Q}^{n \times \ell}$ and $P \in \mathbb{Q}^{n \times s}$ two matrices with full row rank. Assume that $\begin{pmatrix} N \\ P \end{pmatrix}$ has full row rank. One calls* sparsest row basis of N modulo P *any matrix N' which is sparsest and satisfies $N' = CN + VP$ for some invertible matrix C and some matrix V.*

Following Lemma 1 is a rephrasing of [8, Theorem 1]. It is the key ingredient ensuring the greedy approach chosen in [8], which consists in repeatedly reducing the number of nonzeros of some row of M, until it is not possible anymore.

Lemma 1. *Take a full row rank matrix M. The matrix M is not a sparsest row basis of M iff there exists an index i and a row vector v such that $v_i \neq 0$ and $\mathcal{N}(vM) < \mathcal{N}(M_i)$.*

Corollary 3. *Take a full row rank matrix M. If the matrix M is not a spars-est row basis of M, Lemma 1 applies. Moreover, replacing the row M_i by vM yields a sparser row-equivalent matrix. See [8, Algorithm* EnhanceBasis*] for an implementation of Corollary 3.*

Proposition 8. *Take a sparsest basis N' of N modulo P with the same assumptions as in Definition 5. Then there exist matrices P', C', V', G', W' such that $\begin{pmatrix} N' \\ P' \end{pmatrix}$ is a sparsest basis of the matrix $\begin{pmatrix} N \\ P \end{pmatrix}$, with $\begin{pmatrix} N' \\ P' \end{pmatrix} = \begin{pmatrix} C' & V' \\ G' & W' \end{pmatrix} \begin{pmatrix} N \\ P \end{pmatrix}$ where C' is invertible.*

Proof. The existence of C' and V' is given by Definition 5. Consider the matrix $M' = \begin{pmatrix} N' \\ P \end{pmatrix}$. If M' is not a sparsest row basis of $\begin{pmatrix} N \\ P \end{pmatrix}$, it can be made sparser using Corollary 3. Moreover, the row to improve is necessarily not a row of N', since N' is a sparsest row basis of N modulo P. After applying Corollary 3 a certain number of times, one gets a sparsest row basis $\begin{pmatrix} N' \\ P' \end{pmatrix}$ of $\begin{pmatrix} N \\ P \end{pmatrix}$ which proves the existence of P', G' and W'.

Proposition 9. *Take a sparsest basis $\bar{M}$ of a full row rank matrix M. Assume that the rows of $\bar{M}$ and M are sorted by increasing number of nonzeros. Let D be the matrix defined by $D\bar{M} = M$. Then $\mathcal{N}(\bar{M}_i) \leq \mathcal{N}(M_i)$ for any $1 \leq i \leq n$. Moreover, for any $1 \leq i, j \leq n$, if $\mathcal{N}(M_i) < \mathcal{N}(\bar{M}_j)$, then $D_{ij} = 0$.*

Proof. Let us prove the first point and assume $\mathcal{N}(M_i) < \mathcal{N}(\bar{M}_j)$ for some i and j with $D_{ij} \neq 0$. Following ideas from Corollary 3, $\bar{M}$ is not sparsest since $\bar{M}_j$ could be replaced by the sparser row M_i since $M_i = \sum_j D_{ij} \bar{M}_j$ and $D_{ij} \neq 0$.

Let us now prove that $\mathcal{N}(\bar{M}_i) \leq \mathcal{N}(M_i)$ for any $1 \leq i \leq n$. By contradiction, assume that there exists a k such that $\mathcal{N}(\bar{M}_k) > \mathcal{N}(M_k)$ and $\mathcal{N}(\bar{M}_i) = \mathcal{N}(M_i)$ for $i \leq k - 1$. Each row M_i is a linear combination of rows of $\bar{M}$. If all k first rows of M_i were linear combinations of the $k - 1$ rows of $\bar{M}$, then the k first rows would not be linear independent, and M could not have full row rank. Consequently, there exist two indices i, l and a row vector v such that $i \leq k \leq l$, $\bar{M}_i = v\bar{M}$ with $v_l \neq 0$. Since $\mathcal{N}(\bar{M}_l) \geq \mathcal{N}(\bar{M}_k) > \mathcal{N}(M_k) \geq \mathcal{N}(M_i)$, the row $\bar{M}_l$ can be made sparser by replacing it by the sparser row M_i using Lemma 1. This leads to a contradiction since $\bar{M}$ is sparsest.

The new following corollary proves that all sparsest row bases of some fixed matrix share some common structure.

Corollary 4. *Take two sparsest basis $\bar{M}$ and M' of the same matrix M. Assume that the rows of $\bar{M}$ and M' are sorted by increasing number of nonzeros. Let T the matrix defined by $\bar{M} = TM'$. Then for any $1 \leq i \leq n$, one has $\mathcal{N}(\bar{M}_i) = \mathcal{N}(M'_i)$. Moreover, T is a lower block triangular matrix, where the widths of blocks correspond to the width of the blocks of rows of $\bar{M}$ which have the same number of nonzeros.*

Proof. It is a direct consequence of Proposition 9 applied twice: the first time by considering that $\bar{M}$ is a sparsest row basis of M', the second time by considering that M' is a sparsest row basis of $\bar{M}$.

Lemma 2. *Let $H \in \mathbb{Q}^{m \times \ell}$ and $U' = \mathbb{Q}^{t \times m}$. If $U'H = 0$ then* Rank(U') + Rank$(H) \leq m$.

Proof. Consequence of the Rank-nullity theorem applied on the transpose of H

*Proof (*CSBmodulo *is correct).* The fact that the computed matrices $\bar{N}, C$ satisfies $\bar{N} = CN + VP$ for some V is a consequence of the selection strategy in the loop. It is left to the reader. Moreover, the matrix $\bar{N}$ has entries in $\mathbb{Z}$ since it is a submatrix of $\bar{M}$ which also have entries in $\mathbb{Z}$ because it was computed by Algorithm CSB.

The difficult point is to show that $\bar{N}$ is sparsest. To prove that point, we assume that $\mathcal{N}(\bar{N}) > \mathcal{N}(N')$ for some sparsest row basis N' of N modulo P, and show a contradiction. By Proposition 8, there exists a matrix P' such that $\begin{pmatrix} N' \\ P' \end{pmatrix}$ is a sparsest basis of the matrix $\begin{pmatrix} N \\ P \end{pmatrix}$. Let us denote M' the matrix obtained by sorting the rows of $\begin{pmatrix} N' \\ P' \end{pmatrix}$ by increasing number of nonzeros. Let us introduce the indices $s_1 < \ldots < s_n$ such that $M'_{s_i} = N'_i$. By Corollary 4, $\mathcal{N}(\bar{M}_i) = \mathcal{N}(M'_i)$ for any $1 \leq i \leq n$, and there exists an invertible lower block triangular matrix T such that $\bar{M} = TM'$.

For sake of simplicity, one assumes that the number of nonzeros in the rows of $\bar{M}$ are strictly increasing, hence the matrix T is lower triangular with nonzero diagonal elements. Denote by $r_1, \ldots, r_n$ the indices of the rows of $\bar{M}$ which are selected by the loop in Algorithm CSBmodulo to produce the matrix $\bar{N}$ (i.e. $\bar{N}_i = \bar{M}_{r_i}$ for $1 \leq i \leq n$). Since we assumed $\mathcal{N}(\bar{N}) > \mathcal{N}(N')$, then $\mathcal{N}(\bar{N}_k) > \mathcal{N}(N'_k)$ for some k. By taking k minimal, one has $\mathcal{N}(\bar{N}_1) \leq \mathcal{N}(N'_1)$, $\ldots$, $\mathcal{N}(\bar{N}_{k-1}) \leq \mathcal{N}(N'_{k-1})$ and $\mathcal{N}(\bar{N}_k) > \mathcal{N}(N'_k)$. From the inequalities above, one has $r_1 \leq s_1$, $r_2 \leq s_2$, $\ldots$, $r_{k-1} \leq s_{k-1}$ and $r_k > s_k$, which we summarize here:

$$
\begin{array}{ccc}
\bar{M} & T & M' \\[2mm]
\begin{pmatrix} \vdots \\ \bar{M}_{r_1} \\ \vdots \\ \bar{M}_{r_{k-1}} \\ \vdots \\ \vdots \\ \bar{M}_{r_k} \\ \vdots \end{pmatrix}
& = \begin{pmatrix} T_{1,1} & \\ \vdots & \ddots \\ T_{n,1} & \cdots & T_{n,n} \end{pmatrix}
& \begin{pmatrix} \vdots \\ \vdots \\ M'_{s_1} \\ \vdots \\ M'_{s_{k-1}} \\ \vdots \\ M'_{s_k} \\ \vdots \end{pmatrix} .
\end{array}
$$

Among the first s_k rows of $\bar{M}$, there are $k-1$ rows which were selected by the algorithm. As a consequence, $s_k - k + 1$ rows were not selected, implying that each unselected row E_i with $i \leq s_k$ is a linear combination of the previous rows E_j with $j < i$. By storing row-wise those linear combinations above in a matrix U, one gets a $(s_k - k + 1) \times n$ matrix, where columns from indices $s_k + 1$ to n are zero. Moreover, the matrix U has full row rank since U has a echelon form.

Let us write $D = \begin{pmatrix} E\,F \end{pmatrix}$. By definition of U, one has $UD = \begin{pmatrix} UE\,UF \end{pmatrix} = \begin{pmatrix} 0\,UF \end{pmatrix}$. Since $\bar{M} = DM$, then $U\bar{M} = UDM = \begin{pmatrix} 0\,UF \end{pmatrix} \begin{pmatrix} N \\ P \end{pmatrix} = UFP$.

On the other side, since $\bar{M} = TM'$, then $U\bar{M} = UTM'$. Since the columns from indices $s_k + 1$ to n of U are zero, the matrix UT also have columns from indices $s_k + 1$ to n which are zero. Moreover, UT has also full rank.

With notations of Proposition 8 and some easy computations, the product UTM' can be written as $UTHN + JP$ where H has s_k rows including the rows $1, 2, \ldots, k$ of C' and some rows of G', and some matrix J.

Consequently, $U\bar{M} = UFP = UTHN + JP$ which implies $(UF - J)P = (UTH)N$. Since $\begin{pmatrix} N \\ P \end{pmatrix}$ has full row rank, UTH (and also $(UF - J)$) is necessarily the zero matrix.

By Lemma 2, $\mathrm{Rank}(UT) + \mathrm{Rank}(H) \leq s_k$. However, $\mathrm{Rank}(UT) = s_k - k + 1$, and $\mathrm{Rank}(H) \geq k$ since the k rows $C'_1, \ldots, C'_k$ are taken from the invertible matrix C'. Thus $\mathrm{Rank}(UT) + \mathrm{Rank}(H) \geq s_k + 1$ which contradicts $\mathrm{Rank}(UT) + \mathrm{Rank}(H) \leq s_k$. As a consequence, the assumption $\mathcal{N}(\bar{N}) > \mathcal{N}(N')$ leads to a contradiction, so $\mathcal{N}(\bar{N}) \leq \mathcal{N}(N')$ and $\bar{N}$ is indeed sparsest.

References

1. Basu, S., Pollack, R., Roy, M.F.: Algorithms in Real Algebraic Geometry. Algorithms and Computation in Mathematics, vol. 10, 2nd edn. Springer, Heidelberg (2006)
2. Lemaire, F., Ürgüplü, A.: A method for semi-rectifying algebraic and differential systems using scaling type lie point symmetries with linear algebra. In: Proceedings of the 2010 International Symposium on Symbolic and Algebraic Computation, ISSAC 2010, pp. 85–92. ACM, New York (2010)
3. Sedoglavic, A.: Reduction of algebraic parametric systems by rectification of their affine expanded lie symmetries. In: Anai, H., Horimoto, K., Kutsia, T. (eds.) AB 2007. LNCS, vol. 4545, pp. 277–291. Springer, Heidelberg (2007)
4. Hubert, E., Labahn, G.: Scaling invariants and symmetry reduction of dynamical systems. Found. Comput. Math. **13**(4), 479–516 (2013)
5. Fels, M., Olver, P.J.: Moving coframes: II. Regularization and theoretical foundations. Acta Applicandae Math. **55**(2), 127–208 (1999)
6. Olver, P.J.: Applications of Lie Groups to Differential Equations. Graduate Texts in Mathematics, vol. 107, 2nd edn. Springer, New York (1993)
7. Boulier, F., Lemaire, F., Sedoglavic, A., Ürgüplü, A.: Towards an automated reduction method for polynomial ODE models of biochemical reaction systems. Math. Comput. Sci. **2**(3), 443–464 (2009)

8. Lemaire, F., Temperville, A.: On defining and computing "Good" conservation laws. In: Mendes, P., Dada, J.O., Smallbone, K. (eds.) CMSB 2014. LNCS, vol. 8859, pp. 1–19. Springer, Heidelberg (2014)

9. Dumas, J.G., Pernet, C., Sultan, Z.: Computing the rank profile matrix. In: Proceedings of the 2015 International Symposium on Symbolic and Algebraic Computation, ISSAC 2015, pp. 149–156. ACM, New York (2015)

10. Henri, V.: Lois générales de l'action des diastases. Librairie Scientifique A. Hermann, Paris (1903)

11. Michaelis, L., Menten, M.L.: Die Kinetik der Invertinwirkung. Biochemische Zeitschrift **49**, 333–369 (1913)

On the General Analytical Solution
of the Kinematic Cosserat Equations

Dominik L. Michels[1(✉)], Dmitry A. Lyakhov[2], Vladimir P. Gerdt[3],
Zahid Hossain[1,4], Ingmar H. Riedel-Kruse[4], and Andreas G. Weber[5]

[1] Department of Computer Science, Stanford University, 353 Serra Mall,
Stanford, CA 94305, USA
michels@cs.stanford.edu
[2] Visual Computing Center, King Abdullah University of Science and Technology,
Al Khawarizmi Building, Thuwal 23955-6900, Saudi Arabia
dmitry.lyakhov@kaust.edu.sa
[3] Group of Algebraic and Quantum Computations,
Joint Institute for Nuclear Research, Joliot-Curie 6,
141980 Dubna, Moscow Region, Russia
gerdt@jinr.ru
[4] Department of Bioengineering, Stanford University, 318 Campus Drive,
Stanford, CA 94305, USA
{zhossain,ingmar}@stanford.edu
[5] Institute of Computer Science II, University of Bonn, Friedrich-Ebert-Allee 144,
53113 Bonn, Germany
weber@cs.uni-bonn.de

Abstract. Based on a Lie symmetry analysis, we construct a closed
form solution to the kinematic part of the (partial differential) Cosserat
equations describing the mechanical behavior of elastic rods. The solution depends on two arbitrary analytical vector functions and is analytical everywhere except a certain domain of the independent variables in
which one of the arbitrary vector functions satisfies a simple explicitly
given algebraic relation. As our main theoretical result, in addition to
the construction of the solution, we proof its generality. Based on this
observation, a hybrid semi-analytical solver for highly viscous two-way
coupled fluid-rod problems is developed which allows for the interactive
high-fidelity simulations of flagellated microswimmers as a result of a
substantial reduction of the numerical stiffness.

Keywords: Cosserat rods · Differential thomas decomposition ·
Flagellated microswimmers · General analytical solution · Kinematic
equations · Lie symmetry analysis · Stokes flow · Symbolic computation

1 Introduction

Studying the dynamics of nearly one-dimensional structures has various scientific
and industrial applications, for example in biophysics (cf. [11,12] and the references therein) and visual computing (cf. [18]) as well as in civil and mechanical

© Springer International Publishing AG 2016
V.P. Gerdt et al. (Eds.): CASC 2016, LNCS 9890, pp. 367–380, 2016.
DOI: 10.1007/978-3-319-45641-6_24

engineering (cf. [5]), microelectronics and robotics (cf. [7]). In this regard, an appropriate description of the dynamical behavior of flexible one-dimensional structures is provided by the so-called special Cosserat theory of elastic rods (cf. [2], Chap. 8, and the original work [10]). This is a general and geometrically exact dynamical model that takes bending, extension, shear, and torsion into account as well as rod deformations under external forces and torques. In this context, the dynamics of a rod is described by a governing system of twelve first-order nonlinear partial differential equations (PDEs) with a pair of independent variables (s, t) where s is the arc-length and t the time parameter. In this PDE system, the two kinematic vector equations ((9a)–(9b) in [2], Chap. 8) are parameter free and represent the compatibility conditions for four vector functions κ, ω, ν, v in (s, t). Whereas the first vector equation only contains two vector functions κ, ω, the second one contains all four vector functions κ, ω, ν, v. The remaining two vector equations in the governing system are dynamical equations of motion and include two more dependent vector variables $\hat{m}(s, t)$ and $\hat{n}(s, t)$. Moreover, these dynamical equations contain parameters (or parametric functions of s) to characterize the rod and to include the external forces and torques.

Because of its inherent stiffness caused by the different deformation modes of a Cosserat rod, a pure numerical treatment of the full Cosserat PDE system requires the application of specific solvers; see e.g. [15, 17]. In order to reduce the computational overhead caused by the stiffness, we analyzed the Lie symmetries of the first kinematic vector equation ((9a) in [2], Chap. 8) and constructed its general and (locally) analytical solution in [16] which depends on three arbitrary functions in (s, t) and three arbitrary functions in t.

In this contribution we perform a computer algebra-based Lie symmetry analysis to integrate the full kinematic part of the governing Cosserat system based on our previous work in [16]. This allows for the construction of a general analytical solution of this part which depends on six arbitrary functions in (s, t). We prove its generality and apply the obtained analytical solution in order to solve the dynamical part of the governing system. Finally, we prove its practicability by simulating the dynamics of a flagellated microswimmer. To allow for an efficient solution process of the determining equations for the infinitesimal Lie symmetry generators, we make use of the Maple package SADE (cf. [22]) in addition to DESOLV (cf. [8]).

This paper is organized as follows. Section 2 describes the governing PDE system in the special Cosserat theory of rods. In Sect. 3, we show that the functional arbitrariness in the analytical solution to the first kinematic vector equation that we constructed in [16] can be narrowed down to three arbitrary bivariate functions. Our main theoretical result is presented in Sect. 4, in which we construct a general analytical solution to the kinematic part of the governing equations by integrating the Lie equations for a one-parameter subgroup of the Lie symmetry group. Section 5 illustrates the practicability of this approach by realizing a semi-analytical simulation of a flagellated microswimmer. This is based on a combination of the analytical solution of the kinematic part of the Cosserat

PDE and a numerical solution of its dynamical part. Some concluding remarks are given in Sect. 6 and limitations are discussed in Sect. 7.

2 Special Cosserat Theory of Rods

In the context of the special Cosserat theory of rods (cf. [2,7,10,16]), the motion of a rod is defined by a vector-valued function

$$[a, b] \times \mathbb{R} \ni (s, t) \mapsto (\boldsymbol{r}(s, t),\, \boldsymbol{d}_1(s, t),\, \boldsymbol{d}_2(s, t)) \in \mathbb{E}^3\,.$$

Here, t denotes the time and s is the arc-length parameter identifying a *material cross-section* of the rod which consists of all material points whose reference positions are on the plane perpendicular to the rod at s. Moreover, $\boldsymbol{d}_1(s,t)$ and $\boldsymbol{d}_2(s,t)$ are orthonormal vectors, and $\boldsymbol{r}(s,t)$ denotes the position of the material point on the centerline with arc-length parameter s at time t. The Euclidean 3-space is denoted with $\mathbb{E}^3$. The vectors $\boldsymbol{d}_1$, $\boldsymbol{d}_2$, and $\boldsymbol{d}_3 := \boldsymbol{d}_1 \times \boldsymbol{d}_2$ are called *directors* and form a right-handed orthonormal moving frame. The use of the triple $(\boldsymbol{d}_1,\, \boldsymbol{d}_2,\, \boldsymbol{d}_3)$ is natural for the intrinsic description of the rod deformation whereas $\boldsymbol{r}$ describes the motion of the rod relative to the fixed frame $(\mathbf{e}_1, \mathbf{e}_2, \mathbf{e}_3)$. This is illustrated in Fig. 1.

From the orthonormality of the directors follows the existence of so-called *Darboux* and *twist* vector functions $\boldsymbol{\kappa} = \sum_{k=1}^{3} \kappa_k \boldsymbol{d}_k$ and $\boldsymbol{\omega} = \sum_{k=1}^{3} \omega_k \boldsymbol{d}_k$ determined by the kinematic relations

$$\partial_s \boldsymbol{d}_k = \boldsymbol{\kappa} \times \boldsymbol{d}_k\,, \qquad \partial_t \boldsymbol{d}_k = \boldsymbol{\omega} \times \boldsymbol{d}_k\,. \tag{1}$$

The *linear strain* of the rod and the *velocity of the material cross-section* are given by vector functions $\boldsymbol{\nu} := \partial_s \boldsymbol{r} = \sum_{k=1}^{3} \nu_k \boldsymbol{d}_k$ and $\boldsymbol{v} := \partial_t \boldsymbol{r} = \sum_{k=1}^{3} v_k \boldsymbol{d}_k$.

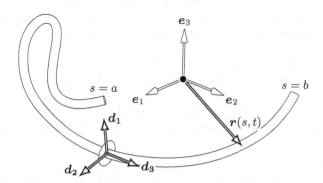

Fig. 1. The vector set $\{\boldsymbol{d}_1, \boldsymbol{d}_2, \boldsymbol{d}_3\}$ forms a right-handed orthonormal basis. The directors $\boldsymbol{d}_1$ and $\boldsymbol{d}_2$ span the local material cross-section, whereas $\boldsymbol{d}_3$ is perpendicular to the cross-section. Note that in the presence of shear deformations $\boldsymbol{d}_3$ is unequal to the tangent $\partial_s \boldsymbol{r}$ of the centerline of the rod.

The components of the *strain variables* κ and ν describe the deformation of the rod: the flexure with respect to the two major axes of the cross-section (κ_1, κ_2), torsion (κ_3), shear (ν_1, ν_2), and extension (ν_3).

The triples

$$\kappa = (\kappa_1, \kappa_2, \kappa_3)\,, \quad \omega = (\omega_1, \omega_2, \omega_3)\,, \quad \nu = (\nu_1, \nu_2, \nu_3)\,, \quad v = (v_1, v_2, v_3) \qquad (2)$$

are functions in (s, t), that satisfy the *compatibility conditions*

$$\partial_t \partial_s \, d_k = \partial_s \partial_t \, d_k \,, \quad \partial_t \partial_s r = \partial_s \partial_t r \,. \qquad (3)$$

The substitution of (1) into the left equation in (3) leads to

$$\tilde{\kappa}_t = \tilde{\omega}_s - \kappa \times \omega \quad \text{with} \quad \tilde{\kappa}_t = \partial_t \sum_{k=1}^{3} \kappa_k \, d_k\,, \quad \tilde{\omega}_s = \partial_s \sum_{k=1}^{3} \omega_k \, d_k\,.$$

On the other hand one obtains, $\tilde{\kappa}_t = \kappa_t + \omega \times \kappa$ and $\tilde{\omega}_s = \omega_s + \kappa \times \omega$ with $\kappa_t = (\partial_t \kappa_1, \partial_t \kappa_2, \partial_t \kappa_3)$ and $\omega_s = (\partial_s \omega_1, \partial_s \omega_2, \partial_s \omega_3)$, and therefore

$$\kappa_t = \omega_s - \omega \times \kappa \,. \qquad (4)$$

Similarly, the second compatibility condition in (3) is equivalent to

$$\nu_t = v_s + \kappa \times v - \omega \times \nu \qquad (5)$$

with $\nu_t = (\partial_t \nu_1, \partial_t \nu_2, \partial_t \nu_3)$ and $v_s = (\partial_s v_1, \partial_s v_2, \partial_s v_3)$.

The first-order PDE system (4) and (5) with independent variables (s, t) and dependent variables (2) forms the kinematic part of the governing Cosserat equations ((9a)–(9b) in [2], Chap. 8). The construction of its general solution is the main theoretical result of this paper.

The remaining part of the governing equations in the special Cosserat theory consists of two vector equations resulting from Newton's laws of motion. For a rod density $\rho(s)$ and cross-section $A(s)$, these equations are given by

$$\rho(s)A(s)\partial_t v = \partial_s n(s, t) + F(s, t)\,,$$
$$\partial_t h(s, t) = \partial_s m(s, t) + \nu(s, t) \times n(s, t) + L(s, t)\,,$$

where $m(s, t) = \sum_{k=1}^{3} m_k(s, t)\, d_k(s, t)$ are the *contact torques*, $n(s, t) = \sum_{k=1}^{3} n_k(s, t)\, d_k(s, t)$ are the *contact forces*, $h(s, t) = \sum_{k=1}^{3} h_k(s, t)\, d_k(s, t)$ are the *angular momenta*, and $F(s, t)$ and $L(s, t)$ are the *external forces* and *torque densities*.

The contact torques $m(s, t)$ and contact forces $n(s, t)$ corresponding to the *internal stresses*, are related to the extension and shear strains $\nu(s, t)$ as well as to the flexure and torsion strains $\kappa(s, t)$ by the *constitutive relations*

$$m(s, t) = \hat{m}\,(\kappa(s, t), \nu(s, t), s)\,, \quad n(s, t) = \hat{n}\,(\kappa(s, t), \nu(s, t), s)\,. \qquad (6)$$

Under certain reasonable assumptions (cf. [2,7,16]) on the structure of the right-hand sides of (6), together with the kinematic relations (4) and (5), it yields to the governing equations (cf. [2], Chap. 8, (9.5a)–(9.5d))

$$
\begin{aligned}
\boldsymbol{\kappa}_t &= \boldsymbol{\omega}_s - \boldsymbol{\omega} \times \boldsymbol{\kappa}, \\
\boldsymbol{\nu}_t &= \boldsymbol{v}_s + \boldsymbol{\kappa} \times \boldsymbol{v} - \boldsymbol{\omega} \times \boldsymbol{\nu}, \\
\rho \boldsymbol{J} \cdot \boldsymbol{\omega}_t &= \hat{\boldsymbol{m}}_s + \boldsymbol{\kappa} \times \hat{\boldsymbol{m}} + \boldsymbol{\nu} \times \hat{\boldsymbol{n}} - \boldsymbol{\omega} \times (\rho \boldsymbol{J} \cdot \boldsymbol{\omega}) + \boldsymbol{L}, \\
\rho A \boldsymbol{v}_t &= \boldsymbol{n}_s + \boldsymbol{\kappa} \times \hat{\boldsymbol{n}} - \boldsymbol{\omega} \times (\rho A \boldsymbol{v}) + \boldsymbol{F},
\end{aligned}
\tag{7}
$$

in which $\boldsymbol{J}$ is the inertia tensor of the cross-section per unit length. The dynamical part of (7) contains parameters characterizing the rod under consideration of $\rho, A, \boldsymbol{J}$ and the external force and torque densities $\boldsymbol{F}$ and $\boldsymbol{L}$, whereas the kinematic part is parameter free.

3   Analytical Form of the Darboux and Twist Functions

In [16], we constructed a general solution to (4) that is the first equation in the PDE system (7). In so doing, we proved that the constructed solution is (locally) analytical and provides the structure of the twist vector function $\boldsymbol{\omega}$ and the Darboux vector function $\boldsymbol{\kappa}$:

$$
\begin{aligned}
\boldsymbol{\omega} =&\, \boldsymbol{f} - \frac{\sin(p)}{p} \, \boldsymbol{p} \times \boldsymbol{f} + \frac{1 - \cos(p)}{p^2} \left(\boldsymbol{p} \, (\boldsymbol{p} \cdot \boldsymbol{f}) - p^2 \, \boldsymbol{f} \right) \\
&+ \boldsymbol{p}_t + \frac{p - \sin(p)}{p^3} \left(\boldsymbol{p} \, (\boldsymbol{p} \cdot \boldsymbol{p}_t) - p^2 \, \boldsymbol{p}_t \right) - \frac{1 - \cos(p)}{p^2} \, \boldsymbol{p} \times \boldsymbol{p}_t, \\
\boldsymbol{\kappa} =&\, \boldsymbol{p}_s + \frac{p - \sin(p)}{p^3} \left(\boldsymbol{p} \, (\boldsymbol{p} \cdot \boldsymbol{p}_s) - p^2 \, \boldsymbol{p}_s \right) - \frac{1 - \cos(p)}{p^2} \, \boldsymbol{p} \times \boldsymbol{p}_s,
\end{aligned}
\tag{8}
$$

where $\boldsymbol{f} := (f_1(t), f_2(t), f_3(t))$ and $\boldsymbol{p} := (p_1(s,t), p_2(s,t), p_3(s,t))$ are arbitrary vector-valued analytical functions, and $p^2 := p_1^2 + p_2^2 + p_3^2$.

It turns out that the functional arbitrariness of $\boldsymbol{f}$ and $\boldsymbol{p}$ is superfluous, and that (8) with $\boldsymbol{f}(t) = 0$ is still a general solution to (4). This fact is formulated in the following proposition.

Proposition 1. *The vector functions $\boldsymbol{\omega}$ and $\boldsymbol{\kappa}$ expressed by*

$$
\boldsymbol{\omega} = \boldsymbol{p}_t + \frac{p - \sin(p)}{p^3} \left(\boldsymbol{p} \, (\boldsymbol{p} \cdot \boldsymbol{p}_t) - p^2 \, \boldsymbol{p}_t \right) - \frac{1 - \cos(p)}{p^2} \, \boldsymbol{p} \times \boldsymbol{p}_t,
\tag{9a}
$$

$$
\boldsymbol{\kappa} = \boldsymbol{p}_s + \frac{p - \sin(p)}{p^3} \left(\boldsymbol{p} \, (\boldsymbol{p} \cdot \boldsymbol{p}_s) - p^2 \, \boldsymbol{p}_s \right) - \frac{1 - \cos(p)}{p^2} \, \boldsymbol{p} \times \boldsymbol{p}_s
\tag{9b}
$$

with an arbitrary analytical vector function $\boldsymbol{p}(s,t)$, are a general analytical solution to (4).

Proof. Let (s_0, t_0) be a fixed point. The right-hand sides of (9a) and (9b) satisfy (4) for an arbitrary vector function $\boldsymbol{p}(s, t)$ analytical in (s_0, t_0). It is an obvious consequence of the fact that (8) is a solution to (4) for arbitrary $\boldsymbol{f}(t)$ analytical in t_0.

Also, the equalities (9a) and (9b) can be transformed into each other with

$$\boldsymbol{\omega}(s,t) \Leftrightarrow \boldsymbol{\kappa}(s,t) \quad \text{and} \quad \partial_s \Leftrightarrow \partial_t \tag{10}$$

reflecting the invariance of (4) under (10). The equalities (9a) and (9b) are linear with respect to the partial derivatives $\boldsymbol{p}_t$ and $\boldsymbol{p}_s$, and their corresponding Jacobians. The determinants of the Jacobian matrices $J_{\boldsymbol{\omega}}(\partial_t p_1, \partial_t p_2, \partial_t p_3)$ and $J_{\boldsymbol{\kappa}}(\partial_s p_1, \partial_s p_2, \partial_s p_3)$ coincide because of the symmetry (10) and read[1]

$$J(\boldsymbol{p}) := \det(J_{\boldsymbol{\omega}}) = \det(J_{\boldsymbol{\kappa}}) = 2\,\frac{\cos(p) - 1}{p^2}. \tag{11}$$

Let $\boldsymbol{\omega}(s,t)$ and $\boldsymbol{\kappa}(s,t)$ be two arbitrary vector functions analytical in (s_0, t_0). We have to show that there is a vector function $\boldsymbol{p}(s,t)$ analytical in (s_0, t_0) satisfying (9a) and (9b). For that, chose real constants a, b, c such that

$$\frac{\cos(\sqrt{a^2 + b^2 + c^2}) - 1}{a^2 + b^2 + c^2} \neq 0$$

and set $\boldsymbol{p}_0 := \{a, b, c\}$. Then (9a) and (9b) are solvable with respect to the partial derivatives of $\boldsymbol{p}_t$ and $\boldsymbol{p}_s$ in a vicinity of (s_0, t_0), and we obtain the first-order PDE system of the form

$$\boldsymbol{p}_t = \boldsymbol{\Phi}(\boldsymbol{\omega}, \boldsymbol{p}), \quad \boldsymbol{p}_s = \boldsymbol{\Phi}(\boldsymbol{\kappa}, \boldsymbol{p}), \tag{12}$$

where the vector function $\boldsymbol{\Phi}$ is linear in its first argument and analytical in $\boldsymbol{p}$ at $\boldsymbol{p}_0$.

Also, the system (12) inherits the symmetry under the swap (10) and is *passive* and *orthonomic* in the sense of the Riquer-Janet theory (cf. [23] and the references therein), since its vector-valued *passivity (integrability) condition*

$$\partial_s \boldsymbol{\Phi}(\boldsymbol{\omega}, \boldsymbol{p}) - \partial_t \boldsymbol{\Phi}(\boldsymbol{\kappa}, \boldsymbol{p}) = 0$$

holds due to symmetry. Therefore, by Riquier's existence theorems [24] that generalize the Cauchy-Kovalevskaya theorem, there is a *unique* solution $\boldsymbol{p}(s,t)$ of (12) analytical in (s_0, t_0) and satisfying $\boldsymbol{p}(s_0, t_0) = \boldsymbol{p}_0$.   □

4 General Solution to the Kinematic Equation System

In this section, we determine a general analytical form of the vector functions $\boldsymbol{\nu}(s,t)$ and $\boldsymbol{v}(s,t)$ in (2) describing the linear strain of a Cosserat rod and its velocity. These functions satisfy the second kinematic equation (5) of the governing PDE system (7) under the condition that the Darboux and the twist

---

[1] The equalities in (11) are easily verifiable with Maple (cf. [16]), Sect. 3.5.

functions, $\kappa(s,t)$ and $\omega(s,t)$, occurring in the last equation, are given by (9a) and (9b) which contain the arbitrary analytical vector function $p(s,t)$.

Similarly, as we carried it out in [16] for the integration of (4), we analyze Lie symmetries (cf. [19] and the references therein) and consider the *infinitesimal generator*

$$\mathcal{X} := \xi^1 \partial_s + \xi^2 \partial_t + \sum_{i=1}^{3} \left(\theta^i \partial_{\omega_i} + \vartheta^i \partial_{\kappa_i} + \phi^i \partial_{\nu_i} + \varphi^i \partial_{v_i} \right) \tag{13}$$

of a Lie group of point symmetry transformations for (4) and (5). The coefficients $\xi^1, \xi^2, \theta^i, \vartheta^j, \phi^m, \varphi^n$ with $i, j, m, n \in \{1, 2, 3\}$ in (13) are functions of the independent and dependent variables.

The *infinitesimal criterion of invariance* of (4) and (5) reads

$$\mathcal{X}^{(pr)} h_1 = \mathcal{X}^{(pr)} h_2 = 0 \quad \text{whenever} \quad h_1 = h_2 = 0, \tag{14}$$

where

$$h_1 := \kappa_t - \omega_s + \omega \times \kappa, \quad h_2 := \nu_t - v_s - \kappa \times v + \omega \times \nu. \tag{15}$$

In addition to those in (13), the *prolonged* infinitesimal symmetry generator $X^{(pr)}$ contains extra terms caused by the presence of the first-order partial derivatives in (4) and (5).

The invariance conditions (14) lead to an overdetermined system of linear PDEs in the coefficients of the infinitesimal generator (13). This *determining* system can be easily computed by any modern computer algebra software (cf. [6]). We make use of the Maple package DESOLV (cf. [8]) which computes the determining system and outputs 138 PDEs.

Since the completion of the determining systems to involution is the most universal algorithmic tool of their analysis (cf. [6,14]), we apply the Maple package JANET (cf. [4]) first and compute a Janet involutive basis (cf. [21]) of 263 elements for the determining system, which took about 80 min of computation time on standard hardware.[2] Then we detected the functional arbitrariness in the general solution of the determining system by means of the differential Hilbert polynomial

$$4s^2 + 18s + 21 = 8 \binom{s+2}{s} + 6 \binom{s+1}{s} + 7 \tag{16}$$

computable by the corresponding routine of the Maple package DIFFEREN-TIALTHOMAS (cf. [3]). It shows that the general solution depends on eight arbitrary functions of (s,t). However, in contrast to the determining system for (4) which is quickly and effectively solvable (cf. [16]) by the routine *pdesolv* built in the package DESOLV, the solution found by this routine to the involutive determining system for (4) and (5) needs around one hour of computation time and has a form which is unsatisfactory for our purposes, since the solution contains

---

[2] The computation time has been measured on a machine with an Intel(R) Xeon E5 with 3.5 GHz and 32 GB DDR-RAM.

nonlocal (integral) dependencies on arbitrary functions. On the other hand, the use of SADE (cf. [22]) leads to a satisfying result. Unlike DESOLV, SADE uses some heuristics to solve simpler equations first in order to simplify the remaining system. In so doing, SADE extends the determining systems with certain integrability conditions for a partial completion to involution. In our case the routine *liesymmetries* of SADE receives components of the vectors in (15) and outputs the set of nine distinct solutions in just a few seconds. The output solution set includes eight arbitrary functions in (s, t) which is in agreement with (16). Each solution represents an infinitesimal symmetry generator (13).

Among the generators, there are three that include an arbitrary vector function, which we denoted by $\boldsymbol{q}(s, t) = (q_1(s, t), q_2(s, t), q_3(s, t))$, with vanishing coefficients $\theta_i, \vartheta_i, i \in \{1, 2, 3\}$. The sum of these generators is given by

$$\mathcal{X}_0 := (-\partial_s q_1 + q_2 \kappa_3 - q_3 \kappa_2) \partial_{\nu_1} + (-\partial_s q_2 + q_3 \kappa_1 - q_1 \kappa_2) \partial_{\nu_2} +$$
$$(-\partial_s q_3 + q_1 \kappa_3 - q_3 \kappa_1) \partial_{\nu_3} + (-\partial_t q_1 + q_2 \omega_3 - q_3 \omega_2) \partial_{\upsilon_1} + \quad (17)$$
$$(-\partial_t q_2 + q_3 \omega_1 - q_1 \omega_2) \partial_{\upsilon_2} + (-\partial_t q_3 + q_1 \omega_3 - q_3 \omega_1) \partial_{\upsilon_3} .$$

It generates a one-parameter Lie symmetry group of point transformations (depending on the arbitrary vector function $\boldsymbol{q}(s, t)$) of the vector functions $\boldsymbol{\nu}(s, t)$ and $\boldsymbol{v}(s, t)$ preserving the equality (5) for fixed $\boldsymbol{\kappa}(s, t)$ and $\boldsymbol{\omega}(s, t)$.

In accordance to Lie's first fundamental theorem (cf. [19]), the symmetry transformations

$$\boldsymbol{\nu} \mapsto \boldsymbol{\nu}'(a), \quad \boldsymbol{v} \mapsto \boldsymbol{v}'(a) \quad \text{with group parameter} \quad a \in \mathbb{R},$$

generated by (17), are solutions to the following differential (Lie) equations whose vector form reads

$$d_a \boldsymbol{\nu}' = \boldsymbol{q} \times \boldsymbol{\kappa} - \boldsymbol{q}_s, \quad d_a \boldsymbol{v}' = \boldsymbol{q} \times \boldsymbol{\omega} - \boldsymbol{q}_t, \quad \boldsymbol{\nu}'(0) = \boldsymbol{\nu}, \quad \boldsymbol{v}'(0) = \boldsymbol{v}. \quad (18)$$

The Eqs. (18) can easily be integrated, and without a loss of generality the group parameter can be absorbed into the arbitrary function q. This gives the following solution[3] to (5):

$$\boldsymbol{\nu} = \boldsymbol{q} \times \boldsymbol{\kappa} - \boldsymbol{q}_s, \quad \boldsymbol{v} = \boldsymbol{q} \times \boldsymbol{\omega} - \boldsymbol{q}_t. \quad (19)$$

Proposition 2. *The vector functions* $\boldsymbol{\omega}(s, t)$, $\boldsymbol{\kappa}(s, t)$, $\boldsymbol{\nu}(s, t)$, *and* $\boldsymbol{v}(s, t)$ *expressed by* (9a)–(9b) *and* (19) *with two arbitrary analytical functions* $\boldsymbol{p}(s, t)$ *and* $\boldsymbol{q}(s, t)$ *form a general analytical solution to* (4) *and* (5).

Proof. The fact that (9a) and (9b) form a general analytical solution to (4) was verified in Proposition 1.

We have to show that, given analytical vector functions $\boldsymbol{\nu}(s, t)$ and $\boldsymbol{v}(s, t)$ satisfying (5) with analytical $\boldsymbol{\omega}(s, t)$ and $\boldsymbol{\kappa}(s, t)$ satisfying (4), there exists an analytical vector function $\boldsymbol{q}(s, t)$ satisfying (19). Consider the last equalities as

---

[3] It is easy to check with Maple that the right-hand sides of (19) satisfy (5) for arbitrary $\boldsymbol{q}(s, t)$ if one takes (9a) and (9b) into account.

a system of first-order PDEs with independent variables (s, t) and a dependent vector variable $\boldsymbol{q}$. According to the argumentation in the proof of Proposition 1, this leads to the fact, that the equations in (19) are invariant under the transformations

$$\boldsymbol{\nu}(s, t) \Leftrightarrow \boldsymbol{v}(s, t), \quad \boldsymbol{\omega}(s, t) \Leftrightarrow \boldsymbol{\kappa}(s, t), \quad \partial_s \Leftrightarrow \partial_t.$$

This symmetry implies the satisfiability of the integrability condition

$$\partial_t(\boldsymbol{q} \times \boldsymbol{\kappa} - \boldsymbol{\nu}) - \partial_s(\boldsymbol{q} \times \boldsymbol{\omega} - \boldsymbol{v}) = 0$$

without any further constraints. Therefore, the system (19) is passive (involutive), and by Riquier's existence theorem, there is a solution $\boldsymbol{q}$ to (19) analytical in a point of analyticity of $\boldsymbol{\omega}, \boldsymbol{\kappa}, \boldsymbol{\nu}, \boldsymbol{v}$. $\square$

5   Simulation of Two-Way Coupled Fluid-Rod Problems

To demonstrate the practical use of the analytical solution to the kinematic Cosserat equations, we combine it with the numerical solution of its dynamical part. The resulting analytical solutions (9a)–(9b) and (19) for the kinematic part of (7) contain two parameterization functions $\boldsymbol{p}(s, t)$ and $\boldsymbol{q}(s, t)$, which can be determined by the numerical integration of the dynamical part of (7). The substitution of the resulting analytical solutions (9a)–(9b) and (19) into the latter two (dynamical) equations of (7), the replacement of the spatial derivatives with central differences, and the replacement of the temporal derivatives according to the numerical scheme of a forward Euler integrator, leads to an explicit expression.[4] Iterating over this recurrence equation allows for the simulation of the dynamics of a rod.

In order to embed this into a scenario close to reality, we consider a flagellated microswimmer. In particular, we simulate the dynamics of a swimming sperm cell, which is of interest in the context of simulations in biology and biophysics. Since such a highly viscous fluid scenario takes place in the low Reynolds number domain, the advection and pressure parts of the Navier-Stokes equations (cf. [13]) can be ignored, such that the resulting so-called *steady Stokes equations* become linear and can be solved analytically. Therefore, numerical errors do not significantly influence the fluid simulation part for which reason this scenario is appropriate for evaluating the practicability of the analytical solution to the kinematic Cosserat equations. The *steady Stokes equations* are given by

$$\mu \Delta \boldsymbol{u} = \nabla p - \boldsymbol{F}, \tag{20}$$

$$\nabla \cdot \boldsymbol{u} = 0, \tag{21}$$

in which μ denotes the fluid viscosity, p the pressure, $\boldsymbol{u}$ the velocity, and $\boldsymbol{F}$ the force. Similar to the fundamental work in [9] we use a regularization in order to develop a suitable integration of (20) and (21). For that, we assume

---

[4] We do not explicitly write out the resulting equations here for brevity. A construction of a hybrid semi-analytical, semi-numerical solver is also described in our recent contribution [17].

$$F(x) = f_0 \, \phi_\epsilon(x - x_0),$$

in which ϕ_ϵ is a smooth and radially symmetric function with $\int \phi_\epsilon(x) \, dx = 1$, is spread over a small ball centered at the point x_0.

Let G_ϵ be the corresponding Green's function, i.e., the solution of $\Delta G_\epsilon(x) = \phi_\epsilon(x)$ and let B_ϵ be the solution of $\Delta B_\epsilon(x) = G_\epsilon(x)$, both in the infinite space bounded for small ϵ. Smooth approximations of G_ϵ and B_ϵ are given by $G(x) = -1/(4\pi \|x\|)$ for $\|x\| > 0$ and $B(x) = -\|x\|/(8\pi)$, the solution of the biharmonic equation $\Delta^2 B(x) = \delta(x)$.

The pressure p satisfies $\Delta p = \nabla \cdot F$, which can be shown by applying the divergence operator on (20) and (21), and is therefore given by $p = f_0 \cdot \nabla G_\epsilon$. Using this, we can rewrite (20) as

$$\mu \Delta u = (f_0 \cdot \nabla) \nabla G_\epsilon - f_0 \phi_\epsilon$$

with its solution

$$\mu u(x) = (f_0 \cdot \nabla) \nabla B_\epsilon(x - x_0) - f_0 G_\epsilon(x - x_0),$$

the so-called *regularized Stokeslet*.

For multiple forces $f_1, \ldots, f_N$ centered at points $x_1, \ldots, x_N$, the pressure p and the velocity u can be computed by superposition. Because G_ϵ and B_ϵ are radially symmetric, we can additionally use $\nabla B_\epsilon(x) = B_\epsilon' x / \|x\|$ and obtain[5]

$$p(x) = \sum_{k=1}^{N} (f_k \cdot (x - x_k)) \frac{G_\epsilon'(\|x - x_k\|)}{\|x - x_k\|}, \tag{22}$$

$$u(x) = \frac{1}{\mu} \sum_{k=1}^{N} \left[f_k \left(\frac{B_\epsilon'(\|x - x_k\|)}{\|x - x_k\|} - G_\epsilon(\|x - x_k\|) \right) \right.$$

$$\left. + (f_k \cdot (x - x_k))(x - x_k) \frac{\|x - x_k\| B_\epsilon''(\|x - x_k\|) - B_\epsilon'(\|x - x_k\|)}{\|x - x_k\|^3} \right]. \tag{23}$$

The flow given by (23) satisfies the incompressibility constraint (21). Because of

$$\Delta G_\epsilon(\|x - x_k\|) = \frac{1}{\|x - x_k\|} (\|x - x_k\| G_\epsilon'(\|x - x_k\|))' = \phi_\epsilon(\|x - x_k\|),$$

the integration of

$$G_\epsilon'(\|x - x_k\|) = \frac{1}{\|x - x_k\|} \int_0^{\|x - x_k\|} s \phi_\epsilon(s) \, ds$$

leads to G_ϵ. Similarly,

$$\frac{1}{\|x - x_k\|} (\|x - x_k\| B_\epsilon'(\|x - x_k\|))' = G_\epsilon(\|x - x_k\|)$$

---

[5] Since at this point, the functions ϕ_ϵ, G_ϵ, and B_ϵ only depend on the norm of their arguments, we change the notation according to this.

leads to the expression

$$B'_\epsilon(\|\boldsymbol{x} - \boldsymbol{x}_k\|) = \frac{1}{\|\boldsymbol{x} - \boldsymbol{x}_k\|} \int_0^{\|\boldsymbol{x} - \boldsymbol{x}_k\|} sG_\epsilon(s)\,\mathrm{d}s$$

to determine B_ϵ. We make use of the specific function

$$\phi_\epsilon(\|\boldsymbol{x}\|) = \frac{15\epsilon^4}{8\pi(\|\boldsymbol{x}\|^2 + \epsilon^2)^{7/2}},$$

which is smooth and radially symmetric.

Up to now, this regularized Stokeslet (22) and (23) allows for the computation of the velocities for given forces. Similarly, we can tread the application of a torque by deriving an analogous *regularized Rodlet*; see e.g. [1]. In the inverse case, the velocity expressions can be rewritten in the form of the equations

$$\boldsymbol{u}(\boldsymbol{x}_i) = \sum_{j=1}^N M_{ij}(\boldsymbol{x}_1, \dots, \boldsymbol{x}_N)\boldsymbol{f}_j$$

for $i \in \{1, \dots, N\}$ which can be transformed into an equation system $\mathsf{U} = \mathsf{M}\,\mathsf{F}$ with a $(3N \times 3N)$-matrix $\mathsf{M} := (M_{ij})_{i,j \in \{1,\dots,N\}}$. Since in general M is not regular, an iterative solver have to be applied.

A flagellated microswimmer can be set up by a rod representing the centerline of the flagellum; see [11]. Additionally, a constant torque perpendicular to the flagellum's base is applied to emulate the rotation of the motor. From forces and torque the velocity field is determined. Repeating this procedure to update the system state iteratively introduces a temporal domain and allows for the dynamical simulation of flagellated microswimmers; see Figs. 2 and 3. Compared to a purely numerical handling of the two-way coupled fluid-rod system, the step size can be increased by four to five orders of magnitude, which leads to an acceleration of four orders of magnitude. This allows for real-time simulations of flagellated microswimmers on a standard desktop computer.[6]

6   Conclusison

We constructed a closed form solution to the kinematic equations (4) and (5) of the governing Cosserat PDE system (7) and proved its generality. The kinematic equations are parameter free whereas the dynamical Cosserat PDEs contain a number of parameters and parametric functions characterizing the rod under consideration of external forces and torques. The solution we found depends on two arbitrary analytical vector functions and is analytical everywhere except at the values of the independent variables (s, t) for which the right-hand side of (11) vanishes. Therefore, the hardness of the numerical integration of the

---

[6] The simulations illustrated in Figs. 2 and 3 can be carried out in real-time on a machine with an Intel(R) Xeon E5 with 3.5 GHz and 32 GB DDR-RAM.

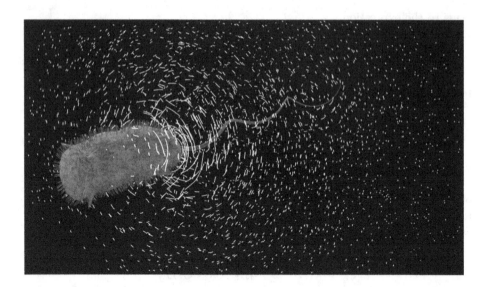

Fig. 2. Simulation of a monotrichous bacteria swimming in a viscous fluid. The rotation of the motor located at the back side of the bacteria's head causes the characteristic motion of the flagellum leading to a movement of the bacteria.

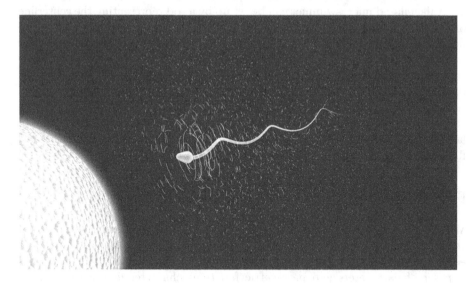

Fig. 3. Simulation of a sperm cell swimming into the direction of an egg. The concentration gradient induced by the egg is linearly coupled with the control of the motor. In contrast to the bacteria in Fig. 2, the flagellum of a sperm cell does not have its motor at its base as simulated here. Instead several motors are distributed along the flagellum (cf. [20]), for which reason this simulation is not fully biologically accurate, but still illustrates the capabilities of the presented approach.

Cosserat system, in particular its stiffness, is substantially reduced by using the exact solution to the kinematic equations. The application of the analytical solution prevents from numerical instabilities and allows for accurate and efficient simulations. This was demonstrated for the two-way coupled fluid-rod scenario of microswimmers, which could efficiently be simulated with an acceleration of four orders of magnitude compared to a purely numerical handling.

7   Limitations

Because of the presence of parameters in the dynamical part of the Cosserat PDEs, the construction of a general closed form solution to this part is hopeless. Even if one specifies all parameters and considers the parametric functions as numerical constants, the exact integration of the dynamical equations is hardly possible. We analyzed Lie symmetries of the kinematic equations extended with one of the dynamical vector equations including all specifications of all parameters and without parametric functions. While the determining equations can be generated in a reasonable time, their completion to involution seems to be practically impossible.

Acknowledgements. This work has been partially supported by the Max Planck Society (FKZ-01IMC01/FKZ-01IM10001), the Russian Foundation for Basic Research (16-01-00080), and a BioX Stanford Interdisciplinary Graduate Fellowship. The reviewers' valuable comments are gratefully acknowledged.

References

1. Ainley, J., Durkin, S., Embid, R., Boindala, P., Cortez, R.: The method of images for regularized stokeslets. J. Comput. Phys. **227**, 4600–4616 (2008)
2. Antman, S.S.: Nonlinear Problems of Elasticity. Applied Mathematical Sciences, vol. 107. Springer, New York (1995)
3. Bächler, T., Gerdt, V., Langer-Hegermann, M., Robertz, D.: Algorithmic Thomas decomposition of algebraic and differential systems. J. Symbolic Comput. **47**, 1233–1266 (2012)
4. Blinkov, Y., Cid, C., Gerdt, V., Plesken, W., Robertz, D.: The Maple package Janet: II. linear partial differential equations. In: Ganzha, V., Mayr, E., Vorozhtsov, E. (eds.) Computer Algebra in Scientific Computing, CASC 2003, pp. 41–54. Springer, Heidelberg (2003)
5. Boyer, F., De Nayer, G., Leroyer, A., Visonneau, M.: Geometrically exact Kirchhoff beam theory: application to cable dynamics. J. Comput. Nonlinear Dyn. **6**(4), 041004 (2011)
6. Butcher, J., Carminati, J., Vu, K.T.: A comparative study of some computer algebra packages which determine the Lie point symmetries of differential equations. Comput. Phys. Commun. **155**, 92–114 (2003)
7. Cao, D.Q., Tucker, R.W.: Nonlinear dynamics of elatic rods using the Cosserat theory: modelling and simulation. Int. J. Solids Struct. **45**, 460–477 (2008)
8. Carminati, J., Vu, K.T.: Symbolic computation and differential equations: Lie symmetries. J. Symb. Comput. **29**, 95–116 (2000)

9. Cortez, R.: The method of the regularized stokeslet. SIAM J. Sci. Comput. **23**(4), 1204–1225 (2001)
10. Cosserat, E., Cosserat, F.: Théorie des corps déformables. Hermann, Paris (1909)
11. Elgeti, J., Winkler, R., Gompper, G.: Physics of microswimmers–single particle motion and collective behavior: a review. Rep. Prog. Phys. **78**(5), 056601 (2015)
12. Goldstein, R.: Green algae as model organisms for biological fluid dynamics. Ann. Rev. Fluid Mech. **47**(1), 343–375 (2015)
13. Granger, R.: Fluid Mechanics. Dover Classics of Science and Mathematics. Courier Corporation, Mineola (1995)
14. Hereman, W.: Review of symbolic software for Lie symmetry analysis. CRC handbook of Lie group analysis of differential equations. In: Ibragimov, N.H. (ed.) New Trends in Theoretical Developments and Computational Methods, pp. 367–413. CRC Press, Boca Raton (1996)
15. Lang, H., Linn, J., Arnold, M.: Multibody dynamics simulation of geometrically exact Cosserat rods. In: Berichte des Fraunhofer ITWM, vol. 209 (2011)
16. Michels, D.L., Lyakhov, D.A., Gerdt, V.P., Sobottka, G.A., Weber, A.G.: Lie symmetry analysis for Cosserat rods. In: Gerdt, V.P., Koepf, W., Seiler, W.M., Vorozhtsov, E.V. (eds.) CASC 2014. LNCS, vol. 8660, pp. 324–334. Springer, Heidelberg (2014)
17. Michels, D., Lyakhov, D., Gerdt, V., Sobottka, G., Weber, A.: On the partial analytical solution to the Kirchhoff equation. In: Gerdt, V., Koepf, W., Seiler, W.M., Vorozhtsov, E.V. (eds.) Computer Algebra in Scientific Computing, CASC 2015, pp. 320–331. Springer, Heidelberg (2015)
18. Michels, D., Mueller, P., Sobottka, G.: A physically based approach to the accurate simulation of stiff fibers and stiff fiber meshes. Comput. Graph. **53B**, 136–146 (2015)
19. Oliveri, F.: Lie symmetries of differential equations: classical results and recent contributions. Symmetry **2**, 658–706 (2010)
20. Riedel-Kruse, I., Hilfinger, A., Howard, J., Jülicher, F.: How molecular motors shape the flagellar beat. HFSP J. **1**(3), 192–208 (2007)
21. Robertz, D.: Formal Algorithmic Elimination for PDEs. Lecture Notes in Mathematics, vol. 2121. Springer, Heidelberg (2014)
22. Filho, R.T.M., Figueiredo, A.: [SADE] a Maple package for the symmetry analysis of differential equations. Comput. Phys. Commun. **182**, 467–476 (2011)
23. Seiler, W.M.: Involution: The Formal Theory of Differential Equations and its Applications in Computer Algebra. Algorithms and Computation in Mathematics. Springer, Heidelberg (2010)
24. Thomas, J.M.: Riquier's existence theorems. Ann. Math. **30**, 285–310 (1929). **30**, 306–311 (1934)

Using Sparse Interpolation in Hensel Lifting

Michael Monagan$^{(\boxtimes)}$ and Baris Tuncer

Department of Mathematics, Simon Fraser University,
Burnaby, BC V5A 1S6, Canada
{mmonagan,ytuncer}@sfu.ca

Abstract. The standard approach to factor a multivariate polynomial in $\mathbb{Z}[x_1, x_2, \ldots, x_n]$ is to factor a univariate image in $\mathbb{Z}[x_1]$ then lift the factors of the image one variable at a time using Hensel lifting to recover the multivariate factors. At each step one must solve a multivariate polynomial Diophantine equation. For polynomials in many variables with many terms we find that solving these multivariate Diophantine equations dominates the factorization time. In this paper we explore the use of sparse interpolation methods, originally introduced by Zippel, to speed this up. We present experimental results in Maple showing that we are able to dramatically speed this up and thereby achieve a good improvement for multivariate polynomial factorization.

1 Introduction

Suppose that we seek to factor a multivariate polynomial $a \in R = \mathbb{Z}[x_1, \ldots, x_n]$ and $a = fg$ with f, g in R and $\gcd(f, g) = 1$. The multivariate Hensel lifting algorithm (MHL) developed by Yun [11] and improved by Wang [9,10] uses a prime number p and an ideal $I = \langle x_2 - \alpha_2, \ldots, x_n - \alpha_n \rangle$ of $\mathbb{Z}_p[x_1, \ldots, x_n]$ where $\alpha_2, \alpha_3, \ldots, \alpha_n \in \mathbb{Z}_p$ is a random evaluation point chosen by the algorithm.

For a given polynomial $h \in R$, let us use the notation

$$h_j := h(x_1, \ldots, x_j, x_{j+1} = \alpha_{j+1}, \ldots, x_n = \alpha_n) \bmod p$$

so that $a_1 = a(x_1, \alpha_2, \ldots, \alpha_n) \bmod p$. The input to MHL is a, I, f_1, g_1 and p such that $a_1 = f_1 g_1$ and $\gcd(f_1, g_1) = 1$ in $\mathbb{Z}_p[x_1]$. The input factorization $a_1 = f_1 g_1$ is obtained by factoring $a(x_1, \alpha_2, \ldots, \alpha_n)$ over the integers. See [2].

Let d_j denote the total degree of a_j with respect to the variables $x_2, \ldots, x_j$ and $I_j = \langle x_2 - \alpha_2, \ldots, x_j - \alpha_j \rangle$ with $j \leq n$. Wang's MHL lifts the factorization $a_1 = f_1 g_1$ variable by variable to $a_j = f_j g_j \in \mathbb{Z}_p[x_1, \ldots, x_j]/I_j^{d_j+1}$. It turns out that $f_n \equiv f \bmod p$ and $g_n \equiv g \bmod p$. For sufficiently large p we recover the factorization of a over $\mathbb{Z}$.

We give a brief description of the jth step of the MHL (assuming that the inputs are monic in the variable x_1, for simplicity) in Algorithm 1 for $j > 1$. For details see [2]. There are two main sub-routines in the design of MHL. The first one is the leading coefficient correction algorithm. The most well-known is the Wang's heuristic leading coefficient algorithm [9] which works well in practice and

© Springer International Publishing AG 2016
V.P. Gerdt et al. (Eds.): CASC 2016, LNCS 9890, pp. 381–400, 2016.
DOI: 10.1007/978-3-319-45641-6_25

is the one Maple currently uses. There are other approaches by Kaltofen [3] and most recently by Lee [4]. In our implementation we use Wang's leading coefficient algorithm. The second main subroutine is the multivariate Diophantine problem (MDP). In MHL, for each j with $j \leq n$, Wang's design of MHL must solve many instances the MDP. In the Maple timings (see Sect. 5), for most of the examples 90% of the time is spent solving MDPs.

In this paper we propose various approaches of sparse interpolation to solve MDP and present the results of our experiments. We will assume that a, f, g are monic in x_1 so as not to complicate the MHL algorithm with leading coefficient correction. In Sect. 2 we define the MDP in detail. In Sect. 3 we show that interpolation is an option to solve the MDP. If the factors to be computed are sparse then the solutions to the MDP are also sparse. We show in Sect. 3.1 how to use Zippel's sparse interpolation to solve the MDP and we describe an improvement to the solution proposed in Sect. 3.2. We have observed that often the evaluation cost is the most expensive part of these algorithms. In Sect. 3.3 we will propose an improvement to the evaluation method used in the sparse interpolation process. Sparse Hensel Lifting (SHL) was first introduced by Zippel [14] and then improved by Kaltofen [3]. In Sect. 4 we show that if we use Wang's leading coefficient correction then Kaltofen's SHL algorithm can be simplified, improved and implemented efficiently. Based on Lemma 1 in Sect. 4 we will propose our SHL organization which is presented as Algorithm 4. Finally in Sect. 5 we will give some timing data to compare our factorization algorithms with Wang's algorithm, which is currently used by Maple.

2   The Multivariate Diophantine Problem (MDP)

Following the notation in Sect. 1, let $u, w, c \in \mathbb{Z}_p[x_1, \ldots, x_j]$ in which u and w are monic polynomials with respect to the variable x_1 with $j \geq 1$ and let $I_j = \langle x_2 - \alpha_2, \ldots, x_j - \alpha_j \rangle$ be an ideal of $\mathbb{Z}_p[x_1, \ldots, x_j]$ with $\alpha_i \in \mathbb{Z}_p$. The MDP consists of finding multivariate polynomials $\sigma, \tau \in \mathbb{Z}_p[x_1, \ldots, x_j]$ that satisfy

$$\sigma u + \tau w = c \mod I_j^{d_j+1}$$

with $\deg_{x_1}(\sigma) < \deg_{x_1}(w)$ where d_j is the maximal total degree of σ and τ with respect to the variables $x_2, \ldots, x_j$ and it is given that

1. $GCD(u, w) \mid c$ and
2. $GCD(u \bmod I_j, w \bmod I_j) = 1$ in $\mathbb{Z}_p[x_1]$.

It can be shown that the solution (σ, τ) exists and is unique provided the second condition is satisfied and that the solution is independent of the choice of the ideal I_j. For $j = 1$ the MDP is in $\mathbb{Z}_p[x_1]$ and can be solved with the extended Euclidean algorithm (see Chap. 2 of [2]).

It can be seen from Algorithm 1 that at step j, there are at most $\max(\deg_{x_j}(f_j), \deg_{x_j}(g_j))$ calls to MDP. To solve the MDP for $j > 1$, Wang uses the same approach as for Hensel Lifting, that is, an ideal-adic approach

---

Algorithm 1. j^{th} step of Multivariate Hensel Lifting for $j > 1$.

Input : $\alpha_j \in \mathbb{Z}_p$, $a_j \in \mathbb{Z}_p[x_1, \ldots, x_j]$, $f_{j-1}, g_{j-1} \in \mathbb{Z}_p[x_1, \ldots, x_{j-1}]$ where a_j, f_{j-1}, g_{j-1} are monic in x_1 and $a_j(x_j = \alpha_j) = f_{j-1}g_{j-1}$.
Output : $f_j, g_j \in \mathbb{Z}_p[x_1, \ldots, x_j]$ such that $a_j = f_j g_j$.

 1: $\sigma_{j0} \leftarrow f_{j-1}, \tau_{j0} \leftarrow g_{j-1}, \sigma_j \leftarrow \sigma_{j0}, \tau_j \leftarrow \tau_{j0}$, *monomial* $\leftarrow 1$
 2: *error* $\leftarrow a_j - f_{j-1}\, g_{j-1}$
 3: **for** i from 1 to $\deg(a_j, x_j)$ **while** *error* $\neq 0$ **do**
 4:     *monomial* $\leftarrow$ *monomial* $\times (x_j - \alpha_j)$
 5:     $c \leftarrow$ coefficient of $(x_j - \alpha_j)^i$ in the Taylor expansion of the *error* about $x_j = \alpha_j$
 6:     **if** $c \neq 0$ **then**
 7:         Solve the MDP $\sigma_{ji}\tau_{j0} + \tau_{ji}\sigma_{j0} = c$ in $\mathbb{Z}_p[x_1, \ldots, x_{j-1}]$ for σ_{ji} and τ_{ji}.
 8:         $(\sigma_j, \tau_j) \leftarrow (\sigma_j + \sigma_{ji} \times$ *monomial*$, \tau_j + \tau_{ji} \times$ *monomial*$)$.
 9:         *error* $\leftarrow a_j - \sigma_j \tau_j$.
10:     **end if**
11: **end for**
12: $f_j \leftarrow \sigma_j$ and $g_j \leftarrow \tau_j$

(see [2]). In general, if $\alpha_k \neq 0$ for $j \leq k$, then an $\langle x_k - \alpha_k \rangle$-adic expansion of the solution is expensive to compute. Since even a sparse solution turns out to be dense in an $\langle x_k - \alpha_k \rangle$-adic expansion, the number of MDP's to be solved significantly increases and hence the time complexity of MHL becomes expensive. In the following sections we will present various approaches to solve the MDP.

3   Solution to the MDP via Interpolation

We consider whether we can interpolate $x_2, \ldots, x_j$ in σ. If $\beta \in \mathbb{Z}_p$ with $\beta \neq \alpha_j$, then we have

$$\sigma(x_j = \beta)u(x_j = \beta) + \tau(x_j = \beta)w(x_j = \beta) = c(x_j = \beta) \bmod I_{j-1}^{d_{j-1}+1}.$$

For $K_j = \langle x_2 - \alpha_2, \ldots, x_{j-1} - \alpha_{j-1}, x_j - \beta \rangle$ and $G_j = \text{GCD}(u \bmod K_j, w \bmod K_j)$, we obtain a unique solution $\sigma(x_1, \ldots, x_{j-1}, \beta)$ iff $G_j = 1$. However it is possible that $G_j \neq 1$. Let $R = \text{res}_{x_1}(u, w)$ be the Sylvester resultant of u and w taken in x_1. Since u, w are monic in x_1 one has

$$G_j \neq 1 \Longleftrightarrow \text{res}_{x_1}(u \bmod K_j, w \bmod K_j) = 0 \Longleftrightarrow R \bmod K_j = 0.$$

Also $\deg(R) \leq \deg(u)\deg(w)$ [1]. Then by the Schwartz-Zippel Lemma [8,13]

$$\text{Prob}(G_j \neq 1) \leq \frac{\deg(u)\deg(w)}{p - 1}.$$

If $\beta \neq \alpha_j$ is chosen at random and p is large, the probability that $G_j = 1$ is high so interpolation is thus an option to solve the MDP. If $G_j \neq 1$, we could choose another β but our implementation does not do this and simply returns FAIL. The bound above for $\text{Prob}(G_j \neq 1)$ is a worst case bound. We note that in [6] we show that the average probability for $\text{Prob}(G_j \neq 1) = 1/(p - 1)$.

3.1   Solution to the MDP via Sparse Interpolation

Following the sparse interpolation idea of Zippel in [12], given a sub-solution $\sigma_j(x_1, \ldots, x_j = \alpha_j)$ for $\alpha_j \in \mathbb{Z}_p$ we use this information to create a sub-solution form σ_f and compute $\sigma_j(x_1, \ldots, x_j = \beta_j)$ for some other random $\beta_j \in \mathbb{Z}_p$ with high probability if p is big. Suppose the form of σ_j is

$$\sigma_f = \sum_{i=1}^{m} c_i(x_2, ..., x_j)x_1^{n_i} \text{ where } c_i = \sum_{k=1}^{t_i} c_{ik}x_2^{\gamma_{2k}} \cdots x_j^{\gamma_{jk}} \text{ with } c_{ik} \in \mathbb{Z}_p\backslash\{0\}.$$

Let $t = \max_{i=1}^{m} t_i$ be the maximum number of terms in the coefficients of σ. In sparse interpolation we obtain each c_{ik} by solving m linear systems of size at most $t \times t$. As explained in [12], each linear system can be solved in $\mathcal{O}(t^2)$ arithmetic operations in $\mathbb{Z}_p$. We then interpolate x_j in σ_j from $\sigma_j(x_1, \ldots, x_{j-1}, \beta_k)$ for $k = 0, \ldots, \deg_{x_j}(\sigma_j)$. Finally we compute $\tau_j = (c_j - \sigma_j u_j)/w_j$.

3.2   First Improvement

The approach introduced in the preceding section solves the interpolation problem based on projection down to $\mathbb{Z}_p[x_1]$. To reduce the cost we tried projecting down to $\mathbb{Z}_p[x_1, x_2]$ because this will likely reduce the number t of evaluation points needed. Let the total degree of σ in x_1, x_2 be bounded by d and let

$$\sigma_f = \sum_{i+k\leq d} c_{ik}(x_3, ..., x_j)x_1^i x_2^k \text{ where } c_{ik} = \sum_{l=0}^{s_{ik}} c_{ikl}x_3^{\gamma_{3l}} \cdots x_j^{\gamma_{jl}} \text{ with } c_{ikl} \in \mathbb{Z}_p\backslash\{0\}.$$

Let $s = \max s_{ik}$ be the maximum number of terms in the coefficients of σ_f. Here we solve $\mathcal{O}(d^2)$ linear systems of size at most $s \times s$. For $s < t$, the complexity of solving the linear systems decreases by a factor of $(t/s)^2$. We also save a factor t/s in the evaluation cost.

To solve the MDP in $\mathbb{Z}_p[x_1, x_2]$ we have implemented an efficient dense bivariate Diophantine solver (BDP) in C. The algorithm incrementally interpolates x_2 in both σ and τ from univariate images in $\mathbb{Z}_p[x_1]$. When σ and τ stabilize we test whether $\sigma(x_1, x_2)u(x_1, x_2) + \tau(x_1, x_2)w(x_1, x_2) = c(x_1, x_2)$ using sufficiently many evaluations to prove the correctness of the solution. The cost is $\mathcal{O}(d^3)$ arithmetic operations in $\mathbb{Z}_p$ where d bounds the total degree of c, u, w, σ and τ in x_1 and x_2. We do not compute τ using division because that would cost $\mathcal{O}(d^4)$ arithmetic operations. This bivariate MDP solving algorithm is presented as algorithm BSDiophant below.

3.3   The Evaluation Cost

In our experiments we found that the sparse interpolation approach we propose reduces the time spent solving MDPs but evaluation becomes the most time dominating part of the factoring algorithm.

Algorithm 2. BSDiophant

Input A big prime p and $u, w, c \in \mathbb{Z}_p[x_1, x_2, \ldots, x_j]$.
Output $(\sigma, \tau) \in \mathbb{Z}_p[x_1, x_2, \ldots, x_j]$ such that $\sigma u + \tau w = c \in \mathbb{Z}_p[x_1, x_2, \ldots, x_j]$ or FAIL.
It returns FAIL if condition 2 (see Sect. 2) is not satisfied for the choice of any β in the
algorithm. This is detected in subroutine BDP.

1: **if** $n = 2$ **then** call BDP to **return** $(\sigma, \tau) \in \mathbb{Z}_p[x_1, x_2]^2$ or FAIL **end if**.
2: Pick $\beta_1 \in \mathbb{Z}_p$ at random
3: $(u_{\beta_1}, w_{\beta_1}, c_{\beta_1}) \leftarrow (u(x_1, \ldots, x_j = \beta_1), w(x_1, \ldots, x_j = \beta_1), c(x_1, \ldots, x_j = \beta_1)$.
4: $(\sigma_1, \tau_1) \leftarrow \text{BSDiophant}(u_{\beta_1}, w_{\beta_1}, c_{\beta_1}, p)$.
5: **if** $\sigma_1 = $ FAIL **then return** FAIL **end if**
6: $k \leftarrow 1$, $\sigma \leftarrow \sigma_1$, $q \leftarrow (x_j - \beta_1)$ and $\sigma_f \leftarrow$ skeleton of σ_1.
7: **repeat**
8:     $h \leftarrow \sigma$
9:     Set $k \leftarrow k + 1$ and pick $\beta_k \in \mathbb{Z}_p$ at random distinct from $\beta_1, \ldots, \beta_{k-1}$
10:     $(u_{\beta_k}, w_{\beta_k}, c_{\beta_k}) \leftarrow (u(x_1, \ldots, x_j = \beta_k), w(x_1, \ldots, x_j = \beta_k), c(x_1, \ldots, x_j = \beta_k)$.
11:     Solve the MDP $\sigma_k u_{\beta_k} + \tau_k w_{\beta_k} = c_{\beta_k}$ using sparse interpolation with σ_f.
12:     **if** $\sigma_k = $ FAIL **then return** FAIL **end if**
13:     Solve $\sigma = h \bmod q$ and $\sigma = \sigma_k \bmod (x_j - \beta_k)$ for $\sigma \in \mathbb{Z}_p[x_1, x_2, \ldots, x_j]$.
14:     $q \leftarrow q \cdot (x_j - \beta_k)$
15: **until** $\sigma = h$ and $w | (c - \sigma u)$
16: Set $\tau \leftarrow (c - \sigma u)/w$ and **return** (σ, τ).

Suppose $f = \sum_{i=1}^{s} c_i X_i Y_i$ where X_i is a monomial in x_1, x_2, Y_i is a monomial
in $x_3, \ldots, x_n$, $0 \neq c_i \in \mathbb{Z}_p$ and we want to compute

$$f_j := f(x_1, x_2, x_3 = \alpha_3^j, \ldots, x_n = \alpha_n^j), \text{ for } j = 1, \ldots, t.$$

To compute f_j efficiently, one way is to pre-compute the powers of α_i's in $(n-2)$
tables and then do the evaluation using tables. We implemented this first. Let
$d_i = \deg(f, x_i)$ and $d = \max_{3 \leq i \leq n} d_i$. For a fixed j, computing the $n-2$ tables
of powers of α_i^j's (i.e. $1, \alpha_i^j, \alpha_i^{2j}, \ldots, \alpha_i^{d_i j}$) costs $\leq (n-2)d$ multiplications. To
evaluate one term $c_i Y_i$ at $(\alpha_3^j, \ldots, \alpha_n^j)$ costs $n-2$ multiplications using the tables.
Then the cost of evaluating f at $(\alpha_3^j, \ldots, \alpha_n^j)$ is $s(n-2)$ multiplications. Hence
the total cost of t evaluations is bounded above by $C_T = s(n-2)t + (n-2)dt = t(n-2)(s+d)$ multiplications using tables.

However when we use sparse interpolation points of the form $(\alpha_3^j, \ldots, \alpha_n^j)$ for
$j = 1, \ldots, t$ we can reduce the evaluation cost by a factor of $(n-2)$ by a simple
organization. As an example suppose

$$f = x_1^{22} + 72x_1^3 x_2^4 x_4 x_5 + 37x_1 x_2^5 x_3^2 x_4 - 92x_1 x_2^5 x_5^2 + 6x_1 x_2^3 x_3 x_4^2$$

and we want to compute $f_j := f(x_1, x_2, \alpha_3^j, \alpha_4^j, \alpha_5^j)$ for $1 \leq j \leq t$. Before combining and sorting, we write the terms of each f_j as

$$f_j = x_1^{22} + 72\alpha_4^j \alpha_5^j x_1^3 x_2^4 + 37(\alpha_3^j)^2 \alpha_4^j x_1 x_2^5 - 92(\alpha_5^j)^2 x_1 x_2^5 + 6\alpha_3^j (\alpha_4^j)^2 x_1 x_2^3$$
$$= x_1^{22} + 72(\alpha_4 \alpha_5)^j x_1^3 x_2^4 + 37(\alpha_3^2 \alpha_4)^j x_1 x_2^5 - 92(\alpha_5^2)^j x_1 x_2^5 + 6(\alpha_3 \alpha_4^2)^j x_1 x_2^3.$$

Now let

$$c^{(0)} := [1, 72, 37, -92, 6] \quad \text{and} \quad \theta := [1, \alpha_4\alpha_5, \alpha_3^2\alpha_4, \alpha_5^2, \alpha_3\alpha_4^2].$$

Then in a for loop $j = 1, \ldots, t$ we can update the coefficient array $c^{(0)}$ by the monomial array θ by defining $c_i^{(j)} = c_i^{(j-1)}\theta_i$ for $1 \leq i \leq s$ so that each iteration computes the coefficient array

$$c^{(j)} = [1, 72(\alpha_4\alpha_5)^j, 37(\alpha_3^2\alpha_4)^j, -92(\alpha_5^2)^j, 6(\alpha_3\alpha_4^2)^j]$$

using $s = \#f$ multiplications in the coefficient field to obtain

$$f_j = x_1^{22} + 72(\alpha_4\alpha_5)^j x_1^3 x_2^4 + 37(\alpha_3^2\alpha_4)^j x_1 x_2^5 - 92(\alpha_5^2)^j x_1 x_2^5 + 6(\alpha_3\alpha_4^2)^j x_1 x_2^3.$$

Then sorting the monomials and combining terms we get

$$f_j = x_1^{22} + 72(\alpha_4\alpha_5)^j x_1^3 x_2^4 + (37(\alpha_3^2\alpha_4)^j - 92(\alpha_5^2)^j)x_1 x_2^5 + 6(\alpha_3\alpha_4^2)^j x_1 x_2^3.$$

Note that sorting is time consuming so it should be done once at the beginning.

   With the organization described above one evaluates Y_i at $(\alpha_3, \ldots, \alpha_n)$ in $(n - 3)$ multiplications using tables. The cost of $n - 2$ tables of powers is $\leq (n - 2)d$. Then at the first step the cost (of creating θ, the monomial array) is $\leq s(n - 3)$. After that the cost of each evaluation is s multiplications. Hence the total cost is bounded above by $C_N = st + s(n - 3) + (n - 2)d$. Compared with $C_T = s(n - 2)t + (n - 2)dt$ the gain is a factor of $(n - 2)$. Roman Pearce implemented this improved evaluation algorithm in C for us in such a way that from Maple, we can obtain the next evaluation using s multiplications.

4   Sparse Hensel Lifting

4.1   On Kaltofen's SHL

Factoring multivariate polynomials via Sparse Hensel Lifting (SHL) uses the same idea of the sparse interpolation [14]. Following the same notation introduced in Sect. 1, at $(j - 1)^{\text{th}}$ step we have $f_{j-1} = x_1^{df} + c_{j1}M_1 + \cdots + c_{jt_j}M_{t_j}$ where t_j is the number of non-zero terms that appear in f_{j-1}, M_k's are the distinct monomials in $x_1, \ldots, x_{j-1}$ and $c_{jk} \in \mathbb{Z}_p$ for $1 \leq k \leq t_j$. Then at the jth step SHL assumes $f_j = x_1^{df} + \Lambda_{j1}M_1 + \cdots + \Lambda_{jt_j}M_{t_j}$ where for $1 \leq k \leq t_j$,

$$\Lambda_{jk} = c_{jk}^{(0)} + c_{jk}^{(1)}(x_j - \alpha_j) + c_{jk}^{(2)}(x_j - \alpha_j)^2 + \cdots + c_{jk}^{(d_{jk})}(x_j - \alpha_j)^{d_{jk}}$$

with $c_{jk}^{(0)} := c_{jk}$ and where $df = \deg_{x_1}(f)$, $d_{jk} = \deg_{x_n}(\Lambda_{jk})$ with $c_{jk}^{(i)} \in \mathbb{Z}_p$ for $0 \leq i \leq d_{jk}$. The assumption is the same for the factor g_{j-1}.

   To recover f_j from f_{j-1} and g_j from g_{j-1}, during the jth step of MHL (see Algorithm 1 above) one starts with $\sigma_{j0} = f_{j-1}$, $\tau_{j0} = g_{j-1}$, then in a for loop starting from $i = 1$ and incrementing it while the error term and its ith Taylor coefficient is non-zero, by solving MDP's $\sigma_{j0}\tau_{ji} + \tau_{j0}\sigma_{ji} = e_j^{(i)}$

for $1 \le i \le \max(\deg_{x_j}(f_j), \deg_{x_j}(g_j))$. After the loop terminates we have $f_j = \sum_{k=0}^{\deg_{x_j}(f_j)} \sigma_{jk}(x_j - \alpha_j)^k$. On the other hand if the assumption of SHL is true then we have also $f_j = x_1^{df} + (\sum_{i=0}^{d_j} c_{j1}^{(i)}(x_j - \alpha_j)^i)M_1 + \cdots + (\sum_{i=0}^{d_j} c_{jt_j}^{(i)}(x_j - \alpha_j)^i)M_{t_j} = x_1^{df} + \sum_{i=0}^{d_j}(c_{j1}^{(i)}M_1 + \cdots + c_{jt_j}^{(i)}M_{t_j})(x_j - \alpha_j)^i$. Similarly for g_j.

Hence we see that if the assumption of SHL is true then the support of each σ_{jk} will be a subset of support of f_{j-1}. Therefore we can use f_{j-1} as the skeleton of the solution of each σ_{jk}. The same is true for τ_{jk}. Although it is not stated explicitly in [3], this is one of the underlying ideas of Kaltofen's SHL (KHL).

In a classical implementation of MHL, at the jth step in the for loop (see Algorithm 1) one gets the monic factors and then after the loop one applies leading coefficient correction. However in [3] leading coefficient correction is also done in the for loop. If we do leading coefficient correction after the for loop, Kaltofen's SHL idea reduces to solve the MDP by assuming for each $1 \le i \le d_j$, $\sigma_{ji} = u_1 M_1 + \cdots + u_{t_j} M_{t_j}$ and $\tau_{ji} = u_{t_j+1} N_1 + \cdots + u_{t_j+r_j} N_{r_j}$ for unknowns u_k and distinct monomials $M_1, \ldots, M_{t_j}$ and $N_1, \ldots, N_{r_j}$ in $x_1, \ldots, x_{j-1}$. Then by equating coefficients of the monomials appearing on the LHS and the RHS in the MDP equation one gets a linear system in the u_k's. By construction this system is homogeneous.

At the jth step of MHL (see Algorithm 1), throughout the loop σ_{j0} and τ_{j0} remain the same. So, if the SHL assumption is true the assumed solution structures of σ_{ji} and τ_{ji} will remain the same on the LHS and only the RHS of the MDP will change. Hence just before the loop it is sufficient to find $r_j + t_j$ linearly independent equations among $\mathcal{O}(r_j t_j)$ linear equations while keeping track of which monomials they correspond. We call this monomial set Mon, construct the corresponding matrix L, and compute L^{-1}. Then in the for loop, for each i, one simply has to compute the Taylor coefficient of $e_j^{(i)}$ of the error, extract the coefficients from it corresponding to each monomial in Mon, form the related vector v, and then compute $w = L^{-1}v$ to recover u_k's. This improvement makes the algorithm faster by a factor of $\deg(a_j, x_j)$.

We present the jth step of KHL in Algorithm 3. We give an example to show explicitly how it works in Appendix KHL.

Our organization of Kaltofen's approach needs no forward translation to $x_j \mapsto x_j + \alpha_j$ and not back translation $x_j + \alpha_j \mapsto x_j$, and also does not need to define the sets $E_{j-1}^{(i)}$ defined in [3]. This simplifies the algorithm.

Let $B = a_j(x_1, \ldots, x_j + \alpha_j, \alpha_{j+1}, \ldots, \alpha_n)$. Note that if we proceed in the way explained in [3] then for each i in the for loop we should compute

$$B - \left(f_j^{(i-1)} + (\sum_{k=1}^{t_j} u_k M_k)x_j^i\right)\left(g_j^{(i-1)} + (\sum_{k=1}^{r_j+t_j} u_k N_k)x_j^i\right) \tag{1}$$

where $f_j^{(i-1)} = \sum_{k=0}^{i-1}\sigma_{jk}(x_j - \alpha_j)^k$, $g_j^{(i-1)} = \sum_{k=0}^{i-1}\tau_{jk}(x_j - \alpha_j)^k$ and then by expanding (1) we need to form a linearly independent system by equating it with the error. Then we should apply back translation $x_j + \alpha_j \mapsto x_j$.

Algorithm 3. j^{th} step of improved Kaltofen's SHL for $j > 2$.

Input : $a_j \in \mathbb{Z}_p[x_1, \ldots, x_j]$, $f_{j-1}, g_{j-1} \in \mathbb{Z}_p[x_1, \ldots, x_{j-1}]$ and $\alpha_j \in \mathbb{Z}_p$ where a_j, f_{j-1}, g_{j-1} are monic in x_1. Also, $a_j(x_1, \ldots, x_{j-1}, x_j = \alpha_j) = f_{j-1}g_{j-1}$.
Let $f_{j-1} = x_1^{df} + c_{j1}M_1 + \cdots + c_{jt_j}M_{t_j}$ and $g_{j-1} = x_1^{dg} + s_{j1}N_1 + \cdots + s_{jr_j}N_{r_j}$ where $M_1, \ldots, M_{t_j}$, and $N_1, \ldots, N_{r_j}$ are monomials in $x_1, \ldots, x_{j-1}$ and $df = \deg_{x_1} f$ and $dg = \deg_{x_1} g$.
Output : $f_j, g_j \in \mathbb{Z}_p[x_1, \ldots, x_j]$ such that $a_j = f_j g_j$
or FAIL (No such factorization exists.)

1: $(\sigma_{j0}, \tau_{j0}) \leftarrow (f_{j-1}, g_{j-1})$.
2: $(\sigma_j, \tau_j) \leftarrow (\sigma_{j0}, \tau_{j0})$.
3: $monomial \leftarrow 1$.
4: Introduce unknowns $u_1, \ldots, u_{r_j + t_j}$ and $D \leftarrow \sigma_{j0}(u_1 N_1 + \cdots + u_{r_j} N_{r_j}) + \tau_{j0}(u_{r_j+1}M_1 + \cdots + u_{t_j}M_{t_j})$
5: Expand D and collect the coefficients of the monomials in $x_1, \ldots, x_{j-1}$. Each coefficient is a homogeneous linear equation in u_k's.
6: Let S be the array of all these homogeneous equations and Mon be the array of monomials such that S_i is the coefficient of Mon_i in the expansion of D.
7: Find $i_1, \ldots, i_{r_j+t_j}$ such that $E = \{S_{i_1}, \ldots, S_{i_{r_j+t_j}}\}$ is a linearly independent set. Do this choosing equations of the form of $c\,u_k$ for some constant c first.
8: **if** no such E exists **then return** FAIL (SHL assumption is wrong) **end if**
9: Construct the $(r_j + t_j) \times (r_j + t_j)$ matrix L corresponding to the set E such that the unknown u_i corresponds to i^{th} column of L
10: Compute L^{-1}.
11: $error \leftarrow a_j - f_{j-1}\,g_{j-1}$
12: **for** i from 1 to $\deg(a_j, x_j)$ **while** $error \neq 0$ **do**
13:     $monomial \leftarrow monomial \times (x_j - \alpha_j)$
14:     $c \leftarrow$ coefficient of $(x_j - \alpha_j)^i$ in the Taylor expansion of the $error$ about $x_j = \alpha_j$
15:     **if** $c \neq 0$ **then**
16:         **for** k from 1 to $r_j + t_j$ **do**
17:             $v_k \leftarrow$ the coefficient of Mon_{i_k} of the polynomial c
18:         **end for**
19:         $w \leftarrow L^{-1}v$
20:         $\sigma_{ji} \leftarrow \sum_{k=1}^{t_j} w_k M_k$ and $\tau_{ji} \leftarrow \sum_{k=1}^{r_j} w_{k+t_j} N_k$.
21:         $(\sigma_j, \tau_j) \leftarrow (\sigma_j + \sigma_{ji} \times monomial, \tau_j + \tau_{ji} \times monomial)$.
22:         $error \leftarrow a_j - \sigma_j \tau_j$.
23:     **end if**
24: **end for**
25: **if** $error \neq 0$ **then return** FAIL **else return** (σ_j, τ_j) **end if**

We have implemented our improved KHL in Maple. The most time consuming step is the step 7 of Algorithm 3 where one has to find $r_j + t_j$ linearly independent equations out of $\mathcal{O}(r_j t_j)$ linear equations and invert the corresponding matrix. The most obvious way to get the linear system is to start with a set of one equation then add new equations to the set, one at time, until the system has full rank $r_j + t_j$.

To implement this we use Maple's `RowReduce` function which performs in-place Gauss elimination on the input mod p Matrix L. This function is implemented in C and optimized. The time complexity is the time complexity of Gauss elimination $\mathcal{O}((r_j + t_j)^3)$ plus the time for the failed cases, which, according to our experiments, is not negligible. In our experiments we have observed that this approach is effective only when the factors are very sparse. According to our experiments in Sect. 5.2, although our improved version of KHL is significantly faster than that described in [3], it is still slower than Wang's algorithm.

4.2 Our SHL Organization

Before explaining our SHL organization we make the following observation:

Lemma 1. *Let* $f \in \mathbb{Z}_p[x_1, \ldots, x_n]$ *and by* Support(f) *we denote the set of monomials present in* f. *Let* α *be a randomly chosen element in* $\mathbb{Z}_p$ *and* $f = \sum_{i=0}^{d_n} b_i(x_1, \ldots, x_{n-1})(x_n - \alpha)^i$ *be the* $(x_n - \alpha)-$*adic expansion of* f, *where* $d_n = \deg_{x_n} f$. *Then for a given* j *with* $0 \le j < d_n$,

$$\text{Prob}(\text{Support}(b_{j+1}) \not\subseteq \text{Support}(b_j)) \le |\text{Support}(b_{j+1})| \frac{d_n - j}{p - d_n + j + 1}.$$

Proof. For simplicity assume that $p > j$, otherwise we will need to introduce Hasse derivatives but the idea will be the same. We have

$$b_j(x_1, \ldots, x_{n-1}) = \frac{1}{j!} \frac{\partial}{\partial x_n^j} f(x_1, \ldots, x_{n-1}, x_n = \alpha).$$

If we write $f \in \mathbb{Z}_p[x_n][x_1, \ldots, x_{n-1}]$ as

$$f = c_1(x_n)M_1 + c_2(x_n)M_2 + \cdots + c_k(x_n)M_k$$

where $M_1, M_2, \ldots, M_k$ are the distinct monomials in $x_1, \ldots, x_{n-1}$ and we denote $\frac{\partial}{\partial x_n^j} c_i(x_n) = c_i^{(j)}(x_n)$ then

$$b_j = \frac{\partial}{\partial x_n^j} f(x_n = \alpha) = c_1^{(j)}(\alpha)M_1 + c_2^{(j)}(\alpha)M_2 + \cdots + c_k^{(j)}(\alpha)M_k.$$

$$b_{j+1} = \frac{\partial}{\partial x_n^{j+1}} f(x_n = \alpha) = c_1^{(j+1)}(\alpha)M_1 + c_2^{(j+1)}(\alpha)M_2 + \cdots + c_k^{(j+1)}(\alpha)M_k.$$

For a given $j > 0$, if $c_i^{(j+1)}(\alpha) \ne 0$, but $c_i^{(j)}(\alpha) = 0$ then $M_i \notin \text{Support}(b_j)$. We need to compute $\text{Prob}(c_i^{(j)}(\alpha) = 0 \,|\, c_i^{(j+1)}(\alpha) \ne 0)$. If A is the event that $c_i^{(j)}(\alpha) = 0$ and B is the event that $c_i^{(j+1)}(\alpha) = 0$ then

$$\text{Prob}(A \,|\, B^c) = \frac{\text{Prob}(A) - \text{Prob}(B)\text{Prob}(A \,|\, B)}{\text{Prob}(B^c)} \le \frac{\text{Prob}(A)}{\text{Prob}(B^c)}.$$

By the Schwartz-Zippel Lemma [8,13]

$$\frac{\text{Prob}(A)}{\text{Prob}(B^c)} \le \frac{\deg_{x_n}(c_i^{(j)}(y))/p}{1 - (\deg_{x_n}(c_i^{(j+1)}(y))/p)} = \frac{(d_n - j)/p}{1 - (d_n - j - 1)/p} = \frac{d_n - j}{p - d_n + j + 1}.$$

$\square$

Lemma 1 shows that for the sparse case, if p is big enough then the probability of Support(b_{j+1}) $\subseteq$ Support(b_j) is high.

Following the notation of Lemma 1 above, for a given $\alpha \in \mathbb{Z}_p$, let us call α unlucky, if Support(b_{j+1}) $\not\subseteq$ Support(b_j) for some $0 \leq j < d_n$. So, for a given f, if $c_i^{(j)}$ has a root but does not have a double root at $x_n = \alpha$, then α is unlucky for b_{j+1}, i.e. Support(b_{j+1}) $\not\subseteq$ Support(b_j): Consider the following example where Support(b_{j+1}) $\not\subseteq$ Support(b_j) for $j = 1, 2$.

$$f := (x_1^6 + x_1^5 + x_1^4)(x_2 - 1)^3 + (x_1^5 + x_1^4 + x_1^3)(x_2 - 1) + x_1^7 + 1 \in \mathbb{Z}_{509}[x_1, x_2].$$

But if we choose another point 301 and compute the $(x_2 - 301)-$adic expansion of $f = \sum_{i=0}^{3} b_i(x_1)(x_2 - 301)^i$ we have

$$b_0 = x_1^7 + 95x_1^6 + 395x_1^5 + 395x_1^4 + 300x_1^3 + 1$$
$$b_1 = 230x_1^6 + 231x_1^5 + 231x_1^4 + x_1^3$$
$$b_2 = 391x_1^6 + 391x_1^5 + 391x_1^4$$
$$b_3 = x_1^6 + x_1^5 + x_1^4$$

and we see that Support(b_{j+1}) $\subseteq$ Support(b_j) for $0 \leq j \leq 2$. In fact for this example $\alpha = 1, 209, -207$ are the only unlucky points as can be seen by considering $f \in \mathbb{Z}_{509}[x_2][x_1]$, that is,

$$f = x_1^7 + (x_2 - 1)^3 x_1^6 + (x_2 - 209)(x_2 - 1)(x_2 + 207)x_1^5$$
$$+ (x_2 - 209)(x_2 - 1)x_1^4 + (x_2 - 1)x_1^3 + 1.$$

Note that these points are unlucky only for b_2. Before we give an upper bound for the number of unlucky points we consider the following example. Let $p = 1021$,

$$f = (x_2 - 841)(x_2 - 414)(x_2 - 15)(x_2 - 277)x_1^9$$
$$+ (x_2 - 339)(x_2 - 761)(x_2 - 752)(x_2 - 345)x_1^7$$

and $f^{(i)} = \frac{\partial}{\partial x_2^i} f(x_1, x_2)$. Then

$$f^{(1)} = 4(x_2 - 384)(x_2 - 230)(x_2 - 291)x_1^9 + 4(x_2 - 441)(x_2 + 127)(x_2 + 453)x_1^7$$
$$f^{(2)} = 12(x_2 - 89)(x_2 - 174)x_1^9 + 12(x_2 - 473)(x_2 - 115)x_1^7$$
$$f^{(3)} = (24x_2 - 93)x_1^9 + (24x_2 + 91)x_1^7 \text{ and}$$
$$f^{(4)} = 24x_1^9 + 24x_1^7.$$

So, the maximum number of unlucky points occurs if each $c_i^{(j)}$ splits for different points, hence $|\text{Support}(f)| \frac{d_n(d_n+1)}{p}$ is an upper bound for the probability of hitting an unlucky point. For a sparse polynomial with 1000 terms, $d_n = 20$, for $p = 2^{31} - 1$, this probability is 0.000097. This observation suggests that we use $\sigma_{i,j-1}$ (or $\tau_{i,j-1}$) as a form of the solution of σ_{ji} (or τ_{ij}).

Back to our discussion on SHL, based on the observation above the j^{th}step ($j > 1$) of our SHL organization is summarized in Algorithm 4. In Appendix SHL we give a concrete example to show how it works.

Algorithm 4. j^{th} step of Sparse Hensel Lifting for $j > 1$.

Input : $a_j \in \mathbb{Z}_p[x_1,\dots,x_j]$, $f_{j-1}, g_{j-1} \in \mathbb{Z}_p[x_1,\dots,x_{j-1}]$ and $\alpha_j \in \mathbb{Z}_p$ where a_j, f_{j-1}, g_{j-1} are monic in x_1. Also, $a_j(x_1,\dots,x_{j-1}, x_j = \alpha_j) = f_{j-1}g_{j-1}$.

Output : $f_j, g_j \in \mathbb{Z}_p[x_1,\dots,x_j]$ such that $a_j = f_j g_j$ or FAIL (No such factorization exists.)

1: **if** $r_j > t_j$ **then** interchange f_{j-1} with g_{j-1} **end if**
2: $(\sigma_{j0}, \tau_{j0}) \leftarrow (f_{j-1}, g_{j-1})$.
3: $(\sigma_j, \tau_j) \leftarrow (\sigma_{j0}, \tau_{j0})$.
4: $monomial \leftarrow 1$.
5: $error \leftarrow a_j - f_{j-1} g_{j-1}$
6: **for** i from 1 to $\deg(a_j, x_j)$ **while** $error \neq 0$ **do**
7:      $monomial \leftarrow monomial \times (x_j - \alpha_j)$
8:      $c \leftarrow$ coefficient of $(x_j - \alpha_j)^i$ in the Taylor expansion of the $error$ about $x_j = \alpha_j$
9:      **if** $c \neq 0$ **then**
10:          $\sigma_g \leftarrow$ skeleton of $\tau_{j,i-1}$
11:          Solve the MDP $\sigma_{j0}\tau_{ji} + \tau_{j0}\sigma_{ji} = c$ for σ_{ji} and τ_{ji} in $\mathbb{Z}_p[x_1,\dots,x_{j-1}]$
12:          using σ_g and our sparse interpolation from Sect. 3.2.
13:          **if** (σ_{ji}, τ_{ji})=FAIL **then**
14:              $(\sigma_{ji}, \tau_{ji}) \leftarrow$ **BSDiophant**$(\sigma_{j0}, \tau_{j0}, c, p)$
15:              **if** (σ_{ji}, τ_{ji})=FAIL **then** restart the factorization with a different ideal
16:          **end if**
17:          **end if**
18:          $(\sigma_j, \tau_j) \leftarrow (\sigma_j + \sigma_{ji} \times monomial, \tau_j + \tau_{ji} \times monomial)$.
19:          $error \leftarrow a_j - \sigma_j\tau_j$.
20:      **end if**
21: **end for**
22: **if** $error \neq 0$ **then return** FAIL (No such factorization exists)
23: **else return** (σ_j, τ_j)
24: **end if**

4.3 Some Remarks on Algorithm 4

Step 8 in the for loop computes the ith Taylor coefficient of the error at $x_j = \alpha_j$. Maple used to compute this using the formula $c = g(x_j = \alpha_j)/i!$ where g is the i'th derivative of $error$ wrt x_j. Instead, Maple now uses the more direct formula $c = \sum_{k=i}^{d} \text{coeff}(error, x_j^k)\alpha_j^{k-i}\binom{k}{i}$ where $d = \deg_{x_j} error$ which is three times faster [7].

At step 10 Algorithm 4 makes the assumption Support$(\tau_{ji}) \subseteq$ Support$(\tau_{j,i-1})$ based on Lemma 1. Note that, if the minimum of the number of the terms of each factor of a_j is $t = \min(\#f_j, \#g_j)$, then at step 11 the probability of the failure of the assumption is $\leq t\frac{d_j - i}{p - d_j - (i-1)} \leq \frac{td_j}{p - 2d_j}$ and its cost is the evaluation cost + cost of a system of linear equation solving which is bounded above by $\mathcal{O}(t^2)$. Another costly operation is the cost of multivariate division, $\sigma_{ji} = (c - \sigma_{j0}\tau_{ji})/\tau_{j0}$, which is hidden in sparse interpolation. If the algorithm fails to compute (σ_{ji}, τ_{ji}) at step 11 then it passes to a safe way at step 14.

Another expensive operation in the Algorithm 4 is the error computation, $error \leftarrow a_j - \sigma_j \tau_j$, in the for loop. To decrease this cost, one of the ideas in [5] can be generalized to MHL. Let $\sigma_j = \sum_{s=0}^{\deg_{x_j} \sigma_j} \sigma_{j,s}(x_j - \alpha_j)^s$ and $\sigma_j^{(i)} = \sum_{s=0}^{i} \sigma_{j,s}(x_j - \alpha_j)^s$ (similarly for τ). One has

$$
\begin{aligned}
e_j^{(i+1)} &= a_j - \sigma_j^{(i)} \tau_j^{(i)} \\
&= a_j - (\sigma_j^{(i-1)} + \sigma_{j,i}(x_j - \alpha_j)^i)(\tau_j^{(i-1)} + \tau_{j,i}(x_j - \alpha_j)^i) \\
&= a_j - \sigma_j^{(i-1)} \tau_j^{(i-1)} - (\sigma_j^{(i-1)} \tau_{j,i} + \tau_j^{(i-1)} \sigma_{j,i})(x_j - \alpha_j)^i \\
&= e_j^{(i)} - U^{(i)}(x_j - \alpha_j)^i
\end{aligned}
$$

where $U^{(i)} := (\sigma_j^{(i-1)} \tau_{j,i} + \tau_j^{(i-1)} \sigma_{j,i})$. Hence in the for loop we have the relation $e_j^{(i+1)} = e_j^{(i)} - (x_j - \alpha_j)^i U^{(i)}$ for a correction term $U^{(i)} \in \mathcal{O}((x_j - \alpha_j)^{i-1})$. Also for $i \geqslant 0$ it is known that $(x_j - \alpha_j)^i$ divides $e_j^{(i)}$. So if we define $c_j^{(i)} := e_j^{(i)}/(x_j - \alpha_j)^i$ then $c_j^{(i)}$ can be computed efficiently using

$$
c_j^{(i+1)} = (c_j^{(i)} - U^{(i)})/(x_j - \alpha_j).
$$

Hence we may compute $c_j^{(i)}$ for $i = 1, 2, \ldots$ until it becomes zero instead of computing $e_j^{(i)}$. According to our experiments, this observation decreases the cost when the number of factors is 2. For more than 2 factors, the generalization of it does not bring a significant advantage. So, in our implementations we only use this update formula when the number of factors is 2.

Also note that, in our SHL organization (Algorithm 4), we use only one of the SHL assumptions and eliminate the recursive step in MHL to compute the skeleton of the solution. In Kaltofen's approach one cannot focus on some subset of the u_k's as we do, since the system of equations are coupled.

5   Some Timing Data

To compare the result of our ideas with Wang's, first we factored the determinants of Toeplitz and Cyclic matrices of different sizes as concrete examples. Note that the factors in these concrete examples are not sparse. Our results are presented in Sect. 5.1. Then we created sparse random polynomials A, B using

$$
x_1^d + \texttt{randpoly}([x_2, .., x_n], \texttt{degree} = d, \texttt{terms} = t)
$$

in Maple and computed $C = AB \in R$. Note that we chose monic factors in x_1 so as not to complicate the algorithm with leading coefficient correction and to have a fair comparison with Maple's factorization algorithm. We used $p = 2^{31} - 1$ and two ideal types for factoring C: ideal type 1: $I = \langle x_2 - 0, x_3 - 0, \cdots, x_n - 0 \rangle$ and ideal type 2: $I = \langle x_2 - \alpha_1, x_3 - \alpha_2, \cdots, x_n - \alpha_n \rangle$ where the α_i's in practice are small. However for sparse Hensel liftings, as explained in Sect. 4, it is important that α_i's should be chosen from a large interval. For these we chose α_i's randomly from $\mathbb{Z}_q - \{0\}$ with $q = 65521$. Our results are presented in Sect. 5.2. In Sect. 5.2 we also included the ideal type 1 case since according to our experiments it is the only case where Wang's algorithm is faster. This is because a sparse polynomial remains sparse for the ideal type 1 and hence the number of MDP's to be solved significantly decreases and the evaluation cost of sparse interpolation becomes dominant which is not the case for Wang's algorithm for the ideal type 1 case. However it is not always possible to use ideal type 1. For example, ideal type 1 cannot be used to factor Cyclic or Toeplitz determinants.

In the tables below all timings are in CPU seconds and are for the Hensel liftings part of the polynomial factorization. They were obtained on an Intel Core i5–4670 CPU running at 3.40 GHz.

tW is the time for Wang's algorithm which Maple currently uses (see [2]),
tUW is the time for Wang's algorithm with the improved Hensel,
tS is the time for Zippel's sparse interpolation from Sect. 3.1,
tBS is the time for the improved sparse interpolation from Sect. 3.2,
tKHL is the time for the Kaltofen's sparse Hensel lifting from Sect. 4.1,
tNBS is the time for the sparse Hensel lifting from Sect. 4.2,
tX(tY) means factoring time tX with tY seconds spent solving MDPs.

5.1 Factoring the Determinants of Cyclic and Toeplitz Matrices

Let C_n denote the $n \times n$ cyclic matrix and let T_n denote the $n \times n$ symmetric Toeplitz matrix below.

$$
C_n = \begin{pmatrix} x_1 & x_2 & \cdots & x_{n-1} & x_n \\ x_n & x_1 & \cdots & x_{n-2} & x_{n-1} \\ \vdots & \vdots & \vdots & \vdots & \vdots \\ x_3 & x_4 & \cdots & x_1 & x_2 \\ x_2 & x_3 & \cdots & x_n & x_1 \end{pmatrix} \text{ and } T_n = \begin{pmatrix} x_1 & x_2 & \cdots & x_{n-1} & x_n \\ x_2 & x_1 & \cdots & x_{n-2} & x_{n-1} \\ & \ddots & \ddots & \ddots & \\ x_{n-1} & x_{n-2} & \cdots & x_1 & x_2 \\ x_n & x_{n-1} & \cdots & x_2 & x_1 \end{pmatrix}
$$

The determinants of C_n and T_n are polynomials in n variables $x_1, x_2, \ldots, x_n$ which factor. For $n > 1$ $\det(T_n)$ has 2 factors and $x_1 + x_2 + \cdots + x_n$ is a factor

Table 1. Timings (CPU seconds) for factoring determinants of $n \times n$ cyclic matrices.

n	tW	tUW	$tKHL$	tS	tBS	$tNBS$
5	0.004 (0.003)	0.014 (0.013)	0.07 (0.068)	0.014 (0.003)	0.015 (0.012)	0.014 (0.012)
7	0.057 (0.054)	0.054 (0.04)	1157.(1157.)	0.018 (0.006)	0.019 (0.014)	0.017 (0.014)
10	0.912 (0.666)	-	-	1.049 (0.823)	0.775 (0.549)	0.434 (0.179)
11	9.437 (8.785)	8.413 (8.107)	∞	0.503 (0.23)	0.505 (0.226)	0.354 (0.071)
12	42.64 (38.38)	-	-	7.705 (4.35)	7.288 (3.913)	4.372 (1.047)
13	258.5 (208.9)	256.5 (208.9)	∞	20.40 (8.936)	20.05 (8.408)	13.78 (1.697)

of C_n. Table 1 presents timings for Hensel liftings in CPU seconds to factor $\det C_n$. For $n = 6, 10, 12$ the number of factors is 3,4 and 6 respectively. For $n = 5, 7, 11, 13$ the number of factors is 2. We didn't implement KHL to factor more than 2 factors. This is why we didn't include the timing for KHL for the case $n = 6$. As can be seen from the data below KHL is not effective for $n \geq 7$. Table 2 presents timings for Hensel liftings in CPU seconds to factor $\det T_n$.

Table 2. Timings for factoring determinants of $n \times n$ symmetric Toeplitz matrices.

n	tW	tUW	$tKHL$	tS	tBS	$tNBS$
5	0.003 (0.002)	0.014 (0.001)	0.02 (0.018)	0.014 (0.014)	0.017 (0.017)	0.015 (0.012)
6	0.016 (0.013)	0.016 (0.005)	0.308 (0.306)	0.04 (0.026)	0.042 (0.031)	0.021 (0.008)
7	0.025 (0.012)	0.044 (0.029)	1157.5(1157.5)	0.031 (0.019)	0.032 (0.019)	0.045 (0.03)
8	0.057 (0.044)	0.072 (0.052)	119.88(119.86)	0.103 (0.086)	0.096 (0.087)	0.059 (0.026)
9	0.167 (0.126)	0.151 (0.123)	486.45(486.41)	0.279 (0.258)	0.194 (0.168)	0.088 (0.06)
10	0.654 (0.461)	0.629 (0.496)	25021.(25021.)	1.389 (1.245)	0.675 (0.531)	0.366 (0.222)
11	2.699 (2.06)	2.538 (2.11)	∞	7.612 (7.109)	2.677 (1.751)	1.133 (0.589)
12	25.93 (18.68)	23.07 (17.95)	∞	69.91 (65.8)	22.08 (15.72)	13.86 (7.579)
13	48.59 (37.43)	47.01 (37.73)	∞	508.3 (495.8)	48.86 (36.11)	32.81 (20.36)

Table 3. The timing table for random data with ideal type 1

$n/d/t$	tW	tUW	tS	tBS
3/35/100	0.11 (0.06)	0.10 (0.06)	0.17 (0.13)	0.07 (0.03)
3/35/500	0.39 (0.16)	0.44 (0.17)	0.60 (0.36)	0.31 (0.08)
5/35/100	0.183 (0.15)	0.18 (0.15)	0.46 (0.43)	0.72 (0.69)
5/35/500	1.42 (0.53)	2.61 (0.51)	5.25 (2.92)	5.05 (2.68)
7/35/100	0.18 (0.16)	0.18 (0.15)	0.79 (0.76)	1.05 (1.02)
7/35/500	1.48 (0.71)	2.36 (0.65)	12.44 (10.43)	7.77 (5.61)

Table 4. The timing table for random data with ideal type 2

$n/d/t$	tW	tUW	$tKHL$	tS	tBS	$tNBS$
3/35/100	2.87 (1.88)	2.14 (1.88)	0.401 (0.046)	0.65 (0.38)	0.38 (0.08)	0.32 (0.04)
3/35/500	5.77 (3.69)	4.30 (3.57)	1.957 (0.057)	1.36 (0.61)	0.90 (0.14)	0.81 (0.05)
5/35/100	88.10 (86.28)	86.45 (85.64)	3.337 (2.551)	6.12 (5.21)	5.04 (4.11)	1.16 (0.36)
5/35/500	472.1 (402.5)	392.2 (370.7)	3732. (3717.)	67.57 (45.98)	48.1 (25.5)	26.0 (4.86)
6/35/100	309.1 (306.3)	323.8 (322.6)	4.383 (3.409)	12.53 (11.42)	9.29 (7.11)	1.49 (0.46)
7/35/100	800.0 (797.0)	829.7 (828.5)	10.22 (9.134)	16.82 (15.15)	10.8 (9.77)	1.58 (0.59)

5.2   Random Data

Table 3 below presents timings for the random data where ideal type 1 is used. For the ideal type 1 case SHL is not used, since the zero evaluation probability is high for the sparse case. Table 4 below presents timings for the random data where ideal type 2 is used. As can be seen KHL is effective only when the factors have 100 terms or less.

6   Conclusion

We have shown that solving the multivariate polynomial diophantine equations in sparse Hensel lifting algorithm can be improved by using sparse interpolation. This leads to an overall improvement in multivariate polynomial factorization. Our experiments show that the improvement is practical.

Appendix KHL

Suppose we seek to factor $a = fg$ where $f = x_1{}^5 + 3\,x_1{}^2x_2x_3{}^2 - 7\,x_1{}^4 - 4\,x_1x_3 + 1$ and $g = x_1{}^5 + x_1{}^2x_2x_3 - 7\,x_3{}^4 - 6$. Let $\alpha_3 = 2$ and $p = 2^{31} - 1$. Before lifting we have a and

$$f^{(0)} := f(x_3 = 2) = x_1{}^5 - 7\,x_1{}^4 + 12\,x_1{}^2x_2 - 8\,x_1 + 1$$
$$g^{(0)} := g(x_3 = 2) = x_1{}^5 + 2\,x_1{}^2x_2 - 118.$$

If the assumption of SHL is true then we assume that $f = \sum_{i=0}^{\deg_{x_3} f} f_i(x_3 - 2)^i$ and $g = \sum_{i=0}^{\deg_{x_3} g} g_i(x_3 - 2)^i$ where each f_i and g_i are in the form

$$f_i = c_1x_1{}^4 + c_2x_1{}^2x_2 + c_3x_1 + c_4 \quad \text{and} \quad g_i = c_5x_1{}^2x_2 + c_6$$

for some unknowns $C = \{c_1, c_2, c_3, c_4, c_5, c_6\}$. First we construct

$$D = \left(x_1{}^5 - 7\,x_1{}^4 + 12\,x_1{}^2x_2 - 8\,x_1 + 1\right)\left(c_5x_1{}^2x_2 + c_6\right)$$
$$+ \left(x_1{}^5 + 2\,x_1{}^2x_2 - 118\right)\left(c_1x_1{}^4 + c_2x_1{}^2x_2 + c_3x_1 + c_4\right).$$

Expanding D we see the system of homogeneous linear equations as coefficients

$$D = c_1x_1{}^9 + (c_2 + c_5)\,x_1{}^7x_2 + (2\,c_1 - 7\,c_5)\,x_1{}^6x_2 + c_3x_1{}^6$$
$$+ (c_4 + c_6)\,x_1{}^5 + (2\,c_2 + 12\,c_5)\,x_1{}^4x_2{}^2 + (-118\,c_1 - 7\,c_6)\,x_1{}^4$$
$$+ (2\,c_3 - 8\,c_5)\,x_1{}^3x_2 + (-118\,c_2 + 2\,c_4 + c_5 + 12\,c_6)\,x_1{}^2x_2$$
$$+ (-118\,c_3 - 8\,c_6)\,x_1 - 118\,c_4 + c_6$$

We need 6 linearly independent equations from these. First we check whether there are single equations. In this example we see that c_1 and c_3 corresponding to monomials $x_1{}^9, x_1{}^6$. Then we go over the equations one by one to get a rank 6 system. In this example we see that equations corresponding to the set $Mon = \{x_1{}^9, x_1{}^6, x_1{}^7 x_2, x_1{}^6 x_2, x_1{}^5, x_1{}^4\}$ are linearly independent. We obtain

$$L = \begin{bmatrix} 1 & 0 & 0 & 0 & 0 & 0 \\ 0 & 0 & 1 & 0 & 0 & 0 \\ 0 & 1 & 0 & 0 & 1 & 0 \\ 2 & 0 & 0 & 0 & -7 & 0 \\ 0 & 0 & 0 & 1 & 0 & 1 \\ -118 & 0 & 0 & 0 & 0 & -7 \end{bmatrix}$$

and compute L^{-1}. In the following $e_3^{(k)}$ denotes the coefficient of $(x_3 - 2)^k$ in the Taylor expansion of the *error* about $x_3 = 2$. Let also $f_0 := f^{(0)}, g_0 := g^{(0)}, f^{(k)} := \sum_{i=0}^{k} f_i (x_3 - 2)^i, g^{(k)} := \sum_{i=0}^{k} g_i (x_3 - 2)^i$.

In Algorithm 3 v is the vector constructed by extracting the coefficients of $e_3^{(k)}$ corresponding to monomials in $Mon = \{x_1{}^9, x_1{}^6, x_1{}^7 x_2, x_1{}^6 x_2, x_1{}^5, x_1{}^4\}$ and $w = L^{-1} v \bmod p$. Now for the loop,

Step $i = 1$: error $= a - f^{(0)} g^{(0)}$

$$e_3^{(1)} = 13 x_1{}^7 x_2 - 7 x_1{}^6 x_2 - 4 x_1{}^6 + 36 x_1{}^4 x_2{}^2 - 224 x_1{}^5$$
$$+1568 x_1{}^4 - 16 x_1{}^3 x_2 - 4103 x_1{}^2 x_2 + 2264 x_1 - 224$$
$$v = \begin{bmatrix} 0 & -4 & 13 & -7 & -224 & 1568 \end{bmatrix}$$
$$w = L^{-1} v = \begin{bmatrix} 0 & 12 & -4 & 0 & 1 & -224 \end{bmatrix}$$
$$f^{(1)} = f^{(0)} + \left(12 x_1{}^2 x_2 - 4 x_1\right)(x_3 - 2)$$
$$= x_1{}^5 - 7 x_1{}^4 + 12 x_1{}^2 x_2 x_3 - 12 x_1{}^2 x_2 - 4 x_1 x_3 + 1$$
$$g^{(1)} = g^{(0)} + \left(x_1{}^2 x_2 - 224\right)(x_3 - 2) = x_1{}^5 + x_1{}^2 x_2 x_3 - 224 x_3 + 330$$

Step $i = 2$: error $= a - f^{(1)} g^{(1)}$

$$e_3^{(2)} = 3 x_1{}^7 x_2 + 6 x_1{}^4 x_2{}^2 - 168 x_1{}^5 + 1176 x_1{}^4 - 2370 x_1{}^2 x_2 + 1344 x_1 - 168$$
$$v = \begin{bmatrix} 0 & 0 & 3 & 0 & -168 & 1176 \end{bmatrix}$$
$$w = L^{-1} v = \begin{bmatrix} 0 & 3 & 0 & 0 & 0 & -168 \end{bmatrix}$$
$$f^{(2)} = f^{(1)} + 3 x_1{}^2 x_2 (x_3 - 2)^2 = x_1{}^5 + 3 x_1{}^2 x_2 x_3{}^2 - 7 x_1{}^4 - 4 x_1 x_3 + 1$$
$$g^{(2)} = g^{(1)} - 168(x_3 - 2)^2 = x_1{}^5 + x_1{}^2 x_2 x_3 - 168 x_3{}^2 + 448 x_3 - 342$$

Step $i = 3$: error $= a - f^{(2)} g^{(2)}$

$$e_3^{(3)} = -56 x_1{}^5 + 392 x_1{}^4 - 672 x_1{}^2 x_2 + 448 x_1 - 56$$
$$v = \begin{bmatrix} 0 & 0 & 0 & 0 & -56 & 392 \end{bmatrix}$$
$$w = L^{-1} v = \begin{bmatrix} 0 & 0 & 0 & 0 & 0 & -56 \end{bmatrix}$$
$$f^{(3)} = f^{(2)} + 0 = x_1{}^5 + 3 x_1{}^2 x_2 x_3{}^2 - 7 x_1{}^4 - 4 x_1 x_3 + 1$$
$$g^{(3)} = g^{(2)} - 56 (x_3 - 2)^3 = x_1{}^5 + x_1{}^2 x_2 x_3 - 56 x_3{}^3 + 168 x_3{}^2 - 224 x_3 + 106$$

At the end of the 3rd iteration we have recovered f actually and so we could obtain $g = a/f$ via trial division and terminate. But let's go further.

Step $i = 4$: error $= a - f^{(3)}g^{(3)}$

$$e_3^{(4)} = -7\,x_1{}^5 + 49\,x_1{}^4 - 84\,x_1{}^2x_2 + 56\,x_1 - 7$$
$$v = \begin{bmatrix} 0\ 0\ 0\ 0\ -7\ 49 \end{bmatrix}$$
$$w = L^{-1}v = \begin{bmatrix} 0\ 0\ 0\ 0\ 0\ -7 \end{bmatrix}$$
$$f^{(4)} = f^{(3)} + 0 = x_1{}^5 + 3\,x_1{}^2x_2x_3{}^2 - 7\,x_1{}^4 - 4\,x_1x_3 + 1$$
$$g^{(4)} = g^{(3)} - 7\,(x_3 - 2)^4 = x_1{}^5 + x_1{}^2x_2x_3 - 7\,x_3{}^4 - 6$$

for $i = 5$, error $= a - f^{(4)}g^{(4)} = 0$ and we have the factors!

Appendix SHL

We give an example of our SHL. Suppose we seek to factor $a = fg$ where

$$f = x_1{}^8 + 2\,x_1x_2{}^2x_4{}^3x_5 + 4\,x_1x_2{}^2x_3{}^3 + 3\,x_1x_2{}^2x_4x_5{}^2 + x_2{}^2x_3x_4 - 5$$
$$g = x_1{}^8 + 3\,x_1{}^2x_2x_3x_4{}^2x_5 + 5\,x_1{}^2x_2x_3{}^2x_4 - 3\,x_4{}^2x_5{}^2 + 4\,x_5$$

Let $\alpha_3 = 1, p = 2^{31} - 1$. Before lifting x_5 we have

$$f^{(0)} := f(x_5 = 1) = x_1{}^8 + 4\,x_1x_2{}^2x_3{}^3 + 2\,x_1x_2{}^2x_4{}^3 + 3\,x_1x_2{}^2x_4 + x_2{}^2x_3x_4 - 5$$
$$g^{(0)} := g(x_5 = 1) = x_1{}^8 + 5\,x_1{}^2x_2x_3{}^2x_4 + 3\,x_1{}^2x_2x_3x_4{}^2 - 3\,x_4{}^2 + 4$$

satisfying $a(x_5 = \alpha_5) = f^{(0)}g^{(0)}$. If the SHL assumption is true then at the first step we assume $f = \sum_{i=0}^{\deg_{x_5} f} f_i(x_5 - 1)^i$ and $g = \sum_{i=0}^{\deg_{x_5} g} g_i(x_5 - 1)^i$ where f_1 and g_1 are in the form

$$f_1 = \left(c_1x_3{}^3 + c_2x_4{}^3 + c_3x_4\right)x_1x_2{}^2 + c_4x_2{}^2x_3x_4 + c_5$$
$$g_1 = \left(c_6x_3{}^2x_4 + c_7x_3x_4{}^2\right)x_1{}^2x_2 + c_8x_4{}^2 + c_9$$

for some unknowns $C = \{c_1, \ldots, c_9\}$. In the following $e_5^{(k)}$ denotes the coefficient of $(x_5 - 1)^k$ in the Taylor expansion of the *error* about $x_5 = 1$. Let also $f_0 := f^{(0)}$, $g_0 := g^{(0)}$, $f^{(k)} := \sum_{i=0}^{k} f_i(x_5 - 1)^i$ and $g^{(k)} := \sum_{i=0}^{k} g_i(x_5 - 1)^i$.

We start by computing the first error term $e_5^{(1)} = a - f^{(0)}g^{(0)}$. We obtain

$$\begin{aligned}
e_5^{(1)} &= 3\,x_1{}^{10}x_2x_3x_4{}^2 + 2\,x_1{}^9x_2{}^2x_4{}^3 + 6\,x_1{}^9x_2{}^2x_4 + 12\,x_1{}^3x_2{}^3x_3{}^4x_4{}^2 - 6\,x_1{}^8x_4{}^2 \\
&\quad + 10\,x_1{}^3x_2{}^3x_3{}^2x_4{}^4 + 12\,x_1{}^3x_2{}^3x_3x_4{}^5 + 30\,x_1{}^3x_2{}^3x_3{}^2x_4{}^2 + 27\,x_1{}^3x_2{}^3x_3x_4{}^3 \\
&\quad + 3\,x_1{}^2x_2{}^3x_3{}^2x_4{}^3 + 4\,x_1{}^8 - 24\,x_1x_2{}^2x_3{}^3x_4{}^2 - 18\,x_1x_2{}^2x_4{}^5 - 15\,x_1{}^2x_2x_3x_4{}^2 \\
&\quad + 16\,x_1x_2{}^2x_3{}^3 - 20\,x_1x_2{}^2x_4{}^3 - 6\,x_2{}^2x_3x_4{}^3 + 36\,x_1x_2{}^2x_4 + 4\,x_2{}^2x_3x_4 + 30\,x_4{}^2 - 20.
\end{aligned}$$

The MDP to be solved is:

$$D := f_0 \left(\left(c_6 x_3{}^2 x_4 + c_7 x_3 x_4{}^2 \right) x_1{}^2 x_2 + c_8 x_4{}^2 + c_9 \right)$$
$$+ g_0 \left(\left(c_1 x_3{}^3 + c_2 x_4{}^3 + c_3 x_4 \right) x_1 x_2{}^2 + c_4 x_2{}^2 x_3 x_4 + c_5 \right) = e_5^{(1)}.$$

Our aim is first to get $\left(\left(c_6 x_3{}^2 x_4 + c_7 x_3 x_4{}^2 \right) x_1{}^2 x_2 + c_8 x_4{}^2 + c_9 \right)$ since it will create a smaller matrix. For sparse interpolation we need 2 evaluations only: we choose $[x_3 = 2, x_4 = 3]$ and $[x_3 = 2^2, x_4 = 3^2]$ and compute $D([x_3 = 2, x_4 = 3])$:

$$\left(x_1{}^8 + 95\, x_1 x_2{}^2 + 6\, x_2{}^2 - 5 \right) \left((12\, c_6 + 18\, c_7)\, x_1{}^2 x_2 + 9\, c_8 + c_9 \right)$$
$$+ \left(x_1{}^8 + 114\, x_1{}^2 x_2 - 23 \right) \left((8\, c_1 + 27\, c_2 + 3\, c_3)\, x_1 x_2{}^2 + 6\, c_4 x_2{}^2 + c_5 \right)$$
$$= 54\, x_1{}^{10} x_2 + 72\, x_1{}^9 x_2{}^2 - 50\, x_1{}^8 + 13338\, x_1{}^3 x_2{}^3 + 324\, x_1{}^2 x_2{}^3 - 270\, x_1{}^2 x_2$$
$$- 6406\, x_1 x_2{}^2 - 300\, x_2{}^2 + 250$$

and $D([x_3 = 4, x_4 = 9])$. Calling BDP to solve these bivariate Diophantine equations we obtain the solutions $[\sigma_1, \tau_1] = [54\, x_1{}^2 x_2 - 50, 72\, x_1 x_2{}^2]$ and $[\sigma_2, \tau_2] = [972\, x_1{}^2 x_2 - 482, 1512\, x_1 x_2{}^2]$. Hence we have

$$(12\, c_6 + 18\, c_7)\, x_1{}^2 x_2 + 9\, c_8 + c_9 = 54\, x_1{}^2 x_2 - 50$$
$$(144\, c_6 + 324\, c_7)\, x_1{}^2 x_2 + 81\, c_8 + c_9 = 972\, x_1{}^2 x_2 - 482$$

Then we solve the Vandermonde linear systems

$$\begin{bmatrix} 12 & 18 \\ 144 & 324 \end{bmatrix} \begin{bmatrix} c_6 \\ c_7 \end{bmatrix} = \begin{bmatrix} 54 \\ 972 \end{bmatrix} \text{ and } \begin{bmatrix} 9 & 1 \\ 81 & 1 \end{bmatrix} \begin{bmatrix} c_8 \\ c_9 \end{bmatrix} = \begin{bmatrix} -50 \\ -482 \end{bmatrix}$$

to obtain $c_6 = 0, c_7 = 3, c_8 = -6, c_9 = 4$. So $g_1 = 3\, x_1{}^2 x_2 x_3 x_4{}^2 - 6\, x_4{}^2 + 4$. Then by division we get $f_1 = (e_5^{(1)} - f_0 g_1)/g_0 = 2\, x_1 x_2 x_4{}^3 + 8\, x_2 x_3{}^4$. Hence

$$f^{(1)} = f_0 + \left(2\, x_1 x_2{}^2 x_4{}^3 + 6\, x_1 x_2{}^2 x_4 \right) (x_5 - 1)$$
$$g^{(1)} = g_0 + \left(3\, x_1{}^2 x_2 x_3 x_4{}^2 - 6\, x_4{}^2 + 4 \right) (x_5 - 1).$$

Note that we use the division step above also as a check for the correctness of the SHL assumption. Since the solution to the MDP is unique, we would have $g_0 \nmid (e_5^{(1)} - f_0 g_1)$, if the assumption was wrong.

Now following Lemma 1 by looking at the monomials of f_1 and g_1, we assume that the form of the f_2 and g_2 are

$$f_2 = c_1 x_1 x_2{}^2 x_4{}^3 + c_2 x_1 x_2{}^2 x_4 + c_3$$
$$g_2 = c_4 x_1{}^2 x_2 x_3 x_4{}^2 + c_5 x_4{}^2 + c_6$$

for some unknowns $C = \{c_1, \ldots, c_6\}$. After computing the next error $a - f^{(1)} g^{(1)}$ we compute $e_5^{(2)}$ and the MDP to be solved is:

$$D := f_0 \left(c_4 x_1{}^2 x_2 x_3 x_4{}^2 + c_5 x_4{}^2 + c_6 \right) + g_0 \left(c_1 x_1 x_2{}^2 x_4{}^3 + c_2 x_1 x_2{}^2 x_4 + c_3 \right) = e_5^{(2)}.$$

We need 2 evaluations again: Choose $[x_3 = 5, x_4 = 6]$ and $[x_3 = 5^2, x_4 = 6^2]$ and compute

$$
\begin{aligned}
D([x_3 = 5, x_4 = 6]) :=& \\
& \left(x_1^8 + 950\, x_1 x_2^2 + 30\, x_2^2 - 5\right)\left(180\, c_4 x_1^2 x_2 + 36\, c_5 + c_6\right) \\
& + \left(x_1^8 + 1290\, x_1^2 x_2 - 104\right)\left(216\, c_1 x_1 x_2^2 + 6\, c_2 x_1 x_2^2 + c_3\right) \\
=& \ 18\, x_1^9 x_2^2 - 108\, x_1^8 + 23220\, x_1^3 x_2^3 - 104472\, x_1 x_2^2 - 3240\, x_2^2 + 540
\end{aligned}
$$

and similarly for $D([x_3 = 25, x_4 = 36])$. Calling BDP we obtain the solutions to these bivariate Diophantine equations $[\sigma_1, \tau_1] = [-108, 18\, x_1 x_2^2]$ and $[\sigma_2, \tau_2] = [-3888, 108\, x_1 x_2^2]$ respectively. Hence we have $180\, c_4 x_1^2 x_2 + 36\, c_5 + c_6 = -108$ and $32400\, c_4 x_1^2 x_2 + 1296\, c_5 + c_6 = -3888$ respectively. Then we solve the Vandermonde linear systems

$$
[180]\,[c_4] = [0] \quad \text{and} \quad \begin{bmatrix} 36 & 1 \\ 1296 & 1 \end{bmatrix} \begin{bmatrix} c_5 \\ c_6 \end{bmatrix} = \begin{bmatrix} -108 \\ -3888 \end{bmatrix}
$$

to obtain $c_4 = 0, c_5 = -3, c_6 = 0$. So $g_2 = -3\, x_4^2$. Then by division we get $f_2 = (e_5^{(2)} - f_0 g_2)/g_0 = 3\, x_1 x_2^2 x_4\, (x_5 - 1)^2$. Hence

$$
\begin{aligned}
f^{(2)} &= f^{(1)} + 3\, x_1 x_2^2 x_4\, (x_5 - 1)^2 \\
&= x_1^8 + 2\, x_1 x_2^2 x_4^3 x_5 + 4\, x_1 x_2^2 x_3^3 + 3\, x_1 x_2^2 x_4 x_5^2 + x_2^2 x_3 x_4 - 5 \\
g^{(2)} &= g^{(1)} + \left(-3\, x_4^2\right)(x_5 - 1)^2 \\
&= x_1^8 + 3\, x_1^2 x_2 x_3 x_4^2 x_5 + 5\, x_1^2 x_2 x_3^2 x_4 - 3\, x_4^2 x_5^2 + 4\, x_5.
\end{aligned}
$$

The next error $e_5^{(3)} = a - f^{(2)} g^{(2)} = 0$ and we have the factors!

We used four evaluations and solved three (2×2) linear systems. For the same problem KHL would need to find 9 linearly independent homogeneous linear equations out of 28 equations first. A natural question is, is it possible for KHL to focus on to some subset of the variables first? The answer is no. The systems of equations constructed by KHL are coupled.

References

1. Cox, D., Little, J., O'Shea, D.: Ideals, Varieties and Algorithms, 3rd edn. Springer, Heidleberg (2007)
2. Geddes, K.O., Czapor, S.R., Labahn, G.: Algorithms for Computer Algebra. Kluwer, Boston (1992)
3. Kaltofen, E.: Sparse Hensel Lifting. In: Caviness, B.F. (ed.) EUROCAL 1985. LNCS, vol. 204, pp. 4–17. Springer, Heidelberg (1985)
4. Lee, M.M.: Factorization of multivariate polynomials. Ph.D. Thesis (2013)
5. Miola, A., Yun, D.Y.Y.: Computational aspects of Hensel-type univariate polynomial greatest common divisor algorithms. In: Proceedings of EUROSAM 1974, pp. 46–54. ACM Press (1974)
6. Monagan, M.B., Tuncer, B.: Some results on counting roots of polynomials and the Sylvester resultant. In: Proceedings of FPSAC 2016. DMTCS (to appear 2016)

7. Monagan, M.B., Pearce, R.: POLY: a new polynomial data structure for Maple 17. In: Feng, R., Lee, W.-S., Sato, Y. (eds.) Computer Mathematics, pp. 325–348. Springer, Heidelberg (2014)
8. Schwartz, J.: Fast probabilistic algorithms for verification of polynomial identities. J. ACM **27**, 701–717 (1980). ACM Press
9. Wang, P.S.: An improved multivariate polynomial factoring algorithm. Math. Comput. **32**, 1215–1231 (1978). AMS
10. Wang, P.S., Rothschild, L.P.: Factoring multivariate polynomials over the integers. Math. Comput. **29**(131), 935–950 (1975). AMS
11. Yun, D.Y.Y.: The Hensel Lemma in algebraic manipulation. Ph.D. Thesis (1974)
12. Zippel, R.: Interpolating polynomials from their values. J. Symbolic Comput. **9**, 375–403 (1990). Academic Press
13. Zippel, R.E.: Probabilistic algorithms for sparse polynomials. In: Proceedings of EUROSAM 1979. LNCS, vol. 72, pp. 216–226. Springer, Heidelberg (1979)
14. Zippel, R.E.: Newton's iteration and the sparse Hensel algorithm. In: Proceedings of SYMSAC 1981, pp. 68–72. ACM Press (1981)

A Survey of Satisfiability Modulo Theory

David Monniaux[1,2]([⊠])

[1] Univ. Grenoble Alpes, VERIMAG, 38000 Grenoble, France
David.Monniaux@imag.fr
[2] CNRS, VERIMAG, 38000 Grenoble, France

Abstract. Satisfiability modulo theory (SMT) consists in testing the satisfiability of first-order formulas over linear integer or real arithmetic, or other theories. In this survey, we explain the combination of propositional satisfiability and decision procedures for conjunctions known as DPLL(T), and the alternative "natural domain" approaches. We also cover quantifiers, Craig interpolants, polynomial arithmetic, and how SMT solvers are used in automated software analysis.

1 Introduction

Satisfiability modulo theory (SMT) solving consists in deciding the satisfiability of a first-order formula with unknowns and relations lying in certain theories. For instance, the following formula has no solution $x, y \in \mathbb{R}$:[1]

$$(x \leq 0 \vee x + y \leq 0) \wedge y \geq 1 \wedge x \geq 1. \tag{1}$$

The formula may contain negations ($\neg$), conjunctions ($\wedge$), disjunctions ($\vee$) and, possibly, quantifiers ($\exists$, $\forall$).

A *SMT-solver* reports whether a formula is satisfiable, and if so, may provide a *model* of this satisfaction; for instance, if one omits $x \geq 1$ in the preceding formula, then its solutions include $(x = 0, y = 1)$. Other possible features include dynamic addition and retraction of constraints, production of proofs and Craig interpolants (Sect. 4.2), and optimization (Sect. 4.3). SMT-solving has major applications in the formal verification of hardware, software, and control systems.

Quantifier-free SMT subsumes Boolean satisfiability (SAT), the canonical NP-complete problem, and certain classes of formulas accepted by SMT-solvers belong to higher complexity classes or are even undecidable. This has not deterred researchers from looking for algorithms that, in practice, solve many

The research leading to these results has received funding from the European Research Council under the European Union's Seventh Framework Programme (FP/2007-2013)/ERC Grant Agreement nr. 306595 "STATOR".

[1] This survey focuses on linear and polynomial numeric constraints over integers and reals. SMT however encompasses theories as diverse as character strings, inductive data structures, bit-vector arithmetic, and ordinary differential equations.

V.P. Gerdt et al. (Eds.): CASC 2016, LNCS 9890, pp. 401–425, 2016.
DOI: 10.1007/978-3-319-45641-6_26

relevant instances at reasonable costs. Care is taken that the worst-case cost does not extend to situations that can be dealt with more cheaply.

Most SMT solvers follow the DPLL(T) framework (Sect. 2.2): a CDCL solver for SAT (Sect. 2.1) is used to traverse the Boolean structure, and conjunctions of atoms from the formula are passed to a solver for the theory. This approach limits the interaction between theory values and Boolean reasoning, which led to the introduction of *natural domain* approaches (Sect. 3). Finally, we shall see in Sect. 4 how to go beyond mere quantifier-free satisfiability testing, by handling quantifiers, providing Craig interpolants, or providing optimal solutions. Let us now first see a few generalities, and how SMT-solving is used in practice.

1.1 Generalities

Consider quantifier-free propositional formulas, that is, formulas constructed from *unknowns* (or *variables*) taking the values "true" ($\mathbf{t}$) and "false" ($\mathbf{f}$) and propositional connectives $\vee$ (or), $\wedge$ (and), $\neg$ (not); $\bar{x}$ shall be short-hand for $\neg x$.[2] A formula is: in *negation normal form* (NNF) if the only $\neg$ connectives are at the leaves of its syntax tree (that is, wrap around unknowns but not larger formulas); a *clause* if it is a disjunction of literals (a literal is an unknown or its negation); in *disjunctive normal form* (DNF) if it is a disjunction of conjunctions of literals; in *conjunctive normal form* (CNF) if it is a conjunction of clauses. If A implies B, then A is *stronger* than B and B *weaker* than A. Uppercase letters (F) shall denote formulas, lowercase letters (x) unknowns, and lowercase bold letters ($\boldsymbol{x}$) vectors of unknowns.

Satisfiability testing consists in deciding whether there exists a *satisfying assignment* (or *solution*) for these unknowns, that is, an assignment making the formula true. For instance, $a = \mathbf{t}, b = \mathbf{t}, c = \mathbf{f}$ is a satisfying assignment for $(a \vee c) \wedge (b \vee c) \wedge (\bar{a} \vee \bar{c})$. In case the formula is satisfiable, a solver is generally expected to provide such a satisfying assignment; in the case it is unsatisfiable, it may be queried for an *unsatisfiable core*, a subset of the conjunction given as input to the solver that is still unsatisfiable.

Satisfiability modulo theory extends propositional satisfiability by having some atomic propositions be predicates from a theory. For instance, $(x > 0 \vee c) \wedge (y > 0 \vee c) \wedge (x \le 0 \vee \bar{c})$ is a formula over linear rational arithmetic (LRA) or linear integer arithmetic (LIA), depending on whether x and y are to be interpreted over the rationals or integers.

Different unknowns may range in different sets; for instance $f(x) \ne f(y) \wedge x = z+1 \wedge z = y-1$ has unknowns $f : \mathbb{Z} \to \mathbb{Z}$ and $x, y, z \in \mathbb{Z}$. This formula is said to be over the combination of *uninterpreted functions and linear integer arithmetic* (UFLIA). In this formula, f is said to be uninterpreted because we give no definition for it; we shall see in Sect. 2.6 that this formula has no satisfying assignment and how to establish this fact automatically.

---

[2] Further propositional connectives, such as exclusive-or, or "let x be e_1 in e_2" constructs may be also considered.

Listing 1.1. Example of SMT-LIB 2 file. Assertions $x \geq 0$, $y \leq 0$, $f(x) \neq f(y)$ and $x + y \leq 0$ are added, then the problem is checked to be unsatisfiable. The last assertion is retracted and replaced by $x + y \leq 1$, the problem becomes satisfiable and a model is requested (see Listing 1.2)

```
(set-logic QF_UFLIA)
(set-option :produce-models true)
(declare-fun x () Int)
(declare-fun y () Int)
(declare-fun f (Int) Int)
(assert (>= x 0))
(assert (>= y 0))
(assert (distinct (f x) (f y)))
(push 1)
(assert (<= (+ x y) 0))
(check-sat)
(pop 1)
(assert (<= (+ x y) 1))
(check-sat)
(get-model)
```

1.2   The SMT-LIB Format and Available Theories

SMT solvers can be used (i) as a library, from an application programming interface, typically from C/C++, Java, Python, or OCaml (ii) as an independent process, from a textual representation, possibly through a bidirectional pipe.

APIs for SMT-solvers are not standardized, though there have been efforts such as JavaSMT[3] to provide a common layer for several solvers. In contrast, much effort has been put into designing and supporting the common SMT-LIB [4] format, a textual representation (Listing 1.1); some solvers support other languages than SMT-LIB, sometimes alongside it. Libraries of benchmark problems, sorted according

Listing 1.2. Z3's answers to the SMT-LIB Listing 1.1

```
unsat
sat
(model
  (define-fun y () Int 1)
  (define-fun x () Int 0)
  (define-fun f ((x!1 Int)) Int
    (ite (= x!1 0) 2
    (ite (= x!1 1) 3
      2)))
)
```

---

[3] https://github.com/sosy-lab/java-smt [42].

to the theories involved and the presence or absence of quantifiers (Table 1), are available in that format. New theories are proposed; for instance, a theory for constraints over IEEE-754 floating-point arithmetic [40] is under evaluation.

Table 1. Categories of formulas in SMT-LIB; e.g. `QF_UFLIA` means quantifier-free combination of uninterpreted functions.

Linear real arithmetic	`LRA`
Linear integer arithmetic	`LIA`
Linear mixed integer and real arithmetic	`LIRA`
Bit-vector arithmetic	`BV`
Nonlinear (polynomial) real arithmetic	`NRA`
Nonlinear (polynomial) integer arithmetic	`NIA`
Nonlinear (polynomial) mixed integer and real arithmetic	`NIRA`
Uninterpreted functions	`UF`
Arrays	`A/AX`
Quantifier-free	`QF_`

Alas, some features, such as quantifier elimination or the extraction of Craig interpolants (Sect. 4.2) do not have standard commands. Furthermore, not all tools implement all operators and commands following the standard.

1.3 Use in Program Analysis Applications

A major use of SMT-solvers is the analysis of software. In most cases (but not always), the solutions of the formula to be tested for satisfiability correspond to execution traces of the software verifying certain desirable or undesirable properties: for instance traces going into error states.

Symbolic Execution. In *symbolic program execution* [44], a program is executed as though operating on symbolic inputs. Along a straight path in the program, the semantics of the instructions and tests encountered accumulate as a *path condition*, expressing the relationship between the final values and the inputs. In case a branching instruction is encountered, the analyzer tests whether either branch may be taken by checking for a solution to the conjunction of the path condition and the guard associated with the branch: branches for which a solution is known not to exist are not retained for the rest of the analysis. The analysis thus explores a tree of possible executions, which in general does not cover all possible executions of the program: this is acceptable in bug-finding applications.

Pure symbolic execution may prove infeasible due to the large number of paths to explore. This is especially true if the program involves loads and writes to memory, due to the *aliasing* conditions to test ("does this read correspond to this write?"). Because of this, often what is done is a mixture of concrete and symbolic execution, dubbed *concolic*: sometimes a non-symbolic value is picked (e.g. memory allocation addresses) for simpler execution. In *whitebox fuzzing*, concolic execution is applied from symbolic values coming from external inputs (files, network communications) so as to reach security hazards [31].

Inductiveness Check and Bounded Model Checking. In some other cases [30,36,37], the formula encodes the full set of executions between two control locations in a program, such that there is no looping construct between these locations: one Boolean variable is added per control location, expressing whether or not the execution goes through that location.

In the Floyd-Hoare approach to proving the correctness of programs (see e.g. [69]), the user is prompted for an inductive invariant for each looping construct: a formula I that holds at loop initiation, and that, if it holds at one loop iteration, holds at the next (*inductiveness*). In other words, there is no execution of the loop guard and loop body that starts in I and ends in $\neg I'$ (I' is I where the variables are renamed in order to express their final, not initial, values). In modern tools, the loop guard and body are turned into a first-order formula that is conjoined with I and $\neg I'$, then checked for unsatisfiability; or equivalently through a *weakest precondition* computation, as in Frama-C [18].

Example 1. Consider the array fill program (assume $n \geq 0$):

```
int t[n];
for(int i=0; i<n; i++) t[i] = 42;
```

In order to prove the postcondition $\forall k \; 0 \leq k < n \Rightarrow t[k] = 42$, one needs the loop invariant

$$I \stackrel{\triangle}{=} (0 \leq i \leq n) \wedge (\forall k \; 0 \leq k < i \Rightarrow t[k] = 42). \tag{2}$$

The inductiveness condition is

$$(I \wedge i < n) \Rightarrow I[i \mapsto i + 1, t \mapsto update(t, i, 42)], \tag{3}$$

where $update(t, i, 42)$ is the array t where i has been replaced by 42, and $I[i \mapsto x]$ is formula I where i has been replaced by x. This condition is checked by showing that the negation of this formula is unsatisfiable — after Skolemization:

$$(0 \leq i \leq n) \wedge (\forall k \; 0 \leq k < i \Rightarrow t[k] = 42) \wedge i < n$$
$$\wedge \; (\neg(0 \leq i + 1 \leq n) \vee (0 \leq k_0 \leq i \wedge update(t, i, 42)[k_0] \neq 42)) . \tag{4}$$

$update(t, i, 42)[k_0]$ expands into $ite(k_0 = i, 42, t[k_0])$ where $ite(a, b, c)$ means "if a then b else c". The universal quantifier is instantiated with $k = k_0$, a new unknown $t_k = t[k]$ is introduced to handle the *uninterpreted function* f (Sect. 2.6) and the resulting problem is solved over linear integer arithmetic (Sect. 2.4).

2 The DPLL(T) Architecture

Most SMT-solvers follow the DPLL(T) architecture: a solver for pure propositional formulas, following the DPLL or CDCL class of algorithms, drives decision
procedures for each theory (e.g. linear arithmetic) by adding or retracting constraints and querying for satisfiability. *DPLL(T) and decision procedures for
many interesting logics are explained in more detail in e.g.* [9,47].

2.1 CDCL Satisfiability Testing

*We shall only give a cursory view of satisfiability testing and refer the reader to
e.g.* [6] *for more in-depth treatment.*

Many algorithms for satisfiability testing for quantifier-free formulas only
accept formulas in conjunctive normal form (conjunction of clauses). Naive conversion into conjunctive normal form, by application of distributivity of $\vee$ over
$\wedge$, incurs an exponential blowup. It is however possible to construct, from any
formula F, a formula F' in CNF but with additional free variables, such that
any satisfying assignment to F can be extended to a satisfying assignment on
F' and any satisfying assignment on F', restricted to the free variables of F, is
a satisfying assignment of F. *Tseitin's encoding* is the simplest way to do so:
to any subformula $e_1 \wedge e_2$ of F, associate a new propositional variable $x_{e_1 \wedge e_2}$
and constrain it such that it is equivalent to $e_1 \wedge e_2$ by clauses $\neg x_{e_1 \wedge e_2} \vee e_1$,
$\neg x_{e_1 \wedge e_2} \vee e_2$, $\neg e_1 \vee \neg e_2 \vee x_{e_1 \wedge e_2}$ (and similarly for $e_1 \vee e_2$).

Example 2. Consider

$$\big((a \wedge \bar{b} \wedge \bar{c}) \vee (b \wedge c \wedge \bar{d})\big) \wedge (\bar{b} \vee \bar{c}). \tag{5}$$

Assign propositional variables to sub-formulas:

$$e \equiv a \wedge \bar{b} \wedge \bar{c} \qquad f \equiv b \wedge c \wedge \bar{d} \qquad g \equiv e \vee f \qquad h \equiv \bar{b} \vee \bar{c} \qquad \phi \equiv g \wedge h; \tag{6}$$

these equivalences are turned into clauses:

$$
\begin{array}{lllll}
\bar{e} \vee a & \bar{e} \vee \bar{b} & \bar{e} \vee \bar{c} & \bar{a} \vee b \vee c \vee e & \\
\bar{f} \vee b & \bar{f} \vee c & \bar{f} \vee d & \bar{b} \vee \bar{c} \vee d \vee f & \\
\bar{e} \vee g & \bar{f} \vee g & \bar{g} \vee e \vee f & & \\
b \vee h & c \vee h & \bar{h} \vee \bar{b} \vee \bar{c} & & \\
\bar{\phi} \vee g & \bar{\phi} \vee h & \bar{g} \vee \bar{h} \vee \phi & \phi. &
\end{array}
\tag{7}
$$

The model $(a, b, c, d) = (\mathbf{t}, \mathbf{f}, \mathbf{f}, \mathbf{t})$ of (5) is extended by $(e, f, g) = (\mathbf{t}, \mathbf{f}, \mathbf{t})$,
producing a model of the system of clauses (7), i.e., the conjunction of these
clauses. Conversely, any model of that system, projected over (a, b, c, d), yields
a model of (5).

Let F' be the conjunction of clauses forming the problem. The Davis–
Putnam–Logemann–Loveland algorithm (DPLL) decides a propositional formula
in CNF (conjunction of clauses) by maintaining a partial assignment of the

variables (that is, an assignment to only some of the variables) and *Boolean constraint propagation*: if we have assigned $a = \mathbf{f}, b = \mathbf{t}$ and we have a clause $a \vee \neg b \vee c$, then we can derive $c = \mathbf{t}$. If an assignment satisfies all clauses, then the algorithm terminates with one solution. If it falsifies at least one clause, then there is no solution for our starting partial assignment (thus no solution at all if our starting partial assignment was empty). If propagation is insufficient to conclude, then the algorithm chooses a variable x and a true value b and extends the assignment with $x = b$; if no solution is found for that assignment, then it *backtracks* and replaces it by $x = \bar{b}$. The solver thus constructs a *search tree*.

The practical performance of the solver depends highly on the heuristics for choosing x and b. Much effort has been put into researching these heuristics, such as *Variable State Independent Decaying Sum* (VSIDS) [55]; understanding why they work well is an active research topic. The Boolean constraint propagation phase must be implemented very efficiently, using data structures that minimize the traversal of irrelevant data (clauses that will not result in further propagation); e.g. the *two watched literals per clause* scheme [49, Sect. 4.5.1.2].

From a run of the DPLL algorithm concluding to unsatisfiability one can extract a *resolution proof* of unsatisfiability. The proof has the form of a tree whose leaves are some of the original clauses of the problem (constituting an unsatisfiable core) and whose inner nodes correspond to the choices made during the search. Each inner node is the application of the *resolution rule*: knowing $C_1 \vee a$ and $C_2 \vee \bar{a}$, where C_1 and C_2 are clauses and a is a choice variable, one can derive $C_1 \vee C_2$, written:

$$\frac{C_1 \vee a \quad C_2 \vee \bar{a}}{C_1 \vee C_2} \quad . \tag{8}$$

Example 3. Consider the system of clauses 7. Boolean clause propagation from unit clause ϕ simplifies $\bar{\phi} \vee g$ and $\bar{\phi} \vee h$ into g and h respectively, and removes clause $\bar{g} \vee \bar{h} \vee \phi$. Since g and h are now $\mathbf{t}$, we can remove clauses $\bar{e} \vee g$ and $\bar{f} \vee g$, $b \vee h$, and $c \vee h$, and simplify $\bar{g} \vee e \vee f$ into $e \vee f$ and $\bar{h} \vee \bar{b} \vee \bar{c}$ into $\bar{b} \vee \bar{c}$:

$$\begin{array}{cccccc} \bar{e} \vee a & \bar{e} \vee \bar{b} & \bar{e} \vee \bar{c} & \bar{a} \vee b \vee c \vee e & \bar{f} \vee b & \\ \bar{f} \vee c & \bar{f} \vee d & \bar{b} \vee \bar{c} \vee d \vee f & e \vee f & \bar{b} \vee \bar{c}. \end{array} \tag{9}$$

The system no longer has unit clauses to propagate and thus must pick a literal, for instance b. By propagation, the system now reaches a contradiction. Since contradiction was reached from assumption b, the converse $\bar{b}$ must be assumed. In fact, it is possible to derive the learned clause $\bar{b}$ by resolution from the set of clauses:

$$\frac{\dfrac{e \vee f \quad \bar{f} \vee c}{e \vee c} \quad \bar{e} \vee \bar{b}}{\dfrac{\bar{b} \vee c \quad \bar{b} \vee \bar{c}}{\bar{b}}} \quad . \tag{10}$$

From any "unsatisfiable" run of a DPLL (even in the CDCL variant, see below) solver, a resolution proof can be extracted. This is a fundamental limitation of that approach, since it is known that for certain families of formulas, such as the *pigeonhole principle* [33], any resolution proof has exponential size in the size of the formula — thus any DPLL/CDCL solver will take exponential time.

Performance was considerably increased by extending DPLL with *clause learning*, yielding constraint-driven clause learning (CDCL) algorithms [49]. In CDCL, when a partial assignments leads by propagation to the falsification of a clause, the deductions made during this propagation are analyzed to obtain a subset of the partial assignment sufficient to entail the falsification of this clause. This subset yields a conjunction $\hat{x}_1 \wedge \cdots \wedge \hat{x}_n$ (where $\hat{x}_i$ is either x_i or $\neg x_i$), such that its conjunction with F' is unsatisfiable. In other words, it yields a clause $\neg \hat{x}_1 \vee \cdots \vee \neg \hat{x}_n$ that is a consequence of F' (in fact, that clause can be obtained by resolution from F'). This clause can thus be conjoined to the problem F' without changing its set of solutions; but *learning* that clause may help cut branches in the search tree early.

Again, the learned clause appears as the root of a resolution proof whose leaves are clauses of the original problem. Since the same learned clause may be used several times, the final proof appears as a directed acyclic graph (DAG, i.e., a tree with shared sub-branches). There exist formulas admitting DAG resolution proofs exponentially shorter than the smallest tree resolution proof [67].

A resolution proof, or a more compact format, may thus be produced during an "unsatisfiable" run. A highly optimized SAT or SMT solver is likely to contain bugs, so it may be desirable to have an independent, simpler, possibly formally verified checker reprocess such as proof [3,8,43].

2.2 DPLL(T)

The most common way to deal with atomic propositions inside satisfiability testing is the so-called DPLL(T) scheme, combining a CDCL satisfiability solver and a decision procedure for conjunctions of propositions from theory T. A quantifier-free formula F over T, say

$$(x \geq 0 \vee 2x + y \geq 1) \wedge (y \geq 0) \wedge (x + y \leq -1), \tag{11}$$

is converted into a propositional formula F' (here $(a \vee b) \wedge c \wedge d$) by replacing each atomic proposition by a propositional variable, using a dictionary (here, $x \geq 0 \mapsto a, 2x + y \geq 1 \mapsto b, y \geq 0 \mapsto c, x + y \leq -1 \mapsto d$) and after conversion to canonical form (so that e.g. $x + y \geq 1$ and $2x + 2y - 2 \geq 0$ are considered the same, and $x + y < 1$ is considered as $\neg(x + y \geq 1)$). F' realizes a *propositional abstraction* of F: any solution of F induces a solution of F', but not all solutions of F' necessarily induce a solution of F.

Consider the solution $a = \mathbf{t}, b = \mathbf{f}, c = \mathbf{t}, d = \mathbf{t}$ of F'; it corresponds to

$$x \geq 0 \wedge \neg(2x + y \geq 1) \wedge y \geq 0 \wedge x + y \leq -1. \tag{12}$$

The inequalities $x \geq 0 \wedge y \geq 0 \wedge x + y \leq -1$ have no common solution; in other words, $\neg(a \wedge c \wedge d)$ is universally true. The *theory clause* $\neg a \vee \neg c \vee \neg d$ can be conjoined to F'. There remains a solution $a = \mathbf{f}, b = \mathbf{t}, c = \mathbf{t}, d = \mathbf{t}$ of F'; but it entails the contradiction $2x + y \geq 1 \wedge y \geq 0 \wedge x + y \leq -1$. The theory clause $\bar{b} \vee \bar{c} \vee \bar{d}$ is then conjoined to F'. Then the propositional problem becomes unsatisfiable, establishing that F has no solution. We have therefore refined the propositional abstraction according to spurious counterexamples.

In current implementations, the propositional solver does not wait until a total satisfying assignment is computed to call the decision procedure for conjunctions of theory formulas. Partial assignments, commonly at each decision point in the DPLL/CDCL algorithm, are tested for satisfiability. In addition, the theory solver may, opportunistically, perform *theory propagation*: if it notices that some asserted constraints imply the truth or falsehood of another known predicate, it can signal it to the SAT solver. The theory solver should be *incremental*, that is, suited for fast addition or retraction of theory constraints, keeping enough internal state to avoid needless recomputation. The SAT solver should be incremental as well, allowing the dynamic addition of clauses.

Multiple theories may be combined, most often by a variant of the Nelson–Oppen approach [47, Chap. 10].

2.3   Linear Real Arithmetic

In the case of linear rational, or equivalently real, arithmetic (LRA), the theory solver is typically implemented using a variant [24,25] of the simplex algorithm [20,63]. The atomic (in)equalities from the formula, put in canonical form, are collected; new variables are introduced for the linear combinations of variables that are not of the form $\pm x$ where x is a variable. For instance, (11) is rewritten as $(x \geq 0 \vee \alpha \geq 1) \wedge (y \geq 0) \wedge (\beta \leq -1)$, together with the system of linear equalities $\alpha = 2x + y$ and $\beta = x + y$.

The simplex algorithm both maintains a tableau and, for each variable, a current valuation and optional lower and upper bounds. At all times, the simplex tableau contains a system of linear equalities equivalent to this system, such that the variables are partitioned into those (*basic variables*) occurring (each alone) on the left side and those occurring on the right side. The non-basic variables are assigned one of their bounds, or at least a value between these bounds. The simplex algorithm tries to fit each basic variable within its bounds; if one does not fit, it makes it non-basic and assigns to it the bound that was exceeded, and selects a formerly non-basic variable to make it basic, through a *pivoting* operation maintaining the equivalence of the system of equalities.

The algorithm stops when either a candidate solution fitting all bounds is found, either one equation in the simplex tableau can be shown to have no solution using interval arithmetic from the bounds of the variables (the interval obtained from the right hand side does not intersect that of the basic variable on the left hand side). A pivot selection ordering is used to ensure that the algorithm always terminates. Theory propagation may be performed by noticing that the current tableau implies that some literals are satisfied.

Example 4. Consider the system

$$\begin{cases} 2 \leq 2x + y \\ -6 \leq 2x - 3y \\ -1000 \leq 2x + 3y \quad \leq 18 \\ -2 \leq -2x + 5y \\ 20 \leq x + y. \end{cases} \tag{13}$$

This system is turned into a system of equations ("tableau") and a system of inequalities on the variables:

$$\begin{cases} a = & 2x +y & 2 \leq a \\ b = & 2x -3y & -6 \leq b \\ c = & 2x \ \ 3y & -1000 \leq c \leq 18 \\ d = & -2x +5y & -2 \leq d \\ e = & x +y & 20 \leq e. \end{cases} \tag{14}$$

The variables on the left of the equal signs are deemed "nonbasic" and those on the right are "basic". The simplex algorithm performs pivoting steps on the tableau, akin to those of Gaussian eliminations, until a tableau such as this one is reached:

$$\begin{cases} e = 7/16c -1/16d \\ a = \ \ 3/4c -1/4d \\ b = \ \ 1/4c -3/4d \\ x = 5/16c -3/16d \\ y = \ \ 1/8c +1/8d. \end{cases} \tag{15}$$

Now consider the first equation ($e =$). By interval analysis, knowing $c \leq 18$ and $d \geq -2$, $-7/16c - 1/16d \leq 8$. Yet $e \geq 20$, thus the system has no solution. These coefficients $7/16$ and $1/16$ can be applied to the original inequalities constraining c and d, with coefficient 1 for that defining e, and the resulting inequalities are summed into a trivially false one:

$$\begin{array}{cccc} 7/16 \ (-2x -3y) & \geq & -7/16 \times 18 \\ 1/16 \ (-2x +5y) & \geq & -1/16 \times 2 \\ 1 \quad x \quad +y & \geq & 20 \\ \hline 0 \quad\quad 0 & \geq & 28. \end{array} \tag{16}$$

By reading nonzero coefficients off the conflicting line of the simplex tableau, one gets a minimal set of contradictory constraints: $d+1$ constraints, corresponding to the nonbasic variable and the basic variables with nonzero multipliers, where d is the dimension of the space. These multipliers may be presented as an *unsatisfiability witness* to an independent proof checker.

Most SMT solvers implement the simplex algorithm using rational arithmetic. In most cases arising from verification problems, rational arithmetic can be performed using machine integers, without need for going into extended precision arithmetic [21]. A common implementation trick is to use a datatype containing a machine-integer (*numerator*, *denominator*) pair or a pointer to an extended precision rational.[4] This approach is however very inefficient in the rare

---

[4] e.g. ZArith https://forge.ocamlcore.org/projects/zarith.

cases where the solver goes a lot into extended precision: the size of numerators and denominators grows fast.

This is why it was proposed to perform linear programming in floating-point arithmetic [11,26,45,58].[5] Because the results of floating-point computations cannot be immediately trusted, some checking is needed. One idea is not to recover floating-point numeric information, but the final partition between basic and nonbasic variables [11,45,58]; once this partition is known, the tableau is uniquely defined and can be computed by plain linear arithmetic — Gaussian elimination, or better algorithms, including multimodular [66, Chap. 7] or p-adic approaches.[6] It is then easy to check the alleged conflicting line, in exact precision.

In some cases, linear arithmetic reasoning may be used to prove the unsatisfiability of polynomial problems. One approach is to expand polynomials and consider all monomials as independent variables (e.g. xy^2 is replaced by a fresh unknown v_{xy^2}). A refinement [50] is to consider lemmas stating that if two polynomials are nonnegative, then so is their product: e.g. $x - 1 \geq 0 \wedge y - 2 \geq 0 \implies v_{xy} - 2x - y + 2 \geq 0$.[7] Because the set of such products has size exponential in the maximal degree, heuristics are used to pick the most promising ones. Experiments have shown this approach to be competitive, even with a rudimentary and sub-optimal connection between linear SMT-solver and nonlinear reasoning.

Some earlier solvers (e.g. CVC3) solver linear real arithmetic by Fourier-Motzkin elimination [29]. This approach is generally not considered efficient, since Fourier-Motzkin elimination tends to generate many redundant constraints, which then may need to be eliminated by linear programming, which defeats the purpose of avoiding using the simplex algorithm.

2.4 Linear Integer Arithmetic

In the case of linear integer arithmetic, the scheme generally used is the same as the one generally used for integer linear programming: the solver first attempts solving the rational relaxation of the problem (nonstrict inequalities are kept, strict inequalities $x < e$ are rewritten as $x \leq e - 1$). If there is no solution over the rationals, there is no integer solution. If a rational solution is found, and has only integral coefficients (say, $(x, y, z) = (0, 1, 2)$), then the problem is decided.

If the proposed solution has non-integral coefficients (say, $(x, y, z) = (\frac{1}{3}, 0, 1)$), then it is excluded by a constraint removing not only that spurious solution but

---

[5] The performance with linear programming solvers meant for large industrial instances was however disappointing [26], due to overhead. Closer integration is needed.

[6] As implemented in e.g. Linbox (http://www.linalg.org/), IML (https://cs.uwaterloo.ca/~astorjoh/iml.html) [12] and SageMath (http://www.sagemath.org/).

[7] One can in fact prove a form of completeness of that approach when the problem contains linear constraints defining a bounded polyhedron, and one nonlinear constraint: if such a problem is unsatisfiable, then this can be proved by going to a sufficiently high degree of products. This follows from Krivine–Handelman's theorem [34,46].

a whole chunk of them. Traditional approaches include (i) *branch-and-bound* [63, Sect. 24.1]: add a lemma excluding one segment of non-integral values of the fractional unknowns (here, $x \leq 0 \lor x \geq 1$); branching is however not guaranteed to terminate in general [45]. (ii) *Gomory cuts* [63, Chap. 23] (iii) *branch-and-cut* [56], a combination of both of the above iv) *cuts from proofs* or *extended branches* [23], which can generate e.g. $x \leq z \lor x \geq z + 1$.

The full integer linear decision procedure can be encapsulated and only export theory lemmas and theory propagation, just as the rational linear procedure, or export the branching lemma to the SMT solver, as a learned clause, so as to allow propositional reasoning over it.

An alternative to linear programming plus branching and/or cuts is Pugh's Omega test [61], which may also be used to simplify constraints. This test is based on Fourier-Motzkin elimination [29], with the twist that, due to divisibility constraints, it may need to enumerate cases up to the least common multiple of the divisors.

2.5   Exponential Behavior Due to Limited Predicate Vocabulary

Example 5. Let $n > 0$ be a constant integer. Let $(t_i)_{0 \leq i \leq n}$, $(x_i)_{0 \leq i < n}$ and $(y_i)_{0 \leq i < n}$ be real unknowns (or rational or integer). Let

$$D_i \overset{\triangle}{=} (x_i - t_i \leq 2) \land (y_i - t_i \leq 3) \land ((t_{i+1} - x_i \leq 3) \lor (t_{i+1} - y_i \leq 2)) , \quad (17)$$

$$P_n \overset{\triangle}{=} \bigwedge_{i=0}^{n-1} D_i \land t_n - t_0 > 5n . \quad (18)$$

These formulas are known as "diamond formulas" since they correspond to paths in a difference graph composed of "diamonds":

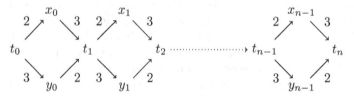

To a human, it is obvious that $D_i \Rightarrow t_{i+1} \leq t_i + 5$ and thus P_n is unsatisfiable. A DPLL(T) solver, however, proceeds by elimination of contradictory conjunctions of atoms from the original formula. Any contradictory conjunction of atoms from P_n must include a conjunction of the form $\bigwedge_{i=0}^{n-1} F_i \land t_n - t_0 > 5n$ where F_i is either $(x_i - t_i \leq 2) \land (t_{i+1} - x_i \leq 3)$ or $(y_i - t_i \leq 3) \land (t_{i+1} - y_i \leq 2)$. There are an exponential number of such conjunctions, and a DPLL(T) solver has to block them by theory lemmas one by one.

In other words, the proof system used by a DPLL(T) solver is sufficient to prove that a "diamond formula" is unsolvable, but needs exponential proofs for doing so. Any pure DPLL(T) solver, whatever its heuristics and implementation, must thereof run in exponential time on this family of formulas. This motivated the study of algorithms capable of inferring lemmas involving new atoms (Sect. 3.2).

Diamond formulas are simplifications of formulas occurring in e.g. worst-case execution time and scheduling applications. The solution proposed in [35] was to pre-compute upper bounds $t_j - t_i \leq B_{ij}$ on the difference of arrival times between i and j (or, equivalently, the total time spent in the program between i and j) and conjoin these bounds to the problems. These bounds are logically implied by the original problem, and thus the set of solutions (valid execution traces with timings) does not change; but the resulting formula isY considerably more tractable. The lemmas $t_j - t_i \leq B_{ij}$ and $t_k - t_j \leq B_{ik}$ allow the solver to avoid exploring many combinations of paths $i \rightarrow j$ and $j \rightarrow k$: for instance, if one searches for a path such that $t_k - t_i \geq 100$, it is known that $t_k - t_j \leq 40$, and the solver explores a path $i \rightarrow j$ such that $t_j - t_i \leq 42$ on this path, then the solver can immediately cut the search without exploring the paths $j \rightarrow k$ in detail.

2.6   Uninterpreted Functions and Arrays

There exists several variants of how to decide uninterpreted functions (UF) in combination with other theories [47, Chap. 4]; we shall expose only one approach here. A quantifier-free formula (e.g. $f(x) \neq f(y) \wedge x = z + 1 \wedge z = y - 1$) is rewritten so that each application of an uninterpreted function is replaced by a fresh variable (e.g. $f_x \neq f_y \wedge x = z + 1 \wedge z = y - 1$), several identical applications getting the same variable. A solution in x, y, z, f_x, f_y is sought. If $x = y$ but not $f_x \neq f_y$ in that solution, the implication $x = y \Rightarrow f_x = f_y$ is conjoined to the problem. Again, this is a counterexample-guided refinement of the theory.

Example 6. $f(x) \neq f(y) \wedge x = z + 1 \wedge z = y - 1$, where $x, y, z \in \mathbb{Z}$ and $f : \mathbb{Z} \rightarrow \mathbb{Z}$, has no solution because $x = z + 1 \wedge z = y - 1$ implies that $x = y$, and it is then impossible that $f(x) \neq f(y)$. One may establish this by solving $f_x \neq f_y \wedge x = z + 1 \wedge z = y - 1$, getting $(x, y, z, f_x, f_y) = (1, 1, 0, 0, 1)$, noticing the conflict between $x = y$ and $f_x \neq f_y$ and conjoining $x = y \Rightarrow f_x = f_y$.

Arrays are "functionally updatable" uninterpreted functions [47, Chap. 7]: $update(f, x_0, y_0)$ is the function mapping $x \neq x_0$ to $f[x]$ and x_0 to y_0.

3   Natural-Domain SMT

In DPLL(T) there is a fundamental difference between propositional and other kinds of unknowns: the second are never dealt with directly during the search process. In contrast, in *natural-domain* SMT, one directly constrains and assigns to numeric unknowns during the search. After initial attempts [17,54], two main directions arose.

3.1   Abstract CDCL (ACDCL)

The DPLL approach is to assign to each unknown (propositional variable) one of $\mathbf{t}$, $\mathbf{f}$, and "undecided" — that is, a non-empty subset of the set of possible values $\{\mathbf{t}, \mathbf{f}\}$. Initially, all variables are assigned to "undecided". Then, the Boolean constraint propagation phase uses each individual clause as a constraint over its literals: if all literals except for one are assigned to $\mathbf{f}$, then the last one gets assigned to $\mathbf{t}$. In other words, information known about some variables leads to information on other variables linked by the same constraint. If the information derived is that some variable cannot be assigned some value ("contradiction"), then it means the problem is unsatisfiable. In most cases, however, a contradiction cannot be derived by only the initial pass of propagation. In that case, the system picks an undecided variable and splits the search between the $\mathbf{t}$ and $\mathbf{f}$ cases. Several splits may be needed, thus the formation of a search tree. If a contradiction is derived in a branch, that branch is closed and the system backtracks to an earlier level.

That approach may be extended to variables lying within an arbitrary domain D, say, the real numbers or the floating-point numbers. The system maintains for each variable an assignment to a subset of D (several types of variables may be used simultaneously, there may therefore be several D), chosen among an *abstract domain*[8] $D^\sharp$ of subsets of D; say, for numeric variables, $D^\sharp$ may be the set of closed intervals of D. Constraints may now constrain variables of different types, and each constraint acts as a propagator of information. For instance, if there is a constraint $x = y + z$, and x is currently assigned the interval $[1, +\infty)$ and y the interval $[4, 10]$, then, applying $z = x - y$, one can derive $x \in [-9, +\infty)$: the current interval for x may thus be refined.

Note that, for soundness, it is not important that the information propagated should be optimally precise, as long as it contains the possible values: in the above example, it would be sound to propagate $x \in [-9.1, +\infty)$ — but unsound to derive $xx \in [-8.99, +\infty)$. In the case of interval propagation for $D = \mathbb{R}$, one sound way to implement it is using floating-point interval arithmetic with directed rounding: the upper bound of an interval is rounded towards $+\infty$, the lower bound towards $-\infty$.

ACDCL also applies clause learning, but in a more general manner than CDCL [10, Sect. 5]. Consider $F \overset{\triangle}{=} y = x \wedge z = x \cdot y \wedge z \leq -1$ and a search context with $x \leq -4$. Then, by interval propagation, $y \leq -4$, and $z \geq 16$, which contradicts $z \leq -1$. CDCL-style clause learning would learn that $x \leq -4$ contradicts F, and thus learn the clause $\neg(x \leq -4) \equiv x > -4$. But there is a weaker reason why such choice of x contradicts F: $x < 0$ is sufficient to ensure contradiction; the solver can exclude a larger part of the search space by learning the clause $\neg(x < 0) \equiv x \geq 0$. Generalizing the reasons for a contradiction is a form of *abduction*. One difficulty is that there may be no weakest generalization expressible in the abstract domain: for instance, the choices $x \geq 10$ and $y \geq 10$ contradict the constraint $x + y < 10$, but $x \geq 0 \wedge y \geq 10$, $x \geq 5 \wedge y \geq 5$

---

[8] Following the terminology of *abstract interpretation*; see [10] for more.

and $x \geq 10 \wedge y \geq 0$ are three incomparable generalizations of the contradiction (leading to three clauses $x < 0 \vee y < 10$ etc.), which are optimal in the sense that if one fixes the interval for x (resp. y), the interval for y (resp. x) is the largest that still ensures contradiction.

3.2   Model-Constructing Satisfiability Calculus (MCSAT)

In DPLL(T) (i) only propositional atoms (including Boolean unknowns) are assigned during the search (ii) the set of atoms considered does not change throughout the search (this may cause exponential behavior, see Sect. 2.5 (iii) when the search process, after assigning $b_1, \ldots, b_n$ concludes that it is impossible to assign a Boolean value to an atom b_{n+1}, it derives a learned clause over a subset of $b_1, \ldots, b_n$ that excludes the current assignment but also, hopefully, many more. In contrast, in *model-constructing satisfiability calculus* (MCSAT) [22], both propositional atoms and numeric unknowns get assigned during the search, and new arithmetic predicates are generated through learning.

Linear Real Arithmetic. Assume variables $x_1, \ldots, x_n$ have been assigned values $v(x_1), \ldots, v(x_n)$ in the current branch of the search, and that two atoms $x_{n+1} \leq a$ and $x_{n+1} \geq b$, where a and b are linear combinations of variables other than x_{n+1}, have been assigned to $\mathbf{t}$, such that $b > a$ in the assignment v; then it is impossible to pick a value for x_{n+1} in that assignment. In fact, it is impossible to pick a value for it in *any* assignment such that $b > a$.

Assignments that conflict for the same reason are eliminated by a *Fourier-Motzkin elimination* [29] elementary step, valid for all $x_1, \ldots, x_{n+1}$:

$$\neg x_{n+1} \leq a \vee \neg x_{n+1} \geq b \vee a \geq b. \tag{19}$$

Example 7. Consider Example 5 with $n = 3$. The solver has clauses $x_i - t_i \leq 2$, $y_i - t_i \leq 3$, $t_{i+1} - x_i \leq 3 \vee t_{i+1} - y_i \leq 2$ for $0 \leq i < 3$, and $t_0 = 0$, $t_3 \geq 16$.

The solver picks $t_0 \mapsto 0$, $t_1 - x_0 \leq 3 \mapsto \mathbf{t}$, $x_0 \mapsto 0$, $t_1 \mapsto 0$, $t_2 - x_1 \leq 3 \mapsto \mathbf{t}$, $x_1 \mapsto 0$, $t_2 \mapsto 0$, $t_3 - x_2 \leq 3 \mapsto \mathbf{t}$, $x_2 \mapsto 0$. But then, there is no way to assign t_3, because of the current assignment $x_2 \mapsto 0$ and the inequalities $t_3 - x_2 \leq 3$ and $t_3 \geq 16$. The solver then learns by Fourier-Motzkin:

$$\neg(t_3 \geq 16) \vee \neg(t_3 - x_2 \leq 3) \vee x_2 \geq 13. \tag{20}$$

which may in fact be immediately simplified by resolution with the original clause $t_3 \geq 16$ to yield $\neg(t_3 - x_2 \leq 3) \vee x_2 \geq 13$. The assignment to x_2 is retracted.

But then, there is no way to assign x_2, because of the current assignment $t_2 \mapsto 0$ and the inequality $x_2 - t_2 \leq 2$. The solver then learns by Fourier-Motzkin:

$$\neg(x_2 \geq 13) \vee \neg(x_2 - t_2 \leq 2) \vee t_2 \geq 11. \tag{21}$$

By resolution, $\neg(t_3 - x_2 \leq 3) \vee t_2 \geq 11$. The truth assignment to $t_3 - x_2 \leq 3$ is retracted.

At this point, the solver has $t_0 \mapsto 0$, $t_1 - x_0 \leq 3 \mapsto \mathbf{t}$, $x_0 \mapsto 0$, $t_1 \mapsto 0$, $t_2 - x_1 \leq 3 \mapsto \mathbf{t}$, $x_1 \mapsto 0$, $t_2 \mapsto 0$, $t_3 - x_2 \leq 3 \mapsto \mathbf{f}$. By similar reasoning in that branch, the solver derives $t_3 - x_2 \leq 3 \vee t_2 \geq 11$. By resolution between the outcomes of both branches, one gets $t_2 \geq 11$.

By similar reasoning, one gets $t_1 \geq 6$ and then $t_0 \geq 1$, but then there is no satisfying assignment to t_0. The problem has no solution.

In contrast to the exponential behavior of DPLL(T) on Example 5, MCSAT has linear behavior: each branch of each individual disjunction is explored only once, and the whole disjunction is then summarized by an extra atom.

The dynamic generation of new atoms by MCSAT, as opposed to DPLL(T), creates two issues. (i) If infinitely many new atoms may be generated, termination is no longer ensured. One can ensure termination by restricting the generation of new atoms to a *finite basis* (this basis of course depends on the original formula); this is the case for instance if the numeric variables $x_1, \ldots, x_n$ are always assigned in the same order, thus the generated new atoms are results of Fourier-Motzkin elimination of x_n, then of x_{n-1} etc. down to x_2.[9] In practice, the interest of being able to choose variable ordering trumps the desire to prove termination. (ii) Since many new atoms and clauses are generated, some garbage collection must be applied, as with learned clauses in a CDCL solver.

Implementation-wise, note that, like a clause in CDCL, a linear inequality is processed only when all variables except for one are assigned. Similar to *two watched literals per clause*, one can apply *two watched variables per inequality*.

Nonlinear Arithmetic (NRA). The MCSAT approach can also be applied to polynomial real arithmetic. Again, the problem is: assuming a set of polynomial constraints over $x_1, \ldots, x_n, x_{n+1}$ have no solution over x_{n+1} for a given valuation $v(x_1), \ldots, v(x_n)$, how can we explain this impossibility by a system of constraints over $x_1, \ldots, x_n$ that excludes $v(x_1), \ldots, v(x_n)$ and hopefully many more?

Jovanović and de Moura [41] proposed applying a modified version of Collin's [15] projection operator in order to perform a partial *cylindrical algebraic decomposition*. In that approach, known as NLSAT, one additional difficulty is that assignments to variables may refer to algebraic reals, and thus the system needs to compute to compute over algebraic reals, including as coefficients to polynomials. It is yet unknown whether this approach could benefit from using other projection operators such as Hong's [39] or McCallum's [51].

4   Beyond Quantifier-Free Decidability

4.1   Quantifiers

Quantifier Elimination by Virtual Substitution. In the case of some theories, such as linear real arithmetic, a finite sequence of instantiations can be

---

[9] Successive applications of Fourier-Motzkin may lead to very large sets of predicates, thus this argument seems of mostly theoretical interest.

produced such that $F \stackrel{\triangle}{=} \forall x\ P(x)$ is equivalent to $\bigwedge_{i=1}^{n} P(v_i)$; note that the v_i are not constants, but functions of the free variables of F, obtained by analyzing the atoms of P. Because this approach amounts to substituting expressions into the quantified variable, it is called *substitution*, or *virtual substitution* if appropriate data structures and algorithms avoid explicit substitution. Examples of substitution-based methods include Cooper's [16] for linear integer arithmetic, Ferrante and Rackoff's [27] and Loos and Weisfpenning's [48] methods for linear real arithmetic.

Example 8. Consider $\forall y\ (y \geq x \Rightarrow y \geq 1)$. Loos and Weisfpenning's method collects the expression to which y is compared (here, x and 1) and then substitutes them into y. For each expression e, one must also substitute $e + \epsilon$ where ϵ is infinitesimal,[10] and also substitute $-\infty$ (equivalently, one can substitute $e - \epsilon$ for each expression, and also $+\infty$). The result is therefore

$$\bigwedge_{e \in \{x, x+\epsilon, 1, 1+\epsilon, -\infty\}} e \geq x \Rightarrow e \geq 1 \tag{22}$$

or, after expansion and simplification, $x \geq 1$.

We thus have *eliminated* the quantifier; by recursion over the structure of a formula and starting at the leaves, we can transform any formula of linear real arithmetic into an equivalent quantifier-free formula.[11]

In these eager approaches, the size of the substitution set may grow quickly (especially for linear integer arithmetic, which may involve enumerating all cases up to the least common multiple of the divisibility constants). For this reason, lazy approaches were proposed where the substitutions are generated from counterexamples, in much the same way that learned lemmas are generated in DPLL(T) [7,60]. For a formula $A(\boldsymbol{x}) \wedge \forall y\ B(\boldsymbol{x}, y)$, the system first solves $A(\boldsymbol{x})$ for a solution $\boldsymbol{x}_0$, then checks whether there exists y such that $\neg B(\boldsymbol{x}_0, y)$; if so, such an y_0 is generalized into one of the possible substitutions $S_1(\boldsymbol{x})$ and the system restarts by solving $A(\boldsymbol{x}) \wedge B(\boldsymbol{x}, S_1(\boldsymbol{x}))$. The process iterates until a solution is found or the substitutions accumulated block all solutions for $\boldsymbol{x}$; termination is ensured because the set of possible symbolic substitutions is finite. Note that a full quantifier elimination is not necessary to produce a solution.

Quantifier Elimination by Projection. In case a quantifier elimination, or *projection*, algorithm, is available for conjunctions of constraints, as happens with linear real arithmetic,[12] one can, given a formula $\exists \boldsymbol{y}\ F(\boldsymbol{x}, \boldsymbol{y})$, find a conjunction

---

[10] $x \geq K + \epsilon$ with K real means $x > K$.

[11] In the case of linear integer arithmetic, we need to enrich the language of the output formula with constraints of divisibility by constants: e.g. $\exists x\ y = 2x$ is equivalent to quantifier-free $2 \mid x$.

[12] This amounts to projection of convex polyhedra, for which there exist algorithms based on conversion to generators (vertices), Fourier-Motzkin elimination and pruning, or parametric linear programming, among others [28].

$C_1 \Rightarrow F$, project C_1 over x as $\pi(C_1)$, conjoin $\neg\pi(C_1)$ to F and repeat the process (generating C_2 etc.) until the F becomes unsatisfiable [57]. $\bigvee_i C_i$ is then equivalent to $\exists y\, F$. Again, this process may be made lazier, for nested quantification in particular [59,60].

Instantiation Heuristics. The addition of quantifiers to theories (such as linear integer arithmetic plus uninterpreted functions) may make them undecidable. This does not however deter designers of SMT solvers from attempting to have them decide as many formulas as possible. A basic approach is quantifier instantiation by E-matching. If a formula in negation normal form contains a subformula $\forall x\ P(x)$, then this formula is replaced by a finite instantiation $\bigwedge_{i=1}^n P(v_i)$. The v_i are extracted from the rest of the formula, possibly guided by counterexamples. This approach is not guaranteed to converge: an infinite sequence of instantiations may be produced for a given quantifier. In the case of *local theories*, one can however prove termination.

4.2   Craig Interpolation

The following conjunction is satisfiable if and only if it is possible to go from a model x_0 of A to a model x_n of B by a sequence of transitions $(\tau_i)_{1 \le i \le n}$:

$$A(x_0) \wedge \tau_1(x_0, x_1) \wedge \cdots \wedge \tau_n(x_{n-1}, x_n) \wedge B(x_n). \tag{23}$$

In program analysis, A typically expresses a precondition, $\neg B$ a postcondition, x_i the variables of the program after i instruction steps, and τ_i the semantics of the i-th instruction in a sequence, and the formula is unsatisfiable if and only if B is always true after executing that sequence of instruction starting from A.

A hand proof of unsatisfiability would often consist in exhibiting predicates $I_1(x_1), \ldots, I_{n-1}(x_{n-1})$, such that, posing $I_0 = A$ and $I_n = B$, for all $0 \le i < n$,

$$\forall x_i, x_{i+1}\ I_i(x_i) \wedge \tau_{i+1}(x_i, x_{i+1}) \Rightarrow I_{i+1}(x_{i+1}), \tag{24}$$

along with proofs of these *local inductiveness* implications.[13]

A SMT-solver, in contrast, produces a monolithic proof of unsatisfiability of (23): it mixes variables from different x_i, that is, in program analysis, from different times of the execution of the program. It is however possible to obtain instead a sequence I_i satisfying (24) by post-processing that proof [13,14,52].

In any theory admitting quantifier elimination, such a sequence must exist:

$$I_{i+1} \equiv \exists x_i\ I_i(x_i) \wedge \tau_{i+1}(x_i, x_{i+1}) \tag{25}$$

defines the strongest sequence of valid interpolants; the weakest is:

$$I_i \equiv \forall x_{i+1}\ \tau_{i+1}(x_i, x_{i+1}) \Rightarrow I_{i+1}(x_{i+1}). \tag{26}$$

---

[13] In program analysis, this corresponds to stating "after the first instruction, the program variables satisfy I_1, but then if one executes the second instruction from I_1, the program variables then satisfy I_2...", and $\{I_i\}\ \tau_i\ \{I_{i+1}\}$ constitute Hoare triples.

The strongest sequence corresponds to computing exactly the sequence of sets of states reachable by τ_1, then $\tau_2 \circ \tau_1$ etc. from A.

Binary interpolation consists in: given A and B, produce I such that

$$\forall x_0, x_1, x_2 \; A(x_0, x_1) \Rightarrow I(x_1) \Rightarrow B(x_1, x_2), \tag{27}$$

in which case, if the theory admits quantifier elimination, $\exists x_0 \; A(x_0, x_1)$ and $\forall x_2 \; B(x_1, x_2)$ are respectively the strongest and weakest interpolant, and any I in between $(\exists x_0 \; A(x_0, x_1) \Rightarrow I \Rightarrow \forall x_2 \; B(x_1, x_2))$ is also an interpolant.

One of the main uses of Craig interpolation in program analysis is to synthesize inductive invariants, for instance by counterexample-guided abstraction refinement in predicate abstraction (CEGAR) [53] or property-guided reachability (PDR). Interpolants obtained by quantifier elimination are too specific (*overfitting*): for instance, strongest interpolants exactly fit the set of states reachable in $1, 2, \ldots$ steps. It has been argued that interpolants likely to be useful as inductive invariants should be "simple" — short formula, with few "magical constants". A variety of approaches have been proposed for getting such interpolants [2,65,68] or to simplify existing interpolants [38].

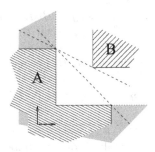

Fig. 1. Binary interpolation in linear arithmetic. The hashed areas represent A and B (Eq. 28) respectively. A possible interpolant I between A and B ($A \Rightarrow I$, $I \Rightarrow \neg B$) is the grey area $x \leq 1 \vee y \leq 1$. The dashed lines define two other possible interpolants, $x + y \leq 5$ and $x + 2y \leq 9$.

Example 9. Consider the interpolation problem $A \Rightarrow I$, $I \Rightarrow \neg B$ (Fig. 1):

$$A_1 \overset{\triangle}{=} x \leq 1 \wedge y \leq 4 \quad A_2 \overset{\triangle}{=} x \leq 4 \wedge y \leq 1$$
$$A \overset{\triangle}{=} A_1 \vee A_2 \qquad B \overset{\triangle}{=} x \geq 3 \wedge y \geq 3. \tag{28}$$

SMTINTERPOL[14] and MATHSAT[15] produce $I \overset{\triangle}{=} x \leq 1 \vee y \leq 1$. This is due to the way these tools produce interpolants from DPLL(T) proofs of unsatisfiability. On this example, a DPLL(T) solver will essentially analyze both branches of

---

[14] SMTINTERPOL 2.1-31-gafd0372-comp.
[15] MATHSAT 5.3.10.

$A_1 \lor A_2$. The first branch yields $A_1 \Rightarrow \neg B$. Finding I_1 such that $A_1 \Rightarrow I_1$ and $I_1 \Rightarrow \neg B$ amounts to finding a separating hyperplane between these two convex polyhedra; $I_1 \overset{\triangle}{=} x \leq 1$ works. Similarly, I_2 such that $A_1 \Rightarrow I_2$ and $I_2 \Rightarrow \neg B$ can be $I_2 \overset{\triangle}{=} y \leq$. $I_1 \lor I_2$ is then produced as interpolant.

Yet, a search for a single separating hyperplane may produce $x + 2y \leq 9$, or $x + y \leq 5$. The second hyperplane may seem preferable according to a criterion limiting the magnitude of integer constants.

It is easy to see that if one can find interpolants for arbitrary conjunctions A, B such that $A \Rightarrow \neg B$, one can find them between arbitrary quantifier-free formulas, by putting them into DNF. Because such a procedure would be needlessly costly due to disjunctive normal forms, the usual approach is to post-process a DPLL(T) proof that $A(\boldsymbol{x}, \boldsymbol{y}) \land B(\boldsymbol{y}, \boldsymbol{z})$ is unsatisfiable [13, 14]. First, interpolants are derived for all theory lemmas: each lemma expresses that a conjunction of atoms from the original formula is unsatisfiable, these atoms can thus be divided into a conjunction α of atoms from A and a conjunction β of atoms from B, and an interpolant I is derived for $\alpha \Rightarrow \neg\beta$. Then, these interpolants are combined following the resolution proof of the solver. This is how the interpolants from Example 9 were produced by the solvers.

The problem is therefore: given $A(\boldsymbol{x}, \boldsymbol{y}) \land B(\boldsymbol{y}, \boldsymbol{z})$ unsatisfiable, where A and B are conjunctions, how do we find $I(\boldsymbol{y})$ such that $A \Rightarrow I$ and $I \Rightarrow \neg B$? If the theory is linear rational arithmetic, this amounts to finding a separating hyperplane between the polyhedra A and B. Let us note

$$A \overset{\triangle}{=} \bigwedge\nolimits_i \boldsymbol{a}''_i \cdot \boldsymbol{x} + \boldsymbol{a}_i \cdot \boldsymbol{y} \geq a'_i, \qquad B \overset{\triangle}{=} \bigwedge\nolimits_j \boldsymbol{b}''_j \cdot \boldsymbol{z} + \boldsymbol{b}_j \cdot \boldsymbol{y} \geq b'_j. \qquad (29)$$

Each a'_i (resp. b'_j) is a pair $(a'^{\mathbb{R}}_i, a'^{\epsilon}_i)$ lexicographically ordered, where $a'^{\mathbb{R}}_i$ is the real part and c'^{ϵ}_i is infinitesimal; all other numbers are assumed to be real. $y \geq (x^{\mathbb{R}}, x^{\epsilon})$ with $x^{\epsilon} > 0$ and $y \in \mathbb{R}$ expresses that $y > x^{\mathbb{R}}$.

Since $A \land B$ is unsatisfiable, by Farkas' lemma, there exists an unsatisfiability witness (λ_i), (μ_j), such that

$$\begin{matrix} \sum_i \lambda_i \boldsymbol{a}''_i = 0 & \sum_j \mu_j \boldsymbol{b}''_j = 0 \\ \sum_i \lambda_i \boldsymbol{a}_i + \sum_j \mu_j \boldsymbol{b}_j = 0 & \sum_i a'_i + \sum_j b'_j > 0 \end{matrix} \qquad (30)$$

Such coefficients can in fact be read off the simplex tableau from the most common way of implementing a DPLL(T) solver for linear real arithmetic, as described in Sect. 2.3. Then the following is a valid interpolant (recall that the right-hand side can contain infinitesimals, leading to $>$):

$$I \overset{\triangle}{=} \sum_i (\lambda_i \boldsymbol{a}_i) \cdot \boldsymbol{y} \geq \sum_i \lambda_i a'_i \qquad (31)$$

For polynomial arithmetic, one approach replaces nonnegative reals by sums-of-squares of polynomials, and Farkas' lemma by Positivstellensatz [19].

Another difficulty is posed for certain theories, for which the solving process involves generating lemmas introducing atoms not present in the original. Consider the approaches for linear integer arithmetic described in Sect. 2.4: except for branch-and-bound, all can generate new constraints involving any of the unknowns, without respecting the original partition of variables. This poses a problem for interpolation: if interpolating for $A(x,y) \wedge B(y,z)$ over linear real arithmetic, we can rely on all atomic propositions being linear inequalities either over x, y or y, z, but here, we have new atomic propositions that can involve both x, z. Special theory-dependent methods are needed to get rid of these new propositions when processing the DPLL(T) proof into an interpolant [13,14].

4.3   Optimization

Instead of finding one solution, one may wish to find a solution that maximizes (or nearly so) some function f.

A simple approach is *binary search*: provided one can get a lower bound l and an upper bound h on the maximum $f(\boldsymbol{x}^*)$, one queries the solver for a solution $\boldsymbol{x}$ such that $f(\boldsymbol{x}) \geq m$, where $m = \frac{l+h}{2}$; if such a solution is found, refine the lower bound $l := f(\boldsymbol{x})$ and restart, otherwise $h := m$ and restart. Proceed until $l = h$. This converges in finite time if f has integer value. This approach has been successfully applied to e.g. worst-case execution time problems [35].

In the case of LRA (resp. LIA), optimization generalizes *linear programming* (resp. *linear integer programming*) to formulas with disjunctions. In fact, linear programming can be applied locally to a polyhedron of solutions: when a DPLL(T) solver finds a solution $\boldsymbol{x}$ of a formula F, it also finds a conjunction C of atoms such that $C \Rightarrow F$; C defines a polyhedron and one can optimize within it, until a local optimum $\boldsymbol{x}^l$. Then one adds the constraint $f(\boldsymbol{x}) > f(\boldsymbol{x}^l)$ and restart; the last $\boldsymbol{x}^l$ found is the optimum (one can also detected unboundedness). This approach can never enumerate the same C (or subsets thereof) twice and thus must terminate. It may, however, scan an exponential number of useless C's; it may be combined with binary search for best effect [64].

5   Conclusion

Considerable progress has been made within the last 15 years on increasingly practical decision procedures for increasingly large classes of formulas, even though worst-case complexity is prohibitive, and sometimes even though the class is undecidable.[16] Major ingredients to that success were (i) lazy generation of lemmas, partial projections or instantiations, guided by counterexamples

---

[16] Worst-case complexity, or completeness in complexity classes, is therefore not always a good indicator of practical performance. Average complexity is difficult to define (one needs to suppose a probability distribution on formulas) and may ill-describe practical use cases: it is well-known that random SAT instances behave unlike industrial examples [1], and same with random linear constraints [58]. For want of better indication, performance is measured on libraries of benchmarks.

(as opposed to eager exhaustive generation, often explosive) (ii) generalization of counterexamples so as to *learn* sufficiently general blocking lemmas (iii) tight integration of propositional and theory-specific reasoning.

Nonlinear arithmetic reasoning (polynomials, or even transcendental functions) is still a very open question. Current approaches in SMT [41] are based on partial cylindrical algebraic decomposition [15]; possibly methods based on critical points [5,32,62] could be investigated as well.

There are several challenges to using computer algebra procedures inside a SMT solver. (i) These procedures may not admit addition or retraction of constraints without recomputation. (ii) They may compute eagerly large sets of formulas (as in conventional cylindrical algebraic decomposition). (iii) They may be very complex and thus likely to contain bugs.[17] Being able to produce independently-checkable proof witnesses would help in this respect.

Acknowledgements. Thanks to the anonymous referees for their careful proofreading.

References

1. Achlioptas, D.: 8. In: [6], pp. 245–270
2. Albarghouthi, A., McMillan, K.L.: Beautiful interpolants. In: Sharygina, N., Veith, H. (eds.) CAV 2013. LNCS, vol. 8044, pp. 313–329. Springer, Heidelberg (2013)
3. Armand, M., Faure, G., Grégoire, B., Keller, C., Théry, L., Werner, B.: A modular integration of SAT/SMT solvers to coq through proof witnesses. In: Jouannaud, J.-P., Shao, Z. (eds.) CPP 2011. LNCS, vol. 7086, pp. 135–150. Springer, Heidelberg (2011)
4. Barrett, C., Fontaine, P., Tinelli, C.: The satisfiability modulo theories library (SMT-LIB). www.SMT-LIB.org (2016)
5. Basu, S., Pollack, R., Roy, M.F.: Algorithms in Real Algebraic Geometry. Springer, Berlin (2006)
6. Handbook of Satisfiability, vol. 185. IOS Press, Amsterdam (2009)
7. Bjørner, N.: Linear quantifier elimination as an abstract decision procedure. In: Giesl, J., Hähnle, R. (eds.) IJCAR 2010. LNCS, vol. 6173, pp. 316–330. Springer, Heidelberg (2010)
8. Böhme, S., Weber, T.: Fast LCF-style proof reconstruction for Z3. In: Kaufmann, M., Paulson, L.C. (eds.) ITP 2010. LNCS, vol. 6172, pp. 179–194. Springer, Heidelberg (2010)
9. Bradley, A.R., Manna, Z.: The Calculus of Computation: Decision Procedures with Applications to Verification. Springer, Berlin (2007)
10. Brain, M., D'Silva, V., Griggio, A., Haller, L., Kroening, D.: Deciding floating-point logic with abstract conflict driven clause learning. Formal Methods Syst. Des. **45**(2), 213–245 (2014)
11. Barbosa, C., de Oliveira, D., Monniaux, D.: Experiments on the feasibility of using a floating-point simplex in an SMT solver. In: Fontaine, P., Schmidt, R.A., Schulz, S. (eds.) Workshop on Practical Aspects of Automated Reasoning (PAAR). EPiC Series, vol. 21, pp. 19–28. Easychair (2012)

---

[17] The author had several computer algebra packages crash or produce wrong results. Perhaps running large libraries of benchmarks would help in finding such bugs.

12. Chen, Z., Storjohann, A.: A BLAS based C library for exact linear algebra on integer matrices. In: Proceedings of the ISSAC 2005, pp. 92–99. ACM, New York (2005)

13. Christ, J.: Interpolation modulo theories. Ph.D. thesis, Albert-Ludwigs-Universität, Freiburg (2015)

14. Christ, J., Hoenicke, J., Nutz, A.: Proof tree preserving interpolation. In: Piterman, N., Smolka, S.A. (eds.) TACAS 2013 (ETAPS 2013). LNCS, vol. 7795, pp. 124–138. Springer, Heidelberg (2013)

15. Collins, G.E.: Quantifier elimination for real closed fields by cylindrical algebraic decomposition. In: Brakhage, H. (ed.) GI-Fachtagung 1975. LNCS, vol. 33, pp. 134–183. Springer, Heidelberg (1975)

16. Cooper, D.C.: Theorem proving in arithmetic without multiplication. In: Meltzer, B., Michie, D. (eds.) Machine Intelligence 7, pp. 91–100. Edinburgh University Press, Edinburgh (1972)

17. Cotton, S.: Natural domain SMT: a preliminary assessment. In: Chatterjee, K., Henzinger, T.A. (eds.) FORMATS 2010. LNCS, vol. 6246, pp. 77–91. Springer, Heidelberg (2010)

18. Cuoq, P., Kirchner, F., Kosmatov, N., Prevosto, V., Signoles, J., Yakobowski, B.: Frama-C. In: Eleftherakis, G., Hinchey, M., Holcombe, M. (eds.) SEFM 2012. LNCS, vol. 7504, pp. 233–247. Springer, Heidelberg (2012)

19. Dai, L., Xia, B., Zhan, N.: Generating non-linear interpolants by semidefinite programming. In: Sharygina, N., Veith, H. (eds.) CAV 2013. LNCS, vol. 8044, pp. 364–380. Springer, Heidelberg (2013)

20. Dantzig, G.B., Thapa, M.N.: Linear Programming 1: Introduction. Springer, New York (1997)

21. de Moura, L.M.: Personal communication

22. de Moura, L., Jovanović, D.: A model-constructing satisfiability calculus. In: Giacobazzi, R., Berdine, J., Mastroeni, I. (eds.) VMCAI 2013. LNCS, vol. 7737, pp. 1–12. Springer, Heidelberg (2013)

23. Dillig, I., Dillig, T., Aiken, A.: Cuts from proofs: a complete and practical technique for solving linear inequalities over integers. Form. Methods Syst. Des. **39**(3), 246–260 (2011)

24. Dutertre, B., de Moura, L.: A fast linear-arithmetic solver for DPLL(T). In: Ball, T., Jones, R.B. (eds.) CAV 2006. LNCS, vol. 4144, pp. 81–94. Springer, Heidelberg (2006)

25. Dutertre, B., de Moura, L.M.: Integrating simplex with DPLL(T). Sri-csl-06-01, SRI International, computer science laboratory (2006)

26. Faure, G., Nieuwenhuis, R., Oliveras, A., Rodríguez-Carbonell, E.: SAT modulo the theory of linear arithmetic: exact, inexact and commercial solvers. In: Kleine Büning, H., Zhao, X. (eds.) SAT 2008. LNCS, vol. 4996, pp. 77–90. Springer, Heidelberg (2008)

27. Ferrante, J., Rackoff, C.: A decision procedure for the first order theory of real addition with order. SIAM J. Comput. **4**(1), 69–76 (1975)

28. Fouilhé, A.: Revisiting the abstract domain of polyhedra: constraints-only representation and formula proof. Ph.D. thesis, Université de Grenoble (2015)

29. Fourier, J.: Histoire de l'acadmie, partie mathmatique. In: Mmoires de l'Acadmie des sciences de l'Institut de France, vol. 7. Gauthier-Villars, xlvij-lv (1827) (1824)

30. Gawlitza, T., Monniaux, D.: Invariant generation through strategy iteration in succinctly represented control flow graphs. Logic. Methods Comput. Sci. 8(3:29), 1–35 (2012)

31. Godefroid, P., Levin, M.Y., Molnar, D.: SAGE: whitebox fuzzing for security testing. Queue 10(1), 20:20–20:27 (2012)

32. Grigor'ev, D.Y., Vorobjov Jr., N.N.: Solving systems of polynomial inequalities in subexponential time. J. Symb. Comput. 5(1–2), 37–64 (1988)

33. Haken, A.: The intractability of resolution. Theor. Comput. Sci. 39, 297–308 (1985)

34. Handelman, D.: Representing polynomials by positive linear functions on compact convex polyhedra. Pacific J. Math. 132(1), 35–62 (1988)

35. Henry, J., Asavoae, M., Monniaux, D., Maiza, C.: How to compute worst-case execution time by optimization modulo theory and a clever encoding of program semantics. In: Zhang, Y., Kulkarni, P. (eds.) Languages, Compilers, Tools and Theory for Embedded Systems (LCTES), pp. 43–52. ACM, New York (2014)

36. Henry, J., Monniaux, D., Moy, M.: PAGAI: a path sensitive static analyzer. In: Jeannet, B. (ed.) Third Workshop on Tools for Automatic Program Analysis (TAPAS 2012). Electronic Notes in Theoretical Computer Science 289, pp. 15–25 (2012)

37. Henry, J., Monniaux, D., Moy, M.: Succinct representations for abstract interpretation. In: Miné, A., Schmidt, D. (eds.) SAS 2012. LNCS, vol. 7460, pp. 283–299. Springer, Heidelberg (2012)

38. Hoder, K., Kovács, L., Voronkov, A.: Playing in the grey area of proofs. In: Field, J., Hicks, M. (eds.) ACM Symposium on Principles of Programming Languages (POPL), pp. 259–272. ACM, New York (2012)

39. Hong, H.: An improvement of the projection operator in cylindrical algebraic decomposition. In: Watanabe, S., Nagata, M. (eds.) Proceedings of the ISSAC 1990, pp. 261–264. ACM, New York (1990)

40. IEEE: IEEE standard for Binary floating-point arithmetic for microprocessor systems. ANSI/IEEE Standard 754–1985 (1985)

41. Jovanović, D., de Moura, L.: Solving non-linear arithmetic. In: Gramlich, B., Miller, D., Sattler, U. (eds.) IJCAR 2012. LNCS, vol. 7364, pp. 339–354. Springer, Heidelberg (2012)

42. Karpenkov, E., Beyer, D., Friedberger, K.: JavaSMT: A unified interface for SMT solvers in Java. In: VSTTE (2016, to appear)

43. Keller, C.: Extended resolution as certificates for propositional logic. In: Blanchette, J.C., Urban, J. (eds.) Proof Exchange for Theorem Proving (PxTP). EPiC Series, vol. 14, pp. 96–109. EasyChair (2013)

44. King, J.C.: Symbolic execution and program testing. Commun. ACM 19(7), 385–394 (1976)

45. King, T., Barrett, C.W., Tinelli, C.: Leveraging linear and mixed integer programming for SMT. In: Formal Methods in Computer-Aided Design, (FMCAD), pp. 139–146. IEEE (2014)

46. Krivine, J.L.: Anneaux préordonnés. J. d'analyse mathématique 12, 307–326 (1964)

47. Kroening, D., Strichman, O.: Decision Procedures. Springer, New York (2008)

48. Loos, R., Weispfenning, V.: Applying linear quantifier elimination. Comput. J. 36(5), 450–462 (1993)

49. Marques-Silva, J.P., Lynce, I., Malik, S.: 4. In: [6], pp. 131–153

50. Maréchal, A., Fouilhé, A., King, T., Monniaux, D., Périn, M.: Polyhedral approximation of multivariate polynomials using Handelman's theorem. In: Jobstmann, B., Leino, K.R.M. (eds.) VMCAI 2016. LNCS, vol. 9583, pp. 166–184. Springer, Berlin (2016)

51. McCallum, S.: An improved projection operation for cylindrical algebraic decomposition. In: Quantifier Elimination and Cylindrical Algebraic Decomposition, pp. 242–268. Springer, Wien (1998)
52. McMillan, K.L.: An interpolating theorem prover. Theor. Comput. Sci. **345**(1), 101–121 (2005)
53. McMillan, K.L.: Lazy abstraction with interpolants. In: Ball, T., Jones, R.B. (eds.) CAV 2006. LNCS, vol. 4144, pp. 123–136. Springer, Heidelberg (2006)
54. McMillan, K.L., Kuehlmann, A., Sagiv, M.: Generalizing DPLL to richer logics. In: Bouajjani, A., Maler, O. (eds.) CAV 2009. LNCS, vol. 5643, pp. 462–476. Springer, Heidelberg (2009)
55. Moskewicz, M.W., Madigan, C.F., Zhao, Y., Zhang, L., Malik, S.: Chaff: Engineering an efficient SAT solver. In: Design Automation Conference (DAC), pp. 530–535. ACM, New York (2001)
56. Mitchell, J.E.: Branch-and-cut algorithms for combinatorial optimization problems. In: Pardalos, P.M., Resende, M.G.C. (eds.) Handbook of Applied Optimization. Oxford University Press, Oxford (2002)
57. Monniaux, D.: A quantifier elimination algorithm for linear real arithmetic. In: Cervesato, I., Veith, H., Voronkov, A. (eds.) LPAR 2008. LNCS (LNAI), vol. 5330, pp. 243–257. Springer, Heidelberg (2008)
58. Monniaux, D.: On using floating-point computations to help an exact linear arithmetic decision procedure. In: Bouajjani, A., Maler, O. (eds.) CAV 2009. LNCS, vol. 5643, pp. 570–583. Springer, Heidelberg (2009)
59. Monniaux, D.: Quantifier elimination by lazy model enumeration. In: Touili, T., Cook, B., Jackson, P. (eds.) CAV 2010. LNCS, vol. 6174, pp. 585–599. Springer, Heidelberg (2010)
60. Phan, A.D., Bjørner, N., Monniaux, D.: Anatomy of alternating quantifier satisfiability (work in progress). In: Fontaine, P., Goel, A. (eds.) 10th International Workshop on Satisfiability Modulo Theories (SMT), pp. 120–130 (2012)
61. Pugh, W.: The Omega test: A fast and practical integer programming algorithm for dependence analysis. In: Supercomputing, pp. 4–13. ACM, New York (1991)
62. Safey El Din, M., Schost, É.: Polar varieties and computation of one point in each connected component of a smooth real algebraic set. In: Proceedings of the ISSAC 2003, pp. 224–231. ACM, New York (2003)
63. Schrijver, A.: Theory of Linear and Integer Programming. Wiley, New York (1998)
64. Sebastiani, R., Tomasi, S.: Optimization in SMT with $\mathcal{LA}(\mathbb{Q})$ cost functions. In: Gramlich, B., Miller, D., Sattler, U. (eds.) IJCAR 2012. LNCS, vol. 7364, pp. 484–498. Springer, Heidelberg (2012)
65. Sharma, R., Nori, A.V., Aiken, A.: Interpolants as classifiers. In: Madhusudan, P., Seshia, S.A. (eds.) CAV 2012. LNCS, vol. 7358, pp. 71–87. Springer, Heidelberg (2012)
66. Stein, W.A.: Modular Forms, a Computational Approach. Graduate Studies in Mathematics, vol. 79. AMS (2007)
67. Tseitin, G.S.: On the complexity of derivation in propositional calculus. In: Siekmann, J.H., Wrightson, G. (eds.) Automation of Reasoning: 2: Classical Papers on Computational Logic 1967–1970, pp. 466–483. Springer, Berlin (1983)
68. Unno, H., Terauchi, T.: Inferring simple solutions to recursion-free horn clauses via sampling. In: Baier, C., Tinelli, C. (eds.) TACAS 2015. LNCS, vol. 9035, pp. 149–163. Springer, Heidelberg (2015)
69. Winskel, G.: The Formal Semantics of Programming Languages: An Introduction. MIT Press, Cambridge (1993)

Quadric Arrangement in Classifying Rigid Motions of a 3D Digital Image

Kacper Pluta[1,2]([⊠]), Guillaume Moroz[5]([⊠]), Yukiko Kenmochi[3],
and Pascal Romon[4]

[1] Université Paris-Est, LIGM, Champs-sur-Marne, France
`kacper.pluta@univ-paris-est.fr`
[2] Université Paris-Est, LAMA, Champs-sur-Marne, France
[3] Université Paris-Est, LIGM, CNRS, ESIEE Paris, Champs-sur-Marne, France
`yukiko.kenmochi@esiee.fr`
[4] Université Paris-Est, LAMA, UPEM, Champs-sur-Marne, France
`pascal.romon@u-pem.fr`
[5] INRIA Nancy-Grand-Est, Project Vegas, Villers-lès-Nancy, France
`guillaume.moroz@inria.fr`

Abstract. Rigid motions are fundamental operations in image processing. While bijective and isometric in $\mathbb{R}^3$, they lose these properties when digitized in $\mathbb{Z}^3$. To understand how the digitization of 3D rigid motions affects the topology and geometry of a chosen image patch, we classify the rigid motions according to their effect on the image patch. This classification can be described by an arrangement of hypersurfaces in the parameter space of 3D rigid motions of dimension six. However, its high dimensionality and the existence of degenerate cases make a direct application of classical techniques, such as cylindrical algebraic decomposition or critical point method, difficult. We show that this problem can be first reduced to computing sample points in an arrangement of quadrics in the 3D parameter space of rotations. Then we recover information about remaining three parameters of translation. We implemented an ad-hoc variant of state-of-the-art algorithms and applied it to an image patch of cardinality 7. This leads to an arrangement of 81 quadrics and we recovered the classification in less than one hour on a machine equipped with 40 cores.

1 Introduction

Rigid motions (*i.e.*, rotations, translations and their compositions) defined on $\mathbb{Z}^3$ are simple yet crucial operations in many image applications (*e.g.*, image registration [33] and motion tracking [32]). However, it is also known that such operations cause geometric and topological defects [18,19,22]. As such alterations happen locally, due to digitization, discrete motion maps have been studied for small image patches, in order to understand such defects at local scale [20,21,23].

For such a local analysis, one wishes to generate all possible images of an image patch under digitized rigid motions. In digital geometry and combinatorics, some complexity analysis of such a problem has been made for some

V.P. Gerdt et al. (Eds.): CASC 2016, LNCS 9890, pp. 426–443, 2016.
DOI: 10.1007/978-3-319-45641-6_27

geometric transformations. The complexities are related to the size of a given image patch in general: $\mathcal{O}(n^3)$ for 2D rotations [2]; $\mathcal{O}(n^9)$ for 2D rigid motions [17] and $\mathcal{O}(n^{18})$ for 2D affine transformations [10], where n stands for a diameter of a subset of an image patch. Later, in this article we show that the theoretical complexity of such a problem for 3D rigid motions is $\mathcal{O}(n^{24})$.

However, there are few algorithms available for generating all the transformed images from a given image patch. Algorithms known to us are: 2D rotations [21]; 3D rotations around a given rational axis [30,31]; 2D rigid motions [17,23] and 2D affine transformations [10]. However, none of them can be applied to 3D rigid motions.

In this article, we reformulate this classification problem on a finite digital image as an arrangement of quadrics, containing many degenerate cases. We then solve the problem by computing all the 3D open cells in this arrangement. The original problem involves a naive decomposition of the six dimensional parameter space of 3D rigid motions, and can be formulated as an arrangement of hypersurfaces given by polynomials of degree two with integer coefficients. Our goal is to compute for each full–dimensional open cell at least one representative point, so-called *sample point*. The state-of-the-art techniques such as cylindrical algebraic decomposition or critical point method [3] are burdened by double exponential [5] and exponential [26] complexity respectively, with respect to the number of variables. Therefore, their direct application to the problem of decomposition of the six dimensional parameter space of 3D digitized rigid motions are practically inefficient. Indeed, high dimensionality and existence of cases such as asymptotic critical values [13]—*e.g.*, a plane orthogonal to a coordinate axis is tangent to a hypersurface in a point at infinity—make a computation of such an arrangement difficult.

In this article, we propose an ad-hoc method as follows. We first show that the problem can be simplified by uncoupling the six parameters of 3D rigid motions to end up with two systems in three variables, and start by studying an arrangement of quadrics in $\mathbb{R}^3$. These two systems correspond to the rotational and translational parameters of rigid motions, respectively. In order to detect all topological changes along one non-generic, chosen direction by sweeping a plane, we compute all critical points including asymptotic critical values in this arrangement of quadrics. Moreover, we compute at least one sample point for each open full dimensional cell in the arrangement – sample points of full dimensional components provide information to generate different images of an image patch under digitized rigid motions. Our strategy is similar to the one proposed by Mourrain *et al.* [15] where the main differences are: we do not use generic directions: we handle asymptotic cases and give new criteria to compute critical values in polynomials of degree two; we compute and store at least one sample point for each full dimensional open cell where Mourrain *et al.* [15] compute full adjacency information for all cells in an arrangement; moreover, we precompute all critical values *a priori* while in the former approach only one type of critical values needs to be computed before the main algorithm. Those sample points are then used to decompose the other three dimensional parameter space. Finally, our implementation is provided together with a numerical experiment for a small image patch.

2   Classifying Rigid Motions of a 3D Digital Image

2.1   Rigid Motions on the 3D Cartesian Grid

Rigid motions on $\mathbb{R}^3$ are bijective isometric maps defined as

$$\left| \begin{aligned} \mathcal{U} : \mathbb{R}^3 &\to \mathbb{R}^3 \\ \mathbf{x} &\mapsto \mathbf{Rx} + \mathbf{t} \end{aligned} \right. \tag{1}$$

where $\mathbf{t} = (t_1, t_2, t_3) \in \mathbb{R}^3$ is a translation vector and $\mathbf{R}$ is a rotation matrix. Let $\mathbf{A}$ be a skew-symmetric matrix

$$\mathbf{A} = \begin{bmatrix} 0 & c & -b \\ -c & 0 & a \\ b & -a & 0 \end{bmatrix}$$

where $a, b, c \in \mathbb{R}$ and $\mathbf{I}$ be the 3×3 identity matrix. Then almost any rotation matrix $\mathbf{R}$ can be obtained by the Cayley transform [4]:

$$\begin{aligned} \mathbf{R} &= (\mathbf{I} - \mathbf{A})(\mathbf{I} + \mathbf{A})^{-1} \\ &= \frac{1}{1 + a^2 + b^2 + c^2} \begin{bmatrix} 1 + a^2 - b^2 - c^2 & 2(ab - c) & 2(b + ac) \\ 2(ab + c) & 1 - a^2 + b^2 - c^2 & 2(bc - a) \\ 2(ac - b) & 2(a + bc) & 1 - a^2 - b^2 + c^2 \end{bmatrix}. \end{aligned} \tag{2}$$

Indeed, rotations by π around any axis can only be obtained by the Cayley transform as a limit: angles of rotation converge to π when a, b, c tend to infinity [29]. In practice, this constraint is negligible and does not affect generality of our study (see the following section which discusses the evolution of an image patch under 3D digitized rigid motions). Using this formula, a rigid motion is parametrized by the six real parameters (a, b, c, t_1, t_2, t_3).

According to Eq. (1), we generally have $\mathcal{U}(\mathbb{Z}^3) \not\subseteq \mathbb{Z}^3$. As a consequence, in order to define digitized rigid motions as maps from $\mathbb{Z}^3$ to $\mathbb{Z}^3$, we combine, as usual, the results of the rotation with a digitization operator

$$\left| \begin{aligned} \mathcal{D} : \mathbb{R}^3 &\to \mathbb{Z}^3 \\ (x_1, x_2, x_3) &\mapsto \left(\lfloor x_1 + \tfrac{1}{2} \rfloor, \lfloor x_2 + \tfrac{1}{2} \rfloor, \lfloor x_3 + \tfrac{1}{2} \rfloor \right) \end{aligned} \right.$$

where $\lfloor s \rfloor$ denotes the largest integer not greater than s. The digitized rigid motion is thus defined by $U = \mathcal{D} \circ \mathcal{U}_{|\mathbb{Z}^3}$. Due to the behavior of $\mathcal{D}$ that maps $\mathbb{R}^3$ onto $\mathbb{Z}^3$, digitized rigid motions are, most of the time, non–bijective. However, some are, and an algorithmic approach to these digitized rotations is given in [24].

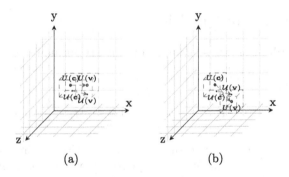

Fig. 1. An example of discontinuity of U. In (a) and (b) the image $\mathcal{U}(\mathbf{c})$ remains within the same unit cube—digitization cell—centered around the origin depicted in blue; thus the image $U(\mathbf{c})$ is the same for the two digitized motions U associated to the continuous motions $\mathcal{U}$ that slightly differ with respect to the parameters. However, the point $\mathbf{v} = \mathbf{c} + (1, 0, 0)^t$, has distinct images $U(\mathbf{v})$ in (a) and (b); in (a), the digitization operator $\mathcal{D}$ sends $\mathcal{U}(\mathbf{v})$ onto the green integer point, while in (b), it sends it onto the red one (Color figure online)

2.2 Image Patch and Its Alterations Under Digitized 3D Rigid Motions

Let us consider a finite set $\mathcal{N} \subset \mathbb{Z}^3$, called an *image patch* whose center $\mathbf{c}$ and radius r of $\mathcal{N}$ are given by $\mathbf{c} = \frac{1}{|\mathcal{N}|} \sum_{\mathbf{v} \in \mathcal{N}} \mathbf{v}$ and $r = \max_{\mathbf{v} \in \mathcal{N}} \|\mathbf{v} - \mathbf{c}\|$, respectively. Note that, in this article, we consider $\mathbf{c}$ as the origin, for simplicity. Next, we express the evolution of such an image patch $\mathcal{N}$ under digitized rigid motions U.

The digitized rigid motions $U = \mathcal{D} \circ \mathcal{U}$ are piecewise constant, and thus non-continuous, which is a consequence of the nature of the digitization operator $\mathcal{D}$. In particular, the image $\mathcal{U}(\mathbf{v})$ of a given point $\mathbf{v}$ may remain constant as the parameters of $\mathcal{U}$ vary, and then suddenly jump from one point of $\mathbb{Z}^3$ to another. In other words, an image patch $\mathcal{N}$ evolves non-continuously, under digitized rigid motions, in accordance with the parameters of $\mathcal{U}$ that underlies U (see Fig. 1). Hereafter, without loss of generality we assume that $\mathcal{U}(\mathbf{c})$ stays in the digitization cell of $\mathbf{c}$, namely $U(\mathbf{c}) = \mathbf{c}$, since translation by an integer vector would not change the geometry of $\mathcal{N}$. Under this assumption we have that $\mathbf{t} \in \left(-\frac{1}{2}, \frac{1}{2}\right)^3$. Moreover, thanks to symmetry (reflections and rotations) we consider only non-negative a, b, c.

Studying the non-continuous evolution of an image patch $\mathcal{N}$ is equivalent to study the discontinuities of $U(\mathbf{v})$ for every $\mathbf{v} \in \mathcal{N} \setminus \{\mathbf{c}\}$, which occur when $\mathcal{U}(\mathbf{v})$ is on the *half-grid* plane, namely a boundary of a digitization cell. This is formulated by

$$\mathbf{R}_i \mathbf{v} + t_i = k_i - \frac{1}{2} \tag{3}$$

where $k_i \in H(\mathcal{N}) = \mathbb{Z} \cap [-r', r'], \mathbf{R}_i$ is the i-th row of the rotation matrix for $i \in \{1, 2, 3\}$ and r' is the longest radius of $\mathcal{U}(\mathcal{N})$ for all $\mathcal{U}$, so that $r' = r + \sqrt{3}$.

3   Arrangement of Quadrics

3.1   The Problem as an Arrangement of Hypersurfaces

For any image patch $\mathcal{N}$, the parameter space

$$\Omega = \left\{ (a, b, c, t_1, t_2, t_3) \in \mathbb{R}^6 \mid a, b, c \geq 0, -\frac{1}{2} < t_i < \frac{1}{2} \text{ for } i = 1, 2, 3 \right\}$$

is partitioned by a set of hypersurfaces given by Eq. (3) into a finite number of connected subsets, namely, 6D open cells whose points induces different rigid motions $\mathcal{U}_{|\mathcal{N}}$ but identical digitized rigid motions $U_{|\mathcal{N}} = \mathcal{D} \circ \mathcal{U}_{|\mathcal{N}}$. For a given image patch $\mathcal{N}$ of radius r, hypersurfaces (3) in Ω are given by the possible combinations of integer 4–tuples (v_1, v_2, v_3, k_i) for $i = 1, 2, 3$ where $\mathbf{v} = (v_1, v_2, v_3) \in \mathcal{N} \setminus \{\mathbf{c}\}$ and $k_i \in H(\mathcal{N})$. Since $|\mathcal{N}| - 1$ is in $\mathcal{O}(r^3)$ and $|H(\mathcal{N})|$ is in $\mathcal{O}(r)$, the number of considerable hypersurfaces is in $\mathcal{O}(r^4)$, and thus in accordance with [8, Theorem 21.1.4] the overall complexity of the arrangement is theoretically bounded by $\mathcal{O}(r^{24})$.

Our goal is to compute for each 6D open cell in Ω at least one representative point, a so-called *sample point*. As the direct application of the cylindrical algebraic decomposition or critical points method to this problem is practically inefficient – due to the high dimensionality and existence of degenerate cases that make computation of the arrangement difficult. Therefore, in the following discussion we develop an indirect but still exact strategy.

3.2   Uncoupling the Parameters

The first idea of our strategy consists in uncoupling the parameters in the six dimensional parameter space Ω. Namely, we show that by considering the differences between the hypersurfaces given in Eq. (3) for different $\mathbf{v} \in \mathcal{N}$ and $\mathbf{k} \in H(\mathcal{N})^3$, we can reduce the problem to the study of an arrangement of surfaces in the (a, b, c)–space, and then lift the solution to the six dimensional space.

Let us consider a rigid motion defined by $\mathbf{R}$ and $\mathbf{t}$. The condition for having $U(\mathbf{v}) = \mathbf{k} = (k_1, k_2, k_3) \in \mathbb{Z}^3$ where $\mathbf{v} \in \mathcal{N}$ is

$$k_i - \frac{1}{2} < \mathbf{R}_i \mathbf{v} + t_i < k_i + \frac{1}{2}$$

for $i = 1, 2, 3$. Equivalently,

$$k_i - \frac{1}{2} - \mathbf{R}_i \mathbf{v} < t_i < k_i + \frac{1}{2} - \mathbf{R}_i \mathbf{v}. \tag{4}$$

Let us call a *configuration* a list of couples $(\mathbf{v}, \mathbf{k})$, which describe how the image patch $\mathcal{N}$ is transformed. This configuration can be described as a function

$$\left| \begin{array}{ll} F : \mathcal{N} & \rightarrow H(\mathcal{N})^3 \\ \mathbf{v} = (v_1, v_2, v_3) & \mapsto \mathbf{k} = (k_1, k_2, k_3). \end{array} \right.$$

We want to ascertain whether a given configuration F arises from some digitized rigid motion U, *i.e.*, corresponds to some parameters a, b, c, t_1, t_2, t_3. Then the inequalities (4) state precisely the necessary and sufficient conditions for the existence of the translation part $\mathbf{t}$ of such a rigid motion, assuming that a, b, c are already known. Let us now remark that all these inequalities can be summed up in *three* inequalities indexed by i:

$$\max_{\mathbf{v} \in \mathcal{N}} \left(F(\mathbf{v})_i - \frac{1}{2} - \mathbf{R}_i \mathbf{v} \right) < \min_{\mathbf{v} \in \mathcal{N}} \left(F(\mathbf{v})_i + \frac{1}{2} - \mathbf{R}_i \mathbf{v} \right). \tag{5}$$

or, equivalently to the following list of inequalities

$$\forall \mathbf{v}, \mathbf{v}' \in \mathcal{N}, \quad F(\mathbf{v}')_i - \frac{1}{2} - \mathbf{R}_i \mathbf{v}' < F(\mathbf{v})_i + \frac{1}{2} - \mathbf{R}_i \mathbf{v}. \tag{6}$$

The key observation is that we have eliminated the variables t_1, t_2, t_3 and have reduced to a subsystem of inequalities in a, b, c. Moreover, due to the rational expression in the Cayley transform (2), we may use the following polynomials of degree 2:

$$q_i[\mathbf{v}, k_i](a, b, c) = (1 + a^2 + b^2 + c^2)(2k_i - 1 - 2\mathbf{R}_i \mathbf{v}), \tag{7}$$

for $i = 1, 2, 3$, namely

$$q_1[\mathbf{v}, k_1](a, b, c) = -2(1 + a^2 - b^2 - c^2)v_1 - 4(ab - c)v_2 - 4(b + ac)v_3 + 2k_1 - 1,$$

$$q_2[\mathbf{v}, k_2](a, b, c) = -4(ab + c)v_1 - 2(1 - a^2 + b^2 - c^2)v_2 - 4(bc - a)v_3 + 2k_2 - 1,$$

$$q_3[\mathbf{v}, k_3](a, b, c) = -4(ac - b)v_1 - 4(a + bc)v_2 - 2(1 - a^2 - b^2 + c^2)v_3 + 2k_3 - 1.$$

Inequality (6) can be rewritten as the quadratic polynomial inequalities

$$\forall \mathbf{v}, \mathbf{v}' \in \mathcal{N}, \quad Q_i[\mathbf{v}, \mathbf{v}', F(\mathbf{v})_i, F(\mathbf{v}')_i](a, b, c) > 0,$$

where

$$Q_i[\mathbf{v}, \mathbf{v}', k_i, k_i'](a, b, c) = q_i[\mathbf{v}, k_i](a, b, c) + 2(1 + a^2 + b^2 + c^2) - q_i[\mathbf{v}', k_i'](a, b, c), \tag{8}$$

for $i = 1, 2, 3$. The set of quadratic polynomials for our problem is then given by $\mathcal{Q} = \{ Q_i[\mathbf{v}, \mathbf{v}', k_i, k_i'](a, b, c) \mid i = 1, 2, 3, \mathbf{v}, \mathbf{v}' \in \mathcal{N}, k_i, k_i' \in H(\mathcal{N}) \}$. Figure 2 illustrates the zero sets of some quadratic polynomials in $\mathcal{Q}$.

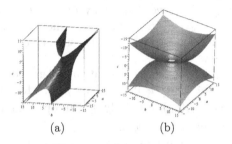

(a)   (b)

Fig. 2. Examples of the zero sets of two quadratic polynomials of $\mathcal{Q}$

4   Computing an Arrangement of Quadrics in 3D

In this section we discuss how to compute the arrangement of quadrics $Q(a,b,c) = 0$ for $Q \in \mathcal{Q}$ given by Eq. (8). Our strategy is similar to one proposed by Mourrain *et al.* [15]. The main differences are that we do not store information about cells different from sample points of full dimensional connected components and we precompute and sort all event points—points which induce changes of topology in an arrangement of quadrics—*a priori*. Moreover, we consider cases such as asymptotic critical values. In short, our method is as follows: Step 1: detect and sort all the events in which topology of an arrangement changes; Step 2: sweep by a plane the set of quadrics along a chosen direction. The sweeping plane stops between two event points and we project quadrics related to them onto the sweeping plane. This reduces to the problem of 2D arrangement of conics for each of such points. After this procedure, for each sample point we recover the translation part of the parameter space of digitized 3D rigid motions. The description of this last part will be given in the next section. Notice that proposed approach could be also applied to solve a similar problem in 2D, *i.e.*, generation of the different images of a 2D image patch under 2D digitized rigid motions – a solution to this problem was already proposed by Ngo *et al.* [17].

4.1   Bifurcation and Critical Values

In [15], the authors show how to describe an arrangement of quadric by sweeping a plane along a generic direction. Using the theory of *generalized critical values* [12,13,25] we will show how to compute a point per open connected component of an arrangement of quadric using a projection along a non-generic direction.

In the following, we consider an arrangement of smooth quadrics $S_i \subset \mathbb{R}^3$ defined by $Q_i(a,b,c) = 0$ for all $Q_i \in \mathcal{Q}$. Note that, if quadric S_i has isolated singularities then without loss of generality one can remove them and work with remaining smooth parts of S_i. We denote by $\mathcal{A}$ the set of maximally connected components of $\mathbb{R}^3 \setminus \bigcup_i S_i$.

Let $\mathcal{C}$ be an open cell of $\mathcal{A}$. We can associate to $\mathcal{C}$ the extremal values $\mathcal{C}_{\inf} = \inf\{a \mid (a,b,c) \in \mathcal{C}\}$ and $\mathcal{C}_{\sup} = \sup\{a \mid (a,b,c) \in \mathcal{C}\}$. We will show in this section that these values are included in a *bifurcation set* (see Definition 1).

In the following, for $i,j,k \in \mathbb{Z}$, we denote by S_i the surface defined by $Q_i(a,b,c) = 0$, by C_{ij} the curve defined by $Q_i = Q_j = 0$ and by P_{ijk} the points defined by $Q_i = Q_j = Q_k = 0$. Furthermore, we assume that S_i are smooth surfaces of dimension two, the C_{ij} are smooth curves of dimension one and $\rho(P_{ijk})$ are finite sets of values.

The projection map on the first coordinate a is denoted by ρ, and its restriction to a submanifold $\mathcal{M} \subset \mathbb{R}^3$ is denoted by $\rho_{|\mathcal{M}}$. Moreover, for $a_0 \in \mathbb{R}$ we denote by $\mathcal{M}_{a_0}$ the set $\rho_{|\mathcal{M}}^{-1}(a_0)$. Similarly, for an open interval $]a_0, b_0[\subset \mathbb{R}$ we denote by $\mathcal{M}_{]a_0,b_0[}$ the set $\rho_{|\mathcal{M}}^{-1}(]a_0, b_0[)$.

We are interested in computing the set of values a above which the topology of the cells of $\mathcal{A}$ change. We will show in Lemma 1 that this set is included in the bifurcation set of the projections on the first axis restricted to different manifolds.

Definition 1. *Let $\mathcal{M}$ be a submanifold of $\mathbb{R}^3$. We call* bifurcation set *of $\rho_{|\mathcal{M}}$ the smallest set $B(\rho_{|\mathcal{M}}) \subset \mathbb{R}$ such that $\rho : \mathcal{M} \setminus \rho^{-1}\left(B(\rho_{|\mathcal{M}})\right) \to \mathbb{R} \setminus B(\rho_{|\mathcal{M}})$ is a locally trivial fibration.*

More specifically, for all $a_0 \in \mathbb{R} \setminus B(\rho_{|\mathcal{M}})$, there exists $\epsilon > 0$ and a homeomorphism

$$\psi : \,]a_0 - \epsilon, a_0 + \epsilon[\, \times \mathcal{M}_{a_0} \to \mathcal{M}_{]a_0 - \epsilon, a_0 + \epsilon[},$$

such that $\rho \circ \psi(x, p) = x$ for all $(x, p) \in \,]a_0 - \epsilon, a_0 + \epsilon[\, \times \mathcal{M}_{a_0}$.

In the following, we will consider the finite set $B \subset \mathbb{R}$ defined as the union of the bifurcation sets of $\rho_{|S_i}$ and $\rho_{|C_{ij}}$ and the projections of P_{ijk}. More precisely, we define:

$$B_i = B(\rho_{|S_i}) \cup \bigcup_{j \neq i} B(\rho_{|C_{ij}}) \cup \bigcup_{j \neq i, k \neq i, j \neq k} \rho(P_{ijk})$$

and $B = \bigcup_i B_i$.

Lemma 1. *Let $\mathcal{C}$ be a maximal open connected cell of $\mathbb{R}^3 \setminus \bigcup_i S_i$. Let β be the smallest value of B such that $\mathcal{C}_{\inf} < \beta$ and let $a_0 \in]\mathcal{C}_{\inf}, \beta[$. Finally, let $\partial \mathcal{C}_{a_0}$ be the boundary of $\mathcal{C}_{a_0}$ and $J_{\mathcal{C}}$ be its edges. More precisely, $J_{\mathcal{C}}$ is the set of indices i such that the intersection of S_i with $\partial \mathcal{C}_{a_0}$ has dimension one. Then $\mathcal{C}_{\inf} \in B_i$ for all $i \in J_{\mathcal{C}}$.*

Proof. Let $i \in J_{\mathcal{C}}$ and let p be a point on $S_i \cap \partial \mathcal{C}_{a_0}$ that does not belong to any surface S_j for $j \neq i$. Let $\alpha \leq \mathcal{C}_{inf}$ be the maximal point of B_i less than a_0. Then $\rho_{|S_i}$ and $\rho_{|C_{ij}}$ are trivial fibrations above $]\alpha, \beta[$ and $\rho(P_{ijk}) \cap]\alpha, \beta[= \emptyset$ for $j \neq i$ and $k \neq i$ different integers. In particular, the points of the curves C_{ij} never cross above $]\alpha, \beta[$. More formally, there exists a continuous function $\phi :]\alpha, \beta[\to S_i$ such that $\phi(a_0) = p$ and $\rho \circ \phi(x) = x$ and $Q_j(\phi(x)) \neq 0$ for all $j \neq i$. Let T_ϵ be the tube defined by $T_\epsilon = \{(a, b, c) \in \mathbb{R}^3 \mid a \in [\mathcal{C}_{\inf}, a_0] \text{ and } \|(a, b, c) - \phi(a)\| < \epsilon\}$. We now prove by contradiction that $\alpha = \mathcal{C}_{\inf}$. If $\alpha < \mathcal{C}_{\inf}$, then there exists a sufficiently small $\epsilon > 0$ such that the respective intersections of T_ϵ with $Q_i < 0$ and $Q_i > 0$ are connected and such that T_ϵ does not intersect any S_j for $j \neq i$. Since $p \in T_\epsilon$, the intersection of T_ϵ with $\mathcal{C}$ is not empty. Moreover, $\mathcal{C}$ is a maximally connected component in the complement of the union of the S_j, such that one of the two connected component of $T_\epsilon \setminus S_i$ is included in $\mathcal{C}$. Thus, the tube T_ϵ intersects $\mathcal{C}_a$ for all $a \in [\mathcal{C}_{\inf}, a_0]$. In particular $\mathcal{C}_{\mathcal{C}_{\inf}}$ is not empty, which is a contradiction with the definition of $\mathcal{C}_{\inf}$. In particular, $\mathcal{C}_{\inf} = \alpha$, which allows us to conclude that $\mathcal{C}_{\inf} \in B_i$. $\square$

Figure 5 illustrates intervals such that the topology of $\mathcal{C}_a$, $a \in]\alpha, \beta[$ remains constant.

For each value $v_0 \in B$, we denote by $J_a \subset \mathbb{N}$ the set of indices i such that $v_0 \in B_i$. Moreover, for a set of indices J, we denote by $\mathcal{A}_J$ the set of maximally open connected components of $\mathbb{R}^3 \setminus \bigcup_{j \in J} S_j$.

Corollary 1. *Let C be a maximal open connected cell of $\mathbb{R}^3 \backslash \bigcup_i S_i$. Let $m > C_{\inf}$ be the smallest value of B greater than $C_{\inf}$. For all $a \in]C_{\inf}, m[$, there exists a cell $C' \in A_{J_{C_{\inf}}}$ such that $C'_a \subset C_a$.*

Proof. According to Lemma 1, $C_{\inf}$ is contained in all B_i such that S_i intersects the border of C_a with dimension one. In particular, one of the cells of $A_{J_{C_{\inf}}} \cap \rho^{-1}(a)$ is included in C_a. $\qquad\square$

From a constructive point of view, the authors of [12] showed that the bifurcation set is included in the union of the critical and asymptotic critical values. More specifically, given a polynomial map $f : M \rightarrow \mathbb{R}$, we have $B(f) \subset K(f) \cup K_\infty(f)$, where $K(f)$ are the critical values of f and K_∞ are its asymptotic critical values. In [15], the authors called the points of $K(\rho_{|S_i})$ events of type A, the points of $K(\rho_{|C_{ij}})$ events of type B and the points $\rho(P_{ijk})$ events of type C. We extended their classification for degenerate projections, and we say that the points of $K_\infty(\rho_{|S_i})$ are of type A_∞ and the points of $K_\infty(\rho_{|C_{ij}})$ are of type B_∞.

From a computational point of view, we recall in the next section how to compute the critical values of types A, B and C. For the types A_∞ and B_∞, we use the results from [12] and simplify them for the case of quadrics.

Finally as described in Subsect. 4.4, our strategy will be to compute the generalized critical values a and for each value, we store also J_a the set of indices i such that either:

- $a \in K(\rho_{|C_i}) \cup K_\infty(\rho_{|C_i})$
- $a \in K(\rho_{|C_{ij}}) \cup K_\infty(\rho_{|C_{ij}})$ for $j \neq i$
- $a \in \rho(P_{ijk})$ for $j \neq i, k \neq i$ and $j \neq k$

This will allow us to reduce the number of quadrics to consider in the intermediate steps of our sweeping plane algorithm.

4.2 Detection of Critical Values

Type A. The first type corresponds to values $s \in K(\rho_{|S_i})$ above which topology of open connected components in A changes. Algebraically, such an event corresponds to a value of $s = a$—called a-critical value—for which there is a solution to the system $Q_i(s, b, c) = \partial_b Q_i(s, b, c) = \partial_c Q_i(s, b, c) = 0$.

Type B. This type corresponds to the case $s \in K(\rho_{|C_{ij}})$. Such an event corresponds to an a-critical values for which there are solutions to the system $Q_i(s, b, c) = Q_j(s, b, c) = (\nabla Q_i \times \nabla Q_j)_1(s, b, c) = 0$.

Type C. There are values $s \in \rho(P_{ijk})$ above which topology of open connected components in A changes. An a-critical value is such that there are solutions to the system $Q_i(s, b, c) = Q_j(s, b, c) = Q_k(s, b, c) = 0$. Note that it can happen that an intersection between three quadrics is a curve. This issue can be solved if a curve projects on a point, thanks to the elimination theory and use of resultants or Gröbner basis, by computing univariate polynomial which vanishes on the projection of the curve [6]. For more information about events of the types A, B and C we refer the reader to [15]. Figure 3 shows examples of events of types A, B and C.

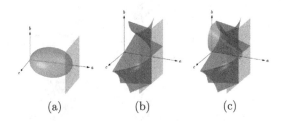

(a)   (b)   (c)

Fig. 3. Example of events of type A—sweeping plane tangent in a point to a quadrics (a), B—curve of intersection of two quadrics lay in a sweeping plane (b) and (c) a point of intersection of three quadrics lays in a sweeping plane. Sets of points, which induce an event are depicted in red and a sweeping plane is depicted in green (Color figure online)

Right now, we are going to discuss the cases of *asymptotic critical values*.

Type A_∞. This type of critical values corresponds to the situation when a plane orthogonal to one of the coordinate axes is tangent to a quadric in a point at infinity (see Fig. 4).

Lemma 2. *Let S be a smooth quadric defined by $Q(a, b, c) = 0$. Denoting by*
$$M(a) \text{ the matrix } \begin{pmatrix} \frac{\partial^2 Q}{\partial b^2} & \frac{\partial^2 Q}{\partial b \partial c} & \frac{\partial Q}{\partial b} \\ \frac{\partial^2 Q}{\partial c \partial b} & \frac{\partial^2 Q}{\partial c^2} & \frac{\partial Q}{\partial c} \end{pmatrix}(a, 0, 0) \text{ that depends only on } a,$$

$$K_\infty(\rho_{|S}) = \{a \mid M(a) \text{ has rank at most } 1\}.$$

Proof. Consider the mapping $f : \mathbb{R}^3 \to \mathbb{R}^2$ such that $(a, b, c) \mapsto (a, Q(a, b, c))$. The definition of K_∞ implies $K_\infty(\rho_{|S}) = K_\infty(f) \cap \mathbb{R} \times \{0\}$. Let $q(a, b, c) = \frac{\max(|\frac{\partial Q}{\partial b}|, |\frac{\partial Q}{\partial c}|)}{\max(|\frac{\partial Q}{\partial a}|, |\frac{\partial Q}{\partial b}|, |\frac{\partial Q}{\partial c}|)}$. Then using [12, Proposition 2.5 and Definition 3.1] with $df = \begin{pmatrix} 1 & 0 & 0 \\ \frac{\partial Q}{\partial a} & \frac{\partial Q}{\partial b} & \frac{\partial Q}{\partial c} \end{pmatrix}$, we have that there exists a sequence $(a_n, b_n, c_n) \in \mathbb{R}^3$ such that $|b_n| + |c_n| \to \infty$ and $a_n \to a$ and $(|b_n| + |c_n|)q(a_n, b_n, c_n) \to 0$. Since $\frac{\partial Q}{\partial a}, \frac{\partial Q}{\partial b}$ and $\frac{\partial Q}{\partial c}$ are linear functions, this implies that in the definition of K_∞, the expression $|b_n| + |c_n|$ divided by the denominator of $q(a_n, b_n, c_n)$ is bounded. In particular the numerator of $q(a_n, b_n, c_n)$ converges toward 0. More specifically, $\frac{\partial Q}{\partial b}$ and $\frac{\partial Q}{\partial c}$ converge toward 0. On the other hand, either $|b_n|$ or $|c_n|$ goes toward infinity. Assume without restriction of generality that $|b_n|$ goes toward infinity. In this case, the function $\frac{\partial^2 Q}{\partial c^2} \frac{\partial Q}{\partial b} - \frac{\partial^2 Q}{\partial b \partial c} \frac{\partial Q}{\partial c}$ is a linear function that depends only on b. Then this function converges toward 0 if and only if the coefficient in front

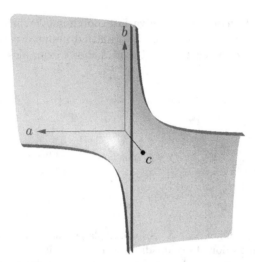

Fig. 4. Example of asymptotic critical value. A line (in blue) is parallel to the b-axis and tangent to a asymptote—red curves laying on the yellow surface—in a point at infinity. For readability only a part of the yellow surface is presented (Color figure online)

of b in this function and its constant coefficient are 0. In particular, if $\frac{\partial^2 Q}{\partial c^2}$ or $\frac{\partial^2 Q}{\partial b \partial c}$ is non-zero, the matrix $M(a)$ has rank 1. If both are 0, then with similar arguments, we can see that $M(a)$ is the null matrix. Thus K_∞ is the set of a such that $M(a)$ has a rank less than or equal 1. $\square$

The algorithm to detect this type of events is as follows. Step 1: we compute $\partial_b Q(a,b,c) = ub + vc + wa + t$ and $\partial_c Q(a,b,c) = u'b + v'c + w'a + t'$, where $u,v,w,t,u',v',w',t' \in \mathbb{Z}$ are coefficients of the corresponding polynomials and $Q \in \mathcal{Q}$. Step 2: let V stands for the cross product of $V_b = (u,v,wa+t)$ and $V_c = (u',v',w'a+t')$, then we solve for a such that all the terms of V are equal 0.

Type B_∞. In this case we are considering the asymptotic critical points of the projection restricted to a curve defined by the intersection of two quadrics $Q_i, Q_j \in \mathcal{Q}$. Using [12, Proposition 4.2], these correspond to the a-coordinate of the sweeping planes that cross the projective closure of the curve at infinity. More formally, we have:

$$K_\infty(\rho_{|C_{ij}}) = \{a \mid \exists (a_n, b_n, c_n) \in C_{ij} \text{ s.t. } |b_n| + |c_n| \to +\infty \text{ and } a_n \to a\}$$

In particular, this set is also the set of values a such that either the projection of C_{ij} on the (a,b)-plane or the projection of C_{ij} on the (a,c)-plane has an asymptote in a.

Using [12, Proposition 4.2], these are the elements of a set of non-properness of the projection map. The properties of this set and algorithms to compute it

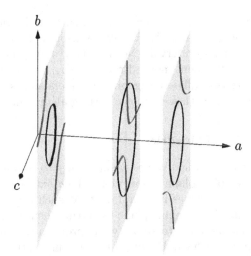

Fig. 5. Visualization of plane sweeping of quadrics. Projection planes at three different midpoints depicted in green. In between the planes we have a-critical values – a values in which topology of an arrangement changes. Conics obtain from quadrics in red and black (Color figure online)

have been studied notably in [7, 11, 14]. In our case the non-properness set of the projection restricted to C_{ij} is the set of a-coordinates of the sweeping planes that cross at infinity the projective closure of C_{ij}.

To detect such a case we apply the following steps. Step 1: we eliminate the c (resp. b) variable, and we denote the corresponding polynomials as $Pb(a, b)$ (resp. $Pc(a, c)$). Step 2: let $Cb(a)$ and $Cc(a)$ stands for head coefficients—coefficients of leading monomials—of $Pb(a, b)$ and $Pc(a, c)$, respectively. The asymptotic critical value for a pair of quadrics happens for $Cb(a) = 0$ or $Cc(a) = 0$.

4.3 Sorting Critical Values

In this section we focus on the representation of a-critical values as real algebraic numbers—roots of univariate polynomials—and operations such as comparison of them, necessary to sort a-critical values.

Similarly to Mourrain *et al.* [15] we represent a real algebraic number α as a pair: an irreducible univariate polynomial $P \in \mathbb{Z}[a]$ such that $P(\alpha) = 0$ and an open isolating interval $]g, h[, g, h \in \mathbb{Q}$, containing α and such that there is no other root of P in this interval. Note that, the isolation of the roots of an irreducible univariate polynomial can be made using Descartes' rule [27].

Let $\alpha = (P,]g, h[)$ and $\beta = (Q,]i, j[)$ such that $P, Q \in \mathbb{Z}[a]$ and $g, h, i, j \in \mathbb{Q}$, stand for two real algebraic numbers. Then we can conclude if $\alpha = \beta$ while checking a sign of GCD of P and Q at an intersecting interval. On the other hand, to conclude if α is bigger than β or β bigger than α we apply a strategy, which consists of refinement of isolating intervals until their disjointness. When two intervals are disjoint then we can compare their bounds and conclude if α is bigger than β (or β bigger than α)[1]. To refine an isolating interval of real roots, one can use *e.g.*, bisection of intervals, Newton interval method [9], [16, Chap. 5] or quadratic interval refinement method proposed by Abbot [1].

Ability to compare two different algebraic numbers allows us to sort a list of events which can be done with well-known sorting algorithm such as quicksort.

4.4 Sweeping a Set of Quadrics

After sorting the set of a-critical values we are ready to compute sample points of open cells. The sweeping plane moves along the a-axis and stops in between two consecutive a-critical values in a *midpoint*. At such a midpoint we project the set of quadrics onto the sweeping plane by setting their a variable to be equal to the midpoint. This allows us to simplify the problem at a midpoint into the arrangement of conics, which can be solved by applying a strategy similar to the one developed so far, such that we compute and sort a set of b-critical values (or c-critical values) in the arrangement of conics and sweep it by a line. One can also apply the critical points method. Note that we found cylindrical algebraic decomposition practically inefficient for such a problem. Figure 5 shows conics for three a-critical values in an arrangement of two quadrics.

The remaining question is which quadrics we should use at each midpoint to miss no open cell. In our approach we use all the quadrics of $\mathfrak{Q}$ for the first midpoint (see Fig. 6). Then for any other midpoint we use only the quadrics, related to the lowermost one from the pair, of a-critical values that bound this midpoint. Indeed, doing so we ensure that at the end of our strategy we collect at least one sample point for each full dimensional open cell thanks to Lemma 1, Corollary 1 and Lemma 2.

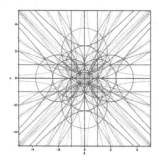

Fig. 6. The first projection of 81 quadrics obtain for the image patch $\mathcal{N} = \{(1,0,0),(0,1,0),(0,0,1),(0,0,0),(-1,0,0),(0,-1,0),(0,0,-1)\}$

---

[1] Our implementation of real algebraic numbers and their comparison can be downloaded from https://github.com/copyme/RigidMotionsMapleTools.

5   Recovering Translation Parameter Values

The algorithm proposed in the previous section gives us the set $\mathcal{R}$ of sample points $(a, b, c) \in \mathbb{Q}^3$, which correspond to the rotation parameters. In this section we discuss how to obtain sample points (t_1, t_2, t_3) of the translation part for each $(a, b, c) \in \mathcal{R}$ and how to generate different images of an image patch under rigid motions.

Let us first note that Eq. (3) under the assumption of $\mathbf{t} \in \left(-\frac{1}{2}, \frac{1}{2}\right)^3$ defines the set of planes in the range $\mathcal{P} = \left(-\frac{1}{2}, \frac{1}{2}\right)^3$ for each $(a, b, c) \in \mathcal{R}$, by setting $\mathbf{v} \in \mathcal{N}$ and $\mathbf{k} \in H(\mathcal{N})^3$. These planes divide $\mathcal{P}$ into cuboidal regions. Figure 7 illustrates an example of such critical planes in $\mathcal{P}$.

To obtain different images of an image patch $\mathcal{N}$ rotated by a given $a, b, c \in \mathbb{Q}$, under translations $(t_1, t_2, t_3) \in \mathcal{P}$, we compute the arrangement of planes in $\mathcal{P}$ which involves sorting of critical planes and finding a midpoint of each cuboidal region bounded by them.

Remark 1. Note that we can have several sample points (a, b, c) inducing the topologically equivalent arrangement of planes (the order of planes is identical). Therefore, to avoid unnecessary calculations we can define a hash function $\mathcal{H}$ which returns a different signature for each sample point (a, b, c) which induce a different order of the critical planes.

To define a hash function $\mathcal{H}$, let $\mathcal{I}$ stands for a collection of indexes of critical planes. Then we define the hash function that returns the sorted indices of $\mathcal{I}$ with respect to the order of critical planes.

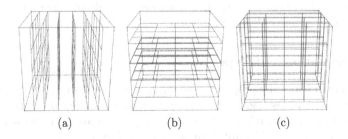

$$(a) \qquad\qquad (b) \qquad\qquad (c)$$

Fig. 7. Visualization of the critical planes for $\mathcal{N} = \{(1,0,0), (0,1,0), (0,0,1), (0,0,0), (-1,0,0), (0,-1,0), (0,0,-1)\}$ and some (a, b, c). For the sake of visibility three types of orthogonal critical planes are presented separately

6   A Case Study

In this section we would like to address some issues of the proposed algorithm while considering a particular image patch $\mathcal{N}$. Hereafter, we consider $\mathcal{N} = \{(1,0,0), (0,1,0), (0,0,1), (0,0,0), (-1,0,0), (0,-1,0), (0,0,-1)\}$.

6.1   Combinatorial Issue

The number of quadrics obtain directly from Eq. (8) for $\mathcal{N}$ is 441. In this section we will show that this number is reduced to 81 by discarding those which are always strictly positive (resp. negative) and ones which are redundant. Note that similar studies remain valid for different image patches.

Let us consider vectors $\mathbf{u}_1 = (1,0,0)$ (resp. $\mathbf{u}_2 = (0,1,0)$, $\mathbf{u}_3 = (0,0,1)$) and $\mathbf{h} = \left(\frac{1}{2}, \frac{1}{2}, \frac{1}{2}\right)$. Then we obtain the following inequality from (6)

$$\mathbf{u}_i \cdot (\mathbf{k}' - \mathbf{h} - \mathbf{R}\mathbf{v}') < \mathbf{u}_i \cdot (\mathbf{k} + \mathbf{h} - \mathbf{R}\mathbf{v}) \tag{9}$$

for $i = 1, 2, 3$ where $\mathbf{v}, \mathbf{v}' \in \mathcal{N}, \mathbf{k}, \mathbf{k}' \in H(\mathcal{N})^3$. This induces

$$k_i - k_i' + 1 - \mathbf{u}_i \cdot \mathbf{R}(\mathbf{v} - \mathbf{v}') > 0, \tag{10}$$

where we know that $K = k_i - k_i' + 1 \in \mathbb{Z} \cap [-1, 3]$. We then consider the following different cases of $\bar{V} = \|\mathbf{v} - \mathbf{v}'\|$:

1. when $\bar{V} = 0$, then there is no $K \in \mathbb{Z} \cap [-1, 3]$ satisfying (10),
2. when $\bar{V} = 1$, then there are 6 different pairs of $(\mathbf{v}, \mathbf{v}')$ and we obtain $K \in \{0\}$,
3. when $\bar{V} = 2$, then there are 6 different pairs of $(\mathbf{v}, \mathbf{v}')$ and we obtain $K \in \{-1, 0, 1\}$,
4. when $\bar{V} = \sqrt{2}$, then there are 12 different pairs of $(\mathbf{v}, \mathbf{v}')$ and we obtain $K \in \{-1, 0, 1\}$.

Therefore, the number of valid quadrics $Q[\mathbf{v}, \mathbf{v}', k_i, k_i']$ for each case is 0 (case 1), 6 (case 2), 18 (case 3) and 36 (case 4). Note that case 2 is included in case 3 up to a constant, as that we can ignore the 6 quadrics. This finally gives us $\frac{18+36}{2} = 27$ quadrics per direction and thus 81 in total.

6.2   Implantation and Experiments

We have implemented the proposed algorithm in Maple 2015 and our code can be downloaded from https://github.com/copyme/RigidMotionsMapleTools. In our implementation we have tried to obtain a good performance. Since the computation of critical values and sample points are not difficult to parallelize, we implemented this part of the algorithm in the Maple Grid framework and we performed tests on a machine equipped with two processors Intel(R) Xeon(R) E5-2680 v2; clocked at 2.8 GHz, with installed 251.717 GiB of memory. After the uncoupling, we obtained 81 quadrics, as predicted. Computation of sample points in such an arrangement took for 20, 15, 10, 5 computational nodes around 16, 19, 22 and 33 min, respectively. Notice that computing and sorting of critical values took around 2 min for 20–10 nodes and around 3 min for 5 nodes. Note that, sorting of critical values was performed on one node. Moreover, computations in such setting need around 21 GiB of memory. Presented real time and memory usage were obtained thanks to the Maple function Usage from the package CodeTools. In our implementation we used for each arrangement of conics the critical point method implemented in RAGlib which returns rational sample points [28]. Note that the average of the number of conics per midpoint is around 11.

Finally, for all the obtained each sample point (a, b, c) we recovered the different images of the image patch under digitized rigid motions which took again around 18, 22, 28 and 42 min for 20, 15, 10 and 5 nodes respectively[2]. Note that the memory usage did not exceed few mebibytes. The computation of different images of the image patch $\mathcal{N}$ consists of calculating for each sample point (a, b, c) an arrangement of planes (see Fig. 7). In such an arrangement each sample point (t_1, t_2, t_3) of full dimensional cell bounded by planes represents the translation part of rigid motions. Using this information we generate an image of the image patch $\mathcal{N}$, by appling to it a digitized rigid motion given by the value of (a, b, c, t_1, t_2, t_3). Note that for different sample points (a, b, c) that belong to the same full dimensional component we observe that planes presented in Fig. 7 move but they do not change their order. Figure 8 shows some images of an image patch for fixed (a, b, c) values and different (t_1, t_2, t_3) values.

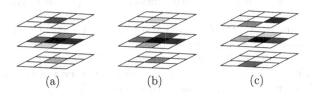

(a)   (b)   (c)

Fig. 8. Visualization of (a) the image patch $\mathcal{N} = \{(1, 0, 0), (0, 1, 0), (0, 0, 1), (0, 0, 0),$ $(-1, 0, 0), (0, -1, 0), (0, 0, -1)\}$ and its images under digitized rigid motions: the rotation given by $a = \frac{330688038827}{274877906944}, b = 7, c = 9$ followed by the translation (b) $\mathbf{t} = \mathbf{0}$ and (c) $\mathbf{t} = (-\frac{35}{100}, \frac{28}{100}, \frac{4}{10})$. For a sake of simplification, $\mathcal{N}$ is presented layer-by-layer. Each point of $\mathcal{N}$ is represented by a colored square with $\mathbf{c}$ in black

7   Conclusions

In this article, we proposed a method to decompose the 6D parameter space of digitized rigid motions for a given 3D image patch. We first uncouple the six parameters of 3D rigid motions to end up with two systems in three variables, and start by studying an arrangement of quadrics in $\mathbb{R}^3$.

Our approach to compute an arrangement of quadrics in 3D is similar to the one proposed by Mourrain et al. [15] where the main differences are: we do not use generic directions: we handle asymptotic cases and give new criteria to compute critical values in polynomials of degree two: we compute and store at least one sample point for each full dimensional open cell where Mourrain et al. [15] compute full adjacency information for all cells in an arrangement; moreover, we precompute all critical values a priori while in the former approach

---

[2] Note that this time is affected by a time needed to read a list of sample points (a, b, c) from a hard drive before the main computations.

only one type of critical values needs to be computed before the main algorithm. Those sample points are then used to decompose the other three dimensional parameter space. We also provided our implementation together with a numerical experiment for some small image patch.

As a part of our future work we would like to use the presented method in a study of topological alteration of $\mathbb{Z}^3$ under 3D digitized rigid motions.

Acknowledgments. This work received funding from the project Singcast (ANR–13–JS02–0006).

References

1. Abbott, J.: Quadratic interval refinement for real roots. Commun. Comput. Algebra **48**(1/187), 3–12 (2014)
2. Amir, A., Kapah, O., Tsur, D.: Faster two-dimensional pattern matching with rotations. Theoret. Comput. Sci. **368**(3), 196–204 (2006)
3. Basu, S., Pollack, R., Roy, M.F.: Algorithms in Real Algebraic Geometry. Springer, Heidelberg (2005)
4. Cayley, A., Forsyth, A.: The Collected Mathematical Papers of Arthur Cayley, vol. 1. The University Press, Cambridge (1898)
5. Collins, G.: Quantifier elimination for real closed fields by cylindrical algebraic decomposition. In: Brakhage, H. (ed.) Automata Theory and Formal Languages. LNCS, vol. 33, pp. 134–183. Springer, Heidelberg (1975)
6. Cox, D., Little, J., O'Shea, D.: Ideals, Varieties and Algorithms. An Introduction to Computational Algebraic Geometry and Commutative Algebra. Springer, New York (1996)
7. El Din, M.S., Schost, E.: Properness defects of projections and computation of atleast one point in each connected component of a real algebraic set. Discrete Comput. Geomet. **32**(3), 417 (2004)
8. Halperin, D.: Arrangements. In: Goodman, J.E., O'Rourke, J. (eds.) Handbook of Discrete and Computational Geometry, 2nd edn., pp. 529–562. Chapman and Hall/CRC (2004)
9. Hansen, E.: Global optimization using interval analysis - the multi-dimensional case. Numerische Mathematik **34**(3), 247–270 (1980)
10. Hundt, C., Liśkiewicz, M.: On the complexity of affine image matching. In: Thomas, W., Weil, P. (eds.) STACS 2007. LNCS, vol. 4393, pp. 284–295. Springer, Heidelberg (2007)
11. Jelonek, Z.: Topological characterization of finite mappings. Bull. Polish Acad. Sci. Math **49**(3), 279–283 (2001)
12. Jelonek, Z., Kurdyka, K.: Quantitative generalized Bertini-Sard theorem for smooth affine varieties. Discrete Comput. Geom. **34**(4), 659–678 (2005)
13. Kurdyka, K., Orro, P., Simon, S., et al.: Semialgebraic Sard theorem for generalized critical values. J. Diff. Geom. **56**(1), 67–92 (2000)
14. Moroz, G.: Properness defects of projection and minimal discriminant variety. J. Symbol. Comput. **46**(10), 1139–1157 (2011)
15. Mourrain, B., Tecourt, J.P., Teillaud, M.: On the computation of an arrangement of quadrics in 3D. Comput. Geom. **30**(2), 145–164 (2005)
16. Neumaier, A.: Interval Methods for Systems of Equations. Encyclopedia of Mathematics and its Applications. Cambridge University Press, Cambridge (1991)

17. Ngo, P., Kenmochi, Y., Passat, N., Talbot, H.: Combinatorial structure of rigid transformations in 2D digital images. Comput. Vis. Image Underst. **117**(4), 393–408 (2013)
18. Ngo, P., Kenmochi, Y., Passat, N., Talbot, H.: Topology-preserving conditions for 2D digital images under rigid transformations. J. Math. Imaging Vis. **49**(2), 418–433 (2014)
19. Ngo, P., Passat, N., Kenmochi, Y., Talbot, H.: Topology-preserving rigid transformation of 2D digital images. IEEE Trans. Image Process. **23**(2), 885–897 (2014)
20. Nouvel, B., Rémila, E.: On colorations induced by discrete rotations. In: Nyström, I., Sanniti di Baja, G., Svensson, S. (eds.) DGCI 2003. LNCS, vol. 2886, pp. 174–183. Springer, Heidelberg (2003)
21. Nouvel, B., Rémila, E.: Configurations induced by discrete rotations: periodicity and quasi-periodicity properties. Discrete Appl. Math. **147**(2–3), 325–343 (2005)
22. Pluta, K., Kenmochi, Y., Passat, N., Talbot, H., Romon, P.: Topological alterations of 3D digital images under rigid transformations. Research report, Université Paris-Est, Laboratoire d'Informatique Gaspard-Monge UMR 8049 (2014). https://hal.archives-ouvertes.fr/hal-01333586
23. Pluta, K., Romon, P., Kenmochi, Y., Passat, N.: Bijective rigid motions of the 2D cartesian grid. In: Normand, N., Guédon, J., Autrusseau, F. (eds.) DGCI 2016. LNCS, vol. 9647, pp. 359–371. Springer, Heidelberg (2016). doi:10.1007/978-3-319-32360-2_28
24. Pluta, K., Romon, P., Kenmochi, Y., Passat, N.: Bijectivity certification of 3D digitized rotations. In: Bac, A., Mari, J. (eds.) CTIC 2016. LNCS, vol. 9667, pp. 30–41. Springer, Heidelberg (2016)
25. Rabier, P.J.: Ehresmann fibrations and Palais-Smale conditions for morphisms of Finsler manifolds. Ann. Math. **146**, 647–691 (1997)
26. Renegar, J.: On the computational complexity and geometry of the first-order theory of the reals. Part I: Introduction. Preliminaries. The geometry of semi-algebraic sets. The decision problem for the existential theory of the reals. J. Symbol. Comput. **13**(3), 255–299 (1992)
27. Rouillier, F., Zimmermann, P.: Efficient isolation of polynomial's real roots. J. Comput. Appl. Math. **162**(1), 33–50 (2004)
28. Safey El Din, M.: Testing sign conditions on a multivariate polynomial and applications. Math. Comput. Sci. **1**(1), 177–207 (2007)
29. Singla, P., Junkins, J.L.: Multi-resolution Methods for Modeling and Control of Dynamical Systems. CRC Press, Boca Raton (2008)
30. Thibault, Y.: Rotations in 2D and 3D discrete spaces. Ph.D. thesis, Université Paris-Est (2010)
31. Thibault, Y., Sugimoto, A., Kenmochi, Y.: 3D discrete rotations using hinge angles. Theoret. Comput. Sci. **412**(15), 1378–1391 (2011)
32. Yilmaz, A., Javed, O., Shah, M.: Object tracking: a survey. ACM Computing Surveys (CSUR) **38**(4), 13 (2006)
33. Zitova, B., Flusser, J.: Image registration methods: a survey. Image Vis. Comput. **21**(11), 977–1000 (2003)

A Lower Bound for Computing Lagrange's Real Root Bound

Swaroop N. Prabhakar and Vikram Sharma[(✉)]

The Institute of Mathematical Sciences CIT Campus,
Taramani, Chennai 600113, India
{npswaroop,vikram}@imsc.res.in

Abstract. In this paper, we study a bound on the real roots of a polynomial by Lagrange. From known results in the literature, it follows that Lagrange's bound is also a bound on the absolute positiveness of a polynomial. A simple $O(n \log n)$ algorithm described in Mehlhorn-Ray (2010) can be used to compute the bound. Our main result is that this is optimal in the real RAM model. Our paper explores the tradeoff between improving the quality of bounds on absolute positiveness and their computational complexity.

Keywords: Real root bounds · Lagrange's bound · Absolute positiveness · Algebraic decision tree · Complexity lower bounds

1 Introduction

Root bounds are functions that operate on univariate polynomials with complex coefficients and compute an upper bound on the absolute value of its roots. The literature contains many root bounds; see, e.g., [16, Chap. 6]. Some of these root bounds (e.g., see van der Sluis [13]), are tight relative to the largest absolute value among all the roots of the polynomial. Often, however, one is interested in the special case of upper bounds on just the positive real roots of a polynomial with real coefficients; for instance, in the continued fraction based algorithms for real root isolation [2,12]. For this special case, the literature contains some bounds [1,4,7,14,15]. In [6], Hong showed that most of the known root bounds are in fact bounds for **absolute positiveness** of a polynomial, i.e., a real number such that the polynomial and all its non-vanishing derivatives are positive for any value greater than this real number. He introduced a new bound and showed that it is tight relative to the threshold of absolute positiveness of the polynomial. The quality of a root bound is defined to be the ratio of the bound with respect to the threshold of absolute positiveness. It was shown in [4] that within a general framework of bounds on absolute positiveness, Hong's bound is nearly optimal, i.e., it is off by a constant factor with respect to the best bound that is possible in this framework. Thus in terms of quality of real root bounds, Hong's bound is nearly the best. However, it was not clear if the quality of the bound was achieved at the cost of the increased effort in computing the bound, since a naive

© Springer International Publishing AG 2016
V.P. Gerdt et al. (Eds.): CASC 2016, LNCS 9890, pp. 444–456, 2016.
DOI: 10.1007/978-3-319-45641-6_28

implementation of Hong's bound has arithmetic cost quadratic in the degree, n, of the polynomial. This computational bottleneck was overcome by Mehlhorn and Ray [9], who gave an $O(n)$ arithmetic cost algorithm to compute Hong's bound for univariate polynomials.

Recently, Collins [5] showed that a real root bound by Lagrange [8] is always better than Hong's bound. It must be noted that the Lagrange's bound had not been covered in the framework proposed in [4]. A simplified derivation of the Lagrange's bound is given in [3,10], and an extension to the complex setting is given in [10]. The improvement is by a constant factor. Given this improvement, one can ask the following questions regarding Lagrange's real root bound:

Q1. Is the bound also a bound on the absolute positiveness of the polynomial?
Q2. Can the bound be computed using $O(n)$ arithmetic operations?

In Theorem 1, we given an affirmative answer to the first question. This result is not very surprising and immediately follows from known results in the literature.

Regarding the second question, we show in Theorem 4 that the answer is negative in the real RAM model [11]. This is done by reducing a certain decision problem in geometry, called the Point-Hull Bijection problem (introduced in Sect. 4), to comparing Lagrange's real root bound with Hong's bound. We then show in Theorem 2 that the complexity of the bijection problem in the real RAM model is $\Omega(n \log n)$. This is done by showing that any algebraic decision tree for deciding the Point-Hull Bijection problem has height roughly $\Omega(n \log n)$. To obtain this lower bound, we derive a lower bound on the "topological complexity" of the bijection problem. Using standard results (see Propositions 1 and 2) these lower bounds translate to lower bounds in various computational models, in particular, the real RAM model. Therefore, the best algorithm to compute Lagrange's real root bound is essentially the $O(n \log n)$ algorithm given in [9, Sect. 3.1]. Our result highlights the tradeoff between obtaining bounds for absolute positiveness that are better in quality than Hong's bound and the arithmetic complexity of computing them. In particular, we show that the constant factor improvement in the quality of Lagrange's real root bound over Hong's bound comes at an increased computational cost. In some sense, therefore, Hong's bound attains the right compromise in this quality-vs-complexity tradeoff.

2   Absolute Positiveness of Lagrange's Real Root Bound

In this section, we will prove that the Lagrange's real root bound [8] is a bound on the absolute positiveness of a polynomial. Let

$$f(x) := x^n - \sum_{k=0}^{n-1} a_k x^k, \tag{1}$$

where $a_k \in \mathbb{R}_{\geq 0}$. Let $R(f)$ be the maximum and $\rho(f)$ be the second maximum in the sequence $|a_k|^{1/(n-k)}$, $k = 0, 1, \ldots, n-1$ (we assume that $n > 1$). **Lagrange's real root bound** of f is defined as

$$L(f) := R(f) + \rho(f). \tag{2}$$

It is known that $L(f)$ is a bound on the positive roots of f [3,5,10]. We show that it is also a bound on the positive roots of its non-vanishing derivatives. First we prove the following result, a variation of the result in [4, Lemma 2.2], which shows that any upper bound on the positive roots of f is a bound on the absolute positiveness of f.

Lemma 1. $L(f)$ *is a bound on the absolute positiveness of* f *defined in* (1).

Proof. The jth derivative of f is given by

$$f^{(j)}(x) = \frac{n!}{(n-j)!}x^{n-j} - \sum_{k=j}^{n-1} \frac{k!}{(k-j)!}a_k x^{k-j}.$$

Taking $n!/(n-j)!$ common from the RHS, we get,

$$f^{(j)}(x) = \frac{n!}{(n-j)!}\left(x^{n-j} - \sum_{k=j}^{n-1} \frac{\frac{k!}{(k-j)!}}{\frac{n!}{(n-j)!}}a_k x^{k-j} \right).$$

Since $\frac{k!}{(k-j)!} < \frac{n!}{(n-j)!}$, we have

$$f^{(j)}(x) > \frac{n!}{(n-j)!}\left(x^{n-j} - \sum_{k=j}^{n-1} a_k x^{k-j} \right), \quad \text{for all } x > 0.$$

So,

$$f^{(j)}(x) > \frac{n!}{(n-j)!}\frac{f(x)}{x^j}, \quad \text{for all } x > 0.$$

Hence, $L(f)$ is a bound on the absolute positiveness of f.   □

Collins [5] used $L(f)$ to improve upon a root bound due to Hong [6]. Consider a general polynomial $f(x):=\sum_{i=0}^{n} a_i x^i \in \mathbb{R}[x]$, where $a_n > 0$. For every $a_i < 0$, define

$$s_i := \operatorname{argmin}\{|a_i/a_j|^{1/(j-i)} : j > i \; a_j > 0\}. \tag{3}$$

Now for each j such that $a_j > 0$, define

$$g_j(x) := a_j x^j + \sum_{s_i=j, a_i<0} a_i x^i.$$

Notice that g_j is in the form given in (1), so $R(g_j)$ and $\rho(g_j)$ are well-defined as the first and the second maximum, respectively, in the sequence $|a_i/a_j|^{1/(j-i)}$. Define $L(g_j)$ as in (2). However, this can be done if g_j has two or more negative coefficients; otherwise, if g_j has exactly one negative a_i, then $L(g_j)$ is taken to be the unique positive root of g_j; if g_j does not have negative coefficients, then $L(g_j):=0$. **Lagrange's Real Root Bound** of f is defined as

$$L(f):= \max_j L(g_j). \tag{4}$$

To compute $L(f)$, we can compute the polynomials g_j. This can be done in $O(n \log n)$ by the algorithm given in [9, Sect. 3.1]. We can then compute $L(g_j)$ in $O(n)$ time over all j. A further linear step of computing $\max_j L(g_j)$ gives us $L(f)$. Our first result is the following:

Theorem 1. *The Lagrange Real Root Bound $L(f)$ is a bound on the absolute positiveness of f.*

Proof. Since every negative monomial $a_i x^i$ has a unique s_i associated with it, we have

$$f(x) = \sum_{a_j > 0} g_j(x).$$

From Lemma 1 and the definition of $L(g_j)$, we know that $L(g_j)$ is a bound on the absolute positiveness of g_j. Hence, from (4), we conclude that $L(f)$ is a bound on the absolute positiveness of f.   □

Collins [5, Theorem 5] showed that $L(f)$ is better than the Hong's bound [6],

$$H(f) := 2 \max_{a_i < 0} \min_{\substack{a_j > 0 \\ j > i}} \left| \frac{a_i}{a_j} \right|^{1/(j-i)}.$$

Mehlhorn and Ray [9] gave an algorithm for computing $H(f)$ in $O(n)$ arithmetic operations. Can a similar algorithm exist for computing $L(f)$? In the following sections, we will answer this question in the negative.

3   Algebraic Decision Trees – Basic Notations and Definitions

Given two positive integers m, d, an (m, d)-**order algebraic decision tree** is a rooted tree T in which every internal node has associated with it a multivariate polynomial in m variables of total degree at most d. The input or domain of the decision tree is $\mathbb{R}^m$. Every internal node u of T has three children labeled "+", "−" and "0". The leaves output either a zero or a one. An algebraic decision tree computes a function from $\mathbb{R}^m$ to $\{0, 1\}$. The value of this function at $\mathbf{p} \in \mathbb{R}^m$ is computed as follows: we evaluate the polynomial associated with the root node of T at $\mathbf{p}$; depending on whether the sign of this evaluation is $-$, 0, or $+$, the computation proceeds recursively from the child of the root node labeled by the corresponding sign; we stop when we reach a leaf and output the value, either zero or one, associated with the leaf. From the description, it follows that the set of points in $\mathbb{R}^m$ that reach a given node in the tree form a semi-algebraic set. It is well known that a semi-algebraic set can be partitioned into connected components. Two points $\mathbf{p}, \mathbf{q} \in \mathbb{R}^m$ are said to be in the same connected component corresponding to a node u of T iff there exists a continuous curve $\gamma : [0, 1] \to \mathbb{R}^m$ such that $\gamma(0) = \mathbf{p}, \gamma(1) = \mathbf{q}$ and for all $t \in [0, 1]$, the point $\gamma(t)$ on the curve satisfies the set of polynomial equalities and inequalities from the root of T to the node u. The measure of complexity in this model is the height of the decision tree T, which counts the number of worst case polynomial evaluations from the root node to a leaf.

We say that an algebraic decision tree T **solves the membership problem** for a set $S \subseteq \mathbb{R}^n$ if it satisfies the following: T outputs 1 on $\mathbf{p} \in \mathbb{R}^m$ iff $\mathbf{p} \in S$. The main idea in showing a lower bound for a membership problem in the algebraic decision tree model is to lower bound the height of T in terms of, $\#S$, the total number of connected components in the set S. We can then use the following fundamental result of Ben-Or [11, p. 102] to obtain a lower bound on the height of T:

Proposition 1. *The height of any (m, d)-order algebraic decision tree T that solves the membership problem for S is $\Omega(\log_d(\#S) - m)$.*

We will crucially use the following fact that relates lower bounds in the algebraic decision tree model with lower bounds in the real RAM model [11, p. 30].

Proposition 2. *A lower bound for a decision problem $\mathcal{A}$ in the algebraic decision tree model implies the same lower bound on $\mathcal{A}$ in the real RAM model.*

4   Lower Bound on a Geometric Problem

Consider the lower hull H of $(n+2)$ points in $\mathbb{R}^2$ such that all the $(n+2)$ points are vertices of the lower hull; note that under this assumption the vertices of H can be ordered in increasing order of x-coordinate; in this paper, we only consider such hulls. From any point $p \in \mathbb{R}^2$ there are two rays that are tangent to the hull H. Of these two rays, the lower ray from p to H is the ray such that direction of the sweep to the other ray is counterclockwise. The **lower tangent** from p to H is the line corresponding to the lower ray from p. Note that p can be on H, in which case the lower tangent is an edge containing p; in particular, if p is a vertex of H, the lower tangent is the edge that has p as the left endpoint. The **point of lower tangency** for p is the left most vertex of H on the lower tangent from p. The definition ensures that the lower tangent is well-defined for all points in the plane.

The **Point-Hull Bijection problem** is the following: For a *fixed H*, given an ordered point set $P = (p_1, \ldots, p_n)$, where $p_i \in \mathbb{R}^2$, such that *all the points in P are to the left of the leftmost vertex of H*, determine if every vertex of H, excluding the leftmost and the rightmost vertex, is a point of lower tangency for some point in P? An ordered point set P that has such a bijection to the vertices of H is called a YES-instance to the problem. All other instances of P are NO-instances; in particular, if P has a point to the right of the leftmost point of H then it is a NO-instance. Since the input is a set of n points in $\mathbb{R}^2$, we take the length of the input to be $2n$.

Known algorithms for computing the points of lower tangency test whether a given point is on one side of a given line or on the line. These tests are equivalent to evaluating a polynomial, and hence these algorithms can be modeled as algebraic

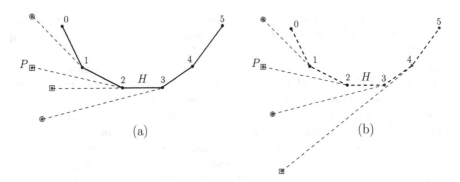

Fig. 1. A point set P shown in blue, hull H and lower tangencies. The points in P_e and P_o are shown circumscribed by boxes and circles, respectively. The figure labelled (a) is a NO-instance, whereas the figure labelled (b) is a YES-instance, to the Point-Hull Bijection problem.

decision trees. So algebraic decision trees solving the Point-Hull bijection problem can be thought of as computing a function from $\mathbb{R}^{2n}$ to the set $\{0, 1\}$. The set of ordered point sets P that are YES-instances to the problem form a connected components in $\mathbb{R}^{2n}$. Two YES-instances are in different connected components iff all continuous paths connecting these two instances contain a NO-instance. We will now derive a lower bound on the number of such components.

Suppose P is an ordered point set that is a YES-instance to the Point-Hull Bijection problem with respect to a given hull H. By enumerating the vertices of H from left to right, starting with 0 to $(n + 1)$, we partition P into two subsets as follows:

$$P_o := \{p_i \in P \mid p_i\text{'s point of lower tangency on H is odd}\}$$

and

$$P_e := \{p_i \in P \mid p_i\text{'s point of lower tangency on H is even}\}.$$

For the ease of exposition, we assume that all the odd indices in P are in P_o and all the even indices are in P_e. We now construct a large set $\mathcal{P}$ of ordered point sets obtained from P such that all these instances are solutions to the Point-Hull Bijection problem. Keeping P_o fixed, we apply a permutation σ to the indices of points in P_e; let P_σ be the ordered point set obtained in this manner from P. Note that the permutation σ only changes the order in which the points from P_e are processed, but P_σ is still a solution to the problem. The set $\mathcal{P}$, therefore, contains $(n/2)!$ many instances that are solutions to the Point-Hull Bijection problem. We are now in a position to derive the following lower bound:

Lemma 2. *There are at least $(n/2)!$ connected components for the Point-Hull Bijection problem.*

Proof. Consider two distinct ordered point sets $P_\sigma, P_{\sigma'} \in \mathcal{P}$. Then we know that there is an even position $2i$ such that $j := \sigma(2i)$ is not the same as $k := \sigma'(2i)$.

In other words, the points $p_j \in P_e$ at the position indexed $2i$ in P_σ and the point $p_k \in P_e$ at the same position in $P_{\sigma'}$ are different (by construction, the points in the odd position are the same in both).

Let ℓ be the vertical line touching the leftmost point of H. Consider a continuous curve $\gamma : [0, 1] \to \mathbb{R}^{2n}$ that connects P_σ and $P_{\sigma'}$. Without loss of generality, we assume that $\gamma(t)$ stays to the left ℓ; otherwise, we obtain a NO-instance to the problem. The component, $\gamma_{2i}(t)$, of $\gamma(t)$ gives us a continuous path between p_j and p_k. Since the points in P are to the left of ℓ, and the lower tangents intersect ℓ in decreasing order of y-coordinates, it follows that the points p_j and p_k are on opposite sides of the lower tangent incident on either the $(j-1)$ or the $(j+1)$ vertex of H. As $\gamma_{2i}(t)$ is a continuous function and is also restricted to the left of ℓ, it intersects one of these tangents. So we have a point set $Q \in \mathbb{R}^{2n}$ on $\gamma(t)$ such that there are two points in Q that have the same lower tangent in H, which means that Q is a NO-instance to the problem. Therefore, P_σ and $P_{\sigma'}$ are in different connected components, and so we have the desired lower bound. For an illustration, see Fig. 2.   □

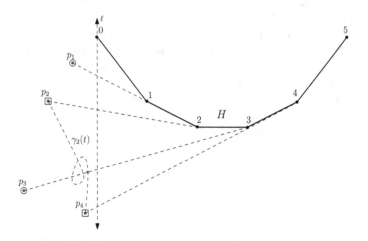

Fig. 2. In the example above $P_\sigma = \{p_1, p_2, p_3, p_4\}$ and $P_{\sigma'} = \{p_1, p_4, p_3, p_2\}$, $j = 2$ and $k = 4$. Now the component $\gamma_2(t)$ is a continuous path in $\mathbb{R}^2$ that takes p_2 to p_4. Clearly, the path intersects the lower tangent of p_3 at the point shown in red. (Color figure online)

Using the lemma above along with Propositions 1 and 2, we obtain the following lower bound.

Theorem 2. *The arithmetic complexity of any algorithm solving the Point-Hull Bijection problem is $\Omega(n \log n)$ in the real RAM model, where $2n$ is the length of the input.*

It must be noted that d does not play a role in the lower bound above, because for a given algebraic decision tree d is fixed and hence $(1/\log d)$ is a constant.

To show the lower bound on algorithms computing $L(f)$, we need a point-hull pair that satisfies certain properties. For a hull H, let MinSlope_H and MaxSlope_H denote the least and the largest slope over the edges of H. We call a point-hull pair (P, H) **nice** if it satisfies the following conditions:

(A1): $\text{MaxSlope}_H < \text{MinSlope}_H + 1$.
(A2): The interval $(\text{MinSlope}_H, \text{MaxSlope}_H]$ contains the slopes of all the lower tangents from P to H.
(A3): The x-coordinates of points in P and H are fixed to $0, \ldots, 2n + 1$.

An example of a nice point-hull pair is given in Fig. 3; assumptions (A1) and (A2) are not restrictive since we can construct instances where these assumptions hold, as shown in the figure.

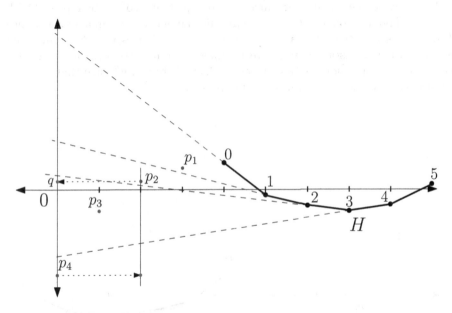

Fig. 3. The vertices labeled 0 to 5 are the vertices of H; the point set P is shown in blue; the red points are obtained by swapping the y-coordinates of p_2 and p_4. Note that q and p_3 have the same point of tangency on H. (Color figure online)

For a nice point-hull pair (P, H), the input is only the ordered set of y-coordinates of the points in P. However, our earlier argument in Lemma 2 breaks down, because we cannot permute points in P_e, since their x-coordinates are fixed and permuting the y-coordinates may yield a NO-instance to the Point-Hull Bijection problem; e.g., in Fig. 3, if we swap the y-coordinates of p_2 and p_4 then the resulting point set is a NO-instance.

For every input $\mathbf{y} \in \mathbb{R}^n$, define the ordered point set

$$P_{\mathbf{y}} := ((0, y_0), \ldots, (n - 1, y_{n-1})).$$

Since the x-coordinates are fixed, we have to count the number of connected components corresponding to $\mathbf{y} \in \mathbb{R}^n$ such that $(P_\mathbf{y}, H)$ is a YES-instance of the Point-Hull Bijection problem.

To create a large number of input instances that are in different connected components we do the following. For $(x_i, y_i) := p_i \in P_e$, we define the following point set

$$Q_i := \{\text{points of intersection of tangents incident on even vertices}$$
$$\text{in } H \text{ with the line } x = x_i\}.$$

For the example shown in Fig. 3, the sets Q_2 and Q_4 are illustrated in Fig. 4. For every $p_i \in P_e$, we have $|Q_i| = n/2$. So p_2 can be replaced with $n/2$ points from Q_2 corresponding to the $n/2$ tangents. However, to maintain a bijection, p_4 has to avoid the tangent on which p_2 is mapped, and so can be replaced with $((n/2)-1)$ points from Q_4. Continuing in this manner, we obtain a YES-instance $P_{\mathbf{y}'}$. The construction gives us $(n/2)!$ such input instances $\mathbf{y}' \in \mathbb{R}^n$. Our claim is that two such instances $\mathbf{y}, \mathbf{y}'$ are in different connected components in $\mathbb{R}^n$, i.e., on every continuous path connecting them there is a $\mathbf{y}''$ such that $P_{\mathbf{y}''}$ is a NO-instance to the Point-Hull Bijection problem.

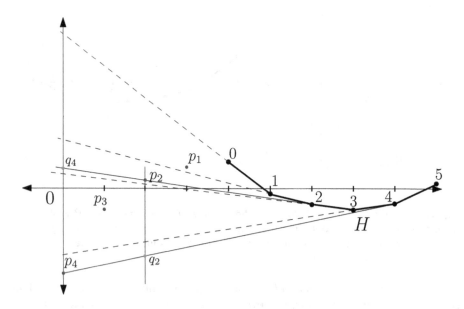

Fig. 4. The sets $Q_2 = \{p_2, q_2\}$ and $Q_4 = \{p_4, q_4\}$ corresponding to the example shown in Fig. 3.

Consider a continuous curve $\gamma : [0, 1] \to \mathbb{R}^n$ connecting $\mathbf{y}$ and $\mathbf{y}'$. There has to be a $p_i \in P_e$ that is mapped to $(x_i, y_i) \in P_\mathbf{y}$ and $(x_i, y_i') \in P_{\mathbf{y}'}$, where $y_i \neq y_i'$. Let $\gamma_i : [0, 1] \to \mathbb{R}$ be the ith component of γ that maps y_i to y_i'. Therefore, in $\mathbb{R}^2$, γ_i takes the point (x_i, y_i) to (x_i, y_i') along the line $x = x_i$. Since (x_i, y_i) and (x_i, y_i') are on two different tangents incident on even vertices in H, and γ_i can only move along the line $x = x_i$, it has to cross a tangent which is incident on an odd vertex of H; e.g., in Fig. 4, the path from p_2 to q_2 keeping the x-coordinate fixed crosses the lower tangent corresponding to p_3. So there is a point $\mathbf{y}'' \in \mathbb{R}^n$ along the path of γ from $\mathbf{y}$ to $\mathbf{y}'$ such that $P_{\mathbf{y}''}$ is a NO-instance to the Point-Hull Bijection problem. Hence $\mathbf{y}$ and $\mathbf{y}'$ are in two different connected components in $\mathbb{R}^n$. Therefore, we apply Propositions 1 and 2, to get the following result.

Theorem 3. *The arithmetic complexity of any algorithm solving the Point-Hull Bijection problem for a nice point-hull pair (P, H) in the real RAM model is $\Omega(n \log n)$, where n is the length of the input.*

5   Lower Bound on Computing Lagrange's Real Root Bound

In this section, we will use Theorem 3 to derive a lower bound on the arithmetic complexity of computing $L(f)$ (recall the definition from (4)). Before we proceed with the derivation, we reinterpret $L(f)$.

Given a polynomial $f(x) := \sum_{i=0}^{n} a_i x^i$, let

$$p_i := (i, \log(1/|a_i|))$$

be the point corresponding to the monomial $a_i x^i$ in f. For $a_i < 0$, define s_i as in (3); recall that s_i is only defined for negative monomials. For a given p_i such that $a_i < 0$, let H_i be the lower hull of the points in the set $\{p_j : j > i, a_j > 0\}$. By definition of s_i we have

$$\left| \frac{a_i}{a_{s_i}} \right|^{\frac{1}{s_i - i}} = \min_{j > i; a_j > 0} \log \left| \frac{a_i}{a_j} \right|^{\frac{1}{j - i}}.$$

This can be interpreted as the slope of the lower tangent from p_i to H_i; note that if $p_j \in H_i$ is the point of lower tangency for p_i then $s_i = j$. For $a_j > 0$, define T_j as the set of lower tangents associated with p_j, i.e.,

$$T_j := \{p_i \in P, \text{such that } s_i = j\}.$$

Let MaxSlope_{1j} and MaxSlope_{2j} be the first and second maximum over the slopes of the lower tangents of the points in T_j; if $|T_j| = 0$, then $\text{MaxSlope}_{1j} = 0$ and if $|T_j| = 1$, then $\text{MaxSlope}_{2j} = 0$. Define

$$\text{MaxSlope} := \max_{j} \{\text{MaxSlope}_{1j}, \text{where } a_j > 0\}. \tag{5}$$

Then we have the following interpretations: For Hong's bound

$$H(f) = 2^{1 + \text{MaxSlope}}, \tag{6}$$

and for Lagrange's real root bound

$$L(f) = \max\left(\max_{j:\,|T_j|=1} 2^{\mathrm{MaxSlope}_{1j}},\ \max_{j:\,|T_j|>1}\left(2^{\mathrm{MaxSlope}_{1j}} + 2^{\mathrm{MaxSlope2j}}\right)\right). \quad (7)$$

Using this interpretation, we will derive a lower bound on computing $L(f)$.

Theorem 4. *An algorithm for computing $L(f)$ for a real polynomial f of degree n requires $\Omega(n\log n)$ arithmetic operations in the real RAM model.*

Proof. The main idea of the proof is to use an algorithm for computing Lagrange's real root bound to decide the Point-Hull Bijection problem for a nice point-hull pair $(P_{\mathbf{y}}, H)$, where $\mathbf{y} \in \mathbb{R}^n$.

Let $(P_{\mathbf{y}}, H)$ be a nice point-hull pair such that

$$P_{\mathbf{y}} = \{(i, a_i) : i \in [0, \ldots, n-1], a_i \in \mathbb{R}\}$$

and

$$H = \{(i, b_i) : i \in [n, \ldots, 2n+1], b_i \in \mathbb{R}\}.$$

From $(P_{\mathbf{y}}, H)$, we construct the following polynomial

$$f(x) := \sum_{(i,b_i)\in H} \frac{x^i}{2^{b_i}} - \sum_{(i,a_i)\in P_{\mathbf{y}}} \frac{x^i}{2^{a_i}}. \quad (8)$$

This reduction from $(P_{\mathbf{y}}, H)$ to f requires $O(n)$ many exponentiation operations.

To decide the Point-Hull Bijection problem for $(P_{\mathbf{y}}, H)$, we do the following: compute $L(f)$ and $H(f)$, for f given in (8). If $2L(f) = H(f)$, we output YES; otherwise, we output NO. We now prove the correctness of this algorithm.

If $(P_{\mathbf{y}}, H)$ is a YES-instance of the Point-Hull Bijection problem, then for all j, such that $a_j > 0$, $|T_j| = 1$. Therefore, from (5), (6) and (7), we obtain that $H(f) = 2L(f)$.

Now we prove the converse: If $(P_{\mathbf{y}}, H)$ is a NO-instance of the Point-Hull Bijection problem then $2L(f) > H(f)$. Let j be an index such that $|T_j| > 1$. Then from the interpretation of $L(f)$ given in (7) we obtain that

$$2L(f) \geq 2(2^{\mathrm{MaxSlope}_{1j}} + 2^{\mathrm{MaxSlope}_{2j}})$$
$$\geq 2^{2+\mathrm{MinSlope}_H}$$
$$> 2^{1+\mathrm{MaxSlope}_H}$$
$$\geq 2^{1+\mathrm{MaxSlope}} = H(f),$$

where the second and fourth inequalities follow from assumption (A2), and the third inequality follows from assumption (A1).

Since $H(f)$ can be computed with $O(n)$ many arithmetic operations, we can decide whether a nice point-hull pair $(P_{\mathbf{y}}, H)$ is a YES-instance in essentially the time taken by the algorithm for computing $L(f)$. From the lower bound in Theorem 3 and the result in [11, p. 29, Proposition 1], we get the desired claim. $\square$

6 Conclusion and Further Directions

In this paper, we show that Lagrange's real root bound $L(f)$ is a bound on the absolute positiveness of a polynomial f. A goal in this line of work is to actually derive a tight bound on the largest positive root f, if one exists. Note that such a bound should be able to detect if f has a positive real root or not. It is clear that any algorithm for isolating real roots can be used to detect existence of a positive real root. In the converse direction, we can ask the following question: Is the problem of deciding whether a polynomial has a positive root at least as hard as isolating its real roots? One way to prove such a statement is to give a reduction from real root isolation that takes sub-quadratic (in the degree) arithmetic cost and makes at most sub-linear calls to detecting positive roots. On the other hand, one can also try to obtain an algorithm with sub-quadratic arithmetic cost for detecting or approximating positive roots.

Another direction to pursue is to generalize $L(f)$ to the multivariate setting. In [6], Hong actually derives a bound on the absolute positiveness of multivariate polynomials. In this setting, the notion of absolute positiveness is the following: A multivariate polynomial $P(x_1, \ldots, x_n)$ with real coefficients is said to be **absolutely positive** from a positive real value B iff P and all its non-zero partial derivatives of arbitrary order are positive for $x_1 \geq B, \ldots, x_n \geq B$. It is natural to derive a version of the Lagrange real root bound for multivariate polynomials and give an algorithm to compute it, similar to the one in [9]. One could then try to generalize the lower bound in Theorem 3 to this more general setting.

Acknowledgement. The authors would like to express their gratitude to Dr. Prashant Batra and the referees for their invaluable comments and suggestions.

References

1. Akritas, A.G., Strzeboński, A., Vigklas, P.: Implementations of a new theorem for computing bounds for positive roots of polynomials. Computing **78**, 355–367 (2006)
2. Akritas, A.: Vincent's theorem in algebraic manipulation. Ph.D. thesis, Operations Research Program, North Carolina State University, Raleigh, North Carolina (1978)
3. Batra, Prashant: On the quality of some root-bounds. In: Kotsireas, Ilias S., Rump, Siegfried M., Yap, Chee K. (eds.) MACIS 2015. LNCS, vol. 9582, pp. 591–595. Springer, Heidelberg (2016). doi:10.1007/978-3-319-32859-1_50
4. Batra, P., Sharma, V.: Bounds on absolute positiveness of multivariate polynomials. J. Symb. Comput. **45**(6), 617–628 (2010)
5. Collins, G.E.: Krandick's proof of Lagrange's real root bound claim. J. Symb. Comput. **70**(C), 106–111 (2015). http://dx.doi.org/10.1016/j.jsc.2014.09.038
6. Hong, H.: Bounds for absolute positiveness of multivariate polynomials. J. Symb. Comput. **25**(5), 571–585 (1998)
7. Kioustelidis, J.: Bounds for the positive roots of polynomials. J. Comput. Appl. Math. **16**, 241–244 (1986)

8. Lagrange, J.L.: Traité de la résolution des équations numériques de tous les degrés, Œuvres de Lagrange, vol. 8, 4th edn. Gauthier-Villars, Paris (1879)
9. Mehlhorn, K., Ray, S.: Faster algorithms for computing Hong's bound on absolute positiveness. J. Symbol. Comput. **45**(6), 677–683 (2010). http://www.science-direct.com/science/article/pii/S0747717110000301
10. Mignotte, M., Ştefănescu, D.: On an Estimation of Polynomial Roots by Lagrange. Prepublication de l'Institut de Recherche Mathématique Avancée, IRMA, Univ. de Louis Pasteur et C.N.R.S. (2002). https://books.google.co.in/books?id=NAd4NAEACAAJ
11. Preparata, F.P., Shamos, M.I.: Computational Geometry: An Introduction. Springer, New York (1985)
12. Sharma, V.: Complexity of real root isolation using continued fractions. Theor. Comput. Sci. **409**(2), 292–310 (2008)
13. van der Sluis, A.: Upper bounds for roots of polynomials. Numer. Math. **15**, 250–262 (1970)
14. Ştefănescu, D.: New bounds for the positive roots of polynomials. J. Univ. Comput. Sci. **11**(12), 2132–2141 (2005)
15. Ştefănescu, D.: A new polynomial bound and its efficiency. In: Gerdt, V.P., Koepf, W., Seiler, W.M., Vorozhtsov, E.V. (eds.) CASC 2015. LNCS, vol. 9301, pp. 457–467. Springer, Switzerland (2015). http://dx.doi.org/10.1007/978-3-319-24021-3_33
16. Yap, C.K.: Fundamental Problems of Algorithmic Algebra. Oxford University Press, Oxford (2000)

Enhancing the Extended Hensel Construction by Using Gröbner Bases

Tateaki Sasaki[1][(✉)] and Daiju Inaba[2]

[1] University of Tsukuba, Tsukuba-shi, Ibaraki 305-8571, Japan
sasaki@math.tsukuba.ac.jp
[2] Japanese Association of Mathematics Certification, Ueno 5-1-1,
Tokyo 110-0005, Japan
d.inaba@su-gaku.net

Abstract. Contrary to that the general Hensel construction (GHC) uses univariate initial Hensel factors, the extended Hensel construction (EHC) uses multivariate initial Hensel factors determined by the Newton polygon of the given multivariate polynomial. In the EHC so far, Moses-Yun's (MY) interpolation functions (see the text) are used for Hensel lifting, but the MY functions often become huge when the degree w.r.t. the main variable is large. In this paper, we propose an algorithm which uses, instead of MY functions, Gröbner bases of two initial factors which are homogeneous w.r.t. the main variable and the total-degree variable for sub-variables. The Hensel factors computed by the EHC are polynomials in the main variable with coefficients of mostly rational functions in sub-variables. We propose a method which converts the rational functions into polynomials by replacing the denominators by system variables. Each of the denominators is determined by the lowest order element of a Gröbner basis. Preliminary experiments show that our new EHC method is much faster than the previous one.

Keywords: Extended Hensel construction · Sparse multivariate polynomial · Singular leading coefficient · Gröbner basis · syzygy

1 Introduction

Let $F(x, \boldsymbol{u})$ be a polynomial in $\mathbb{K}[x, \boldsymbol{u}]$, where $(\boldsymbol{u}) = (u_1, \ldots, u_\ell)$ with $\ell \geq 2$ and $\mathbb{K}$ is a number field (one may consider $\mathbb{K} = \mathbb{Q}$). Let $f_n(\boldsymbol{u})$ be the leading coefficient (LC) of F w.r.t. x. We say that the leading coefficient of F is *singular* (or that F is of *LC-singular*) if $f_n(\boldsymbol{u})$ vanishes at the origin $\boldsymbol{0}$ of $\boldsymbol{u}$: $f_n(\boldsymbol{0}) = 0$.

The Hensel construction is one of the most useful techniques in computer algebra: well known are the Hensel construction of univariate polynomials over $\mathbb{Z}$ and the generalized Hensel construction (GHC) for multivariate polynomials [6,15].

Work supported by Japan Society for Promotion of Science KAKENHI Grant number 15K00005.

© Springer International Publishing AG 2016
V.P. Gerdt et al. (Eds.): CASC 2016, LNCS 9890, pp. 457–472, 2016.
DOI: 10.1007/978-3-319-45641-6_29

In ideal cases, the GHC is used for $F(x, \boldsymbol{u})$ without shifting the origin of $\boldsymbol{u}$. This simple usage fails for polynomials with singular LCs, and we must shift the origin to make the LCs non-singular. Then, another problem arises if F is sparse w.r.t. $\boldsymbol{u}$: origin shifting increases the number of terms very much. This problem is called the *nonzero substitution problem*.

Various attempts have been made to solve the nonzero substitution problem. In 1979, an idea was given by Zippel, based on Schwartz-Zippel's lemma [14,17]. Speaking roughly, the lemma says: *Choose* $(\boldsymbol{r}) = (r_1, \ldots, r_\ell) \in \mathbb{Z}^\ell$ *randomly satisfying* $|r_i| \leq L$ $(1 \leq i \leq \ell)$, *where* L *is enough large. Then, the probability that any nonzero polynomial* $Q(\boldsymbol{u})$ *becomes 0 at* $\boldsymbol{u} = \boldsymbol{r}$ *is enough small.* Zippel's idea is as follows. For unknown polynomial $P(x, \boldsymbol{u}) \in \mathbb{Q}[x, \boldsymbol{u}]$, if $P(x, \boldsymbol{r})$ contains no x^e-term, then interpolate $P(x, \boldsymbol{u})$ by assuming that P has no x^e-term. In [18], Zippel incorporated this algorithm with the GHC for monic polynomials, and showed that the interpolation can be done iteratively by solving a system of polynomial equations by Newton's method. In [5], de Kleine, Monagan and Wittkopf removed the monicness restriction of Zippel's algorithm and developed a very efficient GCD algorithm for sparse polynomials. Kaltofen [4] proposed another algorithm. He factorizes $F(x, \boldsymbol{s})$, with $(\boldsymbol{s}) \in \mathbb{Z}^\ell$, and recovers factors of $f_n(\boldsymbol{u})$ and $F(x, \boldsymbol{u})$ by variable-by-variable Hensel lifting of $F(x, \boldsymbol{u}-\boldsymbol{s})$ performed with Zippel's method so that the expression swell is suppressed.

An essentially new extension of GHC was done by Sasaki and Kako [12,13] in 1993, and they called the construction *extended Hensel construction (EHC)*. The EHC performs the Hensel construction of multivariate polynomials with singular LCs, without the nonzero substitution. The EHC was originally considered to compute "series expansion" of the multivariate algebraic function $x = \phi(\boldsymbol{u})$ defined as a solution of $F(x, \boldsymbol{u}) = 0$, with $\ell \geq 2$, at the *critical point* of $F(x, \boldsymbol{u})$: we say $(\boldsymbol{s}) \in \mathbb{C}^\ell$ a critical point of F if $F(x, \boldsymbol{s})$ has multiple root(s). In 2000, the present authors applied the EHC to factorization of sparse multivariate polynomials with singular LCs, by using polynomials as the initial factors [8]. In 2005, one of the author (D.I) implemented the EHC and showed that the EHC factorizes sparse LC-singular multivariate polynomials, by an order of magnitude faster than the GHC-based algorithm [2]. In 2015, Sanuki and present authors applied the EHC to compute the GCD of sparse LC-singular multivariate polynomials [10], and found that the algorithm is slightly slower than Maple's routine which is based on Zippel's idea.

The conventional algorithm of EHC is very bad for polynomials which are sparse in both x and $\boldsymbol{u}$: the algorithm is based on the so-called "Moses-Yun's" interpolation functions (MY functions) [7], and the computation of MY functions is very heavy; see Sect. 2 for examples. Furthermore, the conventional EHC algorithm has no clever device of handling rational functions in $\boldsymbol{u}$, appearing in the coefficients of Hensel factors. In this paper, we solve these problems on the conventional EHC drastically.

In Sect. 2, we survey briefly the conventional algorithm for EHC, and point out its faults by an example. In Sect. 3, we describe our new algorithm. In particular, we will prove that the denominator factor is determined by the lowest order element of the Gröbner basis. In Sect. 4, we show very good results of experiments for a non-trivial example.

2   A Brief Survey of the EHC

Let polynomial $F(x, \boldsymbol{u}) \in \mathbb{K}[x, \boldsymbol{u}]$ be as follows, where $(\boldsymbol{u}) = (u_1, \ldots, u_\ell)$.

$$F(x, \boldsymbol{u}) = f_n(\boldsymbol{u})x^n + f_{n-1}(\boldsymbol{u})x^{n-1} + \cdots + f_0(\boldsymbol{u}), \qquad f_n(\boldsymbol{u})f_0(\boldsymbol{u}) \neq 0. \quad (2.1)$$

We treat x and $\boldsymbol{u}$ as the main variable and sub-variables, respectively. By $\deg(F), \mathrm{ltm}(F), \mathrm{lc}(F)$, we denote the degree, the leading term, the leading coefficient (LC), respectively, w.r.t. x of F. By $\mathrm{rem}(F, G)$ and $\mathrm{res}(F, G)$, we denote the remainder of F by G and the resultant of F and G, respectively, w.r.t. x. By $\mathrm{cont}(F)$, we denote the content of F w.r.t. x, i.e., $\gcd(f_n(\boldsymbol{u}), \ldots, f_0(\boldsymbol{u}))$, where gcd denotes the greatest common divisor. F is called *primitive* and *of LC-singular* (often called that LC is singular) if $\mathrm{cont}(F) = 1$ and $f_n(0, \ldots, 0) = 0$, respectively. F is called *squarefree* if F has no duplicated root w.r.t. x. Let $T = c\,u_1^{e_1} \cdots u_\ell^{e_\ell}$ be a monomial, with $c \in \mathbb{K}$. By $\mathrm{tdeg}(T)$, we denote the *total-degree* of T: $\mathrm{tdeg}(T) = e_1 + \cdots + e_\ell$. By $\langle G_1, \ldots, G_r \rangle$, we denote the ideal generated by $G_1, \ldots, G_r$.

2.1   Newton Line and Newton Polynomial

We assume that the given polynomial $F(x, \boldsymbol{u})$ is squarefree and primitive. The EHC begins with determination of *Newton polynomial*; the Newton polynomials are unchanged by the EHC, so we define them only for the given polynomial.

Definition 1 (Newton line and Newton polynomial). *Let $F(x, \boldsymbol{u})$ be a polynomial in $\mathbb{K}(\boldsymbol{u})[x]$, where $\mathbb{K}$ is a number field. For each term $cx^d t^e u_1^{e_1} \cdots u_\ell^{e_\ell}$ of $F(x, t\boldsymbol{u})$, where c is a nonzero number, t is the total-degree variable for $\boldsymbol{u}$, and $e = e_1 + \cdots + e_\ell$, plot a dot at the point (d, e) in two-dimensional plane; see Fig. 1. The Newton polygon $\mathcal{N}$ of $F(x, \boldsymbol{u})$ is a convex hull containing all the dots plotted. Let the lower sides of $\mathcal{N}$, traced clockwise, be $\mathcal{N}_1, \ldots, \mathcal{N}_\rho$ which we call Newton lines. For each $i \in \{1, \ldots, \rho\}$, the Newton polynomial $\overline{F}_{\mathcal{N}_i}(x, \boldsymbol{u})$ is the sum of all the terms of $F(x, \boldsymbol{u})$ plotted on $\mathcal{N}_i$. Let n_i be the x-coordinate of the left end of $\mathcal{N}_i$ $(i = 1, \ldots, \rho)$. Then, $\overline{F}_{\mathcal{N}_i}$ can be divided by x^{n_i}. We express $\overline{F}_{\mathcal{N}_i}(x, \boldsymbol{u})/x^{n_i}$ by $F_{\mathcal{N}_i}(x, \boldsymbol{u})$ and call it net Newton polynomial.*

The EHC of $F(x, \boldsymbol{u})$ is performed successively on $\mathcal{N}_1 \to \mathcal{N}_2 \to \cdots \to \mathcal{N}_\rho$. In each construction on $\mathcal{N}_i$ $(i \in \{1, \ldots, \rho\})$, we define a weighting of variable x and $\boldsymbol{u}$ and introduce the order-variable t, depending on the slope λ_i of $\mathcal{N}_i$. Let the coordinate of right end of $\mathcal{N}_1$ be (n_0, ν_0) and the coordinate of left end of $\mathcal{N}_i$ be (n_i, ν_i). Then, the slope of $\mathcal{N}_i$ is given by $\lambda_i = (\nu_{i-1} - \nu_i)/(n_{i-1} - n_i)$. Let $\hat{n}_i$ and $\hat{\nu}_i$ be integers satisfying $\hat{n}_i > 0$, $\hat{\nu}_i/\hat{n}_i = \lambda_i$ and $\gcd(\hat{n}_i, \hat{\nu}_i) = 1$. Then, we define $\mathcal{F}(x, \boldsymbol{u}, t)$ and $\overline{\mathcal{F}}_{\mathcal{N}_i}(x, \boldsymbol{u})$, as follows; the terms of $\overline{\mathcal{F}}_{\mathcal{N}_i}(x, \boldsymbol{u})$ are given weight t^0, hence $\overline{\mathcal{F}}_{\mathcal{N}_i}$ is t-independent.

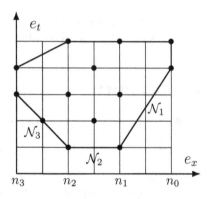

Fig. 1. Newton polygon and Newton lines $\mathcal{N}_3, \mathcal{N}_2, \mathcal{N}_1$

$$\mathcal{F}(x, \boldsymbol{u}, t) \stackrel{\text{def}}{=} t^{\hat{n}_i(\lambda_i n_i - \nu_i)} F(x/t^{\hat{\nu}_i}, t^{\hat{n}_i}\boldsymbol{u}), \tag{2.2}$$

$$\overline{\mathcal{F}}_{\mathcal{N}_i}(x, \boldsymbol{u}) \stackrel{\text{def}}{=} t^{\hat{n}_i(\lambda_i n_i - \nu_i)} \overline{F}_{\mathcal{N}_i}(x/t^{\hat{\nu}_i}, t^{\hat{n}_i}\boldsymbol{u}). \tag{2.3}$$

Note that $F_{\mathcal{N}}(x/t^{\hat{\nu}}, t^{\hat{n}}\boldsymbol{u})$ is $(x/t^{\hat{\nu}}, t^{\hat{n}}\boldsymbol{u})$-homogeneous. The Hensel lifting is done by changing the modulus $\langle t^k \rangle$ as $k = 1 \Rightarrow 2 \Rightarrow 3 \Rightarrow \cdots$.

2.2   EHC Based on MY Interpolation Functions

We briefly survey the conventional EHC which is based on the MY functions. By this survey, we will point out faults of the conventional EHC method and explain the necessity of introducing a new method of EHC suitable for parse multivariate polynomials. Below, we omit the i specifying the index of $\mathcal{N}_i$.

Let $F(x, \boldsymbol{u})$ be a squarefree polynomial having only one Newton line $\mathcal{N}$ of slope λ. Hence, $\deg(F) = \deg(F_{\mathcal{N}}) = n$ and we define $\hat{n}$ and $\hat{\nu}$ as in Sect. 2.1.

First, we factorize the $F_{\mathcal{N}}(x, \boldsymbol{u})$ as follows; since $F_{\mathcal{N}}(x, \boldsymbol{u})$ is a homogeneous polynomial, we can factorize it by treating it to be of $\ell - 1$ sub-variables.

$$\begin{cases} F_{\mathcal{N}}(x, \boldsymbol{u}) = \text{cont}(F_{\mathcal{N}})\, G_1^{(0)}(x, \boldsymbol{u}) \cdots G_r^{(0)}(x, \boldsymbol{u}), \\ \gcd(G_{j_1}^{(0)}, G_{j_2}^{(0)}) = 1 \quad (\forall j_1 \neq j_2). \end{cases} \tag{2.4}$$

The $G_i^{(0)}(x/t^{\hat{\nu}}, t^{\hat{n}}\boldsymbol{u})$ $(\forall i)$ is also $(x/t^{\hat{\nu}}, t^{\hat{n}}\boldsymbol{u})$-homogeneous. Second, for $l = 0, 1, \ldots, n-1$, we compute "Moses-Yun's functions" $A_1^{(l)}(x, \boldsymbol{u}), \cdots, A_r^{(l)}(x, \boldsymbol{u})$ to satisfy

$$\begin{cases} A_1^{(l)}(x, \boldsymbol{u}) \dfrac{F_{\mathcal{N}}(x, \boldsymbol{u})}{G_1^{(0)}(x, \boldsymbol{u})} + \cdots + A_r^{(l)}(x, \boldsymbol{u}) \dfrac{F_{\mathcal{N}}(x, \boldsymbol{u})}{G_r^{(0)}(x, \boldsymbol{u})} = x^l, \\ \deg(A_i^{(l)}) < \deg(G_i^{(0)}) \quad (1 \leq \forall i \leq r). \end{cases} \tag{2.5}$$

Third, we assign the factors of $\text{lc}(F)$ to $F, G_1^{(0)}, \ldots, G_r^{(0)}$ suitably, so that we have $\text{lc}(F) = \text{lc}(G_1^{(0)} \cdots G_r^{(0)})$; see [2] for the assignment. Then, we introduce the

t-order by (2.2), which converts F and $G_i^{(0)}$ ($1 \leq i \leq r$) into $\mathcal{F}$ and $\mathcal{G}_i^{(0)}$, respectively. Finally, using $\mathcal{G}_1^{(0)}, \ldots, \mathcal{G}_r^{(0)}$ as initial factors, we compute $\mathcal{G}_1^{(k)}, \ldots, \mathcal{G}_r^{(k)}$ for $k = 0 \to 1 \to 2 \to \cdots$, by the following formulas.

$$
\begin{cases}
\delta\mathcal{F}^{(k)} \equiv \mathcal{F}(x, \boldsymbol{u}, t) - \mathcal{G}_1^{(k-1)}(x, \boldsymbol{u}, t) \cdots \mathcal{G}_r^{(k-1)}(x, \boldsymbol{u}, t) \pmod{t^{k+1}} \\
\quad = t^{k+\hat{n}(\nu - \lambda n)} [\delta f_{n-1}(\boldsymbol{u}) x^{n-1} + \cdots + \delta f_0(\boldsymbol{u})], \\
\mathcal{G}_i^{(k)} = \mathcal{G}_i^{(k-1)} + t^{k+\gamma_i} \sum_{l=0}^{n-1} A_i^{(l)}(x, \boldsymbol{u}) \delta f_l(\boldsymbol{u}) \quad (i = 1, \ldots, r), \\
\quad \text{where } \gamma_i \text{ is the } t\text{-order of } G_i^{(0)}(x/t^{\hat{\nu}}, t^{\hat{n}}\boldsymbol{u}).
\end{cases}
\tag{2.6}
$$

After the construction, we put $G_i^{(k)}(x, \boldsymbol{u}) = \mathcal{G}_i^{(k)}(x, \boldsymbol{u}, 1)$ ($i = 1, \ldots, r$).

Since the upper equation for $l = 0$ in (2.5) can be rewritten as

$$
A_i^{(0)}(x, \boldsymbol{u}) \frac{F_N(x, \boldsymbol{u})}{G_i^{(0)}(x, \boldsymbol{u})} + B_i^{(0)}(x, \boldsymbol{u}) G_i^{(0)}(x, \boldsymbol{u}) = 1,
\tag{2.7}
$$

we can compute $A_i^{(0)}$ by the extended Euclidean algorithm in $\mathbb{K}(\boldsymbol{u})[x]$. Then, $A_i^{(l)}$ with $l \geq 1$ can be computed as $\text{rem}(x^l A_i^{(0)}, G_i^{(0)})$. By these, we can specify the expression of $A_i^{(l)}$ as follows; for concrete expressions of $A_i^{(l)}$, see [8,10].

Characteristic 1: $A_i^{(l)}(x, \boldsymbol{u})$ is a polynomial in x with coefficients in $\mathbb{K}(\boldsymbol{u})$.
Characteristic 2: The denominators of the rational function coefficients are products of $\text{res}(G_i^{(0)}, F_N/G_i^{(0)})$ and powers of $\text{lc}(G_i^{(0)})$.

2.3 Faults of the Conventional Method

The conventional method is simple mathematically but has the following faults.

Fault 1: Computing MY functions $A_i^{(l)}, B_i^{(l)}$ ($1 \leq i \leq r$; $0 \leq l < \deg(F)$), is based on the extended Euclidean method and quite time-consuming. Furthermore, even if F is sparse w.r.t. x, the sparseness is not utilized at all.
Fault 2: The denominators appearing in MY functions are often too large compared with those appearing in Hensel factors; there occurs cancellation between the denominator and the corresponding numerator.
Fault 3: Handling polynomials with coefficients of rational functions is pretty messy and time-consuming. We want to compute Hensel factors by using polynomial arithmetic as much as possible.

We show actual MS polynomials and Hensel factors by an Example.

Example 1. Let $F = GH + 3x^{10}y^5z^5$, where G and H are as follows.

$$
\begin{aligned}
G &= (x^5y^3 + 2z)(x^5z^3 - 2y) + x^3z^4, \\
H &= (x^5y^3 - 3z)(x^5z^3 + 3y) + x^7y^6.
\end{aligned}
\tag{2.8}
$$

F has only one Newton line and corresponding Newton polynomial is $(x^5y^3 + 2z)(x^5y^3 - 2z)(x^5z^3 + 3y)(x^5z^3 - 2y)$. We set initial polynomials G_0 and H_0 as

$G_0 := (x^5y^3 + 2z)(x^5z^3 - 2y)$ and $H_0 := (x^5y^3 - 3z)(x^5z^3 + 3y)$. Only x^{10}-, x^5- x^0-terms appear in G_0 and H_0, while F contains x^{20}-, x^{17}-, x^{15}-, x^{13}-, x^{12}-, x^{10}-, x^8-, x^7-, x^5-, x^3-, x^0-terms. Therefore, in the EHC of F by the conventional method, we must compute MY functions $A^{(l)}, B^{(l)}$ for 10 different values of l. On the other hand, in the method to be proposed in this paper, we compute a Gröbner basis of $\langle G_0, H_0 \rangle$, and the basis contains polynomials of only x^{10}-, x^5-, x^0-terms.

For reference, we show several of MY functions and Hensel factors computed by the conventional method. We note that the denominators of Hensel factors are factors of $\mathrm{res}(G_0, H_0) = -9765625\, y^{20}z^{20}\, (3y^4 - 2z^4)^5\, (2y^4 - 3z^4)^5$.

$$A^{(0)} = \frac{(y^7z^3 - y^3z^7)x^5 - 2y^8 + 5y^4z^4 - 2z^8}{5\,yz\,(3y^4 - 2z^4)(2y^4 - 3z^4)}, \quad B^{(0)} = \frac{(-y^7z^3 + y^3z^7)x^5 - 3y^8 + 5y^4z^4 - 3z^8}{5\,yz\,(3y^4 - 2z^4)(2y^4 - 3z^4)}$$

$$G^{(14)} = G_0 + x^3z^4 + \frac{18y^5z^5(y^4 - z^4)x^5 + 12y^6z^6}{5\,(3y^4 - 2z^4)(2y^4 - 3z^4)} + \frac{45y^8z^{12}(y^4 - z^4)x^8 + 9y^5z^9(2y^8 - y^4z^4 + 2z^8)x^3}{25\,(3y^4 - 2z^4)^2(2y^4 - 3z^4)^2},$$

$$H^{(14)} = H_0 + x^7y^6 - \frac{18y^5z^5(y^4 - z^4)x^5 + 27y^6z^6}{5\,(3y^4 - 2z^4)(2y^4 - 3z^4)}$$
$$- \frac{45y^8z^{12}(y^4 - z^4)x^8 + 27y^5z^9(9y^8 - 17y^4z^4 + 9z^8)x^3}{25\,(3y^4 - 2z^4)^2(2y^4 - 3z^4)^2}.$$

□

The above computation was done by an EHC package implemented in Mathematica. The above expressions are obtained by applying the GCD and factorization routines many times, as one can easily thinks, hence the computation is very slow and messy. Actual timing data will be given in Sect. 3.4.

3   Use of Gröbner Basis in the EHC

One may think that MY functions $A_1^{(l)}, \dots, A_r^{(l)}$ can be computed efficiently from (2.5), but it is wrong; we compute each $A_i^{(0)}$ by using (2.7), independently from others. Similarly, $G_i^{(k)}$ and $H_i^{(k)}$ are computed independently from others. Therefore, it is enough to discuss the case of two initial factors. In this section, we assume that two initial factors are $G_0, H_0 \in \mathbb{K}[x, \boldsymbol{u}]$. Then, the formulas on EHC become as follows.

$$\begin{cases} \mathcal{F}(x, \boldsymbol{u}, t) \equiv \mathcal{G}^{(k)}(x, \boldsymbol{u}, t)\mathcal{H}^{(k)}(x, \boldsymbol{u}, t) \pmod{t^{k+1}}, \\ t^k\delta F^{(k)} = \delta F^{(k)} \equiv \mathcal{F} - \mathcal{G}^{(k-1)}\mathcal{H}^{(k-1)} \pmod{t^{k+1}}. \end{cases} \tag{3.1}$$

$$\mathcal{G}^{(k)} = \mathcal{G}^{(k-1)} + t^k\delta G^{(k)}, \quad \mathcal{H}^{(k)} = \mathcal{H}^{(k-1)} + t^k\delta H^{(k)}, \quad \mathcal{G}^{(0)} = G_0, \quad \mathcal{H}^{(0)} = H_0. \tag{3.2}$$

$$\delta H^{(k)}G_0 + \delta G^{(k)}H_0 = \delta F^{(k)}, \quad \deg(\delta G^{(k)}) < \deg(G_0), \quad \deg(\delta H^{(k)}) < \deg(H_0). \tag{3.3}$$

Here, we have assumed that the leading coefficient of F has been processed to satisfy the above degree constraints. Note that $\delta G^{(k)}, \delta H^{(k)} \in \mathbb{K}(\boldsymbol{u})[x]$ in general. The leftmost equality in (3.3) suggests us to use the Gröbner basis of ideal $\langle G_0, H_0 \rangle$ for the computation of $\delta G^{(k)}$ and $\delta H^{(k)}$.

Remark 1. *The terms of polynomial $F(x, \boldsymbol{u})$ etc. are weighted depending on the slope of Newton line. However, we compute a Gröbner basis Γ of $\langle G_0.H_0 \rangle$ without the weighting variable t, because the terms of $F_N(x, \boldsymbol{u})$ are weighted to be 0; see (2.3).*

3.1   On a Gröbner Basis of $\langle G_0, H_0 \rangle$

The $x^e u_1^{e_1} \cdots u_\ell^{e_\ell}$ is called the *power product*. Let $\succ$ be a *term-order* which orders the power products in $\mathbb{K}[x, \boldsymbol{u}]$ uniquely, with 0 as the lowest order term. Let $f(\boldsymbol{u})$ be a polynomial in $\mathbb{K}[\boldsymbol{u}]$, expressed as the sum of monomials $c_i T_i$, $1 \leq i \leq m$:

$$f(\boldsymbol{u}) = c_1 T_1 + c_2 T_2 + \cdots + c_m T_m, \quad c_i \in \mathbb{K}, \quad T_1 \succ T_2 \succ \cdots \succ T_m. \quad (3.4)$$

By $\mathrm{htm}(f), \mathrm{hc}(f), \mathrm{hpp}(f), \mathrm{rest}(f)$, we denote the *head term*, the *head coefficient*, the *head power product*, the *rest terms*, respectively, of f: $\mathrm{htm}(f) = c_1 T_1$, $\mathrm{hc}(f) = c_1$, $\mathrm{hpp}(f) = T_1$, $\mathrm{rest}(f) = f - c_1 T_1$. The F is G-irreducible if $\mathrm{hpp}(G)$ does not divide $\mathrm{hpp}(F)$. The most important operations in computing Gröbner basis are the *M-reduction* (Mred) and the *S-polynomial* (Spol) construction: let F and G be polynomials expressed as in (3.4), then $\mathrm{Mred}(F, G) = F - [\mathrm{htm}(F)/\mathrm{htm}(G)]\, G$ if $\mathrm{hpp}(G) \mid \mathrm{hpp}(F)$, and $\mathrm{Spol}(F, G) = [L/\mathrm{htm}(F)]\, F - [L/\mathrm{htm}(G)]\, G$, where $L = \mathrm{lcm}(\mathrm{hpp}(F), \mathrm{hpp}(G))$ with lcm the operation of least common multiple. By $\mathrm{rem}(F, G)$ (we use "rem" again) we denote the result of successive applications of M-reduction of F by G so that $\mathrm{rem}(F, G)$ is G-irreducible. Note that, since we apply the Mred to only the head term, $\mathrm{rest}(\mathrm{rem}(F, G))$ may be G-reducible.

Let Γ be a Gröbner basis of $\langle G_0, H_0 \rangle$ w.r.t. the term order $\succ$ satisfying $x \succ u_1, \ldots, u_\ell$; Γ is not a reduced basis, as we have mentioned just above. Let $\widehat{G}_i$ be an element of Γ, then $\widehat{G}_i$ can be expressed as a linear combination of G_0 and H_0 as

$$\widehat{G}_i = A_i G_0 + B_i H_0, \quad A_i, B_i \in \mathbb{K}[x, \boldsymbol{u}]. \quad (3.5)$$

We call the pair (A_i, B_i) the *syzygy* for $\widehat{G}_i$. We express the syzygy by a polynomial $S_i(\gamma, \eta) = A_i \gamma + B_i \eta$, where γ and η are system variables. Summarizing,

$$\begin{cases} \Gamma = \{\widehat{G}_1, \ldots, \widehat{G}_s\} \subset \mathbb{K}[x, \boldsymbol{u}], \quad \widehat{G}_1 \prec \cdots \prec \widehat{G}_s, \\ \mathcal{S} = \{S_1(\gamma, \eta), \ldots, S_s(\gamma, \eta)\}, \quad S_i(G_0, H_0) = \widehat{G}_i \ (\forall i). \end{cases} \quad (3.6)$$

Lemma 1. *(1) $\widehat{G}_1 \in \mathbb{K}[\boldsymbol{u}]$. (2) If $\widehat{G}_{i \geq 2} \notin \mathbb{K}[\boldsymbol{u}]$ then $\widehat{G}_1 \mid \mathrm{res}(G_0, H_0)$.*

Proof. Since G_0 and H_0 are relatively prime, $R \stackrel{\mathrm{def}}{=} \mathrm{res}(G_0, H_0)$ is not 0. Since $R \in \langle G_0, H_0 \rangle$, R must be M-reducible to 0 by Γ. Therefore, Γ must contain element(s) in $\mathbb{K}[\boldsymbol{u}]$. This proves 1). Claim 2) is obvious from claim 1). □

3.2   Computing Polynomial Parts of $\delta G^{(k)}$ and $\delta H^{(k)}$

Suppose that $\delta F^{(k)} \in \mathbb{K}[x, \boldsymbol{u}]$. M-reducing $\delta F^{(k)}$ by the elements of Γ, we obtain

$$\delta F^{(k)} = \delta \widehat{F}^{(k)} + \delta R^{(k)}, \quad \delta \widehat{F}^{(k)} = q_1 \widehat{G}_1 + \cdots + q_s \widehat{G}_s, \quad q_i \in \mathbb{K}[x, \boldsymbol{u}] \text{ for } \forall i. \quad (3.7)$$

Here, $\delta R^{(k)}$ is in $\mathbb{K}[x, \boldsymbol{u}]$ and Γ-irreducible. We have $\delta \widehat{F}^{(k)} \in \langle G_0, H_0 \rangle$ and $\delta R^{(k)} \notin \langle G_0, H_0 \rangle$. Given $\delta \widehat{F}^{(k)}$, we can compute polynomials $\widehat{h}$ and $\widehat{g}$ in $\mathbb{K}[x, \boldsymbol{u}]$, satisfying $\delta \widehat{F}^{(k)} = \widehat{h} G_0 + \widehat{g} H_0$, as follows.

Method to calculate $\widehat{h}$ and $\widehat{g}$ satisfying $\widehat{h}G_0 + \widehat{g}H_0 = \delta\widehat{F}^{(k)}$: By replacing every $\widehat{G}_i$ in $\delta\widehat{F}^{(k)}$ by $S_i(\gamma, \eta)$, $1 \leq i \leq s$, $\delta\widehat{F}^{(k)}$ becomes a polynomial in $\mathbb{K}[\gamma, \eta, x, \boldsymbol{u}]$. Then, $\widehat{h}$ and $\widehat{g}$ are coefficients of γ and η, respectively, of $\delta\widehat{F}^{(k)}$.

$\square$

Remark 2. *We normalize each $\widehat{G}_i$, hence the syzygy (A_i, B_i) is also multiplied by a normalization constant. The $(q_1, \ldots, q_s)$ in (3.7) is computed as the syzygy for $\delta R^{(k)}$, but we must not normalize $\delta R^{(k)}$ and we must multiply -1 to the syzygy to get the $(q_1, \ldots, q_s)$.*

One may put $\delta G^{(k)} = \widehat{g}$ and $\delta H^{(k)} = \widehat{h}$. This is OK if $\deg(\widehat{g}) < \deg(G_0)$ and $\deg(\widehat{h}) < \deg(H_0)$. If these degree constraints are not satisfied, we must reduce the degrees of $\widehat{g}$ and $\widehat{h}$. For the degree reduction, the next lemma is very useful.

Lemma 2. *Let $\widehat{h}$ and $\widehat{g}$ computed as above be such that $\deg(\widehat{h}) \geq \deg(H_0)$ and $\deg(\widehat{g}) \geq \deg(G_0)$. If there exist $\delta G^{(k)}, \delta H^{(k)} \in \mathbb{K}[x, \boldsymbol{u}]$ satisfying $\deg(\delta G^{(k)}) < \deg(G_0)$, $\deg(\delta H^{(k)}) < \deg(H_0)$ and $\delta\widehat{F}^{(k)} = \delta H^{(k)}G_0 + \delta G^{(k)}H_0$, then we can compute $\delta G^{(k)}$ and $\delta H^{(k)}$ as $\delta G^{(k)} := \mathrm{rem}(\widehat{g}, G_0)$ and $\delta H^{(k)} := \mathrm{rem}(\widehat{h}, H_0)$, where the remainders may be computed either by the division or by Mreds.*

Proof. Since $\delta\widehat{F}^{(k)}$ can be expressed in two ways, we obtain $(\widehat{h} - \delta H^{(k)})G_0 = -(\widehat{g} - \delta G^{(k)})H_0$. Since G_0 and H_0 are relatively prime, this equality implies that $G_0 \mid (\widehat{g} - \delta G^{(k)}))$ and $H_0 \mid (\widehat{h} - \delta H^{(k)}))$. These relations and degree inequalities tell us that $\delta G^{(k)} = \mathrm{rem}(\widehat{g}, G_0)$ and $\delta H^{(k)} = \mathrm{rem}(\widehat{h}, H_0)$. $\square$

Theorem 1 (main theorem 1). *Let $\delta R^{(l)}$ and $\widehat{h}, \widehat{g}$ be defined as above. If $\delta R^{(k)} = 0$ and we can degree-reduce $\widehat{h}, \widehat{g}$ so that we have $\deg(\widehat{h}) < \deg(H_0)$ and $\deg(\widehat{g}) < \deg(G_0)$, then we have $\delta G^{(k)}, \delta H^{(k)} \in \mathbb{K}[x, \boldsymbol{u}]$. If $\delta R^{(k)} \neq 0$ or we cannot degree-reduce $\widehat{h}$ and $\widehat{g}$, then we must introduce rational functions in the coefficients w.r.t. x, of $\delta G^{(k)}$ and $\delta H^{(k)}$.*

Proof We have described the methods of computing $\widehat{h}, \widehat{g}$ and degree-reducing them. Therefore, if there exist $\delta G^{(k)}, \delta H^{(k)}$ in $\mathbb{K}[x, \boldsymbol{u}]$, then the former half of the theorem is valid. If $\delta R^{(k)} \neq 0$ while $\delta G^{(k)}, \delta H^{(k)} \in \mathbb{K}[x, \boldsymbol{u}]$ then (3.3) shows that $\delta R^{(k)}$ can be expressed as $\delta R^{(k)} = \delta F^{(k)} - \delta\widehat{F}^{(k)} = (\delta H^{(k)} - \widehat{h})G_0 + (\delta G^{(k)} - \widehat{g})H_0$. This means that $\delta R^{(k)} \in \langle G^{(0)}, H^{(0)} \rangle$, which is contradictory to that $\delta R^{(k)}$ is Γ-irreducible. The case that the degree-reduction fails will be treated by "forced degree-reduction" to be defined in Sect. 3.3, which shows that $\mathrm{lc}(G_0)$ and $\mathrm{lc}(H_0)$ will enter into the coefficients as the denominator factors. $\square$

3.3   How to Treat the Rational-Function Coefficients

In this subsection, assuming that $\delta R^{(k)} \neq 0$, we first determine a denominator factor, to be denoted by D, then we convert rational functions with denominator D into polynomials by introducing a system variable. We obey the following policy in determining D; with this policy, we need not handle rational-function coefficients superficially.

Our policy for treating rational-function coefficients: We determine a denominator $D \in \mathbb{K}[\boldsymbol{u}]$ so that the product $D \, \delta R^{(k)}$ may become a polynomial in $\langle G_0, H_0 \rangle$, i.e., we may have h and g in $\mathbb{K}[x, \boldsymbol{u}]$ satisfying $D \, \delta R^{(k)} = hG_0 + gH_0$. Then, we replace $1/D$ in $\delta G^{(k)}$ and $\delta H^{(k)}$ by the system variable $\%D$ and convert $\delta G^{(k)}$ and $\delta H^{(k)}$ into polynomials in $\mathbb{K}[\%D, x, \boldsymbol{u}]$. We determine D to be as low order as possible, and we order $\%D$ as $\%D \succ x \succ u_1, \ldots, u_\ell$. □

Method to calculate h and g satisfying $hG_0 + gH_0 = D \, \delta R^{(k)}$: Determine the denominator D as in the above policy, and replace the factor $\widehat{G}_1$ in $D \, \delta R^{(k)}$ by $A_1\gamma + B_1\eta$. Then, h and g are the coefficients of γ and η, respectively, of the resulting expression. Finally, try to degree-reduce the h and g by using Lemma 2. □

Forced degree-reduction of h and g: Consider, for example, to degree-reduce g by G_0. Multiplying $D' = \text{lc}(G_0)/\gcd(\text{lc}(g), \text{lc}(G_0))$ to g, we can eliminate $\text{ltm}(g)$ by G_0. Thus, we compute "pseudo-remainder" of g divided by G_0. Below, by D', we denote the product of terms $\in \mathbb{K}[\boldsymbol{u}]$, multiplied to g and h, in reducing all the high degree terms of h and g. □

Theorem 2. *Let D' be the denominator factor introduced by the forced degree-reduction ($D' = 1$ if no forced degree-reduction is made). First, determining D to be $D = \widehat{G}_1/\gcd(\widehat{G}_1, \delta R^{(k)})$, compute $\delta H^{(k)}, \delta G^{(k)}$ satisfying $D'D \, \delta R^{(k)} = \delta H^{(k)}G_0 + \delta G^{(k)}H_0$, $\deg(\delta H^{(k)}) < \deg(H_0)$ and $\deg(\delta G^{(k)}) < \deg(G_0)$. After this, if $C = \gcd(D'D, \text{cont}(\delta H^{(k)}), \text{cont}(\delta G^{(k)})) \neq 1$, then reset $D'D$ as $D'D := D'D/C$. Then every rational-function coefficient of $\delta G^{(k)}$ and $\delta H^{(k)}$ is expressed as $N_j/D'D$, where $N_j \in \mathbb{K}[\boldsymbol{u}]$.*

Proof. Note that $D \, \delta R^{(k)}$ is a polynomial in $\mathbb{K}[x, \boldsymbol{u}]$, having a factor $\widehat{G}_1$. Hence, We can determine polynomials h and g satisfying $D \, \delta R^{(k)} = hG_0 + gH_0$, and we can determine $\delta H^{(k)}, \delta G^{(k)}$ satisfying $D'D \, \delta R^{(k)} = \delta H^{(k)}G_0 + \delta G^{(k)}H_0$, with degree conditions. Dividing this equality by $D'D/C$, we obtain the theorem. □

Remark 3. *If $\delta F^{(k)}$ contains denominator variables $\%D[1], \ldots$, then we take out coefficients of denominator variables and M-reduce the coefficients by Γ.*

We give a simple choice of D, by assuming that

$$\widehat{G}_1, \ldots, \widehat{G}_\lambda \in \mathbb{K}[\boldsymbol{u}], \quad \widehat{G}_{\lambda+i} \notin \mathbb{K}[\boldsymbol{u}] \quad (\forall i \geq 1). \tag{3.8}$$

Example 2. Let $F(x, y, z)$ be as follows, where x is the main variable.

$$F = (y^2z)\, x^4 + (y^3z^2 + yz^2 + yz)\, x^3 + (y^4z + y^2z^3 + 2y^2z^2 + y - z)\, x^2$$
$$+ (y^3z^2 + y^3z + y^2z + yz^3)\, x + (y^3 - y^2z + yz - z^2) + \underline{3xyz}.$$

Without the last underlined term, F is factorizable. F has two Newton lines of slopes $-1/2$ and 1. The Newton polynomial on the right Newton line is

$$\overline{F}_{\mathcal{N}_1} = (x^2y^2z + xyz + y - z) \times x^2 \overset{\text{def}}{=} G_0H_0, \quad H_0 = x^2.$$

The Gröbner basis of $\langle G_0.H_0 \rangle$ w.r.t. the lexicographic (lex) order is

$$\{\widehat{G}_1,\ldots,\widehat{G}_4\} = \{y^2 - 2yz + z^2,\ xz^2 + y - z,\ xy - xz,\ x^2\},$$

and the corresponding syzygies $S_1,\ldots.S_4$ are as follows.

$$S_1 = (-xyz + y - z)\,\gamma + (xy^3z^2 - y^3z + 2y^2z^2)\,\eta,$$
$$S_2 = (-xz + 1)\,\gamma + (xy^2z^2 - y^2z + yz^2)\,\eta,$$
$$S_3 = x\,\gamma + (-xy^2z - yz)\,\eta, \qquad S_4 = \eta.$$

We note that $\widehat{G}_1 = \mathrm{res}(G_0, H_0)$ in this example.

We show the first 2 steps of the EHC of F with initial factors G_0 and H_0. In this example, Theorem 2 specifies the denominator D to be $y - z$ at $k = 2$, but we dare to choose as $D = \widehat{G}_1 = (y - z)^2$. This is to give an illustrative example of simplification below.

At $k = 1$: $\delta\mathcal{F}^{(1)} = t^1 x^3yz^2 = t^1 (xyz^2 H_0) \Rightarrow \delta G^{(1)} = xyz^2,\ \delta H^{(1)} = 0,$
At $k = 2$: $\delta\mathcal{F}^{(2)} = t^2 (3xyz) = t^2 ((-3y^2z) H_0 + 3 G_0 + \delta R^{(2)}),\ \delta R^{(2)} = -3y + 3z,$
$\quad D_1$: we set $D_1 := \widehat{G}_1 = y^2 - 2yz + z^2,$
$\quad \Rightarrow \delta G^{(2)} = -3y^2z + \%D_1 \times (-3xy^4z^2 + 3xy^3z^3 + 3y^4z - 9y^3z^2 + 6y^2z^3),$
$\quad\ \delta H^{(2)} = 3 + \%D_1 \times (3xy^2z - 3xyz^2 - 3y^2 + 6yz - 3z^2)$

Observing the coefficients of $\%D_1$ in $\delta G^{(2)}$ and $\delta H^{(2)}$, we see that their last three terms, i.e., $(-3y^4z + 9y^3z^2 - 6y^2z^3)$ and $(3y^2 - 6yz + 3z^2)$ can be "simplified" by $\widehat{G}_1$, and that the simplified terms cancel $-3y^2z$ in $\delta G^{(2)}$ and 3 in $\delta H^{(2)}$, respectively. This kind of simplification appears quite frequently. □

If $\lambda \geq 2$ then there may be another candidate of denominator, let it be $D' \in \mathbb{K}[\boldsymbol{u}]$, such that $D' \prec D$. It is not easy to find the simplest one. Below, we will show only one simple choice.

A simple choice of denominator D in the case of $\lambda \geq 2$: Let D be the lowest order element of $\{\widehat{G}_i / \gcd(\widehat{G}_i, C\,\delta R^{(k)}) \mid i = 1, \ldots, \lambda\}$.

3.4 How to Simplify $\delta G^{(k)} \delta H^{(k)}$, etc.

In our method described above, $\delta G^{(k)}$ and $\delta H^{(k)}$ are expressed as $\widehat{g} + g$ and $\widehat{h} + h$, respectively. This seems to be natural. In order to simplify $\widehat{g} + g$ and $\widehat{h} + h$, we have investigated several methods so far, by computing $\widehat{g}$ and g (and $\widehat{h}$ and h) separately. However, we have failed to find a satisfactory one.

On the other hand, we have noticed a very interesting evidence which makes $\delta G^{(k)}$ and $\delta H^{(k)}$ simple. Consider Example 2: $\delta G^{(2)}$ is composed of two polynomials, $-3y^2z$ which came from $\widehat{g}$ and $\%D_1 \times (-3xy^4z^2 + \cdots + 6y^2z^3)$ which came from g. Rewriting $-3y^2z$ as $-\%D_1 \times (3y^2z) \times (y - z)^2$ and adding to the second polynomial, the result becomes a much simpler polynomial. The same is true for $\delta H^{(k)}$. We notice another evidence in Example 1: $G^{(14)}$ and $H^{(14)}$ are expressed very concisely. Therefore, we employ the following policy for the simplification.

Our policy for simplifying $\delta G^{(k)}$ and $\delta H^{(k)}$: Suppose $\delta G^{(k)}$ is given as $\delta G^{(k)} = \%D^e g_e(x, \boldsymbol{u}) + \%D^{e-1} g_{e-1}(x, \boldsymbol{u}) + \cdots + g_0(x, \boldsymbol{u})$, we convert the r.h.s. polynomial as $\%D^e \times \big(g_e(x, \boldsymbol{u}) + D(\boldsymbol{u})g_{e-1}(x, \boldsymbol{u}) + \cdots + D(\boldsymbol{u})^e g_0(x, \boldsymbol{u})\big)$. Here, $D(\boldsymbol{u})$ is a denominator factor which the system variable $\%D$ represents.

We advance this policy further. Since $\widehat{g}$ and g (and $\widehat{h}$ and h, resp.) are added later, we will do as follows.

Our policy for computing $\widehat{h}, \widehat{g}, h, g$: (a) The case of $\delta R^{(k)} = 0$: Compute $\widehat{h}$ and $\widehat{g}$ as describe in Sect. 3.2.
(b) The case of $\delta R^{(k)} \neq 0$: Reset $\delta R^{(k)}$ and $\delta \widehat{F}^{(k)}$ as $\delta R^{(k)} := \delta \widehat{F}^{(k)} + \delta R^{(k)}$ and $\delta \widehat{F}^{(k)} := 0$, then compute h and g as described in Sect. 3.3.

3.5   On Term Order and on Computing Syzygies

In Sect. 3.1, we did not specify the term order on sub-variables $u_1, \ldots, u_\ell$. If we employ the lex order, the computation is often very slow. Therefore, we introduce a term-order named "sub-variable total-degree (stdeg) order", as follows.

$$x^d u_1^{e_1} \cdots u_\ell^{e_\ell} \longleftrightarrow (d, \textstyle\sum_{i=1}^{\ell} e_i, e_1, \ldots, e_\ell). \tag{3.9}$$

In our method, not only the Gröbner basis but also syzygies are inevitable. Method of syzygy computation is simple and well known [1]. However, the conventional method is usually much more expensive than computing the Gröbner basis itself. In the rest of this subsection, we propose an efficient method of syzygy computation.

We explain our method for the case of two initial factors G_0 and H_0. Starting with $P_1 = G_0$ and $P_2 = H_0$, suppose that the Gröbner basis computation generates polynomials as $P_3 \to P_4 \to \cdots \to P_k$. Syzygies are conventionally computed parallel to this computation, as $(A_3, B_3) \to (A_4, B_4) \to \cdots \to (A_k, B_k)$, with initial syzygies $(A_1, B_1) = (1, 0)$ and $(A_2, B_2) = (0, 1)$. However, this method is very inefficient in that, i) most polynomials generated are not included in Γ and ii) syzygies become larger and larger as the computation proceeds, while most computations in final stage are only to convince that $\mathrm{Spol}(P_i, P_j)$ is M-reduced to 0. So, during the Gröbner basis computation, we generate only procedures to generate syzygies, which we call "procedural syzygies (p-syzygies)", and, after finishing the Gröbner basis computation, we convert p-syzygies into actual syzygies.

The computation of Gröbner basis is a sequence of Mreds, Spol constructions and the normalization of polynomials; in this paper, we assume that the normalization makes the polynomial P to satisfy $\mathrm{hc}(P) = 1$. Below, let polynomials F and G be such that $\mathrm{htm}(F) = f_1 S_1$ and $\mathrm{htm}(G) = g_1 T_1$. By $\#_F$ and $\#_G$ we denote the indices given to F and G, respectively; if P is a nonzero polynomial generated i-th and stored in the current intermediate basis then we give i to P as its index. If a polynomial being stored in an intermediate basis is M-reduced later, then the M-reduced polynomial is indexed by a new sequential index. By this, we can obtain actual syzygies by "evaluating" p-syzygies for polynomials from smaller index to larger ones.

The following shows the rules of generating p-syzygies.

"p-syzygies" for Mred, Spol and Normalization:

On Mred(F, G) : (Mred $(\#_F, (0,..,0), 1)$ $(\#_G, S_1/T_1, -f_1/g_1))$,

On Spol(F, G) : (Spol $(\#_F, L/S_1, g_1)$ $(\#_G, L/T_1, -f_1))$,
   where $L := \text{lcm}(S_1, T_1)$,

On Normalization of F : (Nmlz $1/f_1$).

We call the above "$(\#_G, S_1/T_1, -f_1/g_1)$" etc. an "IPC triplet". F will be M-reduced by G_i, G_j, etc., multiply. Then, each time the F is M-reduced, corresponding IPC triplet is appended at the tail of the p-syzygy. The same is true if Spol(F, G) is M-reduced by others. The normalization is made when F is fully M-reduced and the result is not zero. Hence, the "(Nmlz $1/f_1$)" is always appended as the last element of the p-syzygy for F.

We prepare a system array %Syzygy and store the p-syzygy for i-th polynomial in %Syzygy$[i]$. Given p-syzygies in %Syzygy, procedure convPsyz2Asyz converts the p-syzygies into actual syzygies, as follows, where #mx below is the maximum index of nonzero polynomials generated. Procedure IPC2Asyz below converts an IPC-triplet (#, PowP, Coef) into an actual syzygy: it multiplies a monomial 'Coef×PowP' to %Syzygy$[\#]$ (which is already an actual syzygy). Procedures first(l) and rest(l), with l a list of IPC triplets, return the first element of l and the rest of l, respectively.

```
Procedure convPsyz2Asyz(%Syzygy, #mx) ==
     For  i = 3 to #mx do
          { Asyz := evalPsyz(%Syzygy[i]);
            store Asyz to %Syzygy[i] }
Procedure evalPsyz(Psyz) ==
     begin Asyz := IPC2Asyz(first(Psyz));  goto next;
     loop: Asyz := Asyz + IPC2Asyz(first(Psyz));
     next: if (Psyz := rest(Psyz)) ≠ ( ) then goto loop;
          return Asyz;      end
```

Among the #mx polynomials, Γ contains only a part of them, and many syzygies in %Syzygy are needless for polynomials in Γ. Therefore, if we compute syzygies for only polynomials which are necessary to compute syzygies for the elements of Γ, the computation becomes faster. Such a computation is possible if we apply procedure evalPsyz recursively; when we encounter j-th p-syzygy in the execution of evalPsyz(i-th p-syzygy) then we apply evalPsyz to the j-th p-syzygy during the execution.

4   Timing Data and Final Remarks

We have tested our new EHC algorithm by the polynomial given in Example 1 and two related problems. Let $F(x, y, z) \in \mathbb{Q}[x, y, z]$ be as follows, where x is the main variable:

$$F = \{(x^5 y^3 + 2z)(x^5 z^3 - 2y) + x^3 z^4\} \times \{(x^5 z^3 + 3y)(x^5 y^3 - 3z) + x^7 y^6\} + 3x^{10} y^5 z^5. \tag{4.1}$$

F has only one Newton line with Newton polynomial $(x^5 y^3 - 3z)(x^5 y^3 + 2z)(x^5 z^3 + 3y)(x^5 z^3 - 2y)$. The slope of the Newton line is $2/5$, so x, y and z are weighted as $(x, y, z) \rightarrow (x/t^2, t^5 y, t^5 z)$. We choose the initial Hensel factors to be of degree 10 w.r.t. x, hence we have the following three choices.

ChoiceA : $G_0 = (x^5 y^3 + 2z)(x^5 z^3 - 2y)$, $H_0 = (x^5 y^3 - 3z)(x^5 z^3 + 3y)$,
ChoiceB : $G_0 = (x^5 y^3 - 3z)(x^5 z^3 - 2y)$, $H_0 = (x^5 y^3 + 2z)(x^5 z^3 + 3y)$,
ChoiceC : $G_0 = (x^5 y^3 - 3z)(x^5 y^3 + 2z)$, $H_0 = (x^5 z^3 + 3y)(x^5 z^3 - 2y)$.
$$\tag{4.2}$$

The Choice A has been treated in Example 1; the structure of polynomial in (4.1) suggests that the term $x^3 z^4$ will appear in $\delta G^{(4)}$, the term $x^7 y^6$ will appear in $\delta H^{(6)}$, and terms with rational function coefficients will appear in $\delta G^{(10)}$ and $\delta H^{(10)}$. As we will see later, the Hensel factors in Choice C show the most complicated behavior.

Let Γ_A, Γ_B and Γ_C be Gröbner bases in Choices A, B and C, respectively. Γ_A, Γ_B and Γ_C consist of 5, 4 and 6 polynomials, and each of them contains only one polynomial in $\mathbb{Q}[u]$, as follows.

$$\Gamma_A : \widehat{G}_{11} = (1/6)\, yz\, (3y^4 - 2z^4)(2y^4 - 3z^4), \tag{4.3}$$

$$\Gamma_B : \widehat{G}_6 = yz\, (y^4 + z^4), \tag{4.4}$$

$$\Gamma_C : \widehat{G}_5 = (1/6)(3y^4 - 2z^4)(2y^4 - 3z^4)(y^4 + z^4). \tag{4.5}$$

We compare the new EHC algorithm with the old one from the viewpoint of computational time. The new EHC algorithm was implemented on our algebra system named GAL which was developed mainly in Sasaki's Lab., and the computation was done on a computer with Intel(R)-U2300 (1.20GHz), operated by Linux 3.4.100. The old EHC algorithm was implemented on Mathematica, and the computation was done on a computer with Intel(R)-B800 (1.50GHz), operated by MS Windows 7. Each datum in Tables is an average of 100 repetitions of corresponding unit operation (Tables 1 and 2).

It is surprising that the new EHC algorithm is very efficient compared with the old one. One reason for this is that the old EHC algorithm was implemented by using high level commands of Mathematica, while the new EHC algorithm uses various low level procedures of GAL. Another surprise is that the new EHC algorithm spent most time for the factorization of the net Newton polynomial F_N. This slowness of the factorization will be due to that we have applied the

Table 1. Timing data (msec) by old EHC algorithm

Comp. step	Choice A	Choice B	Choice C
F_N & Factri.	see right	133.38	see left
MY-functions	567.07	318.24	472.22
$\delta G^{(4)}, \delta H^{(4)}$	34.630	33.080	35.720
$\delta G^{(6)}, \delta H^{(6)}$	59.280	63.340	67.710
$\delta G^{(8)}, \delta H^{(8)}$	$(\delta F = 0)$	110.14	115.75
$\delta G^{(10)}, \delta H^{(10)}$	99.920	158.65	170.36

Table 2. Timing data (msec) by new EHC algorithm

Comp. step	Choice A	Choice B	Choice C
F_N & Factri.	see right	23.19	see left
Γ &Syzygies	0.670	0.250	0.330
$\delta G^{(4)}, \delta H^{(4)}$	0.018	0.190	0.270
$\delta G^{(6)}, \delta H^{(6)}$	0.175	0.360	0.830
$\delta G^{(8)}, \delta H^{(8)}$	$(\delta F = 0)$	0.400	0.830
$\delta G^{(10)}, \delta H^{(10)}$	0.554	1.160	3.070

GHC-based factorization algorithm for F_N which is sparse both in x and y, z. We are planning to develop algorithms for multivariate GCD and factorization, which are based on the EHC with Gröbner bases.

We explain how the denominator factors were determined.

Choice A: Since $6\widehat{G}_{11} = yz\left(6y^8 - 13y^4z^4 + 6z^8\right)$ and $\delta R^{(10)} = 18y^3z^3$, the denominator factor was determined as $\widehat{G}_{11}/\gcd(\widehat{G}_{11}, \delta R^{(10)}) = (6y^8 - 13y^4z^4 + 6z^8) = (3y^4 - 2z^2)(2y^4 - 3z^4)$.

Choice B: Since $\widehat{G}_6 = yz(y^4 + z^4)$ and $\delta R^{(4)} = -5z^8z^8 - 15x^3yz^5$, we obtained $\widehat{G}_6/\gcd(\widehat{G}_6, \delta R^{(4)}) = y^5 + yz^4$. Then, we found $\gcd(\mathrm{cont}(\delta H^{4)}), \mathrm{cont}(\delta G^{(4)})) = y$, and we finally obtained $y^4 + z^4$ as the denominator factor for $k = 4$. The same is true for $k = 6, 8, 10$.

Choice C: Since $6\widehat{G}_5 = 6y^{12} - 7y^8z^4 - 7y^4z^8 + 6z^{12} = (y^4 + z^4)(3y^4 - 2z^4)(2y^4 - 3z^4)$, we found $\gcd(\widehat{G}_5, \delta R^{(k)}) = 1$ and $\gcd(\mathrm{cont}(\delta H^k), \mathrm{cont}(\delta G^{(k)})) = 6y^8 - 13y^4z^4 + 6z^8$ for $k = 4, 6, 8$, hence we obtained $y^4 + z^4$ as the denominator factor. For $k = 10$, we found $\gcd(\mathrm{cont}(\delta H^{10}), \mathrm{cont}(\delta G^{(10)})) = 1$, so the denominator factor changed to $\widehat{G}_5$.

We see that the denominator factors were determined rather complicatedly, as the theory in Sect. 3 suggests. We note that, for the above experiment, Lemma 2 is enough for the degree reduction, and the forced degree reduction was unnecessary. Our experiments are not enough to judge the superiority of our algorithm, but we did not make enough experiment because of the following reason.

Before the implementation, we feared that the computation of Gröbner bases, in particular the syzygies, is time consuming. We are currently satisfied with the efficiency of our syzygy algorithm for small-sized polynomials; it occupies only a part of the time of whole Gröbner basis computation for small-sized polynomials. However, we have recognized that the syzygy computation becomes heavy in many cases; if we increase the degrees of sub-variables or introduce non-simple numerical coefficients then the syzygies become quite large expressions. Therefore, we have to attain much more enhancements.

After submitting the present paper, we have done such enhancements which improve our algorithm described in Sect. 3 by an order of magnitude in the computation time [11]. The essential idea in the study is that we compute neither a Gröbner basis nor the syzygies for the elements of the basis, but compute only one element in $\mathbb{K}[u]$ and the syzygy for the element, as efficiently as possible. Furthermore, in this study, we have proved a theorem which states that the Gröbner basis Γ of $\langle G_0, H_0 \rangle$ is such that $\Gamma \cap \mathbb{K}[u]$ contains only one element. This theorem makes the discussions in Sect. 3 pretty simple.

References

1. Buchberger, B.: Gröbner bases: an algorithmic methods in polynomial idealtheory. In: Multidimensional Systems Theory, Chapter 6. Reidel Publishing (1985)
2. Inaba, D.: Factorization of multivariate polynomials by extended Hensel construction. ACM SIGSAM Bull. **39**(1), 2–14 (2005)
3. Inaba, D., Sasaki, T.: A numerical study of extended Hensel series. In: Verschede, J., Watt, S.T., (eds.) Proceedings of SNC 2007, pp. 103–109. ACM Press (2007)
4. Kaltofen, E.: Sparse Hensel lifting. In: Caviness, B.F. (ed.) EUROCAL 1985. LNCS, vol. 204, pp. 4–17. Springer, Heidelberg (1985)
5. de Kleine, J., Monagan, M., Wittkopf, A.: Algorithms for the non-monic case of the sparse modular GCD algorithm. In: Proceedings of ISSAC 2005, pp. 124–131 (2005)
6. Musser, D.R.: Algorithms for polynomial factorizations. Ph.D. thesis, University of Wisconsin (1971)
7. Moses, J., Yun, D.Y.Y.: The EZGCD algorithm. In: Proceedings of ACM National Conference, pp. 159–166. ACM (1973)
8. Sasaki, T., Inaba, D.: Hensel construction of $F(x, u_1, \ldots, u_\ell)$, $\ell \geq 2$, at a singular point and its applications. ACM SIGSAM Bull. **34**(1), 9–17 (2000)
9. Sasaki, T., Inaba, D.: A study of Hensel series in general case. In: Moreno Maza, M., (ed.) Proceedings of SNC 2011, pp. 34–43. ACM Press (2011)
10. Sanuki, M., Inaba, D., Sasaki, T.: Computation of GCD of sparse multivariate polynomial by extended Hensel construction. In: Proceedings of SYNASC2015 (Symbolic and Numeric Algorithms for Scientific Computing), pp. 34–41. IEEE Computer Society (2016)
11. Sasaki, T., Inaba, D.: Various enhancements of extended Hensel construction for sparse multivariate polynomials. In: Proceeding of SYNASC 2016 (2016, to appear)
12. Sasaki, T., Kako, F.: Solving multivariate algebraic equation by Hensel construction. Preprint of Univ. Tsukuba, March 1993
13. Sasaki, T., Kako, F.: Solving multivariate algebraic equation by Hensel construction. Japan J. Ind. Appl. Math. **16**(2), 257–285 (1999). (This is almost the same as [12]: the delay of publication is due to very slow reviewing process.)

14. Schwartz, J.T.: Fast probabilistic algorithms for verification of polynomial identities. J. ACM **27**, 701–717 (1980)
15. Wang, P.S., Rothschild, L.P.: Factoring multivariate polynomials over the integers. Math. Comput. **29**, 935–950 (1975)
16. Wang, P.S.: An improved multivariate factoring algorithm. Math. Comput. **32**, 1215–1231 (1978)
17. Zippel, R.: Probabilistic algorithm for sparse polynomials. In: Ng, E.W. (ed.) EUROSAM 1979. LNCS, vol. 72, pp. 216–226. Springer, Heidelberg (1979)
18. Zippel, R.: Newton's iteration and the sparse Hensel lifting (extended abstract). In: Proceedings of SYMSAC 1981, pp. 68–72 (1981)

Symbolic-Numerical Optimization and Realization of the Method of Collocations and Least Residuals for Solving the Navier–Stokes Equations

Vasily P. Shapeev[1,2] and Evgenii V. Vorozhtsov[1(✉)]

[1] Khristianovich Institute of Theoretical and Applied Mechanics,
Russian Academy of Sciences, Novosibirsk 630090, Russia
{shapeev,vorozh}@itam.nsc.ru
[2] Novosibirsk National Research University, Novosibirsk 630090, Russia

Abstract. The computer algebra system (CAS) *Mathematica* has been applied for constructing the optimal iteration processes of the Gauss–Seidel type at the solution of PDE's by the method of collocations and least residuals. The possibilities of the proposed approaches are shown by the examples of the solution of boundary-value problems for the 2D Navier–Stokes equations.

Keywords: Computer algebra system · Symbolic-numerical algorithm · Interface CAS–Fortran · Preconditioner · Krylov subspaces · Multigrid

1 Introduction

In recent decades, rapid development of mathematical methods, first of all, the numerical modelling has taken place. It is used in increasing amount both in conventional domains of physics and technology and in other domains. A wide variety of modelled phenomena and processes, the complexity and peculiarities in arising mathematical problems make increased demands for the properties and capabilities of numerical methods and stimulate a search for the new methods possessing better properties as the previous methods.

Many researchers showed in their works a substantial benefit from using computer algebra systems (CASs) in the process of deriving the formulas of new numerical algorithms, their realization and verification of the corresponding computer codes [1,4,5,24]. In the present work, an emphasis is placed on the demonstration of the efficiency of the CAS application for the optimization of iteration processes for solving the systems of linear algebraic equations (SLAE) arising at the realization of the method of collocations and least residuals (CLR). At the realization of this method, as in other cases, the problem of optimizing the iterative processes of the SLAE solution is important.

The CLR method, which was proposed in [13] and developed further in the subsequent works of other authors, is one of the methods, which enables the efficient solution of PDEs [6,16–19,21–23]. The works [6,19,21–23] have shown the

© Springer International Publishing AG 2016
V.P. Gerdt et al. (Eds.): CASC 2016, LNCS 9890, pp. 473–488, 2016.
DOI: 10.1007/978-3-319-45641-6_30

usefulness of the application of a CAS at the derivation of formulas of the different versions of the CLR method. The versions of the method were constructed, which have enabled the obtaining of the solutions of the 2D and 3D benchmark problems, which are among the most accurate at present [2,20].

Three families of methods for accelerating the iteration processes of solving the SLAEs are the most popular today: the algorithms using Krylov space methods [9,15], multigrid [26], and the preconditioners.

In recent years, a number of researchers have shown a high efficiency of the combined application of different algorithms [10–12,25]. In particular, in [10], the acceleration with factor of 25 was achieved with the aid of a combination of multigrid and GMRES in comparison with a standard multigrid algorithm. Diagonal preconditioners have gained widespread acceptance, see, e.g., [8,14].

In contrast to the previous works [6,7,18,19,21–23], we present in this work for the first time the application of a CAS for formulating the preconditioner and its optimization for the use in the CLR method. Besides, the algorithms of the preconditioner, Krylov subspaces, and of the Fedorenko method have been combined for the first time. Such a combination has been proposed and implemented generally for the first time and has proved to be very efficient and has resulted in a considerable acceleration of the solution of two-dimensional stationary incompressible Navier–Stokes equations.

It was shown previously in [21,23] how one can apply the CAS *Mathematica* for the derivation and verification of the formulae of the CLR method as well as for their translation into the arithmetic operators of the Fortran language. Therefore, these stages of the computer implementation of the symbolic-numerical algorithm of the CLR method are described briefly in the following.

2   Description of the CLR Method

Consider a boundary-value problem for the system of Navier–Stokes equations

$$(\mathbf{V} \cdot \nabla)\mathbf{V} + \nabla p = \frac{1}{\mathrm{Re}} \Delta \mathbf{V} - \mathbf{f}, \quad \mathrm{div}\,\mathbf{V} = 0, \quad (x_1, x_2) \in \Omega, \tag{1}$$

$$\mathbf{V}\big|_{\partial\Omega} = \mathbf{g} \tag{2}$$

in the region Ω with the boundary $\partial\Omega$. In Eq. (1), x_1, x_2 are the Cartesian spatial coordinates, $\mathbf{V} = (v_1(x_1, x_2, v_2(x_1, x_2))$ is the velocity vector; $p = p(x_1, x_2)$ is the pressure, $\mathbf{f} = (f_1, f_2)$ is the given vector function, Re is the Reynolds number, $\Delta = \frac{\partial^2}{\partial x_1^2} + \frac{\partial^2}{\partial x_2^2}$, $(\mathbf{V} \cdot \nabla) = v_1 \frac{\partial}{\partial x_1} + v_2 \frac{\partial}{\partial x_2}$. System (1) is solved under the Dirichlet boundary conditions (2), where $\mathbf{g} = \mathbf{g}(x_1, x_2) = (g_1, g_2)$ is a given vector function. The condition

$$\iint_\Omega p\,dx_1 dx_2 = 0 \tag{3}$$

is imposed on the pressure. It is valid in the absence of the sources and sinks in region Ω [6]. The square

$$\Omega = \{(x_1, x_2), 0 \leq x_i \leq L, i = 1, 2\} \tag{4}$$

is taken in the following as the problem solution region, where $L > 0$ is a given length of the square side. In the given problem (1)–(4), region (4) is discretized by a grid with square cells Ω_{ij}, $i, j = 1, \ldots, I$, $I \geq 1$. It is convenient to introduce the local coordinates y_1 and y_2 in each cell Ω_{ij}. The dependence of local coordinates on global spatial variables x_1 and x_2 is specified by relations $y_m = (x_m - x_{m,i,j})/h$, $m = 1, 2$, where $x_{m,i,j}$ is the value of the coordinate x_m at the center of cell Ω_{ij}, and h is the halved length of the square cell side. Let $\mathbf{u}(y_1, y_2) = (u_1, u_2) = \mathbf{V}(hy_1 + x_{1,i,j}, hy_2 + x_{2,i,j})$, $q(y_1, y_2) = p(hy_1 + x_{1,i,j}, hy_2 + x_{2,i,j})$. The Navier–Stokes equations then take the following form:

$$\Delta u_m - \mathrm{Re}h \left(u_1 \frac{\partial u_m}{\partial y_1} + u_2 \frac{\partial u_m}{\partial y_2} + \frac{\partial q}{\partial y_m} \right) = \mathrm{Re} \cdot h^2 f_m, \quad m = 1, 2; \quad (5)$$

$$\frac{1}{h} \left(\frac{\partial u_1}{\partial y_1} + \frac{\partial u_2}{\partial y_2} \right) = 0, \quad (6)$$

where $\Delta = \frac{\partial^2}{\partial y_1^2} + \frac{\partial^2}{\partial y_2^2}$. The Newton linearization of equations (5) gives

$$\Delta u_m^{s+1} - (\mathrm{Re} \cdot h) \left(u_1^s u_{m,y_1}^{s+1} + u_1^{s+1} u_{m,y_1}^s + u_2^s u_{m,y_2}^{s+1} + u_2^{s+1} u_{m,y_2}^s + q_{y_m}^{s+1} \right) = F_m, \quad (7)$$

where $m = 1, 2$, and s is the number of the iteration over the nonlinearity, $s = 0, 1, 2, \ldots$, u_1^s, u_2^s, q^s is the known approximation to the solution at the sth iteration starting from the chosen initial guess with index $s = 0$, $F_m = \mathrm{Re} \left[h^2 f_m - h \left(u_1^s u_{m,y_1}^s + u_2^s u_{m,y_2}^s \right) \right]$, $u_{m,y_l} = \partial u_m / \partial y_l$, $q_{y_m} = \partial q / \partial y_m$, $l, m = 1, 2$.

The approximate solution in each cell $\Omega_{i,j}$ is sought in the form of a linear combination of the basis vector functions φ_l:

$$(u_1^s, u_2^s, q^s)^T = \sum_l b_{i,j,l}^s \varphi_l, \quad (8)$$

where the superscript T denotes the transposition operation. In the given version of the method, the φ_l are the polynomials. Thus, the approximate solution is a piecewise polynomial. In the given work, the second-degree polynomials in variables y_1, y_2 are employed for the approximation of velocity components, and the first-degree polynomials are used for the pressure approximation. In the chosen space, there are fifteen basis functions in total. Since the coefficients are constant in the continuity equation, which has a simple form, it is easy to satisfy it at the expense of the choice of basis polynomials φ_l. It is not difficult to find that it is required to this end that they satisfy three linear relations. There will finally remain only twelve independent basis polynomials from the original fifteen ones. They are presented in Table 1. One can term their set a solenoidal basis because $\mathrm{div}\, \varphi_l = 0$.

The number of collocation points and their location inside the cell may vary in different versions of the method. In the given work, three versions of the specification of the collocation point coordinates have been implemented. Denote by N_c the number of collocation points inside each cell. In the case when $N_c = 2$, the coordinates of collocation points are as follows: (ω, ω), $(-\omega, \omega)$,

Table 1. The form of basis functions φ_l

l	1	2	3	4	5	6	7	8	9	10	11	12
φ_l	1	y_1	y_2	y_1^2	$-2y_1y_2$	y_2^2	0	0	0	0	0	0
	0	$-y_2$	0	$-2y_1y_2$	y_2^2		0	1	y_1	y_1^2	0	0	0
	0	0	0	0	0		0	0	0	0	1	y_1	y_2

where $0 < \omega < 1$. At $N_c = 4$, the local coordinates of collocation points have the form $(\pm\omega, \pm\omega)$. In the case of $N_c = 8$, the coordinates of collocation points were specified in the following way: the locations of the first four points was the same as at $N_c = 4$, and the coordinates of the next four points were specified by formulas $(\pm\omega, 0)$, $(0, \pm\omega)$.

Substituting (8) as well as the numerical values of the coordinates of each collocation point in (7) we obtain $2N_c$ linear algebraic equations:

$$\sum_{m=1}^{12} a_{\nu,m}^{(1)} \cdot b_m^{s+1} = f_\nu^s, \quad \nu = 1, \ldots, 2N_c. \tag{9}$$

It was proposed in the work [13] to augment the collocation equations by the solution matching conditions at the boundaries between neighboring cells of the spatial computational grid within the framework of the method of collocations and least squares (CLS). The parameters were introduced in the matching conditions in the work [7]. It was shown therein with the aid of numerical experiments that the regions of parameter values at which the global SLAE of the CLS method is well conditioned intersects to a considerable extent with the regions in which one observes the highest accuracy of the approximate solution.

However, the influence of the matching conditions on the condition number of the algebraic system for finding the coefficients affecting the basis functions in the expansions of the solution vector components was not investigated in [7,13]. This investigation is carried out below in Sect. 3, where it is shown that the incorporation of the matching conditions into the SLAE of the CLR method enables a substantial reduction of the condition number of the local SLAEs in internal grid cells by four–five decimal orders of magnitude. This increases the reliability of the CLR method.

By analogy with [6,7,13] let us augment the system of equations of the discrete problem in the Ω_{ij} cell by the conditions of matching with the solutions of the discrete problem, which are taken in all cells adhering to the given cell. We will write these conditions at separate points (called the matching points) on the sides of the Ω_{ij} cell, which are common with its neighboring cells. The matching conditions are taken here in the form

$$h\frac{\partial(u^+)^n}{\partial n} + (u^+)^n = h\frac{\partial(u^-)^n}{\partial n} + (u^-)^n, \tag{10}$$

$$h\frac{\partial(u^+)^\tau}{\partial n} + (u^+)^\tau = h\frac{\partial(u^-)^\tau}{\partial n} + (u^-)^\tau, \tag{11}$$

$$q^+ = q^-. \tag{12}$$

Here $h\frac{\partial}{\partial n} = h\left(n_1\frac{\partial}{\partial x_1} + n_2\frac{\partial}{\partial x_2}\right) = n_1\frac{\partial}{\partial y_1} + n_2\frac{\partial}{\partial y_2}$, $n = (n_1, n_2)$ is the external normal to the side of the Ω_{ij} cell, $(\cdot)^n$, $(\cdot)^\tau$ are the normal and tangential components of the velocity vector with respect to the cell side, u^+, u^- are the limits of the function u as its arguments tend to the matching point from inside and outside the Ω_{ij} cell.

For the uniqueness of the pressure determination in the solution, we either specify its value at a single point of the region or approximate condition (3) by the formula

$$\frac{1}{h}\left(\iint_{\Omega_{i,j}} q\, dy_1 dy_2\right) = \frac{1}{h}\left(-I^* + \iint_{\Omega_{i,j}} q^* dy_1 dy_2\right). \tag{13}$$

Here I^* is the integral over the entire region, which is computed as a sum of the integrals over each cell at the foregoing iteration, q^* is the pressure in a cell from the foregoing iteration.

Denote by N_m the number of matching points for the velocity vector components on the sides of each cell. At $N_m = 4$, the coordinates of these matching points are specified by the formulas $(\pm 1, 0)$, $(0, \pm 1)$. At $N_m = 8$, the coordinates of matching points are as follows: $(\pm 1, -\zeta)$, $(\pm 1, \zeta)$, $(-\zeta, \pm 1)$, $(\zeta, \pm 1)$, where $0 < \zeta < 1$. In the computational examples presented below, the value $\zeta = 1/2$ was used. The matching conditions for pressure (12) are set at four points with coordinates $(\pm 1, 0)$, $(0, \pm 1)$.

Using Eq. (8), we substitute the coordinates of these points in each of three matching conditions (10)–(12). We obtain from the first two conditions $2N_m$ linear algebraic equations for velocity components. The substitution of representation (8) in (12) also yields four linear algebraic (matching) equations.

In the present work, the pressure was specified at the vertex of the $\Omega_{1,1}$ cell or condition (13) was used. If the cell side coincides with the boundary of region Ω, then the boundary conditions are written at the corresponding points instead of the matching conditions for the discrete problem solution: $u_m = g_m$, $m = 1, 2$.

Uniting the equations of collocations, matching, and the equations obtained form the boundary conditions, if the cell Ω_{ij} is the boundary cell, we obtain in each cell a SLAE of the form

$$A_{i,j} \cdot X_{i,j}^{s+1} = f_{i,j}^{s,s+1}, \tag{14}$$

where $X_{i,j}^{s+1} = (b_{i,j,1}^{s+1}, \ldots, b_{i,j,12}^{s+1})^T$. In the versions studied in the present work, system (14) is overdetermined. The symbolic expressions for the coefficients of all equations of SLAE (14) were derived on computer in Fortran form by using symbolic computations with *Mathematica*. At the obtaining of the final form of the formulas for the coefficients of the equations, it is useful to perform the simplifications of the arithmetic expressions of polynomial form to reduce the number of the arithmetic operations needed for their numerical computation. To this end, we employed standard functions of the *Mathematica* system, such as `Simplify` and `FullSimplify` for the simplification of complex symbolic expressions arising at the symbolic stages of the construction of the formulae of the method. Their application enabled a two-three-fold reduction of the length of polynomial expressions.

For the numerical solution of the SLAE of the discrete problem a process was applied, which may be called conventionally the Gauss–Seidel iteration scheme. One global $(s + 1)$th iteration meant that all the cells were considered sequentially in the computational region Ω. In each cell, SLAE (14) was solved by the orthogonal method (of Givens or Householder), and the values known at the solution construction at the $(s+1)$th iteration were taken in the right-hand sides of equations (10)–(12) as the u^- and q^- in a given cell.

3   Preconditioners for the CLR Method

It is necessary to solve in each cell Ω_{ij} the SLAE of the form (14). Let us omit in (14) the superscripts and subscripts for the sake of brevity:

$$\mathrm{A}\boldsymbol{X} = \boldsymbol{f}. \tag{15}$$

The condition number of a rectangular matrix A is calculated by the formula

$$\kappa(\mathrm{A}) = \sqrt{\parallel \mathrm{A}_1 \parallel \cdot \parallel \mathrm{A}_1^{-1} \parallel}, \tag{16}$$

where it is assumed that matrix $\mathrm{A}_1 = \mathrm{A}^T\mathrm{A}$ is non-singular.

We have at first tried the well-known diagonal (Jacobi) preconditioner described in [14], but it has not produced in our case the expected acceleration of the iterations convergence. On the other hand, it is well known that the introduction of parameters in the preconditioner increases its capabilities for a reduction of the condition number because one can then select these parameters from the requirement of the condition number minimization. In our case, we have constructed a preconditioner involving the parameters ξ and η.

The parameter ξ is introduced by multiplying by ξ the both sides of (7):

$$\xi[\Delta u_m^{s+1} - (\mathrm{Re} \cdot h)(u_1^s u_{m,y_1}^{s+1} + u_1^{s+1} u_{m,y_1}^s + u_2^s u_{m,y_2}^{s+1}$$
$$+ u_2^{s+1} u_{m,y_2}^s + q_{y_m}^{s+1})] = \xi F_m, \quad m = 1, 2. \tag{17}$$

The parameter η is introduced in (10) as follows:

$$h\frac{\partial(u^+)^n}{\partial n} + \eta(u^+)^n = h\frac{\partial(u^-)^n}{\partial n} + \eta(u^-)^n. \tag{18}$$

Denote by $\mathrm{A}_{\mathrm{col}}$ the $2N_c \times 12$ matrix of the system obtained from (17) upon substituting the collocation point coordinates. The pressure enters the momentum equation (17) only in the form of the derivatives $\partial q/\partial y_1$, $\partial q/\partial y_2$, therefore, the coefficient affecting b_{10} in the matrix $\mathrm{A}_{\mathrm{col}}$ is equal to zero. Because of this the matrix $\mathrm{A}_{\mathrm{col}}$ is singular. In order to eliminate this singularity we include in $\mathrm{A}_{\mathrm{col}}$ the row corresponding to Eq. (13). This row has the following form: $\{0, 0, 0, 0, 0, 0, 0, 0, 0, h, 0, 0\}$. Denote such an augmented matrix by $\tilde{\mathrm{A}}_{\mathrm{col}}$. This matrix is non-singular. It is seen from (17) that the entries of the matrix $\tilde{\mathrm{A}}_{\mathrm{col}}$ depend on the solution at the foregoing iteration. Therefore, a further investigation of the condition number properties was made on a given grid at the solution of a specific problem as follows. At first we obtained a good approximation of the solution on the grid of 80×80 cells with the aid of the multigrid algorithm. The obtained numerical values of the solution were then used for computing the entries of the matrix $\tilde{A}_{col}$. Thus, the entries of the matrix $\tilde{A}_{col}$ depend on parameters ξ and h. To obtain the matrix $\tilde{A}_{col}$ consisting only of numerical elements the numerical values of the half-step h were specified by the formula $h = 0.5/(2M)$, $M = 20$, 40, 60, 80, 160, 320. Besides, the parameter ξ was varied in the interval from 0.01 to 20. The built-in function of the CAS *Mathematica* Norm[A1, 2] was further used for computing the condition number according to (16). This function calculates the Euclidean norm of a square matrix. The condition number $\kappa(\tilde{A}_{col})$ was found to be independent of the value of the grid half-step h. In the interval $0.01 \leq \xi \leq 20$, the number $\kappa(\tilde{A}_{col})$ varied within the following limits: $4.2255 \cdot 10^6 \leq \kappa(\tilde{A}_{col}) \leq 4.2258 \cdot 10^6$ (here $N_c = 8$). The CAS *Mathematica* printed the following message: "Inverse: luc: Result for inverse of badly conditioned matrix".

Denote by A_{mat} the matrix corresponding to matching conditions (18), (11), and (12). Then one can present the entire matrix A as

$$A = \begin{pmatrix} \tilde{A}_{col} \\ A_{mat} \end{pmatrix}. \tag{19}$$

At a given numerical value of the half-step h, the entries of the matrix A depend on ξ and η. Let $G(\xi, \eta) = \kappa(A(\xi, \eta))$. Denote by ξ_{opt}, η_{opt} the values of the parameters ξ and η, at which the function $G(\xi, \eta)$ reaches its minimum. We describe in the following a numerical algorithm for finding the values ξ_{opt} and η_{opt} in any internal cell Ω_{ij}. To save the CPU time we take only the cell lying at the center of the computational region, and then use the obtained values of ξ_{opt} and η_{opt} in the entire region (4). Therefore, we will call the values ξ_{opt} and η_{opt} *quasi-optimal*.

An attempt was made at the use of the built-in *Mathematica* function NMinimize[...] for finding the minimum of the function $G(\xi, \eta)$ in some given region of the variation of parameters ξ and η. It has turned out, however, that this function requires many hours of the work of the Intel processor with tact frequency of 3.0 GHz. Another built-in function FindArgMin[...], which might also be used for the same purpose, has proved also to be very slow. In this connection, we have implemented the following algorithm of a search for the point (ξ_{opt}, η_{opt}), at which the minimum of the function $G(\xi, \eta)$ is reached. At first some rectangular region $D_{\xi,\eta}^{(1)}$ was specified in the plane (ξ, η), which included those values of the parameters ξ and η, at which the computations by the CLR method with the preconditioner under consideration demonstrated the convergence of the iterative process of the solution obtaining. A rectangular uniform grid of size 10×80 nodes was specified in the region $D_{\xi,\eta}^{(1)}$, where 10 and 80 nodes were taken along the ξ-axis and η-axis, respectively. Then with the aid of of a simple scanning of all nodes (ξ_i, η_j) such a node (ξ_{i_0}, η_{j_0}) was found, in which the value of the function $G(\xi, \eta)$ was minimal. After that, a new region $D_{\xi,\eta}^{(2)}$ was built the sizes of which in each coordinate direction ξ and η were by the factor of $1/2$ smaller than in the case of the region $D_{\xi,\eta}^{(1)}$, but the number of grid nodes in $D_{\xi,\eta}^{(2)}$ was the same as in $D_{\xi,\eta}^{(1)}$. The geometric center of the new region $D_{\xi,\eta}^{(2)}$ was at point (ξ_{i_0}, η_{j_0}), which was found in the region $D_{\xi,\eta}^{(1)}$. This process of the contraction of the regions of the optimum search continued until the values of $(\xi_{\text{opt}}^{(\nu-1)}, \eta_{\text{opt}}^{(\nu-1)})$ and $(\xi_{\text{opt}}^{(\nu)}, \eta_{\text{opt}}^{(\nu)})$, which were found in the regions, respectively, $D_{\xi,\eta}^{(\nu-1)}$ and $D_{\xi,\eta}^{(\nu)}$ $(\nu = 2, 3, \ldots)$, coincided in all three digits of the mantissa of the machine floating-point number. The entire process required no more than three minutes of the work of a desktop computer (Fig. 1). This algorithm was used for elucidating the influence of the value of the grid half-step h, the number of collocation points N_c, and the choice of the norm on the values ξ_{opt} and η_{opt}. It has turned out that the obtained values of ξ_{opt} and η_{opt} do not depend on the value of h. Table 2 presents the values ξ_{opt} and η_{opt} as well as $\kappa(A(\xi_{\text{opt}}, \eta_{\text{opt}}))$ for two different values of the number of collocation points in the cell and of two different norms: the Euclidean norm ($\| \cdot \|_E$) and the Frobenius norm ($\| \cdot \|_F$) for the case of test (25). It is seen that a reduction of N_c affects more significantly the value ξ_{opt} than the value η_{opt}.

However, it was found with the aid of trial computations that a faster convergence was observed at different values of the parameters ξ and η. Therefore, another technique of determining the quasi-optimal parameters ξ_{opt} and η_{opt} of the preconditioner was considered, which is based on the criterion for the iteration process convergence. The iteration process employed for the numerical solution of the SLAE of the discrete problem may be termed the Gauss–Seidel

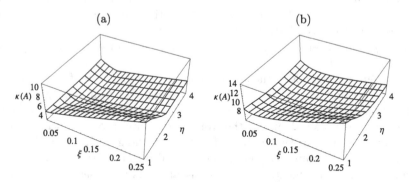

Fig. 1. Surfaces $\kappa(A(\xi, \eta))$ obtained at the use of the Euclidean norm (a) and the Frobenius norm (b), $N_c = 8$

Table 2. Influence of the number of collocation points in the cell on the quasi-optimal values ξ, η for preconditioner

N_c	$(\xi_{\text{opt}})_E$	$(\eta_{\text{opt}})_E$	κ_E	$(\xi_{\text{opt}})_F$	$(\eta_{\text{opt}})_F$	κ_F
4	0.213	1.738	3.83	0.264	1.871	6.74
8	0.156	1.747	3.84	0.189	1.873	6.71

iteration scheme. It is well known that the solution of SLAE (15) with rectangular matrix A by the method of least squares is equivalent to the solution of the SLAE of the form

$$A_1 X = f_1, \tag{20}$$

where $A_1 = A^T A$, $f_1 = A^T f$. Here A_1 is a 12×12 matrix. In the Gauss–Seidel method, the original matrix A_1 in (20) is represented as a sum of two matrices: $A_1 = B + C$, and one considers the iteration process of the form $B X^{n+1} + C X^n = f_1$. We obtain from here

$$X^{n+1} = -B^{-1} C X^n + B^{-1} f_1. \tag{21}$$

Introduce the notation $T = -B^{-1} C$. We can then rewrite (21) in the form

$$X^{n+1} = T X^n + B^{-1} f_1. \tag{22}$$

On the other hand, one can write the Jacobi method for the solution of system (20) also in the form (22), if one sets $T = I - D^{-1} A_1$ and $B = D$, where $D = \mathrm{diag}(\alpha_{ii})_{i=1}^{12}$, and α_{ij} are the entries of the matrix A_1, $i, j = 1, \ldots, 12$. Thus, one can consider the Gauss–Seidel method as a modification of the Jacobi method.

We consider below the problem of optimizing the choice of parameters ξ and η in a particular case when only the values X^n from the foregoing iteration are always used in the right-hand sides of matching conditions. The iteration method for system (22) coincides with the Jacobi method. One can consider this optimization procedure as the first approximation of the solution of the problem of optimizing the parameters of a more general iterative process (21). As is known, the sufficient condition for the convergence of the Jacobi method has the form $\| T(\xi, \eta) \| < 1$. The minimum of $\| T(\xi, \eta) \|$ was found with the aid of the procedure described above at a search for the condition number minimum. It has been shown that there is a ravine in the $\| T(\xi, \eta) \|$ surface, and its bottom is parallel with the ξ-axis. The value $\eta_{opt} = 3.478$ corresponds to this bottom, and ξ may be chosen from the requirement that $\| T(\xi, \eta) \|$ remains small. It is also found that $\| T \|$ slightly exceeds the value 1. But it should be remembered that the process of the Gauss–Seidel type has in fact been implemented, therefore, one can expect that at the values of ξ and η corresponding to the ravine in the $\| T(\xi, \eta) \|$ surface the relation $\| T \| < 1$ will be satisfied. Note that $\kappa(A(\xi, \eta)) = 7.575$ at $\xi = 0.1$, $\eta = 3.5$, thus, it is higher only by a factor of about two than the value of $\kappa(A(\xi, \eta))$, which has been obtained from the requirement of the $\kappa(A)$ minimum.

4   The Krylov and Multigrid Procedures for the CLR Method

In the present work, a version of the Krylov method, which was described in detail in [21,22], was used as the second technique for accelerating the convergence of the iteration process of the SLAE solution. The third technique, which has been used here, is the multigrid technique. The main idea of multigrid is the selective damping of the error harmonics [3,26]. In the CLR method, as in other methods, the number of iterations necessary for reaching the given accuracy of the approximation to the solution depends on the initial guess. As a technique for obtaining a good initial guess for the iterations on the finest grid among the grids used in a multigrid complex we have applied the prolongation operations along the ascending branch of the V-cycle — the computations on a sequence of refining grids. The passage from a coarser grid to a finer grid is made with the aid of the prolongation operators. Let us illustrate the algorithm of the prolongation operation by the example of the velocity component $u_1(y_1, y_2, b_1, \ldots, b_{12})$. Let $h_1 = h$, where h is the half-step of the coarse grid, and let $h_2 = h_1/2$ be the half-step of the fine grid on which one must find the expansion of function u_1 over the basis.

Step 1. Let X_1, X_2 be the global coordinates of the coarse grid cell center. We make the following substitutions into the polynomial expression for u_1:

$$y_l = (x_l - X_l)/h_1, \quad l = 1, 2. \tag{23}$$

As a result, we obtain the polynomial

$$U_1(x_1, x_2, b_1, \ldots, b_{12}) = u_1\left(\frac{x_1 - X_1}{h_1}, \frac{x_2 - X_2}{h_1}, b_1, \ldots, b_{12}\right). \tag{24}$$

Step 2. Let $(\tilde{X}_1, \tilde{X}_2)$ be the global coordinates of the center of any of the four cells of the fine grid, which lie in the coarse grid cell. We make the substitution in (24) $x_l = \tilde{X}_l + \tilde{y}_l \cdot h_2$, $l = 1, 2$. As a result, we obtain the second-degree polynomial $\tilde{U}_1 = P(\tilde{y}_1, \tilde{y}_2, \tilde{b}_1, \ldots, \tilde{b}_{12})$ in variables $\tilde{y}_1, \tilde{y}_2$ with coefficients $\tilde{b}_1, \ldots, \tilde{b}_{12}$. After the collection of terms of similar structure it turns out that the coordinates X_1, X_2 and $\tilde{X}_1, \tilde{X}_2$ enter $\tilde{b}_l$ ($l = 1, \ldots, 12$) only in the form of combinations $\delta x_l = (X_l - \tilde{X}_l)/h_1$. According to (23), the quantity $-\delta x_l = (\tilde{X}_l - X_l)/h_1$ is the local coordinate of the fine grid cell center in the coarse grid cell.

Let us present the expressions for coefficients $\tilde{b}_j$ ($j = 1, \ldots, 12$) of the solution representation in a fine grid cell with the half-step h_2 in terms of the coefficients $b_1, \ldots, b_{12}$ of the solution representation in a cell with the half-step $h_1 = 2h_2$:

$$\tilde{b}_1 = b_1 - b_2\delta x_1 + b_4\delta x_1^2 - (b_3 + 2b_5\delta x_1)\delta x_2 + b_6\delta x_2^2;$$
$$\tilde{b}_2 = \sigma_1(b_2 - 2b_4\delta x_1 + 2b_5\delta x_2); \quad \tilde{b}_3 = \sigma_1[b_3 + 2(b_5\delta x_1 - b_6\delta x_2)];$$
$$\tilde{b}_4 = \sigma_2 b_4; \quad \tilde{b}_5 = \sigma_2 b_5; \quad \tilde{b}_6 = \sigma_2 b_6;$$
$$\tilde{b}_7 = b_7 - b_8\delta x_1 + b_9\delta x_1^2 + \delta x_2(b_2 - 2b_4\delta x_1 + b_5\delta x_2);$$
$$\tilde{b}_8 = \sigma_1(b_8 - 2b_9\delta x_1 + 2b_4\delta x_2);$$
$$\tilde{b}_9 = \sigma_2 b_9; \quad \tilde{b}_{10} = b_{10} - b_{11}\delta x_1 - b_{12}\delta x_2; \quad \tilde{b}_{11} = \sigma_1 b_{11}; \quad \tilde{b}_{12} = \sigma_1 b_{12},$$

where $\sigma_1 = h_2/h_1$, $\sigma_2 = \sigma_1^2$. The analytic expressions for coefficients $\tilde{b}_1, \ldots, \tilde{b}_{12}$ were found efficiently with the aid of the *Mathematica* functions `Expand[...]`,

`Coefficient[...]`, `Simplify[...]`. To reduce the length of obtained coefficients we have applied a number of transformation rules as well as the *Mathematica* function `FullSimplify[...]`. As a result, the length of the final expressions for $\tilde{b}_1, \ldots, \tilde{b}_{12}$ proved to be three times shorter than the length of the original expressions. The Fortran form of the above prolongation operator expressions was produced with the aid of the *Mathematica* functions `ToString[...]` and `FortranForm[...]`.

5 Results of Numerical Experiments

Consider the following exact solution of the Navier–Stokes equations (1):

$$u_1 = \cos(2\pi x_1)\sin(2\pi x_2), \ u_2 = -\sin(2\pi x_1)\cos(2\pi x_2),$$
$$p = \tfrac{1}{2}\left[\cos\left(\tfrac{\pi x_1}{2}\right) + \cos\left(\tfrac{\pi x_2}{2}\right)\right] - \tfrac{2X\sin(\pi X/2)}{\pi}. \tag{25}$$

The expressions f_1 and f_2 in (1) are obtained by substituting (25) in (1). The root-mean-square solution errors were calculated:

$$\delta u(h) = \left[\frac{1}{2M^2}\sum_{i=1}^{M}\sum_{j=1}^{M}\sum_{\nu=1}^{2}(u_{\nu,i,j} - u_{\nu,i,j}^{ex})^2\right]^{\frac{1}{2}}, \ \delta p(h) = \left[\frac{1}{M^2}\sum_{i=1}^{M}\sum_{j=1}^{M}(p_{i,j} - p_{i,j}^{ex})^2\right]^{\frac{1}{2}},$$

where M is the number of cells along each coordinate direction, $u_{i,j}^{ex}$ and $p_{i,j}^{ex}$ are the velocity vector and the pressure according to the exact solution (25). The quantities $u_{i,j}$ and $p_{i,j}$ denote the numerical solution obtained by the CLR method described above. The convergence orders ν_u and ν_p are computed by the well-known formulas [21,23]. Let $b_{i,j,l}^s$, $s = 0, 1, \ldots$ be the values of the coefficients $b_{i,j,l}$ in (8) at the sth iteration. The following condition was used for termination of the iterations over the nonlinearity: $\delta b^{s+1} < \varepsilon$, where $\delta b^{s+1} = \max_{i,j}(\max_{1\le l\le 12}|b_{i,j,l}^{s+1} - b_{i,j,l}^s|)$, and $\varepsilon < h^2$ is a small positive quantity. We will

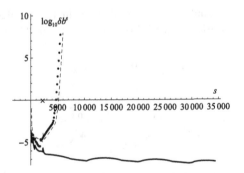

Fig. 2. Error δb^s at the use of different preconditioners: (——) $\xi = \xi_{\mathrm{opt}}$, $\eta = \eta_{\mathrm{opt}}$; (— — —) $\xi = \eta = 1$; ($\cdots\cdots$) the diagonal preconditioner

call the quantity δb^{s+1} the pseudo-error of the approximate solution. A series of computations has been done for the purpose of studying the influence of a specific form of the preconditioner on the convergence of iterations by the CLR method. In this series of runs, the satisfaction of the inequality $\delta b^s < 10^{-9}$ was the criterion for the computation termination. The results are presented in Fig. 2. The cross on the s-axis shows the number of the iteration s, beginning with which the computation within the framework of the multigrid algorithm was carried out on the grid of 80×80 cells. It is seen that in the absence of the preconditioner, when $\xi = 1$ in (17) and $\eta = 1$ in (18), the pseudo-error δb^s starts growing after the passage to the computation on the 80×80 grid.

The diagonal preconditioner of [14] also demonstrates a similar behavior, whereas the preconditioner using (17), (18) with $\xi = \xi_{\mathrm{opt}} = 0.156$, $\eta = \eta_{\mathrm{opt}} = 1.747$ ensures the convergence to $\delta b^s < 10^{-9}$. Therefore, all the computations were done below in this section with the use of the preconditioner of Sect. 3 with $\xi = \xi_{\mathrm{opt}}$, $\eta = \eta_{\mathrm{opt}}$.

Figures 3, 4 and Table 3 present the results of numerical experiments whose aim was to elucidate the influence on the convergence acceleration of the iteration process for solving the Navier–Stokes equations only at the use of the two-parameter preconditioner and the Krylov method. In these computations, it was assumed that the Reynolds number Re $= 1000$ and $L = 0.5$ in (4). These computations were done on the 40×40 grid, $N_c = 8$. The following quasi-optimal values of quantities ξ and η were used according to Table 2: $\xi = \xi_{\mathrm{opt}} = 0.213$,

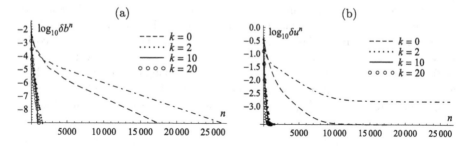

Fig. 3. Influence of the number of the residuals k in Krylov algorithm [22] on the convergence rate of quantities $\log_{10} \delta b^n$ (a) and $\log_{10} \delta u$ (b), where n is the number of iterations

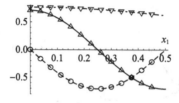

Fig. 4. Comparison of the profiles of approximate solution with the exact one at $x_2 = L/4$

Table 3. Quantities $\delta\mathbf{u}$, δp, ν_u, ν_p on a sequence of grids, Re $=1000$, $L = 0.5$, $N_c = 8$

M	$\delta\mathbf{u}$	δp	ν_u	ν_p
10	2.204E−03	3.087E−03		
20	7.174E−04	9.021E−04	1.62	1.77
40	1.801E−04	2.521E−04	1.99	1.84
80	4.070E−05	9.547E−05	2.15	1.40

$\eta = \eta_{\mathrm{opt}} = 1.738$. The case $k = 0$ (see the caption to Fig. 3) corresponds to a computation without using the algorithm of Krylov subspaces. In this series of runs, Eq. (13) was included in the overdetermined SLAE (14). In the process of iterations by the CLR method, the absolute value of integral (3) dropped from the value of the order $O(10^{-3})$ to the value of the order $O(10^{-12}) - O(10^{-13})$. The number of iterations N_{it} needed for satisfying the inequality $\delta b^n < 10^{-9}$ amounted to 17429, 1577, 969, and 1072, respectively, at $k = 0$, 2, 10, and 20. Thus, the application of the Krylov algorithm at Re $= 1000$ with $k = 10$ has enabled a reduction of the iterations needed for the solution convergence by the factor of 18 in comparison with the case when $k = 0$.

A computation was also done in the absence of the preconditioner that is when $\xi = \eta = 1$, see the dash-dot lines in Fig. 3. In this case, 26384 iterations were required by the CLR method at $k = 0$ to satisfy the inequality $\delta b^n < 10^{-9}$ that is by the factor of 1.5 higher than at the use of the values ξ_{opt} and η_{opt} in the preconditioner. It is seen in Fig. 3(b) that the error δu^n of the converged solution is higher by $10^{0.884}$ times than at the use of the values ξ_{opt} and η_{opt} in the preconditioner. It is seen in Table 3 that at Re $= 1000$, the convergence order ν_u is close to 2, and $\nu_p < \nu_u$ at $M \geq 40$.

Table 4. Influence of the preconditioner and the combination of the Krylov and Fedorenko methods on a sequence of grids of sizes $5 \cdot 2^m \times 5 \cdot 2^m$, $m = 0, \ldots, 4$ on the convergence rate of the CLR method at Re $= 1000$

Method	N_{it}	CPU time, s.	AF	$\delta\mathbf{u}$	δp
$K_{mgr} = 1, k = 0$	851119	464718	1.0	3.587E−05	7.065E−04
$K_{mgr} = 5, k = 0$	1282529	843849	0.55	2.108E−05	4.570E−04
$K_{mgr} = 1, k = 9$	39521	21944	21.18	3.587E−05	7.060E−04
$K_{mgr} = 1, k = 10$	40154	22375	20.77	3.587E−05	7.063E−04
$K_{mgr} = 5, k = 5$	6185	2119	219.31	3.570E−05	6.713E−05
$K_{mgr} = 5, k = 8$	6283	2172	213.96	3.571E−05	6.392E−05
$K_{mgr} = 5, k = 9$	5832	2018	230.29	3.571E−05	5.787E−05
$K_{mgr} = 5, k = 10$	6090	2186	212.59	3.571E−05	5.628E−05

Figure 4 shows the solution obtained by the CLR method by symbols Δ (v_1), $\circ$ (v_2), and ∇ (p); the curves of the exact solution are depicted by the solid, dashed, and dash-dot lines for v_1, v_2, and p, respectively. One can see here a good agreement between the numerical results and the analytic solution.

The computations were done with the use of the ascending branch of the V-cycle, which corresponds to the prolongation operation, for the purpose of elucidating the influence of only multigrid algorithm on the acceleration of convergence in the CLR method. The computations were also done, in which the motion along the ascending branch of the multigrid V-cycle was combined with the acceleration algorithm based on Krylov's subspaces. The results of these computations are presented in Table 4. In this table, K_{mgr} is the number of sequentially used grids in the multigrid complex. If $K_{mgr} = 1$, then this means that only one grid is employed in the computation, and this is the finest grid with the number of cells 80×80. The integer N_{it} is the total number of iterations made on all grids of the complex. The factor AF of the iteration process acceleration as a result of the application of some acceleration technique is calculated as the ratio of the CPU time at $K_{mgr} = 1$, $k = 0$ to the CPU time at the application of a sequence of grids in combination with the application of the multigrid algorithm on each grid ($K_{mgr} > 1$, $k = 0$) or at the application of a sequence of grids combined with the application of the Krylov's algorithm on each grid ($K_{mgr} > 1$, $k > 1$). In all computations shown in Table 4, the quasi-optimal values $\xi_{opt} = 0.08$, $\eta_{opt} = 3.48$ were used in the two-parameter preconditioner. It is seen in Table 4 that the highest convergence acceleration — by the factor of 230 in comparison with the computation only on the finest grid — takes place when five grids are applied in the multigrid algorithm and 9 residuals are applied in the Krylov method.

For a better elucidation of the capabilities of CASs at the construction of the versions of the CLR method, the numerical experiments were done here to solve the benchmark problem of the lid-driven cavity flow on a grid of 320×320 cells. The above-described techniques for accelerating the convergence of iterations were also applied here, and in this case, the value AF $= 162$ was achieved. The obtained results were compared with the most accurate results of other researchers obtained at Re $= 1000$. The obtained results coincided with the results of [2, 6, 20] with the accuracy of $\approx 10^{-3}$.

6   Conclusions

The CAS *Mathematica* has been applied for constructing the optimal iteration processes of the Gauss–Seidel type at the solution of boundary-value problems by the CLR method. A large amount of symbolic computations, which arose at the derivation of the basic formulae of the CLR method, was done efficiently with *Mathematica*. It is very important that the application of CAS has facilitated greatly this work, reduced at all its stages the probability of errors usually introduced by the mathematician at the development of a new algorithm.

Note that the algorithm of a search for optimal parameters of the two-parameter preconditioner, which has been proposed in the present work, can be extended also for the cases of the solution of other PDEs by the CLR method.

References

1. Amodio, P., Blinkov, Y., Gerdt, V., La Scala, R.: On consistency of finite difference approximations to the Navier-Stokes equations. In: Gerdt, V.P., Koepf, W., Mayr, E.W., Vorozhtsov, E.V. (eds.) CASC 2013. LNCS, vol. 8136, pp. 46–60. Springer, Heidelberg (2013)
2. Botella, O., Peyret, R.: Benchmark spectral results on the lid-driven cavity flow. Comput. Fluids **27**, 421–433 (1998)
3. Fedorenko, R.P.: The speed of convergence of one iterative process. USSR Comput. Math. Math. Phys. **4**(3), 227–235 (1964)
4. Ganzha, V.G., Mazurik, S.I., Shapeev, V.P.: Symbolic manipulations on a computer and their application to generation and investigation of difference schemes. In: Caviness, B.F. (ed.) ISSAC 1985 and EUROCAL 1985. LNCS, vol. 204, pp. 335–347. Springer, Heidelberg (1985)
5. Gerdt, V.P., Blinkov, Y.A.: Involution and difference schemes for the Navier–Stokes equations. In: Gerdt, V.P., Mayr, E.W., Vorozhtsov, E.V. (eds.) CASC 2009. LNCS, vol. 5743, pp. 94–105. Springer, Heidelberg (2009)
6. Isaev, V.I., Shapeev, V.P.: High-accuracy versions of the collocations and least squares method for the numerical solution of the Navier-Stokes equations. Comput. Math. Math. Phys. **50**, 1670–1681 (2010)
7. Isaev, V.I., Shapeev, V.P., Eremin, S.A.: Investigation of the properties of the method of collocations and least squares for solving the boundary-value problems for the Poisson equation and the Navier-Stokes equations. Comput. Technol. **12**(3), 1–19 (2007). (in Russian)
8. Jiang, B., Lin, T.L., Povinelli, L.A.: Large-scale computation of incompressible viscous flow by least-squares finite element method. Comput. Meth. Appl. Mech. Engng. **114**(3–4), 213–231 (1994)
9. Krylov, A.N.: On the numerical solution of the equation, which determines in technological questions the frequencies of small oscillations of material systems. Izv. AN SSSR, Otd. matem. i estestv. nauk **4**, 491–539 (in Russian) (1931)
10. Lucas, P., Zuijlen, A.H., Bijl, H.: Fast unsteady flow computations with a Jacobian-free Newton-Krylov algorithm. J. Comp. Phys. **229**(2), 9201–9215 (2010)
11. Nasr-Azadani, M.M., Meiburg, E.: TURBINS: an immersed boundary, Navier-Stokes code for the simulation of gravity and turbidity currents interacting with complex topographies. Comp. Fluids **45**(1), 14–28 (2011)
12. Nickaeen, M., Ouazzi, A., Turek, S.: Newton multigrid least-squares FEM for the V-V-P formulation of the Navier-Stokes equations. J. Comp. Phys. **256**, 416–427 (2014)
13. Plyasunova, A.V., Sleptsov, A.G.: Collocation-grid method of solving the nonlinear parabolic equations on moving grids. Modelirovanie v mekhanike **18**(4), 116–137 (1987). (in Russian)
14. Ramšak, M., Škerget, L.: A subdomain boundary element method for high-Reynolds laminar flow using stream function-vorticity formulation. Int. J. Numer. Meth. Fluids **46**, 815–847 (2004)

15. Saad, Y.: Numerical Methods for Large Eigenvalue Problems. Manchester University Press, Manchester (1991)
16. Semin, L., Shapeev, V.: Constructing the numerical method for Navier–Stokes equations using computer algebra system. In: Ganzha, V.G., Mayr, E.W., Vorozhtsov, E.V. (eds.) CASC 2005. LNCS, vol. 3718, pp. 367–378. Springer, Heidelberg (2005)
17. Semin, L.G., Sleptsov, A.G., Shapeev, V.P.: Collocation and least-squares method for Stokes equations. Comput. Technol. 1(2), 90–98 (1996). (in Russian)
18. Shapeev, V.: Collocation and least residuals method and its applications. EPJ Web Conferences 108, 01009 (2016). doi:10.1051/epjconf/201610801009
19. Shapeev, V.P., Isaev, V.I., Idimeshev, S.V.: The collocations and least squares method: application to numerical solution of the Navier-Stokes equations. In: Eberhardsteiner, J., Böhm, H.J., Rammerstorfer, F.G. (eds.) CD-ROM Proceedings of the 6th ECCOMAS, Sept. 2012, Vienna Univ. of Tech. ISBN: 978-3-9502481-9-7 (2012)
20. Shapeev, A.V., Lin, P.: An asymptotic fitting finite element method with exponential mesh refinement for accurate computation of corner eddies in viscous flows. SIAM J. Sci. Comput. 31, 1874–1900 (2009)
21. Shapeev, V.P., Vorozhtsov, E.V.: CAS application to the construction of the collocations and least residuals method for the solution of 3D Navier–Stokes equations. In: Gerdt, V.P., Koepf, W., Mayr, E.W., Vorozhtsov, E.V. (eds.) CASC 2013. LNCS, vol. 8136, pp. 381–392. Springer, Heidelberg (2013)
22. Shapeev, V.P., Vorozhtsov, E.V., Isaev, V.I., Idimeshev, S.V.: The method of collocations and least residuals for three-dimensional Navier-Stokes equations. Vychislitelnye metody i programmirovanie 14, 306–322 (2013). (in Russian)
23. Shapeev, V.P., Vorozhtsov, E.V.: Symbolic-numeric implementation of the method of collocations and least squaresfor 3D Navier–Stokes equations. In: Gerdt, V.P., Koepf, W., Mayr, E.W., Vorozhtsov, E.V. (eds.) CASC 2012. LNCS, vol. 7442, pp. 321–333. Springer, Heidelberg (2012)
24. Valiullin, A.N., Ganzha, V.G., Meleshko, S.V., Murzin, F.A., Shapeev, V.P., Yanenko, N.N.: Application of symbolic manipulations on a computer for generation and analysis of difference schemes. Preprint Inst. Theor. Appl. Mech. Siberian Branch of the USSR Acad. Sci., Novosibirsk No. 7 (1981). (in Russian)
25. Wang, M., Chen, L.: Multigrid methods for the stokes equations using distributive gauss-seidel relaxations based on the least squares commutator. J. Sci. Comput. 56(2), 409–431 (2013)
26. Wesseling, P.: An Introduction to Multigrid Methods. Wiley, Chichester (1992)

Pruning Algorithms for Pretropisms
of Newton Polytopes

Jeff Sommars$^{(\boxtimes)}$ and Jan Verschelde

Department of Mathematics, Statistics, and Computer Science, University of Illinois
at Chicago, 851 S. Morgan Street (m/c 249), Chicago, IL 60607-7045, USA
{sommars1,janv}@uic.edu

Abstract. Pretropisms are candidates for the leading exponents of Puiseux series that represent positive dimensional solution sets of polynomial systems. We propose a new algorithm to both horizontally and vertically prune the tree of edges of a tuple of Newton polytopes. We provide experimental results with our preliminary implementation in Sage that demonstrates that our algorithm compares favorably to the definitional algorithm.

1 Introduction

Almost all polynomial systems arising in applications are sparse, as few monomials appear with nonzero coefficients, relative to the degree of the polynomials. Polyhedral methods exploit the sparse structure of a polynomial system. In the application of polyhedral methods to compute positive dimensional solution sets of polynomial systems, we look for series developments of the solutions, and in particular we look for Puiseux series [29]. The leading exponents of Puiseux series are called *tropisms*. The *Newton polytope* of a polynomial in several variables is the convex hull of the exponent tuples of the monomials that appear with nonzero coefficient in the polynomial.

In [10], polyhedral methods were defined in tropical algebraic geometry. We refer to [27] for a textbook introduction to tropical algebraic geometry. Our textbook reference for definitions and terminology of polytopes is [39].

Our problem involves the intersection of polyhedral cones. A *normal cone* of a face F of a polytope P is the convex cone generated by all of the facet normals of facets which contain F. The *normal fan* of a polytope P is the union of all of the normal cones of every face of P. Given two fans F_1 and F_2, their *common refinement* $F_1 \wedge F_2$ is defined as

$$F_1 \wedge F_2 = \bigcup_{\substack{C_1 \in F_1 \\ C_2 \in F_2}} C_1 \cap C_2. \tag{1}$$

This material is based upon work supported by the National Science Foundation under Grant No. 1440534.

© Springer International Publishing AG 2016
V.P. Gerdt et al. (Eds.): CASC 2016, LNCS 9890, pp. 489–503, 2016.
DOI: 10.1007/978-3-319-45641-6_31

As the common refinement of two fans is again a fan, the common refinement of three fans F_1, F_2, and F_3 may be computed as $(F_1 \wedge F_2) \wedge F_3$.

Problem Statement. Given the normal fans $(F_1, F_2, \ldots, F_n)$ of the Newton polytopes $(P_1, P_2, \ldots, P_n)$, a *pretropism* is a ray in a cone C,

$$C = C_1 \cap C_2 \cap \cdots \cap C_n \in F_1 \wedge F_2 \wedge \cdots \wedge F_n, \tag{2}$$

where each C_i is the normal cone to some k_i-dimensional face of P_i, for $k_i \geq 1$, for $i = 1, 2, \ldots, n$. Our problem can thus be stated as follows: given a tuple of Newton polytopes, compute all pretropisms.

We say that two pretropisms are equivalent if they are both perpendicular to the same tuples of faces of the Newton polytopes. Modulo this equivalence, there are only a finite number of pretropisms. Ours is a difficult problem because of the dimension restrictions on the cones. In particular, the number of pretropisms can be very small compared to the total number of cones in the common refinement.

Pretropisms are candidates tropisms, but not every pretropism is a tropism, as pretropisms depend only on the Newton polytopes of the system. For polynomial systems with sufficiently generic coefficients, every tropism is also a pretropism. See [9] for an example.

Related Work. A tropical prevariety was introduced in [10] and Gfan [26] is a software system to compute the common refinement of the normal fans of the Newton polytopes. Gfan relies on the reverse search algorithms [4] in cddlib [20].

The problem considered in this paper is a generalization of the problem to compute the mixed volume of a tuple of Newton polytopes, for which pruning methods were first proposed in [15]. Further developments can be found in [21,31], with corresponding free software packages MixedVol [22] and DEMiCS [30]. A recent parallel implementation along with a complexity study appears in [28]. The relationship between triangulations and the mixed subdivisions is explained and nicely illustrated in [13].

The main difference between mixed volume computation and the computation of the tropical prevariety is that in a mixed volume computation the vertices of the polytopes are lifted randomly, thus removing all degeneracies. This lifting gives the powers of an artificial parameter. In contrast, in a Puiseux series development of a space curve, the first variable is typically identified as the parameter and the powers of the first variable in the given polynomials cannot be considered as random.

A practical study on various software packages for exact volume computation of a polytope is described in [12]. Exact algorithms on Newton polytopes are discussed in [18]. The authors of [16] present an experimental study of approximate polytope volume computation. In [17], a polynomial-time algorithm is presented to compute the edge skeleton of a polytope. Computing integer hulls of convex polytopes can be done with polymake [3].

Our contributions and organization of the paper. In this paper we outline two different types of pruning algorithms for the efficient computation of pretropisms. We report on a preliminary implementation in Sage [36] and illustrate the effectiveness on a parallel computer for various benchmark problems. This paper extends the results of our EuroCG paper [35] as well as [34].

2   Pruning Algorithms

2.1   Horizontal and Vertical Pruning Defined

Because we are interested only in those cones of the common refinement that contain rays perpendicular to faces of dimension one or higher, we work with the following modification of (1):

$$F_1 \wedge_1 F_2 = \bigcup_{\substack{C_1 \in F_1, C_1 \perp \text{ edge of } P_1 \\ C_2 \in F_2, C_2 \perp \text{ edge of } P_2}} C_1 \cap C_2. \tag{3}$$

The $\wedge_1$ defines the *vertical pruning* as the replacement of $\wedge$ by $\wedge_1$ in $(F_1 \wedge F_2) \wedge F_3$ so we compute $(F_1 \wedge_1 F_2) \wedge_1 F_3$. Cones in the refinement that do not satisfy the dimension restrictions are pruned away in the computations. Our definition of vertical pruning is currently incomplete, but we will refine it in 2.5 after we have formally defined our algorithms.

The other type of pruning, called *horizontal pruning* is already partially implicitly present in the $\bigcup$ operator of (3), as in a union of sets of cones, every cone is collected only once, even as it may originate as the result of many different cone intersections. With horizontal pruning we remove cones of $F_1 \wedge_1 F_2$ which are contained in larger cones. Formally, we can define this type of pruning via the $\wedge_2$ operator:

$$F_1 \wedge_2 F_2 = \bigcup_{\substack{C \in F_1 \wedge_1 F_2, C \not\subset C' \\ C' \in F_1 \wedge_1 F_2 \setminus \{C\}}} C. \tag{4}$$

2.2   Pseudo Code Definitions of the Algorithms

Algorithm 2 sketches the outline of our algorithm to compute all pretropisms of a set of n polytopes. Along the lines of the gift wrapping algorithm, for every edge of the first polytope we take the plane that contains this edge and consider where this plane touches the second polytope. Algorithm 1 starts exploring the edge skeleton defined by the edges connected to the vertices in this touching plane.

The exploration of the neighboring edges corresponds to tilting the ray r in Algorithm 1, as in rotating a hyperplane in the gift wrapping method. One may wonder why the exploration of the edge skeleton in Algorithm 1 needs to continue after the statement on line 4. This is because the cone C has the potential to intersect many cones in P, particularly if P has small cones and C is large. Furthermore it is reasonable to wonder why we bother checking cone containment when computing the intersection of two cones provides more useful

---

Algorithm 1. Explores the skeleton of edges to find pretropisms of a polytope and a cone.

```
 1: function ExploreEdgeSkeleton(Polytope P, Cone C)
 2:     r := a random ray inside C
 3:     in_r(P) := vertices of P with minimal inner product with r
 4:     EdgesToTest := all edges of P that have vertices in in_r(P)
 5:     Cones := ∅
 6:     TestedEdges := ∅
 7:     while EdgesToTest ≠ ∅ do
 8:         E := pop an edge from EdgesToTest
 9:         C_E := normal cone to E
10:         ShouldAddCone := False
11:         if C_E contains C then
12:             ConeToAdd := C
13:             ShouldAddCone := True
14:         else if C ∩ C_E ≠ {0} then
15:             ConeToAdd := C ∩ C_E
16:             ShouldAddCone := True
17:         end if
18:         if ShouldAddCone then
19:             Cones := Cones ∪ ConeToAdd
20:             Edges := Edges ∪ E
21:             for each neighboring edge e of E do
22:                 if e ∉ TestedEdges then
23:                     EdgesToTest := EdgesToTest∪e
24:                 end if
25:             end for
26:         end if
27:         TestedEdges := TestedEdges ∪ E
28:     end while
29:     return Cones
30: end function
```

---

information. Checking cone containment means checking if each of the generators of C is contained in C_E, which is a far less computationally expensive operation than computing the intersection of two cones.

In the Newton-Puiseux algorithm to compute series expansions, we are interested only in the edges on the lower hull of the Newton polytope, i.e. those edges that have an upward pointing inner normal. For Puiseux for space curves, the expansions are normalized so that the first exponent in the tropism is positive. Algorithm 2 is then easily adjusted so that calls to the edge skeleton computation of Algorithm 1 are made with rays that have a first component that is positive.

2.3   Correctness

To see that these algorithms will do what they claim, we must define an additional term. A *pretropism graph* is the set of edges for a polytope that have normal cones

---

Algorithm 2. Finds pretropisms for a given set of polytopes

1: **function** FINDPRETROPISMS(Polytope P_1, Polytope $P_2, \ldots,$ Polytope P_n)
2:     Cones := set of normal cones to edges in P_1
3:     **for** i := 2 to n **do**
4:         NewCones := $\emptyset$
5:         **for** Cone in Cones **do**
6:             NewCones := NewCones $\cup$ ExploreEdgeSkeleton(P_i, Cone)
7:         **end for**
8:         **for** Cone in NewCones **do**
9:             **if** Cone is contained within another cone in NewCones **then**
10:                 NewCones := NewCones - Cone
11:             **end if**
12:         **end for**
13:         Cones := NewCones
14:     **end for**
15:     Pretropisms := set of generating rays for each cone in Cones
16:     **return** Pretropisms
17: **end function**

---

intersecting a given cone. We will now justify why the cones output by Algorithm 1 correspond to the full set of cones that live on a pretropism graph.

Theorem 1. *Pretropism graphs are connected graphs.*

Proof. Let C be a cone, and let P be a polytope with edges e_1, e_2 such that they are in the pretropism graph of C. Let C_1 be the cone of the intersection of the normal cone of e_1 with C, and let C_2 be the cone of the intersection of the normal cone of e_2 and C. If we can show that there exists a path between e_1 and e_2 that remains in the pretropism graph, then the result will follow.

Let n_1 be a normal to e_1 that is also in C_1 and let n_2 be a normal to e_2 that is also in C_2. Set $n = tn_1 + (1-t)n_2$ where $0 \le t \le 1$. Consider varying t from 0 to 1; this creates the cone C_n, a cone which must lie within C, as both n_1 and n_2 lie in that cone. As n moves from 0 to 1, it will progressively intersect new faces of P that have all of their edges in the pretropism graph. Eventually, this process terminates when we reach e_2, and we have constructed a path from e_1 to e_2. Since a path always exists, we can conclude that pretropism graphs are connected graphs.   □

Since pretropism graphs are connected, Algorithm 1 will find all cones of edges on the pretropism graph. In Algorithm 2, we iteratively explore the edge skeleton of polytope P_i, and use the pruned set of cones to explore P_{i+1}. From this, it is clear that Algorithm 2 will compute the full set of pretropisms.

2.4   Analysis of Computational Complexity

In estimating the cost of our algorithm to compute all pretropisms, we will first consider the case when there are two polytopes. We will take the primitive operation of computing pretropisms to be the number of cone intersections performed, as that number will drive the time required for the algorithm to complete. For a polytope P, denote by $n_e(P)$ its number of edges. The upper bound on the number of primitive operations for two polytopes P_1 and P_2 is the product $n_e(P_1) \times n_e(P_2)$, while the lower bound equals the number of pretropisms.

Denote by $E_{P,e}$ the pretropism graph resting on polytope P corresponding to the ray determined by edge e. Let $n_e(E_{P,e})$ denote the number of edges in $E_{P,e}$.

Proposition 1. *The number of primitive operations in Algorithm 2 on two polytopes P_1 and P_2 is bounded by*

$$\sum_{i=1}^{n_e(P_1)} n_e(E_{P_2,e_i}), \tag{5}$$

where e_i is the i-th edge of P_1.

As $E_{P,e}$ is a subset of the edges of P: $n_e(E_{P,e}) \leq n_e(P)$. Therefore, the bound in (5) is smaller than $n_e(P_1) \times n_e(P_2)$.

To interpret (5), recall that Algorithm 2 takes a ray from inside a normal cone to an edge of the first polytope for the exploration of the edge graph of the second polytope. If we take a simplified view on the second polytopes as a ball, then shining a ray on that ball will illuminate at most half of its surface. If we use the estimate: $n_e(E_{P_2,e_i}) \approx n_e(P_2)/2$, then Algorithm 2 cuts the upper bound on the number of primitive operations in half.

Estimating the cost of the case of n polytopes follows naturally from the cost analysis of the case of 2 polytopes. For n polytopes, the upper bound on the number of primitive operations required is the product $n_e(P_1) \times n_e(P_2) \times \ldots \times n_e(P_n)$.

Proposition 2. *The number of primitive operations in Algorithm 2 on n polytopes $P_1, P_2, \ldots, P_n$ is bounded by*

$$\sum_{i=1}^{n_e(P_1)} \left(\prod_{j=2}^{n} n_e(E_{P_j,e_i}) \right) \tag{6}$$

where e_i is the i-th edge of P_1.

Again, if we use the estimate that $n_e(E_{P_j,e_i}) \approx n_e(P_j)/2$, then Algorithm 2 reduces the upper bound on the number of primitive operations by $\frac{1}{2^{n-1}}$. This estimate depends entirely on the intuition that we are cutting the number of comparisons in half. This estimate may not hold in the case when we have large lineality spaces, and thus have huge input cones. This situation can be partially remedied by sorting the input polytopes from smallest dimension of lineality space to highest. This seeds Algorithm 2 with the smallest possible input cones.

2.5   Horizontal and Vertical Pruning Revisited

The definitional algorithm of pretropism can be interpreted as creating a tree structure. From a root node, connect the cones of P_1. On the next level of the tree, place the cones resulting from intersecting each of the cones of P_2 with the cones of P_1, connecting the new cones with the cone from P_1 that they intersected. It is likely that at this level of the tree there are many cones that are empty. Continue this process creating new levels of cones representing the intersection of the previous level of cones with the next polytope until the nth polytope has been completed. The cones at the nth layer of the tree represent the cones generated by pretropisms.

Our algorithms can be seen to improve on this basic tree structure in two distinct ways. Algorithm 1 reduces the number of comparisons needed through exploring the edge skeletons of the polytopes. Because of this, there are many times that we do not perform cone intersections that will result in 0 dimensional cones. From the perspective of the tree, this is akin to avoiding drawing edges to many 0 dimensional cones; we call this vertically pruning the tree. We horizontally prune the tree through Algorithm 2 which reduces the number of cones necessary to follow for a given level. This is illustrated in Fig. 1. By both horizontally and vertically pruning the tree of cones, we are able to avoid performing many unnecessary cone intersections. We will demonstrate the benefits of pruning experimentally in the sections to come.

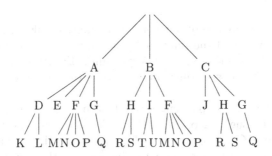

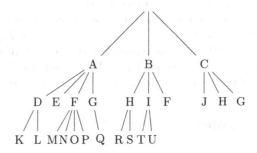

Fig. 1. Nodes A, B, C represent cones to P_1. Nodes D, E, F, and G represent intersections of cone A with cones to P_2, etc. Nodes K and L represent intersections of cone D with cones to P_3, etc. Duplicate nodes are removed from the second tree at the bottom.

3   Implementation

Our algorithm takes as input a modified version of the data structure output by the gift wrapping algorithm to compute convex hulls. It conceptually exploits the connectivity between vertices, edges, and facets, but only requires the edge skeleton of the polytope. To accomplish this, we created edge objects that had vertices, references to their neighboring edges, and the normal cone of the edge. When polytopes were not full dimensional, we included the generating rays of the lineality space when we created the normal cones. This has the negative effect of increasing the size of the cone, but is essential for the algorithm to work.

3.1   Code

We developed a high level version of Algorithm 2 in Sage [36], using its modules for lattice polytopes [32], and polyhedral cones [11]. To compute the intersections of cones, Sage uses PPL [6]. Our preliminary code is available at https://github. com/sommars/GiftWrap.

3.2   Parallelism

To improve the performance of our core algorithms, we also implemented high level parallelism. We used the built in Python queue structure to create a job queue of cones in lines 5 through 7 of Algorithm 2. Each call to the Algorithm 1 is done independently on a distinct process, using the computers resources more efficiently.

Additionally, we have implemented parallelism in checking the cone containments in lines 8 through 12 of Algorithm 1. To check if cone C_1 is contained within another cone C_2, it requires checking if each of the linear equations and inequalities of C_1 is or is not restricted by each of the linear equations and inequalities of C_2. However, there is much overlap between cones, with many distinct cones sharing some of the same linear equations or inequalities. Because of this, we optimized by creating a lookup table to avoid performing duplicate calculations. However, with large benchmark problems, creation of this table becomes prohibitively slow. To amend this problem, we parallelized the creation of the table, with distinct processes performing distinct calculations.

4   Generic Experiments

To test the algorithms, we generate $n - 1$ simplices spanned by integer points with coordinates uniformly generated within the range of 0 to 30. This input corresponds to considering systems of $n - 1$ sparse Laurent polynomials in n variables with $n + 1$ monomials per equation. We can compare with the mixed volume computation if we add one extra linear equation to the Laurent polynomial system. Then the mixed volume of the n-tuple will give the sum of the degrees of all the curves represented by the Puiseux series. Assuming generic choices for the coefficients, the degrees of the curves can be computed directly from the tropisms, as used in [38] and applied in [1,2].

Denoting by $MV(P)$ the mixed volume of an n-tuple P of Newton polytopes:

$$MV(P) = \sum_{\mathbf{v}} \left(\max_{i=1}^{n} v_i - \min \left(\min_{i=1}^{n} v_i, 0 \right) \right) \tag{7}$$

where the sum ranges over all tropisms $\mathbf{v}$.

All computations were done on a 2.6 GHz Intel Xeon E5-2670 processor in a Red Hat Linux workstation with 128 GB RAM using 32 threads. When performing generic tests, the program did not perform any cone containment tests because no cone can be contained in another cone in this case.

4.1 Benchmarking

Table 4 shows a comparison between the two distinct methods of computing pretropisms. The mixed volume was computed with the version of MixedVol [22], available in PHCpack [37] since version 2.3.13. For systems with generic coefficients, the mixed volume equals the number of isolated solutions [7]. While a fast multicore workstation Table 1 can compute millions of solutions, a true supercomputer will be needed in the case of billions of solutions. For larger dimensions, the new pruning method dominates the method suggested by the definition of pretropism.

Table 1. Comparisons between the definitional and our pruning method, for randomly generated generic simplices. Timings are listed in seconds.

n	Definitional	Pruning	#Pretropisms	Mixed volume
3	0.008	0.20	7	319
4	0.11	0.42	18	7,384
5	1.33	0.76	58	152,054
6	13.03	2.75	171	4,305,758
7	243.88	20.17	614	91,381,325
8	2054.11	220.14	1,878	2,097,221,068

4.2 Number of Cone Intersections

Another way that the definitional algorithm can be compared to our new algorithm is through comparing the number of cone intersections required for each algorithm. Table 2 contains a comparison of these numbers. A large number of trials were performed at each dimension so we could conclude statistically if our mean number of intersections differed from the number of intersections required by the cone intersection algorithm. To test this hypothesis, we performed t-tests using the statistical software package R [33]. For every dimension from 3 to 8, we were able to reject the null hypothesis that they had the same mean and we were able to conclude that the new algorithm has a lower mean number of intersections ($p < 2 \times 10^{-16}$ for every test). We had estimated the cost to be an improvement by a factor of $\frac{1}{2^{n-1}}$, but experimentally we found a greater improvement as can be seen in Table 2.

Table 2. Average number of cone intersections required for each algorithm, comparing the definitional algorithm with our pruning algorithm for generic inputs. The second to last column contains the ratio predicted by our cost estimate and the final column contains the actual ratio.

n	Definitional	Pruning	Predicted ratio	Actual ratio
3	36	29	0.5	0.72
4	1,000	288	0.25	0.288
5	50,625	2,424	0.125	0.0478
6	4,084,101	18,479	0.0625	0.00452
7	481,890,304	145,134	0.03125	0.000301
8	78,364,164,096	1,150,386	0.015625	0.0000147

4.3   Comparison with Gfan

In the generic case, our code is competitive with Gfan. Table 3 contains timing comparisons, with input polynomials determined as they were previously determined; the timings in the Gfan column were obtained by running the current version 0.5 of Gfan [25].

Table 3. Comparisons between Gfan and our implementation, for dimensions 3 through 8. Timings are listed in seconds.

n	Gfan	Pruning
3	0.036	0.12
4	0.23	0.25
5	2.03	0.80
6	23.49	10.73
7	299.32	49.53
8	3,764.83	540.32

5   Benchmark Polynomial Systems

Many of the classic mixed volume benchmark problems like Katsura-n, Chandra-n, eco-n, and Noonberg-n are inappropriate to use as benchmark systems for computing pretropisms. A good testing system needs to have a positive dimensional solution set as well as being a system that can be scaled up in size. The aforementioned mixed volume benchmark problems all lack positive dimensional solution sets, so we did not perform tests on them. We have found the cyclic n-roots problem to be the most interesting system that fulfills both criteria, as there are a variety of sizes of solution sets within them and the difficulty of computing pretropisms increases slowly. We also provide experimental data for the n-vortex and the n-body problem, but these problems quickly become uncomputable with our prototype Sage implementation.

5.1   Cyclic-n Experiments

The cyclic n-roots problem asks for the solutions of a polynomial system, commonly formulated as

$$\begin{cases} x_0 + x_1 + \cdots + x_{n-1} = 0 \\ i = 2, 3, 4, \ldots, n-1 : \sum_{j=0}^{n-1} \prod_{k=j}^{j+i-1} x_{k \bmod n} = 0 \\ x_0 x_1 x_2 \cdots x_{n-1} - 1 = 0. \end{cases} \tag{8}$$

This problem is important in the study of biunimodular vectors, a notion that traces back to Gauss, as stated in [19]. In [5], Backelin showed that if n has a divisor that is a square, i.e. if d^2 divides n for $d \geq 2$, then there are infinitely many cyclic n-roots. The conjecture of Björck and Saffari [8], [19, Conjecture 1.1] is that if n is not divisible by a square, then the set of cyclic n-roots is finite. As shown in [1], the result of Backelin can be recovered by polyhedral methods.

Instead of directly calculating the pretropisms of the Newton polytopes of the cyclic n-root problem, we chose to calculate pretropisms of the reduced cyclic n-root problem. This reformulation [14] is obtained by performing the substitution $x_i = \frac{y_i}{y_0}$ for $i = 0 \ldots n-1$. Clearing the denominator of each equation leaves the first $n-1$ equations as polynomials in $y_1, \ldots y_{n-1}$. We compute pretropisms of the Newton polytopes of these $n-1$ equations because they yield meaningful sets of pretropisms. Calculating with the reduced cyclic n-roots problem has the benefit of removing much of the symmetry present in the standard cyclic n-roots problem, as well as decreasing the ambient dimension by one. Unlike the standard cyclic n-roots problem, some of the polytopes of the reduced cyclic n-roots problem are full dimensional, which leads to calculation speed ups. A simple transformation can be performed on the pretropisms we calculate of reduced cyclic n-root problem to convert them to the pretropisms of cyclic n-root problem, so calculating the pretropisms of reduced cyclic n-roots problem is equivalent to calculating the pretropisms of the cyclic n-roots problem.

Table 4 shows how our implementation scales with time. As with the generic case, our implementation shows great improvement over the definitional algorithm as n becomes larger. For $n > 8$, the definitional algorithm was too inefficient to terminate in the time allotted.

Table 4. Comparisons between the definitional and our pruning method for reduced cyclic-n. Timings are listed in seconds

n	Definitional	Pruning	#Pretropisms	Mixed Volume
4	0.02	0.62	2	4
5	0.43	1.04	0	14
6	17.90	1.56	8	26
7	301.26	2.57	28	132
8	33681.66	9.43	94	320
9		44.97	276	1224
10		978.67	712	3594

Just as we surpassed our estimates of the expected number of cone intersections in the generic case, we also surpassed our estimated ratio in the case of reduced cyclic-n. Table 5 contains experimental results.

Table 5. Number of cone intersections required for each algorithm, comparing the definitional algorithm with our pruning algorithm for reduced cyclic-n. The second to last column contains the ratio predicted by our cost estimate and the final column contains the actual ratio

n	Definitional	Pruning	Predicted ratio	Actual ratio
4	120	44	0.25	0.36
5	1850	210	0.125	0.113
6	63,981	2,040	0.0625	0.0318
7	989,751	6,272	0.03125	0.00634
8	58,155,904	39,808	0.015625	0.000684
9		198,300	0.0078125	
10		1,933,147	0.00390625	

5.2 n-body and n-vortex

The n-body problem [23] is a classical problem from celestial dynamics that states that the acceleration due to Newtonian gravity can be found by solving a system of equations (9). These equations can be turned into a polynomial system by clearing the denominators.

$$\ddot{x}_j = \sum_{i \neq j} \frac{m_i(x_i - x_j)}{r_{ij}^3} \qquad 1 \leq j \leq n \tag{9}$$

The n-vortex problem [24] arose from a generalization of a problem from fluid dynamics that attempted to model vortex filaments (10). Again, these equations can be turned into polynomials through clearing denominators.

$$V_i = I \sum_{i \neq j} \frac{\Gamma_j}{z_i - z_j} \qquad 1 \leq j \leq n \tag{10}$$

Table 6 displays experimental results for both the n-body problem and the n-vortex problem. We expect to be able to compute higher n for these benchmark problems when we develop a compiled version of this code.

Table 6. Experimental results of our new algorithm. Timings are in seconds. The last column gives the number of cone intersections.

System	n	Pruning time	#Pretropisms	#Intersections
n-body	3	0.62	4	121
	4	5.07	57	25,379
	5	13,111.42	2,908	18,711,101
n-vortex	3	0.71	4	87
	4	2.93	25	10,595
	5	1457.48	569	5,021,659

6   Conclusion

To compute all pretropisms of a Laurent polynomial system, we propose to exploit the connectivity of edge skeletons to prune the tree of edges of the tuple of Newton polytopes. The horizontal and vertical pruning concepts we introduce are innovations that reduce the computational complexity of the problem. Our first high level implementation in Sage provides practical evidence that shows that our new pruning method is better than the definitional method with a variety of types of polynomial systems.

References

1. Adrovic, D., Verschelde, J.: Computing Puiseux series for algebraic surfaces. In: van der Hoeven, J., van Hoeij, M. (eds.) Proceedings of the 37th International Symposium on Symbolic and Algebraic Computation (ISSAC 2012), pp. 20–27. ACM (2012)

2. Adrovic, D., Verschelde, J.: Polyhedral methods for space curves exploiting symmetry applied to the cyclic n-roots problem. In: Gerdt, V.P., Koepf, W., Mayr, E.W., Vorozhtsov, E.V. (eds.) CASC 2013. LNCS, vol. 8136, pp. 10–29. Springer, Heidelberg (2013)

3. Assarf, B., Gawrilow, E., Herr, K., Joswig, M., Lorenz, B., Paffenholz, A., Rehn, T.: Computing convex hulls and counting integer points with polymake. arXiv:1408.4653v2

4. Avis, D., Fukuda, K.: A pivoting algorithm for convex hulls and vertex enumeration of arrangements and polyhedra. Discrete Comput. Geom. **8**(3), 295–313 (1992)

5. Backelin, J.: Square multiples n give infinitely many cyclic n-roots. Reports, Matematiska Institutionen 8, Stockholms universitet (1989)

6. Bagnara, R., Hill, P., Zaffanella, E.: The Parma Polyhedral Library: toward a complete set of numerical abstractions for the analysis and verification of hardware and software systems. Sci. Comput. Program. **72**(1–2), 3–21 (2008)

7. Bernshteĭn, D.: The number of roots of a system of equations. Funct. Anal. Appl. **9**(3), 183–185 (1975)

8. Bjöck, G., Saffari, B.: New classes of finite unimodular sequences with unimodular Fourier transforms. Circulant Hadamard matrices with complex entries. C.R. Acad. Sci. Paris Série I **320**, 319–324 (1995)

9. Bliss, N., Verschelde, J.: Computing all space curve solutions of polynomial systems by polyhedral methods. In: Gerdt, V.P., Koepf, W., Seiler, W.M., Vorozhtsov, E.V. (eds.) CASC 2016. LNCS, vol. 9890, pp. 73–86. Springer, Heidelberg (2016)

10. Bogart, T., Jensen, A., Speyer, D., Sturmfels, B., Thomas, R.: Computing tropical varieties. J. Symbolic Comput. **42**(1), 54–73 (2007)

11. Braun, V., Hampton, M.: `Polyhedra` module of Sage, The Sage Development Team (2011)

12. Büeler, B., Enge, A., Fukuda, K.: Exact volume computation for polytopes:a practical study. In: Kalai, G., Ziegler, G. (eds.) Polytopes -Combinatorics and Computation, DMV Seminar, vol. 29, pp. 131–154. Springer, Heidelberg (2000)

13. De Loera, J., Rambau, J., Santos, F.: Triangulations, Structures for Algorithms and Applications. Algorithms and Computation in Mathematics, vol. 25. Springer, Heidelberg (2010)

14. Emiris, I.: Sparse Elimination and Applications in Kinematics. Ph.D. thesis, University of California at Berkeley, Berkeley (1994)

15. Emiris, I., Canny, J.: Efficient incremental algorithms for the sparse resultant and the mixed volume. J. Symbolic Comput. **20**(2), 117–149 (1995)

16. Emiris, I., Fisikopoulos, V.: Efficient random-walk methods for approximating polytope volume. In: Proceedings of the Thirtieth Annual Symposium on Computational Geometry (SoCG 2014), pp. 318–327. ACM (2014)

17. Emiris, I., Fisikopoulos, V., Gärtner, B.: Efficient edge-skeleton computation for polytopes defined by oracles. J. Symbolic Comput. **73**, 139–152 (2016)

18. Emiris, I., Fisikopoulos, V., Konaxis, C.: Exact and approximate algorithms for resultant polytopes. In: Proceedings of the 28th European Workshop on Computational Geometry (EuroCG 2012) (2012)

19. Führ, H., Rzeszotnik, Z.: On biunimodular vectors for unitary matrices. Linear Algebra Appl. **484**, 86–129 (2015)

20. Fukuda, K., Prodon, A.: Double description method revisited. In: Deza, M., Manoussakis, I., Euler, R. (eds.) CCS 1995. LNCS, vol. 1120, pp. 91–111. Springer, Heidelberg (1996)

21. Gao, T., Li, T.: Mixed volume computation for semi-mixed systems. Discrete Comput. Geom. **29**(2), 257–277 (2003)

22. Gao, T., Li, T., Wu, M.: Algorithm 846: MixedVol: a software package for mixed-volume computation. ACM Trans. Math. Softw. **31**(4), 555–560 (2005)
23. Hampton, M., Jensen, A.: Finiteness of relative equilibria in the planar generalized n-body problem with fixed subconfigurations. J. Geom. Mech. **7**(1), 35–42 (2015)
24. Hampton, M., Moeckel, R.: Finiteness of stationary configurations of the four-vortex problem. Trans. Am. Math. Soc. **361**(3), 1317–1332 (2009)
25. Jensen, A.: Gfan, a software system for Gröbner fans and tropical varieties. http:// home.imf.au.dk/jensen/software/gfan/gfan.html
26. Jensen, A.: Computing Gröbner fans and tropical varieties in Gfan. In: Stillman, M., Takayama, N., Verschelde, J. (eds.) Software for Algebraic Geometry. The IMA Volumes in Mathematics and its Applications, vol. 148, pp. 33–46. Springer, Heidelberg (2008)
27. Maclagan, D., Sturmfels, B.: Introduction to Tropical Geometry, Graduate Studies in Mathematics, vol. 161. American Mathematical Society, Providence (2015)
28. Malajovich, G.: Computing mixed volume and all mixed cells in quermass-integral time, to appear in Found. Comput. Math. http://dx.doi.org/10.1007/ s10208-016-9320-1
29. Maurer, J.: Puiseux expansion for space curves. Manuscripta Math. **32**, 91–100 (1980)
30. Mizutani, T., Takeda, A.: DEMiCs: a software package for computing the mixed volume via dynamic enumeration of all mixed cells. In: Stillman, M., Takayama, N., Verschelde, J. (eds.) Software for Algebraic Geometry. The IMA Volumes in Mathematics and Its Applications, vol. 148, pp. 59–79. Springer, New York (2008)
31. Mizutani, T., Takeda, A., Kojima, M.: Dynamic enumeration of all mixed cells. Discrete Comput. Geom. **37**(3), 351–367 (2007)
32. Novoseltsev, A.: lattice_polytope module of Sage, The Sage Development Team (2011)
33. R Development Core Team: R: A Language and Environment for Statistical Computing. R Foundation for Statistical Computing, Vienna, Austria (2008). http:// www.R-project.org, ISBN 3-900051-07-0
34. Sommars, J., Verschelde, J.: Exact gift wrapping to prune the tree of edges of Newton polytopes to compute pretropisms. arXiv:1512.01594
35. Sommars, J., Verschelde, J.: Computing pretropisms for the cyclic n-roots problem. In: 32nd European Workshop on Computational Geometry (EuroCG 2016), pp. 235–238 (2016)
36. Stein, W., et al.: Sage Mathematics Software (Version 6.9). The Sage Development Team (2015). http://www.sagemath.org
37. Verschelde, J.: Algorithm 795: PHCpack: a general-purpose solver for polynomial systems by homotopy continuation. ACM Trans. Math. Softw. **25**(2), 251–276 (1999)
38. Verschelde, J.: Polyhedral methods in numerical algebraic geometry. In: Bates, D., Besana, G., Di Rocco, S., Wampler, C. (eds.) Interactions of Classical and Numerical Algebraic Geometry, Contemporary Mathematics, vol. 496, pp. 243–263. AMS (2009)
39. Ziegler, G.: Lectures on Polytopes, Graduate Texts in Mathematics, vol. 152. Springer, New York (1995)

Computational Aspects of a Bound of Lagrange

Doru Ştefănescu[(✉)]

University of Bucharest, Bucharest, Romania
stef@rms.unibuc.ro

Abstract. We consider the bound $R + \rho$ of Lagrange and we obtain some improvements of it. We also discuss the efficiency of this bound of Lagrange and of its refinements.

1 Introduction

During the last decades, there were discovered new bounds for the positive roots of univariate polynomials with real coefficients. However, there exists an old bound of Lagrange that can be useful in applications, this is $R + \rho$ bound:

Theorem 1 (Lagrange, 1767). *Let F be a nonconstant monic polynomial of degree n over $\mathbb{R}$ and let $\{a_j \, ; \, j \in J\}$ be the set of its negative coefficients. Then an upper bound for the positive real roots of F is given by the sum of the largest and the second largest numbers in the set*

$$\left\{ \sqrt[j]{|a_j|} \, ; \, j \in J \right\}.$$

The bound $R + \rho$ can be used also for bounding the roots of univariate polynomials with complex coefficients, see [9]:

Theorem 2. *Let $F(X) = X^d + a_1 X^{d-1} + \cdots + a_{d-1} X + a_d$ be a polynomial with complex coefficients, with $d > 0$ and $a_d \neq 0$ and*

$$|a_{i_1}|^{1/i_1} \geq |a_{i_2}|^{1/i_2} \geq \cdots \geq |a_{i_d}|^{1/i_d}.$$

We put $R = |a_{i_1}|^{1/i_1}$, $\rho = |a_{i_2}|^{1/i_2}$. The number $R + \rho$ is an upper bound for the absolute values of the roots of the polynomial F .

Lagrange has not given a proof, and just to the beginning of XXIth century, his result was not used in rootbounding. However, during the XIXth and XXth centuries, Theorem 1 was considered at least three times. In 1842, *Nouvelles Annales de Mathématiques* proposed its statement as a problem, and A. Pury gave a proof which seems to be forgotten until very recently. The result was stated and proved in a paper of E.C. Westerfield [15], and he considered it as a new bound of him [17]. We note that Westerfield ignored the memory of Lagrange. The result of Lagrange was explicitly cited in the book of A.M. Ostrowski [12], p. 125.

© Springer International Publishing AG 2016
V.P. Gerdt et al. (Eds.): CASC 2016, LNCS 9890, pp. 504–511, 2016.
DOI: 10.1007/978-3-319-45641-6_32

As the bound $R + \rho$ was considered unproved it was considered in a technical paper by M. Mignotte and D. Ştefănescu [9], where they gave two proofs. Also G.E. Collins and W. Krandick thought that the result was unproved and gave a proof, see [4]. The proof Krandick–Collins was recently simplified by P. Batra [2]. On the other hand, the implementation of the bound $R + \rho$ was discussed by Akritas-Strebonski-Vigklas [1].

Remark. From Theorem 2 — the complex version of the theorem of Lagrange — it is easy to deduce as corollary the bound of Fujiwara [5].

$$2 \cdot \max_{i=1}^{d} |a_i|^{1/i},$$

which is one of the best for the absolute values of the roots of complex polynomials.

2   The Method of Cauchy

Nowadays we have more approaches for proving the result of Lagrange. One of them is the method of Cauchy, considered explicitly by Mignotte–Ştefănescu [9]. We note that the method of Cauchy (1829, [3]) was used indirectly also by other authors, for example, by Pury [13].

Theorem 3 (Cauchy, 1829). *Let $F(X) = X^n + a_1 X^{n-1} + \cdots + a_{n-1} X + a_n \in \mathbb{C}[X]$ be nonconstant and let σ be the unique positive real root of the polynomial*

$$G(X) = X^n - |a_1| X^{n-1} - \cdots - |a_{n-1}| X - |a_n|.$$

Then any number surpassing σ is a bound for the moduli of the roots of F.

The basic idea is to observe that if we consider a nonconstant polynomial $F(X) = X^n + a_1 X^{n-1} + \cdots + a_{n-1} X + a_n$ with complex coefficients, it is sufficient to prove that $R + \rho$ is an upper bound for the unique root of the polynomial $X^n - |a_1| X^{n-1} - \cdots - |a_{n-1}| X - |a_n|$, where R and ρ are the two largest numbers in the set $\left\{ |a_k|^{1/k}; 1 \le k \le n \right\}$. This implies the same bound for the roots of F.

We note that Collins–Krandick and Batra follow the strategy proposed by Lagrange in [7], namely to use the device of Newton for finding the largest root of a polynomial.

3   Refinements of the Bound of Lagrange

If we compare the bound $R + \rho$ with other bounds we note that it seems to be one of the best. So it is natural to ask if it can be improved. Such a refinement was obtained by Mignotte–Ştefănescu in [10]:

Theorem 4. *Let* $F(X) = X^d + a_1 X^{d-1} + \cdots + a_{d-1}X + a_d$ *be a polynomial with complex coefficients, with* $d > 0$ *and* $a_d \neq 0$. *Suppose that* $|a_{i_1}|^{1/i_1} \geq |a_{i_2}|^{1/i_2} \geq \cdots \geq |a_{i_d}|^{1/i_d}$ *and put* $R = |a_{i_1}|^{1/i_1}$, $\rho = |a_{i_2}|^{1/i_2}$, $i_1 = j$ *and*

$$
C_j = \begin{cases} \dfrac{R + \rho + \sqrt{R^2 - 2R\rho + 5\rho^2}}{2} & \text{if } j = 1, \\ (R^{j-1} + R^{j-2}\rho + \cdots + \rho^{j-1})^{1/(j-1)} & \text{for } j = 2, \ldots, d. \end{cases}
$$

Then for any complex root z *of* F *we have*

$$
|z| \leq \max \left\{ C_j, \frac{R + \rho + \sqrt{R^2 - 2R\rho + 5\rho^2}}{2} \right\}.
$$

3.1   A Cubic Polynomial Related to the Bound of Lagrange

We observe that for $j = 2$, Theorem 4 gives exactly the bound $R + \rho$ of Lagrange. We shall give another bound for this case. More precisely, we shall prove that a bound for the positive roots of the polynomial F is given by the largest positive root of a particular cubic polynomial. Then we shall prove that this bound is smaller than the bound of Lagrange and we shall give estimates for it.

Proposition 1. *Let* $F(X) = X^d + a_1 X^{d-1} + \cdots + a_{d-1}X + a_d$ *be a polynomial with complex coefficients such that*

$$
|a_2|^{1/2} := R > \rho := |a_{i_2}|^{1/i_2} \geq \cdots \geq |a_{i_d}|^{1/i_d}.
$$

Then the largest positive root of the polynomial

$$
H(X) = X^3 - 2\rho X^2 - (R^2 - \rho^2)X + \rho(R^2 - \rho^2)
$$

is an upper bound for the absolute values of the roots of the polynomial F.

Proof. We use the method of Cauchy, so we associate the polynomial

$$
G(X) = X^d - |a_1|X^{d-1} - \cdots - |a_{d-1}|X - |a_d|.
$$

By Theorem 3 of Cauchy we search an upper bound for the positive root of G. We have

$$
G(x) \geq g(x) := x^d - R^2 x^{d-2} + r^2 x^{d-2} - \sum_{k=1}^{d} \rho^k x^{d-k}.
$$

We consider $h(x) := (x - \rho)g(x)$. We have

$$
h(x) = (x - \rho)\left(x^d - (R^2 - \rho^2)x^{d-2} - \sum_{k=1}^{d} \rho^k x^{d-k} \right)
$$

$$= x^{d+1} - (R^2 - \rho^2)x^{d-1} - \sum_{k=1}^{d} \rho^k\, x^{d-k+1} - \rho\, x^d + \rho(R^2 - \rho^2)x^{d-2}$$

$$+ \sum_{k=0}^{d} \rho^{k+1}\, x^{d-k}$$

$$= x^{d+1} - (R^2 - \rho^2)x^{d-1} - \sum_{s=0}^{d-1} \rho^{s+1}\, x^{d-s} - \rho\, x^d + \rho(R^2 - \rho^2)x^{d-2}$$

$$+ \sum_{s=1}^{d} \rho^{s+1}\, x^{d-s}$$

$$= x^{d+1} - 2\rho\, x^d - (R^2 - \rho^2)x^{d-1} + \rho(R^2 - \rho^2)x^{d-2} + \rho^{d+1}$$

$$> x^{d+1} - 2\rho\, x^d - (R^2 - \rho^2)x^{d-1} + \rho(R^2 - \rho^2)x^{d-2}$$

$$= \left(x^3 - 2\rho\, x^2 - (R^2 - \rho^2)x + \rho(R^2 - \rho^2)\right) x^{d-2}$$

$$= H(x)\, x^{d-2},$$

with $H(x) = x^3 - 2\rho x^2 - (R^2 - \rho^2)x + \rho(R^2 - \rho^2)$.   □

Estimates for the Largest Positive Root of the Polynomial H.

Lemma 1. *The polynomial H has three real roots and its largest positive root is smaller than $R + \rho$.*

Proof. We observe that

$$H(-R) \quad = -\rho(R^2 + R\rho + \rho^2) \quad < 0,$$

$$H(0) \qquad = \rho(R^2 - \rho^2) \qquad\qquad > 0,$$

$$H(R) \qquad = -\rho(R^2 - R\rho + \rho^2) < 0,$$

$$H(R + \rho) = \rho(R^2 - \rho^2) \qquad\qquad > 0,$$

It follows that the polynomial H has three real roots, namely in the intervals $(-R, 0)$, $(0, R)$ and $(R, R + \rho)$.

We observe that the largest of these roots lies in the interval $(R, R + \rho)$, so it is positive and smaller than $R + \rho$.   □

Proposition 2. *The largest positive root of the polynomial*

$$H(x) = x^3 - 2\rho x^2 - (R^2 - \rho^2)x + \rho(R^2 - \rho^2)$$

lies in the interval $(R + \rho/2, R + \rho)$.

Proof. We observe that

$$H(R + \frac{\rho}{2}) = -\frac{\rho^2}{4}\left(R + \frac{7\rho}{2}\right) < 0,$$

and

$$H(R + \rho) = \rho(R^2 - \rho^2) > 0.$$

So the largest positive root of the polynomial H lies in the interval $(R + \rho/2, R + \rho)$. □

3.2   Experiments

We keep the notation from Proposition 2. We search for smaller intervals than that from the previous Proposition.

We first consider

$$H(R + \frac{3}{4}\rho) = \frac{\rho}{2}\left(R^2 - \frac{5}{8}R\rho - \frac{61}{32}\rho^2\right)$$

and we note that $H(R + \frac{3}{4}\rho) > 0$ if $R/\rho > 1.73$.

We also have

$$H(R + \frac{3}{5}\rho) = \frac{\rho}{5}\left(R^2 - \frac{8}{5}R\rho - \frac{113}{25}\rho^2\right),$$

so $H(R + \frac{3}{5}\rho)$ is positive for $R/\rho > 3.072$.

3.3   Another Polynomial

We can obtain another improvements of the bound of Lagrange by considering another polynomial which gives upper bounds for the roots of the F.

We first give a proof shorter than those in [9] of the result of Lagrange:

Proposition 3. *Let* $F(X) = X^d + a_1 X^{d-1} + \cdots + a_{d-1}X + a_d$ *be a polynomial with complex coefficients, with* $d > 0$ *and* $a_d \neq 0$ *and*

$$|a_{i_1}|^{1/i_1} \geq |a_{i_2}|^{1/i_2} \geq \cdots \geq |a_{i_d}|^{1/i_d}.$$

We put $R = |a_{i_1}|^{1/i_1}$, $\rho = |a_{i_2}|^{1/i_2}$ *and* $j = i_1$. *Then the largest positive root of the polynomial*

$$u(X) = (X - 2\rho)X^j - (R^j - \rho^j)(X - \rho)$$

is an upper bound for the absolute values of the roots of the polynomial F.

Proof. We first note, by Descartes's rule of signs, that the polynomial u has an unique positive root.

By Theorem 3 of Cauchy, it is sufficient to find an estimation for the unique positive root of the associated polynomial

$$G(X) = X^d - |a_1|X^{d-1} - \cdots - |a_{d-1}|X - |a_d|.$$

We observe that $|a_i| \leq \rho^i$ for all $i = 1, 2, \ldots, d$, $i \neq j$. Then, for $x > \rho$, we have

$$f(x) \geq x^d - (R^j - \rho^j)x^{d-j} - \sum_{k=1}^{d} \rho^k x^{d-k}.$$

We put

$$g(X) = X^d - (R^j - \rho^j)X^{d-j} - \sum_{k=1}^{d} \rho^k X^{d-k}, \quad h(X) = (X - \rho)g(X)$$

and we observe that

$$h(X) = (X - \rho)X^{d+1} - (R^j - \rho^j)(X - \rho)X^{d-j} - \sum_{s=0}^{d-1} \rho^{s+1}X^{d-s} + \sum_{s=1}^{d} \rho^{s+1}X^{d-s}$$

$$= (X - 2\rho)X^d - (X - \rho)(R^j - \rho^j)X^{d-j} + \rho^{d+1}.$$

For $x > 0$ we have

$$h(x) = \left((x - 2\rho)x^j - (x - \rho)(R^j - \rho^j)\right)x^{d-j} + \rho^{d+1}.$$

We put $u(X) = (X - 2\rho)X^j - (X - \rho)(R^j - \rho^j)$ and we notice that, if, for $x > 0$, we have $u(x) > 0$, then $h(x) > 0$. $\qquad\square$

We search intervals containing the largest positive root of the polynomial $u(X) = (X - 2\rho)X^j - (X - \rho)(R^j - \rho^j)$.

Lemma 2. *We have $u(R + \rho) > 0$.*

Proof. We have the inequalities

$$u(R + \rho) = (R - \rho)(R + \rho)^j - R(R^j - \rho^j)$$

$$= (R - \rho)\left(\sum_{k=0}^{j} \binom{j}{k} R^{k-j}\rho^k - (R^j + R^{j-1}\rho + \cdots + R\rho^{j-1})\right)$$

$$= (R - \rho)\left((jR^{j-1}\rho + \frac{j(j-1)}{2}R^{j-2}\rho^2 + \cdots + jR\rho^{j-1})\right.$$

$$- (R^{j-1}\rho + R^{j-2}\rho^2 + \cdots + R^2\rho^{j-2} + R\rho^{j-1}))$$

$$> 0. \qquad\qquad\qquad\qquad\qquad\qquad\qquad\qquad\qquad\square$$

We observe that

$$u(R + \frac{\rho}{2}) = (R + \frac{\rho}{2})^j(R - \frac{3\rho}{2}) - (R - \frac{\rho}{2})(R^j - \rho^j).$$

Lemma 3. *If $R < \frac{3}{2}\rho$ we have*

$$u\left(R + \frac{\rho}{2}\right) < 0 \,.$$

Proof. In this case we have

$$u\left(R + \frac{\rho}{2}\right) = \left(R - \frac{3}{2}\rho\right)\left(R + \frac{\rho}{2}\right)^j - \left(R - \frac{\rho}{2}\right)(R^j - \rho^j) < 0 \,. \qquad \square$$

Corollary 1. *If $R < \frac{3}{2}\rho$, a bound for the positive roots of the polynomial F lies in the interval $\left(R + \frac{\rho}{2}, R + \rho\right)$.*

4   Comparisons and Numerical Results

We consider the bounds given by Theorems 2, 4, and Proposition 3. We denote by $H(P)$ the bound of H. Hong [6] and by $A(P)$ the threshold of positiveness of P:

$$H(P) = 2 \cdot \max_{\substack{i \\ a_i < 0}} \min_{\substack{j > i \\ a_j > 0}} \left(\frac{|a_i|}{a_j}\right)^{\frac{1}{d_i - d_j}} \,,$$

the *threshold* of positiveness of an univariate polynomial P with real coefficients is

$$A(P) = \inf \{D(P) \,; \, D(P) \in \{\text{bounds for positiveness of } P\}\} \,.$$

Example 1. We consider the polynomial $P(X) = X^5 - 2X^4 + 2X^3 - 3X^2 - 2X - 2$. We have $R = 2$, $\rho = 1.442$. We obtain the following bounds

Th 2	Th 4	H(P)	Prop 3	A(P)
3.442	3.189	4.000	3.189	2.069

The best result is given by Theorem 4. It can be obtained also by finding the largest positive root of the polynomial u from Proposition 3.

Example 2. Let $P(X) = X^7 - X^6 - 9X^5 - 8X^4 - 3X^3 + 12X + 1$. We have $R = 3$ and $\rho = 2$. In this case $j = 2$, so we will consider also the cubic polynomial H from Proposition 1. We have

Th 2	Th 4	H(P)	Prop 3	Prop 1	A(P)
5.0	5.0	6.0	4.613	4.613	3.884

We observe that both Propositions 1 and 3 give the best bound. In fact, in this case, the polynomials H and u coincide.

d	R	ρ	Th 2	Th 4	H(P)	Prop 3	Prop 1	A(P)
6	2	1.259	3.259	2.498	4	2.498	—	1.233
7	1.414	1.189	2.603	2.603	2.828	2.378	2.378	1.166
8	1.104	1.080	2.184	2.172	2.208	2.172	—	1.082

Example 3. We consider the family of polynomials $P_d(X) = X^d - 2X^5 + 3X^4 - 2X^3 - 1$, where $d \geq 6$, discussed in [14]. Note that for $d = 7$, we have $j = 2$, and we can apply Proposition 1.

Conclusions. The bound of Lagrange gives good results with respect to known efficient bounds for polynomials with real coefficients. We proved that it can be improved. Because of the Theorem of Cauchy such results give also accurate bounds for absolute values of the roots of univariate polynomials with complex coefficients.

References

1. Akritas, A.G., Strzeboński, A.W., Vigklas, P.S.: Lagrange's bound on the values of the positive roots of polynomials (private communication)
2. Batra, P.: On a proof of Lagrange's bound for real roots (private communication)
3. Cauchy, A.-L.: Exercices de Mathématiques, t. 4, Paris (1829)
4. Collins, G.E.: Krandick's proof of Lagrange's real root bound claim. J. Symb. Comp. **70**, 106–111 (2015)
5. Fujiwara, M.: Über die obere Schranke des absoluten Betrages der Wurzeln einer algebraischen Gleichung. Tôhoku Math. J. **10**, 167–171 (1916)
6. Hong, H.: Bounds for absolute positiveness of multivariate polynomials. J. Symb. Comp. **25**, 571–585 (1998)
7. Lagrange, J.-L.: Traité de la résolution des équations numériques, Paris (Reprinted in Œuvres, t. VIII, Gauthier-Villars, Paris (1879)) (1798)
8. Mignotte, M., Ştefănescu, D.: Polynomials - An Algorithmic Approach. Springer, Singapore (1999)
9. Mignotte, M., Ştefănescu, D.: On an estimation of polynomial roots by Lagrange. Prépubl. IRMA, 25/2002, 17 pag., Strasbourg (2002)
10. Mignotte, M., Ştefănescu, D.: On the bound $R + \rho$ of Lagrange (available from the authors on request)
11. Problème 6. Limite de Lagrange. Nouv. Annales Math. 1, 58 (1842)
12. Ostrowski, A.M.: Solutions of Equations and Systems of Equations. Academic Press, New York (1960)
13. Pury, A.: Solution du problème 6. Limite de Lagrange. Nouv. Annales Math. **1**, 243–244 (1842)
14. Ştefănescu, D.: A new polynomial bound and its efficiency. In: Gerdt, V.P., Koepf, W., Seiler, W.M., Vorozhtsov, E.V. (eds.) CASC 2015. LNCS, vol. 9301, pp. 457–467. Springer, Heidelberg (2015). doi:10.1007/978-3-319-24021-3_33
15. Westerfield, E.C.: New bounds for the roots of an algebraic equation. Amer. Math. Monthly **38**, 30–35 (1931)
16. Westerfield, E.C.: A new bound for the zeros of polynomials. Amer. Math. Monthly **40**, 18–23 (1933)
17. Westerfiled, E.C.: New bounds for the roots of an algebraic equation. Amer. Math. Monthly **38**, 30–35 (1931)

Author Index

Printed in the United States
By Bookmasters